Geomathematically Oriented Potential Theory

PURE AND APPLIED MATHEMATICS

A Program of Monographs, Textbooks, and Lecture Notes

MONOGRAPHS AND TEXTBOOKS IN PURE AND APPLIED MATHEMATICS

Recent Titles

Applications to Fluid Structure Interactions (2006)

Alfred Geroldinger and Franz Halter-Koch, Non-Unique Factorizations: Algebraic, Combinatorial and Analytic Theory (2006)

Kevin J. Hastings, Introduction to the Mathematics of Operations Research with *Mathematica*®, Second Edition (2006)

Robert Carlson, A Concrete Introduction to Real Analysis (2006)

John Dauns and Yiqiang Zhou, Classes of Modules (2006)

N. K. Govil, H. N. Mhaskar, Ram N. Mohapatra, Zuhair Nashed, and J. Szabados, Frontiers in Interpolation and Approximation (2006)

Luca Lorenzi and Marcello Bertoldi, Analytical Methods for Markov Semigroups (2006)

M. A. Al-Gwaiz and S. A. Elsanousi, Elements of Real Analysis (2006)

Theodore G. Faticoni, Direct Sum Decompositions of Torsion-Free Finite Rank Groups (2007)

R. Sivaramakrishnan, Certain Number-Theoretic Episodes in Algebra (2006)

Aderemi Kuku, Representation Theory and Higher Algebraic K-Theory (2006)

Robert Piziak and P. L. Odell, Matrix Theory: From Generalized Inverses to Jordan Form (2007)

Norman L. Johnson, Vikram Jha, and Mauro Biliotti, Handbook of Finite Translation Planes (2007)

Lieven Le Bruyn, Noncommutative Geometry and Cayley-smooth Orders (2008)

Fritz Schwarz, Algorithmic Lie Theory for Solving Ordinary Differential Equations (2008)

Jane Cronin, Ordinary Differential Equations: Introduction and Qualitative Theory, Third Edition (2008)

Su Gao, Invariant Descriptive Set Theory (2009)

Christopher Apelian and Steve Surace, Real and Complex Analysis (2010)

Norman L. Johnson, Combinatorics of Spreads and Parallelisms (2010)

Lawrence Narici and Edward Beckenstein, Topological Vector Spaces, Second Edition (2010)

Moshe Sniedovich, Dynamic Programming: Foundations and Principles, Second Edition (2010)

Drumi D. Bainov and Snezhana G. Hristova, Differential Equations with Maxima (2011)

Willi Freeden, Metaharmonic Lattice Point Theory (2011)

Murray R. Bremner, Lattice Basis Reduction: An Introduction to the LLL Algorithm and Its Applications (2011)

Clifford Bergman, Universal Algebra: Fundamentals and Selected Topics (2011)

A. A. Martynyuk and Yu. A. Martynyuk-Chernienko, Uncertain Dynamical Systems: Stability and Motion Control (2012)

Washek F. Pfeffer, The Divergence Theorem and Sets of Finite Perimeter (2012)

Willi Freeden and Christian Gerhards, Geomathematically Oriented Potential Theory (2013)

Geomathematically Oriented Potential Theory

Willi Freeden
University of Kaiserslautern
Germany

Christian Gerhards
University of Kaiserslautern
Germany

CRC Press
Taylor & Francis Group
Boca Raton London New York

CRC Press is an imprint of the
Taylor & Francis Group, an **informa** business
A CHAPMAN & HALL BOOK

CRC Press
Taylor & Francis Group
6000 Broken Sound Parkway NW, Suite 300
Boca Raton, FL 33487-2742

Printed in the United States of America on acid-free paper
Version Date: 2012918

International Standard Book Number: 978-1-4398-9542-9 (Hardback)

Library of Congress Cataloging-in-Publication Data

Freeden, W.
Geomathematically oriented potential theory / Willi Freeden, Christian Gerhards.
pages cm. -- (Pure and applied mathematics)
Includes bibliographical references and index.
ISBN 978-1-4398-9542-9 (hardback)
1. Gravitational potential. 2. Geomagnetism--Mathematics. 3. Potential theory (Physics) I. Gerhards, Christian, 1982- II. Title.

QC178.F66 2013
515'.96--dc23 2012026783

Visit the Taylor & Francis Web site at
http://www.taylorandfrancis.com

and the CRC Press Web site at
http://www.crcpress.com

Contents

Preface

In the year 1782, P.S. Laplace showed that the Newtonian potential obeys a certain partial differential equation. Laplace's equation – as it is called today – has become a central differential equation in arguably all physical geosciences because of the wide range of phenomena that it characterizes. It is important in gravitation, electromagnetism, and fluid dynamics because it describes the behavior of gravitational, magnetic, and fluid potentials. The theory of potentials concerned with the Laplace equation is known as *potential theory*. It is the subject of this textbook, but with particular emphasis on the Earth's gravitational and magnetic field. Our understanding of potential theory, however, does not intend to explain the nature of gravitation or magnetism. This is the task of physics; it requires a detailed study of Newton's and Einstein's theories of gravitation and Maxwell's theory to magnetism as well as modern developments in these fields. Our work rather intends to gain some insight on how gravitation and magnetism can be geomathematically handled for the relevant observables and on how the resulting geopotential problems can be solved in a systematic, mathematically rigorous framework involving three-dimensional Euclidean as well as spherical concepts.

Considering the huge literature in geosciences on the gravitational and the magnetic field, it might be asked why a new textbook on potential theory is needed at this time. Indeed, classical potential theory is a well-understood and frequently documented area in mathematics. It deals with harmonic functions, maximum/minimum principles, single and multipole expansions, Green functions, integral representations of potentials, boundary-value problems, volume and surface integral equation methods, etc. The essential texts on potential theory published during the last century are still available today, notably those of O.D. Kellogg [1967], L.L. Helms [1969], and R.J. Blakely [1996]. These books deal thoroughly with the fundamentals of potential theory in three-dimensional Euclidean space, more precisely in the interior and/or exterior of "potato-like" bodies such as a ball, ellipsoid, geoid, and the actual Earth. Even more, multiscale methods for boundary-value and inverse problems (cf. W. Freeden, V. Michel [2004]) have been studied in more detail.

Nevertheless, we believe that this book on geomathematically oriented potential theory will fill a significant gap. As mathematicians interested in any kind of Earth-related gravitational and magnetic data and processes, we immediately found ourselves involved with the Laplace operator in specific application to functions defined on surfaces in three-dimensional Euclidean space. It seemed to us, and we also find it true today, that no single textbook is available covering topics of potential theory with respect to the "space Laplacian" together with the "surface Laplacian", i.e., the Beltrami operator. This work attempts to fill the gap by exploring and presenting the principles of *surface potential theory*, particularly for the sphere, in addition to those in the Euclidean space $\mathbb{R}^3$. In doing so we are led by the observation that the Earth's surface is an almost perfect sphere. Deviations from its spherical shape are less than 0.4% of its radius. All level (equipotential) surfaces are nearly spherical. Moreover, almost all modern satellite missions providing gravitational or magnetic data sets collect their data on nearly spherical orbits. Consequently, spherical methods and tools play an important part in the mathematical treatment of Earth's gravitational and magnetic field.

At the present time, the use of spherical harmonics is a well-established technique in all geosciences for the purpose of representing spherical fields. However, spherical harmonics are polynomials of global nature. Orthogonal expansions in terms of spherical harmonics are well suited to resolve the trend of a signal, while their application to obtain high resolution is critical, particularly in models where local data and boundary information come into play. These situations indispensably require the development of a potential theory related to the Beltrami operator. Like the Laplacian, the Beltrami operator is the divergence of a gradient. In turn, the essential properties known from *space potential theory* can be formulated in close similarity for the Beltrami operator, leading to a *surface potential theory* including maximum/minimum principles, integral expressions for potentials, Green functions, integral equation methods, Runge procedures, etc. As a matter of fact, looking at the contents of space as well as surface potential theory, we realize that most of the material can be presented in parallel for the Laplace and the Beltrami operator. However, there are also essential differences that are caused by the angular nature of the Beltrami operator, the characteristic logarithmic singularity of its Green function, and the fact that the exterior space of a regular region on the sphere remains bounded (in contrast to the Euclidean case $\mathbb{R}^3$).

All in all, this book breaks new mathematical ground in dealing generically with potential theoretic aspects of gravitation and geomagnetism. Our work shows a canonical subdivision into three parts. The first two chapters give all notational material and background that are needed throughout the remaining work. The next three chapters build the foundation of potential theory in Euclidean space $\mathbb{R}^3$ and its application to gravitation and geomagnetism. The three chapters on space potential theory in $\mathbb{R}^3$ are followed by three chapters on surface potential theory on the unit sphere Ω and corresponding applications. As far as the notation is concerned, the authors had to find their own way, thereby adopting autochthonous nomenclature of geophysics and physical geodesy to a certain extent.

The work covers a two-semester graduate course in the teaching cycle of geomathematics, but it can as well be used as a reference for researchers and aims at presenting the state of the art in three-dimensional Euclidean spatial as well as two-dimensional spherical potential theory. The seeds of the work began in graduate-level classes of the first author on potential theory at the University of Kaiserslautern. The final concept of the textbook, however, is a reflection of discussions and interactions with several members involved in the course "Geomathematics" at the University of Kaiserslautern, especially M. Schreiner, H. Nutz, M. Gutting, T. Fehlinger, K. Wolf, C. Blick, S. Möhringer, and S. Eberle. The second part of the book is essentially influenced by the PhD thesis of the second author finished in early spring 2011. We have the pleasure of thanking our students A. Hunt, S. Nelles, and P. Easwaran for many hints and remarks. Furthermore, our gratitude to C. Korb has to be expressed for her continuous support in handling the typing job.

The cover illustration includes a model of the geoid (i.e., the equipotential surface of the Earth at sea level). The "geoidal potato" constitutes a typical geophysically relevant regular region as discussed in this work. We are obliged to R. Haagmans, head of Earth Surfaces and Interior Section, Mission Science Division, ESA-ESTEC, Noordwijk, the Netherlands, for providing us with the image (ESA ID number: SEMLXEOA90).

Finally, the authors thank Bob Stern and Marsha Hecht, Taylor & Francis, for the interest in the book and the cooperative work.

W. Freeden, C. Gerhards
Kaiserslautern

About the Authors

Willi Freeden: Studies in mathematics, geography, and philosophy at the RWTH Aachen, 1971 Diplom in mathematics, 1972 Staatsexamen in mathematics and geography, 1975 PhD in mathematics, 1979 Habilitation in mathematics, 1981/1982 visiting research professor at The Ohio State University, Columbus (Department of Geodetic Science and Surveying), 1984 professor of mathematics at the RWTH Aachen (Institute of Pure and Applied Mathematics), 1989 professor of technomathematics (industrial mathematics), 1994 head of the Geomathematics Group, 2002–2006 vice-president for Research and Technology at the University of Kaiserslautern, 2009 editor in chief of the *International Journal on Geomathematics* (GEM), 2010 editor of the *Handbook of Geomathematics*, member of the editorial board of seven international journals.

Christian Gerhards: Studies in mathematics at the RWTH Aachen, 2008 Diplom in mathematics, studies in geomathematics at the University of Kaiserslautern, 2011 PhD in mathematics, 2012/13 visiting postdoc researcher (funded by the DAAD postdoc fellowship D/11/46067) at the University of New South Wales, Sydney, Australia (Department of Mathematics and Statistics).

List of Symbols

$\mathbb{N}_0$ ... set of non–negative integers
$\mathbb{N}$ ... set of positive integers
$\mathbb{Z}$ ... set of integers
$\mathbb{R}$ ... set of real numbers
$\mathbb{C}$ ... set of complex numbers
$\Re(s)$ ... real part of $s \in \mathbb{C}$
$\Im(s)$ ... imaginary part of $s \in \mathbb{C}$
$\mathbb{R}^3$ ... three-dimensional Euclidean space
x, y, z ... elements of $\mathbb{R}^3$
$x \cdot y$ ... scalar product of vectors in $\mathbb{R}^3$
$x \otimes y$ ... tensor product in $\mathbb{R}^3$
$x \wedge y$ ... vector product in $\mathbb{R}^3$
$|x|$ ... Euclidean norm of $x \in \mathbb{R}^3$
$\varepsilon^i,\ i = 1, 2, 3$ ... canonical orthonormal basis vectors in $\mathbb{R}^3$
ε^{ijk} ... alternator (Levi–Cività alternating symbol)
$\delta_{i,j}$ ... Kronecker symbol
$\mathrm{C}^{(k)}, \mathrm{L}^p$ ... classes of scalar functions
$\mathrm{c}^{(k)}, \mathrm{l}^p$ ... classes of vector functions
$\mathbf{c}^{(k)}, \mathbf{l}^p$ ... classes of tensor functions
F, G ... scalar-valued functions
f, g ... vector-valued functions
$\mathbf{f}, \mathbf{g}$ ... tensor-valued functions
$\mathbf{i}$ ... identity tensor in $\mathbb{R}^{3\times 3}$
$\mathbf{t}$ ... orthogonal matrix in $\mathbb{R}^{3\times 3}$
$\mathbf{t}^T$ ... transpose of the matrix $\mathbf{t}$
$\det \mathbf{t}$ ... determinant of the matrix $\mathbf{t}$
$F|_M$ ... restriction of the function F to M
$\{\ldots\}$ ... set of elements
$\emptyset$ ... empty set
$\in$ ... element
$\cup$ ... union
$\cap$ ... intersection
$\subset$ ... is contained in
$\Subset$ ... is totally contained in
$\oplus$ ... orthogonal sum
$\lfloor x \rfloor$ ... the largest integer $\leq x$

$\lceil x \rceil$	the smallest integer $\geq x$
$o, \mathcal{O}$	Landau symbols
$\mathcal{G}$	regular region in $\mathbb{R}^3$
$\overline{\mathcal{G}} = \mathcal{G} \cup \partial\mathcal{G}$	closure of $\mathcal{G}$
$\partial\mathcal{G}$	boundary of $\mathcal{G}$
$\mathcal{G}^c$	open complement of $\mathcal{G}$ in $\mathbb{R}^3$
$\|\|\mathcal{G}\|\|$	volume of the regular region $\mathcal{G}$ in $\mathbb{R}^3$
$\alpha(x)$	solid angle at x subtended by $\partial\mathcal{G}$
$\mathcal{B}_R(y)$	ball in $\mathbb{R}^3$ with radius R around y
$\mathcal{B}_{\rho,R}(y)$	spherical shell with radii ρ and R around y
Ω	unit sphere in $\mathbb{R}^3$ around the origin
$\Omega_R(y)$	sphere in $\mathbb{R}^3$ with radius R around y
r, t, φ	spherical coordinates in $\mathbb{R}^3$
ε^r, ε^t, ε^φ	orthonormal spherical basis vectors in $\mathbb{R}^3$
ξ, η, ζ	elements of Ω
Γ	regular region on Ω
$\overline{\Gamma} = \Gamma \cup \partial\Gamma$	closure of Γ on Ω
$\partial\Gamma$	boundary curve of Γ on Ω
Γ^c	open complement of Γ in Ω
$\|\|\Gamma\|\|$	"volume" of a regular region Γ on Ω
$\alpha(\xi)$	solid angle at ξ subtended by $\partial\Gamma$
$\Gamma_\rho(\xi)$	spherical cap with radius ρ around $\xi \in \Omega$
$\mathcal{C}_{\rho,R}(\Gamma)$	conical shell with radii ρ and R
∇	gradient
L	curl gradient
$\nabla\cdot$	divergence
$\mathrm{L}\cdot$	curl divergence
Δ	Laplace operator
∇^*	surface gradient on Ω
L^*	surface curl gradient on Ω
$\nabla^*\cdot$	surface divergence on Ω
$\mathrm{L}^*\cdot$	surface curl divergence on Ω
Δ^*	Beltrami operator on Ω
$o^{(i)}, \tilde{o}^{(i)}$, $i = 1, 2, 3$	spherical Helmholtz operators on Ω
dV	volume element
$d\omega$	surface element
$d\sigma$	line element

Introduction

Several potentials play a major role in the course of this book. Most of them have a canonical representation in the Euclidean context of $\mathbb{R}^3$ as well as in the spherical framework on the unit sphere Ω. In this introduction, we present some prominent examples of potentials and their relation to geophysical problems.

Euclidean Case

Let $\mathcal{G} \subset \mathbb{R}^3$ be a region. A real-valued function U is called *harmonic* on $\mathcal{G}$ if U is of class $\mathrm{C}^{(2)}(\mathcal{G})$ and satisfies the Laplace equation

$$\Delta U(x) = \left(\frac{\partial}{\partial x_1}\right)^2 U(x) + \left(\frac{\partial}{\partial x_2}\right)^2 U(x) + \left(\frac{\partial}{\partial x_3}\right)^2 U(x) = 0 \qquad (0.1)$$

for all $x = (x_1, x_2, x_3)^T \in \mathcal{G}$. Some important examples of harmonic functions should be listed in the classical nomenclature of geophysics.

(a) *Potential of a mass point:* According to Newton's Law of Gravitation two points with masses M_x, M_y attract each other with a force given by

$$-\frac{G}{4\pi}\frac{M_x M_y}{|x-y|^3}(x-y), \quad x, y \in \mathbb{R}^3,\ x \neq y. \qquad (0.2)$$

The force is directed along the line connecting the two points x, y (cf. Figure 1). The constant G denotes Newton's gravitational constant (note that G will be assumed to be equal to one in the theoretical part of our later work, but not in numerical applications).

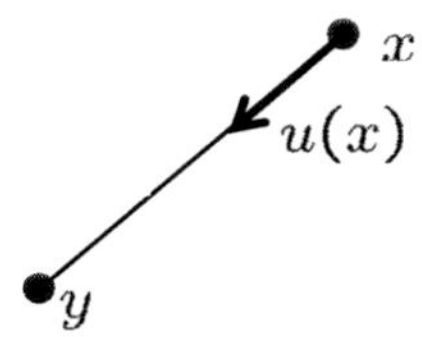

FIGURE 1
Gravitation between point masses.

Although the masses M_x, M_y attract each other in a symmetric way, it is convenient to call one of them the *attracting mass* and the other one the *attracted mass.* Conventionally, the attracted mass is set equal to unity and the attracting mass is denoted by M:

$$u(x) = -\frac{G}{4\pi}\frac{M}{|x-y|^3}(x-y), \quad x \in \mathbb{R}^3 \backslash \{y\}. \tag{0.3}$$

The formula (0.3) expresses the force exerted by the mass M on a unit mass located at the distance $|x-y|$ from M. Obviously, the intensity $|u(x)|$ of the force $u(x)$ is given by

$$|u(x)| = \frac{G}{4\pi}\frac{M}{|x-y|^2}, \quad x \in \mathbb{R}^3 \backslash \{y\}. \tag{0.4}$$

The scalar function U defined by

$$U(x) = GM \; G(\Delta; |x-y|), \quad x \in \mathbb{R}^3 \backslash \{y\}, \tag{0.5}$$

is called the *potential of gravitation,* where we have used the abbreviation

$$G(\Delta; |x-y|) = \frac{1}{4\pi}\frac{1}{|x-y|}, \quad x \in \mathbb{R}^3 \backslash \{y\}. \tag{0.6}$$

The force vector $u(x)$ is the gradient vector of the scalar $U(x)$. In other words,

$$u(x) = \nabla U(x), \quad x \in \mathbb{R}^3 \backslash \{y\}. \tag{0.7}$$

Calculating the second derivatives of U, it readily follows that

$$\Delta U(x) = 0, \quad x \in \mathbb{R}^3 \backslash \{y\}. \tag{0.8}$$

(b) *Potential of a finite mass point system:* The potential for N points x_i with masses M_i, $i = 1, \ldots, N$, is the sum of the individual contributions

$$U(x) = G \sum_{i=1}^{N} M_i \; G(\Delta; |x-y_i|), \quad x \in \mathbb{R}^3 \backslash \{y_1, \ldots, y_N\}. \tag{0.9}$$

Clearly we have

$$\Delta U(x) = 0, \quad x \in \mathbb{R}^3 \backslash \{y_1, \ldots, y_N\}. \tag{0.10}$$

(c) *Potential of a volume:* The point masses are distributed continuously over a (regular) body $\mathcal{G} \subset \mathbb{R}^3$, with density F (i.e., $F = \frac{dM}{dV}$, where dV is the element of volume and dM is the element of mass). Then the discrete sum (0.9) becomes a continuous sum, i.e., an integral over the body $\mathcal{G}$:

$$U(x) = G \int_{\mathcal{G}} F(y) G(\Delta; |x-y|) \, dV(y) = G \int_{\mathcal{G}} G(\Delta; |x-y|) \, dM(y). \tag{0.11}$$

Needless to say

$$\Delta U(x) = 0, \quad x \in \mathbb{R}^3 \backslash \overline{\mathcal{G}}, \tag{0.12}$$

where $\overline{\mathcal{G}} = \mathcal{G} \cup \partial\mathcal{G}$ denotes the closure of $\mathcal{G}$. Note that U is defined on the whole space $\mathbb{R}^3$, however, $\Delta U(x)$ may not be obtained simply by interchanging the Laplace operator and the integral over $\mathcal{G}$ for all points x inside $\mathcal{G}$. At infinity, the potential behaves like

$$|U(x)| = \mathcal{O}\left(\frac{1}{|x|}\right), \qquad |\nabla U(x)| = \mathcal{O}\left(\frac{1}{|x|^2}\right), \qquad |x| \to \infty, \tag{0.13}$$

uniformly with respect to all directions.

(d) *Potential of a single surface layer:* The attracting masses are assumed to form a layer on a certain closed surface, e.g., $\partial\mathcal{G}$, with thickness zero and (continuous) density F (i.e., $F = \frac{dM}{d\omega}$, where $d\omega$ is the element of surface)

$$U(x) = G \int_{\partial\mathcal{G}} F(y) G(\Delta; |x-y|)\, d\omega(y) = G \int_{\partial\mathcal{G}} G(\Delta; |x-y|)\, dM(y). \tag{0.14}$$

This definition is fictitious, but it is nevertheless of great theoretical significance. The integral (0.14) also exists for all $x \in \partial\mathcal{G}$. At infinity, the potential of a single surface layer behaves in the same way as a volume potential. U is harmonic in the inner space $\mathcal{G}$ and the outer space $\mathcal{G}^c = \mathbb{R}^3 \backslash \overline{\mathcal{G}}$.

(e) *Potential of a double surface layer:* A double-layer on a surface $\partial\mathcal{G}$ can be generated by two single layers separated by a small distance τ. Every pair of points $\left(y - \frac{\tau}{2}\nu(y), y + \frac{\tau}{2}\nu(y)\right)$ for $y \in \partial\mathcal{G}$ and sufficiently small values τ defines, via the Taylor formula (in linearized sense for $\tau \to 0+$), the potential of a dipole

$$U(x) = GM\, \nu(y) \cdot \nabla_y G(\Delta; |x-y|), \quad x \in \mathbb{R}^3 \backslash \{y\}, \tag{0.15}$$

with dipole moment M. Integrating over all dipoles on $\partial\mathcal{G}$, we are led to the potential of a double layer

$$\begin{aligned} U(x) &= G \int_{\partial\mathcal{G}} \frac{\partial}{\partial\nu(y)} G(\Delta; |x-y|)\, dM(y) \qquad (0.16) \\ &= G \int_{\partial\mathcal{G}} F(y)\, \nu(y) \cdot \nabla_y G(\Delta; |x-y|)\, d\omega(y). \end{aligned}$$

An easy calculation yields

$$\frac{\partial}{\partial\nu(y)} \frac{1}{|x-y|} = \frac{\nu(y) \cdot (x-y)}{|x-y|^3}. \tag{0.17}$$

Consequently, as far as the boundary surface $\partial\mathcal{G}$ shows a certain amount of smoothness, the existence of the surface integral (0.16) on $\partial\mathcal{G}$ is evident. At infinity, the potential of a double-layer behaves like $\mathcal{O}(|x|^{-2})$, $|x| \to \infty$. U is harmonic in the inner space $\mathcal{G}$ and the outer space $\mathcal{G}^c = \mathbb{R}^3 \backslash \overline{\mathcal{G}}$.

The double-layer potential (0.16) must be sharply distinguished from the single-layer potential (0.14). Common to both is the fact that they vanish at infinity and satisfy the Laplace equation in the inner and outer space of $\partial\mathcal{G}$. On the surface $\partial\mathcal{G}$ itself, however, they have a completely different nature, and it is this difference that makes these fictitious potentials mathematically useful. Indeed, layer potentials help us to handle inner and outer boundary-value problems by surface integral equations over $\partial\mathcal{G}$. The essential tool for their solution is the Fredholm theory.

The (Newton) volume potential (0.11) appears from different points of view in the context of geomathematics. For example, *gravimetry* denotes the ill-posed inverse problem of determining the mass density F inside $\partial\mathcal{G}$ from the knowledge of U on and outside $\partial\mathcal{G}$. But also the determination of the potential U itself from discrete data in the whole Euclidean space $\mathbb{R}^3$ or certain subsets is of great interest, e.g., for the determination of the geoid or other equipotential surfaces. Terrestrial measurements of the gravitational force intensity $|u(x)| = |\nabla U(x)|$ typically lead to an *oblique derivative boundary-value problem for the Earth's gravitational potential* U. The obliqueness is a result of the fact that the real Earth's surface does not coincide with the geoidal surface (except over certain parts on oceans). Satellite measurements on orbits yield vectorial and/or tensorial derivatives of first and second order. Depending on the type of measurements, modern satellite problems for the determination of the Earth's external gravitational potential are categorized as *satellite-to-satellite tracking* (SST) problems (i.e., ∇U is derivable from orbit deviations) or *satellite gravity gradiometry* (SGG) problems ($\nabla \otimes \nabla U$ is available by gradiometer measurements on the orbit).

The Earth's gravitational field is not the only geoscientifically relevant quantity where potential theory plays a major role. In source-free (regular) regions $\mathcal{G} \subset \mathbb{R}^3$, the Earth's magnetic field satisfies the homogeneous pre-Maxwell equations

$$\nabla \wedge b(x) = 0, \quad x \in \mathcal{G}, \tag{0.18}$$
$$\nabla \cdot b(x) = 0, \quad x \in \mathcal{G}. \tag{0.19}$$

Mathematically, this implies the existence of a harmonic function U such that $b(x) = \nabla U(x),\ x \in \mathcal{G}$. Thus, similar techniques as for the gravitational potential can be applied leading to vectorial problems in magnetic field modeling. However, in non-source-free regions, for which $\nabla \wedge b(x) \neq 0$, $x \in \mathcal{G}$, the magnetic field cannot be described solely by a scalar harmonic potential U. In this case, the Helmholtz decomposition

$$b(x) = \nabla U(x) + \nabla \wedge v(x), \quad x \in \mathcal{G}, \tag{0.20}$$

requires an additional vector potential v. The Mie decomposition

$$b(x) = \nabla \wedge \mathrm{L}P_b(x) + \mathrm{L}Q_b(x), \quad x \in \mathcal{G}, \tag{0.21}$$

manages on two scalar (however, not necessarily harmonic) functions.

Spherical Case

Geomathematical features in the Euclidean context of $\mathbb{R}^3$ often reduce to the sphere (being a good approximation of the Earth's surface or satellite orbits). However, many geophysical problems do not only relate to the sphere as the boundary surface that determines the behavior of the observable in the entire three-dimensional region $\mathcal{G}$, but they form problems intrinsic to a sphere. One should be aware that these spherical problems do not evolve in such a canonical manner as, e.g., for the gravitational potential in the Euclidean framework of $\mathbb{R}^3$. Yet, spherical operators like the surface gradient ∇^* connect, e.g., the Earth's disturbing potential T to the deflections of the vertical Θ via the surface gradient equation

$$\nabla^*_\xi T(R\xi) = -\frac{GM}{R}\Theta(R\xi), \quad \xi \in \Omega, \tag{0.22}$$

where R is the mean Earth's radius and Ω denotes the unit sphere in $\mathbb{R}^3$.

In geomagnetism, the radial ionospheric current density J_1 at a spherical satellite orbit Ω_r of mean radius r is related to the toroidal scalar Q_b of the induced magnetic field b by the Beltrami operator Δ^* in the form of the differential equation

$$\frac{1}{r}\Delta^*_\xi Q_b(r\xi) = \mu_0 J_1(r\xi), \quad \xi \in \Omega \tag{0.23}$$

(note that the vacuum permeability μ_0 is simply set to one for the theoretical considerations in this book). Different from the Euclidean case, we typically do not deal with potentials U that are harmonic (with respect to the Beltrami operator) on the sphere, i.e.,

$$\Delta^* U(\xi) = 0, \quad \xi \in \Omega, \tag{0.24}$$

but we are more often confronted with problems like (0.23) having a non-vanishing right-hand side. The arising conceptual tools, however, are similar to the Euclidean settings, in particular when working on (regular) subregions Γ of the sphere Ω. The bridging relation between the spatial and spherical approach becomes especially apparent by observing the connection between the Beltrami operator and the Laplace operator in terms of the spherical coordinates $x = r\xi, r = |x|, \xi = \frac{x}{|x|}$, i.e.,

$$\Delta_x = \frac{1}{r^2}\frac{\partial}{\partial r}r^2\frac{\partial}{\partial r} + \frac{1}{r^2}\Delta^*_\xi. \tag{0.25}$$

The Beltrami operator Δ^* represents the spherical (i.e., tangential) part of the Laplace operator Δ. Consequently, important potentials can be handled in a broad similarity to the Euclidean case:

(a) *Potential of a "volume":* Having a function F defined on a (regular) region $\Gamma \subset \Omega$, we are able to handle the integral

$$U(\xi) = \int_\Gamma F(\eta) G(\Delta^*; \xi \cdot \eta)\, d\omega(\eta), \quad \xi \in \Omega, \tag{0.26}$$

in parallel to the volume potential in $\mathbb{R}^3$. Obviously, the integral (0.26) is not a volume but a surface integral over $\Gamma \subset \Omega$. Nonetheless, it takes over the role of volume potentials known from the Euclidean setting.

$$G(\Delta^*;\xi\cdot\eta) = \frac{1}{4\pi}\ln(1-\xi\cdot\eta) + \frac{1}{4\pi}(1-\ln(2)), \quad \xi\in\Omega\setminus\{\eta\}, \tag{0.27}$$

denotes the *fundamental solution for the Beltrami operator* Δ^*, being the counterpart to the *fundamental solution* (0.6) *for the Laplace operator* Δ in $\mathbb{R}^3$, i.e.,

$$G(\Delta;|x-y|) = \frac{1}{4\pi}\frac{1}{|x-y|}, \quad x\in\mathbb{R}^3\backslash\{y\}. \tag{0.28}$$

From the integral representation (0.26) it can easily be derived that

$$\Delta^* U(\xi) = -\frac{1}{4\pi}\int_\Gamma F(\eta)\,d\omega(\eta), \quad \xi\in\Omega\backslash\overline{\Gamma}. \tag{0.29}$$

At this point, we already realize a major difference in comparison to the Euclidean concept: the potential (0.26) is not harmonic in the sense that a homogeneous Beltrami differential equation is satisfied in the exterior $\Gamma^c = \Omega\backslash\overline{\Gamma}$. Furthermore, the periodic structure of the sphere makes decay conditions at infinity obsolete (note that the exterior Γ^c again is a bounded (regular) region like Γ itself).

(b) *Potential of a single "surface" layer:* Letting $\partial\Gamma$ be the boundary of a (regular) region $\Gamma\subset\Omega$, the single-layer potential correspondingly is defined by

$$U(\xi) = \int_{\partial\Gamma} F(\eta)\; G(\Delta^*;\xi\cdot\eta)\; d\sigma(y), \quad \xi\in\Omega, \tag{0.30}$$

where $d\sigma$ denotes the line element (thus, we are concerned with a curve and not with a surface integral; nevertheless, U takes over the role of surface potentials in the Euclidean context). As in the case of (0.26), the single-layer potential is not harmonic (with respect to the Beltrami operator).

(c) *Potential of a double "surface" layer:* The double-layer potential is defined in the following way:

$$U(\xi) = \int_{\partial\Gamma} F(\eta)\; \frac{\partial}{\partial\nu(\eta)} G(\Delta^*;\xi\cdot\eta)\; d\sigma(y), \quad \xi\in\Omega. \tag{0.31}$$

Due to the additional normal derivative applied to the η-variable of the fundamental solution, this potential is actually harmonic (with respect to the Beltrami operator) in $\Omega\setminus\partial\Gamma$. More precisely, U satisfies the equation

$$\Delta^* U(\xi) = 0, \quad \xi\in\Omega\setminus\partial\Gamma. \tag{0.32}$$

Aside from this striking difference in comparison to the single-layer potential, the nature of (0.31) is analogous in its behavior to the corresponding potential in the Euclidean framework when approaching the boundary $\partial\Gamma$.

All in all, the spherical potentials have a somewhat more artificial origin than those in the Euclidean framework of $\mathbb{R}^3$, but they have the same important meaning for the solution of various problems arising in gravitation and geomagnetism. The major differences between the spatial and the spherical setting arise on the one hand from the reduced dimension (implying a logarithmic singularity of the fundamental solution of the Beltrami operator instead of a singularity of type $|\cdot|^{-1}$ for the Laplacian in $\mathbb{R}^3$), and on the other hand from specific characteristics of the sphere (leading to the non-harmonicity of the single-layer potential and the lack of decay conditions at infinity).

Layout

The geomathematically oriented background to be realized in this book demands a specific layout. The content is handled in parallel by a column-by-column subdivision into space and sphere, thereby documenting in a formal way that a variety of topics and features of the book can be treated in far-reaching similarity.

Content	**Space**	**Sphere**
Notational background	Chapter 1	Chapter 2
Conceptional background	Chapter 3	Chapter 6
Geomathematical methods and applications	Chapter 4 (Earth's gravitation) Chapter 5 (geomagnetism)	Chapter 7 (Earth's gravitation) Chapter 8 (geomagnetism)

More concretely, as visualized in our scheme, the column "Space" starts with the introduction of necessary notation and settings in the Euclidean space $\mathbb{R}^3$ (Chapter 1). It is followed by Chapter 3 dealing with the general potential theoretic concepts (existence/uniqueness for boundary-value problems, limit/jump relations for layer potentials, approximation methods) in $\mathbb{R}^3$. Their specific applications to Earth's gravitation and geomagnetism are elaborated in Chapters 4 and 5, respectively.

In close analogy, the column "Sphere" first provides the notational background and settings on the sphere (Chapter 2). Subsequently, Chapter 6 treats the corresponding general potential theoretic background. The specific applications in spherically reflected Earth's gravitation and geomagnetism are worked out in Chapters 7 and 8, respectively.

Part I

Preliminaries

1

Three-Dimensional Euclidean Space $\mathbb{R}^3$

CONTENTS

1.1 Basic Notation

The letters $\mathbb{N}$, $\mathbb{N}_0$, $\mathbb{Z}$, $\mathbb{R}$, and $\mathbb{C}$ denote the sets of positive integers, non-negative integers, integers, real numbers, and complex numbers, respectively.

Let us use $x, y, \ldots$ to represent the elements of Euclidean space $\mathbb{R}^3$. For all $x \in \mathbb{R}^3$, $x = (x_1, x_2, x_3)^T$, different from the origin, we have

$$x = r\xi, \qquad r = |x| = \sqrt{x_1^2 + x_2^2 + x_3^2}, \tag{1.1}$$

where $\xi = \frac{x}{|x|}$ is the uniquely determined directional unit vector of $x \in \mathbb{R}^3$. $\mathcal{B}_R(x)$ designates the *(open) ball in* $\mathbb{R}^3$ *with center* x *and radius* R:

$$\mathcal{B}_R(x) = \{y \in \mathbb{R}^3 : |x - y| < R\}. \tag{1.2}$$

The *sphere of radius* R *around* x (i.e., the boundary $\partial\mathcal{B}_R(x)$ of the ball $\mathcal{B}_R(x)$) is denoted by $\Omega_R(x)$:

$$\Omega_R(x) = \partial\mathcal{B}_R(x) = \{y \in \mathbb{R}^3 : |x - y| = R\}. \tag{1.3}$$

Throughout this work, the *unit sphere* $\Omega_1(0)$ *around the origin* is denoted simply by Ω, while Ω_R denotes the *sphere with radius* R *around the origin.*

1.1.1 Vectors and Tensors

If the vectors $\varepsilon^1 = (1,0,0)^T$, $\varepsilon^2 = (0,1,0)^T$, $\varepsilon^3 = (0,0,1)^T$ form the canonical orthonormal basis in $\mathbb{R}^3$, we may represent the points $x \in \mathbb{R}^3$ in Cartesian coordinates by

$$x = \sum_{i=1}^{3} (x \cdot \varepsilon^i)\varepsilon^i = \sum_{i=1}^{3} x_i \varepsilon^i. \tag{1.4}$$

The *inner (scalar), vector,* and *tensor (dyadic) products* of two elements $x, y \in \mathbb{R}^3$ are defined by

$$x \cdot y = x^{\mathrm{T}} y = \sum_{i=1}^{3} x_i y_i, \tag{1.5}$$

$$x \wedge y = (x_2 y_3 - x_3 y_2, x_3 y_1 - x_1 y_3, x_1 y_2 - x_2 y_1)^{\mathrm{T}}, \tag{1.6}$$

$$x \otimes y = x y^{\mathrm{T}} = \begin{pmatrix} x_1 y_1 & x_1 y_2 & x_1 y_3 \\ x_2 y_1 & x_2 y_2 & x_2 y_3 \\ x_3 y_1 & x_3 y_2 & x_3 y_3 \end{pmatrix}, \tag{1.7}$$

respectively. With the *alternator* (Levi–Cività alternating symbol)

$$\varepsilon^{ijk} = \begin{cases} 1 & \text{if } (i,j,k) \text{ is an even permutation of } (1,2,3), \\ -1 & \text{if } (i,j,k) \text{ is an odd permutation of } (1,2,3), \\ 0 & \text{if } (i,j,k) \text{ is not a permutation of } (1,2,3), \end{cases} \tag{1.8}$$

we obtain

$$(x \wedge y) \cdot \varepsilon^i = \sum_{j=1}^{3} \sum_{k=1}^{3} \varepsilon^{ijk} x_j y_k. \tag{1.9}$$

Moreover, we have $\sum_{i=1}^{3} \varepsilon^{ijk} \varepsilon^{ipq} = \delta_{j,p}\delta_{k,q} - \delta_{j,q}\delta_{k,p}$, where $\delta_{i,j}$ is the *Kronecker delta*:

$$\delta_{i,j} = \begin{cases} 0 & \text{if } i \neq j, \\ 1 & \text{if } i = j. \end{cases} \tag{1.10}$$

As usual, a *tensor* $\mathbf{x} \in \mathbb{R}^{3\times 3}$ *of rank 2* is understood to be a linear mapping that assigns to each $x \in \mathbb{R}^3$ a vector $y \in \mathbb{R}^3$: $y = \mathbf{x}x$. The (Cartesian) components $\mathbf{x}_{ij}$ of $\mathbf{x}$ are defined by $\mathbf{x}_{ij} = \varepsilon^i \cdot (\mathbf{x}\varepsilon^j)$, so that $y = \mathbf{x}x$ is equivalent to

$$y_i = y \cdot \varepsilon^i = \sum_{j=1}^{3} \mathbf{x}_{ij} (x \cdot \varepsilon^j) = \sum_{j=1}^{3} \mathbf{x}_{ij} x_j. \tag{1.11}$$

We write $\mathbf{x}^{\mathrm{T}}$ for the *transpose* of $\mathbf{x}$. It is the unique tensor satisfying $(\mathbf{x}y) \cdot x = y \cdot (\mathbf{x}^{\mathrm{T}} x)$ for all $x, y \in \mathbb{R}^3$. Moreover, we write $\mathrm{tr}(\mathbf{x})$ for the *trace* of $\mathbf{x}$ and $\det(\mathbf{x})$ for the *determinant* of $\mathbf{x}$. We call $\mathbf{x}$ *symmetric* if $\mathbf{x} = \mathbf{x}^{\mathrm{T}}$, and *skew* if $\mathbf{x} = -\mathbf{x}^{\mathrm{T}}$. Every tensor $\mathbf{x}$ admits the unique decomposition $\mathbf{x} = \mathrm{sym}(\mathbf{x}) + \mathrm{skw}(\mathbf{x})$, where $\mathrm{sym}(\mathbf{x})$ is symmetric and $\mathrm{skw}(\mathbf{x})$ is skew. More explicitly, $\mathrm{sym}(\mathbf{x}) = \frac{1}{2}\left(\mathbf{x} + \mathbf{x}^{\mathrm{T}}\right)$, $\mathrm{skw}(\mathbf{x}) = \frac{1}{2}\left(\mathbf{x} - \mathbf{x}^{\mathrm{T}}\right)$. We call $\mathrm{sym}(\mathbf{x})$ the symmetric part and $\mathrm{skw}(\mathbf{x})$ the skew part of $\mathbf{x}$.

The tensor (dyadic) product $x \otimes y$ of two elements $x, y \in \mathbb{R}^3$, as given in (1.7), is the tensor that assigns to each $z \in \mathbb{R}^3$ the vector $(y \cdot z)x$. More explicitly, for every $z \in \mathbb{R}^3$,

$$(x \otimes y)z = (y \cdot z)x. \tag{1.12}$$

The *inner product* $\mathbf{x} \cdot \mathbf{y}$ of two tensors $\mathbf{x}, \mathbf{y} \in \mathbb{R}^{3\times 3}$ of rank 2 is defined by

$$\mathbf{x} \cdot \mathbf{y} = \mathrm{tr}(\mathbf{x}^{\mathrm{T}}\mathbf{y}) = \sum_{i=1}^{3}\sum_{j=1}^{3} \mathbf{x}_{ij}\mathbf{y}_{ij}, \tag{1.13}$$

while $|\mathbf{x}| = (\mathbf{x} \cdot \mathbf{x})^{\frac{1}{2}}$ is called the *norm* of $\mathbf{x} \in \mathbb{R}^{3\times 3}$. Given any tensor $\mathbf{x}$ and any pair x, $y \in \mathbb{R}^3$, we have $x \cdot (\mathbf{x}y) = \mathbf{x} \cdot (x \otimes y)$ and $(x \otimes y)\mathbf{x} = x \otimes (\mathbf{x}^{\mathrm{T}}y)$. For $x, y, w, z \in \mathbb{R}^3$, we get $(x \otimes y)(w \otimes z) = (y \cdot w)(x \otimes z)$. It is easy to see that

$$(\varepsilon^i \otimes \varepsilon^j) \cdot (\varepsilon^k \otimes \varepsilon^l) = \delta_{i,k}\, \delta_{j,l}, \tag{1.14}$$

so that the nine tensors $\varepsilon^i \otimes \varepsilon^j$, $i, j = 1, 2, 3$, are orthonormal. Moreover, it follows that $\mathbf{x} \in \mathbb{R}^{3\times 3}$ is representable in the form

$$\mathbf{x} = \sum_{i=1}^{3}\sum_{j=1}^{3} \mathbf{x}_{ij}(\varepsilon^i \otimes \varepsilon^j). \tag{1.15}$$

In particular, the identity tensor $\mathbf{i}$ is given by $\mathbf{i} = \sum_{i=1}^{3} \varepsilon^i \otimes \varepsilon^i$. Furthermore, it is not hard to see that

$$\mathrm{tr}(x \otimes y) = x \cdot y \tag{1.16}$$

for $x, y \in \mathbb{R}^3$ and

$$\mathbf{x} \cdot (\mathbf{y}\mathbf{z}) = \left(\mathbf{y}^{\mathrm{T}}\mathbf{x}\right) \cdot \mathbf{z} = \left(\mathbf{x}\mathbf{z}^{\mathrm{T}}\right) \cdot \mathbf{y} \tag{1.17}$$

for $\mathbf{x}, \mathbf{y}, \mathbf{z} \in \mathbb{R}^{3\times 3}$.

1.1.2 Differential Operators and Function Spaces

If $\mathcal{G}$ is a set of points in $\mathbb{R}^3$, then $\partial\mathcal{G}$ denotes its *boundary*. The set $\overline{\mathcal{G}} = \mathcal{G} \cup \partial\mathcal{G}$ is called the *closure* of $\mathcal{G}$. The *complement* of $\overline{\mathcal{G}}$ in $\mathbb{R}^3$ is denoted by $\mathcal{G}^c$. A set $\mathcal{G} \subset \mathbb{R}^3$ is called a *region* if and only if it is open and connected. An open set $\mathcal{A} \subset \mathbb{R}^3$ is said to be *totally contained* in the open set $\mathcal{G} \subset \mathbb{R}^3$ (in brief, $\mathcal{A} \Subset \mathcal{G}$), if $\mathcal{A} \subset \mathcal{G}$ and $\mathrm{dist}(\overline{\mathcal{A}}, \partial\mathcal{G}) > 0$. An open set $\mathcal{A} \subset \mathbb{R}^3$ is said to be *compactly contained* in the open set $\mathcal{G} \subset \mathbb{R}^3$, if $\mathcal{A}$ is bounded and $\mathcal{A} \Subset \mathcal{G}$.

Concerning the general nomenclature, we use the following scheme for functions:

upper-case letters F, G:	scalar functions,
lower-case letters f, g:	vector fields,
lower-case boldface letters $\mathbf{f}, \mathbf{g}$:	tensor fields (of rank 2).

By a scalar, vector, or tensor function (field) on a region $\mathcal{G} \subset \mathbb{R}^3$ we mean a function that assigns to each point of $\mathcal{G}$ a scalar, vector, or tensor function value, respectively. Unless otherwise specified, all fields are assumed to be real valued throughout this book. The restriction of a scalar-valued function F, vector-valued function f, or tensor-valued function $\mathbf{f}$ to a subset M of its domain is denoted by $F|_M$, $f|_M$, or $\mathbf{f}|_M$, respectively. For a set L of functions, we define $\mathrm{L}|_M = \{F|_M : F \in \mathrm{L}\}$. We call

$$\nabla F(x) = \left(\frac{\partial}{\partial x_1} F(x), \frac{\partial}{\partial x_2} F(x), \frac{\partial}{\partial x_3} F(x) \right)^T \tag{1.18}$$

the *gradient* of a function $F : \mathcal{G} \to \mathbb{R}$. The *partial derivatives* of F at $x \in \mathcal{G}$ are designated by

$$\frac{\partial}{\partial x_i} F(x) = F_{|i}(x) = (\nabla F(x)) \cdot \varepsilon^i, \quad i = 1, 2, 3. \tag{1.19}$$

The scalar function $F : \mathcal{G} \to \mathbb{R}$, the vector function $f : \mathcal{G} \to \mathbb{R}^3$, and the tensor function $\mathbf{f} : \mathcal{G} \to \mathbb{R}^{3\times 3}$ are members of class $\mathrm{C}^{(1)}(\mathcal{G})$, $\mathrm{c}^{(1)}(\mathcal{G})$, and $\mathbf{c}^{(1)}(\mathcal{G})$, respectively, if $F, f, \mathbf{f}$, are differentiable at every point x of $\mathcal{G}$ and $\nabla F, \nabla f, \nabla \mathbf{f}$ are continuous on $\mathcal{G}$. The n-fold application of the gradient to F, f, $\mathbf{f}$ is denoted by $\nabla^{(n)} F$, $\nabla^{(n)} f$, $\nabla^{(n)} \mathbf{f}$, respectively (where $n \in \mathbb{N}_0$). We say that $F, f, \mathbf{f}$ are of class $\mathrm{C}^{(n)}(\mathcal{G}), \mathrm{c}^{(n)}(\mathcal{G}), \mathbf{c}^{(n)}(\mathcal{G})$, respectively, if $\nabla^{(n)} F$, $\nabla^{(n)} f$,$\nabla^{(n)} \mathbf{f}$ exist and are continuous in $\mathcal{G}$. Obviously, the gradient of a differentiable scalar field is a vector field, while the gradient of a differentiable vector field is a tensor field (of rank 2), the gradient of a tensor field (of rank 2) is a tensor field (of rank 3), etc. Instead of ∇f for a vector field f, we often use the equivalent notation

$$\nabla f = \nabla \otimes f \tag{1.20}$$

to indicate that the gradient of a vector field is nothing more than the tensor product of a vector field with the vectorial gradient operator. Moreover, for twice differentiable scalar fields F, we get

$$\nabla^{(2)} f = \nabla \otimes \nabla F. \tag{1.21}$$

We say that F is of class $\mathrm{C}^{(n)}(\overline{\mathcal{G}})$ if F is of class $\mathrm{C}^{(n)}(\mathcal{G})$ and, for each $k = 0, \ldots, n$, the k-th derivative $\nabla^{(k)} F$ has a continuous extension to $\overline{\mathcal{G}}$. Now, let $\mathcal{G}$ be a bounded subregion of $\mathbb{R}^3$. A function F on $\overline{\mathcal{G}}$ is said to be *μ-Hölder continuous* on $\overline{\mathcal{G}}$, if there exists a constant $C > 0$ such that

$$|F(x) - F(y)| \leq C\, |x - y|^\mu \tag{1.22}$$

holds for all $x, y \in \overline{\mathcal{G}}$. By $\mathrm{C}^{(0,\mu)}(\overline{\mathcal{G}})$ we denote the space of all μ-Hölder continuous functions on $\overline{\mathcal{G}}$. If $\mathcal{G}$ is unbounded, then we mean by $F \in \mathrm{C}^{(0,\mu)}(\overline{\mathcal{G}})$ that F is bounded and satisfies the inequality (1.22). Clearly, if F is of class

$\mathrm{C}^{(0,\mu)}(\overline{\mathcal{G}}), \mu \in (0,1]$, then F is uniformly continuous on $\overline{\mathcal{G}}$. The Hölder space $\mathrm{C}^{(0,\mu)}(\overline{\mathcal{G}})$ is a normed space equipped with

$$\|F\|_{\mathrm{C}^{(0)}(\overline{\mathcal{G}})} = \sup_{x\in\overline{\mathcal{G}}} |F(x)| \tag{1.23}$$

and a Banach space equipped with

$$\|F\|_{\mathrm{C}^{(0,\mu)}(\overline{\mathcal{G}})} = \sup_{x\in\overline{\mathcal{G}}} |F(x)| + \sup_{\substack{x,y\in\overline{\mathcal{G}}\\ x\neq y}} \frac{|F(x)-F(y)|}{|x-y|^{\mu}}. \tag{1.24}$$

$F \in \mathrm{C}^{(n,\mu)}(\overline{\mathcal{G}})$ means that F is n-times differentiable in $\overline{\mathcal{G}}$ and the n-th derivative is a member of the space $\mathrm{C}^{(0,\mu)}(\overline{\mathcal{G}})$. In the same way, definitions can be given for the vectorial and tensorial cases. A field $F : \mathcal{G} \to \mathbb{R}$ is *analytic* on $\mathcal{G}$ if, given any point $x \in \mathcal{G}$, F can be represented by a power series in some neighborhood of x. Of course, if F is analytic in $\mathcal{G}$, then F is of class $\mathrm{C}^{(\infty)}(\mathcal{G})$.

The space of scalar functions $F : \mathcal{G} \to \mathbb{R}$ that satisfy

$$\|F\|_{\mathrm{L}^p(\mathcal{G})} = \left(\int_{\mathcal{G}} |F(y)|^p dV(y)\right)^{\frac{1}{p}} < \infty, \tag{1.25}$$

is denoted by $\mathrm{L}^p(\mathcal{G})$, $1 \leq p < \infty$. Together with the norm $\|\cdot\|_{\mathrm{L}^p(\mathcal{G})}$, these spaces form Banach spaces. Even more, for the special case $p = 2$, $\mathrm{L}^2(\mathcal{G})$ denotes a Hilbert space with the inner product

$$(F,G)_{\mathrm{L}^2(\mathcal{G})} = \int_{\mathcal{G}} F(y)G(y)dV(y), \tag{1.26}$$

for $F, G \in \mathrm{L}^2(\mathcal{G})$. The vector and tensor spaces $\mathrm{l}^p(\mathcal{G})$ and $\mathbf{l}^p(\mathcal{G})$, $1 \leq p < \infty$, respectively, can be defined in analogy to the scalar case.

Let $f : \mathcal{G} \to \mathbb{R}^3$ be a vector field, and suppose that f is differentiable at a point $x \in \mathcal{G}$. Then the *divergence* of f at $x \in \mathcal{G}$ is the scalar value

$$\nabla \cdot f(x) = \mathrm{tr}(\nabla f)(x) = \sum_{i=1}^{3} \frac{\partial}{\partial x_i} f_i(x), \tag{1.27}$$

assuming that $f = \sum_{i=1}^3 f_i \varepsilon^i$. The partial derivatives of f at $x \in \mathcal{G}$ are given by

$$\frac{\partial}{\partial x_j} f_i(x) = f_{i|j}(x) = \varepsilon^i \cdot \left((\nabla f(x))\varepsilon^j\right), \quad i,j = 1,2,3. \tag{1.28}$$

The *curl* of f at $x \in \mathcal{G}$, denoted by $\nabla \wedge f(x)$, is defined via

$$(\nabla \wedge f(x)) \cdot \varepsilon^i = \sum_{j=1}^{3}\sum_{k=1}^{3} \varepsilon^{ijk} \frac{\partial}{\partial x_j} f_k(x), \tag{1.29}$$

for $i = 1, 2, 3$. Furthermore, the curl of f in x is the unique vector with the

Symbol	Differential Operator
∇	Gradient
L	Curl gradient
$\nabla\wedge$	Curl
$\nabla\cdot$	Divergence
L$\cdot$	Curl divergence
Δ	Laplace operator

TABLE 1.1
Important differential operators.

property $\left(\nabla_x f(x) - (\nabla_x f(x))^{\mathrm{T}}\right)y = (\nabla_x \wedge f(x)) \wedge y$, for every $y \in \mathbb{R}^3$. The so-called *curl gradient* L is the operator acting on differentiable scalar fields $F : \mathcal{G} \to \mathbb{R}$ via

$$\mathrm{L}_x F(x) = x \wedge \nabla_x F(x). \tag{1.30}$$

Let $\mathbf{f} : \mathcal{G} \to \mathbb{R}^{3\times 3}$ be a tensor field of second order, and suppose that $\mathbf{f}$ is differentiable at $x \in \mathcal{G}$. Then the tensor field $\mathbf{f}^{\mathrm{T}} : x \mapsto (\mathbf{f}(x))^{\mathrm{T}}$, $x \in \mathcal{G}$, is also differentiable at the point $x \in \mathcal{G}$. The *divergence* of $\mathbf{f}$ at x, denoted by $\nabla \cdot \mathbf{f}(x)$, is the unique vector with the property

$$(\nabla_x \cdot \mathbf{f}(x)) \cdot y = \nabla_x \cdot \left(\mathbf{f}^{\mathrm{T}}(x) y\right) \tag{1.31}$$

for every vector $y \in \mathbb{R}^3$. In other words, the divergence of a tensor field can be regarded as the row-wise application of the already known divergence for vector fields. In the same manner, we define the *curl* of $\mathbf{f}$ at x, denoted by $\nabla \wedge \mathbf{f}(x)$, to be the unique tensor with the property

$$(\nabla_x \wedge \mathbf{f}(x))\, y = \nabla_x \wedge \left(\mathbf{f}^{\mathrm{T}}(x) y\right) \tag{1.32}$$

for every vector $y \in \mathbb{R}^3$. The *partial derivatives* of $\mathbf{f}$ at $x \in \mathcal{G}$ are given by

$$\frac{\partial}{\partial x_k} \mathbf{f}_{ij}(x) = \mathbf{f}_{ij|k}(x) = \varepsilon^i \cdot \left((\nabla \mathbf{f}(x))\, \varepsilon^k \right) \varepsilon^j \quad i,j,k = 1,2,3. \tag{1.33}$$

Let $F : \mathcal{G} \to \mathbb{R}$ be a twice differentiable scalar field. Then we introduce the *Laplace operator (Laplacian)* of F at $x \in \mathcal{G}$ by

$$\Delta F(x) = \nabla \cdot (\nabla F(x)) = \sum_{i=1}^{3} \left(\frac{\partial}{\partial x_i} \right)^2 F(x). \tag{1.34}$$

Analogously, we define the Laplacian of a twice differentiable vector field $f : \mathcal{G} \to \mathbb{R}^3$ as the uniquely determined vector field satisfying $(\Delta_x f(x)) \cdot y = \Delta_x (f(x) \cdot y)$ for any $y \in \mathbb{R}^3$. Finally, the Laplacian $\Delta \mathbf{f}(x)$ of a twice differentiable tensor field $\mathbf{f}$ is the unique tensor (of rank 2) with the property

$(\Delta_x \mathbf{f}(x))\, y = \Delta_x\, (\mathbf{f}(x)y)$ for any $y \in \mathbb{R}^3$. In other words, the Laplace operator applied to vector and tensor fields is nothing more than the componentwise application of the already known Laplace operator for scalar fields (see Table 1.1 for a list of differential operators).

The Laplacian Δ is invariant under orthogonal transformations. More concretely, if $\mathbf{t} \in \mathbb{R}^{3\times 3}$ is an orthogonal matrix with $\det(\mathbf{t}) = 1$ (i.e., $\mathbf{t}$ is a rotation matrix) and $y = \mathbf{t}(x-a)$, $a \in \mathbb{R}^3$ fixed, then we formally have

$$\Delta_x = \sum_{i=1}^{3} \frac{\partial^2}{\partial x_i^2} = \sum_{i=1}^{3} \frac{\partial^2}{\partial y_i^2} = \Delta_y.$$

This property is called *spherical symmetry of the Laplacian.*

Let $\mathcal{G} \subset \mathbb{R}^3$ be an open set, $a \in \mathbb{R}^3 \backslash \{0\}$, and $U : \mathcal{G} \to \mathbb{R}$. The limit $\lim_{\delta\to 0+} \frac{1}{\delta}(U(x+\delta a) - U(x))$ (if it exists) is called the *directional derivative* of U with respect to a at $x \in \mathcal{G}$. It is designated by $\frac{\partial U}{\partial a}$. If U is differentiable in $x \in \mathcal{G}$, then the directional derivative exists for all $a \in \mathbb{R}^3\backslash\{0\}$, and we have

$$\frac{\partial U}{\partial a}(x) = \sum_{i=1}^{3} a_i \frac{\partial U}{\partial x_i}(x) = a \cdot \nabla_x U(x). \tag{1.35}$$

In particular, if $x \in \mathbb{R}^3\backslash\{0\}$ and $a = \xi = \frac{x}{|x|}$, then

$$\frac{\partial U}{\partial \xi}(x) = \frac{x}{|x|} \cdot \nabla_x U(x) \tag{1.36}$$

is called the *radial derivative* in $x \in \mathbb{R}^3$. By convention, we let

$$\frac{\partial}{\partial r} = \xi \cdot \nabla_x = \frac{x}{|x|} \cdot \nabla_x. \tag{1.37}$$

Next we are interested in introducing the *Beltrami operator* in Cartesian coordinates of $\mathbb{R}^3$. For that purpose we start from

$$\begin{aligned} \left(x_k \frac{\partial}{\partial x_i} - x_i \frac{\partial}{\partial x_k}\right)^2 &= \left(x_k \frac{\partial}{\partial x_i} - x_i \frac{\partial}{\partial x_k}\right)\left(x_k \frac{\partial}{\partial x_i} - x_i \frac{\partial}{\partial x_k}\right) \\ &= x_k \delta_{i,k} \frac{\partial}{\partial x_i} + x_k^2 \left(\frac{\partial}{\partial x_i}\right)^2 - x_i \frac{\partial}{\partial x_i} - x_i x_k \frac{\partial^2}{\partial x_k \partial x_i} \\ &\quad + x_i \delta_{i,k} \frac{\partial}{\partial x_k} + x_i^2 \left(\frac{\partial}{\partial x_k}\right)^2 - x_k \frac{\partial}{\partial x_k} - x_i x_k \frac{\partial^2}{\partial x_i \partial x_k}. \end{aligned} \tag{1.38}$$

Introducing polar coordinates $x = r\xi$, $r = |x|$, $\xi = \frac{x}{|x|}$, $x \neq 0$, we see that

$$\left(\frac{\partial}{\partial r}\right)^2 = \left(\frac{x}{|x|} \cdot \nabla\right)\left(\frac{x}{|x|} \cdot \nabla\right) = \frac{x}{|x|} \cdot (\nabla \otimes \nabla)\, \frac{x}{|x|} = \sum_{i=1}^{3}\sum_{k=1}^{3} \frac{x_i x_k}{|x|^2} \frac{\partial^2}{\partial x_i \partial x_k}. \tag{1.39}$$

An easy calculation yields the identity

$$\sum_{i=1}^{3}\sum_{k=1}^{3}\left(x_k\frac{\partial}{\partial x_i}-x_i\frac{\partial}{\partial x_k}\right)^2 = 2r^2\Delta_x - 4r\frac{\partial}{\partial r} - 2r^2\left(\frac{\partial}{\partial r}\right)^2. \tag{1.40}$$

For $x \neq 0$, the Laplace operator therefore admits the representation

$$\begin{aligned}\Delta_x &= \frac{x}{|x|}\cdot(\nabla\otimes\nabla)\,\frac{x}{|x|}+\frac{2}{|x|}\frac{x}{|x|}\cdot\nabla+\frac{1}{2|x|^2}\sum_{i=1}^{3}\sum_{k=1}^{3}\left(x_k\frac{\partial}{\partial x_i}-x_i\frac{\partial}{\partial x_k}\right)^2\\ &= \left(\frac{\partial}{\partial r}\right)^2+\frac{2}{r}\frac{\partial}{\partial r}+\frac{1}{2r^2}\sum_{i=1}^{3}\sum_{k=1}^{3}\left(x_k\frac{\partial}{\partial x_i}-x_i\frac{\partial}{\partial x_k}\right)^2. \end{aligned}\tag{1.41}$$

Definition 1.1. The operator

$$\Delta^* = \frac{1}{2}\sum_{i=1}^{3}\sum_{k=1}^{3}\left(x_k\frac{\partial}{\partial x_i}-x_i\frac{\partial}{\partial x_k}\right)^2 \tag{1.42}$$

is called the *Beltrami operator in* $\mathbb{R}^3$.

Remark 1.2. If $F \in \mathrm{C}^{(1)}((0,R))$, $R>0$, then

$$\left(x_k\frac{\partial}{\partial x_i}-x_i\frac{\partial}{\partial x_k}\right)F(|x|)=0 \tag{1.43}$$

is valid for all $x \in \mathcal{B}_R(0)\backslash\{0\}$ (cf. Exercise 1.5). Consequently, for $F \in \mathrm{C}^{(2)}((0,R))$, $R>0$,

$$\Delta_x F(|x|) = F''(|x|)+\frac{2}{|x|}F'(|x|), \tag{1.44}$$

where F' and F'' denote the first- and second-order derivative of F. If $F \in \mathrm{C}^{(2)}\,(\mathcal{B}_R(0)\backslash\{0\})$ with $F(x) = F(\frac{x}{|x|})$ for all $x \in \mathcal{B}_R(0)\backslash\{0\}$, then (cf. (1.37))

$$\frac{\partial}{\partial r}F(x)=\left(\frac{x}{|x|}\cdot\nabla_x\right)F(x)=0, \tag{1.45}$$

and we have

$$\Delta_x F(x)=\frac{1}{|x|^2}\Delta_x^*F(x)=\frac{1}{2|x|^2}\sum_{i=1}^{3}\sum_{k=1}^{3}\left(x_k\frac{\partial}{\partial x_i}-x_i\frac{\partial}{\partial x_k}\right)^2F(x). \tag{1.46}$$

Remark 1.3. The Beltrami operator Δ_x^* can be formally represented as the square of the vector product of x and ∇_x, i.e.,

$$\Delta_x^* = (x\wedge\nabla_x)\cdot(x\wedge\nabla_x) = \mathrm{L}_x\cdot\mathrm{L}_x. \tag{1.47}$$

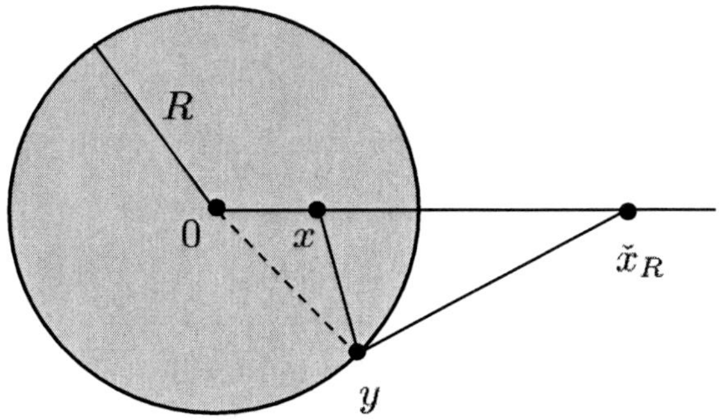

FIGURE 1.1
The inversion $x \mapsto \breve{x}_R$ with respect to the sphere Ω_R.

1.1.3 Kelvin Transform

The mapping

$$x \mapsto \breve{x}_R = \left(\frac{R}{|x|}\right)^2 x, \qquad x \in \overline{\mathcal{B}_R(0)}\backslash\{0\}, \tag{1.48}$$

transforms $\mathcal{B}_R(0)\backslash\{0\}$ into $\mathbb{R}^3\backslash\overline{\mathcal{B}_R(0)}$ and $\Omega_R = \partial\mathcal{B}_R(0)$ onto itself. Referring to Figure 1.1, we observe that the two triangles with edges $(\breve{x}_R, y, 0)$ and $(x, y, 0)$ are similar whenever $y \in \Omega_R$. Furthermore, the ratios $\frac{|x|}{|y|}$ and $\frac{|y|}{|\breve{x}_R|}$ are equal, provided that $y \in \partial\Omega_R$.

Simple calculations yield that, on the one hand, for $x = |x|\xi$, $\xi \in \Omega$, and $y = |y|\eta$, $R = |y|$, $\eta \in \Omega$, we have

$$|x-y|^2 = x^2 + y^2 - 2x\cdot y = |x|^2 + R^2 - 2|x|R\ \xi\cdot\eta. \tag{1.49}$$

On the other hand, we see that

$$\begin{aligned}\left(\frac{|x|}{R}\right)^2 \left|\frac{R^2}{|x|^2}x - y\right|^2 &= \frac{|x|^2}{R^2}\left(\frac{R^4}{|x|^4}|x|^2 + R^2 - 2\frac{R^2}{|x|^2}x\cdot y\right) \\ &= |x|^2 + R^2 - 2|x|R\ \xi\cdot\eta.\end{aligned} \tag{1.50}$$

As a consequence, we get

Lemma 1.4. *For all* $y \in \Omega_R$ *and* $x \in \mathcal{B}_R(0)$,

$$|x-y| = \frac{|x|}{R}|\breve{x}_R - y|. \tag{1.51}$$

After these preparations concerning the inversion of points with respect to a sphere Ω_R, $R > 0$, we discuss the *Kelvin transform*. For simplicity, we first choose $R = 1$, i.e., we restrict the inversion to the unit sphere Ω.

Theorem 1.5. *Assume that U is of class $\mathrm{C}^{(2)}(\mathcal{G})$, $\mathcal{G} \subset \mathbb{R}^3\backslash\{0\}$ open. Let $\breve{\mathcal{G}}$ be the image of $\mathcal{G}$ under the inversion $x \mapsto \breve{x} = |x|^{-2}x$, and denote by $\breve{U} = \mathcal{K}[U] : \breve{\mathcal{G}} \to \mathbb{R}$, with*

$$\breve{U}(x) = \mathcal{K}[U](x) = \frac{1}{|x|} U(\breve{x}), \tag{1.52}$$

the Kelvin transform of U. *Then*

$$\Delta \breve{U}(x) = \frac{1}{|x|^5} \Delta U(\breve{x}), \quad x \in \breve{\mathcal{G}}. \tag{1.53}$$

Proof. From $|\breve{x}| = \breve{r}$ and $|x| = r$ it follows that $r\breve{r} = 1$. It is not hard to deduce that

$$\begin{aligned} \frac{\partial U}{\partial r}(\breve{x}) &= \sum_{j=1}^{3} \frac{\partial U}{\partial \breve{x}_j}\left(\frac{x}{|x|^2}\right) \frac{\partial}{\partial r}\left(\frac{x_j}{|x|^2}\right) \\ &= \sum_{j=1}^{3} \frac{\partial U}{\partial \breve{x}_j}\left(\frac{x}{|x|^2}\right) \frac{x_j}{|x|} \frac{\partial}{\partial r}\frac{1}{r} \\ &= -\frac{1}{r^2}\frac{\partial}{\partial \breve{r}} U(\breve{x}). \end{aligned} \tag{1.54}$$

Hence we find

$$\frac{\partial}{\partial r}\breve{U}(x) = \frac{\partial}{\partial r}\frac{1}{r} U\left(\frac{x}{|x|^2}\right) = -\frac{1}{r^2}U(\breve{x}) - \frac{1}{r^3}\frac{\partial}{\partial \breve{r}}U(\breve{x}) \tag{1.55}$$

and

$$\begin{aligned} \frac{\partial^2}{\partial r^2}\breve{U}(x) &= 2\frac{1}{r^3}U(\breve{x}) + \frac{1}{r^4}\frac{\partial}{\partial \breve{r}}U(\breve{x}) \\ &\quad + 3\frac{1}{r^4}\frac{\partial}{\partial \breve{r}}U(\breve{x}) + \frac{1}{r^5}\frac{\partial^2}{\partial \breve{r}^2}U(\breve{x}). \end{aligned} \tag{1.56}$$

Furthermore we obtain

$$\begin{aligned} \left(x_k \frac{\partial}{\partial x_i} - x_i \frac{\partial}{\partial x_k}\right) U(\breve{x}) &= \sum_{j=1}^{3} \frac{\partial U}{\partial \breve{x}_j}(\breve{x}) \left(x_k \frac{\partial}{\partial x_i} - x_i \frac{\partial}{\partial x_k}\right) \breve{x}_j \\ &= \sum_{j=1}^{3} \frac{\partial U}{\partial \breve{x}_j}(\breve{x}) \frac{1}{|x|^2} \left(x_k \frac{\partial}{\partial x_i} - x_i \frac{\partial}{\partial x_k}\right) x_j \\ &= \left(\breve{x}_k \frac{\partial}{\partial \breve{x}_i} - \breve{x}_i \frac{\partial}{\partial \breve{x}_k}\right) U(\breve{x}). \end{aligned} \tag{1.57}$$

Therefore we get

$$\left(x_k \frac{\partial}{\partial x_i} - x_i \frac{\partial}{\partial x_k}\right) \breve{U}(x) = \frac{1}{|x|} \left(\breve{x}_k \frac{\partial}{\partial \breve{x}_i} - \breve{x}_i \frac{\partial}{\partial \breve{x}_k}\right) U(\breve{x}) \tag{1.58}$$

such that

$$\begin{aligned}
&\left(x_k \frac{\partial}{\partial x_i} - x_i \frac{\partial}{\partial x_k}\right)^2 \breve{U}(x) \qquad (1.59)\\
&= \left(x_k \frac{\partial}{\partial x_i} - x_i \frac{\partial}{\partial x_k}\right)\left(\frac{1}{|x|}\left(\breve{x}_k \frac{\partial}{\partial \breve{x}_i} - \breve{x}_i \frac{\partial}{\partial \breve{x}_k}\right) U(\breve{x})\right)\\
&= \frac{1}{|x|}\left(\breve{x}_k \frac{\partial}{\partial \breve{x}_i} - \breve{x}_i \frac{\partial}{\partial \breve{x}_k}\right)^2 U(\breve{x})\\
&= \frac{1}{|x|^3}\frac{1}{|\breve{x}|^2}\left(\breve{x}_k \frac{\partial}{\partial \breve{x}_i} - \breve{x}_i \frac{\partial}{\partial \breve{x}_k}\right)^2 U(\breve{x})
\end{aligned}$$

(note that $|\breve{x}||x| = 1$). Observing the representation (1.41) of the Laplace operator we see that

$$\begin{aligned}
&\Delta_x \breve{U}(x) \qquad (1.60)\\
&= \frac{\partial^2}{\partial r^2}\breve{U}(x) + \frac{2}{r}\frac{\partial}{\partial r}\breve{U}(x) + \frac{1}{2r^2}\sum_{i=1}^{3}\sum_{k=1}^{3}\left(x_k \frac{\partial}{\partial x_i} - x_i \frac{\partial}{\partial x_k}\right)^2 \breve{U}(x)\\
&= \frac{1}{r^5}\left(\frac{\partial^2}{\partial \breve{r}^2}U(\breve{x}) + \frac{2}{\breve{r}}\frac{\partial}{\partial \breve{r}}U(\breve{x}) + \frac{1}{2\breve{r}^2}\sum_{i=1}^{3}\sum_{k=1}^{3}\left(\breve{x}_k \frac{\partial}{\partial \breve{x}_i} - \breve{x}_i \frac{\partial}{\partial \breve{x}_k}\right)^2 U(\breve{x})\right).
\end{aligned}$$

Thus we finally arrive at the formula

$$\Delta \breve{U}(x) = \frac{1}{|x|^5}\Delta_{\breve{x}} U(\breve{x}) = \frac{1}{|x|^5}(\Delta U)\left(\frac{x}{|x|^2}\right), \qquad (1.61)$$

which completes the proof. □

Corollary 1.6. *Assume that U is of class $\mathrm{C}^{(2)}(\mathcal{G}), \mathcal{G} \subset \mathbb{R}^3\backslash\{0\}$ open. Let $\breve{\mathcal{G}}_R$ be the image of $\mathcal{G}$ under the inversion $x \mapsto \breve{x}_R = R^2|x|^{-2}x$, and denote by $\breve{U} = \mathcal{K}^R[U]\colon \breve{\mathcal{G}}_R \to \mathbb{R}$, with*

$$\breve{U}(x) = \mathcal{K}^R[U](x) = \frac{R}{|x|}\, U\left(\breve{x}_R\right), \qquad (1.62)$$

the Kelvin transform of U with respect to $\Omega_R = \partial\mathcal{B}_R(0)$. Then

$$\Delta \breve{U}(x) = \left(\frac{R}{|x|}\right)^5 \Delta U\left(\breve{x}_R\right), \quad x \in \breve{\mathcal{G}}_R. \qquad (1.63)$$

Proof. We let $W(\breve{x}) = \frac{1}{|\breve{x}|}U\left(\frac{\breve{x}}{|\breve{x}|}\right)$. Then we have $\breve{U}(x) = \frac{1}{R}W\left(\frac{x}{R^2}\right)$, such that

$$\Delta \breve{U}(x) = \frac{1}{R^5}\Delta W\left(\frac{x}{R^2}\right) = \frac{1}{R^5}\left|\frac{x}{R^2}\right|^{-5}\Delta U\left(\left(\frac{R}{|x|}\right)^2 x\right) \qquad (1.64)$$

follows by standard manipulations. □

1.2 Integral Theorems

We begin with the definition of a *regular region* in the three-dimensional Euclidean space $\mathbb{R}^3$. Its geometric character is of basic significance in our geomathematically oriented potential theory. The most typical regions in our approach are, e.g., ball, ellipsoid, geoid, (actual) real Earth.

1.2.1 Regular Region

Our considerations start with the definition of a regular region in the Euclidean space $\mathbb{R}^3$.

Definition 1.7. A region $\mathcal{G} \subset \mathbb{R}^3$ is called *regular* if it satisfies the following properties:

(i) $\partial\mathcal{G}$ divides the three-dimensional Euclidean space $\mathbb{R}^3$ into the bounded, simply connected region $\mathcal{G}$ *(inner space)* and the unbounded region $\mathcal{G}^c$ *(outer space)* defined by $\mathcal{G}^c = \mathbb{R}^3 \backslash \overline{\mathcal{G}}$,

(ii) $\mathcal{G}$ contains the origin,

(iii) $\partial\mathcal{G}$ is a closed and compact surface free of double points,

(iv) there exist (universal) constants $M, \delta > 0$ such that, for every $x \in \partial\mathcal{G}$, the set $\partial\mathcal{G} \cap \mathcal{B}_\delta(x)$ can be represented in a local *(tangential-normal) coordinate system* of Cartesian coordinates associated with x (cf. Figure 1.2). More precisely, any $y \in \partial\mathcal{G} \cap \mathcal{B}_\delta(x)$ can be represented in the form $(u_1, u_2, u_3)^T$ with

$$u_3 = F(u), \quad u = (u_1, u_2)^T \in \mathcal{U}, \tag{1.65}$$

where $\mathcal{U}$ is a subset of the tangent plane of $\partial\mathcal{G}$ at x (i.e., the ε^3-axis coincides with the unit normal vector $\nu(x)$ in $x \in \partial\mathcal{G}$, pointing into the outer space $\mathcal{G}^c = \mathbb{R}^3 \backslash \overline{\mathcal{G}}$) and $F : \mathcal{U} \to \mathbb{R}$ is a twice continuously differentiable function satisfying

$$|F(u)| \leq M|u|^2, \tag{1.66}$$

$$|\nabla F(u)| \leq M|u|, \tag{1.67}$$

for all $u = (u_1, u_2)^T \in \mathcal{U}$.

In the tangential-normal system, the point $x \in \partial\mathcal{G}$ represents the origin, i.e., it has the coordinates $(0,0,0)^T$, so that $F(0,0) = 0$. Furthermore, by rotating the coordinate system, the plane $u_3 = 0$ can be transferred into the tangential plane at x such that the first partial derivatives of F vanish at the origin. Via Taylor's formula, this observation explains the estimates (1.66) and (1.67).

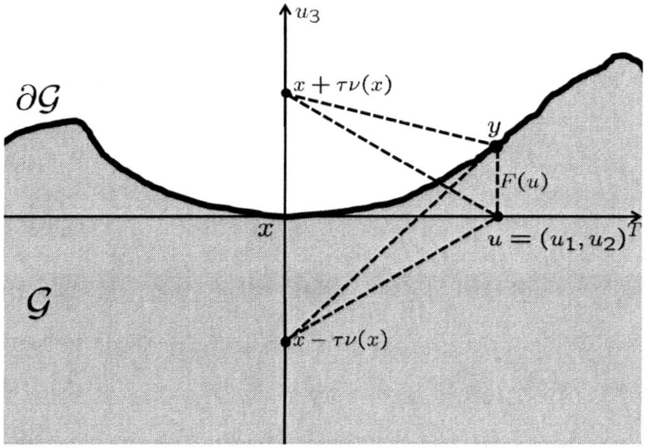

FIGURE 1.2
Sectional representation of a tangential-normal system (with $\nu(x) = \varepsilon^3$) illustrating the geometric relations (1.68).

1.2.2 Basic Estimates

Let $\mathcal{G} \subset \mathbb{R}^3$ be a regular region in the sense of Definition 1.7. Then, from (1.65), we are able to associate a point $y \in \partial\mathcal{G}$ to $y = (u_1, u_2, F(u_1, u_2))^T$, with x the as origin of the local tangential-normal system. In the nomenclature of the local tangential-normal system (see Figure 1.2) we are able to write

$$|x \pm \tau\nu(x) - y| = \left(u_1^2 + u_2^2 \ + \ (\tau \mp F(u_1, u_2))^2\right)^{\frac{1}{2}}. \tag{1.68}$$

Furthermore, we find

$$\begin{aligned} |\nu(x) \cdot (y - x)| &= \left|\varepsilon^3 \cdot \left(u_1\varepsilon^1 + u_2\varepsilon^2 + F(u_1, u_2)\varepsilon^3\right)\right| \\ &= |F(u_1, u_2)| \\ &\leq M(u_1^2 + u_2^2) \\ &\leq M(u_1^2 + u_2^2 + F(u_1, u_2)^2) \\ &\leq M|y - x|^2. \end{aligned} \tag{1.69}$$

Observing the representation

$$\nu(y) \ = \ \frac{1}{(1 + |\nabla F(u_1, u_2)|^2)^{\frac{1}{2}}} \left(\varepsilon^3 - F_{|1}(u_1, u_2)\varepsilon^1 - F_{|2}(u_1, u_2)\varepsilon^2\right) \tag{1.70}$$

we find

$$
\begin{aligned}
&|\nu(x) - \nu(y)| && (1.71)\\
&= \frac{\left|F_{|1}(u_1,u_2)\varepsilon^1 + F_{|2}(u_1,u_2)\varepsilon^2 + \left((1+|\nabla F(u_1,u_2)|^2)^{\frac{1}{2}} - 1\right)\varepsilon^3\right|}{(1+|\nabla F(u_1,u_2)|^2)^{\frac{1}{2}}}\\
&\leq \frac{\left(|\nabla F(u_1,u_2)|^2 + \frac{1}{4}|\nabla F(u_1,u_2)|^4\right)^{\frac{1}{2}}}{(1+|\nabla F(u_1,u_2)|^2)^{\frac{1}{2}}}\\
&\leq |\nabla F(u_1,u_2)|\\
&\leq M\ |x-y|,
\end{aligned}
$$

where we have used the well-known inequality $(1+\alpha)^{\frac{1}{2}} - 1 \leq \frac{\alpha}{2}$ for $\alpha > 0$. Summarizing, we are able to formulate

Lemma 1.8. *Let $\mathcal{G}$ be a regular region (with M, δ specified by Definition 1.7). Then, for every $x \in \partial\mathcal{G}$ and all $y \in \partial\mathcal{G} \cap \mathcal{B}_\delta(x)$,*

$$
\begin{aligned}
|\nu(x) - \nu(y)| &\leq M|x-y|,\\
|\nu(x) \cdot (x-y)| &\leq M|x-y|^2.
\end{aligned}
$$

Next we are interested in verifying an additional estimate.

Lemma 1.9. *For every $x \in \partial\mathcal{G}$ and all $y \in \partial\mathcal{G} \cap \mathcal{B}_\delta(x)$ (with δ, M specified by Definition 1.7) we have*

$$
|\nu(y) \cdot (y-x)| \leq 2M|x-y|^2. \tag{1.72}
$$

Proof. For $x \in \partial\mathcal{G}$ and $y \in \partial\mathcal{G} \cap \mathcal{B}_\delta(x)$ we find, by virtue of (1.71) and Lemma 1.8,

$$
\begin{aligned}
|\nu(y) \cdot (y-x)| &= |(\nu(y) - \nu(x)) \cdot (y-x) + \nu(x) \cdot (y-x)| \qquad (1.73)\\
&\leq |\nu(y) - \nu(x)|\ |y-x| + M|x-y|^2\\
&\leq 2M|x-y|^2.
\end{aligned}
$$

This yields the desired result of Lemma 1.9. □

In our geomathematically oriented approach to potential theory, bounding spheres for a regular region $\mathcal{G}$ play a particular role. To be more concrete, given a regular region $\mathcal{G} \subset \mathbb{R}^3$, then there exist positive constants $\underline{R}, \overline{R}$ with

$$
\underline{R} < \inf_{x \in \partial\mathcal{G}} |x| \leq \sup_{x \in \partial\mathcal{G}} |x| < \overline{R} \tag{1.74}
$$

such that $\mathcal{B}_{\underline{R}}(0) \Subset \mathcal{G} \Subset \mathcal{B}_{\overline{R}}(0)$ (cf. Figure 1.3). $\Omega_{\underline{R}}$ is called an *inner Runge sphere for $\mathcal{G}$*, while $\Omega_{\overline{R}}$ is called an *outer Runge sphere for $\mathcal{G}$* (note that, in physical geodesy, $\Omega_{\underline{R}}$ is called a *Bjerhammar sphere* for $\mathcal{G}$, while $\Omega_{\overline{R}}$ is

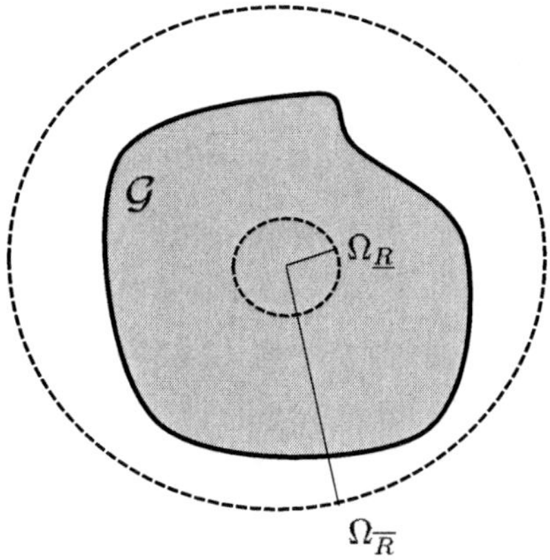

FIGURE 1.3
Sectional illustration of a regular region $\mathcal{G} \subset \mathbb{R}^3$ (together with bounding Runge spheres).

called a *Brillouin sphere* for $\mathcal{G}$). Clearly, under the assumption (1.74), we have $\mathbb{R}^3 \backslash \overline{\mathcal{B}_{\overline{R}}(0)} \Subset \mathcal{G}^c = \mathbb{R}^3 \backslash \overline{\mathcal{G}} \Subset \mathbb{R}^3 \backslash \overline{\mathcal{B}_{\underline{R}}(0)}$.

1.2.3 Interior Green Formulas

In the following, we recapitulate the classical *Gauss theorem* in $\mathbb{R}^3$.

Theorem 1.10 (Gauss Theorem). *Let $\mathcal{G}$ be a regular region, and suppose that F is of class $\mathrm{C}^{(1)}(\mathcal{G}) \cap \mathrm{C}^{(0)}(\overline{\mathcal{G}})$ and f of class $\mathrm{C}^{(1)}(\mathcal{G}) \cap \mathrm{C}^{(0)}(\overline{\mathcal{G}})$. Then*

$$\int_{\partial\mathcal{G}} F(y)\nu(y)\; d\omega(y) = \int_{\mathcal{G}} \nabla_y F(y)\; dV(y), \tag{1.75}$$

$$\int_{\partial\mathcal{G}} f(y)\cdot\nu(y)\; d\omega(y) = \int_{\mathcal{G}} \nabla_y \cdot f(y)\; dV(y), \tag{1.76}$$

$$\int_{\partial\mathcal{G}} \nu(y)\wedge f(y)\; d\omega(y) = \int_{\mathcal{G}} \nabla_y \wedge f(y)\; dV(y), \tag{1.77}$$

provided that the integrand on the right-hand side is Lebesgue-integrable on $\mathcal{G}$ and $\nu : \partial\mathcal{G} \to \mathbb{R}^3$ is the (unit) normal field pointing into the exterior of $\mathcal{G}$.

Remark 1.11. The identities (1.76) and (1.77) are valid for all vector fields, whatever their physical meaning. Of special interest is the case (1.76) in which f may be understood to be the velocity vector of an incompressible fluid. Inside the surface $\partial\mathcal{G}$, there may be sources in which the fluid is generated or sinks in which the fluid is annihilated. The divergence $\nabla \cdot f$ measures the strength of the sources and sinks. The volume integral $\int_{\mathcal{G}} \nabla \cdot f(y)\; dV(y)$ is the total amount of the fluid generated in unit time. The surface integral

$\int_{\partial\mathcal{G}} f(y) \cdot \nu(y)\, d\omega(y)$ is the total amount of fluid flowing in unit time across the surface $\partial\mathcal{G}$. Therefore, the Gauss formula expresses a balance equation, namely the evident fact that both integrals are equal.

Next we come to the *interior Green formulas* for regular regions $\mathcal{G} \subset \mathbb{R}^3$. Suppose that $f = \nabla F$, where $F \in \mathrm{C}^{(1)}(\overline{\mathcal{G}}) \cap \mathrm{C}^{(2)}(\mathcal{G})$, i.e., $F : \overline{\mathcal{G}} \to \mathbb{R}$ is continuously differentiable in $\overline{\mathcal{G}}$ and $F|_{\mathcal{G}}$ is twice continuously differentiable in $\mathcal{G}$. Let ΔF be Lebesgue-integrable in $\mathcal{G}$. Then we obtain from the Gauss theorem (1.76)

$$\int_{\partial\mathcal{G}} \frac{\partial F}{\partial \nu}(y)\, d\omega(y) = \int_{\mathcal{G}} \Delta F(y)\, dV(y), \tag{1.78}$$

where, as always, $\frac{\partial}{\partial\nu} = \nu \cdot \nabla$ denotes the derivative in the direction of the outer (unit) normal field ν. Consequently, the harmonicity of F in $\mathcal{G}$, i.e., $\Delta F = 0$ in $\mathcal{G}$, implies the identity

$$\int_{\partial\mathcal{G}} \frac{\partial F}{\partial \nu}(y)\, d\omega(y) = 0. \tag{1.79}$$

Under the special choice $f = F\, \nabla G$, the Gauss Theorem yields

Theorem 1.12 (Interior First Green Theorem). *Suppose that $\mathcal{G} \subset \mathbb{R}^3$ is a regular region. For $F \in \mathrm{C}^{(1)}(\overline{\mathcal{G}})$, $G \in \mathrm{C}^{(1)}(\overline{\mathcal{G}}) \cap \mathrm{C}^{(2)}(\mathcal{G})$, with ΔG Lebesgue-integrable on $\mathcal{G}$, we have*

$$\int_{\mathcal{G}} (F(y)\Delta G(y) + \nabla F(y) \cdot \nabla G(y))\, dV(y) = \int_{\partial\mathcal{G}} F(y) \frac{\partial G}{\partial \nu}(y)\, d\omega(y). \tag{1.80}$$

Taking $f = F\, \nabla G - G\, \nabla F$, we finally obtain

Theorem 1.13 (Interior Second Green Theorem). *Suppose that $\mathcal{G} \in \mathbb{R}^3$ is a regular region. For $F, G \in \mathrm{C}^{(1)}(\overline{\mathcal{G}}) \cap \mathrm{C}^{(2)}(\mathcal{G})$, with $\Delta F, \Delta G$ Lebesgue-integrable on $\mathcal{G}$ we have*

$$\begin{aligned} &\int_{\mathcal{G}} (G(y)\Delta F(y) - F(y)\Delta G(y))\, dV(y) \\ &= \int_{\partial\mathcal{G}} \left(G(y)\frac{\partial F}{\partial \nu}(y) - F(y)\frac{\partial G}{\partial \nu}(y) \right) d\omega(y). \end{aligned} \tag{1.81}$$

1.3 Exercises

Exercise 1.1. Prove the following relations for $w, x, y, z \in \mathbb{R}^3$ and $\mathbf{x} \in \mathbb{R}^{3\times 3}$:

$$\begin{aligned}(x \otimes y)w &= (y \cdot w)\ x, && (1.82)\\ x \cdot (\mathbf{x}y) &= \mathbf{x} \cdot (x \otimes y), && (1.83)\\ (x \otimes y)(w \otimes z) &= (y \cdot w)(x \otimes z), && (1.84)\\ \mathrm{tr}(x \otimes y) &= x \cdot y\ . && (1.85)\end{aligned}$$

Exercise 1.2. Verify that the identities

$$\begin{aligned}\nabla(UV) &= U\nabla V + V\nabla U, && (1.86)\\ \nabla \cdot (Uv) &= U\nabla \cdot v + \nabla U \cdot v, && (1.87)\\ \nabla \wedge (\nabla U) &= 0, && (1.88)\\ \nabla \wedge (\Delta u) &= \Delta(\nabla \wedge u), && (1.89)\\ \nabla \cdot (\nabla \wedge u) &= 0, && (1.90)\\ \nabla \wedge \nabla \wedge u &= \nabla(\nabla \cdot u) - \Delta u. && (1.91)\end{aligned}$$

are valid for all sufficiently often differentiable functions U, V and vector fields u, v.

Exercise 1.3 (Hobson's Formula)**.** Prove by induction that

$$\left(\frac{\partial}{\partial x_i}\right)^n \frac{1}{|x|} = (-1)^n \frac{1}{2^n n!} \frac{1}{|x|^{2n+1}} \left(\sum_{k=0}^{\lfloor \frac{n}{2} \rfloor} (-1)^k \frac{n!(2n-2k)!}{(n-k)!k!} |x|^{2k} \Delta^k \right) x_i^n \tag{1.92}$$

is valid for all $n \in \mathbb{N}_0$ and all $x \in \mathbb{R}^3 \backslash \{0\}$.

Exercise 1.4. Verify that the identity

$$\Delta_x (x \cdot y)^n = n(n-1)|y|^2 (x \cdot y)^{n-2} \tag{1.93}$$

holds true for all integers $n \geq 2$ and all $x, y \in \mathbb{R}^3$.

Exercise 1.5. Prove the following assertion: If $F \in \mathrm{C}^{(1)}((0, R))$, $R > 0$, then

$$\Delta_x^* F(|x|) = 0 \tag{1.94}$$

is valid for all $x \in \mathcal{B}_R(0)\backslash\{0\}$.

Exercise 1.6. Let $\mathcal{G}$ be a regular region. Suppose that $f : \overline{\mathcal{G}} \to \mathbb{R}$ is continuous on $\overline{\mathcal{G}}$ and continuously differentiable in $\mathcal{G}$, i.e., $f \in \mathrm{c}^{(0)}(\overline{\mathcal{G}}) \cap \mathrm{c}^{(1)}(\mathcal{G})$. Prove that

$$\int_{\partial\mathcal{G}} \nu(y) \wedge f(y)\ d\omega(y) = \int_{\mathcal{G}} \nabla_y \wedge f(y)\ dV(y), \tag{1.95}$$

provided that the integrand on the right side is (Lebesgue) integrable on $\mathcal{G}$ and $\nu : \partial\mathcal{G} \to \mathbb{R}^3$ is the (unit) normal field pointing into $\mathcal{G}$.

Exercise 1.7. Let $\mathcal{G} \subset \mathbb{R}^3$ be a regular region. Show that

$$\int_{\mathcal{G}} |\nabla U(y)|^2 \, dV(y) = \frac{1}{2} \int_{\partial\mathcal{G}} \nu(y) \cdot \nabla_y \left(U(y)\right)^2 \, d\omega(y) \tag{1.96}$$

for all $U \in \mathrm{C}^{(1)}(\overline{\mathcal{G}}) \cap \mathrm{C}^{(2)}(\mathcal{G})$ with $\Delta U = 0$ in $\mathcal{G}$.

2

Two–Dimensional Sphere Ω

CONTENTS

2.1 Basic Notation

We start with the general settings for the sphere. Much of the notation is adapted to the Euclidean concept in Chapter 1.

2.1.1 Spherical Setting

A *spherical cap* with radius $\rho \in (0, 2]$ and center $\xi \in \Omega$ is a subregion of the unit sphere Ω defined by

$$\Gamma_\rho(\xi) = \{\eta \in \Omega : 1 - \xi \cdot \eta < \rho\}. \tag{2.1}$$

For the choice $\rho = 2$, the spherical cap coincides with the punctured sphere $\Omega \setminus \{\xi\}$. If the underlying spherical cap is meant with respect to a sphere of radius $R > 0$ and center $x \in \Omega_R$, we still write $\Gamma_\rho(x) = \{y \in \Omega_R : R^2 - x \cdot y < \rho\}$, where ρ can be chosen from the interval $(0, 2R^2]$ (the fact $x \in \Omega_R$ already indicates that $\Gamma_\rho(x)$ is meant with respect to Ω_R). While $\mathcal{B}_{R_0,R_1}(y) = \{x \in \mathbb{R}^3 : R_0 < |x - y| < R_1\}$, $0 \leq R_0 < R_1$, denotes a *spherical shell* around the center $y \in \mathbb{R}^3$, the set

$$\mathcal{C}_{R_0,R_1}(\Gamma) = \left\{x \in \mathbb{R}^3 : \frac{x}{|x|} \in \Gamma,\ R_0 < |x| < R_1\right\} \tag{2.2}$$

is called a *conical shell* if Γ is a subregion of the sphere Ω. It is always meant with the origin as its center. As already known, any vector $x \in \mathbb{R}^3 \setminus \{0\}$ can be represented by $x = r\xi$, with $r = |x|$ and $\xi = \frac{x}{|x|} \in \Omega$ (generally, Greek letters ξ, η, ζ denote vectors of the unit sphere). Using spherical coordinates, we write

$$x(r, \varphi, t) = r\xi = r \begin{pmatrix} \sqrt{1-t^2}\cos\varphi \\ \sqrt{1-t^2}\sin\varphi \\ t \end{pmatrix}, \quad r > 0,\ \varphi \in [0, 2\pi),\ t \in [-1, 1], \tag{2.3}$$

where φ denotes longitude, $\theta \in [0, \pi]$ colatitude, and $t = \cos(\theta)$ polar distance. As mentioned before, the vectors ε^1, ε^2, ε^3 denote the canonical Cartesian basis vectors. Beside this basis, the typical orthonormal basis in a spherical framework is given by the moving triad

$$\varepsilon^r(\varphi, t) = \begin{pmatrix} \sqrt{1-t^2}\cos\varphi \\ \sqrt{1-t^2}\sin\varphi \\ t \end{pmatrix}, \tag{2.4}$$

$$\varepsilon^\varphi(\varphi, t) = \begin{pmatrix} -\sin\varphi \\ \cos\varphi \\ 0 \end{pmatrix}, \qquad \varepsilon^t(\varphi, t) = \begin{pmatrix} -t\cos\varphi \\ -t\sin\varphi \\ \sqrt{1-t^2} \end{pmatrix}.$$

While $\xi = \varepsilon^r$ has radial direction, ε^φ and ε^t are always tangential to the sphere (note that $\varepsilon^\varphi \wedge \varepsilon^t = \varepsilon^r$).

2.1.2 Function Spaces

A function $F : \Omega \to \mathbb{R}$ possessing k continuous derivatives on the unit sphere Ω is said to be of class $\mathrm{C}^{(k)}(\Omega)$, $k \in \mathbb{N}$. $\mathrm{C}^{(0)}(\Omega)$ is the space of continuous scalar-valued functions on Ω, which is a complete normed space endowed with the norm $\|F\|_{\mathrm{C}^{(0)}(\Omega)} = \sup_{\xi\in\Omega} |F(\xi)|$. The set of scalar functions $F : \Omega \to \mathbb{R}$ that are (Lebesgue-)measurable, and for which

$$\|F\|_{\mathrm{L}^p(\Omega)} = \left(\int_\Omega |F(\eta)|^p \; d\omega(\eta) \right)^{\frac{1}{p}} < \infty, \tag{2.5}$$

is known as $\mathrm{L}^p(\Omega)$, $1 \le p < \infty$. Clearly, $\mathrm{L}^p(\Omega) \subset \mathrm{L}^q(\Omega)$ for $1 \le q \le p$. $\mathrm{L}^2(\Omega)$ is a Hilbert space with respect to the inner product $(\cdot,\cdot)_{\mathrm{L}^2(\Omega)}$ defined by

$$(F,G)_{\mathrm{L}^2(\Omega)} = \int_\Omega F(\eta)G(\eta) \; d\omega(\eta), \quad F,G \in \mathrm{L}^2(\Omega). \tag{2.6}$$

Equipped with this inner product, $\mathrm{C}^{(0)}(\Omega)$ is a pre-Hilbert space, and $\mathrm{L}^2(\Omega)$ is its completion with respect to the $\mathrm{L}^2(\Omega)$-norm, i.e.,

$$\mathrm{L}^2(\Omega) = \overline{\mathrm{C}^{(0)}(\Omega)}^{\|\cdot\|_{\mathrm{L}^2(\Omega)}}. \tag{2.7}$$

For each $F \in \mathrm{C}^{(0)}(\Omega)$, we have the norm estimate

$$\|F\|_{\mathrm{L}^2(\Omega)} \le \sqrt{4\pi} \; \|F\|_{\mathrm{C}^{(0)}(\Omega)}. \tag{2.8}$$

This estimate is a direct consequence of the fact that the surface area of the unit sphere is equal to $\|\Omega\| = \int_\Omega d\omega(\eta) = 4\pi$.

A function $G : \Omega \to \mathbb{R}$ is called a *ξ-zonal function* on Ω (or radial basis function) if a function $\tilde{G} : [-1,1] \to \mathbb{R}$ exists such that $G(\eta) = \tilde{G}(\xi\cdot\eta)$, $\eta \in \Omega$. The set of all ξ-zonal functions is isomorphic to the set of functions $\tilde{G} : [-1,1] \to \mathbb{R}$. This allows us to interpret $\mathrm{C}^{(0)}([-1,1])$ and $\mathrm{L}^p([-1,1])$ as subspaces of $\mathrm{C}^{(0)}(\Omega)$ and $\mathrm{L}^p(\Omega)$, respectively, where the norms are defined correspondingly, i.e.,

$$\|\tilde{G}\|_{\mathrm{C}^{(0)}([-1,1])} = \|\tilde{G}(\varepsilon^3\cdot)\|_{\mathrm{C}^{(0)}(\Omega)} \tag{2.9}$$

and (cf. Exercise 2.1)

$$\begin{aligned} \|\tilde{G}\|_{\mathrm{L}^p([-1,1])} &= \left(2\pi \int_{-1}^{1} |\tilde{G}(t)|^p \; dt \right)^{\frac{1}{p}} \\ &= \left(\int_\Omega \tilde{G}(\varepsilon^3\cdot\eta)|^p \; d\omega(\eta) \right)^{\frac{1}{p}} = \|\tilde{G}(\varepsilon^3\cdot)\|_{\mathrm{L}^p(\Omega)}. \end{aligned} \tag{2.10}$$

Using the Cartesian basis $\varepsilon^1, \varepsilon^2, \varepsilon^3$ of $\mathbb{R}^3$, we may write any vector field $f : \Omega \to \mathbb{R}^3$ in the form $f = \sum_{i=1}^3 f_i\varepsilon^i$, where the component functions f_i are given by $f_i(\xi) = f(\xi)\cdot\varepsilon^i$, $\xi \in \Omega$.

The space $\mathrm{c}^{(k)}(\Omega)$, $k \in \mathbb{N}$, consists of all k-times continuously differentiable vector fields on Ω. The space of continuous vector fields is denoted by $\mathrm{c}^{(0)}(\Omega)$. It is complete with respect to the norm $\|f\|_{\mathrm{c}^{(0)}(\Omega)} = \sup_{\xi\in\Omega} |f(\xi)|$. By $\mathrm{l}^p(\Omega)$,

$1 \leq p < \infty$, we mean the space of all (Lebesgue-)measurable vector fields $F : \Omega \to \mathbb{R}^3$ with

$$\|f\|_{\mathrm{l}^p(\Omega)} = \left(\int_\Omega |f(\xi)|^p \; d\omega(\xi) \right)^{\frac{1}{p}} < \infty. \tag{2.11}$$

Equipped with the inner product

$$(f,g)_{\mathrm{l}^2(\Omega)} = \int_\Omega f(\xi) \cdot g(\xi) \; d\omega(\xi), \quad f,g \in \mathrm{l}^2(\Omega), \tag{2.12}$$

$\mathrm{l}^2(\Omega)$ is a Hilbert space and the completion of $\mathrm{c}^{(0)}(\Omega)$ with respect to the corresponding norm, i.e.,

$$\overline{\mathrm{c}^{(0)}(\Omega)}^{\|\cdot\|_{\mathrm{l}^2(\Omega)}} = \mathrm{l}^2(\Omega). \tag{2.13}$$

For all $f \in \mathrm{c}^{(0)}(\Omega)$, we have the norm estimate

$$\|f\|_{\mathrm{l}^2(\Omega)} \leq \sqrt{4\pi} \; \|f\|_{\mathrm{c}^{(0)}(\Omega)}. \tag{2.14}$$

In order to separate vector fields into their tangential and radial parts, we introduce the projection operators p_{rad} and p_{tan} by

$$p_{\mathrm{rad}}[f](\xi) = (f(\xi) \cdot \xi) \; \xi, \tag{2.15}$$
$$p_{\mathrm{tan}}[f](\xi) = f(\xi) - p_{\mathrm{rad}}[f](\xi), \tag{2.16}$$

for $\xi \in \Omega$ and $f \in \mathrm{l}^2(\Omega)$. For brevity, we often write f_{tan} and f_{rad} instead of $p_{\mathrm{tan}}[f]$ and $p_{\mathrm{rad}}[f]$, respectively. Furthermore, we define

$$\mathrm{l}^2_{\mathrm{rad}}(\Omega) = \{ f \in \mathrm{l}^2(\Omega) : f = p_{\mathrm{rad}}[f] \}, \tag{2.17}$$
$$\mathrm{l}^2_{\mathrm{tan}}(\Omega) = \{ f \in \mathrm{l}^2(\Omega) : f = p_{\mathrm{tan}}[f] \}. \tag{2.18}$$

We say $f \in \mathrm{l}^2(\Omega)$ is *radial* if $f = p_{\mathrm{rad}}[f]$ and *tangential* if $f = p_{\mathrm{tan}}[f]$. Clearly, we have the orthogonal decomposition $\mathrm{l}^2(\Omega) = \mathrm{l}^2_{\mathrm{rad}}(\Omega) \oplus \mathrm{l}^2_{\mathrm{tan}}(\Omega)$. The spaces $\mathrm{c}^{(k)}_{\mathrm{rad}}(\Omega)$ and $\mathrm{c}^{(k)}_{\mathrm{tan}}(\Omega)$, $k \in \mathbb{N}_0$, are defined in the same fashion.

Finally, we turn to tensor fields (of rank 2) $\mathbf{f} : \Omega \to \mathbb{R}^{3\times 3}$. Any such tensor field can be represented as $\mathbf{f} = \sum_{i=1}^{3} \sum_{k=1}^{3} \mathbf{f}_{i,k} (\varepsilon^i \otimes \varepsilon^k)$. The component functions are given by $\mathbf{f}_{i,k}(\xi) = (\varepsilon^i)^T \mathbf{f}(\xi) \varepsilon^k$, for $\xi \in \Omega$. As usual, the identity tensor is defined by

$$\mathbf{i} = \sum_{i=1}^{3} \varepsilon^i \otimes \varepsilon^i. \tag{2.19}$$

Its projection onto the tangential components at a point $\xi \in \Omega$ defines the surface identity tensor field

$$\mathbf{i}_{\mathrm{tan}}(\xi) = \mathbf{i} - \xi \otimes \xi, \quad \xi \in \Omega. \tag{2.20}$$

Moreover, we introduce the surface rotation tensor field

$$\mathbf{j}_{\tan}(\xi) = \sum_{i=1}^{3} (\xi \wedge \varepsilon^i) \otimes \varepsilon^i, \quad \xi \in \Omega, \tag{2.21}$$

and obtain

$$\begin{array}{rclrcl} \mathbf{i}_{\tan}(\xi)\,\xi & = & 0, & \mathbf{j}_{\tan}(\xi)\,\xi & = & 0, \\ \mathbf{i}_{\tan}(\xi)\,x_\xi & = & x_\xi, & \mathbf{j}_{\tan}(\xi)\,x_\xi & = & \xi \wedge x_\xi, \end{array} \tag{2.22}$$

for $\xi \in \Omega$ and $x_\xi \in \mathbb{R}^3$ with $x_\xi \cdot \xi = 0$. The function spaces $\mathbf{c}^{(k)}(\Omega)$, $k \in \mathbb{N}_0$, and $\mathbf{l}^p(\Omega)$, $1 \leq p < \infty$, can be defined analogously to the scalar and vector case.

The same settings as on the entire sphere can be introduced for subregions $\Gamma \subset \Omega$ of the sphere. For the norms of the spaces $\mathrm{L}^p(\Gamma), \mathrm{l}^p(\Gamma), \mathbf{l}^p(\Gamma)$, the integration simply needs to be restricted to Γ. $\mathrm{C}^{(k)}(\Gamma), \mathrm{c}^{(k)}(\Gamma), \mathbf{c}^{(k)}(\Gamma)$ denote the spaces of k-times continuously differentiable scalar, vector, and tensor fields on Γ, respectively. If they are k-times continuously differentiable up to the boundary, i.e., on the closure $\overline{\Gamma} = \Gamma \cup \partial\Gamma$, we write $\mathrm{C}^{(k)}(\overline{\Gamma}), \mathrm{c}^{(k)}(\overline{\Gamma}), \mathbf{c}^{(k)}(\overline{\Gamma})$. If the boundary curve $\partial\Gamma$ is sufficiently smooth, we are also able to deal with the spaces $\mathrm{C}^{(k)}(\partial\Gamma), \mathrm{c}^{(k)}(\partial\Gamma), \mathbf{c}^{(k)}(\partial\Gamma)$. The norm in $\mathrm{L}^p(\partial\Gamma)$ is given by

$$\|F\|_{\mathrm{L}^p(\partial\Gamma)} = \left(\int_{\partial\Gamma} |F(\eta)|^p \, d\sigma(\eta) \right)^{\frac{1}{p}}, \quad F \in \mathrm{L}^p(\partial\Gamma), \tag{2.23}$$

where $d\sigma$ denotes the line element. $\mathrm{L}^2(\partial\Gamma)$ is a Hilbert space with the corresponding inner product $(\cdot,\cdot)_{\mathrm{L}^2(\partial\Gamma)}$. The analogous definitions hold true for $\mathrm{l}^p(\partial\Gamma)$ and $\mathbf{l}^p(\partial\Gamma)$.

2.1.3 Differential Operators

The gradient field ∇ can be decomposed into a radial and a tangential component. The *surface gradient* ∇^* denotes the tangential part of the gradient ∇. More explicitly,

$$\nabla = \varepsilon^r \frac{\partial}{\partial r} + \frac{1}{r} \nabla^*, \tag{2.24}$$

where $\xi = \varepsilon^r$ denotes the radial direction. The *surface curl gradient* L^* is given by

$$\mathrm{L}^*_\xi F(\xi) = \xi \wedge \nabla^*_\xi F(\xi), \quad \xi \in \Omega, \tag{2.25}$$

for F of class $\mathrm{C}^{(1)}(\Omega)$. As usual, the index ξ denotes the variable on which the operator acts. It is often omitted if it is obvious which variable is meant. According to its definition, $\mathrm{L}^* F$ is a tangential vector field perpendicular to $\nabla^* F$. Elementary calculations in spherical coordinates (2.4) show that

$$\nabla^* = \varepsilon^\varphi \frac{1}{\sqrt{1-t^2}} \frac{\partial}{\partial \varphi} + \varepsilon^t \sqrt{1-t^2} \frac{\partial}{\partial t}, \tag{2.26}$$

$$\mathrm{L}^* = -\varepsilon^\varphi \sqrt{1-t^2} \frac{\partial}{\partial t} + \varepsilon^t \frac{1}{\sqrt{1-t^2}} \frac{\partial}{\partial \varphi}\ . \tag{2.27}$$

Symbol	Differential Operator
∇^*	Surface gradient
L^*	Surface curl gradient
$\nabla^*\cdot$	Surface divergence
$\mathrm{L}^*\cdot$	Surface curl divergence
Δ^*	Beltrami operator

TABLE 2.1
Important spherical differential operators on the sphere.

The *surface divergence* and the *surface curl divergence* of a vectorial function $f = \sum_{i=1}^{3} f_i \varepsilon^i$ of class $\mathrm{c}^{(1)}(\Omega)$ are defined as

$$\nabla^* \cdot f = \sum_{i=1}^{3} \varepsilon^i \cdot \nabla^* f_i, \qquad \mathrm{L}^* \cdot f = \sum_{i=1}^{3} \varepsilon^i \cdot \mathrm{L}^* f_i, \tag{2.28}$$

respectively. Less frequently used, but still of some importance, are the operators

$$\nabla^* \wedge f = \sum_{i=1}^{3} (\nabla^* f_i) \wedge \varepsilon^i, \qquad \mathrm{L}^* \wedge f = \sum_{i=1}^{3} (\mathrm{L}^* f_i) \wedge \varepsilon^i, \tag{2.29}$$

and

$$\nabla^* \otimes f = \sum_{i=1}^{3} (\nabla^* f_i) \otimes \varepsilon^i, \qquad \mathrm{L}^* \otimes f = \sum_{i=1}^{3} (\mathrm{L}^* f_i) \otimes \varepsilon^i. \tag{2.30}$$

$\nabla^* \otimes f$ can also be understood as the surface gradient $\nabla^* f$ of a vector field f of class $\mathrm{c}^{(1)}(\Omega)$. The Laplace operator $\Delta = \nabla \cdot \nabla$ and the *Beltrami operator* $\Delta^* = \nabla^* \cdot \nabla^*$ are related by

$$\Delta = \frac{1}{r^2} \frac{\partial}{\partial r} r^2 \frac{\partial}{\partial r} + \frac{1}{r^2} \Delta^*, \tag{2.31}$$

where the representation in spherical coordinates reads

$$\Delta^* = \frac{\partial}{\partial t} \left(1 - t^2\right) \frac{\partial}{\partial t} + \frac{1}{1 - t^2} \left(\frac{\partial}{\partial \varphi}\right)^2. \tag{2.32}$$

The Beltrami operator as given in (2.32) is just the restriction to the sphere Ω of the operator introduced in Definition 1.1 in Cartesian coordinates of $\mathbb{R}^3$.

Throughout this book, the operators ∇^*, L^*, Δ^* (see Table 2.1) are always used in coordinate-free representation, thereby avoiding any singularity at the poles. The following properties can be shown

$$\begin{aligned} \nabla^* \cdot \nabla^* F &= \mathrm{L}^* \cdot \mathrm{L}^* F = \Delta^* F, && (2.33)\\ \nabla^* \cdot \mathrm{L}^* F &= \mathrm{L}^* \cdot \nabla^* F = 0, && (2.34)\\ (\nabla^* F) \cdot (\mathrm{L}^* F) &= 0, && (2.35) \end{aligned}$$

for $F \in \mathrm{C}^{(1)}(\Omega)$, and

$$\begin{aligned} \nabla^*_\xi \cdot \xi &= 2, && (2.36)\\ \mathrm{L}^*_\xi \cdot \xi &= 0, && (2.37)\\ \nabla^*_\xi \otimes \xi &= \mathbf{i}_{\mathrm{tan}}(\xi), && (2.38)\\ \mathrm{L}^*_\xi \otimes \xi &= \mathbf{j}_{\mathrm{tan}}(\xi), && (2.39) \end{aligned}$$

for $\xi \in \Omega$. Moreover, we have

$$\begin{aligned} \nabla^* \cdot (F\, f) &= (\nabla^* F) \cdot f + F(\nabla^* \cdot f), && (2.40)\\ \nabla^* \wedge (F\, f) &= (\nabla^* F) \wedge f + F(\nabla^* \wedge f), && (2.41)\\ \nabla^* \otimes (F\, f) &= (\nabla^* F) \otimes f + F(\nabla^* \otimes f), && (2.42) \end{aligned}$$

for $F \in \mathrm{C}^{(1)}(\Omega)$ and $f \in \mathrm{c}^{(1)}(\Omega)$. If F is of class $\mathrm{C}^{(1)}([-1,1])$ and if $F' \in \mathrm{C}^{(0)}([-1,1])$ is its (one-dimensional) derivative, then

$$\begin{aligned} \nabla^*_\xi F(\xi \cdot \eta) &= F'(\xi \cdot \eta)(\eta - (\xi \cdot \eta)\xi), && (2.43)\\ \mathrm{L}^*_\xi F(\xi \cdot \eta) &= F'(\xi \cdot \eta)(\xi \wedge \eta), && (2.44) \end{aligned}$$

whereas, for $F \in \mathrm{C}^{(2)}([-1,1])$,

$$\Delta^*_\xi F(\xi \cdot \eta) = -2(\xi \cdot \eta)F'(\xi \cdot \eta) + (1 - (\xi \cdot \eta)^2)F''(\xi \cdot \eta), \qquad (2.45)$$

for $\xi, \eta \in \Omega$. Thus, the spherical differential operators have particularly simple representations when applied to zonal functions. Furthermore, we find

$$\begin{aligned} \Delta^*_\xi F(\xi \cdot \eta) &= \Delta^*_\eta F(\xi \cdot \eta), && (2.46)\\ \mathrm{L}^*_\xi F(\xi \cdot \eta) &= -\mathrm{L}^*_\eta F(\xi \cdot \eta) && (2.47) \end{aligned}$$

but $\nabla^*_\xi F(\xi \cdot \eta) \neq \pm \nabla^*_\eta F(\xi \cdot \eta)$.

2.2 Integral Theorems

In order to state the well-known integral theorems of Gauss and Stokes, we first need the notion of a regular region on the sphere. The definition is given in close similarity to the definition of a regular region in the three-dimensional Euclidean space $\mathbb{R}^3$ (cf. Definition 1.7). However, in contrast to the Euclidean settings, the open complement of a regular region on the unit sphere is again a (bounded) regular region.

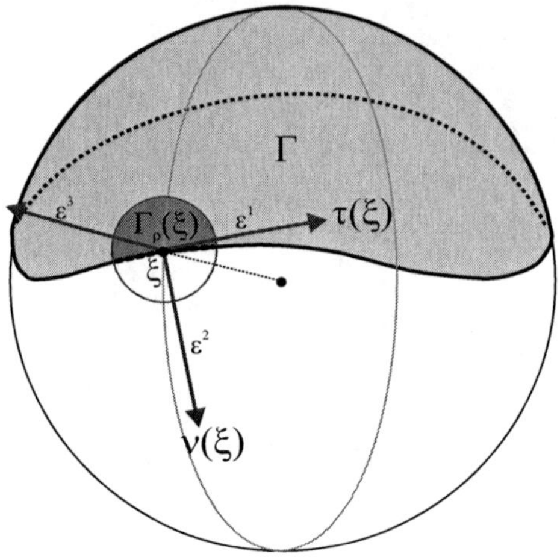

FIGURE 2.1
Illustration of the local (tangential-normal) coordinate system.

2.2.1 Regular Regions

Definition 2.1. A region $\Gamma \subset \Omega$ is called a *regular region (on the sphere)* if it satisfies the following properties:

(i) The boundary curve $\partial\Gamma$ divides Ω into two disjoint simply connected regions, namely Γ and its open complement $\Gamma^c = \Omega \setminus \overline{\Gamma}$,

(ii) $\partial\Gamma$ is a closed and compact curve, free of double points,

(iii) there exist (universal) constants $M, \rho > 0$ such that, for every point $\xi \in \partial\Gamma$, the set $\partial\Gamma \cap \Gamma_\rho(\xi)$ can be represented in a local (tangential-normal) coordinate system in the form

$$(\eta_2, \eta_3)^T = \gamma(\eta_1), \quad \eta_1 \in (-\tilde{\rho}, \tilde{\rho}), \tag{2.48}$$

for $\eta = (\eta_1, \eta_2, \eta_3)^T \in \partial\Gamma \cap \Gamma_\rho(\xi)$ and a fixed $\tilde{\rho} \in (0, \sqrt{2\rho})$. The parameterization $\gamma : [-\tilde{\rho}, \tilde{\rho}] \to \mathbb{R}^2$ is twice continuously differentiable and satisfies

$$|\gamma(t)| \leq Mt^2, \quad t \in (-\tilde{\rho}, \tilde{\rho}), \tag{2.49}$$

$$\left|\frac{d}{dt}\gamma(t)\right| \leq Mt, \quad t \in (-\tilde{\rho}, \tilde{\rho}). \tag{2.50}$$

The *local (tangential-normal) coordinate system with center* $\xi \in \Omega$ is chosen such that ξ coincides with the origin, the ε^1-axis coincides with the tangential unit vector $\tau(\xi)$, the ε^2-axis with the normal unit vector $\nu(\xi)$, and the ε^3-axis with the radial direction ξ (cf. Figure 2.1).

$\tau(\xi)$ denotes the positively oriented unit tangential vector of the boundary curve $\partial\Gamma$ at the point $\xi \in \partial\Gamma$. The unit normal vector $\nu(\xi)$ points into the exterior of Γ and is perpendicular to $\tau(\xi)$ and ξ (i.e., $\nu(\xi)$ is perpendicular to the boundary curve $\partial\Gamma$ in the point $\xi \in \partial\Gamma$ but tangential to the sphere Ω). The *normal derivative* is defined as

$$\frac{\partial}{\partial\nu}F(\xi) = \nu(\xi)\cdot\nabla^*_\xi F(\xi) \left(= \tau(\xi)\cdot \mathrm{L}^*_\xi F(\xi)\right), \quad \xi \in \partial\Gamma, \tag{2.51}$$

for a function F of class $\mathrm{C}^{(1)}(\overline{\Gamma})$. Although the notation of the normal ν is the same for regular regions in $\mathbb{R}^3$ and regular regions on the sphere, it should always be clear from the context which case is meant.

Remark 2.2. The properties from Definition 2.1 imply that a constant $M > 0$ exists such that

$$|\nu(\xi) - \nu(\eta)| \le M|\xi-\eta|, \quad \xi,\eta \in \partial\Gamma, \tag{2.52}$$

$$|\nu(\xi)\cdot\eta| \le M|\xi-\eta|^2, \quad \xi,\eta \in \partial\Gamma. \tag{2.53}$$

Next, we give some basic estimates for regular regions.

Proposition 2.3. *Let* $\Gamma \subset \Omega$ *be a regular region. Then there exists a constant* $\tau_0 \in (0,1)$ *such that*

$$\left|\frac{\xi+\tau\nu(\xi)}{\sqrt{1+\tau^2}} - \eta\right| \ge \frac{1}{2\sqrt{2}}|\xi-\eta|, \tag{2.54}$$

for $\xi,\eta \in \partial\Gamma$ *and* $|\tau| < \tau_0$.

Proof. Due to the regularity assumptions on Γ, one can find a $\tau_0 \in (0,1)$ such that

$$\left|\frac{\xi+\tau\nu(\xi)}{\sqrt{1+\tau^2}} - \eta\right| \ge \left|\frac{\xi+\tau\nu(\xi)}{\sqrt{1+\tau^2}} - \xi\right| = \sqrt{2\left(1-\frac{1}{\sqrt{1+\tau^2}}\right)}, \tag{2.55}$$

for $\xi,\eta \in \partial\Gamma$ and any $|\tau| < \tau_0$. Furthermore, we have

$$\left|\frac{\xi+\tau\nu(\xi)}{\sqrt{1+\tau^2}} - \eta\right| = \left|\xi-\eta+\left(\frac{1}{\sqrt{1+\tau^2}}-1\right)\xi + \frac{\tau\nu(\xi)}{\sqrt{1+\tau^2}}\right|, \tag{2.56}$$

and

$$\left|\left(\frac{1}{\sqrt{1+\tau^2}}-1\right)\xi + \frac{\tau\nu(\xi)}{\sqrt{1+\tau^2}}\right| = \sqrt{\left(\frac{1}{\sqrt{1+\tau^2}}-1\right)^2 + \frac{\tau^2}{1+\tau^2}}, \tag{2.57}$$

for $\xi,\eta \in \partial\Gamma$. Let now $|\xi-\eta| \ge 2\left(\left(\frac{1}{\sqrt{1+\tau^2}}-1\right)^2 + \frac{\tau^2}{1+\tau^2}\right)^{\frac{1}{2}}$, then we find from (2.56), by use of the triangle inequality, that

$$\left|\frac{\xi+\tau\nu(\xi)}{\sqrt{1+\tau^2}} - \eta\right| \ge \frac{1}{2}|\xi-\eta|. \tag{2.58}$$

We observe that

$$\sqrt{\left(\frac{1}{\sqrt{1+\tau^2}}-1\right)^2+\frac{\tau^2}{1+\tau^2}} \leq \sqrt{2}\frac{\tau}{\sqrt{1+\tau^2}}. \tag{2.59}$$

For the case $|\xi-\eta| < 2\left(\left(\frac{1}{\sqrt{1+\tau^2}}-1\right)^2+\frac{\tau^2}{1+\tau^2}\right)^{\frac{1}{2}}$, we then get, by use of (2.55),

$$\begin{aligned}\left|\frac{\xi+\tau\nu(\xi)}{\sqrt{1+\tau^2}}-\eta\right| &\geq \sqrt{2\left(1-\frac{1}{\sqrt{1+\tau^2}}\right)}\frac{|\xi-\eta|}{2\sqrt{2}\frac{\tau}{\sqrt{1+\tau^2}}} \\ &= \frac{1}{2}\frac{\sqrt{1+\tau^2-\sqrt{1+\tau^2}}}{\tau}|\xi-\eta|.\end{aligned} \tag{2.60}$$

Estimating the right-hand side with respect to τ, we find

$$\left|\frac{\xi+\tau\nu(\xi)}{\sqrt{1+\tau^2}}-\eta\right| \geq \frac{1}{2\sqrt{2}}|\xi-\eta|, \tag{2.61}$$

which proves the assertion, together with (2.58). □

Proposition 2.4. *Let $\Gamma \subset \Omega$ be a regular subregion. Then there exists a constant $\tau_0 \in (0,1)$ such that*

$$\left|\frac{\xi+\tau\nu(\xi)}{\sqrt{1+\tau^2}}-\eta\right| \geq \frac{1}{5}\left|\xi-\frac{\eta+\tau\nu(\eta)}{\sqrt{1+\tau^2}}\right|, \tag{2.62}$$

for $\xi,\eta \in \partial\Gamma$ and $|\tau| < \tau_0$.

Proof. Using Proposition 2.3 and $\left|\frac{\xi+\tau\nu(\xi)}{\sqrt{1+\tau^2}}-\eta\right| \geq \left|\frac{\eta+\tau\nu(\eta)}{\sqrt{1+\tau^2}}-\eta\right|$ for $|\tau| < \tau_0$ and $\xi,\eta \in \partial\Gamma$, we are able to deduce that

$$\begin{aligned}\frac{\left|\xi-\frac{\eta+\tau\nu(\eta)}{\sqrt{1+\tau^2}}\right|}{\left|\frac{\xi+\tau\nu(\xi)}{\sqrt{1+\tau^2}}-\eta\right|} &= \frac{1}{\sqrt{1+\tau^2}}\frac{\left|\sqrt{1+\tau^2}(\xi-\eta)+(\sqrt{1+\tau^2}-1)\eta+\tau\nu(\eta)\right|}{\left|\frac{\xi+\tau\nu(\xi)}{\sqrt{1+\tau^2}}-\eta\right|} \\ &\leq 2\sqrt{2}\frac{\left|\sqrt{1+\tau^2}(\xi-\eta)\right|}{|\xi-\eta|}+\frac{\left|(\sqrt{1+\tau^2}-1)\eta+\tau\nu(\eta)\right|}{\left|(\sqrt{1+\tau^2}-1)\eta-\tau\nu(\eta)\right|} \\ &= 2\sqrt{2}\sqrt{1+\tau^2}+1 \leq 5,\end{aligned} \tag{2.63}$$

which proves the statement of Proposition 2.4. □

2.2.2 Green Formulas

The following spherical versions of the theorems of Gauss and Stokes are well-known and can be found in any textbook on differential geometry.

Theorem 2.5 (Surface Theorems of Gauss and Stokes). *Let $\Gamma \subset \Omega$ be a regular region. If $f \in \mathrm{c}^{(1)}(\overline{\Gamma})$ is tangential, i.e., $\xi \cdot f(\xi) = 0$ for $\xi \in \overline{\Gamma}$, then*

$$\int_\Gamma \nabla^*_\eta \cdot f(\eta) d\omega(\eta) = \int_{\partial\Gamma} \nu(\eta) \cdot f(\eta) d\sigma(\eta), \tag{2.64}$$

$$\int_\Gamma \mathrm{L}^*_\eta \cdot f(\eta) d\omega(\eta) = \int_{\partial\Gamma} \tau(\eta) \cdot f(\eta) d\sigma(\eta). \tag{2.65}$$

It is important to point out the assumption of f being tangential in Theorem 2.5. This causes an additional term in *Green's formulas* involving ∇^*, but it does not affect those for L^*, which is due to the fact that $\nabla^*_\xi \cdot \xi = 2$ but $\mathrm{L}^*_\xi \cdot \xi = 0$, $\xi \in \Omega$. Whenever we refer to Green's formulas (on the sphere) in this book, we mean the corresponding assertion of one of the following formulas.

Lemma 2.6. *Let $\Gamma \subset \Omega$ be a regular region, and suppose that F, G are of class $\mathrm{C}^{(1)}(\overline{\Gamma})$. Then*

$$\begin{aligned} &\int_\Gamma G(\eta)\nabla^*_\eta F(\eta) d\omega(\eta) + \int_\Gamma F(\eta)\nabla^*_\eta G(\eta) d\omega(\eta) \\ &\quad = \int_{\partial\Gamma} \nu(\eta) \left(F(\eta)G(\eta)\right) d\sigma(\eta) + 2\int_\Gamma \eta \left(F(\eta)G(\eta)\right) d\omega(\eta), \end{aligned} \tag{2.66}$$

$$\begin{aligned} &\int_\Gamma G(\eta)\mathrm{L}^*_\eta F(\eta) d\omega(\eta) + \int_\Gamma F(\eta)\mathrm{L}^*_\eta G(\eta) d\omega(\eta) \\ &\quad = \int_{\partial\Gamma} \tau(\eta) \left(F(\eta)G(\eta)\right) d\sigma(\eta). \end{aligned} \tag{2.67}$$

Proof. It is clear that

$$\begin{aligned} \int_\Gamma G(\eta)\nabla^*_\eta F(\eta) d\omega(\eta) &= \sum_{i=1}^3 \varepsilon^i \int_\Gamma \left(\varepsilon^i G(\eta)\right) \cdot \nabla^*_\eta F(\eta) d\omega(\eta) \\ &= \sum_{i=1}^3 \varepsilon^i \int_\Gamma \left((\varepsilon^i - (\varepsilon^i \cdot \eta)\eta)G(\eta)\right) \cdot \nabla^*_\eta F(\eta) d\omega(\eta). \end{aligned} \tag{2.68}$$

Now, both factors in the integral of the last row are tangential vector fields.

Observing that $g \cdot \nabla^* F = -F\nabla^* \cdot g + \nabla^* \cdot (gF)$, we obtain from Theorem 2.5

$$\begin{aligned}
\int_\Gamma G(\eta)\nabla^*_\eta F(\eta) d\omega(\eta) &= -\sum_{i=1}^{3} \varepsilon^i \int_\Gamma F(\eta)\nabla^*_\eta \cdot \left((\varepsilon^i - (\varepsilon^i \cdot \eta)\eta)G(\eta)\right) d\omega(\eta) \\
&\quad +\sum_{i=1}^{3} \varepsilon^i \int_\Gamma \nabla^*_\eta \cdot \left((\varepsilon^i - (\varepsilon^i \cdot \eta)\eta)G(\eta)F(\eta)\right) d\omega(\eta) \\
&= -\sum_{i=1}^{3} \varepsilon^i \int_\Gamma F(\eta)\nabla^*_\eta \cdot \left((\varepsilon^i - (\varepsilon^i \cdot \eta)\eta)G(\eta)\right) d\omega(\eta) \\
&\quad +\sum_{i=1}^{3} \varepsilon^i \int_{\partial\Gamma} F(\eta)\nu(\eta) \cdot \left((\varepsilon^i - (\varepsilon^i \cdot \eta)\eta)G(\eta)\right) d\sigma(\eta) \\
&= -\sum_{i=1}^{3} \varepsilon^i \int_\Gamma F(\eta)\varepsilon^i \cdot \nabla^*_\eta G(\eta) d\omega(\eta) \qquad (2.69) \\
&\quad +\sum_{i=1}^{3} \varepsilon^i \int_\Gamma 2(\varepsilon^i \cdot \eta)F(\eta)G(\eta) d\omega(\eta) \\
&\quad +\sum_{i=1}^{3} \varepsilon^i \int_{\partial\Gamma} (\nu(\eta) \cdot \varepsilon^i)F(\eta)G(\eta) d\sigma(\eta) \\
&= -\int_\Gamma F(\eta)\nabla^*_\eta G(\eta) d\omega(\eta) + 2\int_\Gamma \eta(F(\eta)G(\eta)) d\omega(\eta) \\
&\quad +\int_{\partial\Gamma} \nu(\eta)(F(\eta)G(\eta)) d\sigma(\eta),
\end{aligned}$$

which provides the desired statement. Analogously, we are able to achieve the corresponding formula for the surface curl gradient. However, since $\mathrm{L}^*_\xi \cdot \xi = 0$, $\xi \in \Omega$, we notice that the term with the pre-factor 2 vanishes in this case. □

In the same manner, we obtain the following versions.

Lemma 2.7. *Let $\Gamma \subset \Omega$ be a regular region. Furthermore, if F is of class $\mathrm{C}^{(1)}(\overline{\Gamma})$ and f of class $\mathrm{c}^{(1)}(\overline{\Gamma})$, then*

$$\begin{aligned}
&\int_\Gamma f(\eta) \cdot \nabla^*_\eta F(\eta) d\omega(\eta) + \int_\Gamma F(\eta)\nabla^*_\eta \cdot f(\eta) d\omega(\eta) \qquad (2.70) \\
&\quad = \int_{\partial\Gamma} \nu(\eta) \cdot (F(\eta)f(\eta))\, d\sigma(\eta) + 2\int_\Gamma \eta \cdot (F(\eta)f(\eta))\, d\omega(\eta),
\end{aligned}$$

$$\begin{aligned}
&\int_\Gamma f(\eta) \cdot \mathrm{L}^*_\eta F(\eta) d\omega(\eta) + \int_\Gamma F(\eta)\mathrm{L}^*_\eta \cdot f(\eta) d\omega(\eta) \qquad (2.71) \\
&\quad = \int_{\partial\Gamma} \tau(\eta) \cdot (F(\eta)f(\eta))\, d\sigma(\eta).
\end{aligned}$$

For functions f, g *of class* $\mathrm{c}^{(1)}(\overline{\Gamma})$, *we have*

$$\int_\Gamma f(\eta)\left(\nabla^*_\eta \cdot g(\eta)\right)d\omega(\eta) + \int_\Gamma \left(\nabla^*_\eta \otimes f(\eta)\right)^T g(\eta)d\omega(\eta) \tag{2.72}$$
$$= \int_{\partial\Gamma} \left(f(\eta)\otimes\nu(\eta)\right) g(\eta)d\sigma(\eta) + 2\int_\Gamma \left(f(\eta)\otimes\eta\right) g(\eta)d\omega(\eta),$$

$$\int_\Gamma f(\eta)\left(\mathrm{L}^*_\eta \cdot g(\eta)\right)d\omega(\eta) + \int_\Gamma \left(\mathrm{L}^*_\eta \otimes f(\eta)\right)^T g(\eta)d\omega(\eta) \tag{2.73}$$
$$= \int_{\partial\Gamma} \left(f(\eta)\otimes\tau(\eta)\right) g(\eta)d\sigma(\eta).$$

Lemma 2.8. *Let* $\Gamma \subset \Omega$ *be a regular region. If* F *is of class* $\mathrm{C}^{(1)}(\overline{\Gamma})$ *and* f *of class* $\mathrm{c}^{(1)}(\overline{\Gamma})$, *then*

$$\int_\Gamma \left(\nabla^*_\eta F(\eta)\right)\wedge f(\eta)d\omega(\eta) + \int_\Gamma F(\eta)\nabla^*_\eta \wedge f(\eta)d\omega(\eta) \tag{2.74}$$
$$= \int_{\partial\Gamma} \nu(\eta)\wedge\left(F(\eta)f(\eta)\right)d\sigma(\eta) + 2\int_\Gamma \eta\wedge\left(F(\eta)f(\eta)\right)d\omega(\eta),$$

$$\int_\Gamma \left(\mathrm{L}^*_\eta F(\eta)\right)\wedge f(\eta)d\omega(\eta) + \int_\Gamma F(\eta)\mathrm{L}^*_\eta \wedge f(\eta)d\omega(\eta) \tag{2.75}$$
$$= \int_{\partial\Gamma} \tau(\eta)\wedge\left(F(\eta)f(\eta)\right)d\sigma(\eta).$$

Lemma 2.9. *Let* $\Gamma \subset \Omega$ *be a regular region, and suppose that* F, H *are of class* $\mathrm{C}^{(2)}(\overline{\Gamma})$. *Then*

$$\int_\Gamma F(\eta)\Delta^*_\eta H(\eta)d\omega(\eta) - \int_\Gamma H(\eta)\Delta^*_\eta F(\eta)d\omega(\eta) \tag{2.76}$$
$$= \int_{\partial\Gamma} F(\eta)\frac{\partial}{\partial\nu(\eta)}H(\eta)d\sigma(\eta) - \int_{\partial\Gamma} H(\eta)\frac{\partial}{\partial\nu(\eta)}F(\eta)d\sigma(\eta).$$

Remark 2.10. The aforementioned statements hold as well for the entire sphere Ω instead of a subregion Γ, observing that no boundary integrals occur. For functions F of class $\mathrm{C}^{(1)}(\Omega)$ and tangential vector fields f of class $\mathrm{c}^{(1)}_{\mathrm{tan}}(\Omega)$, this especially implies

$$\int_\Omega f(\eta)\cdot\nabla^*_\eta F(\eta)\,d\omega(\eta) = -\int_\Omega F(\eta)\nabla^*_\eta\cdot f(\eta)\,d\omega(\eta), \tag{2.77}$$
$$\int_\Omega f(\eta)\cdot\mathrm{L}^*_\eta F(\eta)\,d\omega(\eta) = -\int_\Omega F(\eta)\mathrm{L}^*_\eta\cdot f(\eta)\,d\omega(\eta), \tag{2.78}$$

and

$$\int_\Omega \nabla^*_\eta\cdot f(\eta)d\omega(\eta) = \int_\Omega \mathrm{L}^*_\eta\cdot f(\eta)d\omega(\eta) = 0. \tag{2.79}$$

2.3 (Scalar) Spherical Harmonics

Spherical harmonics are the functions most commonly used in geosciences to represent scalar fields on a spherical surface $\Omega \subset \mathbb{R}^3$. They are used extensively in the gravitational and magnetic applications involving Laplace's equation. The introduction of (scalar) spherical harmonics and the derivation of some important properties occupies us in this section. More detailed information can be found, e.g., W. Freeden, T. Gervens, M. Schreiner [1998], W. Freeden, M. Schreiner [2009], E.W. Hobson [1955], N.N. Lebedev [1973], C. Müller [1966], and C. Müller [1998].

2.3.1 Basic Properties

Definition 2.11. Let $H_n : \mathbb{R}^3 \to \mathbb{R}$ be a homogeneous and harmonic polynomial of degree $n \in \mathbb{N}_0$, i.e., $H_n(\lambda x) = \lambda^n H_n(x)$, $\lambda \in \mathbb{R}$, and $\Delta H_n(x) = 0$, $x \in \mathbb{R}^3$. Then, the restriction

$$Y_n = H_n|_\Omega \tag{2.80}$$

is called a *(scalar) spherical harmonic (of degree n)*. The space of all spherical harmonics of degree n is denoted by $\mathrm{Harm}_n(\Omega)$.

The spherical harmonics of degree n form a space of dimension $2n+1$, i.e.,

$$\dim(\mathrm{Harm}_n(\Omega)) = 2n+1, \quad n \in \mathbb{N}_0. \tag{2.81}$$

Using the standard method of separation by spherical coordinates and observing the homogeneity, we have $H_n(x) = r^n Y_n(\xi)$, for $r = |x|$, $\xi = \frac{x}{|x|} \in \Omega$. From the identity $\left(\frac{1}{r^2}\frac{d}{dr}r^2\frac{d}{dr}\right) r^n = n(n+1)r^{n-2}$ it follows, in connection with (2.31) and the harmonicity of H_n, that

$$0 = \Delta H_n(x) = r^{n-2}n(n+1)Y_n(\xi) + r^{n-2}\Delta^*_\xi Y_n(\xi). \tag{2.82}$$

Lemma 2.12. *Any spherical harmonic $Y_n \in \mathrm{Harm}_n(\Omega)$, $n \in \mathbb{N}_0$, is an infinitely often differentiable eigenfunction of the Beltrami operator corresponding to the eigenvalue $-n(n+1)$. More explicitly,*

$$\Delta^* Y_n(\xi) = -n(n+1)Y_n(\xi), \qquad \xi \in \Omega. \tag{2.83}$$

Conversely, every infinitely often differentiable eigenfunction of the Beltrami operator with respect to the eigenvalue $-n(n+1)$ constitutes a spherical harmonic of degree n.

Using Green's formulas for the Beltrami operator, the above relation implies that spherical harmonics of different degrees are orthogonal with respect to the $\mathrm{L}^2(\Omega)$-inner product, i.e.,

$$(Y_n, Y_m)_{\mathrm{L}^2(\Omega)} = \int_\Omega Y_n(\xi) Y_m(\xi) \, d\omega(\xi) = 0, \qquad n \neq m. \tag{2.84}$$

Remark 2.13. The Gram-Schmidt method allows the orthonormalization of any set of linearly independent spherical harmonics of degree n with respect to the $\mathrm{L}^2(\Omega)$-inner product. By convention, any set

$$\{Y_{n,k}\}_{k=1,\ldots,2n+1} \subset \mathrm{Harm}_n(\Omega) \tag{2.85}$$

denotes an orthonormal basis of $\mathrm{Harm}_n(\Omega)$.

2.3.2 Legendre Polynomials

The Legendre polynomials are one-dimensional orthogonal polynomials that are of great importance when treating spherical harmonics. In the following, we restrict ourselves only to the very basic properties required in the course of this book. For more details, there is a large amount of literature available (e.g., W. Freeden, T. Gervens, M. Schreiner [1998], W. Freeden, M. Schreiner [2009], N.N. Lebedev [1973], C. Müller [1966], C. Müller [1998], and G. Szegö [1939]). Further properties can also be found in the exercises.

Definition 2.14. A polynomial $P_n : [-1,1] \to \mathbb{R}$ of degree $n \in \mathbb{N}_0$ is called a *Legendre polynomial (of degree n)* if the following properties are satisfied:

(i) $\int_{-1}^{1} P_n(t) P_m(t) dt = 0, \quad n \neq m,$

(ii) $P_n(1) = 1.$

The Legendre polynomials are uniquely determined by Definition 2.14. They have the explicit representation

$$P_n(t) = \sum_{k=0}^{\lfloor \frac{n}{2} \rfloor} (-1)^k \frac{(2n-2k)!}{2^n (n-2k)!(n-k)!k!} t^{n-2k}, \qquad t \in [-1,1]. \tag{2.86}$$

For an illustration, see Figure 2.2. Additionally, they are infinitely often differentiable eigenfunctions of the Legendre operator $\frac{d}{dt}(1-t^2)\frac{d}{dt}$ with respect to the eigenvalues $-n(n+1)$, i.e.,

$$\left(\frac{d}{dt}(1-t^2)\frac{d}{dt} \right) P_n(t) = -n(n+1) P_n(t), \qquad t \in [-1,1]. \tag{2.87}$$

The zonal function $\eta \mapsto P_n(\xi \cdot \eta)$, $\eta \in \Omega$, is the only normalized (i.e., $P_n(\xi \cdot \xi) = 1$) spherical harmonic of degree n that is invariant with respect to orthogonal transformations that leave $\xi \in \Omega$ fixed (cf. Exercise 2.8). A consequence of this fact is the addition theorem, which states the close relation of (univariate) Legendre polynomials to spherical harmonics.

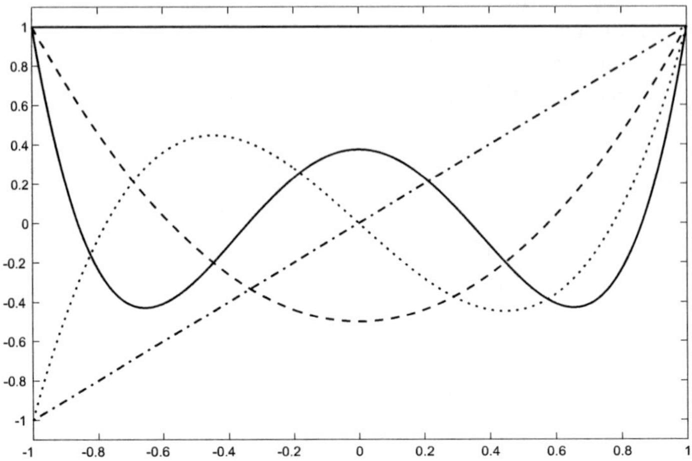

FIGURE 2.2
Illustration of the Legendre polynomials $P_n(t)$, $n = 0, \ldots, 4$.

Theorem 2.15 (Addition Theorem). *For $n \in \mathbb{N}_0$ and $\xi, \eta \in \Omega$, we have*

$$\sum_{k=1}^{2n+1} Y_{n,k}(\xi) Y_{n,k}(\eta) = \frac{2n+1}{4\pi} P_n(\xi \cdot \eta). \tag{2.88}$$

As a direct consequence, estimates for Legendre polynomials and spherical harmonics are derivable by standard arguments.

Corollary 2.16. *For $n \in \mathbb{N}_0$ and $k = 1, \ldots, 2n+1$, we have*

$$|P_n(t)| \leq 1, \quad t \in [-1, 1] \tag{2.89}$$

and

$$|Y_{n,k}(\xi)| \leq \sqrt{\frac{2n+1}{4\pi}}, \quad \xi \in \Omega. \tag{2.90}$$

Proof. We apply the Cauchy–Schwarz inequality and the addition theorem twice to obtain

$$\begin{aligned} \frac{2n+1}{4\pi} |P_n(\xi \cdot \eta)| &= \left| \sum_{k=1}^{2n+1} Y_{n,k}(\xi) Y_{n,k}(\eta) \right| \\ &\leq \left(\sum_{k=1}^{2n+1} |Y_{n,k}(\xi)|^2 \right)^{\frac{1}{2}} \left(\sum_{k=1}^{2n+1} |Y_{n,k}(\eta)|^2 \right)^{\frac{1}{2}} \\ &= \frac{2n+1}{4\pi} P_n(1), \end{aligned} \tag{2.91}$$

for $\xi, \eta \in \Omega$. Due to $P_n(1) = 1$, this is the desired assertion for the Legendre polynomials when setting $t = \xi \cdot \eta \in [-1,1]$. The assertion for spherical harmonics follows from the same estimate by the particular choice $\xi = \eta$. $\square$

The connection between the orthogonal invariance of the sphere and the addition theorem is established by the *Funk–Hecke formula*

$$\int_\Omega F(\xi \cdot \eta) Y_n(\eta)\, d\omega(\eta) = F^\wedge(n) Y_n(\xi), \quad \xi \in \Omega, \tag{2.92}$$

for $F \in \mathrm{L}^1([-1,1])$, $Y_n \in \mathrm{Harm}_n$, where

$$F^\wedge(n) = 2\pi \int_{-1}^{1} F(t) P_n(t) dt \tag{2.93}$$

denotes the *Legendre coefficient (of degree n)* of the function F. By virtue of the closure and completeness of the Legendre polynomials in $\mathrm{L}^2([-1,1])$ we get

$$\lim_{N\to\infty} \left\| F - \sum_{n=0}^{N} \frac{2n+1}{4\pi} F^\wedge(n) P_n \right\|_{\mathrm{L}^2([-1,1])} = 0, \tag{2.94}$$

provided that F is of class $\mathrm{L}^2([-1,1])$. In order to develop a similar representation on the unit sphere in terms of spherical harmonics, we need further properties of the Legendre polynomials. In a first step, we derive a closed representation for the generating series expansion of the Legendre polynomials.

Lemma 2.17. *For* $t \in [-1,1]$ *and* $h \in (-1,1)$, *we have*

$$\sum_{n=0}^{\infty} P_n(t) h^n = \frac{1}{\sqrt{1+h^2-2ht}}. \tag{2.95}$$

Proof. The power series

$$\phi(h) = \sum_{n=0}^{\infty} P_n(t) h^n, \quad t \in [-1,1], \tag{2.96}$$

is absolutely and uniformly convergent for all h with $|h| \le h_0$, and $h_0 \in [0,1)$ fixed, since $|P_n(t)| \le 1$, $t \in [-1,1]$, by Corollary 2.16. Differentiating with respect to h, using Exercise 2.5, and comparing the coefficients according to (2.96) leads to

$$(1+h^2-2ht)\phi'(h) = (t-h)\phi(h). \tag{2.97}$$

This differential equation is uniquely solvable under the initial condition $\phi(0) = 1$. Since it is not hard to show that $\phi(h) = (1+h^2-2ht)^{-\frac{1}{2}}$ also solves this initial value problem, we end up with the desired generating series expansion. $\square$

Remark 2.18. Of special importance later on is the fundamental solution for the Laplace operator $G(\Delta; |x-y|) = \frac{1}{4\pi}|x-y|^{-1}$, $x, y \in \mathbb{R}^3$, $x \neq y$. This function relates to the gravitational potential between a mass point y and a point in free space x (see our explanation in the introduction). An easy manipulation yields

$$\frac{1}{|x-y|} = \frac{1}{|y|}\left(1 + \left(\frac{|x|}{|y|}\right)^2 - 2\frac{|x|}{|y|}\xi \cdot \eta\right)^{-\frac{1}{2}}, \tag{2.98}$$

where $x, y \in \mathbb{R}^3$, $|x| < |y|$, and $\xi = \frac{x}{|x|}$, $\eta = \frac{y}{|y|}$. With $t = \xi \cdot \eta$ and $h = \frac{|x|}{|y|}$, Lemma 2.17 implies the series expansion

$$\frac{1}{|x-y|} = \frac{1}{|y|}\sum_{n=0}^{\infty}\left(\frac{|x|}{|y|}\right)^n P_n(\xi \cdot \eta). \tag{2.99}$$

Moreover, further calculations in connection with Hobson's formula (cf. Exercise 1.3) show that

$$\frac{1}{|x-y|} = \sum_{n=0}^{\infty}\frac{(-1)^n}{n!}|x|^n(\xi \cdot \nabla_y)^n\frac{1}{|y|}, \tag{2.100}$$

where

$$\frac{(-1)^n}{n!}(\xi \cdot \nabla_y)^n\frac{1}{|y|} = \frac{P_n(\xi \cdot \eta)}{|y|^{n+1}}, \qquad n \in \mathbb{N}_0. \tag{2.101}$$

Identity (2.101) is known as *Maxwell's representation formula.* As $y \mapsto |y|^{-1}$, $y \neq 0$, is (apart from a multiplicative constant) the fundamental solution for the Laplace operator, this representation tells us that the Legendre polynomials may be obtained by repeated differentiation of the fundamental solution in the radial direction of ξ. The potential on the right-hand side may be regarded as the potential of a pole of order n with the axis ξ at the origin.

Lemma 2.19. *For all $t \in [-1,1]$ and $h \in (-1,1)$, we have*

$$\frac{1-h^2}{(1+h^2-2ht)^{\frac{3}{2}}} = \sum_{n=0}^{\infty}(2n+1)h^n P_n(t). \tag{2.102}$$

Proof. The power series in Lemma 2.17 can be differentiated with respect to $h \in (-1,1)$. It follows that

$$-\frac{h-t}{(1+h^2-2ht)^{\frac{3}{2}}} = \sum_{n=1}^{\infty} nP_n(t)h^{n-1}. \tag{2.103}$$

Furthermore, it is easy to see that

$$\frac{1}{\sqrt{1+h^2-2ht}} - \frac{2h^2-2ht}{(1+h^2-2ht)^{\frac{3}{2}}} = \frac{1-h^2}{(1+h^2-2ht)^{\frac{3}{2}}}. \tag{2.104}$$

Thus, combining (2.103) and Lemma 2.17, we obtain the desired representation (2.102). □

2.3.3 Closure and Completeness

We are now prepared to develop the Abel–Poisson integral formula.

Theorem 2.20 (Abel–Poisson Integral Formula). *Let* F *be of class* $\mathrm{C}^{(0)}(\Omega)$. *Then*

$$\lim_{h\to 1-}\sup_{\xi\in\Omega}\left|\frac{1}{4\pi}\int_\Omega \frac{(1-h^2)}{(1+h^2-2h(\xi\cdot\eta))^{\frac{3}{2}}}F(\eta)\,d\omega(\eta)-F(\xi)\right|=0. \tag{2.105}$$

Proof. Lemma 2.19 enables us to conclude that

$$\int_{-1}^{1}\frac{1-h^2}{(1+h^2-2ht)^{\frac{3}{2}}}\,dt=\sum_{n=0}^{\infty}(2n+1)h^n\int_{-1}^{1}P_n(t)\,dt \tag{2.106}$$

for all $h\in(-1,1)$. Since $\int_{-1}^{1}P_n(t)P_0(t)\,dt=0$, $n\geq 1$, we have

$$\frac{1}{4\pi}\int_\Omega\frac{1-h^2}{(1+h^2-2h(\xi\cdot\eta))^{\frac{3}{2}}}d\omega(\eta)=\frac{1}{2}\int_{-1}^{1}\frac{1-h^2}{(1+h^2-2ht)^{\frac{3}{2}}}\,dt=1. \tag{2.107}$$

Observing the identity (2.107), we get

$$\begin{aligned}&\frac{1}{4\pi}\int_\Omega\frac{(1-h^2)}{(1+h^2-2h(\xi\cdot\eta))^{\frac{3}{2}}}F(\eta)d\omega(\eta)-F(\xi)\\ &\quad=\frac{1}{4\pi}\int_\Omega\frac{(1-h^2)(F(\eta)-F(\xi))}{(1+h^2-2h(\xi\cdot\eta))^{\frac{3}{2}}}\,d\omega(\eta).\end{aligned} \tag{2.108}$$

For $h\in\left[\frac{1}{2},1\right)$, we split the integral into two parts, namely

$$\int_\Omega\ldots\,d\omega(\eta)=\int_{\substack{\eta\in\Omega\\ -1\leq\xi\cdot\eta\leq 1-\sqrt[3]{1-h}}}\ldots\,d\omega(\eta)+\int_{\substack{\eta\in\Omega\\ 1-\sqrt[3]{1-h}\leq\xi\cdot\eta\leq 1}}\ldots\,d\omega(\eta). \tag{2.109}$$

On the one hand, we find

$$1+h^2-2ht=(1-h)^2+2h(1-t)\geq 2h\sqrt[3]{1-h} \tag{2.110}$$

and, consequently,

$$\frac{1-h^2}{(1+h^2-2ht)^{\frac{3}{2}}}\leq\frac{1-h^2}{(2h\sqrt[3]{1-h})^{\frac{3}{2}}}=\frac{1+h}{(2h)^{\frac{3}{2}}}\frac{1-h}{\sqrt{1-h}}\leq 2\sqrt{1-h}, \tag{2.111}$$

provided that $t \in [-1, 1-\sqrt[3]{1-h}]$. This leads us to the estimate

$$\left| \int\limits_{\substack{\eta\in\Omega \\ -1\leq\xi\cdot\eta\leq 1-\sqrt[3]{1-h}}} \frac{(1-h^2)(F(\xi)-F(\eta))}{(1+h^2-2h(\xi\cdot\eta))^{\frac{3}{2}}}\, d\omega(\eta) \right| \tag{2.112}$$

$$\leq 2\|F\|_{\mathrm{C}^{(0)}(\Omega)} \int\limits_{\substack{\eta\in\Omega \\ -1\leq\xi\cdot\eta\leq 1-\sqrt[3]{1-h}}} \frac{1-h^2}{(1+h^2-2h(\xi\cdot\eta))^{\frac{3}{2}}}\, d\omega(\eta)$$

$$= 4\pi\|F\|_{\mathrm{C}^{(0)}(\Omega)} \int_{-1}^{1-\sqrt[3]{1-h}} \frac{1-h^2}{(1+h^2-2ht)^{\frac{3}{2}}}\, dt$$

$$\leq 16\pi\|F\|_{\mathrm{C}^{(0)}(\Omega)}\sqrt{1-h}.$$

On the other hand, F is uniformly continuous on Ω. Thus, there exists a positive function $\mu : [-1,1] \mapsto \mathbb{R}$ with $\lim_{h\to 1-}\mu(h) = 0$ such that the estimate $|F(\xi)-F(\eta)| \leq \mu(h)$ holds for all $\eta \in \Omega$ satisfying $1-\sqrt[3]{1-h} \leq \xi\cdot\eta \leq 1$. Consequently, in connection with (2.107), we are able to deduce that

$$\left| \int\limits_{\substack{\eta\in\Omega \\ 1-\sqrt[3]{1-h}\leq\xi\cdot\eta\leq 1}} \frac{(1-h^2)(F(\xi)-F(\eta))}{(1+h^2-2h(\xi\cdot\eta))^{\frac{3}{2}}}\, d\omega(\eta) \right| \leq 4\pi\mu(h). \tag{2.113}$$

Letting h tend towards 1, we obtain the desired result. □

Combining Theorem 2.20, Lemma 2.19, and Theorem 2.15, we get the Abel–Poisson summability of a Fourier series expansion.

Theorem 2.21. *Let F be of class $\mathrm{C}^{(0)}(\Omega)$. Then the series*

$$\sum_{n=0}^{\infty} h^n \sum_{k=1}^{2n+1} F^\wedge(n,k)Y_{n,k}(\xi) \tag{2.114}$$

converges uniformly with respect to all $\xi \in \Omega$, for fixed $|h| \leq h_0$, $h_0 \in [0,1)$.

$$F^\wedge(n,k) = \int_\Omega F(\eta)Y_{n,k}(\eta)d\omega(\eta) \tag{2.115}$$

denotes the Fourier coefficient (of degree n and order k) *of F. Furthermore, understood in a pointwise sense, we have*

$$\lim_{h\to 1-} \sum_{n=0}^{\infty} h^n \sum_{k=1}^{2n+1} F^\wedge(n,k)Y_{n,k}(\xi) \;=\; F(\xi), \quad \xi\in\Omega. \tag{2.116}$$

Theorem 2.21 enables us to prove the closure of the system of spherical harmonics in $\mathrm{C}^{(0)}(\Omega)$ with respect to $\|\cdot\|_{\mathrm{C}^{(0)}(\Omega)}$.

Theorem 2.22 (Closure in $\mathrm{C}^{(0)}(\Omega)$). *The system $\{Y_{n,k}\}_{n\in\mathbb{N}_0,k=1,\ldots,2n+1}$ is closed in $\mathrm{C}^{(0)}(\Omega)$. That is, for any given $\varepsilon > 0$ and each $F \in \mathrm{C}^{(0)}(\Omega)$, there exist coefficients $a_{n,k} \in \mathbb{R}$ such that*

$$\left\| F - \sum_{n=0}^{N}\sum_{k=1}^{2n+1} a_{n,k} Y_{n,k} \right\|_{\mathrm{C}^{(0)}(\Omega)} \leq \varepsilon. \tag{2.117}$$

Proof. Let $F \in \mathrm{C}^{(0)}(\Omega)$. Theorem 2.21 implies that, for any given $\varepsilon > 0$, there exists a real number $h \in [0,1)$ such that

$$\sup_{\xi\in\Omega}\left| F(\xi) - \sum_{n=0}^{\infty} h^n \sum_{k=1}^{2n+1} F^{\wedge}(n,k) Y_{n,k}(\xi) \right| \leq \frac{\varepsilon}{2}. \tag{2.118}$$

Furthermore, there exists an index $N = N(\varepsilon)$ such that

$$\sup_{\xi\in\Omega}\left| \sum_{n=0}^{\infty} h^n \sum_{k=1}^{2n+1} F^{\wedge}(n,k) Y_{n,k}(\xi) - \sum_{n=0}^{N} h^n \sum_{k=1}^{2n+1} F^{\wedge}(n,k) Y_{n,k}(\xi) \right| \leq \frac{\varepsilon}{2}. \tag{2.119}$$

But this means that

$$\sup_{\xi\in\Omega}\left| F(\xi) - \sum_{n=0}^{N}\sum_{k=1}^{2n+1} h^n F^{\wedge}(n,k) Y_{n,k}(\xi) \right| \leq \varepsilon, \tag{2.120}$$

which proves Theorem 2.22, by choosing $a_{n,k} = h^n F^{\wedge}(n,k)$. □

Finally, we are interested in closure and completeness in the Hilbert space $\mathrm{L}^2(\Omega)$ with respect to the norm $\|\cdot\|_{\mathrm{L}^2(\Omega)}$. The next auxiliary corollary follows immediately from Theorem 2.22 by using the norm estimate (2.8).

Corollary 2.23. *The system $\{Y_{n,k}\}_{n\in\mathbb{N}_0,k=1,\ldots,2n+1}$ is closed in the space $\mathrm{C}^{(0)}(\Omega)$ with respect to $\|\cdot\|_{\mathrm{L}^2(\Omega)}$. That is, for any given $\varepsilon > 0$ and any given $F \in \mathrm{C}^{(0)}(\Omega)$, there exist coefficients $b_{n,k} \in \mathbb{R}$ such that*

$$\left\| F - \sum_{n=0}^{N}\sum_{k=1}^{2n+1} b_{n,k} Y_{n,k} \right\|_{\mathrm{L}^2(\Omega)} \leq \varepsilon. \tag{2.121}$$

Theorem 2.24 (Closure in $\mathrm{L}^2(\Omega)$). *The system $\{Y_{n,k}\}_{n\in\mathbb{N}_0,k=1,\ldots,2n+1}$ is closed in the space $\mathrm{L}^2(\Omega)$ with respect to $\|\cdot\|_{\mathrm{L}^2(\Omega)}$. That is, for any given $\varepsilon > 0$ and each $F \in \mathrm{L}^2(\Omega)$, there exist coefficients $b_{n,k} \in \mathbb{R}$ such that*

$$\left\| F - \sum_{n=0}^{N}\sum_{k=1}^{2n+1} b_{n,k} Y_{n,k} \right\|_{\mathrm{L}^2(\Omega)} \leq \varepsilon. \tag{2.122}$$

Proof. $\mathrm{C}^{(0)}(\Omega)$ is dense in $\mathrm{L}^2(\Omega)$. Thus, for every $F \in \mathrm{L}^2(\Omega)$ and $\varepsilon > 0$ there exists a function $\tilde{F} \in \mathrm{C}^{(0)}(\Omega)$ with $\|F - \tilde{F}\|_{\mathrm{L}^2(\Omega)} \leq \frac{\varepsilon}{2}$. By Corollary 2.23, the function $\tilde{F}$ admits an approximation by finite linear combinations of spherical harmonics with accuracy $\frac{\varepsilon}{2}$. Therefore, the proof of the closure is clear. □

In a Hilbert space, the closure property of a function system is equivalent to the *completeness property* (see, e.g., H.W. Alt [2006] and P.J. Davis [1963]). In our case, completeness means: If $F \in \mathrm{L}^2(\Omega)$ has vanishing Fourier coefficients

$$F^\wedge(n,k) = \int_\Omega F(\eta) Y_{n,k}(\eta) d\omega(\eta) = 0, \tag{2.123}$$

for all $n \in \mathbb{N}_0, k = 1, \ldots, 2n+1$, then $F = 0$ in $\mathrm{L}^2(\Omega)$-sense. In other words, F is uniquely determined by its Fourier coefficients. Furthermore, we can state the spherical analogon to (2.94), i.e.,

$$\lim_{N\to\infty} \left\| F - \sum_{n=0}^{N} \sum_{k=1}^{2n+1} F^\wedge(n,k) Y_{n,k} \right\|_{\mathrm{L}^2(\Omega)} = 0, \tag{2.124}$$

for F of class $\mathrm{L}^2(\Omega)$.

2.3.4 Inner/Outer Harmonics

In the following, we are interested in polynomial solutions of the Laplace equation. To this end, we consider the sphere Ω_R around the origin with radius $R > 0$. By virtue of the isomorphism $\xi \mapsto R\xi$, $\xi \in \Omega$, we can assume a function $F : \Omega_R \to \mathbb{R}$ to be reduced to the unit sphere Ω. Obviously, an $\mathrm{L}^2(\Omega)$-orthonormal system of spherical harmonics forms an $\mathrm{L}^2(\Omega_R)$-orthogonal system. More explicitly,

$$(Y_{n,k}, Y_{p,q})_{\mathrm{L}^2(\Omega_R)} = \int_{\Omega_R} Y_{n,k}\left(\frac{y}{|y|}\right) Y_{p,q}\left(\frac{y}{|y|}\right) d\omega(y) = R^2 \delta_{n,p}\delta_{k,q}. \tag{2.125}$$

Introducing the system

$$Y_{n,k}^R(x) = \frac{1}{R} Y_{n,k}\left(\frac{x}{|x|}\right), \quad x \in \Omega_R, \tag{2.126}$$

we get an orthonormal basis $\{Y_{n,k}^R\}_{n\in\mathbb{N}_0, k=1,\ldots,2n+1}$ of the space $\mathrm{L}^2(\Omega_R)$.

Definition 2.25 (Inner/Outer Harmonics).

(a) The functions

$$H_{n,k}^R(x) = \left(\frac{|x|}{R}\right)^n Y_{n,k}^R(x), \qquad x \in \mathbb{R}^3, \tag{2.127}$$

for $n \in \mathbb{N}_0$, $k = 1, \ldots, 2n+1$, are called *inner harmonics (of degree n and order k).*

(b) The functions

$$H^R_{-n-1,k}(x) = \left(\frac{R}{|x|}\right)^{n+1} Y^R_{n,k}(x), \qquad x \in \mathbb{R}^3\backslash\{0\}, \tag{2.128}$$

for $n \in \mathbb{N}_0$, $k = 1, \ldots, 2n+1$, are called *outer harmonics (of degree n and order k)*.

Remark 2.26. It is not difficult to see that the inner harmonics satisfy the following properties:

(i) $H^R_{n,k}$ is of class $\mathrm{C}^{(\infty)}(\mathbb{R}^3)$,

(ii) $H^R_{n,k}$ satisfies $\Delta H^R_{n,k}(x) = 0$, $x \in \mathbb{R}^3$,

(iii) $H^R_{n,k}|_{\Omega_R} = Y^R_{n,k}$,

(iv) $\left(H^R_{n,k}, H^R_{p,q}\right)_{\mathrm{L}^2(\Omega_R)} = \delta_{n,p}\delta_{k,q}$.

Analogously, the outer harmonics represent those functions that are harmonic in $\mathbb{R}^3 \setminus \{0\}$ and regular at infinity, and that coincide with the spherical harmonics $Y^R_{n,k}$ on the sphere Ω_R. More precisely,

(i) $H^R_{-n-1,k}$ is of class $\mathrm{C}^{(\infty)}(\mathbb{R}^3\backslash\{0\})$,

(ii) $H^R_{-n-1,k}$ satisfies $\Delta H^R_{-n-1,k}(x) = 0$, $x \in \mathbb{R}^3\backslash\{0\}$,

(iii) H^R_{-n-1} is regular at infinity,

(iv) $H^R_{-n-1,k}|_{\Omega_R} = Y^R_{n,k}$,

(v) $\left(H^R_{-n-1,k}, H^R_{-p-1,q}\right)_{\mathrm{L}^2(\Omega_R)} = \delta_{n,p}\, \delta_{k,q}$.

Furthermore, it should be noted that an inner harmonic $H^R_{n,k}$ is related to its corresponding outer harmonic $H^R_{-n-1,k}$ in the following way:

$$H^R_{-n-1,k}(x) = \left(\frac{R}{|x|}\right)^{2n+1} H^R_{n,k}(x) = \frac{R}{|x|} H^R_{n,k}\left(\frac{R^2}{|x|^2}x\right), \tag{2.129}$$

for $x \in \mathbb{R}^3 \setminus \{0\}$. In other words, the outer harmonics are obtainable by the Kelvin transform $\mathcal{K}^R$ from their inner counterparts, and vice versa. More precisely,

$$H^R_{-n-1,k}(x) = \mathcal{K}^R\left[H^R_{n,k}\right](x), \tag{2.130}$$

$$H^R_{n,k}(x) = \mathcal{K}^R\left[H^R_{-n-1,k}\right](x), \tag{2.131}$$

for $x \in \mathbb{R}^3 \setminus \{0\}$.

2.4 (Scalar) Circular Harmonics

The previous considerations have been undertaken with respect to two-dimensional spheres embedded in the Euclidean space $\mathbb{R}^3$. This can be generalized to $(q-1)$-dimensional spheres embedded in the Euclidean space $\mathbb{R}^q$, for $q \in \mathbb{N}$ (see, e.g., W. Freeden [2011] and C. Müller [1998]). In this section, we give a brief overview of the case $q = 2$. We use some slightly modified notation:

$$\Omega_R^{(2)} = \{x \in \mathbb{R}^2 : |x| = R\} \tag{2.132}$$

denotes the circle of radius $R > 0$ in $\mathbb{R}^2$, and $\Omega^{(2)}$ stands for the unit circle. As before, $\Omega_R\big(= \Omega_R^{(3)}\big)$ still denotes the sphere of radius $R > 0$ in $\mathbb{R}^3$, and $\Omega\big(= \Omega^{(3)}\big)$ stands for the unit sphere (note that, except in this subsection, we generally omit the index (3) indicating the case $q = 3$). The disc with radius $R > 0$ and center x in $\mathbb{R}^2$ is denoted by

$$\mathcal{B}_R^{(2)}(x) = \{y \in \mathbb{R}^2 : |x - y| < R\}, \tag{2.133}$$

while the ball with radius $R > 0$ and center x in $\mathbb{R}^3$ is still designated by $\mathcal{B}_R(x)\ \big(= \mathcal{B}_R^{(3)}(x)\big)$.

Actually, the case $q = 2$ is essentially a more abstract reformulation of well-known results for the trigonometric sine and cosine functions. The main motivation for this section on $q = 2$ is to transfer concepts that are used for boundary-value problems in the Euclidean space $\mathbb{R}^3$ to boundary-value problems on the unit sphere $\Omega^{(3)}$ (the boundary of a spherical cap is nothing more than a circle).

2.4.1 Basic Properties

Definition 2.27. Let $H_n : \mathbb{R}^2 \to \mathbb{R}$ be a homogeneous and harmonic polynomial of degree $n \in \mathbb{N}_0$, i.e., $H_n(\lambda x) = \lambda^n H_n(x)$, $\lambda \in \mathbb{R}$, and $\Delta H_n(x) = 0$, $x \in \mathbb{R}^2$. Then, the restriction

$$Y_n(2;\cdot) = H_n|_{\Omega^{(2)}} \tag{2.134}$$

is called a *(scalar) circular harmonic (of degree n)*. The space of all circular harmonics of degree n is denoted by $\mathrm{Harm}_n(\Omega^{(2)})$. The notation $Y_n(2;\cdot)$ is again used to distinguish the case $q = 2$ from the case $q = 3$ where we use $Y_n\big(= Y_n(3;\cdot)\big)$.

The circular harmonics of degree n form a space of dimension

$$\dim(\mathrm{Harm}_n(\Omega^{(2)})) = \begin{cases} 1, & n = 0, \\ 2, & n \geq 1. \end{cases} \tag{2.135}$$

Similar to the spherical harmonics of higher dimensions, any $Y_n(2;\cdot) \in$

$\mathrm{Harm}_n(\Omega^{(2)})$ is an infinitely often differentiable eigenfunction of the Beltrami operator $\Delta^*_{(2)}$ (acting on the circle $\Omega^{(2)}$). More precisely,

$$\Delta^*_{(2)} Y_n(2;\xi) = -n^2 Y_n(2;\xi), \quad \xi \in \Omega^{(2)}. \tag{2.136}$$

Note that, for the case $q = 2$, the Beltrami operator $\Delta^*_{(2)}$ satisfies

$$\Delta_{(2)} = \frac{1}{r}\frac{\partial}{\partial r} r \frac{\partial}{\partial r} + \frac{1}{r^2}\Delta^*_{(2)}, \tag{2.137}$$

where $\Delta_{(2)}$ denotes the Laplace operator in $\mathbb{R}^2$. Using a representation in polar coordinates (i.e., $x = r\xi = (r\cos(\varphi), r\sin(\varphi))^T$, $r > 0$, $\varphi \in [0, 2\pi)$), for any vector $x \in \mathbb{R}^2 \setminus \{0\}$), we have

$$\Delta^*_{(2)} = \left(\frac{\partial}{\partial \varphi}\right)^2. \tag{2.138}$$

Circular harmonics of different degrees are orthogonal with respect to the $\mathrm{L}^2(\Omega^{(2)})$-inner product, i.e.,

$$\left(Y_n(2;\cdot), Y_m(2;\cdot)\right)_{\mathrm{L}^2(\Omega^{(2)})} = \int_{\Omega^{(2)}} Y_n(2;\eta) Y_m(2;\eta)\, d\sigma(\eta) = 0, \quad n \neq m. \tag{2.139}$$

Remark 2.28. Throughout the course of this book, any set

$$\{Y_{n,k}(2;\cdot)\}_{k=1,2} \subset \mathrm{Harm}_n(\Omega^{(2)}) \tag{2.140}$$

denotes an orthonormal basis of $\mathrm{Harm}_n(\Omega^{(2)})$, $n \in \mathbb{N}$, with respect to the $\mathrm{L}^2(\Omega^{(2)})$-inner product. For $n = 0$, we set $Y_{0,1}(2;\xi) = (2\pi)^{-\frac{1}{2}}$, $\xi \in \Omega^{(2)}$. It can be easily seen that an orthonormal basis of $\mathrm{Harm}_n(\Omega^{(2)})$ is given by

$$Y_{n,1}(2;\xi) = \frac{1}{\sqrt{\pi}}\cos(n\varphi), \qquad Y_{n,2}(2;\xi) = \frac{1}{\sqrt{\pi}}\sin(n\varphi), \tag{2.141}$$

for $n \in \mathbb{N}$, $\xi = (\cos(\varphi), \sin(\varphi))^T$, and $\varphi \in [0, 2\pi)$.

2.4.2 Chebyshev Polynomials

For the case $q = 2$, the Chebyshev polynomials take over the role of the Legendre polynomials $P_n (= P_n(3;\cdot))$. For more details, see, e.g., W. Freeden [2011], N.N. Lebedev [1973], C. Müller [1998], and G. Szegö [1939].

Definition 2.29. A polynomial $P_n(2;\cdot) : [-1,1] \to \mathbb{R}$ of degree $n \in \mathbb{N}_0$ is called a *Chebyshev polynomial (of degree n)* if it satisfies the properties

(i) $\int_{-1}^{1} P_n(2;t) P_m(2;t) \frac{1}{\sqrt{1-t^2}}\, dt = 0, \quad n \neq m,$

(ii) $P_n(2;1) = 1.$

The Chebyshev polynomials are uniquely determined by Definition 2.29 and have the explicit representation

$$P_n(2;t) = \frac{n}{2}\sum_{k=0}^{\lfloor \frac{n}{2} \rfloor}(-1)^k \frac{(n-1-k)!}{2^{2k-n}(n-2k)!k!}t^{n-2k} \tag{2.142}$$

$$= \cos(n \arccos(t)), \tag{2.143}$$

for $t \in [-1,1]$. Furthermore, they satisfy the differential equation

$$\left(\sqrt{1-t^2}\frac{d}{dt}(1-t^2)\frac{d}{dt}\right) P_n(2;t) = -n^2 P_n(2;t), \qquad t \in [-1,1]. \tag{2.144}$$

Similar to the Legendre polynomials, we obtain an addition theorem connecting the Chebychev polynomials to the circular harmonics.

Theorem 2.30 (Addition Theorem). *For $n \in \mathbb{N}$, we have*

$$\sum_{k=1}^{2} Y_{n,k}(2;\xi)Y_{n,k}(2;\eta) = \frac{1}{\pi}P_n(2;\xi\cdot\eta), \quad \xi,\eta \in \Omega^{(2)}. \tag{2.145}$$

Theorem 2.30 is just an abstract version of well-known trigonometric addition theorems like $\cos(n\varphi)\cos(n\theta) + \sin(n\varphi)\sin(n\theta) = \cos(n(\varphi-\theta))$.

Corollary 2.31. *For $n \in \mathbb{N}_0$, we have*

$$|P_n(2;t)| \leq 1, \quad t \in [-1,1]. \tag{2.146}$$

For $n \in \mathbb{N}$, $k = 1,2$, or $n = 0$, $k = 1$, we get

$$|Y_{n,k}(2;\xi)| \leq \frac{1}{\sqrt{\pi}}, \quad \xi \in \Omega^{(2)}. \tag{2.147}$$

It is possible to derive closed representations for different generating series of the Chebychev polynomials.

Lemma 2.32. *For $t \in [-1,1]$ and $h \in (-1,1)$, we have*

$$\sum_{n=0}^{\infty} P_n(2;t)h^n = \frac{1-ht}{1+h^2-2ht}, \tag{2.148}$$

$$\sum_{n=1}^{\infty} \frac{1}{n}P_n(2;t)h^n = -\frac{1}{2}\ln(1+h^2-2ht). \tag{2.149}$$

While (2.148) is the canonical counterpart of Lemma 2.17, Equation (2.149) states the relevant expression to achieve a *multipole expansion* of the fundamental solution for the Laplace operator in $\mathbb{R}^2$. More precisely, we obtain

$$\ln(|x-y|) = \ln(|y|) + \ln\left(1 + \left(\frac{|x|}{|y|}\right)^2 - 2(\xi\cdot\eta)\frac{|x|}{|y|}\right) \tag{2.150}$$

$$= \ln(|y|) - \sum_{n=1}^{\infty}\frac{2}{n}\left(\frac{|x|}{|y|}\right)^n P_n(2;\xi\cdot\eta),$$

for $x, y \in \mathbb{R}^2$, $|x| < |y|$, and $\xi = \frac{x}{|x|}$, $\eta = \frac{y}{|y|}$. We conclude this subsection with the closure and completeness theorem of the circular harmonics.

Theorem 2.33 (Closure in $\mathrm{C}^{(0)}(\Omega^{(2)})$ and $\mathrm{L}^2(\Omega^{(2)})$). *The system of circular harmonics* $\{Y_{0,1}(2;\cdot)\} \oplus \{Y_{n,k}(2;\cdot)\}_{n\in\mathbb{N},k=1,2}$ *is closed in* $\mathrm{C}^{(0)}(\Omega^{(2)})$ *(with respect to* $\|\cdot\|_{\mathrm{C}^{(0)}(\Omega^{(2)})}$ *and* $\|\cdot\|_{\mathrm{L}^2(\Omega^{(2)})}$*) and in* $\mathrm{L}^2(\Omega^{(2)})$ *(with respect to* $\|\cdot\|_{\mathrm{L}^2(\Omega^{(2)})}$*).*

2.4.3 Stereographic Projection

Now that we have treated circular harmonics on $\Omega^{(2)}$ embedded in the Euclidean plane $\mathbb{R}^2$, our purpose is to transfer the results to circles embedded in the sphere $\Omega^{(3)}$. An essential tool is the stereographic projection (see Figure 2.3).

Definition 2.34. The mapping $p_{\mathrm{stereo}} : \Omega^{(3)} \setminus \{-\tilde{\xi}\} \to \mathbb{R}^2$, for $\tilde{\xi} \in \Omega^{(3)}$ fixed, is called a *stereographic projection (with respect to* $\tilde{\xi}$*)* if

$$p_{\mathrm{stereo}}(\xi) = \left(\frac{2\xi\cdot(\mathbf{t}\varepsilon^1)}{1+\xi\cdot\tilde{\xi}}, \frac{2\xi\cdot(\mathbf{t}\varepsilon^2)}{1+\xi\cdot\tilde{\xi}} \right)^T, \quad \xi \in \Omega^{(3)} \setminus \{-\tilde{\xi}\}, \tag{2.151}$$

where $\mathbf{t} \in \mathbb{R}^{3\times 3}$ denotes the orthogonal transformation satisfying $\mathbf{t}\varepsilon^3 = \tilde{\xi}$. An additional index $p_{\mathrm{stereo}} = p_{\tilde{\xi},\mathrm{stereo}}$, indicating the dependence on $\tilde{\xi}$, is usually omitted. Furthermore, let $\Gamma \subset \Omega^{(3)} \setminus \{-\tilde{\xi}\}$ be a regular region and $\mathcal{G} = p_{\mathrm{stereo}}(\Gamma) \subset \mathbb{R}^2$ its image in the plane. Then the stereographic projection $p_{\mathrm{stereo}}[F] : \Gamma \to \mathbb{R}$ of a function $F : \mathcal{G} \to \mathbb{R}$ is defined by

$$p_{\mathrm{stereo}}[F](\xi) = F(p_{\mathrm{stereo}}(\xi)), \quad \xi \in \Gamma. \tag{2.152}$$

An interesting connection between the Laplace operator and the Beltrami operator is given in the following lemma.

Lemma 2.35. *Let* $\Gamma \subset \Omega^{(3)} \setminus \{-\tilde{\xi}\}$*, with* $\tilde{\xi} \in \Omega^{(3)}$ *fixed, be a regular region, and* $\mathcal{G} = p_{\mathrm{stereo}}(\Gamma) \subset \mathbb{R}^2$ *its image in the plane. If* F *is of class* $\mathrm{C}^{(2)}(\mathcal{G})$*, then*

$$\frac{(1+\xi\cdot\tilde{\xi})^2}{4} \Delta^* p_{\mathrm{stereo}}[F](\xi) = \Delta F(x)|_{x=p_{\mathrm{stereo}}(\xi)}, \quad \xi \in \Gamma. \tag{2.153}$$

Lemma 2.35 implies that the stereographic projection of a function that is harmonic with respect to the Laplace operator is harmonic with respect to the Beltrami operator. This observation allows us to transfer some results for harmonic functions from the Euclidean setting in the plane $\mathbb{R}^2$ to the sphere $\Omega^{(3)}$, and vice versa.

Definition 2.36. A *circular harmonic (of degree* n *and order* k*) on the boundary of a spherical cap* $\Gamma_\rho(\tilde{\xi}) \subset \Omega^{(3)}$ with radius $\rho \in (0,2)$ and center $\tilde{\xi} \in \Omega^{(3)}$ is defined as

$$Y_{n,k}^{\rho,\mathrm{stereo}}(\xi) = p_{\mathrm{stereo}}\left[Y_{n,k}^{\rho^{\frac{1}{4}}(2-\rho)^{\frac{1}{4}}}(2;\cdot) \right](\xi), \quad \xi \in \partial\Gamma_\rho(\tilde{\xi}). \tag{2.154}$$

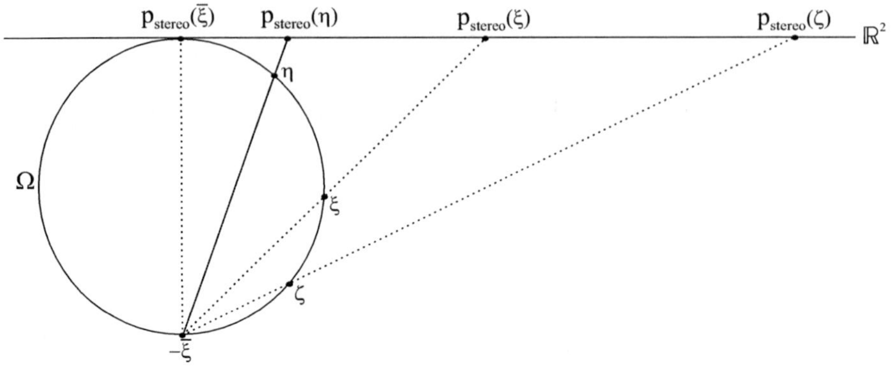

FIGURE 2.3
Two-dimensional illustration of the stereographic projection.

By $Y^R_{n,k}(2;x) = R^{-1}Y_{n,k}(2;\frac{x}{|x|})$, $x \in \Omega^{(2)}_R$, we denote a set of orthonormalized circular harmonics on the circle of radius $R > 0$.

Remark 2.37. The functions $Y^{\rho,\text{stereo}}_{n,k}$ form an orthonormal system with respect to the $\mathrm{L}^2(\partial\Gamma_\rho(\tilde{\xi}))$-inner product. Furthermore, observing that the stereographic projection does not affect the longitude of a point on the sphere $\Omega^{(3)}$ (at least for the choice $\tilde{\xi} = \varepsilon^3$), it can be seen that $Y^{\rho,\text{stereo}}_{n,k}$ and $Y^{\rho^{\frac{1}{4}}(2-\rho)^{\frac{1}{4}}}_{n,k}(2;\cdot)$ have the same representation when using spherical coordinates and polar coordinates, respectively. More precisely, for $n \in \mathbb{N}$,

$$Y^{\rho,\text{stereo}}_{n,1}(\xi) = \frac{\cos(n\varphi)}{\pi^{\frac{1}{2}}\rho^{\frac{1}{4}}(2-\rho)^{\frac{1}{4}}}, \qquad Y^{\rho,\text{stereo}}_{n,2}(\xi) = \frac{\sin(n\varphi)}{\pi^{\frac{1}{2}}\rho^{\frac{1}{4}}(2-\rho)^{\frac{1}{4}}}, \tag{2.155}$$

where $\xi = \left(\rho^{\frac{1}{2}}(2-\rho)^{\frac{1}{2}}\cos(\varphi), \rho^{\frac{1}{2}}(2-\rho)^{\frac{1}{2}}\sin(\varphi), 1-\rho\right)^T \in \partial\Gamma_\rho(\varepsilon^3)$, $\varphi \in [0, 2\pi)$.

2.4.4 Inner/Outer Harmonics

First, we stick to the plane $\mathbb{R}^2$. Here, inner and outer harmonics can be defined according to Definition 2.38. Only for the outer harmonics there occurs a minor modification due to the reduced dimension for the case $q = 2$.

Definition 2.38 (Circular Inner/Outer Harmonics).

(a) The functions

$$H^R_{n,k}(2;x) = \left(\frac{|x|}{R}\right)^n Y^R_{n,k}(2;x), \qquad x \in \mathbb{R}^2, \tag{2.156}$$

for $n \in \mathbb{N}$, $k = 1, 2$, or $n = 0$, $k = 1$, are called *circular inner harmonics (of degree n and order k).*

(b) The functions

$$H^R_{-n,k}(2;x) = \left(\frac{R}{|x|}\right)^n Y^R_{n,k}(2;x), \qquad x \in \mathbb{R}^2\backslash\{0\}, \tag{2.157}$$

for $n \in \mathbb{N}$, $k = 1, 2$, or $n = 0$, $k = 1$, are called *circular outer harmonics (of degree n and order k).*

Remark 2.39. Similar to the already known versions of inner harmonics, it can easily be seen that the following properties hold true:

(i) $H^R_{n,k}(2;\cdot)$ is of class $\mathrm{C}^{(\infty)}(\mathbb{R}^2)$,

(ii) $H^R_{n,k}(2;\cdot)$ satisfies $\Delta H^R_{n,k}(2;x) = 0$, $x \in \mathbb{R}^2$,

(iii) $H^R_{n,k}(2;\cdot)\big|_{\Omega^{(2)}_R} = Y^R_{n,k}(2;\cdot)$,

(iv) $\left(H^R_{n,k}(2;\cdot), H^R_{p,q}(2;\cdot)\right)_{\mathrm{L}^2(\Omega^{(2)}_R)} = \delta_{n,p}\delta_{k,q}.$

For the circular outer harmonics, we obtain:

(i) $H^R_{-n,k}(2;\cdot)$ is of class $\mathrm{C}^{(\infty)}(\mathbb{R}^2\backslash\{0\})$,

(ii) $H^R_{-n,k}(2;\cdot)$ satisfies $\Delta H^R_{-n,k}(2;x) = 0$, $x \in \mathbb{R}^2\backslash\{0\}$,

(iii) $H^R_{-n,k}(2;\cdot)$ is regular at infinity,

(iv) $H^R_{-n,k}(2;\cdot)\big|_{\Omega^{(2)}_R} = Y^R_{n,k}(2;\cdot)$,

(v) $\left(H^R_{-n,k}(2;\cdot), H^R_{-p,q}(2;\cdot)\right)_{\mathrm{L}^2(\Omega^{(2)}_R)} = \delta_{n,p}\,\delta_{k,q}.$

What we are interested in for later considerations are not the circular inner and outer harmonics in the plane $\mathbb{R}^2$, but their spherical counterparts on the sphere $\Omega^{(3)}$. This can be achieved by a simple application of the stereographic projection.

Definition 2.40 (Inner/Outer Harmonics on the Sphere)**.** The following definitions are meant with respect to the spherical cap $\Gamma_\rho(\tilde{\xi})$ with radius $\rho \in (0,2)$ and center $\tilde{\xi} \in \Omega^{(3)}$.

(a) The functions

$$H^{\rho,\mathrm{stereo}}_{n,k}(\xi) = p_{\mathrm{stereo}}\left[H^{\rho^{\frac{1}{4}}(2-\rho)^{\frac{1}{4}}}_{n,k}(2;\cdot)\right](\xi), \qquad \xi \in \Omega^{(3)} \setminus \{-\tilde{\xi}\}, \tag{2.158}$$

for $n \in \mathbb{N}$, $k = 1, 2$, or $n = 0$, $k = 1$, are called *inner harmonics (of degree n and order k) with respect to the spherical cap $\Gamma_\rho(\tilde{\xi})$* .

(b) The functions

$$H^{\rho,\mathrm{stereo}}_{-n,k}(\xi) = p_{\mathrm{stereo}}\left[H^{\rho^{\frac{1}{4}}(2-\rho)^{\frac{1}{4}}}_{-n,k}(2;\cdot)\right](\xi), \qquad \xi \in \Omega^{(3)} \setminus \{\tilde{\xi}\}, \quad (2.159)$$

for $n \in \mathbb{N}$, $k = 1, 2$, or $n = 0$, $k = 1$, are called *outer harmonics (of degree n and order k) with respect to the spherical cap $\Gamma_\rho(\tilde{\xi})$* .

Since the stereographic projection depends on the center $\tilde{\xi}$ of the spherical cap, the inner and outer harmonics also depend on the choice of $\tilde{\xi}$. For brevity, we usually do not indicate this in the notation.

Remark 2.41. As usual, the inner harmonics with respect to the cap $\Gamma_\rho(\tilde{\xi})$ satisfy

(i) $H^{\rho,\mathrm{stereo}}_{n,k}$ is of class $\mathrm{C}^{(\infty)}(\Omega^{(3)} \setminus \{-\tilde{\xi}\})$,

(ii) $H^{\rho,\mathrm{stereo}}_{n,k}$ satisfies $\Delta^* H^{\rho,\mathrm{stereo}}_{n,k}(\xi) = 0$, $\xi \in \Omega^{(3)} \setminus \{-\tilde{\xi}\}$,

(iii) $H^{\rho,\mathrm{stereo}}_{n,k}\big|_{\partial\Gamma_\rho(\tilde{\xi})} = Y^{\rho,\mathrm{stereo}}_{n,k}$,

(iv) $\left(H^{\rho,\mathrm{stereo}}_{n,k}, H^{\rho,\mathrm{stereo}}_{p,q}\right)_{\mathrm{L}^2(\partial\Gamma_\rho(\tilde{\xi}))} = \delta_{n,p}\,\delta_{k,q}$.

The properties for the outer harmonics with respect to the cap $\Gamma_\rho(\tilde{\xi})$, however, show slightly modified properties in comparison to the Euclidean cases we have treated up to now:

(i) $H^{\rho,\mathrm{stereo}}_{-n,k}$ is of class $\mathrm{C}^{(\infty)}(\Omega^{(3)} \setminus \{\tilde{\xi}\})$,

(ii) $H^{\rho,\mathrm{stereo}}_{-n,k}$ satisfies $\Delta^* H^{\rho,\mathrm{stereo}}_{-n,k}(\xi) = 0$, $\xi \in \Omega^{(3)} \setminus \{\tilde{\xi}\}$,

(iii) $H^{\rho,\mathrm{stereo}}_{-n,k}\big|_{\partial\Gamma_\rho(\tilde{\xi})} = Y^{\rho,\mathrm{stereo}}_{n,k}$,

(iv) $\left(H^{\rho,\mathrm{stereo}}_{-n,k}, H^{\rho,\mathrm{stereo}}_{-p,q}\right)_{\mathrm{L}^2(\partial\Gamma_\rho(\tilde{\xi}))} = \delta_{n,p}\,\delta_{k,q}$.

While there is an actual difference between inner and outer harmonics in the Euclidean framework of $\mathbb{R}^2$ and $\mathbb{R}^3$, because the exterior of a regular region is unbounded, this is not true in the spherical context. Consequently, the inner harmonics $H^{\rho,\mathrm{stereo}}_{n,k}$ for the spherical cap $\Gamma_\rho(\tilde{\xi})$ coincide with outer harmonics $H^{2-\rho,\mathrm{stereo}}_{-n,k}$ for the spherical cap $\Gamma_{2-\rho}(-\tilde{\xi})$.

Of special interest to us is the multipole representation (2.150). The addition theorem implies

$$\ln(|x-y|) \;=\; \ln(|y|) - \pi R^2 \sum_{n=1}^{\infty}\sum_{k=1}^{2} \frac{2}{n} H^R_{n,k}(2;x) H^R_{-n,k}(2;y), \quad (2.160)$$

for $x, y \in \mathbb{R}^2$, $|x| < |y|$, and some fixed $R > 0$. Applying the stereographic projection to (2.160), we obtain the following spherical version.

Lemma 2.42. *Let $\tilde{\xi} \in \Omega^{(3)}$ and $\rho \in (0,2)$. Then we have*

$$\begin{aligned} \ln(1 - \xi \cdot \eta) &= -\ln(2) + \ln(1 + \xi \cdot \tilde{\xi}) + \ln(1 - \tilde{\xi} \cdot \eta) \\ &\quad - \sqrt{\rho(2-\rho)}\pi \sum_{n=1}^{\infty} \sum_{k=1}^{2} \frac{2}{n} H_{n,k}^{\rho,\text{stereo}}(\xi) H_{-n,k}^{\rho,\text{stereo}}(\eta), \end{aligned} \tag{2.161}$$

for $\xi \in \Omega^{(3)} \setminus \{-\tilde{\xi}\}$, $\eta \in \Omega^{(3)} \setminus \{\tilde{\xi}\}$, and $|p_{\text{stereo}}(\xi)| < |p_{\text{stereo}}(\eta)|$.

2.5 Vector Spherical Harmonics

Analogous to the scalar case, we want to derive complete orthonormal function systems for the vectorial space $\mathrm{l}^2(\Omega)$. Overviews of this topic are given, e.g., in W. Freeden, T. Gervens, M. Schreiner [1998], and W. Freeden, M. Schreiner [2009] (for more detailed information, we refer to further references therein). We restrict ourselves to aspects that are of importance in the course of this book.

2.5.1 Radial-Tangential System

As is well known, the *spherical Helmholtz decomposition* of a vector field f of class $\mathrm{c}^{(1)}(\Omega)$ is given by

$$f(\xi) = \xi F_1(\xi) + \nabla^* F_2(\xi) + \mathrm{L}^* F_3(\xi), \quad \xi \in \Omega, \tag{2.162}$$

where F_1 is of class $\mathrm{C}^{(1)}(\Omega)$, and F_2, F_3 are scalar fields of class $\mathrm{C}^{(2)}(\Omega)$. Since any of the scalar fields F_1, F_2, F_3 can be expanded in terms of (scalar) spherical harmonics, the upcoming Definition 2.43 of vector spherical harmonics is straightforward by introducing the *Helmholtz operators*

$$o_\xi^{(1)} F(\xi) = \xi F(\xi), \tag{2.163}$$

$$o_\xi^{(2)} F(\xi) = \nabla_\xi^* F(\xi), \tag{2.164}$$

$$o_\xi^{(3)} F(\xi) = \mathrm{L}_\xi^* F(\xi), \tag{2.165}$$

for $\xi \in \Omega$ and sufficiently smooth scalar functions $F : \Omega \to \mathbb{R}$. The adjoint operators $O^{(i)}$ in the sense of

$$(o^{(i)} F, f)_{\mathrm{l}^2(\Omega)} = (F, O^{(i)} f)_{\mathrm{L}^2(\Omega)}, \tag{2.166}$$

for $F : \Omega \to \mathbb{R}$ and $f : \Omega \to \mathbb{R}^3$ sufficiently smooth, are given by

$$O_\xi^{(1)} f(\xi) = \xi \cdot f(\xi), \tag{2.167}$$

$$O_\xi^{(2)} f(\xi) = -\nabla_\xi^* \cdot p_{\tan}[f](\xi), \tag{2.168}$$

$$O_\xi^{(3)} f(\xi) = -\mathrm{L}_\xi^* \cdot p_{\tan}[f](\xi), \tag{2.169}$$

for $\xi \in \Omega$. This can easily be seen from Green's formulas on the unit sphere.

Definition 2.43. For $i = 1, 2, 3$, we denote by

$$y_{n,k}^{(i)} = \left(\mu_n^{(i)}\right)^{-\frac{1}{2}} o^{(i)} Y_{n,k}, \quad n \in \mathbb{N}_{0_i}, k = 1, \dots 2n+1 \tag{2.170}$$

a *vector spherical harmonic (of type i, degree n, and order k)*, where

$$0_i = \begin{cases} 0, & i = 1, \\ 1, & i = 2, 3, \end{cases} \tag{2.171}$$

such that $\mathbb{N}_{0_i} = \mathbb{N}_0$, for $i = 1$, and $\mathbb{N}_{0_i} = \mathbb{N}$, for $i = 2, 3$. The normalization factor $\mu_n^{(i)}$ is given by

$$\mu_n^{(i)} = \begin{cases} 1, & i = 1, \\ n(n+1), & i = 2, 3. \end{cases} \tag{2.172}$$

By $\mathrm{harm}_n^{(i)}(\Omega)$ we denote the space of all vector spherical harmonics of type i and degree n. Furthermore, we set $\mathrm{harm}_0(\Omega) = \mathrm{harm}_0^{(1)}(\Omega)$ and $\mathrm{harm}_n(\Omega) = \bigoplus_{i=1}^{3} \mathrm{harm}_n^{(i)}(\Omega)$, $n \in \mathbb{N}$.

The notation 0_i takes into account the fact that $o_\xi^{(i)} Y_{0,1}(\xi) = 0$, for $\xi \in \Omega$ and $i = 2, 3$. Since $\{Y_{n,k}\}_{n \in \mathbb{N}_0, k=1,\dots,2n+1}$ forms an orthonormal system in $\mathrm{L}^2(\Omega)$, the nature of the spherical differential operators ∇^*, L^* implies the orthonormality of $\{y_{n,k}^{(i)}\}_{i=1,2,3,\, n \in \mathbb{N}_{0_i}, k=1,\dots,2n+1}$ in $\mathrm{l}^2(\Omega)$. More precisely,

$$\int_\Omega y_{n,k}^{(i)}(\xi) \cdot y_{p,q}^{(j)}(\xi)\, d\omega(\xi) = \delta_{i,j} \delta_{n,p} \delta_{k,q}, \tag{2.173}$$

which actually justifies the notation $\bigoplus_{i=1}^{3} \mathrm{harm}_n^{(i)}(\Omega)$ in Definition 2.43. The closure (and completeness) in the Hilbert space $\mathrm{l}^2(\Omega)$ follows from the Helmholtz decomposition (2.162) and the closure (and completeness) of $\{Y_{n,k}\}_{n \in \mathbb{N}_0, k=1,\dots,2n+1}$ in $\mathrm{L}^2(\Omega)$.

Theorem 2.44 (Closure in $\mathrm{l}^2(\Omega)$). *The system* $\{y_{n,k}^{(i)}\}_{i=1,2,3,\, n \in \mathbb{N}_{0_i}, k=1,\dots,2n+1}$ *is closed and complete in the space* $\mathrm{l}^2(\Omega)$ *with respect to* $\|\cdot\|_{\mathrm{l}^2(\Omega)}$. *More precisely,*

$$\lim_{N \to \infty} \left\| f - \sum_{i=1}^{3} \sum_{n=0}^{N} \sum_{k=1}^{2n+1} (f^{(i)})^\wedge(n,k) y_{n,k}^{(i)} \right\|_{\mathrm{l}^2(\Omega)} = 0, \tag{2.174}$$

provided that f *is of class* $\mathrm{l}^2(\Omega)$, *where*

$$(f^{(i)})^\wedge(n,k) = \int_\Omega f(\eta) \cdot y_{n,k}^{(i)}(\eta) d\omega(\eta) \tag{2.175}$$

denotes the corresponding Fourier coefficient (of type i, degree n, and order k) .

For convenience, we introduce the spaces

$$\mathrm{l}^2_{(i)}(\Omega) = \overline{\{o^{(i)}F : F \in \mathrm{C}^{(\infty)}(\Omega)\}}^{\|\cdot\|_{\mathrm{l}^2(\Omega)}}, \quad i = 1,2,3, \tag{2.176}$$

and obtain

$$\mathrm{l}^2(\Omega) = \bigoplus_{i=1}^{3} \mathrm{l}^2_{(i)}(\Omega). \tag{2.177}$$

To span the space $\mathrm{l}^2_{\mathrm{tan}}(\Omega)$ of tangential functions, we need $\mathrm{l}^2_{(2)}(\Omega)$ and $\mathrm{l}^2_{(3)}(\Omega)$. The space $\mathrm{l}^2_{(1)}(\Omega)$ contains only functions directed in radial direction.

Besides completeness, many other properties of the scalar harmonics can be transferred to the vectorial setting. One property that we need later on is a vectorial counterpart to the *Funk-Hecke formula* (2.92):

$$\int_\Omega F(\xi\cdot\eta)o^{(1)}_\eta Y_n(\eta)d\omega(\eta) = F^\wedge_{(1,1)}(n)o^{(1)}_\xi Y_n(\xi) + F^\wedge_{(1,2)}(n)o^{(2)}_\xi Y_n(\xi), \tag{2.178}$$

$$\int_\Omega F(\xi\cdot\eta)o^{(2)}_\eta Y_n(\eta)d\omega(\eta) = F^\wedge_{(2,1)}(n)o^{(1)}_\xi Y_n(\xi) + F^\wedge_{(2,2)}(n)o^{(2)}_\xi Y_n(\xi), \tag{2.179}$$

$$\int_\Omega F(\xi\cdot\eta)o^{(3)}_\eta Y_n(\eta)d\omega(\eta) = F^\wedge_{(3,1)}(n)o^{(3)}_\xi Y_n(\xi), \tag{2.180}$$

for $\xi \in \Omega$, and functions $F \in \mathrm{L}^1([-1,1])$, $Y_n \in \mathrm{Harm}_n(\Omega)$. The coefficients $F^\wedge_{(i,j)}(n)$ are given by

$$F^\wedge_{(1,1)}(n) = \frac{1}{2n+1}((n+1)F^\wedge(n+1) + nF^\wedge(n-1)), \tag{2.181}$$

$$F^\wedge_{(1,2)}(n) = \frac{1}{2n+1}(F^\wedge(n-1) - F^\wedge(n+1)), \tag{2.182}$$

$$F^\wedge_{(2,1)}(n) = \frac{n(n+1)}{2n+1}(F^\wedge(n-1) - F^\wedge(n+1)), \tag{2.183}$$

$$F^\wedge_{(2,2)}(n) = \frac{1}{2n+1}(nF^\wedge(n+1) + (n+1)F^\wedge(n-1)), \tag{2.184}$$

$$F^\wedge_{(3,1)}(n) = F^\wedge(n), \tag{2.185}$$

for $n \in \mathbb{N}$, where $F^\wedge(n) = 2\pi\int_{-1}^{1} F(t)P_n(t)dt$ denotes the already introduced Legendre coefficient.

Different from the (scalar) spherical harmonics, the vector spherical harmonics of type 1 and 2 are not eigenfunctions of the Beltrami operator Δ^* (acting componentwise on vectorial functions). This is clarified in more detail by the following lemma.

Lemma 2.45.

(a) Let $F : \Omega \to \mathbb{R}$ be sufficiently smooth. Then

$$\Delta^* o^{(1)}F = o^{(1)}(\Delta^* - 2)F + 2o^{(2)}F, \tag{2.186}$$

$$\Delta^* o^{(2)}F = -2o^{(1)}\Delta^*F + o^{(2)}\Delta^*F, \tag{2.187}$$

$$\Delta^* o^{(3)}F = o^{(3)}\Delta^*F. \tag{2.188}$$

(b) Let $f : \Omega \to \mathbb{R}^3$ be sufficiently smooth. Then

$$O^{(1)}\Delta^* f = (\Delta^* - 2)O^{(1)} f + 2O^{(2)} f, \tag{2.189}$$
$$O^{(2)}\Delta^* f = -2\Delta^* O^{(1)} f + \Delta^* O^{(2)} f, \tag{2.190}$$
$$O^{(3)}\Delta^* f = \Delta^* O^{(3)} f. \tag{2.191}$$

Proof. We only prove (2.186); the other relations follow in a similar manner. From Exercise 2.2, we know that

$$\begin{aligned} \Delta^*_\xi o^{(1)}_\xi F(\xi) &= \Delta^*_\xi (\xi F(\xi)) \\ &= \nabla^*_\xi \left(\nabla^*_\xi \cdot (\xi F(\xi))\right) - \nabla^*_\xi \wedge \left(\nabla^*_\xi \wedge (\xi F(\xi))\right) \\ &\quad + \xi \wedge \left(\nabla^*_\xi \wedge (\xi F(\xi))\right) - \xi \left(\nabla^*_\xi \cdot (\xi F(\xi))\right), \end{aligned} \tag{2.192}$$

for $\xi \in \Omega$. Next, we observe $\nabla^*_\xi \cdot (\xi F(\xi)) = 2F(\xi)$ and

$$\xi \wedge \left(\nabla^*_\xi \wedge (\xi F(\xi))\right) = -\xi \wedge \left(\xi \wedge \left(\nabla^*_\xi F(\xi)\right)\right) = \nabla^*_\xi F(\xi) = o^{(2)}_\xi F(\xi). \tag{2.193}$$

By letting $g(x) = \nabla^*_\xi F(\xi)$, for $\xi = \frac{x}{|x|}$, we obtain a continuously differentiable function on $\mathbb{R}^3 \setminus \{0\}$ that does not depend on the radial distance $r = |x|$. Using standard relations for the differential operators in $\mathbb{R}^3$ (see Chapter 1), we are able to deduce that

$$\begin{aligned} \nabla^*_\xi \wedge \left(\nabla^*_\xi \wedge (\xi F(\xi))\right) &= -\nabla^*_\xi \wedge \left(\xi \wedge (\nabla^*_\xi F(\xi))\right) \\ &= -\nabla_x \wedge \left(\frac{x}{|x|} \wedge g(x))\right)\Big|_{x=\xi} \\ &= -\frac{x}{|x|}(\nabla_x \cdot g(x))\Big|_{x=\xi} + g(x)\left(\nabla_x \cdot \frac{x}{|x|}\right)\Big|_{x=\xi} \\ &\quad - (g(x) \cdot \nabla_x)\frac{x}{|x|}\Big|_{x=\xi} + \left(\frac{x}{|x|} \cdot \nabla_x\right) g(x)\Big|_{x=\xi}. \end{aligned} \tag{2.194}$$

Making use of the fact that g is tangential and that $g(x) = g(\xi)$, $\xi = \frac{x}{|x|}$, some basic but lengthy calculations lead to

$$\begin{aligned} \nabla^*_\xi \wedge \left(\nabla^*_\xi \wedge (\xi F(\xi))\right) &= -\xi \nabla^*_\xi \cdot g(\xi) + 2g(\xi) - g(\xi) \\ &= -o^{(1)}_\xi \Delta^*_\xi F(\xi) + o^{(2)}_\xi F(\xi), \end{aligned} \tag{2.195}$$

for $\xi \in \Omega$. Inserting all previous results into (2.192), we end up with the desired relation (2.186). □

Lemma 2.45 motivates a vectorial Beltrami operator $\boldsymbol{\Delta}^*$, which possesses the vector spherical harmonics as eigenfunctions.

Lemma 2.46. *Let the* vectorial Beltrami operator *be defined by*

$$\boldsymbol{\Delta}^* = p_{\mathrm{rad}}(\Delta^* + 2)p_{\mathrm{rad}} + p_{\mathrm{tan}}\Delta^* p_{\mathrm{tan}}. \tag{2.196}$$

Then, any vector spherical harmonic $y_n \in \mathrm{harm}_n(\Omega)$*,* $n \in \mathbb{N}_0$*, is an infinitely often differentiable eigenfunction of the vectorial Beltrami operator corresponding to the eigenvalue* $-n(n+1)$*. More explicitly,*

$$\boldsymbol{\Delta}^* y_n(\xi) = -n(n+1)y_n(\xi), \quad \xi \in \Omega. \tag{2.197}$$

2.5.2 Eigenfunction System

Next, we are concerned with a set of vector spherical harmonics that additionally constitute eigenfunctions of the (scalar) Beltrami operator. This set does not separate anymore into tangential and radial components but it is of special importance in geomagnetism (cf. Chapter 8) and quantum mechanics (see, e.g., A.R. Edmonds [1957]). Our considerations start with the operators (*alternative Helmholtz operators*)

$$\tilde{o}^{(1)} = o^{(1)}\left(\mathrm{D}+\frac{1}{2}\right) - o^{(2)}, \tag{2.198}$$

$$\tilde{o}^{(2)} = o^{(1)}\left(\mathrm{D}-\frac{1}{2}\right) + o^{(2)}, \tag{2.199}$$

$$\tilde{o}^{(3)} = o^{(3)}, \tag{2.200}$$

and the associated adjoint operators

$$\tilde{O}^{(1)} = \left(\mathrm{D}+\frac{1}{2}\right)O^{(1)} - O^{(2)}, \tag{2.201}$$

$$\tilde{O}^{(2)} = \left(\mathrm{D}-\frac{1}{2}\right)O^{(1)} + O^{(2)}, \tag{2.202}$$

$$\tilde{O}^{(3)} = O^{(3)}. \tag{2.203}$$

The pseudodifferential operator D is given by

$$\mathrm{D} = \left(-\Delta^* + \frac{1}{4}\right)^{\frac{1}{2}}, \tag{2.204}$$

such that

$$\mathrm{D}Y_{n,k} = \left(n+\frac{1}{2}\right)Y_{n,k}, \quad n \in \mathbb{N}_0,\ k = 1,\ldots,2n+1, \tag{2.205}$$

or equivalently

$$\left(\mathrm{D}-\frac{1}{2}\right)Y_{n,k} = nY_{n,k}, \quad n \in \mathbb{N}_0,\ k = 1,\ldots,2n+1. \tag{2.206}$$

To make the notion of a pseudodifferential operator like D a bit more transparent, we first discuss Sobolev spaces.

Definition 2.47. The *Sobolev space* $\mathrm{H}_s(\Omega)$, $s \in \mathbb{R}$, is the completion of the space of all functions F of class $\mathrm{C}^{(\infty)}(\Omega)$ with respect to the norm $\|\cdot\|_{\mathrm{H}_s(\Omega)}$ induced by the inner product

$$(F,G)_{\mathrm{H}_s(\Omega)} = \sum_{n=0}^{\infty}\sum_{k=1}^{2n+1}\left(n+\frac{1}{2}\right)^{2s}(F,Y_{n,k})_{\mathrm{L}^2(\Omega)}(G,Y_{n,k})_{\mathrm{L}^2(\Omega)}. \quad (2.207)$$

In brief, $\mathrm{H}_s(\Omega) = \overline{\mathrm{C}^{(\infty)}(\Omega)}^{\|\cdot\|_{\mathrm{H}_s(\Omega)}}$.

Definition 2.48. Let $\{\Lambda^\wedge(n)\}_{n\in\mathbb{N}_0}$ be a sequence of real numbers satisfying

$$\lim_{n\to\infty}\frac{|\Lambda^\wedge(n)|}{(n+\frac{1}{2})^t} = K, \quad (2.208)$$

for a constant $K \neq 0$ and some $t \in \mathbb{R}$. Then the operator $\Lambda : \mathrm{H}_s(\Omega) \to \mathrm{H}_{s-t}(\Omega)$ defined by

$$\Lambda F = \sum_{n=0}^{\infty}\sum_{k=1}^{2n+1}\Lambda^\wedge(n)(F,Y_{n,k})_{\mathrm{L}^2(\Omega)}Y_{n,k}, \quad F\in\mathrm{H}_s(\Omega), \quad (2.209)$$

is called a *pseudodifferential operator of order* t. $\{\Lambda^\wedge(n)\}_{n\in\mathbb{N}_0}$ is called the *symbol of* Λ.

From Definition 2.48 we see that the Beltrami operator is a pseudodifferential operator of order $t = 2$ with symbol $(\Delta^*)^\wedge(n) = -n(n+1)$, while D is a pseudodifferential operator of order $t = 1$ with symbol $\mathrm{D}^\wedge(n) = n+\frac{1}{2}$. Later on, the inverse

$$\mathrm{D}^{-1} = \left(-\Delta^* + \frac{1}{4}\right)^{-\frac{1}{2}} \quad (2.210)$$

will be of some importance. It obviously has the symbol $(n+\frac{1}{2})^{-1}$, such that

$$\mathrm{D}^{-1}Y_{n,k} = \frac{1}{n+\frac{1}{2}}Y_{n,k}, \quad n\in\mathbb{N}_0,\ k=1,\ldots,2n+1. \quad (2.211)$$

Consequently, D^{-1} is a pseudodifferential operator of order $t=-1$ that turns out to be an integral operator.

Lemma 2.49. *Let* F *be of class* $\mathrm{C}^{(0)}(\Omega)$. *Then*

$$\mathrm{D}^{-1}F(\xi) = \frac{1}{4\pi}\int_\Omega S(\xi\cdot\eta)F(\eta)\,d\omega(\eta), \quad \xi\in\Omega, \quad (2.212)$$

with the kernel

$$S(\xi\cdot\eta) = \frac{\sqrt{2}}{\sqrt{1-\xi\cdot\eta}}, \quad \xi,\eta\in\Omega,\ 1-\xi\cdot\eta>0. \quad (2.213)$$

Proof. From (2.98) and the addition theorem, we get

$$\begin{aligned}
\frac{1}{4\pi}\int_\Omega \frac{\sqrt{2}}{\sqrt{1-\xi\cdot\eta}}\, F(\eta)d\omega(\eta) &= \frac{1}{2\pi}\int_\Omega \frac{1}{|\xi-\eta|}F(\eta)d\omega(\eta) \qquad (2.214)\\
&= \frac{1}{2\pi}\int_\Omega F(\eta)\sum_{n=0}^{\infty}\sum_{k=1}^{2n+1}\frac{4\pi}{2n+1}Y_{n,k}(\xi)Y_{n,k}(\eta)d\omega(\eta)\\
&= \sum_{n=0}^{\infty}\sum_{k=1}^{2n+1}\frac{1}{n+\frac{1}{2}}(F,Y_{n,k})_{\mathrm{L}^2(\Omega)}Y_{n,k}(\xi),
\end{aligned}$$

for $\xi\in\Omega$. By (2.211), the last line can be understood as the spectral representation of D^{-1}, so that the desired statement holds true. □

Remark 2.50. Since the fundamental solution of the Laplace operator is directly connected to the convolution kernel S by

$$G(\Delta;|\xi-\eta|) = \frac{1}{4\pi}\frac{1}{|\xi-\eta|} = \frac{1}{8\pi}S(\xi\cdot\eta), \quad \xi,\eta\in\Omega,\ 1-\xi\cdot\eta>0, \qquad (2.215)$$

the operator D^{-1} is also called the *(spherical) single-layer operator*. It should, however, not be confused with the layer potentials used in Chapters 3 and 6. The kernel S is commonly called *(spherical) single-layer kernel*. Furthermore, we note that the operator D^{-1} actually maps $\mathrm{C}^{(k)}(\Omega)$ into $\mathrm{C}^{(k)}(\Omega)$, $k\in\mathbb{N}_0$, which is important for later considerations.

Next, we come to the already announced definition of the alternative system of vector spherical harmonics.

Definition 2.51. For $i=1,2,3$, we denote by

$$\tilde{y}^{(i)}_{n,k} = \left(\tilde{\mu}^{(i)}_n\right)^{-\frac{1}{2}}\tilde{o}^{(i)}Y_{n,k}, \quad n\in\mathbb{N}_{0_i}, k=1,\ldots 2n+1 \qquad (2.216)$$

a *vector spherical harmonic (of type* i*, degree* n*, and order* k*)*, where the normalization factor $\tilde{\mu}^{(i)}_n$ is given by

$$\tilde{\mu}^{(i)}_n = \begin{cases} (n+1)(2n+1), & i=1,\\ n(2n+1), & i=2,\\ n(n+1), & i=3.\end{cases} \qquad (2.217)$$

By $\widetilde{\mathrm{harm}}^{(i)}_n(\Omega)$ we denote the space of all vector spherical harmonics of type i and degree n. Furthermore, we set $\mathrm{harm}_0(\Omega)=\widetilde{\mathrm{harm}}_0(\Omega)=\widetilde{\mathrm{harm}}^{(1)}_0(\Omega)$ and $\mathrm{harm}_n(\Omega)=\widetilde{\mathrm{harm}}_n(\Omega)=\bigoplus_{i=1}^3\widetilde{\mathrm{harm}}^{(i)}_n(\Omega)$, for $n\in\mathbb{N}$.

Observing the previous definition, we are able to express the set (2.216) of vector spherical harmonics by

$$\tilde{y}^{(1)}_{n,k} = \sqrt{\frac{n+1}{2n+1}} y^{(1)}_{n,k} - \sqrt{\frac{n}{2n+1}} y^{(2)}_{n,k}, \tag{2.218}$$

$$\tilde{y}^{(2)}_{n,k} = \sqrt{\frac{n}{2n+1}} y^{(1)}_{n,k} + \sqrt{\frac{n+1}{2n+1}} y^{(2)}_{n,k}, \tag{2.219}$$

$$\tilde{y}^{(3)}_{n,k} = y^{(3)}_{n,k}. \tag{2.220}$$

Using Lemma 2.45 and observing the known framework for (scalar) spherical harmonics, it is possible to show that $\tilde{y}^{(i)}_{n,k}$ is an eigenfunction of the (scalar) Beltrami operator.

Lemma 2.52. *Any vector spherical harmonic $\tilde{y}^{(i)}_{n,k}$, $i = 1, 2, 3$, $n \in \mathbb{N}_{0_i}$, $k = 1, \ldots, 2n + 1$, is an infinitely often differentiable eigenfunction of the Beltrami operator. More explicitly,*

$$\Delta^* \tilde{y}^{(1)}_{n,k}(\xi) = -(n+1)(n+2)\tilde{y}^{(1)}_{n,k}(\xi), \tag{2.221}$$

$$\Delta^* \tilde{y}^{(2)}_{n,k}(\xi) = -n(n-1)\tilde{y}^{(2)}_{n,k}(\xi), \tag{2.222}$$

$$\Delta^* \tilde{y}^{(3)}_{n,k}(\xi) = -n(n+1)\tilde{y}^{(3)}_{n,k}(\xi), \tag{2.223}$$

for $\xi \in \Omega$.

Furthermore, these vector spherical harmonics again form an orthonormal basis in $\mathrm{l}^2(\Omega)$. The spaces $\tilde{\mathrm{l}}^2_{(i)}(\Omega)$, $i = 1, 2, 3$, can analogously be defined by (2.176).

Theorem 2.53 (Closure in $\mathrm{l}^2(\Omega)$). *The system $\{\tilde{y}^{(i)}_{n,k}\}_{i=1,2,3,\, n\in\mathbb{N}_{0_i},k=1,\ldots,2n+1}$ is closed and complete in the space $\mathrm{l}^2(\Omega)$ with respect to $\|\cdot\|_{\mathrm{l}^2(\Omega)}$. More precisely,*

$$\lim_{N\to\infty} \left\| f - \sum_{i=1}^{3}\sum_{n=0}^{N}\sum_{k=1}^{2n+1} (\tilde{f}^{(i)})^\wedge(n,k)\tilde{y}^{(i)}_{n,k} \right\|_{\mathrm{l}^2(\Omega)} = 0, \tag{2.224}$$

for f of class $l^2(\Omega)$, where

$$(\tilde{f}^{(i)})^\wedge(n,k) = \int_\Omega f(\eta)\cdot \tilde{y}^{(i)}_{n,k}(\eta)\, d\omega(\eta) \tag{2.225}$$

denotes the Fourier coefficient (of type i, degree n, and order k) .

An important property of $\tilde{y}^{(i)}_{n,k}$, $i = 1, 2$, is the relationship to the (scalar) inner and outer harmonics. As a matter of fact, this feature makes the alternative system of vector spherical harmonics especially suitable for vectorial boundary-value problems.

Lemma 2.54. *Let* $R > 0$ *and* $\xi = \frac{x}{|x|}$, $r = |x|$. *Then we have*

$$\begin{aligned} \nabla_x H_{n,k}^R(x) &= \frac{1}{R^2}\left(\frac{r}{R}\right)^{n-1} (\tilde{\mu}_n^{(2)})^{\frac{1}{2}} \tilde{y}_{n,k}^{(2)}(\xi), \quad x \in \overline{\Omega_R^{int}}, \\ -\nabla_x H_{-n-1,k}^R(x) &= \frac{1}{R^2}\left(\frac{R}{r}\right)^{n+2} (\tilde{\mu}_n^{(1)})^{\frac{1}{2}} \tilde{y}_{n,k}^{(1)}(\xi), \quad x \in \overline{\Omega_R^{ext}}, \end{aligned}$$

for $n \in \mathbb{N}_{0_i}$, $i = 1, 2$, *and* $k = 1, \ldots, 2n+1$.

2.5.3 Vector Inner/Outer Harmonics

We conclude this section by introducing the vectorial counterparts to the (scalar) inner and outer harmonics. The system

$$\tilde{y}_{n,k}^{(i);R}(x) = \frac{1}{R}\tilde{y}_{n,k}^{(i)}\left(\frac{x}{|x|}\right), \quad x \in \Omega_R, \tag{2.226}$$

constitutes an orthonormal basis $\{\tilde{y}_{n,k}^{(i);R}\}_{i=1,2,3,\, n\in\mathbb{N}_{0_i},\, k=1,\ldots,2n+1}$ of the space $\mathrm{l}^2(\Omega_R)$. Once more, it should be emphasized that the vector spherical harmonics $\tilde{y}_{n,k}^{(i);R}$, and not $y_{n,k}^{(i);R}$, are used to generate the vector inner and outer harmonics.

Definition 2.55 (Vector Inner/Outer Harmonics).

(a) The vector fields

$$h_{n,k}^{(1);R}(x) = \left(\frac{|x|}{R}\right)^{n+1} \tilde{y}_{n,k}^{(1);R}(x), \quad x \in \mathbb{R}^3, \tag{2.227}$$

$$h_{n,k}^{(2);R}(x) = \left(\frac{|x|}{R}\right)^{n-1} \tilde{y}_{n,k}^{(2);R}(x), \quad x \in \mathbb{R}^3, \tag{2.228}$$

$$h_{n,k}^{(3);R}(x) = \left(\frac{|x|}{R}\right)^{n} \tilde{y}_{n,k}^{(3);R}(x), \quad x \in \mathbb{R}^3, \tag{2.229}$$

for $n \in \mathbb{N}_{0_i}$, $i = 1, 2, 3$, and $k = 1, \ldots, 2n+1$, are called *vector inner harmonics (of type i, degree n and order k).*

(b) The vector fields

$$h_{-n-1,k}^{(1);R}(x) = \left(\frac{R}{|x|}\right)^{n+2} \tilde{y}_{n,k}^{(1);R}(x), \quad x \in \mathbb{R}^3\backslash\{0\}, \tag{2.230}$$

$$h_{-n-1,k}^{(2);R}(x) = \left(\frac{R}{|x|}\right)^{n} \tilde{y}_{n,k}^{(2);R}(x), \quad x \in \mathbb{R}^3\backslash\{0\}, \tag{2.231}$$

$$h_{-n-1,k}^{(3);R}(x) = \left(\frac{R}{|x|}\right)^{n+1} \tilde{y}_{n,k}^{(3);R}(x), \quad x \in \mathbb{R}^3\backslash\{0\}, \tag{2.232}$$

for $n \in \mathbb{N}_{0_i}$, $i = 1, 2, 3$, and $k = 1, \ldots, 2n+1$, are called *vector outer harmonics (of type i, degree n and order k).*

Remark 2.56. Analgous to the scalar case of Subsection 2.3.4, the vector inner harmonics fulfill the conditions

(i) $h^{(i);R}_{n,k}$ is of class $c^{(\infty)}(\mathbb{R}^3)$, for $i = 1, 2, 3$,

(ii) $h^{(i);R}_{n,k}$ satisfies $\Delta h^{(i);R}_{n,k}(x) = 0$, $x \in \mathbb{R}^3$, for $i = 1, 2, 3$,

(iii) $h^{(i);R}_{n,k}\big|_{\Omega_R} = \tilde{y}^{(i);R}_{n,k}$, for $i = 1, 2, 3$,

(iv) $\left(h^{(i);R}_{n,k}, h^{(j);R}_{p,q}\right)_{l^2(\Omega_R)} = \delta_{i,j}\delta_{n,p}\delta_{k,q}$.

Correspondingly, the vector outer harmonics satisfy the following properties:

(i) $h^{(i);R}_{-n-1,k}$ is of class $c^{(\infty)}(\mathbb{R}^3\backslash\{0\})$, for $i = 1, 2, 3$,

(ii) $h^{(i);R}_{-n-1,k}$ satisfies $\Delta h^{(i);R}_{-n-1,k}(x) = 0$, $x \in \mathbb{R}^3\backslash\{0\}$, for $i = 1, 2, 3$,

(iii) $h^{(i);R}_{-n-1,k} = \mathcal{O}(|x|^{-2})$, for $i = 1, 2, 3$,

(iv) $h^{(i);R}_{-n-1,k}\big|_{\Omega_R} = \tilde{y}^{(i);R}_{n,k}$, for $i = 1, 2, 3$,

(v) $\left(h^{(i);R}_{-n-1,k}, h^{(j);R}_{-p-1,q}\right)_{l^2(\Omega_R)} = \delta_{i,j}\delta_{n,p}\delta_{k,q}$.

Furthermore, the vector inner and outer harmonics are connected via the Kelvin transform in the same manner as the (scalar) inner and outer harmonics (see (2.130) and (2.131)).

Finally, we notice that (cf. Lemma 2.54)

$$\nabla H^R_{-n-1,k} = -\frac{1}{R}\left(\tilde{\mu}^{(1)}_n\right)^{\frac{1}{2}} h^{(1);R}_{-n-1,k} = -\frac{1}{R}\sqrt{(n+1)(2n+1)}\, h^{(1);R}_{-n-1,k}. \tag{2.233}$$

This enables us to write the gradient field ∇F of a harmonic function $F : \mathbb{R}^3\backslash\mathcal{B}_R(0) \to \mathbb{R}$, given as the series expansion

$$F = \sum_{n=0}^{\infty}\sum_{k=1}^{2n+1} F^{\wedge_{L^2(\Omega_R)}}(n,k) H^R_{-n-1,k} \tag{2.234}$$

with $F|_{\Omega_R} \in L^2(\Omega_R)$, in the form

$$\nabla F = -\frac{1}{R}\sum_{n=0}^{\infty}\sum_{k=1}^{2n+1} F^{\wedge_{L^2(\Omega_R)}}(n,k)\left(\tilde{\mu}^{(1)}_n\right)^{\frac{1}{2}} h^{(1);R}_{-n-1,k}. \tag{2.235}$$

By the integral

$$F^{\wedge_{L^2(\Omega_R)}}(n,k) = \int_{\Omega_R} F(y) Y^R_{n,k}(y)\, d\omega(y), \tag{2.236}$$

we mean the Fourier coefficient with respect to $L^2(\Omega_R)$. If we just write $F^\wedge(n,k)$, we always mean the Fourier coefficient with respect to $L^2(\Omega)$. Moreover, it is easily seen that

$$\int_{\Omega_r} \nabla F(y) \cdot h^{(1);r}_{-n-1,k}(y)\, d\omega(y) = -\left(\frac{R}{r}\right)^{n+1} \frac{\left(\tilde{\mu}_n^{(1)}\right)^{\frac{1}{2}}}{r} F^{\wedge_{L^2(\Omega_R)}}(n,k), \quad (2.237)$$

for $r > R$. Therefore, we obtain the following reformulation of (2.234), which plays a particular role in the modeling of the SST problem (cf. Section 4.2):

$$F = \sum_{n=0}^{\infty} \sum_{k=1}^{2n+1} \left(\nabla F, h^{(1);r}_{-n-1,k}\right)_{l^2(\Omega_r)} \frac{r}{\left(\tilde{\mu}_n^{(1)}\right)^{\frac{1}{2}}} \left(\frac{r}{R}\right)^{n+1} H^R_{-n-1,k}. \quad (2.238)$$

Indeed, the last formula expresses the gravitational potential on $\mathbb{R}^3 \backslash \mathcal{B}_R(0)$ in terms of the gravitational gradient on the satellite orbit Ω_r. The vector outer harmonics are essential tools. The equality on Ω_R is understood in $L^2(\Omega_R)$-sense, while the convergence on each $\overline{\mathcal{K}}$ with $\mathcal{K} \Subset \mathbb{R}^3 \backslash \overline{\mathcal{B}_R(0)}$ is understood in uniform sense.

2.6 Tensor Spherical Harmonics

By $\mathbf{l}^2(\Omega)$ we denote the Hilbert space of square-integrable tensor fields $\mathbf{f} : \Omega \to \mathbb{R}^{3\times 3}$, equipped with the inner product

$$(\mathbf{f}, \mathbf{g})_{\mathbf{l}^2(\Omega)} = \int_\Omega \mathbf{f}(\eta) \cdot \mathbf{g}(\eta)\, d\omega(\eta), \quad \mathbf{f}, \mathbf{g} \in \mathbf{l}^2(\Omega), \quad (2.239)$$

and the associated norm $\|\cdot\|_{\mathbf{l}^2(\Omega)}$. Note that the space $\mathbf{l}^2(\Omega)$ is the completion of $\mathbf{c}^{(0)}(\Omega)$ with respect to $\|\cdot\|_{\mathbf{l}^2(\Omega)}$. Analogous to the vectorial case, tensor spherical harmonics can be defined with the help of certain spherical operators. They form a useful tool in solid Earth physics and satellite technology. In this section, we just give a basic introduction that suffices to present some applications in Chapter 4.

2.6.1 Radial-Tangential System

The operators $\mathbf{o}^{(i,k)} : C^{(\infty)}(\Omega) \to \mathbf{c}^{(\infty)}(\Omega)$, $i,k = 1,2,3$, transform scalar functions into tensor fields (cf. W. Freeden, M. Schreiner [2009]):

$$\begin{aligned} \mathbf{o}^{(1,1)} F(\xi) &= \xi \otimes \xi F(\xi), && (2.240)\\ \mathbf{o}^{(1,2)} F(\xi) &= \xi \otimes \nabla^*_\xi F(\xi), && (2.241)\\ \mathbf{o}^{(1,3)} F(\xi) &= \xi \otimes L^*_\xi F(\xi), && (2.242) \end{aligned}$$

$$\begin{aligned}
\mathbf{o}^{(2,1)}F(\xi) &= (\nabla_\xi^* F(\xi)) \otimes \xi, & (2.243)\\
\mathbf{o}^{(3,1)}F(\xi) &= (\mathrm{L}_\xi^* F(\xi)) \otimes \xi, & (2.244)\\
\mathbf{o}^{(2,2)}F(\xi) &= \mathbf{i}_{\mathrm{tan}}(\xi)F(\xi), & (2.245)\\
\mathbf{o}^{(2,3)}F(\xi) &= \left(\nabla_\xi^* \otimes \nabla_\xi^* - \mathrm{L}_\xi^* \otimes \mathrm{L}_\xi^*\right) F(\xi) + 2(\nabla_\xi^* F(\xi)) \otimes \xi, & (2.246)\\
\mathbf{o}^{(3,2)}F(\xi) &= \left(\nabla_\xi^* \otimes \mathrm{L}_\xi^* + \mathrm{L}_\xi^* \otimes \nabla_\xi^*\right) F(\xi) + 2(\mathrm{L}_\xi^* F(\xi)) \otimes \xi, & (2.247)\\
\mathbf{o}^{(3,3)}F(\xi) &= \mathbf{j}_{\mathrm{tan}}(\xi)F(\xi), & (2.248)
\end{aligned}$$

for $F \in \mathrm{C}^{(2)}(\Omega)$, $\xi \in \Omega$. Note that the tensors $\mathbf{i}_{\mathrm{tan}}$ and $\mathbf{j}_{\mathrm{tan}}$ are the surface identity tensor and the surface rotation tensor, respectively, as defined in (2.20) and (2.21). The adjoint operators $O^{(i,k)}$ to $\mathbf{o}^{(i,k)}$ satisfying

$$\left(\mathbf{o}^{(i,k)}F, \mathbf{f}\right)_{\mathrm{l}^2(\Omega)} = \left(F, O^{(i,k)}\mathbf{f}\right)_{\mathrm{L}^2(\Omega)} \tag{2.249}$$

for $F \in \mathrm{C}^{(2)}(\Omega)$ and $\mathbf{f} \in \mathbf{c}^{(2)}(\Omega)$ are given by

$$\begin{aligned}
O^{(1,1)}\mathbf{f}(\xi) &= \xi^T \mathbf{f}(\xi)\xi, & (2.250)\\
O^{(1,2)}\mathbf{f}(\xi) &= -\nabla_\xi^* \cdot p_{\mathrm{tan}}\left[\xi^T \mathbf{f}(\xi)\right], & (2.251)\\
O^{(1,3)}\mathbf{f}(\xi) &= -\mathrm{L}_\xi^* \cdot p_{\mathrm{tan}}\left[\xi^T \mathbf{f}(\xi)\right], & (2.252)\\
O^{(2,1)}\mathbf{f}(\xi) &= -\nabla_\xi^* \cdot p_{\mathrm{tan}}\left[\mathbf{f}(\xi)\xi\right], & (2.253)\\
O^{(3,1)}\mathbf{f}(\xi) &= -\mathrm{L}_\xi^* \cdot p_{\mathrm{tan}}\left[\mathbf{f}(\xi)\xi\right], & (2.254)\\
O^{(2,2)}\mathbf{f}(\xi) &= \mathbf{i}_{\mathrm{tan}}(\xi) \cdot \mathbf{f}(\xi), & (2.255)\\
O^{(2,3)}\mathbf{f}(\xi) &= \nabla_\xi^* \cdot p_{\mathrm{tan}}\left[\nabla_\xi^* \cdot \mathbf{p}_{\mathrm{tan},*}[\mathbf{f}](\xi)\right] - \mathrm{L}_\xi^* \cdot p_{\mathrm{tan}}\left[\mathrm{L}_\xi^* \cdot \mathbf{p}_{\mathrm{tan},*}[\mathbf{f}](\xi)\right] & \\
&\quad -2\nabla_\xi^* \cdot p_{\mathrm{tan}}\left[\mathbf{f}(\xi)\xi\right], & (2.256)\\
O^{(3,2)}\mathbf{f}(\xi) &= \mathrm{L}_\xi^* \cdot p_{\mathrm{tan}}\left[\nabla_\xi^* \cdot \mathbf{p}_{\mathrm{tan},*}[\mathbf{f}](\xi)\right] + \nabla_\xi^* \cdot p_{\mathrm{tan}}\left[\mathrm{L}_\xi^* \cdot \mathbf{p}_{\mathrm{tan},*}[\mathbf{f}](\xi)\right] & \\
&\quad -2\mathrm{L}_\xi^* \cdot p_{\mathrm{tan}}\left[\mathbf{f}(\xi)\xi\right], & (2.257)\\
O^{(3,3)}\mathbf{f}(\xi) &= \mathbf{j}_{\mathrm{tan}}(\xi) \cdot \mathbf{f}(\xi), & (2.258)
\end{aligned}$$

where

$$\begin{aligned}
p_{\mathrm{tan}}[f](\xi) &= f(\xi) - (\xi \cdot f(\xi))\xi, \quad \xi \in \Omega, & (2.259)\\
\mathbf{p}_{\mathrm{tan},*}[\mathbf{f}](\xi) &= \mathbf{f}(\xi) - \xi \otimes ((\mathbf{f}(\xi))^T \xi), \quad \xi \in \Omega. & (2.260)
\end{aligned}$$

With the help of the operators $\mathbf{o}^{(i,k)}$, we are able to introduce a set of tensor spherical harmonics.

Definition 2.57. For $i, k \in \{1, 2, 3\}$, we denote by

$$\mathbf{y}_{n,m}^{(i,k)} = \left(\mu_n^{(i,k)}\right)^{-1/2} \mathbf{o}^{(i,k)} Y_{n,m}, \quad n \in \mathbb{N}_{0_{i,k}}, m = 1, \ldots, 2n+1, \tag{2.261}$$

a *tensor spherical harmonic (of type (i,k), degree n, and order m)*. The normalization constants $\mu_n^{(i,k)}$ are given by

$$\mu_n^{(i,k)} = \begin{cases} 1, & (i,k) = (1,1), \\ 2, & (i,k) \in \{(2,2),(3,3)\}, \\ n(n+1), & (i,k) \in \{(1,2),(1,3),(2,1),(3,1)\}, \\ 2n(n+1)(n(n+1)-2), & (i,k) \in \{(2,3),(3,2)\}. \end{cases} \tag{2.262}$$

For brevity, we set

$$0_{i,k} = \begin{cases} 0, & (i,k) \in \{(1,1),(2,2),(3,3)\}, \\ 1, & (i,k) \in \{(1,2),(1,3),(2,1),(3,1)\}, \\ 2, & (i,k) \in \{(2,3),(3,2)\}. \end{cases} \tag{2.263}$$

By $\mathbf{harm}_n^{(i,k)}(\Omega)$ we denote the space of all tensor spherical harmonics of degree n and type (i,k), and by $\mathbf{harm}_n(\Omega)$ the space of all tensor spherical harmonics of degree n.

As already seen for the scalar and vectorial case, the tensor spherical harmonics form a closed and complete $\mathbf{l}^2(\Omega)$-orthonormal set of functions.

Theorem 2.58 (Closure). *The system* $\{\mathbf{y}_{n,m}^{(i,k)}\}_{i,k=1,2,3,\ n\in\mathbb{N}_{0_{i,k}},\ m=1,\ldots,2n+1}$ *is closed and complete in* $\mathbf{l}^2(\Omega)$ *with respect to* $\|\cdot\|_{\mathbf{l}^2(\Omega)}$. *More precisely,*

$$\lim_{N\to\infty} \left\| \mathbf{f} - \sum_{i=1}^{3}\sum_{k=1}^{3}\sum_{n=0_{i,k}}^{N}\sum_{m=1}^{2n+1} \left(\mathbf{f}^{(i,k)}\right)^{\wedge}(n,m)\,\mathbf{y}_{n,m}^{(i,k)} \right\|_{\mathbf{l}^2(\Omega)} = 0, \tag{2.264}$$

for $\mathbf{f}$ *of class* $\mathbf{l}^2(\Omega)$, *where the Fourier coefficients are given by*

$$\left(\mathbf{f}^{(i,k)}\right)^{\wedge}(n,m) = \int_\Omega \mathbf{f}(\eta)\cdot \mathbf{y}_{n,m}^{(i,k)}(\eta)\,d\omega(\eta). \tag{2.265}$$

2.6.2 Eigenfunction System

The tensor spherical harmonics $\mathbf{y}_{n,m}^{(i,k)}$ concentrate on the fact that the decomposition into normal and tangential tensor fields is fulfilled. A disadvantage of this set of tensor spherical harmonics is that these functions are not eigenfunctions of the (scalar) Beltrami operator. To overcome this problem, we introduce the operators $\tilde{\mathbf{o}}^{(i,k)} : \mathrm{C}^{(\infty)}(\Omega) \to \mathbf{c}^{(\infty)}(\Omega)$, $i,k = 1,2,3$, based on the operators $\mathbf{o}^{(i,k)}$ that have been used for the radial-tangential system:

$$\begin{pmatrix} \tilde{\mathbf{o}}^{(1,1)}F \\ \tilde{\mathbf{o}}^{(1,2)}F \\ \tilde{\mathbf{o}}^{(2,1)}F \\ \tilde{\mathbf{o}}^{(2,2)}F \\ \tilde{\mathbf{o}}^{(3,3)}F \end{pmatrix} = \mathbf{a}_{\mathrm{D}} \begin{pmatrix} F \\ F \\ F \\ F \\ F \end{pmatrix}, \qquad \begin{pmatrix} \tilde{\mathbf{o}}^{(1,3)}F \\ \tilde{\mathbf{o}}^{(2,3)}F \\ \tilde{\mathbf{o}}^{(3,1)}F \\ \tilde{\mathbf{o}}^{(3,2)}F \end{pmatrix} = \mathbf{b}_{\mathrm{D}} \begin{pmatrix} F \\ F \\ F \\ F \end{pmatrix}, \tag{2.266}$$

where the tensorial operators $\mathbf{a}_{\mathrm{D}}$ and $\mathbf{b}_{\mathrm{D}}$ are defined via

$$\mathbf{a}_{\mathrm{D}} = \begin{pmatrix} \mathbf{o}^{(1,1)}(\mathrm{D}+\frac{1}{2})(\mathrm{D}+\frac{3}{2}) & -\mathbf{o}^{(1,2)}(\mathrm{D}+\frac{3}{2}) & -\mathbf{o}^{(2,1)}(\mathrm{D}+\frac{3}{2}) & -\frac{1}{2}o^{(2,2)}(\mathrm{D}+\frac{3}{2})(\mathrm{D}+\frac{1}{2}) & \frac{1}{2}\mathbf{o}^{(2,3)} \\ \mathbf{o}^{(1,1)}(\mathrm{D}-\frac{1}{2})^2 & \mathbf{o}^{(1,2)}(\mathrm{D}-\frac{1}{2}) & -\mathbf{o}^{(2,1)}(\mathrm{D}-\frac{3}{2}) & -\frac{1}{2}\mathbf{o}^{(2,2)}(\mathrm{D}-\frac{1}{2})(\mathrm{D}-\frac{3}{2}) & -\frac{1}{2}\mathbf{o}^{(2,3)} \\ \mathbf{o}^{(1,1)}(\mathrm{D}+\frac{1}{2})^2 & -\mathbf{o}^{(1,2)}(\mathrm{D}+\frac{1}{2}) & \mathbf{o}^{(2,1)}(\mathrm{D}+\frac{3}{2}) & \frac{1}{2}\mathbf{o}^{(2,2)}(\mathrm{D}+\frac{3}{2})(D+\frac{1}{2}) & -\frac{1}{2}\mathbf{o}^{(2,3)} \\ \mathbf{o}^{(1,1)}(\mathrm{D}-\frac{1}{2})(\mathrm{D}-\frac{3}{2}) & \mathbf{o}^{(1,2)}(\mathrm{D}-\frac{3}{2}) & \mathbf{o}^{(2,1)}(\mathrm{D}-\frac{3}{2}) & -\frac{1}{2}\mathbf{o}^{(2,2)}(\mathrm{D}-\frac{1}{2})(\mathrm{D}-\frac{3}{2}) & \frac{1}{2}\mathbf{o}^{(2,3)} \\ 0 & 0 & \mathbf{o}^{(2,1)} & -\frac{1}{2}\mathbf{o}^{(2,2)}(\mathrm{D}-\frac{1}{2})(D+\frac{1}{2}) & -\frac{1}{2}\mathbf{o}^{(2,3)} \end{pmatrix},$$

$$\mathbf{b}_{\mathrm{D}} = \begin{pmatrix} \mathbf{o}^{(1,3)}(\mathrm{D}+\frac{1}{2}) & \mathbf{o}^{(3,1)} & -\frac{1}{2}\mathbf{o}^{(3,2)} & -\frac{1}{2}\mathbf{o}^{(3,3)}(\mathrm{D}-\frac{1}{2})(\mathrm{D}+\frac{1}{2}) \\ \mathbf{o}^{(1,3)}(\mathrm{D}-\frac{1}{2}) & -\mathbf{o}^{(3,1)} & \frac{1}{2}\mathbf{o}^{(3,2)} & \frac{1}{2}\mathbf{o}^{(3,3)}(\mathrm{D}-\frac{1}{2})(\mathrm{D}+\frac{1}{2}) \\ 0 & \mathbf{o}^{(3,1)}(\mathrm{D}+\frac{3}{2}) & -\frac{1}{2}\mathbf{o}^{(3,2)} & \frac{1}{2}\mathbf{o}^{(3,3)}(\mathrm{D}+\frac{3}{2})(\mathrm{D}+\frac{1}{2}) \\ 0 & \mathbf{o}^{(3,1)}(\mathrm{D}-\frac{3}{2}) & \frac{1}{2}\mathbf{o}^{(3,2)} & -\frac{1}{2}\mathbf{o}^{(3,3)}(\mathrm{D}-\frac{1}{2})(\mathrm{D}-\frac{3}{2}) \end{pmatrix}, \tag{2.267}$$

and the function F is of class $\mathrm{C}^{(\infty)}(\Omega)$. The pseudodifferential operator D has already been introduced in (2.204) and (2.205). The adjoint operators $\tilde{O}^{(i,k)} : \mathbf{c}^{(\infty)}(\Omega) \to \mathrm{C}^{(\infty)}(\Omega)$, $i, k \in \{1, 2, 3\}$, to the operators $\tilde{\mathbf{o}}^{(i,k)}$ satisfying the relation

$$(\tilde{\mathbf{o}}^{(i,k)}G, \mathbf{f})_{\mathrm{l}^2(\Omega)} = (G, \tilde{O}^{(i,k)}\mathbf{f})_{\mathrm{L}^2(\Omega)}, \tag{2.268}$$

for $\mathbf{f} \in \mathbf{c}^{(\infty)}(\Omega)$, $G \in \mathrm{C}^{(\infty)}(\Omega)$, are easily obtainable as follows:

$$\begin{pmatrix} \tilde{O}^{(1,1)}\mathbf{f} \\ \tilde{O}^{(1,2)}\mathbf{f} \\ \tilde{O}^{(2,1)}\mathbf{f} \\ \tilde{O}^{(2,2)}\mathbf{f} \\ \tilde{O}^{(3,3)}\mathbf{f} \end{pmatrix} = \mathbf{a}_{\mathrm{D}}^{*} \begin{pmatrix} \mathbf{f} \\ \mathbf{f} \\ \mathbf{f} \\ \mathbf{f} \\ \mathbf{f} \end{pmatrix}, \qquad \begin{pmatrix} \tilde{O}^{(1,3)}\mathbf{f} \\ \tilde{O}^{(2,3)}\mathbf{f} \\ \tilde{O}^{(3,1)}\mathbf{f} \\ \tilde{O}^{(3,2)}\mathbf{f} \end{pmatrix} = \mathbf{b}_{\mathrm{D}}^{*} \begin{pmatrix} \mathbf{f} \\ \mathbf{f} \\ \mathbf{f} \\ \mathbf{f} \end{pmatrix}, \tag{2.269}$$

where $\mathbf{a}_{\mathrm{D}}^{*}$ and $\mathbf{b}_{\mathrm{D}}^{*}$ denote the adjoint tensors to $\mathbf{a}_{\mathrm{D}}$ and $\mathbf{b}_{\mathrm{D}}$, respectively. After these preliminaries, we are now able to introduce an alternative set of tensor spherical harmonics.

Definition 2.59. For $i, k \in \{1, 2, 3\}$, we denote by

$$\tilde{\mathbf{y}}_{n,m}^{(i,k)} = \left(\tilde{\mu}_n^{(i,k)}\right)^{-1/2} \tilde{\mathbf{o}}^{(i,k)} Y_{n,m}, \quad n \in \mathbb{N}_{\tilde{0}_{i,k}}, m = 1, \ldots, 2n+1, \tag{2.270}$$

a *tensor spherical harmonic (of type (i, k), degree n, and order m)*. The normalization constants $\tilde{\mu}_n^{(i,k)}$ are given by

$$\tilde{\mu}_n^{(1,1)} = (n+2)(n+1)(2n+3)(2n+1), \tag{2.271}$$
$$\tilde{\mu}_n^{(1,2)} = 3n^4, \tag{2.272}$$
$$\tilde{\mu}_n^{(1,3)} = n(n+1)^2(2n+1), \tag{2.273}$$
$$\tilde{\mu}_n^{(2,1)} = (n+1)^2(2n+3)(2n+1), \tag{2.274}$$
$$\tilde{\mu}_n^{(2,2)} = n(n-1)(2n+1)(2n-1), \tag{2.275}$$
$$\tilde{\mu}_n^{(2,3)} = n^2(n+1)^2, \tag{2.276}$$
$$\tilde{\mu}_n^{(3,1)} = n^2(n+1)(2n+1), \tag{2.277}$$
$$\tilde{\mu}_n^{(3,2)} = n(n+1)^2(2n+1), \tag{2.278}$$
$$\tilde{\mu}_n^{(3,3)} = n^2(n-1)(2n+1). \tag{2.279}$$

For brevity, we set

$$\tilde{0}_{i,k} = \begin{cases} 0, & (i,k) \in \{(1,1),(2,1),(3,1)\}, \\ 1, & (i,k) \in \{(1,2),(1,3),(2,3),(3,3)\}, \\ 2, & (i,k) \in \{(2,2),(3,2)\}. \end{cases} \tag{2.280}$$

According to this construction, the tensor spherical harmonics from Definition 2.59 are eigenfunctions of the Beltrami operator.

Theorem 2.60. *Any tensor spherical harmonic $\tilde{\mathbf{y}}^{(i,k)}_{n,m}$ is an infinitely often differential eigenfunction of the (scalar) Beltrami operator. More precisely,*

$$\begin{aligned}
\Delta^*_\xi \tilde{\mathbf{y}}^{(1,1)}_{n,m} &= -(n+2)(n+3)\tilde{\mathbf{y}}^{(1,1)}_{n,m}, && (2.281)\\
\Delta^*_\xi \tilde{\mathbf{y}}^{(1,2)}_{n,m} &= -n(n+1)\tilde{\mathbf{y}}^{(1,2)}_{n,m}, && (2.282)\\
\Delta^*_\xi \tilde{\mathbf{y}}^{(2,1)}_{n,m} &= -n(n+1)\tilde{\mathbf{y}}^{(2,1)}_{n,m}, && (2.283)\\
\Delta^*_\xi \tilde{\mathbf{y}}^{(2,2)}_{n,m} &= -(n-1)(n-2)\tilde{\mathbf{y}}^{(2,2)}_{n,m}, && (2.284)\\
\Delta^*_\xi \tilde{\mathbf{y}}^{(3,3)}_{n,m} &= -n(n+1)\tilde{\mathbf{y}}^{(3,3)}_{n,m}, && (2.285)\\
\Delta^*_\xi \tilde{\mathbf{y}}^{(1,3)}_{n,m} &= -(n+1)(n+2)\tilde{\mathbf{y}}^{(1,3)}_{n,m}, && (2.286)\\
\Delta^*_\xi \tilde{\mathbf{y}}^{(2,3)}_{n,m} &= -n(n-1)\tilde{\mathbf{y}}^{(2,3)}_{n,m}, && (2.287)\\
\Delta^*_\xi \tilde{\mathbf{y}}^{(3,1)}_{n,m} &= -(n+1)(n+2)\tilde{\mathbf{y}}^{(3,1)}_{n,m}, && (2.288)\\
\Delta^*_\xi \tilde{\mathbf{y}}^{(3,2)}_{n,m} &= -n(n-1)\tilde{\mathbf{y}}^{(3,2)}_{n,m}, && (2.289)
\end{aligned}$$

where the application of the Beltrami operator is understood componentwise.

Theorem 2.61 (Closure). *The system $\{\tilde{\mathbf{y}}^{(i,k)}_{n,m}\}_{i,k=1,2,3,\ n\in\mathbb{N}_{\tilde{0}_{i,k}},\ m=1,\ldots,2n+1}$ is closed and complete in $\mathbf{l}^2(\Omega)$ with respect to $\|\cdot\|_{\mathbf{l}^2(\Omega)}$. More precisely,*

$$\lim_{N\to\infty} \left\| \mathbf{f} - \sum_{i=1}^{3}\sum_{k=1}^{3}\sum_{n=\tilde{0}_{i,k}}^{N}\sum_{m=1}^{2n+1} \left(\tilde{\mathbf{f}}^{(i,k)}\right)^{\wedge}(n,m)\,\tilde{\mathbf{y}}^{(i,k)}_{n,m} \right\|_{\mathbf{l}^2(\Omega)} = 0, \tag{2.290}$$

for $\mathbf{f}$ of class $\mathbf{l}^2(\Omega)$, where the Fourier coefficients of $\mathbf{f}$ are given by

$$\left(\tilde{\mathbf{f}}^{(i,k)}\right)^{\wedge}(n,m) = \int_\Omega \mathbf{f}(\eta)\cdot\tilde{\mathbf{y}}^{(i,k)}_{n,m}(\eta)\,d\omega(\eta). \tag{2.291}$$

More details on the entire topic of tensor spherical harmonics, including addition theorems and Funk–Hecke formulas, can be found, e.g., in W. Freeden, T. Gervens, M. Schreiner [1998] and W. Freeden, M. Schreiner [2009].

2.6.3 Tensor Outer Harmonics

Next, we introduce an associated class of tensor inner/outer harmonics, using tensor spherical harmonics from Definition 2.59. For simplicity, we restrict ourselves to the geomathematically important case of *tensor outer harmonics.*

Definition 2.62 (Tensor Outer Harmonics). The tensor fields

$$\mathbf{h}^{R;(1,1)}_{-n-1,m}(x) = \frac{1}{R}\left(\frac{R}{|x|}\right)^{n+3} \tilde{\mathbf{y}}^{(1,1)}_{n,m}\left(\frac{x}{|x|}\right), \quad x \in \mathbb{R}^3\backslash\{0\}, \tag{2.292}$$

$$\mathbf{h}^{R;(1,2)}_{-n-1,m}(x) = \frac{1}{R}\left(\frac{R}{|x|}\right)^{n+1} \tilde{\mathbf{y}}^{(1,2)}_{n,m}\left(\frac{x}{|x|}\right), \quad x \in \mathbb{R}^3\backslash\{0\}, \tag{2.293}$$

$$\mathbf{h}^{R;(1,3)}_{-n-1,m}(x) = \frac{1}{R}\left(\frac{R}{|x|}\right)^{n+2} \tilde{\mathbf{y}}^{(1,3)}_{n,m}\left(\frac{x}{|x|}\right), \quad x \in \mathbb{R}^3\backslash\{0\}, \tag{2.294}$$

$$\mathbf{h}^{R;(2,1)}_{-n-1,m}(x) = \frac{1}{R}\left(\frac{R}{|x|}\right)^{n+1} \tilde{\mathbf{y}}^{(2,1)}_{n,m}\left(\frac{x}{|x|}\right), \quad x \in \mathbb{R}^3\backslash\{0\}, \tag{2.295}$$

$$\mathbf{h}^{R;(2,2)}_{-n-1,m}(x) = \frac{1}{R}\left(\frac{R}{|x|}\right)^{n-1} \tilde{\mathbf{y}}^{(2,2)}_{n,m}\left(\frac{x}{|x|}\right), \quad x \in \mathbb{R}^3\backslash\{0\}, \tag{2.296}$$

$$\mathbf{h}^{R;(2,3)}_{-n-1,m}(x) = \frac{1}{R}\left(\frac{R}{|x|}\right)^{n} \tilde{\mathbf{y}}^{(2,3)}_{n,m}\left(\frac{x}{|x|}\right), \quad x \in \mathbb{R}^3\backslash\{0\}, \tag{2.297}$$

$$\mathbf{h}^{R;(3,1)}_{-n-1,m}(x) = \frac{1}{R}\left(\frac{R}{|x|}\right)^{n+2} \tilde{\mathbf{y}}^{(3,1)}_{n,m}\left(\frac{x}{|x|}\right), \quad x \in \mathbb{R}^3\backslash\{0\}, \tag{2.298}$$

$$\mathbf{h}^{R;(3,2)}_{-n-1,m}(x) = \frac{1}{R}\left(\frac{R}{|x|}\right)^{n} \tilde{\mathbf{y}}^{(3,2)}_{n,m}\left(\frac{x}{|x|}\right), \quad x \in \mathbb{R}^3\backslash\{0\}, \tag{2.299}$$

$$\mathbf{h}^{R;(3,3)}_{-n-1,m}(x) = \frac{1}{R}\left(\frac{R}{|x|}\right)^{n+1} \tilde{\mathbf{y}}^{(3,3)}_{n,m}\left(\frac{x}{|x|}\right), \quad x \in \mathbb{R}^3\backslash\{0\}, \tag{2.300}$$

for $n \in \mathbb{N}_{\tilde{0}_{i,k}}$, $m = 1, ..., 2n+1$, are called *tensor outer harmonics (of type (i,k), degree n, and order m)*.

Remark 2.63. It is not difficult to show that the following properties are satisfied:

(i) $\mathbf{h}^{R;(i,k)}_{-n-1,m}$ is of class $\mathbf{c}^{(\infty)}(\mathbb{R}^3\backslash\{0\})$, for $i,k \in \{1,2,3\}$,

(ii) $\Delta_x \mathbf{h}^{R;(i,k)}_{-n-1,m}(x) = 0$, for $x \in \mathbb{R}^3\backslash\{0\}$ and $i,k \in \{1,2,3\}$,

(iii) $\mathbf{h}^{R;(i,k)}_{-n-1,m}\big|_{\Omega_R} = \frac{1}{R}\tilde{\mathbf{y}}^{(i,k)}_{n,m}$, for $i,k \in \{1,2,3\}$,

(iv) $\left|\mathbf{h}^{R;(i,k)}_{n,m}(x)\right| = \mathcal{O}\left(|x|^{-3}\right)$, $|x| \to \infty$, for $i,k \in \{1,2,3\}$,

(v) $\left(\mathbf{h}^{R;(i,k)}_{-n-1,m}, \mathbf{h}^{R;(p,q)}_{-l-1,s}\right)_{\mathbf{l}^2(\Omega_R)} = \delta_{i,p}\delta_{k,q}\delta_{n,l}\delta_{m,s}$.

Finally, we discuss the interrelation between scalar outer harmonics and their Hesse tensor in more detail. In spherical coordinates, we have as always that $x = r\xi$, $r = |x|$, $\xi = \frac{x}{|x|}$, and get (see W. Freeden, M. Schreiner [2009])

$$\begin{aligned}
&(\nabla_x \otimes \nabla_x) H^R_{-n-1,m}(x) \qquad (2.301)\\
&= \xi\otimes\xi\frac{\partial^2}{\partial r^2}H^R_{-n-1,m}(x) + \frac{\partial}{\partial r}\xi\otimes\frac{1}{r}\nabla^*_\xi H^R_{-n-1,m}(x)\\
&\quad +\frac{1}{r}\nabla^*_\xi\otimes\xi\frac{\partial}{\partial r}H^R_{-n-1,m}(x) + \frac{1}{r}\nabla^*_\xi\otimes\frac{1}{r}\nabla^*_\xi H^R_{-n-1,m}(x).\\
&= \xi\otimes\xi\frac{\partial^2}{\partial r^2}H^R_{-n-1,m}(x) - \frac{1}{r^2}\xi\otimes\nabla^*_\xi H_{-n-1,m}(x)\\
&\quad +\frac{\partial}{\partial r}\xi\otimes\nabla^*_\xi\frac{\partial}{\partial r}H^R_{-n-1,m}(x) + \frac{1}{r}\mathbf{i}_{\mathrm{tan}}(\xi)\frac{\partial}{\partial r}H^R_{-n-1,m}(x)\\
&\quad +\frac{1}{r}\nabla^*_\xi\left(\frac{\partial}{\partial r}H^R_{-n-1,m}(x)\right)\otimes\xi + \frac{1}{r^2}\nabla^*_\xi\otimes\nabla^*_\xi H^R_{-n-1,m}(x).
\end{aligned}$$

For $F \in \mathrm{C}^{(2)}(\Omega)$, observing the identity

$$\left(\nabla^*_\xi\otimes\nabla^*_\xi + \mathrm{L}^*_\xi\otimes\mathrm{L}^*_\xi\right)F(\xi) = \mathbf{i}_{\mathrm{tan}}(\xi)\Delta^*_\xi F(\xi), \quad \xi\in\Omega, \qquad (2.302)$$

we find

$$\frac{1}{2}\mathbf{o}^{(2,3)}_\xi F(\xi) = \nabla^*_\xi\otimes\nabla^*_\xi F(\xi) + (\nabla^*_\xi F(\xi))\otimes\xi - \frac{1}{2}\mathbf{i}_{\mathrm{tan}}\Delta^*_\xi F(\xi), \quad \xi\in\Omega, \ (2.303)$$

which yields

$$\begin{aligned}
&(\nabla_x \otimes \nabla_x) H^R_{-n-1,m}(x) \qquad (2.304)\\
&= \mathbf{o}^{(1,1)}_\xi\left(\frac{\partial^2}{\partial r^2}H^R_{-n-1,m}(x)\right) + \mathbf{o}^{(2,3)}_\xi\left(\frac{1}{2r^2}H^R_{-n-1,m}(x)\right)\\
&\quad +\mathbf{o}^{(1,2)}_\xi\left(\frac{1}{r}\frac{\partial}{\partial r}H^R_{-n-1,m}(x) - \frac{1}{r^2}H^R_{-n-1,m}(x)\right)\\
&\quad +\mathbf{o}^{(2,1)}_\xi\left(\frac{1}{r}\frac{\partial}{\partial r}H^R_{-n-1,m}(x) - \frac{1}{r^2}H^R_{-n-1,m}(x)\right)\\
&\quad +\mathbf{o}^{(2,2)}_\xi\left(\frac{1}{2}{}_\xi\Delta^*_\xi H^R_{-n-1,m}(x) + \frac{1}{r}\frac{\partial}{\partial r}H^R_{-n-1,m}(x)\right).
\end{aligned}$$

Some elementary calculations using (2.266) show that

$$\begin{aligned}
\tilde{\mathbf{o}}^{(1,1)}Y_{n,m} &= (n+1)(n+2)\mathbf{o}^{(1,1)}Y_{n,m} - (n+2)\mathbf{o}^{(1,2)}Y_{n,m} \qquad (2.305)\\
&\quad -(n+2)\mathbf{o}^{(2,1)}Y_{n,m} - \frac{1}{2}(n+2)(n+1)\mathbf{o}^{(2,2)}Y_{n,m}\\
&\quad +\frac{1}{2}\mathbf{o}^{(2,3)}Y_{n,m}.
\end{aligned}$$

Consequently, we are finally able to verify the relation

$$(\nabla_x \otimes \nabla_x) H^R_{-n-1,m}(x) = \frac{1}{R^2}\left(\tilde{\mu}^{(1,1)}_n\right)^{\frac{1}{2}}\mathbf{h}^{R;(1,1)}_{-n-1,m}(x), \qquad (2.306)$$

$n \in \mathbb{N}_0, m = 1, \ldots, 2n+1$. The last identity enables us to deduce that any harmonic function F of the form (2.234) satisfies a *Meissl relation* of the type

$$\int_{\Omega_r} (\nabla_y \otimes \nabla_y F(y)) \cdot \mathbf{h}^{r;(1,1)}_{-n-1,m}(y)\, d\omega(y) = \left(\frac{R}{r}\right)^n \frac{\left(\tilde{\mu}_n^{(1,1)}\right)^{\frac{1}{2}}}{r^2} F^{\wedge_{\mathrm{L}^2(\Omega_R)}}(n,m) \tag{2.307}$$

for $r > R$. This leads us to the outer harmonic expansion

$$F = \sum_{n=0}^{\infty} \sum_{m=1}^{2n+1} \left(\frac{r}{R}\right)^n r^2 \left(\tilde{\mu}_n^{(1,1)}\right)^{-\frac{1}{2}} \left(\nabla \otimes \nabla F, \mathbf{h}^{r;(1,1)}_{-n-1,m}\right)_{\mathbf{l}^2(\Omega_r)} H^R_{-n-1,m}. \tag{2.308}$$

The formula (2.308) is extremely suitable in the determination of the scalar gravitational potential on a spherical Earth Ω_R from tensorial $(\nabla \otimes \nabla F)$-data on the spherical orbit Ω_r (compare the considerations on SGG problems in Section 4.2). It expresses the gravitational potential F by means of the gravitational tensor $\nabla \otimes \nabla F$ on the sphere Ω_S in terms of a spherical harmonic expansion, where the convergence of the series (2.308) is understood in a uniform sense on every $\overline{\mathcal{K}}$ with $\mathcal{K} \Subset \mathbb{R}^3 \backslash \overline{\mathcal{B}_R(0)}$. Even more, the convergence on Ω_R can be understood in the $\mathrm{L}^2(\Omega_R)$-topology. In other words, potential theoretically reflected concepts exclusively use the tensor outer harmonics of type $(1,1)$ specified by (2.292).

2.7 Exercises

Exercise 2.1. Let F be of class $\mathrm{L}^1([-1,1])$. Show that for any fixed $\xi \in \Omega$,

$$\int_{\Omega} F(\xi \cdot \eta) d\omega(\eta) = 2\pi \int_{-1}^{1} F(t) dt. \tag{2.309}$$

Exercise 2.2. Prove the following rules for spherical differential operators:

(a) For F of class $\mathrm{C}^{(1)}(\Omega)$, we have

$$\mathrm{L}^*_\xi F(\xi) = \xi \wedge \nabla^*_\xi F(\xi) = -\nabla^*_\xi \wedge (\xi F(\xi)), \quad \xi \in \Omega \tag{2.310}$$

$$\nabla^*_\xi F(\xi) = -\xi \wedge \mathrm{L}^*_\xi F(\xi), \quad \xi \in \Omega. \tag{2.311}$$

Additionally, if F is of class $\mathrm{C}^{(1)}(\mathbb{R}^3)$, we have

$$\mathrm{L}_x F(x)|_{x=\xi} = \mathrm{L}^*_\xi F(\xi), \quad \xi \in \Omega, \tag{2.312}$$

where $\mathrm{L}_x F(x) = x \wedge \nabla_x F(x)$, for $x \in \mathbb{R}^3$, is the Euclidean counterpart to the spherical operator L^*.

(b) Let f be a vector field of class $c^{(2)}(\Omega)$. Then

$$\begin{aligned} \Delta^*_\xi f(\xi) &= \nabla^*_\xi \left(\nabla^*_\xi \cdot f(\xi)\right) - \nabla^*_\xi \wedge \left(\nabla^*_\xi \wedge f(\xi)\right) \\ &\quad + \xi \wedge \left(\nabla^*_\xi \wedge f(\xi)\right) - \xi \left(\nabla^*_\xi \cdot f(\xi)\right), \end{aligned} \tag{2.313}$$

for $\xi \in \Omega$. (Hint: you may use the formula $\Delta f = \nabla(\nabla \cdot f) - \nabla \wedge (\nabla \wedge f)$ for a vector field f of class $c^{(2)}(\mathbb{R}^3)$.)

Exercise 2.3. Prove that every infinitely often differentiable eigenfunction of the Beltrami operator with respect to the eigenvalue $-n(n+1)$ constitutes a spherical harmonic of degree n.

Exercise 2.4. Prove the Rodriguez formula for Legendre polynomials, i.e.,

$$P_n(t) = \frac{1}{2^n n!} \left(\frac{d}{dt}\right)^n \left(t^2 - 1\right)^n, \tag{2.314}$$

for $t \in [-1, 1]$ and $n \in \mathbb{N}_0$. Derive the representation

$$P_n(t) = \sum_{k=0}^{\lfloor \frac{n}{2} \rfloor} (-1)^k \frac{(2n-2k)!}{2^n (n-2k)!(n-k)!k!} t^{n-2k}. \tag{2.315}$$

Exercise 2.5. Prove the following recursion formulas for Legendre polynomials:

$$(n+1)P_{n+1}(t) = (2n+1)tP_n(t) - nP_{n-1}(t), \tag{2.316}$$

$$(t^2 - 1)\frac{d}{dt}P_n(t) = ntP_n(t) - nP_{n-1}(t), \tag{2.317}$$

$$(2n+1)P_n(t) = \frac{d}{dt}P_{n+1}(t) - \frac{d}{dt}P_{n-1}(t), \tag{2.318}$$

for $t \in [-1, 1]$ and $n \in \mathbb{N}$.

Exercise 2.6. The associated Legendre polynomial of degree n and order m is defined by

$$P_{n,m}(t) = (1 - t^2)^{\frac{m}{2}} \left(\frac{d}{dt}\right)^m P_n(t), \quad t \in [-1, 1]. \tag{2.319}$$

(a) Show that the associated Legendre polynomials satisfy the differential equation

$$\left(\frac{d}{dt}(1 - t^2)\frac{d}{dt}\right) P_{n,m}(t) = -\left(n(n+1) - \frac{m^2}{1 - t^2}\right) P_{n,m}(t),$$

for $t \in (-1, 1)$, $n \in \mathbb{N}_0$, and $m = 0, \ldots, n$.

(b) Prove that

$$L_{n,m}(\xi) = \begin{cases} P_{n,|m|}(t)\cos(|m|\varphi), & n \in \mathbb{N}_0,\ m = -n, \ldots, 0, \\ P_{n,m}(t)\sin(m\varphi), & n \in \mathbb{N}_0,\ m = 1, \ldots, n, \end{cases} \tag{2.320}$$

for $\xi = (\sqrt{1-t^2}\cos(\varphi), \sqrt{1-t^2}\sin(\varphi), t)^T \in \Omega$, are eigenfunctions of the Beltrami operator with respect to the eigenvalue $-n(n+1)$ (i.e., $L_{n,m}$ provides a specific system of spherical harmonics of degree n). Furthermore, show that $\{L_{n,m}\}_{m=-n,\ldots,n}$ forms an orthogonal basis of $\mathrm{Harm}_n(\Omega)$ with respect to the inner product $(\cdot,\cdot)_{\mathrm{L}^2(\Omega)}$.

Exercise 2.7. Let $H_n : \mathbb{R}^3 \to \mathbb{R}$ be a homogeneous and harmonic polynomial of degree n. Show that H_n can be expressed as

$$H_n(x) = \sum_{k=0}^{n} x_3^k A_{n-k}(x_1, x_2), \quad x = (x_1, x_2, x_3)^T \in \mathbb{R}^3, \tag{2.321}$$

where the $A_{n-k} : \mathbb{R}^2 \to \mathbb{R}$ are homogeneous polynomials of degree $n-k$ that are recursively defined via

$$A_{n-k-2}(x) = -\frac{1}{(k+1)(k+2)}\left(\left(\frac{\partial}{\partial x_1}\right)^2 + \left(\frac{\partial}{\partial x_2}\right)^2\right) A_{n-k}(x), \tag{2.322}$$

for $x = (x_1, x_2)^T \in \mathbb{R}^2$ and $k = 0, \ldots, n-2$. Verify that $\dim(\mathrm{Harm}_n(\Omega)) = 2n+1$.

Exercise 2.8. Let $H_n : \mathbb{R}^3 \to \mathbb{R}$ be a homogeneous and harmonic polynomial of degree n with the following properties:

(i) $H_n(\mathbf{t}x) = H_n(x)$, for $x \in \mathbb{R}^3$ and all orthogonal transformations $\mathbf{t} \in \mathbb{R}^{3\times 3}$ that leave ε^3 invariant (i.e., $\mathbf{t}\varepsilon^3 = \varepsilon^3$),

(ii) $H_n(\varepsilon^3) = 1$.

Show that H_n is uniquely determined by these properties. Moreover, use this fact to show that $\eta \mapsto P_n(\eta \cdot \varepsilon^3)$ is the only normalized (in the sense of (ii)) spherical harmonic of degree n that is invariant with respect to all orthogonal transformations that leave ε^3 fixed.

Exercise 2.9. Prove the Funk–Hecke formula, i.e.,

$$\int_\Omega F(\xi\cdot\eta)P_n(\zeta\cdot\eta)\, d\omega(\eta) = F^\wedge(n)P_n(\xi\cdot\zeta), \quad \xi,\zeta \in \Omega, \tag{2.323}$$

for $F \in \mathrm{L}^1([-1,1])$ and $n \in \mathbb{N}_0$, where

$$F^\wedge(n) = 2\pi \int_{-1}^{1} F(t)P_n(t)dt. \tag{2.324}$$

Use the addition theorem to derive

$$\int_\Omega F(\xi \cdot \eta) Y_n(\eta)\, d\omega(\eta) = F^\wedge(n) Y_n(\xi), \quad \xi \in \Omega, \tag{2.325}$$

for $Y_n \in \mathrm{Harm}_n(\Omega)$.

Exercise 2.10. Prove the following recursion formulas for Chebychev polynomials:

$$\begin{aligned} P_{n+1}(2;t) &= 2tP_n(2;t) - P_{n-1}(2;t), & (2.326)\\ (t^2-1)\frac{d}{dt}P_n(2;t) &= ntP_n(2;t) - nP_{n-1}(2;t), & (2.327)\\ 2(n^2-1)P_n(2;t) &= (n-1)\frac{d}{dt}P_{n+1}(2;t) - (n+1)\frac{d}{dt}P_{n-1}(2;t), & (2.328) \end{aligned}$$

for $t \in [-1,1]$ and $n \in \mathbb{N}$.

Exercise 2.11. Calculate an explicit representation for the inner and outer harmonics $H_{n,k}^{\rho,\mathrm{stereo}}$ and $H_{-n,k}^{\rho,\mathrm{stereo}}$ on the unit sphere.

Exercise 2.12. We define the differential operators

$$\begin{aligned} k_n^{(1)}F(x) &= \left((2n+1)x - |x|^2\nabla_x\right)F(x), \quad x \in \mathbb{R}^3, & (2.329)\\ k_n^{(2)}F(x) &= \nabla_x F(x), \quad x \in \mathbb{R}^3, & (2.330)\\ k_n^{(3)}F(x) &= \mathrm{L}_x F(x), \quad x \in \mathbb{R}^3, & (2.331) \end{aligned}$$

for sufficiently smooth $F : \mathbb{R}^3 \to \mathbb{R}$. Show that $k_n^{(i)} H_{n,k}^R$ is a harmonic vector field such that

$$k_n^{(i)} H_{n,k}^R\big|_\Omega = \tilde{o}^{(i)} H_{n,k}^R\big|_\Omega, \quad i = 1,2,3. \tag{2.332}$$

As usual, $H_{n,k}^R$ denotes the inner harmonic of degree n and order k. More generally, we have

$$\begin{aligned} k_n^{(1)}(r^n Y_n(\xi)) &= (n+1)r^{n+1}o_\xi^{(1)}Y_n(\xi) - r^{n+1}o_\xi^{(2)}Y_n(\xi), & (2.333)\\ k_n^{(2)}(r^n Y_n(\xi)) &= nr^{n-1}o_\xi^{(1)}Y_n(\xi) - r^{n-1}o_\xi^{(2)}Y_n(\xi), & (2.334)\\ k_n^{(3)}(r^n Y_n(\xi)) &= r^n o_\xi^{(3)}Y_n(\xi), & (2.335) \end{aligned}$$

for $x = r\xi$ with $\xi \in \Omega$, $r > 0$, and Y_n of class $\mathrm{Harm}_n(\Omega)$.

Exercise 2.13. Show that

$$\begin{aligned} \nabla_x \wedge (k_n^{(1)}F(x)) &= -(2n+3)k_n^{(3)}F(x), & (2.336)\\ \nabla_x \wedge (k_n^{(2)}F(x)) &= 0, & (2.337)\\ \nabla_x \wedge (k_n^{(3)}F(x)) &= x\Delta_x F(x) - 2k_n^{(2)}F(x) - (x\cdot\nabla_x)k_n^{(2)}F(x) & (2.338)\\ &= o_\xi^{(1)}\left(\frac{1}{r}\Delta_\xi^* F(r\xi)\right) + o_\xi^{(2)}\left(-\frac{1}{r}\frac{\partial}{\partial r}(rF(r\xi))\right), \end{aligned}$$

for $x \in \mathbb{R}^3$, $\xi = \frac{x}{|x|}$, $r = |x|$, and sufficiently smooth functions $F : \mathbb{R}^3 \to \mathbb{R}$.

Part II

Potential Theory in the Euclidean Space $\mathbb{R}^3$

3

Basic Concepts

CONTENTS

3.1 Background Material

In the following, we collect some basic material well known from classical potential theory in the Euclidean space $\mathbb{R}^3$. First we have a closer look at the already mentioned fundamental solution of the Laplace operator. Observing its specific properties, we are able to formulate the third interior Green formula. Mean value theorems and a maximum/minimum principle are the canonical consequences. Harmonic functions are recognized to be analytic in their harmonicity domain. The Kelvin transform enables us to study harmonic functions that are regular at infinity. Keeping the regularity at infinity in mind, we are finally led to exterior Green formulas. The third exterior Green formula is formulated analogously to its interior counterpart, thereby observing the regularity at infinity.

3.1.1 Fundamental Solution of the Laplace Operator

We begin with the introduction of the fundamental solution (in geophysical and geodetic applications sometimes also called the *singularity function* or *single-pole function*).

Definition 3.1. The function $G(\Delta; \cdot) : (0, \infty) \to \mathbb{R}$ given by

$$G(\Delta; r) = \frac{1}{4\pi r}, \quad r \in (0, \infty), \tag{3.1}$$

is called the *fundamental solution of the Laplace operator in* $\mathbb{R}^3$.

$G(\Delta; |x-y|), x \neq y$, possesses the following interesting properties, which are listed for later use:

Symmetry: $G(\Delta; |x-y|), x \neq y$, defines a function that depends only on the distance $r = |x-y|$ of x and y, i.e., a radial basis function in $\mathbb{R}^3$.

Differential Equation: $G(\Delta; |x-y|),\ x \in \mathbb{R}^3 \backslash \{y\}$, is infinitely often differentiable in the variable x. From the identities

$$\frac{\partial}{\partial x_i} G(\Delta; |x-y|) = G'(\Delta; |x-y|) \frac{x_i - y_i}{|x-y|}, \tag{3.2}$$

$$\left(\frac{\partial}{\partial x_i}\right)^2 G(\Delta; |x-y|) = G''(\Delta; |x-y|) \frac{(x_i - y_i)^2}{|x-y|^2} + G'(\Delta; |x-y|) \left(\frac{1}{|x-y|} - \frac{(x_i-y_i)^2}{|x-y|^3} \right) \tag{3.3}$$

we easily obtain

$$\Delta_x G(\Delta; |x-y|) = 0,\ x \in \mathbb{R}^3 \backslash \{y\}. \tag{3.4}$$

By $G'(\Delta; |x-y|)$ and $G''(\Delta; |x-y|)$ we mean the first- and second-order one-dimensional derivatives at $r = |x-y|$, respectively.

Estimates: An elementary calculation yields

$$-\nabla_x G(\Delta; |x-y|) = \nabla_y G(\Delta; |x-y|). \tag{3.5}$$

Moreover, it is not difficult to show that

$$\left| \frac{\partial}{\partial x_i \partial x_j} G(\Delta; |x-y|) \right| \leq \; 64\pi^2 \; (G(\Delta; |x-y|))^3, \; x \in \mathbb{R}^3 \backslash \{y\}. \tag{3.6}$$

For $r \to 0+$, we have (cf. Exercise 3.1)

$$\int_{\Omega_r(x)} |G(\Delta; |x-y|)| \; d\omega(y) \; = \; \mathcal{O}(r) \tag{3.7}$$

and

$$\int_{\Omega_r(x)} \left| \frac{\partial}{\partial x_i} G(\Delta; |x-y|) \right| \; d\omega(y) \; = \; \mathcal{O}(1), \quad i = 1, 2, 3. \tag{3.8}$$

3.1.2 Interior Third Green Formula

Lemma 3.2. *For a continuous function F on $\mathcal{B}_\rho(x)$, $\rho > 0, x \in \mathbb{R}^3$, we have*

$$\lim_{r \to 0+} \int_{\Omega_r(x)} F(y) \; \frac{x-y}{|x-y|} \cdot \nabla_y G(\Delta; |x-y|) \; d\omega(y) \; = \; F(x), \tag{3.9}$$

$$\lim_{r \to 0+} \int_{\Omega_r(x)} F(y) \; G(\Delta; |x-y|) \; d\omega(y) \; = \; 0. \tag{3.10}$$

Proof. Because of the continuity of the function F in each ball $\overline{\mathcal{B}_r(x)}$, $r < \rho$, we find

$$\left| \int_{\Omega_r(x)} F(y) \frac{1}{4\pi|x-y|} \; d\omega(x) \right| \leq \frac{C}{4\pi} \int_{\Omega_r(x)} \frac{1}{|x-y|} \; d\omega(y) = \frac{C}{4\pi} 4\pi r \tag{3.11}$$

for some positive constant C. Hence, the limit relation (3.10) is valid.

If we apply the mean value theorem of multidimensional analysis to the integral in (3.9), we obtain

$$\begin{aligned} \int_{\Omega_r(x)} F(y) \; \frac{x-y}{|x-y|} \cdot \nabla_y \frac{1}{4\pi|x-y|} \; d\omega(y) &= \frac{1}{4\pi r^2} \int_{\Omega_r(x)} F(y) \; d\omega(y) \\ &= \frac{1}{4\pi r^2} 4\pi r^2 F(y_r) \end{aligned} \tag{3.12}$$

for certain points $y_r \in \Omega_r(x)$. The limit $r \to 0+$ implies $y_r \to x$, such that the continuity of F on $\overline{\mathcal{B}_r(x)}$, $r < \rho$, yields $\lim_{r \to 0+} F(y_r) = F(x)$. This is the desired result. □

From Lemma 3.2, in combination with the Second Green Theorem, we are able to prove

Theorem 3.3 (Interior Third Green Theorem). *Let $\mathcal{G} \subset \mathbb{R}^3$ be a regular region. Suppose that U is of class $\mathrm{C}^{(1)}(\overline{\mathcal{G}}) \cap \mathrm{C}^{(2)}(\mathcal{G})$, with ΔU being Lebesgue-integrable on $\mathcal{G}$. Then*

$$\int_{\partial\mathcal{G}} \left(G(\Delta; |x-y|) \frac{\partial}{\partial\nu(y)} U(y) - U(y) \frac{\partial}{\partial\nu(y)} G(\Delta; |x-y|) \right) d\omega(y) \tag{3.13}$$
$$- \int_{\mathcal{G}} G(\Delta; |x-y|) \Delta_y U(y) \, dV(y) = \begin{cases} U(x), & x \in \mathcal{G}, \\ \frac{1}{2} U(x), & x \in \partial\mathcal{G}, \\ 0, & x \in \mathcal{G}^c. \end{cases}$$

Proof. We first consider the case $x \in \mathcal{G}$. For every (sufficiently small) $r > 0$, the Second Green Theorem (Theorem 1.13) tells us that

$$\begin{aligned} &- \int_{\mathcal{G}\setminus\overline{\mathcal{B}_r(x)}} G(\Delta; |x-y|) \Delta U(y) \, dV(y) \\ &= \int_{\partial\mathcal{G}} \left(U(y) \frac{\partial}{\partial\nu(y)} G(\Delta; |x-y|) - G(\Delta; |x-y|) \frac{\partial}{\partial\nu(y)} U(y) \right) d\omega(y) \\ &\quad + \int_{\Omega_r(x)\cap\mathcal{G}} \left(U(y) \frac{\partial}{\partial\nu(y)} G(\Delta; |x-y|) - G(\Delta; |x-y|) \frac{\partial}{\partial\nu(y)} U(y) \right) d\omega(y). \end{aligned} \tag{3.14}$$

For $r \to 0+$, the assertion of Theorem 3.3 follows immediately from the integral relations of Lemma 3.2. For $x \in \mathcal{G}^c$, a similar argument holds true, except we find $\Omega_r(x) \cap \mathcal{G} = \emptyset$ for sufficiently small $r > 0$. Thus, the second integral on the right-hand side of (3.14) vanishes as r tends to zero. Finally, if $x \in \partial\mathcal{G}$, the surface area of $\Omega_r(x) \cap \mathcal{G}$ behaves like $2\pi r^2$ for $r \to 0+$ (and not $4\pi r^2$, as would be the case for the entire sphere $\Omega_r(x)$). This causes the pre-factor $\frac{1}{2}$ on the right-hand side of (3.13). □

As a special case, we obtain for continuously differentiable functions U in $\overline{\mathcal{G}}$ which are harmonic in $\mathcal{G}$ the so-called *Interior Fundamental Theorem.*

Corollary 3.4. *Suppose that U is of class $\mathrm{C}^{(1)}(\overline{\mathcal{G}}) \cap \mathrm{C}^{(2)}(\mathcal{G})$ with $\Delta U = 0$ on $\mathcal{G}$. Then*

$$\int_{\partial\mathcal{G}} \left(G(\Delta; |x-y|) \frac{\partial U}{\partial\nu(y)}(y) - U(y) \frac{\partial}{\partial\nu(y)} G(\Delta; |x-y|) \right) d\omega(y) \tag{3.15}$$
$$= \begin{cases} U(x), & x \in \mathcal{G}, \\ \frac{1}{2} U(x), & x \in \partial\mathcal{G}, \\ 0, & x \in \mathcal{G}^c. \end{cases}$$

Letting $U = 1$ in $\overline{\mathcal{G}}$, we obviously find, in connection with (3.1) and Corollary 3.4, the following representation of the solid angle α.

Lemma 3.5 (Solid Angle). *Let $\mathcal{G} \subset \mathbb{R}^3$ be a regular region. Then the solid angle $\alpha(x)$, subtended by the boundary $\partial\mathcal{G}$ at the point $x \in \mathbb{R}^3$, given by*

$$\alpha(x) = \begin{cases} 4\pi, & x \in \mathcal{G}, \\ 2\pi, & x \in \partial\mathcal{G}, \\ 0, & x \in \mathcal{G}^c, \end{cases} \tag{3.16}$$

can be expressed in the integral form

$$\alpha(x) \quad = \quad -\int_{\partial\mathcal{G}} \frac{\partial}{\partial\nu(y)} \frac{1}{|x-y|} \, d\omega(y). \tag{3.17}$$

Remark 3.6. Of course, the interior Green theorems remain valid for translations $\mathcal{G}+\{x\}$ with $x \in \mathbb{R}^3$ and $\mathcal{G} \subset \mathbb{R}^3$ regular. Furthermore, the interior Green theorems holds true, e.g., for unions, intersections, and differences of two regular regions. In doing so, Theorem 3.3 may be extended to certain regions $\mathcal{G}$ with non-smooth boundaries (for further details on the Green theorems, the reader is referred to textbooks on vector analysis).

3.1.3 Mean Value Theorems

Next, we present some basic material on harmonic functions in the Euclidean space $\mathbb{R}^3$. A more comprehensive treatment of classical potential theory can be found in any standard textbook, e.g., N.M. Günter [1957], L.L. Helms [1969], O.D. Kellogg [1967], V.D. Kupradze [1965], R. Leis [1967], E. Martensen [1968], W. Walter [1971], S.G. Michlin [1975], E. Wienholtz et al. [2009].

Our point of departure is to verify that continuous functions that satisfy a Mean Value Property are harmonic.

Theorem 3.7. *Let $\mathcal{G} \subset \mathbb{R}^3$ be a regular region as introduced by Definition 1.7. Then the following statements are equivalent:*

(a) $U : \mathcal{G} \to \mathbb{R}$ *is* harmonic *in $\mathcal{G}$, i.e., $U \in \mathrm{C}^{(2)}(\mathcal{G})$ and $\Delta U = 0$ in $\mathcal{G}$,*

(b) $U : \mathcal{G} \to \mathbb{R}$ *possesses the* Mean Value Property *on $\mathcal{G}$, i.e., U is of class $\mathrm{C}^{(0)}(\mathcal{G})$ and, for all $x \in \mathcal{G}$ and $r > 0$ with $\mathcal{B}_r(x) \Subset \mathcal{G}$,*

$$U(x) = \frac{1}{4\pi r^2} \int_{\Omega_r(x)} U(y) \, d\omega(y), \tag{3.18}$$

(c) *U is of class $\mathrm{C}^{(0)}(\overline{\mathcal{G}})$ and for all $r > 0$ with $\mathcal{B}_r(x) \Subset \mathcal{G}$*

$$\int_{\mathcal{B}_r(x)} (U(x) - U(y)) \, dV(y) = 0. \tag{3.19}$$

Proof. "(a) $\Rightarrow$ (b)": U is harmonic in $\mathcal{G}$. The property (b) immediately follows from the Interior Third Green Theorem, i.e.,

$$U(x) = \frac{1}{4\pi} \int_{\Omega_r(x)} \left(\frac{1}{r} \frac{\partial U}{\partial \nu}(y) - \frac{1}{r^2} U(y) \right) \, d\omega(y), \tag{3.20}$$

in connection with (1.79).

"(b) $\Rightarrow$ (c)": From (3.18) we are led to

$$4\pi \int_0^r U(x) \; s^2 \; ds = \int_0^r \left(\int_{\Omega_s(x)} U(y) \; d\omega(y) \right) \; ds. \tag{3.21}$$

Hence it follows that

$$\frac{4\pi r^3}{3} U(x) = \int_{\mathcal{B}_r(x)} U(y) \; dV(y). \tag{3.22}$$

Clearly, $\|\mathcal{B}_r(x)\| = \frac{4\pi r^3}{3}$ is the volume of the ball $\mathcal{B}_r(x)$ such that

$$\int_{\mathcal{B}_r(x)} (U(x) - U(y)) \; dV(y) = 0 \tag{3.23}$$

holds true for all $\mathcal{B}_r(x) \Subset \mathcal{G}$.

"(c) $\Rightarrow$ (b)": Differentiation yields

$$\begin{aligned} 0 &= \frac{d}{dr} \int_{\mathcal{B}_r(x)} (U(x) - U(y)) \; dV(y) \\ &= \int_{\Omega_r(x)} (U(x) - U(y)) \; d\omega(y) \\ &= 4\pi r^2 U(x) - \int_{\Omega_r(x)} U(y) \; d\omega(y), \end{aligned} \tag{3.24}$$

remembering that $\partial \mathcal{B}_r(x) = \Omega_r(x)$.

"(b) $\Rightarrow$ (a)": First, we prove a remarkable auxiliary result, namely that U is of class $\mathrm{C}^{(\infty)}(\mathcal{G})$. For that purpose we assume that U is a member of class $\mathrm{C}^{(l)}(\mathcal{G})$ (note that this assertion is valid for the initial step $l = 0$ of the induction). From a standard argument of multidimensional analysis it follows in connection with (c) (which is implied by (b)) that

$$\begin{aligned} \left(\frac{\partial}{\partial x_k} \right)^l U(x) &= \frac{3}{4\pi r^3} \left(\frac{\partial}{\partial x_k} \right)^l \int_{\mathcal{B}_r(x)} U(y) \; dV(y) \\ &= \frac{3}{4\pi r^3} \left(\frac{\partial}{\partial x_k} \right)^l \int_{\mathcal{B}_r(0)} U(y + x) \; dV(y) \\ &= \frac{3}{4\pi r^3} \int_{\mathcal{B}_r(0)} \left(\frac{\partial}{\partial x_k} \right)^l U(y + x) \; dV(y), \end{aligned} \tag{3.25}$$

for $x \in \mathcal{G}$, $k \in \{1,2,3\}$, and $r > 0$ sufficiently small. The integrand of the last integral is continuous on $\mathcal{G}$. It can directly be seen that

$$\int_{\mathcal{B}_r(0)} \left(\frac{\partial}{\partial x_k}\right)^l U(y+x)\, dV(y) = \int_{\mathcal{B}_r(0)} \left(\frac{\partial}{\partial y_k}\right)^l U(y+x)\, dV(y). \tag{3.26}$$

In other words, for $k \in \{1,2,3\}$, we have

$$\left(\frac{\partial}{\partial x_k}\right)^l U(x) = \frac{3}{4\pi r^3} \int_{\mathcal{B}_r(x)} \left(\frac{\partial}{\partial y_k}\right)^l U(y)\, dV(y). \tag{3.27}$$

Since for any $x \in \mathcal{G}$ a ball $B_r(x) \subset \mathcal{G}$ can be found, it follows from (3.27) that $\left(\frac{\partial}{\partial x_k}\right)^l U \in \mathrm{C}^{(1)}(\mathcal{G})$. Thus, we find $U \in \mathrm{C}^{(l+1)}(\mathcal{G})$. By inductively continuing this process, we are finally led to the conclusion that U is a member of class $\mathrm{C}^{(\infty)}(\mathcal{G})$.

Now, as we have seen that U is of class $\mathrm{C}^{(\infty)}(\mathcal{G})$, we are able to form the Laplace derivative ΔU. For sufficiently small r, it follows from the Gauss Theorem (cf. (1.76)) that

$$\begin{aligned} \int_{\mathcal{B}_r(x)} \Delta U(y)\, dV(y) &= \int_{\Omega_r(x)} \nu(y) \cdot \nabla U(y)\, d\omega(y) \\ &= r^2 \int_{\Omega} (\nabla U)(x + r\eta) \cdot \eta\, d\omega(\eta). \end{aligned} \tag{3.28}$$

Remember that $\Omega = \Omega_1(0)$. Introducing the auxiliary expression

$$H(r) = \int_{\Omega} U(x + r\eta)\, d\omega(\eta), \tag{3.29}$$

we obtain (by aid of Exercise 3.4) the derivative

$$H'(r) = \int_{\Omega} (\nabla U)(x + r\eta) \cdot \eta\, d\omega(\eta). \tag{3.30}$$

In connection with (3.28), we therefore have

$$H'(r) = \frac{1}{r^2} \int_{\mathcal{B}_r(x)} \Delta U(y)\, dV(y). \tag{3.31}$$

Integrating over an interval $[s, r]$ provides

$$H(r) - H(s) = \int_s^r \frac{1}{t^2} \left(\int_{\mathcal{B}_t(x)} \Delta U(y)\, dV(y) \right) dt. \tag{3.32}$$

Property (b), in connection with (3.29), leads us to

$$\lim_{s \to 0+} H(s) = \int_{\Omega} U(x)\, d\omega(\eta) = 4\pi U(x). \tag{3.33}$$

Combining (3.32) and (3.33) we therefore obtain

$$U(x) = \frac{1}{4\pi} \int_{\Omega} U(x+r\eta)\, d\omega(\eta) - \frac{1}{4\pi} \int_0^r \frac{1}{t^2} \left(\int_{\mathcal{B}_t(x)} \Delta U(y)\, dV(y) \right) dt. \quad (3.34)$$

Suppose now that there exists a point $z \in \mathcal{G}$ with $\Delta U(z) \neq 0$. Without loss of generality, we assume $\Delta U(z) > 0$. Because of the continuity of ΔU in the neighborhood of z, there exists a ball $\mathcal{B}_r(z) \Subset \mathcal{G}$ such that $\Delta U(x) > 0$ for all $x \in \overline{\mathcal{B}_r(z)}$. This means

$$0 < \int_{\mathcal{B}_\rho(z)} \Delta U(y)\, dV(y) \quad (3.35)$$

for all $\rho \in (0, r)$. As an immediate consequence we obtain from (3.34)

$$U(z) < \frac{1}{4\pi r^2} \int_{\Omega_r} U(x)\, d\omega(x). \quad (3.36)$$

This, however, is a contradiction to the Mean Value Property (3.18). Altogether, the equivalencies listed in Theorem 3.7 are verified. □

As a particularly important consequence of the proof of Theorem 3.7 we recapitulate:

Corollary 3.8. *A harmonic function U on a regular region $\mathcal{G}$ is of class $\mathrm{C}^{(\infty)}(\mathcal{G})$. Moreover,*

$$\Delta \left(\frac{\partial}{\partial x_j} \right) U(x) = \frac{\partial}{\partial x_j} \Delta U(x) = 0, \quad x \in \mathcal{G}, \quad (3.37)$$

for all $j = 1, 2, 3$. Especially,

$$\Delta(\nabla U) \;=\; \nabla(\Delta U) \;=\; 0. \quad (3.38)$$

Our results enable us to derive the so-called *Harnack Theorem.*

Theorem 3.9. *Let $\mathcal{G} \subset \mathbb{R}^3$ be a regular region. Suppose that $\{U_n\}_{n\in\mathbb{N}_0}$ is a sequence of harmonic functions that converges in a locally uniform sense, i.e., for every $\mathcal{K} \Subset \mathcal{G}$, $\{U_n\}_{n\in\mathbb{N}_0}$ converges uniformly on $\overline{\mathcal{K}}$. Then $\{U_n\}_{n\in\mathbb{N}_0}$ converges to a limit function U that is harmonic on $\mathcal{G}$. Furthermore, every sequence*

$$\left\{ \left(\frac{\partial}{\partial x_j} \right)^l U_n \right\}_{n\in\mathbb{N}_0}, \quad l \in \mathbb{N}_0,\ j \in \{1,2,3\},$$

converges uniformly to $\left\{ \left(\frac{\partial}{\partial x_j} \right)^l U \right\}$ on every $\overline{\mathcal{K}}$ with $\mathcal{K} \Subset \mathcal{G}$.

Proof. Based on a standard argument of analysis about the uniform convergence of continuous functions, we know that $\{U_n\}_{n\in\mathbb{N}_0}$ has a continuous limit function U. From Theorem 3.7 we are able to deduce that for all $\mathcal{B}_r(x) \Subset \mathcal{G}$, the identity

$$\int_{\mathcal{B}_r(x)} (U_n(y) - U_n(x)) \; dV(y) = 0 \tag{3.39}$$

holds true for every n. By virtue of the uniform convergence, we therefore have

$$\int_{\mathcal{B}_r(x)} (U(y) - U(x)) \; dV(y) = 0. \tag{3.40}$$

Consequently, Theorem 3.7 informs us that U is of class $\mathrm{C}^{(\infty)}(\mathcal{G})$ and harmonic on $\mathcal{G}$. Next, we consider a sequence $\{(\partial/\partial x_j)^l U_n\}_{n\in\mathbb{N}_0}, j \in \{1,2,3\}$, converging uniformly on every set $\overline{\mathcal{K}'}$ with $\mathcal{K}' \Subset \mathcal{K} \Subset \mathcal{G}$. In connection with the Mean Value Property, we obtain

$$\left(\frac{\partial}{\partial x_j}\right)^l U_n(x) = \frac{3}{4\pi r^3}\int_{\mathcal{B}_r(x)} \left(\frac{\partial}{\partial y_j}\right)^l U_n(y) \, dV(y), \tag{3.41}$$

for all $x \in \mathcal{K}'$ and $j \in \{1,2,3\}$. Moreover, for $k \in \{1,2,3\}$,

$$\frac{\partial}{\partial x_k}\left(\frac{\partial}{\partial x_j}\right)^l U_n(x) = \frac{3}{4\pi r^3}\int_{\Omega_r(x)} \left(\frac{\partial}{\partial y_j}\right)^l U_n(y)\, \nu(y)\cdot \varepsilon^k \; d\omega(y). \tag{3.42}$$

The uniform convergence of $\{(\partial/\partial x_j)^l U_n\}_{n\in\mathbb{N}_0}, j \in \{1,2,3\}$, in $\overline{\mathcal{K}}$ implies the uniform convergence of $\{\partial/\partial x_k \,(\partial/\partial x_j)^l U_n\}_{n\in\mathbb{N}_0}$ in $\overline{\mathcal{K}'}$, $k \in \{1,2,3\}$. Consequently, $\{(\partial/\partial x_j)^{l+1} U_n\}_{n\in\mathbb{N}_0}$ is uniformly convergent. Therefore, Theorem 3.9 follows by already known arguments. □

3.1.4 Maximum/Minimum Principle

A central result in the theory of harmonic functions is the Maximum/Minimum Principle. The Mean Value Property is an essential tool.

Theorem 3.10 (Maximum/Minimum Principle). *Let $\mathcal{G} \subset \mathbb{R}^3$ be a regular region. Suppose that U is harmonic in $\mathcal{G}$ and non-constant. Then U does not reach its minimum or maximum in $\mathcal{G}$. If, in addition, U is of class $\mathrm{C}^{(0)}(\overline{\mathcal{G}})$, then U reaches its minimum and maximum in $\overline{\mathcal{G}}$, and the extremal points lie on $\partial\mathcal{G}$. More precisely,*

$$\sup_{x\in\overline{\mathcal{G}}} |U(x)| \leq \sup_{x\in\partial\mathcal{G}} |U(x)|. \tag{3.43}$$

Proof. Contrary to the assertion, we assume that U reaches its maximum at $x_0 \in \mathcal{G}$, i.e., $U(x_0) = \max_{x\in\overline{\mathcal{G}}} |U(x)| = \overline{M}$ (without loss of generality, we only

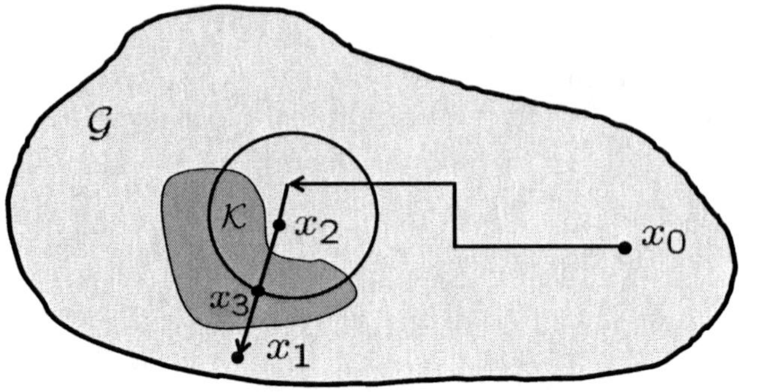

FIGURE 3.1
Auxiliary figure illustrating the proof of the maximum principle.

deal with the maximum). Since U is non-constant, there exists a point $x_1 \in \mathcal{G}$ with $U(x_1) < \overline{M}$. We connect the points x_0 and x_1 by a continuous piecewise linear curve lying completely inside $\mathcal{G}$ (see Figure 3.1).

Let x_2 be the first point on the way back from x_1 to x_0 on the specified continuous, piecewise linear curve satisfying $U(x_2) = \overline{M}$ (note that $x_2 = x_0$ is allowed). We consider a ball $\mathcal{B}_r(x_2) \Subset \mathcal{G}$, which does not contain the point $x_1 \in \mathcal{G}$. The bounding sphere $\Omega_r(x_2)$ intersects the curve at $x_3 \in \mathcal{G}$ for which we know $U(x_3) < \overline{M}$. Since U is continuous in $\mathcal{G}$, there exists an open neighborhood $\mathcal{K}$ of x_3 with $U(x) < \overline{M}$ for all $x \in \mathcal{K}$. Under these circumstances, the Mean Value Property tells us that

$$\begin{aligned} \overline{M} &= U(x_2) = \frac{3}{4\pi r^3} \int_{\mathcal{B}_r(x_2)} U(y)\, dV(y) \qquad (3.44) \\ &= \frac{3}{4\pi r^3} \left(\int_{\mathcal{B}_r(x_2)\cap\mathcal{K}} U(y)\, dV(y) + \int_{\mathcal{B}_r(x_2)\setminus\mathcal{K}} U(y)\, dV(y) \right) \\ &< \overline{M}. \end{aligned}$$

Obviously, the relation (3.44) yields a contradiction. □

A direct consequence of the Maximum/Minimum Principle is the following *stability theorem.*

Theorem 3.11. *Let $\mathcal{G} \subset \mathbb{R}^3$ be a regular region, and suppose that U, V are of class $\mathrm{C}^{(0)}(\overline{\mathcal{G}}) \cap \mathrm{C}^{(2)}(\mathcal{G})$ and harmonic in $\mathcal{G}$. Let ε be an arbitrary positive number. If*

$$\sup_{x\in\partial\mathcal{G}} |U(x) - V(x)| \le \varepsilon, \qquad (3.45)$$

then

$$\sup_{x\in\overline{\mathcal{G}}} |U(x) - V(x)| \le \varepsilon. \qquad (3.46)$$

Proof. For all $x \in \partial\mathcal{G}$ we have $-\varepsilon \leq U(x) - V(x) \leq \varepsilon$. Moreover, $U - V$ is of class $\mathrm{C}^{(0)}(\overline{\mathcal{G}}) \cap \mathrm{C}^{(2)}(\mathcal{G})$ and harmonic in $\mathcal{G}$. By virtue of the Maximum/Minimum Principle, we therefore find $-\varepsilon \leq U(x) - V(x) \leq \varepsilon$ for all $x \in \overline{\mathcal{G}}$. □

3.1.5 Real Analyticity

The following a priori estimate for first partial derivatives of harmonic functions does not depend on the specific character of the harmonic function, it is dependent only on global bounds $\underline{M}, \overline{M}$.

Theorem 3.12. *Let $\mathcal{G} \subset \mathbb{R}^3$ be a regular region. Suppose that U is harmonic in $\mathcal{G}$ such that $\underline{M} \leq U(x) \leq \overline{M}$ for all $x \in \mathcal{G}$. Then*

$$\left| \frac{\partial U}{\partial x_i}(x) \right| \leq \frac{3}{2\ \mathrm{dist}(x, \partial\mathcal{G})} (\overline{M} - \underline{M}), \quad i \in \{1,2,3\} \tag{3.47}$$

holds true for all $x \in \mathcal{G}$.

Proof. We set $Q(x) = U(x) - \frac{1}{2}(\underline{M} + \overline{M})$ for all $x \in \mathcal{G}$. Then we readily see that $|Q(x)| \leq \frac{1}{2}(\overline{M} - \underline{M})$. Clearly, the first derivatives of Q coincide with the derivatives of U. We know that $\frac{\partial Q}{\partial x_j}$, $j \in \{1,2,3\}$, is harmonic in $\mathcal{G}$. From the Mean Value Property we therefore get

$$\frac{\partial Q}{\partial x_j}(x) = \frac{3}{4\pi r^3} \int_{\mathcal{B}_r(x)} \frac{\partial Q}{\partial y_j}(y)\ dV(y), \tag{3.48}$$

provided that $\mathcal{B}_r(x) \Subset \mathcal{G}$. Hence, from the Gauss Theorem, it follows that

$$\begin{aligned} \left| \frac{\partial U}{\partial x_j}(x) \right| &= \left| \frac{\partial Q}{\partial x_j}(x) \right| = \frac{3}{4\pi r^3} \left| \int_{\Omega_r(x)} Q(y) \frac{(x-y) \cdot \varepsilon^j}{|x-y|}\ d\omega(y) \right| \\ &\leq \frac{3}{4\pi r^3} \int_{\Omega_r(x)} \frac{\overline{M} - \underline{M}}{2}\ d\omega(y) = \frac{3}{r} \frac{\overline{M} - \underline{M}}{2}. \end{aligned} \tag{3.49}$$

This is the announced result of Theorem 3.12. □

Next we deal with the (real) analyticity of harmonic functions in $\mathcal{G}$.

Theorem 3.13. *Let $\mathcal{G} \subset \mathbb{R}^3$ be a regular region. Then the estimate*

$$\left| \left(\frac{\partial}{\partial x_j} \right)^l U(x) \right| \leq \frac{3^l l!\ e^{l-1}}{r^l} \sup_{y \in \mathcal{B}_r(x)} |U(y)|, \quad l \in \mathbb{N},\ j \in \{1,2,3\}, \tag{3.50}$$

holds for all $x \in \mathcal{G}$, $0 < r < \mathrm{dist}(x, \partial\mathcal{G})$ and all harmonic functions U on $\mathcal{G}$.

Proof. A harmonic function is infinitely often differentiable, and all derivatives are again harmonic. For $l = 1$ the estimate is clear from Theorem 3.12. We prove the estimate of Theorem 3.13 by induction. Assuming that the assertion is true for l, the point of departure is

$$\left| \left(\frac{\partial}{\partial x_k} \right)^{l+1} U(x) \right| = \left| \left(\frac{\partial}{\partial x_k} \right)^{l} \frac{\partial}{\partial x_k} U(x) \right| \leq \frac{3^l l! \ e^{l-1}}{s^l} \sup_{y \in \mathcal{B}_s(x)} \left| \frac{\partial}{\partial x_k} U(y) \right| , \tag{3.51}$$

where $s = \frac{l}{l+1} r, \quad 0 < r < \operatorname{dist}(x, \partial \mathcal{G})$ (note that this choice of s becomes obvious if we first let $s = (1-\beta) r$ and then take β in appropriate adaptation). From Theorem 3.12 we get

$$\sup_{y \in \mathcal{B}_s(x)} \left| \frac{\partial}{\partial x_k} U(y) \right| \leq \frac{3}{r-s} \sup_{y \in \mathcal{B}_r(x)} |U(y)| . \tag{3.52}$$

Moreover, it is not difficult to see that

$$\frac{3^l l! \ e^{l-1} \cdot 3}{s^l (r-s)} = \frac{3^{l+1} (l+1)! \ e^l}{r^{l+1}} \frac{\left(1 + \frac{1}{l}\right)^l}{e} . \tag{3.53}$$

Since $\left(1 + \frac{1}{l}\right)^l$ converges (monotonically increasing) to the limit e, the induction is verified. □

From Theorem 3.13 we immediately obtain

Corollary 3.14. *Let U be harmonic in $\mathcal{G} \subset \mathbb{R}^3$. Suppose that $x \in \mathcal{B}_s(x_0) \subset \mathcal{B}_r(x_0)$ with $\mathcal{B}_r(x_0) \Subset \mathcal{G}$. Then*

$$\left| \left(\frac{\partial}{\partial x_j} \right)^{l} U(x) \right| \leq \frac{3^l \ l! \ e^{l-1}}{(r-s)^l} \sup_{y \in \mathcal{B}_r(x_0)} |U(y)|, \quad j \in \{1,2,3\}. \tag{3.54}$$

Proof. For $x \in \mathcal{B}_s(x_0)$ we have $\mathcal{B}_{r-s}(x) \subset \mathcal{B}_r(x_0)$. □

In the following, we are interested in the proof of

Lemma 3.15. *Let U be of class $\mathrm{C}^{(\infty)}(\mathcal{B}_s(x_0))$. Assume that there exist constants $C > 0$ and $M > 0$ such that*

$$\left| \left(\frac{\partial}{\partial x_j} \right)^{l} U(x) \right| \leq M C^l \ l!, \quad j \in \{1,2,3\}, \tag{3.55}$$

for all $x \in \mathcal{B}_s(x_0)$ and $l \in \mathbb{N}_0$. Then U is representable as a (real) power series.

Proof. We start from Taylor's formula:

$$U(x_0 + h) = \sum_{j=0}^{m-1} \frac{1}{j!} \left((h \cdot \nabla)^j U \right) (x_0) + \frac{1}{m!} \left((h \cdot \nabla)^m U \right) (x_0 + \theta h) \tag{3.56}$$

for $h \in \mathbb{R}^3$ with $|h| < s$ and $\theta \in (0,1)$. We have to estimate the remainder term. To this end, we observe that $h \cdot \nabla^m = \sum_{i=1}^{3} h_i \frac{\partial}{\partial x_i}$. In the multi-index notation, i.e., using $\alpha = (\alpha_1, \alpha_2, \alpha_3)^T \in \mathbb{N}_0^3$, we find

$$\left| \frac{1}{m} \left((h \cdot \nabla)^m U \right) (x_0 + \theta h) \right| \leq C^l M \left(m! \sum_{\alpha_1+\alpha_2+\alpha_3=m} \frac{|h_1|^{\alpha_1} |h_2|^{\alpha_2} |h_3|^{\alpha_3}}{\alpha_1! \alpha_2! \alpha_3!} \right). \tag{3.57}$$

The binomial formula yields

$$m! \sum_{\alpha_1+\alpha_2+\alpha_3=m} \frac{|h_1|^{\alpha_1} |h_2|^{\alpha_2} |h_3|^{\alpha_3}}{\alpha_1! \alpha_2! \alpha_3!} = \left(\sum_{i=1}^{3} |h_i| \right)^m . \tag{3.58}$$

By the Cauchy–Schwarz inequality, the right side of (3.58) can be estimated by $(\sqrt{3}|h|)^m$. Consequently, for $|h| < \frac{1}{\sqrt{3}C}$, the remainder term $\frac{1}{m!} \left((h \cdot \nabla)^m U \right) (x_0 + \theta h)$ converges to zero if $m \to \infty$. □

Now we are prepared to establish the (real) analyticity.

Theorem 3.16 (Analyticity). *Let $\mathcal{G} \subset \mathbb{R}^3$ be a regular region. Suppose that U is harmonic on $\mathcal{G}$. Then U is (real) analytic, i.e., for $x_0 \in \mathcal{G}$ there exists $\rho > 0$ such that*

$$U(x_0 + h) = \sum_{j=0}^{\infty} \frac{1}{j!} ((h \cdot \nabla)^j U)(x_0) \tag{3.59}$$

for all $h \in \mathbb{R}^3$ with $|h| < \rho$.

Proof. It is known that U is of class $\mathrm{C}^{(\infty)}(\mathcal{G})$. Furthermore, for $r > 0$ with $\overline{\mathcal{B}_r(x_0)} \Subset \mathcal{G}$, there exists a constant $C > 0$ such that $|U(x)| \leq C$, for $x \in \mathcal{B}_r(x_0)$. Combining Corollary 3.14 with $s = \frac{r}{2}$ and Lemma 3.15 with $C = 6\frac{e}{r}$, $M = \frac{C}{e}$, we arrive at the desired result. □

3.1.6 Regularity at Infinity

The Newton (volume) potential (0.10) extended over $\mathcal{G}$ is harmonic in the exterior $\mathcal{G}^c = \mathbb{R}^3 \backslash \overline{\mathcal{G}}$. This is the reason why potential theory under geoscientifically relevant aspects essentially aims at concepts in the outer space of a regular region. The treatment of the outer space in the Euclidean space $\mathbb{R}^3$, however, includes the discussion at infinity. Consequently, Green's integral theorems must be formulated under geophysically relevant conditions imposed on harmonic functions at infinity. Mathematically, the *regularity at infinity* can be deduced appropriately via the Kelvin transform by a transition from functions harmonic in the inner space to their counterparts in outer space, and vice versa. In order to explain this assertion in more detail, we remember the fact that a function U that is harmonic in $\mathcal{G}^c$ has a harmonic Kelvin transform in every punctured neighborhood around the origin. Thus it remains to discuss *harmonicity at the origin.* In this context the notion of a removable singularity is of particular interest.

Definition 3.17. Suppose that U satisfies the Laplace equation in the punctured neighborhood $\mathcal{U}\backslash\{x_0\} \subset \mathbb{R}^3$. The point $x_0 \in \mathbb{R}^3$ is called a *removable singularity* of U if U can be extended continuously in x_0 so that the resulting extension is harmonic in all of $\mathcal{U}$.

Obviously, the point $x_0 \in \mathbb{R}$ is not a removable singularity for $G(\Delta; |x_0 - \cdot|)$. This observation suggests that for a point x_0 to be a removable singularity, the behavior of U in the neighborhood of x_0 should be better than that of the fundamental solution.

Lemma 3.18. *Let x_0 be a point of a regular region $\mathcal{G} \subset \mathbb{R}^3$. Assume that U is harmonic in $\mathcal{G}\backslash\{x_0\}$ and*

$$\frac{U(x)}{G(\Delta; |x - x_0|)} = o(1), \quad |x - x_0| \to 0. \tag{3.60}$$

Then x_0 is a removable singularity for U, i.e., the continuous extension of U in x_0 is harmonic in $\mathcal{G}$.

Proof. Let V be a harmonic function in $\mathcal{B}_\rho(x_0) \subset \mathcal{G}$ satisfying the boundary values $V|_{\Omega_\rho(x_0)} = U|_{\Omega_\rho(x_0)}$ (such a function can be constructed by the Abel–Poisson formula). Now, regard $W = V - U$. Our goal is to show $W = 0$ in $\overline{\mathcal{G}}\backslash\{x_0\}$. For some arbitrary $\varepsilon \in (0, \rho)$, we set

$$C_\varepsilon = \sup_{x \in \Omega_\varepsilon(x_0)} |V(x) - U(x)|. \tag{3.61}$$

From (3.60) it follows that, for every $\delta \in (0, 1)$ fixed, there exists $\varepsilon_0 \in (0, \rho)$ such that $C_\varepsilon \leq \delta\varepsilon^{-1}$ holds for all $\varepsilon \leq \varepsilon_0$. The functions $H^\pm$, given by

$$H^\pm(x) = \frac{C_\varepsilon \varepsilon}{|x - x_0|} \mp W(x), \tag{3.62}$$

are harmonic in the spherical shell $\mathcal{B}_{\varepsilon,\rho}(x_0)$. Furthermore, both functions are non-negative on $\Omega_\rho(x_0)$, since $W = 0$ on $\Omega_\rho(x_0)$. On the sphere $\Omega_\varepsilon(x_0)$, we obtain

$$H^\pm\big|_{\Omega_\varepsilon(x_0)} = C_\varepsilon \mp W\big|_{\Omega_\varepsilon(x_0)} \geq 0. \tag{3.63}$$

Therefore, the Maximum/Minimum Principle shows us that $H^\pm(x) \geq 0$ is valid for all $x \in \mathcal{B}_{\varepsilon,\rho}(x_0)$. In other words,

$$|W(x)| \leq \frac{C_\varepsilon \varepsilon}{|x - x_0|} \leq \frac{\delta}{|x - x_0|}, \quad x \in \mathcal{B}_{\varepsilon,\rho}(x_0). \tag{3.64}$$

This proves Lemma 3.18. □

A generalization of Lemma 3.18 is Lemma 3.19 (cf. E. DiBenedetto [1995]). Choosing the function $U - C\,G(\Delta; |\cdot - x_0|)$ instead of U, for some constant $C \in \mathbb{R}$, it is actually a direct consequence of Lemma 3.18.

Lemma 3.19. *Let x_0 be a point in the regular region $\mathcal{G} \subset \mathbb{R}^3$. If U is harmonic in $\mathcal{G}\backslash\{x_0\}$ and*

$$\lim_{|x-x_0|\to 0} \frac{U(x)}{G(\Delta; |x-x_0|)} = C \tag{3.65}$$

for some $C \in \mathbb{R}$, then

$$U(x) = C\ G(\Delta; |x-x_0|) + V(x), \tag{3.66}$$

where V is harmonic in $\mathcal{G}$.

In general, Lemma 3.18 shows us that U can be replaced by a harmonic function V in any neighborhood of the origin if U increases more weakly (in o-sense) than the fundamental solution. This leads us to

Lemma 3.20. *If U is harmonic in $\mathcal{G}^c$ and $|U(x)| = o(1)$, $|x| \to \infty$ (uniformly with respect to all directions $x/|x|$), then the Kelvin transform $\breve{U} = \mathcal{K}[U]$ is harmonic in any neighborhood around the origin.*

Proof. It suffices to verify $\breve{U}(x) = o(|x|^{-1})$ for $|x| \to 0$. Then we get the desired assertion by Lemma 3.18. The required asymptotic relation immediately follows from the observation that $U(\breve{x}) \to 0$, for $|\breve{x}| \to \infty$, implies $\breve{U}(x)|x| = U(\breve{x}) \to 0$, $|x| \to 0$. □

The next statement can be regarded as a conversion of Lemma 3.20.

Lemma 3.21. *Let U be harmonic in $\mathcal{G}^c$ such that the Kelvin transform $\breve{U}$ is harmonic in any neighborhood around the origin. Then the asymptotic relations $|U(x)| = \mathcal{O}(|x|^{-1})$ and $|\nabla U(x)| = \mathcal{O}(|x|^{-2})$ hold true for $|x| \to \infty$ uniformly with respect to all directions $\frac{x}{|x|}$.*

Proof. Since $\breve{U}$ is a harmonic function in any neighborhood of the origin, $\breve{U}$ and its first-order partial derivatives are bounded in these neighborhoods. We know that U is the Kelvin transform of $\breve{U}$, i.e.,

$$U(x) = \frac{1}{|x|}\ \breve{U}\left(\frac{x}{|x|^2}\right), \tag{3.67}$$

and

$$\frac{\partial U}{\partial x_i}(x) = -\frac{x_i}{|x|^3}\breve{U}(\breve{x}) + \frac{1}{|x|}\left(\frac{1}{|x|^2}\frac{\partial \breve{U}}{\partial x_i}(\breve{x}) - \sum_{j=1}^{3}\frac{2x_ix_j}{|x|^4}\frac{\partial \breve{U}}{\partial x_j}(\breve{x})\right), \tag{3.68}$$

for $x \in \mathcal{G}^c$ and $\breve{x} = \frac{x}{|x|^2} \in \mathcal{G}$. The boundedness of $\breve{U}$ finally guarantees the assertion of Theorem 3.21. □

Combining Lemma 3.20 and Lemma 3.21, we are led to the following conclusion.

Corollary 3.22. *If U is harmonic in $\mathcal{G}^c$ and U converges to zero as $|x| \to \infty$, uniformly in all directions, then $|U(x)| = \mathcal{O}(|x|^{-1})$ and $|\nabla U(x)| = \mathcal{O}(|x|^{-2})$, $|x| \to \infty$.*

Corollary 3.22 leads us to the definition of the *regularity at infinity.*

Definition 3.23 (Regularity at Infinity)**.** A function $U : \mathcal{G}^c \to \mathbb{R}$ is called *regular at infinity* if U satisfies the asymptotic relations $|U(x)| = \mathcal{O}(|x|^{-1})$ and $|\nabla U(x)| = \mathcal{O}(|x|^{-2})$, $|x| \to \infty$, uniformly with respect to all directions $\frac{x}{|x|}$.

Remark 3.24. Definition 3.23 is also used for non-harmonic functions $U : \mathcal{G}^c \to \mathbb{R}$ (for example, in geomagnetic theory). But it should be noted that, if the regularity at infinity is satisfied for a harmonic function U in $\mathcal{G}^c$, then its Kelvin transform $\breve{U}$ is harmonic in any neighborhood around the origin.

3.1.7 Exterior Third Green Formula

Now we are prepared to discuss exterior versions of Green's identities involving harmonic functions that are regular at infinity. All these identities can be obtained by first considering the auxiliary set $\mathcal{G}^c \cap \mathcal{B}_R(0)$ (with R sufficiently large such that $\mathcal{G} \Subset \mathcal{B}_R(0)$) and then letting R tend to infinity (note that $\mathcal{G}^c \cap \mathcal{B}_R(0)$, as the intersection of the regular region $\mathcal{B}_R(0)$ and the exterior of the regular region $\mathcal{G}$, allows the application of the interior Green formulas).

Theorem 3.25 (Exterior First Green Theorem)**.** *Let F be a function of class $\mathrm{C}^{(2)}(\mathcal{G}^c) \cap \mathrm{C}^{(1)}(\overline{\mathcal{G}^c})$ such that F is harmonic in $\mathcal{G}^c$ and regular at infinity. Suppose that the function $H \in \mathrm{C}^{(1)}(\overline{\mathcal{G}^c})$ satisfies the asymptotic relations*

$$|F(y)\nabla H(y)| = o\left(\frac{1}{|y|^2}\right) \tag{3.69}$$

and

$$|\nabla F(y) \cdot \nabla H(y)| = \mathcal{O}\left(\frac{1}{|y|^{3+\varepsilon}}\right), \quad \varepsilon > 0, \tag{3.70}$$

for $|y| \to \infty$. Then

$$\int_{\mathcal{G}^c} \nabla F(y) \cdot \nabla H(y) \, dV(y) = \int_{\partial\mathcal{G}} F(y) \frac{\partial H}{\partial \nu}(y) \, d\omega(y), \tag{3.71}$$

where ν is the outer unit normal field to $\mathcal{G}^c$, i.e., the inner unit normal field to $\mathcal{G}$.

Proof. As already announced, we consider the set $\mathcal{G}^c \cap \mathcal{B}_R(0)$ with $R > \sup_{y \in \partial\mathcal{G}} |y|$ sufficiently large and apply the Interior First Green Theorem to $\overline{\mathcal{G}^c \cap \mathcal{B}_R(0)}$. Then, we get

$$\int\limits_{\mathcal{G}^c \cap \mathcal{B}_R(0)} \nabla F(y) \cdot \nabla H(y) \, dV(y) = \int_{\partial \mathcal{G}} F(y) \frac{\partial H}{\partial \nu}(y) \, d\omega(y) \tag{3.72}$$

$$+ \int_{\Omega_R} F(y) \frac{\partial H}{\partial \nu}(y) \, d\omega(y).$$

Observing the estimates (3.69) and (3.70) for $R \to \infty$, we obtain the desired result. □

In the same way we are able to verify the following theorems.

Theorem 3.26 (Exterior Second Green Theorem). *Let the functions $F, G \in \mathrm{C}^{(1)}(\overline{\mathcal{G}^c}) \cap \mathrm{C}^{(2)}(\mathcal{G}^c)$ be harmonic in $\mathcal{G}^c$ and regular at infinity. Then*

$$\int_{\partial \mathcal{G}} \left(F(y) \frac{\partial}{\partial \nu} H(y) - H(y) \frac{\partial}{\partial \nu} F(y) \right) \, d\omega(y) = 0. \tag{3.73}$$

Theorem 3.27 (Exterior Third Green Theorem). *Let the function U be of class $\mathrm{C}^{(1)}(\overline{\mathcal{G}^c}) \cap \mathrm{C}^{(2)}(\mathcal{G}^c)$ such that U is harmonic in $\mathcal{G}^c$ and regular at infinity. Then*

$$\int_{\partial \mathcal{G}} \left(G(\Delta; |x-y|) \frac{\partial}{\partial \nu(y)} U(y) - U(y) \frac{\partial}{\partial \nu(y)} G(\Delta; |x-y|) \right) \, d\omega(y)$$

$$= \begin{cases} U(x), & x \in \mathcal{G}^c, \\ \frac{1}{2} U(x), & x \in \partial \mathcal{G}, \\ 0, & x \in \mathcal{G}, \end{cases} \tag{3.74}$$

where ν is the outer unit normal field to $\mathcal{G}^c$, i.e., the inner unit normal field to $\mathcal{G}$.

3.2 Volume Potentials

We already know that the Newton (volume) integral over a regular region $\mathcal{G}$ corresponding to a mass density distribution F satisfies the Laplace equation in the outer space $\mathcal{G}^c = \mathbb{R}^3 \backslash \overline{\mathcal{G}}$. Clearly, this property is an immediate consequence of the harmonicity of the fundamental solution for the Laplace operator.

Theorem 3.28. *Let $F : \overline{\mathcal{G}} \to \mathbb{R}$ be an integrable, bounded function. Then*

$$U(x) = \int_{\mathcal{G}} F(y) \, G(\Delta; |x-y|) \, dV(y), \quad x \in \mathcal{G}^c, \tag{3.75}$$

satisfies

$$\Delta_x \int_{\mathcal{G}} F(y) \, G(\Delta; |x-y|) \, dV(y) = 0 \tag{3.76}$$

for all $x \in \mathcal{G}^c$, i.e., U is harmonic in $\mathcal{G}^c$.

3.2.1 Differentiability

Next, we are interested in showing that the Newton integral in the inner space satisfies the Poisson equation at least under some canonical conditions on the mass density function.

Theorem 3.29. *Let $F : \overline{\mathcal{G}} \to \mathbb{R}$ be of class $\mathrm{C}^{(0)}(\overline{\mathcal{G}})$. Then U, as defined by (3.75), is of class $\mathrm{C}^{(1)}(\overline{\mathcal{G}})$. Furthermore, we have*

$$\nabla_x U(x) = \int_{\mathcal{G}} F(y) \; \nabla_x G(\Delta; |x-y|) \; dV(y), \; x \in \overline{\mathcal{G}}. \tag{3.77}$$

Proof. By one-dimensional Taylor linearization we obtain

$$\frac{1}{\sqrt{u}} = \frac{1}{\sqrt{u_0}} - \frac{1}{2u_0^{\frac{3}{2}}}(u-u_0) + \frac{3}{8}\frac{1}{(u_0+\theta(u-u_0))^{\frac{5}{2}}}(u-u_0)^2 \tag{3.78}$$

for some $\theta \in (0,1)$. Setting $u = r^2$ and $u_0 = \rho^2$ we therefore find

$$\frac{1}{r} = \frac{1}{2\rho}\left(3 - \frac{r^2}{\rho^2}\right) + \frac{3}{8}\frac{1}{(\rho^2+\theta(r^2-\rho^2))^{\frac{5}{2}}}(r^2-\rho^2)^2. \tag{3.79}$$

In other words, by letting $r = |x-y|$ we are able to *regularize* the fundamental solution of the Laplace operator

$$G(\Delta; r) = \frac{1}{4\pi r}, \qquad r > 0, \tag{3.80}$$

by

$$^{l}G^{\rho}(\Delta; r) = \begin{cases} \dfrac{1}{8\pi\rho}\left(3 - \dfrac{1}{\rho^2}r^2\right), & r \leq \rho, \\[2ex] \dfrac{1}{4\pi r}, & r > \rho, \end{cases} \tag{3.81}$$

such that $^{l}G^{\rho}(\Delta; r)$ is continuously differentiable for all $r \geq 0$ (note that an upper left index "l" typically indicates a regularization by linearization). Obviously, $G(\Delta; r) = {}^{l}G^{\rho}(\Delta; r)$ for all $r > \rho$. Consequently,

$$G(\Delta; |x-y|) = \frac{1}{4\pi}\frac{1}{|x-y|}, \qquad |x-y| \neq 0, \tag{3.82}$$

admits a regularization of the form

$$^{l}G^{\rho}(\Delta; |x-y|) = \begin{cases} \dfrac{1}{8\pi\rho}\left(3 - \dfrac{1}{\rho^2}|x-y|^2\right), & |x-y| \leq \rho, \\[2ex] \dfrac{1}{4\pi|x-y|}, & \rho < |x-y|. \end{cases} \tag{3.83}$$

For brevity, we set

$$U(x) = \int_{\mathcal{G}} F(y)\ G(\Delta; |x-y|)\ dV(y) \tag{3.84}$$

and

$${}^{l}U^{\rho}(x) = \int_{\mathcal{G}} F(y)\ {}^{l}G^{\rho}(\Delta; |x-y|)\ dV(y). \tag{3.85}$$

The integrands of U and ${}^{l}U^{\rho}$ differ only in the ball $\mathcal{B}_\rho(x)$ around the point x with radius ρ. Moreover, the function $F : \overline{\mathcal{G}} \to \mathbb{R}$ is supposed to be continuous on $\overline{\mathcal{G}}$. Hence, it is uniformly bounded on $\overline{\mathcal{G}}$. This shows us that

$$\begin{aligned} \left| U(x) - {}^{l}U^{\rho}(x) \right| &= \mathcal{O}\left(\int_{\mathcal{B}_\rho(x)} |G(\Delta; |x-y|) - {}^{l}G^{\rho}(\Delta; |x-y|)|\ dV(y) \right) \\ &= \mathcal{O}(\rho^2). \end{aligned} \tag{3.86}$$

Therefore, U is of class $\mathrm{C}^{(0)}(\overline{\mathcal{G}})$ as the limit of a uniformly convergent sequence of continuous functions on $\overline{\mathcal{G}}$. Furthermore, we let

$$u(x) = \int_{\mathcal{G}} F(y)\ \nabla_x G(\Delta; |x-y|)\ dV(y) \tag{3.87}$$

and

$${}^{l}u^{\rho}(x) = \int_{\mathcal{G}} F(y)\ \nabla_x {}^{l}G^{\rho}(\Delta; |x-y|)\ dV(y). \tag{3.88}$$

Because of $|\nabla_x G(\Delta; |x-y|)| = \mathcal{O}(|x-y|^{-2})$, the integrals u and ${}^{l}u^{\rho}$ exist for all $x \in \overline{\mathcal{G}}$. It is not hard to see that

$$\sup_{x \in \overline{\mathcal{G}}} \left| u(x) - {}^{l}u^{\rho}(x) \right| = \sup_{x \in \overline{\mathcal{G}}} |u(x) - \nabla_x {}^{l}U^{\rho}(x)| = \mathcal{O}(\rho). \tag{3.89}$$

Consequently, u is a continuous vector field on $\overline{\mathcal{G}}$. Moreover, as the relation (3.89) holds uniformly on $\overline{\mathcal{G}}$, we obtain

$$u(x) = \nabla U(x) = \int_{\mathcal{G}} F(y)\ \nabla_x G(\Delta; |x-y|)\ dV(y). \tag{3.90}$$

This is the desired result. □

3.2.2 Poisson Differential Equation

Next, we come to the Poisson equation under the assumption of Hölder continuity for the function F on $\overline{\mathcal{G}}$.

Theorem 3.30. *If F is of class $C^{(0,\mu)}(\overline{\mathcal{G}})$, $\mu \in (0,1]$, then the Poisson differential equation*

$$\Delta_x \int_{\mathcal{G}} F(y)\ G(\Delta; |x-y|)\ dV(y) = -F(x) \tag{3.91}$$

holds true for all $x \in \mathcal{G}$.

Proof. The Taylor linearization of $s^{-\frac{3}{2}}$ is $s_0^{\frac{3}{2}} - \frac{3}{2}s_0^{-\frac{5}{2}}(s-s_0)$. Hence, by letting $s = r^2$ and $s_0 = \rho^2$, we are able to replace r^{-3} by $\frac{1}{2\rho^3}(5 - \frac{3}{\rho^2}r^2)$. Consequently, as regularization of

$$Z(\Delta; |x-y|) = \frac{1}{4\pi|x-y|^3}, \qquad |x-y| \neq 0, \tag{3.92}$$

we introduce

$$^{l}Z^{\rho}(\Delta; |x-y|) = \begin{cases} \dfrac{1}{8\pi\rho^3}\left(5 - \dfrac{3}{\rho^2}|x-y|^2\right), & |x-y| \leq \rho \\[2ex] \dfrac{1}{4\pi|x-y|^3}, & \rho < |x-y|. \end{cases} \tag{3.93}$$

$^{l}Z^{\rho}(\Delta; \cdot)$ is continuously differentiable for all $r \geq 0$. Moreover, by already known arguments, it can be shown (cf. Theorem 3.29) that the vector field

$$^{l}z^{\rho}(x) = -\int_{\mathcal{G}} F(y)\, ^{l}Z^{\rho}(\Delta; |x-y|)(x-y)\ dV(y) \tag{3.94}$$

converges uniformly on $\overline{\mathcal{G}}$ to the limit field

$$u(x) = \nabla U(x) = \int_{\mathcal{G}} F(y)\ \nabla_x G(\Delta; |x-y|)\ dV(y). \tag{3.95}$$

For all $x \in \mathcal{B}_\rho(x)$ we obtain by a simple calculation (cf. Exercise 3.10)

$$\nabla_x \cdot ((x-y)\, ^{l}Z^{\rho}(\Delta; |x-y|)) = \frac{15}{8\pi}\left(\frac{1}{\rho^3} - \frac{|x-y|^2}{\rho^5}\right). \tag{3.96}$$

Furthermore, we find

$$\int_{\mathcal{B}_\rho(x)} \nabla_x \cdot \left((x-y)\, ^{l}Z^{\rho}(\Delta; |x-y|)\right)\ dV(y) = 1. \tag{3.97}$$

Hence, we obtain

$$\begin{aligned} &\nabla_x \cdot {^{l}z^{\rho}}(x) \qquad\qquad (3.98) \\ &= \ -\nabla_x \cdot \int_{\mathcal{G}} F(y)\, ^{l}Z^{\rho}(\Delta; |x-y|)(x-y)\ dV(y) \\ &= \ -\int_{\mathcal{B}_\rho(x)} F(y)\nabla_x \cdot \left(^{l}Z^{\rho}(\Delta; |x-y|)(x-y)\right)\ dV(y) \\ &= \ -F(x) + \int_{\mathcal{B}_\rho(x)} (F(x) - F(y))\nabla_x \cdot \left(^{l}Z^{\rho}(\Delta|x-y|)(x-y)\right)\ dV(y). \end{aligned}$$

The μ-Hölder continuity of F guarantees the estimate

$$\sup_{x\in\overline{\mathcal{G}}} \left|\nabla_x \cdot {}^l z^\rho(x) \ + \ F(x)\right| = \mathcal{O}(\rho^\mu) \tag{3.99}$$

uniformly with respect to $x \in \overline{\mathcal{G}}$. In an analogous way, we are able to show that the first partial derivatives of (3.94) converge uniformly to continuous limit fields. Again, the uniform convergence shows us that ∇U is differentiable in $\overline{\mathcal{G}}$, and we have

$$\nabla_x \cdot u(x) \ = \Delta U(x) \ = \ \Delta_x \int_{\mathcal{G}} F(y) G(\Delta; |x-y|) \, dV(y) = -F(x), \quad x \in \mathcal{G}, \tag{3.100}$$

as required. □

Remark 3.31. In Theorem 3.30, the assumption of μ-Hölder continuity of F, $\mu \in (0,1]$, is needed for its proof. Indeed, H. Petrini [1900] (for more details see also C. Müller [1969], E. Wienholtz et al. [2009]) showed that the μ-Hölder continuity of F, $\mu \in (0,1]$, is necessary to imply the second continuous differentiability of the Newton volume potential.

3.3 Surface Potentials

Let $\mathcal{G} \subset \mathbb{R}^3$ be a regular region as introduced by Definition 1.7. Suppose that F is a continuous function on the boundary surface $\partial\mathcal{G}$. Then the function defined by

$$x \mapsto \int_{\partial\mathcal{G}} F(y) \left(\frac{\partial}{\partial\nu(y)}\right)^{n-1} G(\Delta; |x-y|) \, d\omega(y), \quad x \in \mathbb{R}^3\backslash\partial\mathcal{G}, \ n \in \mathbb{N}, \tag{3.101}$$

is called a *potential of the n-th layer on $\partial\mathcal{G}$*. It is infinitely often differentiable in $\mathbb{R}^3\backslash\partial\mathcal{G}$ and harmonic in $\mathcal{G}$ as well as in $\mathcal{G}^c$. In addition, the integral (3.101) is regular at infinity.

For $n = 1$, the function

$$x \mapsto \int_{\partial\mathcal{G}} F(y) G(\Delta; |x-y|) \, d\omega(y), \ x \in \mathbb{R}^3 \tag{3.102}$$

is called the *potential of the single-layer* on $\partial\mathcal{G}$, while the function (by letting $n = 2$)

$$x \mapsto \int_{\partial\mathcal{G}} F(y) \frac{\partial}{\partial\nu(y)} G(\Delta; |x-y|) \, d\omega(y), \ x \in \mathbb{R}^3 \tag{3.103}$$

is called the *potential of the double-layer* on $\partial\mathcal{G}$. The function

$$x \mapsto \int_{\partial\mathcal{G}} F(y) \frac{\partial}{\partial\nu(x)} G(\Delta; |x-y|) \, d\omega(y), \; x \in \mathbb{R}^3 \tag{3.104}$$

is called the *normal derivative of the potential of the single-layer on* $\partial\mathcal{G}$.

Note that the integrals (3.102), (3.103), and (3.104) also exist as improper integrals on $\partial\mathcal{G}$. This follows from the basic estimates for normal fields on $\partial\mathcal{G}$ (see Lemma 1.8 and Lemma 1.9). However, the improper integrals on $\partial\mathcal{G}$ are not necessarily the limits of the regular integrals when approaching $\partial\mathcal{G}$ from the inside (i.e., from within $\mathcal{G}$) or from the outside (i.e., from within $\mathcal{G}^c$). In fact, from classical potential theory (see, for example, O.D. Kellogg [1967], N.M. Günter [1957], J. Schauder [1931] and the references therein) the pointwise limit behavior of the surface potentials is known. Roughly speaking, the potential of a single layer with continuous density does not show any discontinuity as we pass from the outside (inside) to the inside (outside), whereas its normal derivative shows a discontinuity. The difference from the outside (inside) to the inside (outside) is (except from a multiplicative constant) the functional value of the density function at the point on $\partial\mathcal{G}$ under consideration. The situation is different for the potential of a double layer with a continuous density. The transition of the potential of a double layer with continuous density is discontinuous through the boundary $\partial\mathcal{G}$, while its normal derivative is continuous (understood symmetrically in the direction of the normal). The discontinuity of the potential of a double layer is again characterized (except from a constant factor) by the functional value of the density at the point on $\partial\mathcal{G}$ under consideration.

In the following, we are not only interested in formulating the limit and jump relations in a pointwise sense, but also in a uniform sense, more explicitly, in the $\mathrm{C}^{(0)}(\partial\mathcal{G})$- as well as $\mathrm{L}^2(\partial\mathcal{G})$-topology. For this purpose we first have to collect some preliminaries.

3.3.1 Preparatory Estimates

Let $\partial\mathcal{G}$ be the boundary of a regular region $\mathcal{G} \subset \mathbb{R}^3$. The surface

$$\partial\mathcal{G}(\tau) = \left\{ x \in \mathbb{R}^3 : x = y + \tau\nu(y), y \in \partial\mathcal{G} \right\} \tag{3.105}$$

generates a *parallel surface* (at the distance $|\tau|$), which is exterior to $\partial\mathcal{G}$ for $\tau > 0$ and interior for $\tau < 0$ (cf. Figure 3.2).

$$\mathcal{G}(\tau) = \begin{cases} \mathcal{G} \cup \{y + \sigma\nu(y) : y \in \partial\mathcal{G}, 0 \leq \sigma < \tau\}, & \tau > 0 \\ \mathcal{G} \backslash \{y + \sigma\nu(y) : y \in \partial\mathcal{G}, \tau \leq \sigma < 0\}, & \tau < 0 \end{cases} \tag{3.106}$$

is the interior space of $\partial\mathcal{G}(\tau)$. The mapping $x \mapsto x + \tau\nu(x)$, $x \in \partial\mathcal{G}$ (together with the observation that the normal field of $\partial\mathcal{G}(\tau)$ coincides with the normal field of $\partial\mathcal{G}$ as long as $|\tau|$ is sufficiently small) enables us to verify that $\mathcal{G}(\tau)$

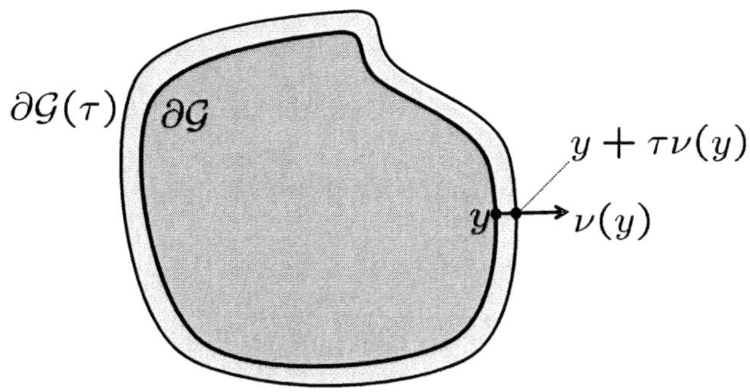

FIGURE 3.2
Exterior parallel surface $\partial\mathcal{G}(\tau)$ to the surface $\partial\mathcal{G}$.

also is a regular region (more details on the regularity of parallel surfaces can be found in C. Müller [1966]).

Lemma 3.32. *Suppose that* $|\tau| < \frac{1}{4M}$. *For every* $x \in \partial\mathcal{G}$ *and* $y \in \partial\mathcal{G} \cap \mathcal{B}_\delta(x)$, *with* δ, M *specified by Definition 1.7, we have*

$$\sqrt{\frac{1}{2}}\sqrt{\tau^2 + |x-y|^2} \le |x + \tau\nu(x) - y| \le \sqrt{\frac{3}{2}}\sqrt{\tau^2 + |x-y|^2}. \tag{3.107}$$

Proof. For $x \in \partial\mathcal{G}$ and $y \in \partial\mathcal{G} \cap \mathcal{B}_\delta(x)$, the relation (3.107) follows from

$$|x + \tau\nu(x) - y| = \sqrt{|x-y|^2 + \tau^2 + 2\tau\nu(x)\cdot(x-y)}, \tag{3.108}$$

since

$$\begin{aligned} |x + \tau\nu(x) - y| &\le \sqrt{|x-y|^2 + \tau^2 + 2M|\tau|\,|x-y|^2} \\ &= \sqrt{\tau^2 + (1 + 2M|\tau|)|x-y|^2} \\ &\le \sqrt{\frac{3}{2}\tau^2 + \frac{3}{2}|x-y|^2} \end{aligned} \tag{3.109}$$

and

$$\begin{aligned} |x + \tau\nu(x) - y| &\ge \sqrt{\tau^2 + (1 - 2|\tau|M)|x-y|^2} \\ &\ge \sqrt{\frac{1}{2}}\sqrt{\tau^2 + |x-y|^2}, \end{aligned} \tag{3.110}$$

by use of the estimate $|\tau| < \frac{1}{4M}$. □

Lemma 3.33. *For $k \in \mathbb{N}$, under the assumptions of Lemma 3.32, we have*

$$\begin{aligned} & \left| |x+\tau\nu(x)-y|^k - |x-\tau\nu(x)-y|^k \right| \\ & \quad \leq \; 2^{\frac{4-k}{2}} 3^{\frac{k-1}{2}} M \; k|\tau| \; |x-y|^2 \left(\tau^2 + |x-y|^2\right)^{\frac{k-2}{2}} . \end{aligned}$$

In particular,

$$||x+\tau\nu(x)-y| - |x-\tau\nu(x)-y|| \leq 3M|\tau| \frac{|x-y|^2}{\sqrt{\tau^2+|x-y|^2}} \tag{3.111}$$

and

$$\left| |x+\tau\nu(x)-y|^3 - |x-\tau\nu(x)-y|^3 \right| \leq 13M|\tau| \; |x-y|^2 \sqrt{\tau^2+|x-y|^2}. \tag{3.112}$$

Proof. We easily see that

$$|x+\tau\nu(x)-y| - |x-\tau\nu(x)-y| = \frac{4\tau\nu(x)\cdot(x-y)}{|x+\tau\nu(x)-y| + |x-\tau\nu(x)-y|}. \tag{3.113}$$

Hence, it follows that

$$||x+\tau\nu(x)-y| - |x-\tau\nu(x)-y|| \leq 2\sqrt{2}M|\tau| \frac{|x-y|^2}{\sqrt{\tau^2+|x-y|^2}}. \tag{3.114}$$

For $k \in \mathbb{N}$ we know that

$$\begin{aligned} |\alpha^k - \beta^k| &= \left| (\alpha-\beta) \left(\sum_{i=1}^{k-1} \alpha^i \beta^{k-1-i} \right) \right| \\ &\leq k \max\{|\alpha|,|\beta|\}^{k-1} \; |\alpha-\beta|. \end{aligned} \tag{3.115}$$

Letting $\alpha = |x+\tau\nu(x)-y|$ and $\beta = |x-\tau\nu(x)-y|$ we obtain the desired result. $\square$

For later use, we need some estimates for integrals over $\partial\mathcal{G} \cap \mathcal{B}_\delta(x)$ of the type

$$\int_{\partial\mathcal{G}\cap\mathcal{B}_\delta(x)} \frac{|x-y|^p}{(\tau^2+|x-y|^2)^q} \, d\omega(y), \qquad p,q \geq 0. \tag{3.116}$$

In order to discuss (3.116) for $\tau \to 0+$, we go back to the local tangential-normal coordinate systems in Definition 1.7:

$$\left| \int\limits_{\partial\mathcal{G}\cap\mathcal{B}_\delta(x)} \frac{|x-y|^p}{(\tau^2+|x-y|^2)^q}\, d\omega(y) \right| \tag{3.117}$$

$$\leq \int\limits_{u_1^2+u_2^2\leq\delta^2} \frac{(u_1^2+u_2^2+F(u_1,u_2)^2)^{\frac{p}{2}}}{(\tau^2+u_1^2+u_2^2)^q} \sqrt{1+|\nabla F(u_1,u_2)|^2}\, du_1\, du_2$$

$$\leq (1+M^2\delta^2)^{\frac{1+p}{2}} \int\limits_{u_1^2+u_2^2\leq\delta^2} \frac{(u_1^2+u_2^2)^{\frac{p}{2}}}{(\tau^2+u_1^2+u_2^2)^q}\, du_1 du_2$$

$$\leq 2\pi(1+M^2\delta^2)^{\frac{1+p}{2}} \int_0^\delta \frac{\rho^{p+1}}{(\tau^2+\rho^2)^q}\, d\rho.$$

The last integrals enable us to control the local behavior of the integrals (3.116). We only mention some special cases.

Lemma 3.34. *Under the conditions of Definition 1.7 we have for $\tau \to 0+$,*

$$\int\limits_{\partial\mathcal{G}\cap\mathcal{B}_\delta(x)} \frac{|x-y|^p}{(\tau^2+|x-y|^2)^q}\, d\omega(y) = \begin{cases} \mathcal{O}(1), & q=\frac{1}{2}, p=0 \\ o(\frac{1}{\tau}), & q=1, p=0 \\ \mathcal{O}(\frac{1}{\tau}), & q=\frac{3}{2}, p=0 \\ \mathcal{O}(\ln\frac{1}{\tau}), & q=\frac{3}{2}, p=1. \end{cases}$$

For the evaluation of the integrals over $\partial\mathcal{G}\backslash\overline{\mathcal{B}_\delta(x)}$, i.e., the remaining integrals, it is helpful to assume $|\tau| < \frac{\delta}{2}$. Because of the relation $|x-y| \geq \delta$, this leads to the following result.

Lemma 3.35. *If $x \in \partial\mathcal{G}$ and $y \in \partial\mathcal{G}\backslash\overline{\mathcal{B}_\delta(x)}$, with δ specified by Definition 1.7, then*

$$|x \pm \tau\nu(x) - y| \geq ||x-y| - |\tau|| \geq \frac{\delta}{2}. \tag{3.118}$$

Lemma 3.36. *If $x \in \partial\mathcal{G}$ and $y \in \partial\mathcal{G}\backslash\overline{\mathcal{B}_\delta(x)}$, then*

$$||x+\tau\nu(x)-y| - |x-\tau\nu(x)-y|| \leq \frac{4\tau}{\delta}|x-y|, \tag{3.119}$$

$$\left||x+\tau\nu(x)-y|^3 - |x-\tau\nu(x)-y|^3\right| \leq \frac{12\tau}{\delta}|x-y|\left(|x-y|+\frac{\delta}{2}\right)^2, \tag{3.120}$$

$$\left||x+\tau\nu(x)-y|^5 - |x-\tau\nu(x)-y|^5\right| \leq \frac{20\tau}{\delta}|x-y|\left(|x-y|+\frac{\delta}{2}\right)^4, \tag{3.121}$$

provided that $|\tau| < \min\left(\frac{\delta}{2}, \frac{1}{4M}\right)$, with δ, M specified by Definition 1.7.

Additionally we are confronted with the problem of estimating expressions of the form $|y+\tau\nu(y)-x|$.

Lemma 3.37. *Suppose that $|\tau| < \min\left(\frac{\delta}{2}, \frac{1}{8M}\right)$, with δ, M specified by Definition 1.7. If $x \in \partial\mathcal{G}$ and $y \in \partial\mathcal{G} \cap \mathcal{B}_\delta(x)$, then*

$$\sqrt{\frac{1}{2}}\sqrt{\tau^2 + |x-y|^2} \le |y + \tau\nu(y) - x| \le \sqrt{\frac{3}{2}}\sqrt{\tau^2 + |x-y|^2} \tag{3.122}$$

and

$$\begin{aligned} ||y + \tau\nu(y) - x| - |y - \tau\nu(y) - x|| &\le 3M|\tau|\,|x-y|^2, & (3.123)\\ \left||y + \tau\nu(y) - x|^3 - |y - \tau\nu(y) - x|^3\right| &\le 13M|\tau|\,|x-y|^2\sqrt{\tau^2 + |x-y|^2}. & (3.124)\end{aligned}$$

Finally, we deal with the distance between two parallel surfaces close to the surface $\partial\mathcal{G}$.

Lemma 3.38. *Let $\mathcal{G}$ be a regular region. Then*

$$\inf_{x,y\in\partial\mathcal{G}} |x + \tau\nu(x) - (y + \sigma\nu(y))| = |\tau - \sigma| \tag{3.125}$$

provided that $|\tau|, |\sigma| < \delta$, with δ specified by Definition 1.7.

Proof. At first we see that, for $x, y \in \partial\mathcal{G}$,

$$|x + \tau\nu(x) - (y + \sigma\nu(y))|^2 = |(x-y) + (\tau\nu(x) - \sigma\nu(y))|^2. \tag{3.126}$$

This leads us to the identity

$$\begin{aligned} |x + \tau\nu(x) - (y + \sigma\nu(y))|^2 &= |x-y|^2 + |\tau\nu(x) - \sigma\nu(y)|^2 & (3.127)\\ &\quad +2\tau\nu(x)\cdot(x-y) - 2\sigma\nu(y)\cdot(x-y). \end{aligned}$$

Furthermore, we obtain

$$|\tau\nu(x) - \sigma\nu(y)|^2 = \tau^2 + \sigma^2 - 2\tau\sigma\ \nu(x)\cdot\nu(y) \tag{3.128}$$

and

$$\nu(x)\cdot\nu(y) = 1 - \frac{1}{2}|\nu(x) - \nu(y)|^2. \tag{3.129}$$

In connection with (1.71) we find

$$|x + \tau\nu(x) - (y + \sigma\nu(y))|^2 \ge (\tau - \sigma)^2 + \gamma(\sigma,\tau)|x-y|^2, \tag{3.130}$$

where the expression

$$\gamma(\sigma,\tau) = 1 - 2(|\tau| + |\sigma|)M - |\tau|\,|\sigma|M^2 \tag{3.131}$$

(with $|\tau|$, $|\sigma| < \delta$) is positive. But this shows us that

$$|x + \tau\nu(x) - (y + \sigma\nu(y))| \ge |\tau - \sigma|. \tag{3.132}$$

From the identity

$$|x + \tau\nu(x) - (x + \sigma\nu(x))| = |\tau - \sigma|, \tag{3.133}$$

we finally obtain the desired result. □

3.3.2 Limit and Jump Relations in $\mathrm{C}^{(0)}$-Topology

As already pointed out, the question arises whether the values of the improper integral (3.102) on the surface $\partial\mathcal{G}$ are the limits of the values of the proper integrals (3.102) from the inside and outside, respectively.

In order to give an answer we first start with the discussion of the limit behavior of (3.102) at $\partial\mathcal{G}$ as a two-sided limit in the direction of the normal.

Lemma 3.39 (Two-Sided Jump Relation of the Single-Layer Potential)**.** *Let $\partial\mathcal{G}$ be the boundary of a regular region $\mathcal{G} \subset \mathbb{R}^3$. Suppose that the function $F : \partial\mathcal{G} \to \mathbb{R}$ is continuous. Then*

$$\lim_{\tau\to 0+} \sup_{x\in\partial\mathcal{G}} \left| \int_{\partial\mathcal{G}} F(y) \left(\frac{1}{|x+\tau\nu(x)-y|} - \frac{1}{|x-\tau\nu(x)-y|} \right) d\omega(y) \right| = 0. \tag{3.134}$$

Proof. By splitting the integral over $\partial\mathcal{G}$ into the two parts $\partial\mathcal{G} \cap \mathcal{B}_\delta(x)$ and $\partial\mathcal{G}\backslash\overline{\mathcal{B}_\delta(x)}$ we get for $|\tau| < \min(\frac{\delta}{2}, \frac{1}{4M})$ (with δ, M as specified in Definition 1.7)

$$\begin{aligned} &\left| \int_{\partial\mathcal{G}} F(y) \left(\frac{1}{|x+\tau\nu(x)-y|} - \frac{1}{|x-\tau\nu(x)-y|} \right) d\omega(y) \right| \qquad (3.135)\\ &\quad\le \|F\|_{\mathrm{C}^{(0)}(\partial\mathcal{G})} \left(\int_{\partial\mathcal{G}\backslash\overline{\mathcal{B}_\delta(x)}} \left| \frac{1}{|x+\tau\nu(x)-y|} - \frac{1}{|x-\tau\nu(x)-y|} \right| d\omega(y) \right.\\ &\qquad\qquad \left. + \int_{\partial\mathcal{G}\cap\mathcal{B}_\delta(x)} \left| \frac{1}{|x+\tau\nu(x)-y|} - \frac{1}{|x-\tau\nu(x)-y|} \right| d\omega(y) \right). \end{aligned}$$

For the integral over $\partial\mathcal{G}\backslash\overline{\mathcal{B}_\delta(x)}$, Lemma 3.35 shows us that the relation $|x \pm \tau\nu(x) - y| \geq ||x-y| - |\tau|| \geq \frac{\delta}{2}$ is valid. Hence, we have

$$\begin{aligned} \left| \frac{1}{|x+\tau\nu(x)-y|} - \frac{1}{|x-\tau\nu(x)-y|} \right| &= \frac{||x-\tau\nu(x)-y| - |x+\tau\nu(x)-y||}{|x+\tau\nu(x)-y||x-\tau\nu(x)-y|}\\ &\le \frac{2|\tau|}{\left(\frac{\delta}{2}\right)^2}. \end{aligned} \tag{3.136}$$

In other words, for $|\tau| \to 0$,

$$\int_{\partial\mathcal{G}\backslash\overline{\mathcal{B}_\delta(x)}} \left| \frac{1}{|x+\tau\nu(x)-y|} - \frac{1}{|x-\tau\nu(x)-y|} \right| d\omega(y) = \mathcal{O}\left(|\tau|\right). \tag{3.137}$$

For the integral over $\partial\mathcal{G}\backslash\overline{\mathcal{B}_\delta(x)}$, we find

$$\begin{aligned} &\int_{\partial\mathcal{G}\cap\mathcal{B}_\delta(x)} \left| \frac{1}{|x+\tau\nu(x)-y|} - \frac{1}{|x-\tau\nu(x)-y|} \right| d\omega(y) \qquad (3.138)\\ &\quad= \mathcal{O}\left(\int_{\partial\mathcal{G}\cap\mathcal{B}_\delta(x)} \frac{|\tau|}{|x-y|^2+\tau^2}\, d\omega(y) \right). \end{aligned}$$

In accordance with Lemma 3.34 we are therefore able to conclude that

$$\int_{\partial\mathcal{G}\cap\mathcal{B}_\delta(x)} \left| \frac{1}{|x+\tau\nu(x)-y|} - \frac{1}{|x-\tau\nu(x)-y|} \right| d\omega(y) = |\tau| o\left(\frac{1}{|\tau|}\right). \quad (3.139)$$

Summarizing our results (3.137), (3.139) we are therefore led to the asymptotic relation

$$\sup_{x\in\partial\mathcal{G}} \left| \int_{\partial\mathcal{G}} F(y) \left(\frac{1}{|x+\tau\nu(x)-y|} - \frac{1}{|x-\tau\nu(x)-y|} \right) d\omega(y) \right| = o(1), \quad (3.140)$$

for $|\tau| \to 0$. All in all, this is the desired result of Lemma 3.39. □

Next we deal with *the one-sided limit relation for the single-layer potential.*

Lemma 3.40. *Under the assumptions of Lemma 3.39,*

$$\lim_{\tau\to 0+} \sup_{x\in\partial\mathcal{G}} \left| \int_{\partial\mathcal{G}} F(y) \left(\frac{1}{|x\pm\tau\nu(x)-y|} - \frac{1}{|x-y|} \right) d\omega(y) \right| = 0. \quad (3.141)$$

Proof. We start from $0 < \tau < \min(\frac{\delta}{2}, \frac{1}{4M})$. The usual splitting of $\partial\mathcal{G}$ yields

$$\begin{aligned} &\left| \int_{\partial\mathcal{G}} F(y) \left(\frac{1}{|x\pm\tau\nu(x)-y|} - \frac{1}{|x-y|} \right) d\omega(y) \right| \quad (3.142) \\ &\leq \|F\|_{\mathrm{C}^{(0)}(\partial\mathcal{G})} \left(\int_{\partial\mathcal{G}\setminus\overline{\mathcal{B}_\delta(x)}} \left| \frac{1}{|x\pm\tau\nu(x)-y|} - \frac{1}{|x-y|} \right| d\omega(y) \right. \\ &\qquad \left. + \int_{\partial\mathcal{G}\cap\mathcal{B}_\delta(x)} \left| \frac{1}{|x\pm\tau\nu(x)-y|} - \frac{1}{|x-y|} \right| d\omega(y) \right). \end{aligned}$$

For the integral over $\partial\mathcal{G}\setminus\overline{\mathcal{B}_\delta(x)}$, it is easily seen that

$$\left| \frac{1}{|x\pm\tau\nu(x)-y|} - \frac{1}{|x-y|} \right| = \left| \frac{|x-y| - |x\pm\tau\nu(x)-y|}{|x\pm\tau\nu(x)-y||x-y|} \right| \leq \frac{2|\tau|}{\delta^2}. \quad (3.143)$$

Hence,

$$\int\limits_{\partial\mathcal{G}\setminus\overline{\mathcal{B}_\delta(x)}} \left| \frac{1}{|x\pm\tau\nu(x)-y|} - \frac{1}{|x-y|} \right| d\omega(y) = \mathcal{O}\left(|\tau|\right), \ \tau \to 0+ . \quad (3.144)$$

Turning to $\partial\mathcal{G}\cap\mathcal{B}_\delta(x)$, we obtain

$$|x\pm\tau\nu(x)-y| \; |x-y| = ||x-y|^2 \pm \tau\nu(x)\cdot(x-y)|. \quad (3.145)$$

The triangle inequality yields by aid of Lemma 1.8

$$||x-y|^2 \pm \tau\nu(x)\cdot(x-y)| \geq \left||x-y|^2 - |\tau||\nu(x)\cdot(x-y)|\right|. \quad (3.146)$$

Therefore, we are able to deduce in the usual way, by introducing local coordinates, that

$$\int_{\partial\mathcal{G}\cap\mathcal{B}_\delta(x)} \left| \frac{1}{|x \pm \tau\nu(x) - y|} - \frac{1}{|x-y|} \right| d\omega(y) = o(1), \ \tau \to 0+. \qquad (3.147)$$

Altogether, we obtain from (3.144) and (3.147)

$$\sup_{x\in\partial\mathcal{G}} \int_{\partial\mathcal{G}} \left| \frac{1}{|x \pm \tau\nu(x) - y|} - \frac{1}{|x-y|} \right| d\omega(y) = o(1), \ \tau \to 0+, \qquad (3.148)$$

which guarantees Lemma 3.40. □

The existence of the *potential of the double-layer* (3.103) as a proper integral is clear for $\mathbb{R}^3\setminus\partial\mathcal{G}$. Moreover, the existence of (3.103) as an improper integral for all $x \in \partial\mathcal{G}$ becomes obvious from Lemma 1.8. In order to investigate the limit behavior of (3.103) when approaching $\partial\mathcal{G}$ from the inside and/or the outside, we are led to *one-sided limit relations.*

Lemma 3.41. *Under the assumptions of Lemma 3.39, the following relations hold true:*

(a)
$$\lim_{\tau\to 0+} \sup_{x\in\partial\mathcal{G}} \left| \int_{\partial\mathcal{G}} F(y) \left(\frac{\partial}{\partial\nu(y)} \frac{1}{|x+\tau\nu(x)-y|} - \frac{\partial}{\partial\nu(y)} \frac{1}{|x-y|} \right) d\omega(y) - 2\pi F(x) \right| = 0. \qquad (3.149)$$

(b)
$$\lim_{\tau\to 0+} \sup_{x\in\partial\mathcal{G}} \left| \int_{\partial\mathcal{G}} F(y) \left(\frac{\partial}{\partial\nu(y)} \frac{1}{|x-\tau\nu(x)-y|} - \frac{\partial}{\partial\nu(y)} \frac{1}{|x-y|} \right) d\omega(y) + 2\pi F(x) \right| = 0. \qquad (3.150)$$

Proof. Concerning (a), we suppose that $0 < \tau \leq \frac{\sigma}{2} < \min\left(\frac{\delta}{2}, \frac{1}{4M}\right)$. Then, in connection with Lemma 3.5, we find by the usual splitting of $\partial\mathcal{G}$ into the two parts $\partial\mathcal{G} \cap \mathcal{B}_\sigma(x)$ and $\partial\mathcal{G}\setminus\overline{\mathcal{B}_\sigma(x)}$ that

$$\left| \int_{\partial\mathcal{G}} F(y) \frac{\partial}{\partial\nu(y)} \frac{1}{|x+\tau\nu(x)-y|} d\omega(y) - \int_{\partial\mathcal{G}} F(y) \frac{\partial}{\partial\nu(y)} \frac{1}{|x-y|} d\omega(y) - F(x) \int_{\partial\mathcal{G}} \frac{\partial}{\partial\nu(y)} \frac{1}{|x-y|} d\omega(y) \right| \qquad (3.151)$$

$$\leq \int\limits_{\partial\mathcal{G}\cap\mathcal{B}_\sigma(x)} |F(y)-F(x)| \left| \frac{\partial}{\partial\nu(y)} \frac{1}{|x+\tau\nu(x)-y|} \right| d\omega(y)$$

$$+ |F(x)| \left| \int\limits_{\partial\mathcal{G}} \frac{\partial}{\partial\nu(y)} \frac{1}{|x+\tau\nu(x)-y|} \, d\omega(y) \right|$$

$$+ \left| \int\limits_{\partial\mathcal{G}\setminus\overline{\mathcal{B}_\sigma(x)}} F(x) \left(\frac{\partial}{\partial\nu(y)} \frac{1}{|x-y|} - \frac{\partial}{\partial\nu(y)} \frac{1}{|x+\tau\nu(x)-y|} \right) d\omega(y) \right|$$

$$+ \int\limits_{\partial\mathcal{G}\cap\mathcal{B}_\sigma(x)} |F(x)-F(y)| \left| \frac{\partial}{\partial\nu(y)} \frac{1}{|x-y|} \right| d\omega(y)$$

$$+ \left| \int\limits_{\partial\mathcal{G}\setminus\overline{\mathcal{B}_\sigma(x)}} F(y) \left(\frac{\partial}{\partial\nu(y)} \frac{1}{|x+\tau\nu(x)-y|} - \frac{\partial}{\partial\nu(y)} \frac{1}{|x-y|} \right) d\omega(y) \right| .$$

Because of (1.79), the second integral on the right-hand side of (3.151) vanishes (note that $x+\tau\nu(x)$ is an element of $\mathcal{G}^c$). By use of the modulus of continuity $\mu(\sigma;F)$ and the estimates in Lemma 3.36, the remaining right-hand side of (3.151) can be estimated as follows:

$$\mu(\sigma;F) \int_{\partial\mathcal{G}\cap\mathcal{B}_\sigma(x)} \left| \frac{\partial}{\partial\nu(y)} \frac{1}{|x+\tau\nu(x)-y|} \right| d\omega(y) \tag{3.152}$$
$$+ \quad \mathcal{O}\left(\frac{\tau}{\sigma^3}\right) + \mu(\sigma;F) \int_{\partial\mathcal{G}\cap\mathcal{B}_\sigma(x)} \left| \frac{\partial}{\partial\nu(y)} \frac{1}{|x-y|} \right| d\omega(y).$$

All integrals in (3.152) are bounded for small $\frac{\sigma}{2} < \min(\frac{\delta}{2}, \frac{1}{4M})$. Choosing $\sigma = \tau^{\frac{1}{4}}$ (i.e., $\frac{\sigma}{2} < \min(\frac{\delta}{2}, \frac{1}{4M}, \frac{1}{2\sqrt[3]{2}})$) we are led to the assertion (a).

The proof of the relation (b) follows by the same arguments as in (a), thereby observing that $x-\tau\nu(x)$ is a member of $\mathcal{G}$, so that the corresponding second integral on the right side of (3.151) becomes -4π (cf. Lemma 3.5). □

Next we derive *two-sided jump relations for the double-layer potential.*

Theorem 3.42. *Under the assumption of Lemma 3.39, the following statements are valid:*

(a)
$$\lim_{\tau\to 0+} \sup_{x\in\partial\mathcal{G}} \left| \int_{\partial\mathcal{G}} F(y) \left(\frac{\partial}{\partial\nu(y)} \frac{1}{|x+\tau\nu(x)-y|} \right.\right. \tag{3.153}$$
$$\left.\left. - \frac{\partial}{\partial\nu(y)} \frac{1}{|x-\tau\nu(x)-y|} \right) d\omega(y) - 4\pi F(x) \right| = 0.$$

(b) There exists a constant $C > 0$ such that, for all $0 < \tau < \min(\frac{\delta}{2}, \frac{1}{4M})$,

$$\sup_{F \in \mathrm{C}^{(0)}(\partial\mathcal{G})} \sup_{x \in \partial\mathcal{G}} \left| \int_{\partial\mathcal{G}} F(y) \left(\frac{\partial}{\partial\nu(y)} \frac{1}{|x + \tau\nu(x) - y|} \right.\right. \tag{3.154}$$

$$\left.\left. - \frac{\partial}{\partial\nu(y)} \frac{1}{|x - \tau\nu(x) - y|} \right) d\omega(y) - 4\pi F(x) \right| \leq C \|F\|_{\mathrm{C}^{(0)}(\partial\mathcal{G})}.$$

Proof. Lemma 3.41, in connection with the triangle inequality, yields in the usual way that, for $x \in \partial\mathcal{G}$,

$$\begin{aligned} &\left| \int_{\partial\mathcal{G}} F(y) \left(\frac{\partial}{\partial\nu(y)} \frac{1}{|x + \tau\nu(x) - y|} - \frac{\partial}{\partial\nu(y)} \frac{1}{|x - \tau\nu(x) - y|} \right) d\omega(y) - 4\pi F(x) \right| \\ \leq\ & \mu(\sigma; F)) \int_{\partial\mathcal{G} \cap \mathcal{B}_\sigma(x)} \left| \frac{\partial}{\partial\nu(y)} \left(\frac{1}{|x + \tau\nu(x) - y|} - \frac{1}{|x - \tau\nu(x) - y|} \right) \right| d\omega(y) \\ & + 2\mu(\sigma; F) \int_{\partial\mathcal{G} \setminus \overline{\mathcal{B}_\sigma(x)}} \left| \frac{\partial}{\partial\nu(y)} \frac{1}{|x - y|} \right| d\omega(y) \\ & + \mathcal{O}\left(\|F\|_{\mathrm{C}^{(0)}(\partial\mathcal{G})} \, \|\partial\mathcal{G}\| \, \frac{\tau}{\sigma^3} \right). \end{aligned} \tag{3.155}$$

The inequality $\mu(\sigma; F) \leq 2\|F\|_{\mathrm{C}^{(0)}(\partial\mathcal{G})}$ for the modulus of continuity yields the desired proof of (b). In addition, letting $\sigma = \tau^{\frac{1}{4}}$, the proof of (a) follows from (3.155). However, part (a) is already a direct consequence of Lemma 3.41. □

After the jump relation for the potential of the double-layer, we discuss *one-sided limit relations of the normal derivative of the single-layer potential.*

Lemma 3.43. *Under the assumptions of Lemma 3.39, the following relations are valid:*

(a)
$$\lim_{\tau \to 0+} \sup_{x \in \partial\mathcal{G}} \left| \int_{\partial\mathcal{G}} F(y) \left(\frac{\partial}{\partial\tau} \frac{1}{|x + \tau\nu(x) - y|} - \frac{\partial}{\partial\nu(x)} \frac{1}{|x - y|} \right) d\omega(y) + 2\pi F(x) \right| = 0, \tag{3.156}$$

(b)
$$\lim_{\tau \to 0+} \sup_{x \in \partial\mathcal{G}} \left| \int_{\partial\mathcal{G}} F(y) \left(\frac{\partial}{\partial\tau} \frac{1}{|x - \tau\nu(x) - y|} - \frac{\partial}{\partial\nu(x)} \frac{1}{|x - y|} \right) d\omega(y) - 2\pi F(x) \right| = 0. \tag{3.157}$$

Proof. We only prove case (a). An easy calculation shows that

$$\begin{aligned} &\int_{\partial\mathcal{G}} F(y) \left(\frac{\partial}{\partial\tau} \frac{1}{|x + \tau\nu(x) - y|} - \frac{\partial}{\partial\nu(x)} \frac{1}{|x - y|} \right) d\omega(y) \\ =\ & - \int_{\partial\mathcal{G}} F(y) \, \nu(x) \cdot \left(\frac{x + \tau\nu(x) - y}{|x + \tau\nu(x) - y|^3} - \frac{x - y}{|x - y|^3} \right) d\omega(y) \\ =\ & - \int_{\partial\mathcal{G}} F(y) \, \nu(y) \cdot \left(\frac{x + \tau\nu(x) - y}{|x + \tau\nu(x) - y|^3} - \frac{x - y}{|x - y|^3} \right) d\omega(y) \\ & - \int_{\partial\mathcal{G}} F(y) \, (\nu(x) - \nu(y)) \cdot \left(\frac{x + \tau\nu(x) - y}{|x + \tau\nu(x) - y|^3} - \frac{x - y}{|x - y|^3} \right) d\omega(y). \end{aligned} \tag{3.158}$$

The limit behavior of the integral

$$-\int_{\partial\mathcal{G}} F(y) \left(\frac{\partial}{\partial\nu(y)} \frac{1}{|x+\tau\nu(x)-y|} - \frac{\partial}{\partial\nu(y)} \frac{1}{|x-y|} \right) d\omega(y) \tag{3.159}$$

is known from the treatment of the potential of the double-layer. Therefore it remains to guarantee that the second integral does not contribute to the limit as τ tends to 0. For that purpose we split $\partial\mathcal{G}$ in the usual way into $\partial\mathcal{G}\cap\mathcal{B}_\sigma(x)$ and $\partial\mathcal{G}\setminus\overline{\mathcal{B}_\sigma(x)}$. The essential step toward Lemma 3.43 is provided by

$$\int_{\partial\mathcal{G}\cap\mathcal{B}_\sigma(x)} F(y)\ (\nu(x)-\nu(y)) \cdot \left(\frac{x+\tau\nu(x)-y}{|x+\tau\nu(x)-y|^3} - \frac{x-y}{|x-y|^3} \right) d\omega(y)$$
$$= \mathcal{O}\left(\int_0^\sigma \frac{\rho^2}{\rho^2+\tau^2} d\rho \right), \tag{3.160}$$

where the integral occurring in the last term is of order $\mathcal{O}(\tau)$ (choosing, e.g., $\sigma=\tau$). □

The *two-sided jump relations of the normal derivative of the single-layer potential* follow in a canonical manner.

Theorem 3.44. *Under the assumptions of Lemma 3.39 the following relations hold true:*

(a)
$$\lim_{\tau\to 0+} \sup_{x\in\partial\mathcal{G}} \left| \int_{\partial\mathcal{G}} F(x) \left(\frac{\partial}{\partial\tau} \frac{1}{|x+\tau\nu(x)-y|} - \frac{\partial}{\partial\tau} \frac{1}{|x-\tau\nu(x)-y|} \right) d\omega(y) + 4\pi F(x) \right| = 0. \tag{3.161}$$

(b) *There exists a constant $C>0$ such that for all $0<\tau<\min(\frac{\delta}{2},\frac{1}{8M})$,*

$$\sup_{F\in \mathrm{C}^{(0)}(\partial\mathcal{G})} \sup_{x\in\partial\mathcal{G}} \left| \int_{\partial\mathcal{G}} F(y) \left(\frac{\partial}{\partial\tau} \frac{1}{|x+\tau\nu(x)-y|} - \frac{\partial}{\partial\tau} \frac{1}{|x-\tau\nu(x)-y|} \right) d\omega(y) - 4\pi F(x) \right| \le C\|F\|_{\mathrm{C}^{(0)}(\partial\mathcal{G})}. \tag{3.162}$$

Proof. First, we observe that

$$\frac{\partial}{\partial\tau} \frac{1}{|x\pm\tau\nu(x)-y|} = \frac{\nu(x)\cdot(y\pm\tau\nu(y)-x)}{|y+\tau\nu(y)-x|^3}. \tag{3.163}$$

For $0<\tau\le\frac{\sigma}{2}<\min(\frac{\delta}{2},\frac{1}{8M})$, with δ, M as specified by Definition 1.7, we are able to deduce that

$$\begin{aligned}
&\left| \int_{\partial\mathcal{G}} F(y) \left(\frac{\nu(x)\cdot(y+\tau\nu(y)-x)}{|y+\tau\nu(y)-x|^3} - \frac{\nu(x)\cdot(y-\tau\nu(y)-x)}{|y-\tau\nu(y)-x|^3} \right) d\omega(y) - 4\pi F(x) \right| \\
&\leq 4\pi \|F\|_{C^{(0)}(\partial\mathcal{G})} \\
&\quad + \left| \int_{\partial\mathcal{G}} F(y) \left(\frac{\nu(x)\cdot(y+\tau\nu(y)-x)}{|y+\tau\nu(y)-x|^3} - \frac{\nu(x)\cdot(y-\tau\nu(y)-x)}{|y-\tau\nu(y)-x|^3} \right) d\omega(y) \right| \\
&\leq 4\pi \|F\|_{C^{(0)}(\partial\mathcal{G})} \\
&\quad + \|F\|_{C^{(0)}(\partial\mathcal{G})} \int_{\partial\mathcal{G}} \left| \frac{\nu(x)\cdot(y+\tau\nu(y)-x)}{|y+\tau\nu(y)-x|^3} - \frac{\nu(x)\cdot(y-\tau\nu(y)-x)}{|y-\tau\nu(y)-x|^3} \right| d\omega(y) \\
&\leq 4\pi \|F\|_{C^{(0)}(\partial\mathcal{G})} \\
&\quad + \|F\|_{C^{(0)}(\partial\mathcal{G})} \int_{\partial\mathcal{G}} \left| \frac{(y+\tau\nu(y)-x)\,|y-\tau\nu(y)-x|^3}{|y+\tau\nu(y)-x|^3|y-\tau\nu(y)-x|^3} \right. \qquad (3.164) \\
&\qquad\qquad \left. - \frac{(y-\tau\nu(y)-x)\,|y+\tau\nu(y)-x|^3}{|y+\tau\nu(y)-x|^3|y-\tau\nu(y)-x|^3} \right| d\omega(y).
\end{aligned}$$

In the usual way we split the last integral over $\partial\mathcal{G}$ on the right side of (3.164) into the two parts $\partial\mathcal{G} \cap \mathcal{B}_\sigma(x)$ and $\partial\mathcal{G}\backslash\overline{\mathcal{B}_\sigma(x)}$. We consider the denominator separately for the two parts in the splitting. More explicitly, for the denominator of the second part we are able to use Lemma 3.35 and determine that $\int_{\partial\mathcal{G}\backslash\overline{\mathcal{B}_\sigma(x)}} \ldots d\omega(y)$ is bounded. For the denominator of the first part of this splitting, considerations similar to Lemma 3.34 can be applied. We obtain an estimate of the order

$$\mathcal{O}\left(\int\limits_{\partial\mathcal{G}\cap\mathcal{B}_\sigma(x)} \frac{|\tau|\ |y-x| + \tau^2}{(|y-x|^2+\tau^2)^2}\, d\omega(y) \right). \tag{3.165}$$

Local coordinates show that the integral in (3.165) is of the order $\mathcal{O}(1)$, more concretely,

$$\int_{\partial\mathcal{G}\cap\mathcal{B}_\sigma(x)} \frac{|\tau||y-x| + \tau^2}{(|y-x|^2+\tau^2)^2}\, d\omega(y) = \mathcal{O}\left(\tau^2 \left(\frac{1}{\tau^2} - \frac{1}{\sigma^2+\tau^2} \right) \right). \tag{3.166}$$

This guarantees (b). Part (a) is a direct consequence of Lemma 3.43. □

Finally, we come to the *jump relations for the normal derivative of the double-layer potential.*

Lemma 3.45. *Under the assumption of Lemma 3.39, the following relation holds true:*

$$\lim_{\tau\to 0+} \sup_{x\in\partial\mathcal{G}} \left| \int_{\partial\mathcal{G}} F(y) \frac{\partial}{\partial\tau}\frac{\partial}{\partial\nu(y)} \left(\frac{1}{|x+\tau\nu(x)-y|} - \frac{1}{|x-\tau\nu(x)-y|} \right) d\omega(y) \right| = 0. \tag{3.167}$$

Proof. We discuss the limit relation for the density $\tilde{F}(y) = F(y) - F(x)$ instead of $F(y)$. We have to estimate

$$\begin{aligned} &\int_{\partial\mathcal{G}} \tilde{F}(y) \frac{\partial}{\partial\tau} \frac{\partial}{\partial\nu(y)} \left(\frac{1}{|x+\tau\nu(x)-y|} - \frac{1}{|x-\tau\nu(x)-y|} \right) d\omega(y) \qquad (3.168) \\ &= -\int_{\partial\mathcal{G}} \tilde{F}(y)\,(\nu(x)\cdot\nu(y)) \left(\frac{1}{|x+\tau\nu(x)-y|^3} - \frac{1}{|x-\tau\nu(x)-y|^3} \right) d\omega(y) \\ &\quad +3\int_{\partial\mathcal{G}} \tilde{F}(y) \left(\frac{\nu(x)\cdot(x+\tau\nu(x)-y)\ \nu(y)\cdot(x+\tau\nu(x)-y)}{|x+\tau\nu(x)-y|^5} \right. \\ &\qquad \left. - \frac{\nu(x)\cdot(x-\tau\nu(x)-y)\ \nu(y)\cdot(x-\tau\nu(x)-y)}{|x-\tau\nu(x)-y|^5} \right) d\omega(y). \end{aligned}$$

We split $\partial\mathcal{G}$ in the usual way. Using Lemma 3.32 and Lemma 3.33, we are able to show that

$$\begin{aligned} &\left| \int_{\partial\mathcal{G}\cap\mathcal{B}_\sigma(x)} \tilde{F}(y)\,(\nu(x)\cdot\nu(y)) \left(\frac{1}{|x+\tau\nu(x)-y|^3} - \frac{1}{|x-\tau\nu(x)-y|^3} \right) d\omega(y) \right| \\ &= \mathcal{O}\left(\sup_{z\in\partial\mathcal{G}\cap\mathcal{B}_\sigma(x)} \left|\tilde{F}(z)\right| \int_{\mathcal{G}\cap\mathcal{B}_\sigma(x)} \frac{\tau}{(|x-y|^2+\tau^2)^{\frac{3}{2}}}\, d\omega(y) \right). \qquad (3.169) \end{aligned}$$

For $\tau \leq \frac{\sigma}{2}$ we get, from Lemma 3.35 and Lemma 3.36, that

$$\begin{aligned} &\left| \int_{\partial\mathcal{G}\setminus\overline{\mathcal{B}_\sigma(x)}} \tilde{F}(y)(\nu(x)\cdot\nu(y)) \left(\frac{1}{|x+\tau\nu(x)-y|^3} - \frac{1}{|x-\tau\nu(x)-y|^3} \right) d\omega(y) \right| \\ &\qquad = \mathcal{O}\left(\frac{\tau}{\sigma^4}\right). \qquad (3.170) \end{aligned}$$

Thus, observing the arbitrary choice of $\sigma > 0$, Lemma 3.34 and the estimates (3.169), (3.170) imply that the first integral on the right-hand side of (3.168) vanishes as τ tends to zero (at this point it becomes important that we have used $\tilde{F}$ and not F in order to guarantee that the term $\sup_{z\in\partial\mathcal{G}\cap\mathcal{B}_\sigma(x)} |\tilde{F}(z)|$ in (3.169) gets small for small $\sigma > 0$). Analogous estimates can be obtained for the second integral on the right-hand side of (3.168). □

We conclude our investigations concerning limit and jump relations with more detailed information about the behavior of the single- as well as the double-layer potential on the boundary surface $\partial\mathcal{G}$ itself. From the literature (see, e.g., C. Müller [1969], D. Colton, R. Kress [1983]) we know the following:

Lemma 3.46 (Hölder Continuity of the Layer Potentials on $\partial\mathcal{G}$).

(a) *The single-layer potential with continuous density F on $\partial\mathcal{G}$ represents a μ-Hölder continuous function on $\mathbb{R}^3$ such that*

$$\left\| \int_{\partial\mathcal{G}} F(y) \frac{1}{|\cdot - y|}\, d\omega(y) \right\|_{\mathrm{C}^{(0,\mu)}(\mathbb{R}^3)} \leq \mathrm{C}_\mu \, \|F\|_{\mathrm{C}^{(0)}(\partial\mathcal{G})} \qquad (3.171)$$

for $\mu \in (0,1)$ and some constant C_μ depending on μ and $\partial\mathcal{G}$.

(b) The double-layer potential (3.103) with density F of class $\mathrm{C}^{(0,\mu)}(\partial\mathcal{G})$, $\mu \in (0,1)$, is μ-Hölder continuous in $\overline{\mathcal{G}}$ and $\overline{\mathcal{G}^c}$ such that

$$\left\| \int_{\partial\mathcal{G}} F(y) \frac{\partial}{\partial\nu(y)} \frac{1}{|x-y|} \, d\omega(y) \right\|_{\mathrm{C}^{(0,\mu)}(\overline{\mathcal{G}})} \leq \mathrm{C}_\mu \|F\|_{\mathrm{C}^{(0,\mu)}(\partial\mathcal{G})} \tag{3.172}$$

and

$$\left\| \int_{\partial\mathcal{G}} F(y) \frac{\partial}{\partial\nu(y)} \frac{1}{|\cdot - y|} \, d\omega(y) \right\|_{\mathrm{C}^{(0,\mu)}(\overline{\mathcal{G}^c})} \leq \mathrm{C}_\mu \|F\|_{\mathrm{C}^{(0,\mu)}(\partial\mathcal{G})}, \tag{3.173}$$

where C_μ is some constant depending on μ and $\partial\mathcal{G}$.

For our approach, it suffices to prove the μ-Hölder continuity of the double-layer potential on $\partial\mathcal{G}$, for some $\mu \in (0, \frac{1}{4}]$ and an integrable and bounded function on $\partial\mathcal{G}$.

Lemma 3.47 (Hölder Continuity of the Double-Layer Potential on $\partial\mathcal{G}$). *Let $\mathcal{G} \subset \mathbb{R}^3$ be a regular region. If the density $F : \partial\mathcal{G} \to \mathbb{R}$ is integrable and bounded, then the double-layer potential (3.103) is of class $\mathrm{C}^{(0,\mu)}(\partial\mathcal{G})$, $\mu \in (0, \frac{1}{4}]$.*

Proof. For two points $x_1, x_2 \in \partial\mathcal{G}$ we set $\rho = |x_1 - x_2|$. For sufficiently small $\tau > 0$ with $\rho < \frac{\tau}{2}$ we consider

$$\begin{aligned} &\left| \int_{\partial\mathcal{G}} F(y) \left(\frac{\partial}{\partial\nu(y)} \frac{1}{|x_1-y|} - \frac{\partial}{\partial\nu(y)} \frac{1}{|x_2-y|} \right) d\omega(y) \right| \\ &\leq \left| \int_{\partial\mathcal{G}\setminus\overline{\mathcal{B}_\tau(x_1)}} F(y) \left(\frac{\partial}{\partial\nu(y)} \frac{1}{|x_1-y|} - \frac{\partial}{\partial\nu(y)} \frac{1}{|x_2-y|} \right) d\omega(y) \right| \\ &\quad + \left| \int_{\partial\mathcal{G}\cap\mathcal{B}_\tau(x_1)} F(y) \frac{\partial}{\partial\nu(y)} \frac{1}{|x_1-y|} \, d\omega(y) \right| \\ &\quad + \left| \int_{\partial\mathcal{G}\cap\mathcal{B}_\tau(x_1)} F(y) \frac{\partial}{\partial\nu(y)} \frac{1}{|x_2-y|} \, d\omega(y) \right|. \end{aligned} \tag{3.174}$$

We apply the mean value theorem of multivariate analysis to the first integral on the right side of (3.174). In connection with the boundedness of the density function, we obtain

$$\begin{aligned} &\left| \int_{\partial\mathcal{G}\setminus\overline{\mathcal{B}_\tau(x_1)}} F(y) \left(\frac{\partial}{\partial\nu(y)} \frac{1}{|x_1-y|} - \frac{\partial}{\partial\nu(y)} \frac{1}{|x_2-y|} \right) d\omega(y) \right| \\ &= \mathcal{O}\left(\rho \int_{\partial\mathcal{G}\setminus\overline{\mathcal{B}_\tau(x_1)}} \left| \nabla_z \frac{\partial}{\partial\nu(y)} \frac{1}{|z-y|} \right| d\omega(y) \right), \end{aligned} \tag{3.175}$$

where z is located in $\partial\mathcal{G} \cap \mathcal{B}_{\frac{\tau}{2}}(x_1)$. Thus it is clear that (3.175) behaves like

$O(\rho\tau^{-3})$. The second and third integral of the right side of (3.174) are of order $O(\tau)$. Thus it follows, for $\tau \to 0+$, that

$$\int_{\partial\mathcal{G}} F(y) \left(\frac{\partial}{\partial\nu(y)} \frac{1}{|x_1 - y|} - \frac{\partial}{\partial\nu(y)} \frac{1}{|x_2 - y|} \right) d\omega(y) = \mathcal{O}\left(\frac{\rho}{\tau^3} + \tau \right). \quad (3.176)$$

Therefore, letting $\rho = \tau^4$, we obtain for all $\rho < (\frac{1}{2})^{\frac{4}{3}}$ that

$$\left| \int_{\partial\mathcal{G}} F(y) \left(\frac{\partial}{\partial\nu(y)} \frac{1}{|x_1 - y|} - \frac{\partial}{\partial\nu(y)} \frac{1}{|x_2 - y|} \right) d\omega(y) \right| \leq C\, |x_1 - x_2|^{\frac{1}{4}}. \quad (3.177)$$

This guarantees Lemma 3.47. □

Remark 3.48. The proof of Lemma 3.47 only uses the uniform boundedness of the density function F on $\partial\mathcal{G}$. This is the reason why we are able to conclude that the equiboundedness of a sequence $\{F_n\}_{n\in\mathbb{N}}$, i.e., $\sup_{x\in\partial\mathcal{G}}|F_n(x)| \leq C$ for all $n \in \mathbb{N}$, implies the equicontinuity of the corresponding sequence of double-layer potentials, i.e., for arbitrary $\varepsilon > 0$ there exists a (sufficiently small) $\delta > 0$ such that for all $n \in \mathbb{N}$ and all $|x_1 - x_2| \leq \delta$ the estimate

$$\left| \int_{\partial\mathcal{G}} F_n(y) \left(\frac{\partial}{\partial\nu(y)} \frac{1}{|x_1 - y|} - \frac{\partial}{\partial\nu(y)} \frac{1}{|x_2 - y|} \right) d\omega(y) \right| \leq \varepsilon \quad (3.178)$$

holds true. Together with the Theorem of Arzéla–Ascoli, this aspect leads to the complete continuity of the potential operators under consideration. Thus, it is of special significance for the solution of the (Fredholm) integral equations derived for boundary-value problems (as presented in Subsection 3.4.4).

3.3.3 Limit and Jump Relations in L^2-Topology

Next, we are interested in extending the $\mathrm{C}^{(0)}(\partial\mathcal{G})$-limit and jump relations to the $\mathrm{L}^2(\partial\mathcal{G})$-nomenclature. An essential tool for the transfer of the limit relations from the $\mathrm{C}^{(0)}(\partial\mathcal{G})$- into the $\mathrm{L}^2(\partial\mathcal{G})$-framework is their reformulation by means of potential operators.

We remember that, for $\tau \neq \sigma$ with $|\tau|, |\sigma|$ sufficiently small, the functions

$$(x, y) \mapsto G(\Delta; |x + \tau\nu(x) - (y + \sigma\nu(y))|), \quad (x, y) \in \partial\mathcal{G} \times \partial\mathcal{G}, \quad (3.179)$$

are continuous. Thus, the *potential operators* $P(\tau, \sigma)$ defined by

$$P(\tau, \sigma)[F](x) = \int_{\partial\mathcal{G}} F(y) G(\Delta; |x + \tau\nu(x) - (y + \sigma\nu(y))|)\, d\omega(y) \quad (3.180)$$

form mappings from $\mathrm{L}^2(\partial\mathcal{G})$ into $\mathrm{C}^{(0)}(\partial\mathcal{G})$ and are continuous with respect to $\|\cdot\|_{\mathrm{C}^{(0)}(\partial\mathcal{G})}$. Formally, for sufficiently small $|\tau| > 0$, and $\sigma = 0$, we obtain the *potential operator* $P(\tau, 0) : \mathrm{L}^2(\partial\mathcal{G}) \to \mathrm{C}^{(0)}(\partial\mathcal{G})$ given by

$$P(\tau, 0)[F](x) = \int_{\partial\mathcal{G}} F(y) G(\Delta; |x + \tau\nu(x) - y|)\, d\omega(y). \quad (3.181)$$

This is *the operator of the single-layer potential on $\partial\mathcal{G}$ for values on $\partial\mathcal{G}(\tau)$*. Moreover, $P_{|2}(\tau,0) : \mathrm{L}^2(\partial\mathcal{G}) \to \mathrm{C}^{(0)}(\partial\mathcal{G})$ given by

$$\begin{aligned} P_{|2}(\tau,0)[F](x) &= \frac{\partial}{\partial\sigma}P(\tau,\sigma)[F](x)\Big|_{\sigma=0} \\ &= \int_{\partial\mathcal{G}} F(y)\frac{\partial}{\partial\nu(y)}G(\Delta;|x+\tau\nu(x)-y|)\,d\omega(y) \\ &= \frac{1}{4\pi}\int_{\partial\mathcal{G}} F(y)\frac{\nu(y)\cdot(x+\tau\nu(x)-y)}{|x+\tau\nu(x)-y|^3}\,d\omega(y) \end{aligned} \tag{3.182}$$

is *the operator of the double-layer potential on $\partial\mathcal{G}$ for values on $\partial\mathcal{G}(\tau)$* (note that the notation $P_{|i}$ indicates differentiation with respect to the i-th variable). Analogously, we introduce the operator $P_{|1}(\tau,0) : \mathrm{L}^2(\partial\mathcal{G}) \to \mathrm{C}^{(0)}(\partial\mathcal{G})$ by

$$\begin{aligned} P_{|1}(\tau,0)[F](x) &= \frac{\partial}{\partial\tau}P(\tau,\sigma)[F](x)\Big|_{\sigma=0} \\ &= \int_{\partial\mathcal{G}} F(y)\frac{\partial}{\partial\tau}G(\Delta;|x-\tau\nu(x)-y|)\,d\omega(y) \\ &= -\frac{1}{4\pi}\int_{\partial\mathcal{G}} F(y)\frac{\nu(x)\cdot(x+\tau\nu(x)-y)}{|x+\tau\nu(x)-y|^3}\,d\omega(y). \end{aligned} \tag{3.183}$$

$P_{|1}(\tau,0)$ is the *operator of the normal derivative of the single-layer potential on $\partial\mathcal{G}$ for values on $\partial\mathcal{G}(\tau)$*. Moreover, $P_{|2|1}(\tau,0) : \mathrm{L}^2(\partial\mathcal{G}) \to \mathrm{C}^{(0)}(\partial\mathcal{G})$ given by

$$\begin{aligned} &P_{|2|1}(\tau,0)[F](x) \\ &= \left(\frac{\partial^2}{\partial\tau\partial\sigma}P(\tau,\sigma)[F](x)\right)\Big|_{\sigma=0} \\ &= \int_{\partial\mathcal{G}} F(y)\frac{\partial}{\partial\tau}\frac{\partial}{\partial\nu(y)}G(\Delta;|x-\tau\nu(x)-y|)\,d\omega(y) \\ &= \frac{1}{4\pi}\int_{\partial\mathcal{G}} F(y)\frac{\nu(x)\cdot(x+\tau\nu(x)-y)\;\nu(y)\cdot(x+\tau\nu(x)-y)}{|x+\tau\nu(x)-y|^3}\,d\omega(y) \end{aligned} \tag{3.184}$$

is the *operator of the normal derivative of the double-layer potential for values on $\partial\mathcal{G}(\tau)$*. We collect the jump relations from Subsection 3.3.2 in the nomenclature of the potential operators.

Theorem 3.49. *Let $\mathcal{G} \subset \mathbb{R}^3$ be a regular region. Suppose that F is of class $\mathrm{C}^{(0)}(\partial\mathcal{G})$. Then*

(a) $\lim_{\tau\to 0+} \|P(\tau,0)[F] - P(-\tau,0)[F]\|_{\mathrm{C}^{(0)}(\partial\mathcal{G})} = 0,$

(b) $\lim_{\tau\to 0+} \left\|P_{|1}(\tau,0)[F] - P_{|1}(-\tau,0)[F] + F\right\|_{\mathrm{C}^{(0)}(\partial\mathcal{G})} = 0,$

(c) $\lim_{\tau\to 0+} \left\|P_{|2}(\tau,0)[F] - P_{|2}(-\tau,0)[F] - F\right\|_{\mathrm{C}^{(0)}(\partial\mathcal{G})} = 0,$

(d) $\lim_{\tau\to 0+} \left\|P_{|2|1}(\tau,0)[F] - P_{|2|1}(-\tau,0)[F]\right\|_{\mathrm{C}^{(0)}(\partial\mathcal{G})} = 0.$

The limit relations can, of course, be formulated in such a potential operator notation as well.

Theorem 3.50. *Let $\mathcal{G} \subset \mathbb{R}^3$ be a regular region. Suppose that F is of class $\mathrm{C}^{(0)}(\partial\mathcal{G})$. Then*

(a) $\lim_{\tau\to 0+} \|P(\pm\tau,0)[F] - P(0,0)[F]\|_{\mathrm{C}^{(0)}(\partial\mathcal{G})} = 0,$

(b) $\lim_{\tau\to 0+} \left\|P_{|1}(\pm\tau,0)[F] - P_{|1}(0,0)[F] \pm \frac{1}{2}F\right\|_{\mathrm{C}^{(0)}(\partial\mathcal{G})} = 0,$

(c) $\lim_{\tau\to 0+} \left\|P_{|2}(\pm\tau,0)[F] - P_{|2}(0,0)[F] \mp \frac{1}{2}F\right\|_{\mathrm{C}^{(0)}(\partial\mathcal{G})} = 0,$

Next, we consider potentials on parallel surfaces $\partial\mathcal{G}(\sigma)$ for values on $\partial\mathcal{G}$. For $F \in \mathrm{L}^2(\partial\mathcal{G})$ and for sufficiently small positive σ, the operator $P(0,\sigma) : \mathrm{L}^2(\partial\mathcal{G}) \to \mathrm{C}^{(0)}(\partial\mathcal{G})$ given by

$$P(0,\sigma)[F](x) = \int_{\partial\mathcal{G}} F(y) G(\Delta; |x-(y+\sigma\nu(y))|)\, d\omega(y) \tag{3.185}$$

is the *operator of the single-layer potential on $\partial\mathcal{G}(\sigma)$ for values on $\partial\mathcal{G}$.* $P_{|2}(0,\sigma) : \mathrm{L}^2(\partial\mathcal{G}) \to \mathrm{C}^{(0)}(\partial\mathcal{G})$ given by

$$P_{|2}(0,\sigma)[F](x) = \int_{\partial\mathcal{G}} F(y) \frac{\partial}{\partial\sigma} G(\Delta; |x-(y+\sigma\nu(y))|)\, d\omega(y) \tag{3.186}$$

is the *operator of the double-layer potential on $\partial\mathcal{G}(\sigma)$ for values on $\partial\mathcal{G}$.* The operator $P_{|1}(0,\sigma) : \mathrm{L}^2(\partial\mathcal{G}) \to \mathrm{C}^{(0)}(\partial\mathcal{G})$ given by

$$P_{|1}(0,\sigma)[F](x) = \int_{\partial\mathcal{G}} F(y) \frac{\partial}{\partial\nu(x)} G(\Delta; |x-(y+\sigma\nu(y))|)\, d\omega(y) \tag{3.187}$$

is the *operator of the normal derivative of the single-layer potential on $\partial\mathcal{G}(\sigma)$ for values on $\partial\mathcal{G}$.* Finally, $P_{|2|1}(0,\sigma) : \mathrm{L}^2(\partial\mathcal{G}) \to \mathrm{C}^{(0)}(\partial\mathcal{G})$ given by

$$P_{|2|1}(0,\sigma)[F](x) = \int_{\partial\mathcal{G}} F(y) \frac{\partial}{\partial\nu(x)} \frac{\partial}{\partial\sigma} G(\Delta; |x-(y+\sigma\nu(y))|)\, d\omega(y) \tag{3.188}$$

is the *operator of the normal derivative of the double-layer potential on $\partial\mathcal{G}(\sigma)$ for values on $\partial\mathcal{G}$.*

Mixed limit relations for differences of both types of potential operators are listed in the next lemma.

Lemma 3.51. *Let $\mathcal{G} \subset \mathbb{R}^3$ be a regular surface. Suppose that F is of class $\mathrm{C}^{(0)}(\partial\mathcal{G})$. Then*

(a) $\lim_{\tau\to 0+} \|P(\pm\tau,0)[F] - P(0,\mp\tau)[F]\|_{\mathrm{C}^{(0)}(\partial\mathcal{G})} = 0,$

(b) $\lim_{\tau\to 0+} \left\|P_{|2}(\pm\tau,0)[F] - P_{|2}(0,\mp\tau)[F]\right\|_{\mathrm{C}^{(0)}(\partial\mathcal{G})} = 0,$

(c) $\lim_{\tau \to 0+} \left\| P_{|1}(\pm\tau, 0)[F] - P_{|1}(0, \mp\tau)[F] \right\|_{\mathrm{C}^{(0)}(\partial\mathcal{G})} = 0,$

(d) $\lim_{\tau \to 0+} \left\| P_{|2|1}(\pm\tau, 0)[F] - P_{|2|1}(0, \mp\tau)[F] \right\|_{\mathrm{C}^{(0)}(\partial\mathcal{G})} = 0.$

Proof. All relations can be proved in an elementary way, but the justification is rather technical. For brevity, we restrict ourselves to the limit relation (a): For $0 < \tau \leq \frac{\sigma}{2} < \min(\frac{\delta}{2}, \frac{1}{4M})$, with δ, M as specified by Definition 1.7, and $F \in \mathrm{C}^{(0)}(\partial\mathcal{G})$ we get

$$\begin{aligned} & \|P(\tau,0)[F] - P(0,-\tau)[F]\|_{\mathrm{C}^{(0)}(\partial\mathcal{G})} \\ = & \sup_{x \in \partial\mathcal{G}} |P(\tau,0)[F](x) - P(0,-\tau)[F](x)| \\ = & \sup_{x \in \partial\mathcal{G}} \left| \frac{1}{4\pi} \int_{\partial\mathcal{G}} F(y) \left(\frac{1}{|x + \tau\nu(x) - y|} - \frac{1}{|x - y + \tau\nu(y)|} \right) d\omega(y) \right| . \end{aligned} \tag{3.189}$$

The surface $\partial\mathcal{G}$ is split into the parts $\partial\mathcal{G} \cap \mathcal{B}_\sigma(x)$ and $\partial\mathcal{G} \setminus \overline{\mathcal{B}_\sigma(x)}$ in the usual way. In connection with Lemma 1.9, this yields

$$\begin{aligned} & \sup_{x \in \partial\mathcal{G}} \left| \int_{\partial\mathcal{G} \cap \mathcal{B}_\sigma(x)} F(y) \left(\frac{1}{|x + \tau\nu(x) - y|} - \frac{1}{|x - y + \tau\nu(y)|} \right) d\omega(y) \right| \\ \leq & \|F\|_{\mathrm{C}^{(0)}(\partial\mathcal{G})} \sup_{x \in \partial\mathcal{G}} \left| \int_{\partial\mathcal{G} \cap \mathcal{B}_\sigma(x)} \frac{\tau M |x - y|}{|x + \tau\nu(x) - y| \, |x - y + \tau\nu(y)|} d\omega(y) \right| \\ \leq & \|F\|_{\mathrm{C}^{(0)}(\partial\mathcal{G})} \frac{M\sigma}{2} \sup_{x \in \partial\mathcal{G}} \left| \int_{\partial\mathcal{G} \cap \mathcal{B}_\sigma(x)} \frac{1}{|x + \tau\nu(x) - y|} d\omega(y) \right| . \end{aligned} \tag{3.190}$$

It follows that

$$\sup_{x \in \partial\mathcal{G}} \left| \int_{\partial\mathcal{G} \cap \mathcal{B}_\sigma(x)} F(y) \left(\frac{1}{|x + \tau\nu(x) - y|} - \frac{1}{|x - y + \tau\nu(y)|} \right) d\omega(y) \right| = \mathcal{O}(\sigma^2). \tag{3.191}$$

By virtue of Lemma 1.9 and the estimate $|x - y - \tau\nu(x)| \geq \frac{\sigma}{2}$, we obtain in the usual manner that

$$\begin{aligned} & \sup_{x \in \partial\mathcal{G}} \left| \int_{\partial\mathcal{G} \setminus \overline{\mathcal{B}_\sigma(x)}} F(y) \left(\frac{1}{|x + \tau\nu(x) - y|} - \frac{1}{|x - y + \tau\nu(y)|} \right) d\omega(y) \right| \\ \leq & \|F\|_{\mathrm{C}^{(0)}(\partial\mathcal{G})} \sup_{x \in \partial\mathcal{G}} \int_{\partial\mathcal{G} \setminus \overline{\mathcal{B}_\sigma(x)}} \frac{3\tau|\nu(y) - \nu(x)|}{|x + \tau\nu(x) - y|^2} d\omega(y), \end{aligned} \tag{3.192}$$

where

$$\sup_{x \in \partial\mathcal{G}} \int_{\partial\mathcal{G} \setminus \overline{\mathcal{B}_\sigma(x)}} \frac{\tau|\nu(y) - \nu(x)|}{|x + \tau\nu(x) - y|^2} d\omega(y) = \mathcal{O}\left(\frac{\tau}{\sigma^2} \right) . \tag{3.193}$$

Altogether, we have

$$\|P(\tau,0)[F] - P(0,-\tau)[F]\|_{\mathrm{C}^{(0)}(\partial\mathcal{G})} = \mathcal{O}(\sigma^2) + \mathcal{O}\left(\frac{\tau}{\sigma^2}\right). \tag{3.194}$$

Letting $\sigma = \tau^{\frac{1}{3}}$ we get for $\tau < \min\left(\frac{\delta}{2}, \frac{1}{4M}, \frac{1}{2\sqrt{2}}\right)$,

$$\|P(\tau,0)[F] - P(0,-\tau)[F]\|_{\mathrm{C}^{(0)}(\partial\mathcal{G})} = \mathcal{O}\left(\tau^{\frac{1}{3}}\right). \tag{3.195}$$

This is the desired result. □

By aid of Lemma 3.49, we are led to the following jump relations.

Theorem 3.52. *Let $\mathcal{G} \subset \mathbb{R}^3$ be a regular region. Suppose that F is of class $\mathrm{C}^{(0)}(\partial\mathcal{G})$. Then*

(a) $\lim_{\tau\to 0+} \|P(0,\tau)[F] - P(0,-\tau)[F]\|_{\mathrm{C}^{(0)}(\partial\mathcal{G})} = 0,$

(b) $\lim_{\tau\to 0+} \left\|P_{|1}(0,\tau)[F] - P_{|1}(0,-\tau)[F] - F\right\|_{\mathrm{C}^{(0)}(\partial\mathcal{G})} = 0,$

(c) $\lim_{\tau\to 0+} \left\|P_{|2}(0,\tau)[F] - P_{|2}(0,-\tau)[F] + F\right\|_{\mathrm{C}^{(0)}(\partial\mathcal{G})} = 0,$

(d) $\lim_{\tau\to 0+} \left\|P_{|2|1}(0,\tau)[F] - P_{|2|1}(0,-\tau)[F]\right\|_{\mathrm{C}^{(0)}(\partial\mathcal{G})} = 0.$

Proof. All assertions can be easily obtained by use of the triangle inequality. We restrict ourselves to (a): For $F \in \mathrm{C}^{(0)}(\partial\mathcal{G})$, we obviously have

$$\begin{aligned} & \|P(0,\tau)[F] - P(0,-\tau)[F]\|_{\mathrm{C}^{(0)}(\partial\mathcal{G})} \\ \leq\ & \|P(0,\tau)[F] - P(-\tau,0)[F]\|_{\mathrm{C}^{(0)}(\partial\mathcal{G})} + \|P(\tau,0)[F] - P(0,-\tau)[F]\|_{\mathrm{C}^{(0)}(\partial\mathcal{G})} \\ & + \|P(-\tau,0)[F] - P(\tau,0)[F]\|_{\mathrm{C}^{(0)}(\partial\mathcal{G})}. \end{aligned} \tag{3.196}$$

We already know that the three terms on the right side of (3.197) tend to zero with $\tau \to 0+$. This proves (a) of Theorem 3.52. □

The operator $P(\tau,\sigma)^*$ satisfying

$$(F, P(\tau,\sigma)[G])_{\mathrm{L}^2(\partial\mathcal{G})} = (P(\tau,\sigma)^*[F], G)_{\mathrm{L}^2(\partial\mathcal{G})}, \tag{3.197}$$

for all $F, G \in \mathrm{L}^2(\partial\mathcal{G})$, is called the *adjoint operator* of $P(\tau,\sigma)$ with respect to $(\cdot,\cdot)_{\mathrm{L}^2(\partial\mathcal{G})}$. According to Fubini's theorem, it follows that

$$\begin{aligned} & (F, P(\tau,\sigma)[G])_{\mathrm{L}^2(\partial\mathcal{G})} \\ =\ & \frac{1}{4\pi}\int_{\partial\mathcal{G}} F(x)\left(\int_{\partial\mathcal{G}} \frac{G(y)}{|x+\tau\nu(x) - (y+\sigma\nu(y))|}\, d\omega(y)\right) d\omega(x) \\ =\ & \frac{1}{4\pi}\int_{\partial\mathcal{G}} G(y)\left(\int_{\partial\mathcal{G}} \frac{F(x)}{|x+\tau\nu(x) - (y+\sigma\nu(y))|}\, d\omega(x)\right) d\omega(y) \\ =\ & (P(\sigma,\tau)[F], G)_{\mathrm{L}^2(\partial\mathcal{G})}, \end{aligned} \tag{3.198}$$

i.e.,

$$P(\tau,\sigma)^* = P(\sigma,\tau). \tag{3.199}$$

By comparison we thus have

$$P(\tau,0)^*[F](x) = \frac{1}{4\pi}\int_{\partial\mathcal{G}} F(y)\frac{1}{|y+\tau\nu(y)-x|}\,d\omega(y). \tag{3.200}$$

Analogously, we obtain expressions of $P_{|1}(\tau,0)^*$ and $P_{|2}(\tau,0)^*$

$$\begin{aligned} &P_{|1}(\tau,0)^*[F](x) \\ &\quad = -\frac{1}{4\pi}\int_{\partial\mathcal{G}} F(y)\frac{\nu(y)\cdot(y+\tau\nu(y)-x)}{|y+\tau\nu(y)-x|^3}\,d\omega(y), \end{aligned} \tag{3.201}$$

$$\begin{aligned} &P_{|2}(\tau,0)^*[F](x) \\ &\quad = \frac{1}{4\pi}\int_{\partial\mathcal{G}} F(y)\frac{\nu(x)\cdot(y+\tau\nu(y)-x)}{|y+\tau\nu(y)-x|^3}\,d\omega(y), \end{aligned} \tag{3.202}$$

$$\begin{aligned} &P_{|2|1}(\tau,0)^*[F](x) \\ &\quad = \frac{1}{4\pi}\int_{\partial\mathcal{G}} F(y)\frac{\nu(y)\cdot(y+\tau\nu(y)-x)\nu(x)\cdot(y+\tau\nu(y)-x)}{|y+\tau\nu(y)-x|^3}\,d\omega(y). \end{aligned} \tag{3.203}$$

Comparing the adjoints with the potential operators on parallel surfaces $\partial\mathcal{G}(\tau)$ for values on $\partial\mathcal{G}$, we find

Lemma 3.53. *Let $\mathcal{G}\subset\mathbb{R}^3$ be a regular surface. Then*

$$\begin{aligned} P(\pm\tau,0)^* &= P(0,\pm\tau), & P_{|1}(\pm\tau,0)^* &= P_{|2}(0,\pm\tau), \\ P_{|2}(\pm\tau,0)^* &= P_{|1}(0,\pm\tau), & P_{|2|1}(\pm\tau,0)^* &= P_{|2|1}(0,\pm\tau) \end{aligned}$$

and

$$\begin{aligned} P(0,\pm\tau)^* &= P(\pm\tau,0), & P_{|1}(0,\pm\tau)^* &= P_{|2}(\pm\tau,0), \\ P_{|2}(0,\pm\tau)^* &= P_{|1}(\pm\tau,0), & P_{|2|1}(0,\pm\tau)^* &= P_{|2|1}(\pm\tau,0), \end{aligned}$$

where all potential operators form mappings from $\mathrm{C}^{(0)}(\partial\mathcal{G})$ to $\mathrm{C}^{(0)}(\partial\mathcal{G})$.

In other words, we obtain the following *jump relations for adjoint potential operators.*

Theorem 3.54. *Let $\mathcal{G}\subset\mathbb{R}^3$ be a regular region. Suppose that F is of class $\mathrm{C}^{(0)}(\partial\mathcal{G})$. Then*

(a) $\lim_{\tau\to0+}\|P^*(\tau,0)[F]-P^*(-\tau,0)[F]\|_{\mathrm{C}^{(0)}(\partial\mathcal{G})}=0,$

(b) $\lim_{\tau\to0+}\|P^*_{|1}(\tau,0)[F]-P^*_{|1}(-\tau,0)[F]+F\|_{\mathrm{C}^{(0)}(\partial\mathcal{G})}=0,$

(c) $\lim_{\tau\to0+}\|P^*_{|2}(\tau,0)[F]-P^*_{|2}(-\tau,0)[F]-F\|_{\mathrm{C}^{(0)}(\partial\mathcal{G})}=0,$

(d) $\lim_{\tau\to0+}\|P^*_{|2|1}(\tau,0)[F]-P^*_{|2|1}(-\tau,0)[F]\|_{\mathrm{C}^{(0)}(\partial\mathcal{G})}=0.$

If $\tau = \sigma = 0$, we know that the kernels of the potentials have weak singularities. The integrals formally defined by

$$P(0,0)[F](x) = \frac{1}{4\pi}\int_{\partial\mathcal{G}} F(y)\frac{1}{|x-y|}\, d\omega(y), \tag{3.204}$$

$$P_{|2}(0,0)[F](x) = \frac{1}{4\pi}\int_{\partial\mathcal{G}} F(y)\frac{\partial}{\partial\nu(y)}\frac{1}{|x-y|}\, d\omega(y), \tag{3.205}$$

$$P_{|1}(0,0)[F](x) = \frac{1}{4\pi}\frac{\partial}{\partial\nu(x)}\int_{\partial\mathcal{G}} F(y)\frac{1}{|x-y|}\, d\omega(y), \tag{3.206}$$

however, exist and define linear bounded operators in $\mathrm{L}^2(\partial\mathcal{G})$. Furthermore,

$$P_{|1}(0,0)^* = P_{|2}(0,0), \qquad P_{|2}(0,0)^* = P_{|1}(0,0). \tag{3.207}$$

Applying the one-sided limit relations from Theorem 3.50 in connection with Lemma 3.53, i.e., the relations of the adjoint potential operators, we are led to

Lemma 3.55. *Suppose that F is of class $\mathrm{C}^{(0)}(\partial\mathcal{G})$. Then*

(a) $\lim_{\tau\to 0+} \|P^*(\pm\tau,0) - P^*(0,0)[F]\|_{\mathrm{C}^{(0)}(\partial\mathcal{G})} = 0,$

(b) $\lim_{\tau\to 0+} \|P^*_{|1}(\pm\tau,0) - P^*_{|1}(0,0)[F] \pm \frac{1}{2}F\|_{\mathrm{C}^{(0)}(\partial\mathcal{G})} = 0,$

(c) $\lim_{\tau\to 0+} \|P^*_{|2}(\pm\tau,0)[F] - P^*_{|2}(0,0)[F] \mp \frac{1}{2}F\|_{\mathrm{C}^{(0)}(\partial\mathcal{G})} = 0.$

All in all, the potential operators enable us to give a concise formulation of the limit and jump relations in the Euclidean space $\mathbb{R}^3$: Suppose that, for all sufficiently small values $\tau > 0$, the limit operators $L_i^{\pm}(\tau)$, $i = 1,2,3$, are defined by

$$L_1^{\pm}(\tau) = P(\pm\tau,0) - P(0,0), \tag{3.208}$$

$$L_2^{\pm}(\tau) = P_{|1}(\pm\tau,0) - P_{|1}(0,0) \pm \frac{1}{2}I, \tag{3.209}$$

$$L_3^{\pm}(\tau) = P_{|2}(\pm\tau,0) - P_{|2}(0,0) \mp \frac{1}{2}I, \tag{3.210}$$

while the jump operators $J_i(\tau)$, $i = 1,2,3,4$, are given in the form

$$\mathrm{J}_1(\tau) = P(\tau,0) - P(-\tau,0), \tag{3.211}$$

$$\mathrm{J}_2(\tau) = P_{|1}(\tau,0) - P_{|1}(-\tau,0) + I, \tag{3.212}$$

$$\mathrm{J}_3(\tau) = P_{|2}(\tau,0) - P_{|2}(-\tau,0) - I, \tag{3.213}$$

$$\mathrm{J}_4(\tau) = P_{|2|1}(\tau,0) - P_{|2|1}(-\tau,0), \tag{3.214}$$

where I denotes the identity operator. Then, for $F \in \mathrm{C}^{(0)}(\partial\mathcal{G})$, we have

$$\begin{aligned} &\lim_{\tau\to 0+} \|L_i^{\pm}(\tau)[F]\|_{\mathrm{C}^{(0)}(\partial\mathcal{G})} = 0, && \lim_{\tau\to 0+} \|J_i(\tau)[F]\|_{\mathrm{C}^{(0)}(\partial\mathcal{G})} = 0,\\ &\lim_{\tau\to 0+} \|L_i^{\pm}(\tau)^*[F]\|_{\mathrm{C}^{(0)}(\partial\mathcal{G})} = 0, && \lim_{\tau\to 0+} \|J_i(\tau)^*[F]\|_{\mathrm{C}^{(0)}(\partial\mathcal{G})} = 0. \end{aligned} \tag{3.215}$$

The relations (3.215) can be generalized to the Hilbert space $\mathrm{L}^2(\partial\mathcal{G})$ (see W. Freeden [1980], W. Freeden, H. Kersten [1981]).

Theorem 3.56. *For all $F \in \mathrm{L}^2(\partial\mathcal{G})$, we have*

$$\begin{array}{ll} \lim_{\tau\to 0+} \|\mathrm{L}_i^{\pm}(\tau)[F]\|_{\mathrm{L}^2(\partial\mathcal{G})} = 0, & \lim_{\tau\to 0+} \|J_i(\tau)[F]\|_{\mathrm{L}^2(\partial\mathcal{G})} = 0, \\ \lim_{\tau\to 0+} \|\mathrm{L}_i^{\pm}(\tau)^*[F]\|_{\mathrm{L}^2(\partial\mathcal{G})} = 0, & \lim_{\tau\to 0+} \|J_i(\tau)^*[F]\|_{\mathrm{L}^2(\partial\mathcal{G})} = 0. \end{array} \tag{3.216}$$

Proof. Denote by $T(\tau)$ one of the operators $\mathrm{L}_i^{\pm}(\tau), i = 1,2,3$, or $J_i(\tau), i = 1,2,3,4$. Then, by virtue of the norm estimate,

$$\|F\|_{\mathrm{L}^2(\partial\mathcal{G})} \leq \sqrt{\|\partial\mathcal{G}\|}\,\|F\|_{\mathrm{C}^{(0)}(\partial\mathcal{G})}, \tag{3.217}$$

with $\|\partial\mathcal{G}\| = \int_{\partial\mathcal{G}} d\omega$ being the surface area of $\partial\mathcal{G}$, we obtain

$$\lim_{\tau\to 0+} \|T(\tau)[F]\|_{\mathrm{L}^2(\partial\mathcal{G})} = 0, \ \lim_{\tau\to 0+} \|T(\tau)^*[F]\|_{\mathrm{L}^2(\partial\mathcal{G})} = 0, \tag{3.218}$$

for all $F \in \mathrm{C}(\partial\mathcal{G})$. Therefore, there exists a constant $\mathrm{C}(F) > 0$ such that

$$\|T(\tau)[F]\|_{\mathrm{L}^2(\partial\mathcal{G})} \leq C(F), \quad \|T(\tau)^*[F]\|_{\mathrm{L}^2(\partial\mathcal{G})} \leq C(F), \tag{3.219}$$

for all $\tau \leq \tau_0$ (and τ_0 a sufficiently small constant). The uniform boundedness principle of functional analysis (see, e.g., H.W. Alt [2006]) then shows us that there exists a constant $M > 0$ such that

$$\|T(\tau)\|_{\mathrm{C}^{(0)}(\partial\mathcal{G})\to\mathrm{L}^2(\partial\mathcal{G})} \leq M, \quad \|T(\tau)^*\|_{\mathrm{C}^{(0)}(\partial\mathcal{G})\to\mathrm{L}^2(\partial\mathcal{G})} \leq M \tag{3.220}$$

for all $\tau \leq \tau_0$, where $\|T(\tau)\|_{X\to Y}$ denotes the operator norm $\sup_{F\in X} \frac{\|T(\tau)F\|_Y}{\|F\|_X}$, with $X = \mathrm{C}^{(0)}(\partial\mathcal{G})$ and $Y = \mathrm{L}^2(\partial\mathcal{G})$. In the same way we get

$$\|T(\tau)\|_{\mathrm{C}^{(0)}(\partial\mathcal{G})\to\mathrm{C}^{(0)}(\partial\mathcal{G})} \leq M, \quad \|T(\tau)^*\|_{\mathrm{C}^{(0)}(\partial\mathcal{G})\to\mathrm{C}^{(0)}(\partial\mathcal{G})} \leq M \tag{3.221}$$

for all $\tau \leq \tau_0$. The operators $(T(\tau)^*T(\tau))$ are self-adjoint, and their restrictions to the Banach space $\mathrm{C}^{(0)}(\partial\mathcal{G})$ are continuous. We now modify a technique due to P.D. Lax [1954] in the formulation by C. Müller [1969]. According to the Cauchy–Schwarz inequality, we get for $F \in \mathrm{C}^{(0)}(\partial\mathcal{G})$

$$\begin{aligned} \|T(\tau)[F]\|^2_{\mathrm{L}^2(\partial\mathcal{G})} &= (T(\tau)[F], T(\tau)[F])_{\mathrm{L}^2(\partial\mathcal{G})} \\ &= (F, (T(\tau)^*T(\tau))[F])_{\mathrm{L}^2(\partial\mathcal{G})} \\ &\leq \|F\|_{\mathrm{L}^2(\partial\mathcal{G})} \|(T(\tau)^*T(\tau))[F]\|_{\mathrm{L}^2(\partial\mathcal{G})}. \end{aligned} \tag{3.222}$$

Consequently, it follows that

$$\begin{aligned} \|T(\tau)[F]\|^{2^2}_{\mathrm{L}^2(\partial\mathcal{G})} &\leq \|F\|^2_{\mathrm{L}^2(\partial\mathcal{G})} \|T(\tau)^*T(\tau)[F]\|^2_{\mathrm{L}^2(\partial\mathcal{G})} \\ &\leq \|F\|^2_{\mathrm{L}^2(\partial\mathcal{G})} \|F\|_{\mathrm{L}^2(\partial\mathcal{G})} \|(T(\tau)^*T(\tau))^2[F]\|_{\mathrm{L}^2(\partial\mathcal{G})}. \end{aligned} \tag{3.223}$$

Iterating this procedure we find by induction

$$\|T(\tau)[F]\|^{2^n}_{\mathrm{L}^2(\partial\mathcal{G})} \leq \|F\|^{2^n-1}_{\mathrm{L}^2(\partial\mathcal{G})} \|(T(\tau)^*T(\tau))^{2^{n-1}}[F]\|_{\mathrm{L}^2(\partial\mathcal{G})} \tag{3.224}$$

for all positive integers n. Due to the norm estimate (3.217) and the boundedness of the operators $T(\tau), T(\tau)^*$ for all $\tau \leq \tau_0$, there exists a positive constant K such that

$$\|T(\tau)[F]\|^{2^n}_{\mathrm{L}^2(\partial\mathcal{G})} \leq \sqrt{\|\partial\mathcal{G}\|}\ K^{2^n} \|F\|^{2^n-1}_{\mathrm{L}^2(\partial\mathcal{G})} \|F\|_{\mathrm{C}^{(0)}(\partial\mathcal{G})}. \tag{3.225}$$

Therefore, for positive integers n and all $F \in \mathrm{C}^{(0)}(\partial\mathcal{G})$ with $F \neq 0$, we find

$$\frac{\|T(\tau)[F]\|_{\mathrm{L}^2(\partial\mathcal{G})}}{\|F\|_{\mathrm{L}^2(\partial\mathcal{G})}} \leq K \left(\frac{\sqrt{\|\partial\mathcal{G}\|}\ \|F\|_{\mathrm{C}^{(0)}(\partial\mathcal{G})}}{\|F\|_{\mathrm{L}^2(\partial\mathcal{G})}} \right)^{2^{-n}}. \tag{3.226}$$

Observing $\lim_{n\to\infty} \|F\|^{2^{-n}}_{\mathrm{C}^{(0)}(\partial\mathcal{G})} \|F\|^{-2^{-n}}_{\mathrm{L}^2(\partial\mathcal{G})} = 1$, we are therefore led to

$$\|T(\tau)F\|_{\mathrm{L}^2(\partial\mathcal{G})} \leq K\ \|F\|_{\mathrm{L}^2(\partial\mathcal{G})}, \quad \|T(\tau)^*F\|_{\mathrm{L}^2(\partial\mathcal{G})} \leq K\ \|F\|_{\mathrm{L}^2(\partial\mathcal{G})} \tag{3.227}$$

for all F of class $\mathrm{C}^{(0)}(\partial\mathcal{G})$ and $0 < \tau \leq \tau_0$. Since $\mathrm{C}^{(0)}(\partial\mathcal{G})$ is dense in $\mathrm{L}^2(\partial\mathcal{G})$, the Hahn–Banach theorem (see, e.g., L.W. Kantorowitsch, G. Akilow [1964], K. Yoshida [1980]) informs us that both operators $T(\tau) : \mathrm{L}^2(\partial\mathcal{G}) \to \mathrm{L}^2(\partial\mathcal{G})$ as well as $T(\tau)^* : \mathrm{L}^2(\partial\mathcal{G}) \to \mathrm{L}^2(\partial\mathcal{G})$ are bounded by K. Choosing F_ε of class $\mathrm{C}^{(0)}(\partial\mathcal{G})$ corresponding to $F \in \mathrm{L}^2(\partial\mathcal{G})$ such that $\|F - F_\varepsilon\|_{\mathrm{L}^2(\partial\mathcal{G})} \leq \varepsilon$, we finally arrive at the estimate

$$\begin{aligned} \lim_{\tau\to 0+} \|T(\tau)F\|_{\mathrm{L}^2(\partial\mathcal{G})} &\leq \lim_{\tau\to 0+} \|T(\tau)[F_\varepsilon]\|_{\mathrm{L}^2(\partial\mathcal{G})} + \lim_{\tau\to 0+} \|T(\tau)[F - F_\varepsilon]\|_{\mathrm{L}^2(\partial\mathcal{G})} \\ &\leq \lim_{\tau\to 0+} \|T(\tau)[F_\varepsilon]\|_{\mathrm{L}^2(\partial\mathcal{G})} + K\ \|F - F_\varepsilon\|_{\mathrm{L}^2(\partial\mathcal{G})} \\ &\leq K\ \varepsilon. \end{aligned} \tag{3.228}$$

Since $\varepsilon > 0$ can be chosen arbitrarily small, this immediately yields Theorem 3.56. □

3.4 Boundary-Value Problems

The task of solving a boundary-value problem for the Laplace equation $\Delta U = 0$ from given data on the boundary $\partial\mathcal{G}$ of a regular region $\mathcal{G}$ arises in many applications (e.g., gravitation, magnetics, solid Earth mechanics, electromagnetism). Of particular importance is the Dirichlet (resp. Neumann) boundary-value problem, i.e., the determination of U from given potential values (resp. normal derivatives) on the boundary. Finding the solution in the interior/exterior space of a geoscientifically relevant boundary (such as, e.g., sphere, ellipsoid, geoidal surface, (actual) Earth's surface) is of significance in all geosciences.

3.4.1 Formulation and Uniqueness

We begin with the formulation of the four classical boundary-value problems.

Interior Dirichlet Problem (IDP): Suppose that a function F of class $C^{(0)}(\partial\mathcal{G})$ is given. We are looking for a function $V : \overline{\mathcal{G}} \to \mathbb{R}$ satisfying the following conditions:

(i) V is of class $C^{(2)}(\mathcal{G}) \cap C^{(0)}(\overline{\mathcal{G}})$,

(ii) V satisfies Laplace's equation $\Delta V = 0$ in $\mathcal{G}$,

(iii) $V^-|_{\partial\mathcal{G}} = F$ (i.e., $V^-(x) = \lim_{\tau\to 0+} V(x - \tau\nu(x)) = F(x),\ x \in \partial\mathcal{G}$), where ν is the normal field directed into the exterior of $\mathcal{G}$).

Interior Neumann Problem (INP): Suppose that $F \in C^{(0)}(\partial\mathcal{G})$, with $\int_{\partial\mathcal{G}} F(x)\ d\omega(x) = 0$, is given. We are looking for a function $V : \overline{\mathcal{G}} \to \mathbb{R}$ satisfying the following conditions:

(i) V is of class $C^{(2)}(\mathcal{G}) \cap C^{(1)}(\overline{\mathcal{G}})$,

(ii) V satisfies Laplace's equation $\Delta V = 0$ in $\mathcal{G}$,

(iii) $\left.\frac{\partial V^-}{\partial\nu}\right|_{\partial\mathcal{G}} = F$ (i.e., $\frac{\partial V^-}{\partial\nu}(x) = \lim_{\tau\to 0+} \nu(x)\cdot(\nabla V)(x - \tau\nu(x)) = F(x), x \in \partial\mathcal{G}$, where ν is the normal field directed into the exterior of $\mathcal{G}$).

Exterior Dirichlet Problem (EDP): Suppose that $F \in C^{(0)}(\partial\mathcal{G})$ is given. We are looking for a function $V : \overline{\mathcal{G}^c} \to \mathbb{R}$ satisfying the following conditions:

(i) V is of class $C^{(2)}(\mathcal{G}^c) \cap C^{(0)}(\overline{\mathcal{G}^c})$,

(ii) V satisfies Laplace's equation $\Delta V = 0$ in $\mathcal{G}^c$,

(iii) V is regular at infinity,

(iv) $V^+|_{\partial\mathcal{G}} = F$ (i.e., $V^+(x) = \lim_{\tau\to 0+} V(x + \tau\nu(x)) = F(x), x \in \partial\mathcal{G}$, where ν is the normal field directed into the exterior of $\mathcal{G}$).

Exterior Neumann Problem (ENP): Suppose that $F \in C^{(0)}(\partial\mathcal{G})$ is given. We are looking for a function $V : \overline{\mathcal{G}^c} \to \mathbb{R}$ satisfying the following conditions:

(i) V is of class $C^{(2)}(\mathcal{G}^c) \cap C^{(1)}(\overline{\mathcal{G}^c})$,

(ii) V satisfies Laplace's equation $\Delta V = 0$ in $\mathcal{G}^c$,

(iii) V is regular at infinity,

(iv) $\left.\frac{\partial V^+}{\partial\nu}\right|_{\partial\mathcal{G}} = F$ (i.e., $\frac{\partial V^+}{\partial\nu}(x) = \lim_{\tau\to 0+} \nu(x)\cdot(\nabla V)(x + \tau\nu(x)) = F(x), x \in \partial\mathcal{G}$, where ν is the normal field directed into the exterior of $\mathcal{G}$).

Remark 3.57. The indices "+" and "−" denote whether we approach the boundary $\partial\mathcal{G}$ from the inside or the outside of $\mathcal{G}$. When it is obvious that we are referring to the exterior or the interior case, we often just write $V|_{\partial\mathcal{G}}$ or $\frac{\partial V}{\partial\nu}|_{\partial\mathcal{G}}$. Then the normal ν is the actual outer normal, i.e., it is directed into the exterior of $\mathcal{G}$ in case of interior problems in $\mathcal{G}$, and it is directed into the exterior of $\mathcal{G}^c$ (i.e., the interior of $\mathcal{G}$) in case of exterior problems in $\mathcal{G}^c$.

In the following, the *well-posedness* of the boundary-value problems, i.e., *uniqueness, existence, and continuous dependence of the boundary data,* are discussed in more detail. We begin with the uniqueness.

Lemma 3.58. *A solution of* (IDP) *is uniquely determined.*

Proof. Let V and $\tilde{V}$ satisfy (IDP). Then the difference function $D : \overline{\mathcal{G}} \to \mathbb{R}$, $D = V - \tilde{V}$, is of class $\mathrm{C}^{(2)}(\mathcal{G}) \cap \mathrm{C}^{(0)}(\overline{\mathcal{G}})$, satisfies Laplace's equation $\Delta D = 0$ in $\mathcal{G}$, and $D^-|_{\partial\mathcal{G}} = 0$. Under these assumptions the Maximum/Minimum Principle (see Theorem 3.10) tells us that $\sup_{x\in\overline{\mathcal{G}}} |D(x)| \leq \sup_{x\in\partial\mathcal{G}} |D(x)| = 0$. But this means $D = 0$ in $\overline{\mathcal{G}}$. □

Lemma 3.59. *A solution of* (INP) *is uniquely determined (up to an additive constant).*

Proof. Clearly, the difference $D : \overline{\mathcal{G}} \to \mathbb{R}$ of two solutions $V, \tilde{V}$ is of class $\mathrm{C}^{(2)}(\mathcal{G}) \cap \mathrm{C}^{(1)}(\overline{\mathcal{G}})$ and satisfies Laplace's equation $\Delta D = 0$ in $\mathcal{G}$. Moreover, we have $\frac{\partial D^-}{\partial\nu}|_{\partial\mathcal{G}} = 0$. Under these properties the Interior First Green Theorem (cf. Theorem 1.12) shows that

$$\int_{\mathcal{G}} |\nabla D(x)|^2 \, dV(x) = \int_{\partial\mathcal{G}} D(x) \underbrace{\frac{\partial D}{\partial\nu}(x)}_{=0} \, d\omega(x) = 0. \tag{3.229}$$

Thus, $\nabla D = 0$ in $\mathcal{G}$ such that $D =$const. on $\mathcal{G}$. In other words, together with V, the function $\tilde{V} = V + C$ is also a solution of (INP). □

Lemma 3.60. *A solution of* (EDP) *is uniquely determined.*

Proof. We consider the bounded region $\mathcal{G}^c \cap \mathcal{B}_R(0)$ (for R sufficiently large such that $\sup_{x\in\overline{\mathcal{G}}}|x| < R$). Then, from the Maximum/Minimum Principle (cf. Theorem 3.10), we obtain for the difference D of two solutions of (EDP) that $\sup_{x\in\overline{\mathcal{G}^c\cap\mathcal{B}_R(0)}} |D(x)| \leq \sup_{x\in\partial\mathcal{G}\cup\Omega_R} |D(x)|$. Now, we know $D^+|_{\partial\mathcal{G}} = 0$ and, because of the regularity at infinity, we have $\sup_{x\in\Omega_R} |D(x)| \to 0$ for $R \to \infty$. This implies the uniqueness. □

Lemma 3.61. *A solution of* (ENP) *is uniquely determined.*

Proof. For the difference D of two solutions of (ENP) it follows that

$$\int\limits_{\mathcal{G}^c\cap\mathcal{B}_R(0)} |\nabla D(x)|^2 \, dV(x) = \int\limits_{\partial\mathcal{G}} D(x) \underbrace{\frac{\partial D}{\partial\nu}(x)}_{=0} \, d\omega(x) + \int\limits_{\Omega_R} D(x) \frac{\partial D}{\partial\nu}(x) \, d\omega(x). \tag{3.230}$$

The regularity of D at infinity yields

$$\int_{\Omega_R} D(x) \frac{\partial D}{\partial \nu}(x) \, d\omega(x) \;=\; o(1), \qquad R \to \infty. \tag{3.231}$$

Consequently, $\int_{\mathcal{G}^c} |\nabla D(x)|^2 \, dV \;=\; 0$ so that $D \;=\;$ const. in $\overline{\mathcal{G}^c}$. Thus, the regularity at infinity gives $D = 0$ in $\overline{\mathcal{G}^c}$. □

Already at this stage, the Maximum/Minimum Principle enables us to guarantee the *continuous dependence of boundary data of the Dirichlet type.*

Lemma 3.62. *Let $V, \tilde{V} : \overline{\mathcal{G}} \to \mathbb{R}$ satisfy the inner Dirichlet problem* (IDP) *with $V^-|_{\partial\mathcal{G}} = F$ and $\tilde{V}^-|_{\partial\mathcal{G}} = \tilde{F}$, respectively, such that*

$$\sup_{x \in \partial\mathcal{G}} |F(x) - \tilde{F}(x)| \leq \varepsilon \tag{3.232}$$

for some $\varepsilon > 0$. Then,

$$\sup_{x \in \overline{\mathcal{G}}} |V(x) - \tilde{V}(x)| \leq \sup_{x \in \partial\mathcal{G}} |F(x) - \tilde{F}(x)| \leq \varepsilon. \tag{3.233}$$

Lemma 3.63. *Let $V, \tilde{V} : \overline{\mathcal{G}^c} \to \mathbb{R}$ satisfy the exterior Dirichlet problem* (EDP) *with $V|_{\partial\mathcal{G}} = F$ and $\tilde{V}|_{\partial\mathcal{G}} = \tilde{F}$, respectively, such that*

$$\sup_{x \in \partial\mathcal{G}} |F(x) - \tilde{F}(x)| \leq \varepsilon \tag{3.234}$$

for some $\varepsilon > 0$. Then, because of the regularity at infinity,

$$\sup_{x \in \overline{\mathcal{G}^c}} |V(x) - \tilde{V}(x)| \leq \sup_{x \in \partial\mathcal{G}} |F(x) - \tilde{F}(x)| \leq \varepsilon. \tag{3.235}$$

3.4.2 Boundary-Value Problems for a Ball

Next, we come to boundary-value problems corresponding to spherical boundaries. The explicit formulas for exterior problems play a significant role in geophysical and geodetic reality. First, we deal with a method of solving the exterior Dirichlet problem (EDP) by use of a so-called Green function. In a subsequent step, we apply the Kelvin transform to solve the interior Dirichlet problem (IDP). Second, we are concerned with the Neumann boundary-value problem corresponding to spherical boundaries.

Dirichlet Problem

Definition 3.64. $G^{(\mathrm{EDP})}(\Delta; \cdot, \cdot)$ is called *Green's function for $\mathbb{R}^3 \backslash \overline{\mathcal{B}_R(0)}$ with respect to the exterior Dirichlet problem* (EDP) if the following conditions are valid:

(i) for every $x \in \mathbb{R}^3 \backslash \overline{\mathcal{B}_R(0)}$,

$$y \mapsto \Phi(x,y) \;=\; G^{(\mathrm{EDP})}(\Delta;x,y) - G(\Delta;|x-y|) \tag{3.236}$$

is continuously differentiable on $\mathbb{R}^3 \backslash \mathcal{B}_R(0)$, twice continuously differentiable in $\mathbb{R}^3 \backslash \overline{\mathcal{B}_R(0)}$, harmonic in $\mathbb{R}^3 \backslash \overline{\mathcal{B}_R(0)}$, and regular at infinity, and

(ii) for every $x \in \mathbb{R}^3 \backslash \overline{\mathcal{B}_R(0)}$,

$$G^{(\mathrm{EDP})}(\Delta;x,y) = 0, \quad y \in \Omega_R. \tag{3.237}$$

Clearly, for every $x \in \mathbb{R}^3 \backslash \overline{\mathcal{B}_R(0)}$, Green's function for $\mathbb{R}^3 \backslash \overline{\mathcal{B}_R(0)}$ with respect to the exterior Dirichlet problem (EDP) is uniquely determined. The importance of the Green function is based on the following fact: Suppose that a function $U : \mathbb{R}^3 \backslash \mathcal{B}_R(0) \to \mathbb{R}$ is continuously differentiable on $\mathbb{R}^3 \backslash \mathcal{B}_R(0)$, twice continuously differentiable in $\mathbb{R}^3 \backslash \overline{\mathcal{B}_R(0)}$, harmonic in $\mathbb{R}^3 \backslash \overline{\mathcal{B}_R(0)}$, and regular at infinity. Then, from Theorem 3.27, we obtain for every $x \in \mathbb{R}^3 \backslash \overline{\mathcal{B}_R(0)}$,

$$U(x) = \int_{\Omega_R} \left(G(\Delta;|x-y|) \frac{\partial U}{\partial \nu}(y) - U(y) \frac{\partial}{\partial \nu(y)} \left(G(\Delta;|x-y|) \right) \right) d\omega(y), \tag{3.238}$$

where ν is the normal to Ω_R pointing into $\mathcal{B}_R(0)$. Moreover, from the Exterior Second Green Theorem, we are able to deduce that

$$\begin{aligned} 0 \;&=\; \int_{\mathbb{R}^3 \backslash \overline{\mathcal{B}_R(0)}} \left(\underbrace{\Delta_y U(y)}_{=0} \Phi(x,y) - U(y) \underbrace{\Delta_y \Phi(x,y)}_{=0} \right) dV(y) \\ &=\; \int_{\Omega_R} \left(\Phi(x,y) \frac{\partial U}{\partial \nu}(y) - U(y) \frac{\partial}{\partial \nu(y)} \Phi(x,y) \right) d\omega(y) \end{aligned} \tag{3.239}$$

holds for every $x \in \mathbb{R}^3 \backslash \overline{\mathcal{B}_R(0)}$. Therefore, summing up (3.238) and (3.239), we find

$$U(x) = \int_{\Omega_R} \left(\underbrace{G^{(\mathrm{EDP})}(\Delta;x,y)}_{=0} \frac{\partial U}{\partial \nu}(y) - U(y) \frac{\partial}{\partial \nu(y)} G^{(\mathrm{EDP})}(\Delta;x,y) \right) d\omega(y), \tag{3.240}$$

where ν is the unit normal field on Ω_R pointing into $\mathcal{B}_R(0)$. This leads us to the following statement: Let $U : \mathbb{R}^3 \backslash \mathcal{B}_R(0) \to \mathbb{R}$ satisfy the aforementioned assumptions with $U|_{\Omega_R} = F, F \in \mathrm{C}^{(0)}(\Omega_R)$. Then U allows the integral representation

$$U(x) \;= - \int_{\Omega_R} F(y) \frac{\partial}{\partial \nu(y)} G^{(\mathrm{EDP})}(\Delta;x,y)\, d\omega(y). \tag{3.241}$$

To achieve an explicit construction of Green's function for $\mathbb{R}^3 \backslash \overline{\mathcal{B}_R(0)}$ with respect to the exterior Dirichlet problem (EDP), we have to find, for every $x \in \mathbb{R}^3 \backslash \overline{\mathcal{B}_R(0)}$, the function $y \mapsto \Phi(x,y)$ satisfying the following properties:

(i) $\Phi(x,\cdot)$ is continuously differentiable on $\mathbb{R}^3\backslash\mathcal{B}_R(0)$, twice continuously differentiable in $\mathbb{R}^3\backslash\overline{\mathcal{B}_R(0)}$, harmonic in $\mathbb{R}^3\backslash\overline{\mathcal{B}_R(0)}$, and regular at infinity,

(ii) On the sphere Ω_R, $\Phi(x,\cdot)$ coincides with $-G(\Delta;|x-\cdot|)$.

For every $x \in \mathbb{R}^3\backslash\overline{\mathcal{B}_R(0)}$, this problem is again an exterior Dirichlet problem (EDP). It can be solved by observing that, for all $y \in \Omega_R$, the identity (1.51) holds true. Consequently, it follows that

$$G^{(\mathrm{EDP})}(\Delta;x,y) = G(\Delta;|x-y|) - \frac{R}{|x|}G\left(\Delta;\left|\frac{R^2}{|x|^2}x - y\right|\right) \tag{3.242}$$

represents Green's function for $\mathbb{R}^3\backslash\overline{\mathcal{B}_R(0)}$ with respect to the exterior Dirichlet problem (EDP). In connection with (3.241), our approach yields

$$U(x) = -\int_{\Omega_R} F(y)\frac{\partial}{\partial\nu(y)}\left(\frac{1}{4\pi}\left(\frac{1}{|x-y|} - \frac{1}{\frac{|x|}{R}\left|\frac{R^2}{|x|^2}x - y\right|}\right)\right)\,d\omega(y). \tag{3.243}$$

We have to calculate the (inner) normal derivative

$$\frac{\partial}{\partial\nu(y)}G^{(\mathrm{EDP})}(\Delta;x,y) = \nu(y)\cdot\nabla_y G^{(\mathrm{EDP})}(\Delta;x,y) = -\frac{y}{R}\cdot\nabla_y G^{(\mathrm{EDP})}(\Delta;x,y). \tag{3.244}$$

An easy manipulation yields, with $\breve{x}_R = \frac{R^2}{|x|^2}x$,

$$\nabla_y\left(\frac{1}{4\pi}\frac{1}{|x-y|} - \frac{1}{4\pi}\frac{1}{\frac{|x|}{R}|\breve{x}_R - y|}\right) = \frac{\frac{|x|^2}{R^2}y - y}{4\pi|x-y|^3} = \frac{1}{4\pi R^2}\frac{|x|^2-R^2}{|x-y|^3}y. \tag{3.245}$$

Combining (3.244) and (3.245) we obtain the so-called *Abel–Poisson formula*

$$U(x) = \frac{1}{4\pi R}\int_{\Omega_R} F(y)\,\frac{|x|^2-R^2}{|x-y|^3}\,d\omega(y). \tag{3.246}$$

All in all, the Abel–Poisson formula (3.246) solves the exterior Dirichlet problem for functions $U:\mathbb{R}^3\backslash\mathcal{B}_R(0)\to\mathbb{R}$ that are continuously differentiable on $\mathbb{R}^3\backslash\mathcal{B}_R(0)$, twice continuously differentiable in $\mathbb{R}^3\backslash\overline{\mathcal{B}_R(0)}$, harmonic in $\mathbb{R}^3\backslash\overline{\mathcal{B}_R(0)}$, and regular at infinity, such that $U|_{\Omega_R} = F, F \in \mathrm{C}^{(0)}(\Omega_R)$. From Theorem 2.20, we know that

$$F(R\xi) = \lim_{r\to R+}\frac{1}{4\pi R}\int_{\Omega_R} F(R\eta)\,\frac{r^2-R^2}{(r^2+R^2-2Rr(\xi\cdot\eta))^{\frac{3}{2}}}\,d\omega(R\eta) \tag{3.247}$$

holds true for all $F \in \mathrm{C}^{(0)}(\Omega_R)$. Consequently, even though (3.246) has been derived for solutions U of class $\mathrm{C}^{(2)}(\mathbb{R}^3\backslash\overline{\mathcal{B}_R(0)})\cap\mathrm{C}^{(1)}(\mathbb{R}^3\backslash\mathcal{B}_R(0))$, this formula gives the unique solution of the Dirichlet problem corresponding to continuous boundary values on the sphere Ω_R.

Theorem 3.65 (Exterior Abel–Poisson Formula). *Let F be of class $\mathrm{C}^{(0)}(\Omega_R)$. Then the function $U : \mathbb{R}^3\backslash\overline{\mathcal{B}_R(0)} \to \mathbb{R}$ given by*

$$U(x) = \int_{\Omega_R} F(y)\ D(x,y)\ d\omega(y), \quad x \in \mathbb{R}^3\backslash\overline{\mathcal{B}_R(0)}, \tag{3.248}$$

with the exterior Abel–Poisson kernel function

$$D(x,y) = \frac{1}{4\pi R}\frac{|x|^2 - R^2}{|x-y|^3} \tag{3.249}$$

is the unique solution of the exterior Dirichlet boundary-value problem (EDP)*:*

(i) U is continuous in $\mathbb{R}^3\backslash\mathcal{B}_R(0)$, twice continuously differentiable in $\mathbb{R}^3\backslash\overline{\mathcal{B}_R(0)}$, harmonic on $\mathbb{R}^3\backslash\overline{\mathcal{B}_R(0)}$, i.e., $\Delta U = 0$ in $\mathbb{R}^3\backslash\overline{\mathcal{B}_R(0)}$, and regular at infinity, and

(ii) $U^+|_{\Omega_R} = F$.

Remark 3.66. By use of the spherical harmonic expansion for the Abel–Poisson kernel (cf. Lemma 2.19), U can be represented by a Fourier series expansion in terms of outer harmonics

$$U = \sum_{n=0}^{\infty}\sum_{k=1}^{2n+1} F^{\wedge_{\mathrm{L}^2(\Omega_R)}}(n,k) H^R_{-n-1,k}\ , \tag{3.250}$$

where the Fourier coefficients are given by

$$F^{\wedge_{\mathrm{L}^2(\Omega_R)}}(n,k) = \int_{\Omega_R} F(y) H^R_{-n-1,k}(y)\ d\omega(y), \tag{3.251}$$

$n \in \mathbb{N}_0$, $k = 1, \ldots, 2n+1$, and the series expansion is absolutely and uniformly convergent on each set $\overline{\mathcal{K}}$ with $\mathcal{K} \Subset \mathbb{R}^3\backslash\overline{\mathcal{B}_R(0)}$.

Next we turn to the interior Dirichlet problem corresponding to continuous boundary values on the sphere Ω_R.

Theorem 3.67 (Interior Abel–Poisson Formula). *Let F be of class $\mathrm{C}^{(0)}(\Omega_R)$. Then the function $V : \mathcal{B}_R(0) \to \mathbb{R}$ given by*

$$V(x) = \int_{\Omega_R} F(y)\ D(x,y)\ d\omega(y), \quad x \in \mathcal{B}_R(0) \tag{3.252}$$

with the interior Abel–Poisson kernel function

$$D(x,y) = \frac{1}{4\pi R}\frac{R^2 - |x|^2}{|x-y|^3}, \tag{3.253}$$

is the unique solution of the interior Dirichlet boundary-value problem (IDP):

(i) *V is continuous in $\overline{\mathcal{B}_R(0)}$, twice continuously differentiable in $\mathcal{B}_R(0)$, and harmonic in $\mathcal{B}_R(0)$, i.e., $\Delta V = 0$ in $\mathcal{B}_R(0)$,*

(ii) *$V^-|_{\Omega_R} = F$.*

Proof. In order to verify the solution (3.252) of the Dirichlet problem for $\mathcal{B}_R(0)$ we use the Kelvin transform (see Theorem 1.5). We follow a two-step strategy: First we apply the Kelvin transform and introduce the new (unknown) function $V : \mathcal{B}_R(0)\backslash\{0\} \to \mathbb{R}$ defined by

$$V(x) = \frac{R}{|x|} U(\breve{x}_R). \tag{3.254}$$

We know from Lemma 3.18 that the singularity in the origin is removable. Second, we solve the corresponding exterior Dirichlet problem for U corresponding to the boundary data $U(x) = F(x)$, $x \in \Omega_R$, by means of the known exterior Abel–Poisson formula (cf. Theorem 3.65). Inserting this representation into (3.254), we obtain the desired expression (3.252). □

Setting $F = 1$ in (3.252) we find

$$\frac{1}{4\pi R} \int_{\Omega_R} \frac{R^2 - |x|^2}{|x-y|^3} \, d\omega(y) = 1, \quad x \in \mathcal{B}_R(0). \tag{3.255}$$

Remark 3.68. The solution V to (IDP) can be represented by a Fourier series expansion in terms of inner harmonics. More precisely,

$$V = \sum_{n=0}^{\infty} \sum_{k=1}^{2n+1} F^{\wedge_{\mathrm{L}^2(\Omega_R)}}(n,k) \; H^R_{n,k} \,, \tag{3.256}$$

where the Fourier coefficients are given by

$$F^{\wedge_{\mathrm{L}^2(\Omega_R)}}(n,k) = \int_{\Omega_R} F(y) H^R_{n,k}(y) \, d\omega(y), \tag{3.257}$$

$n \in \mathbb{N}_0$, $k = 1, \ldots, 2n+1$. The series expansion is absolutely and uniformly convergent on each set $\overline{\mathcal{K}}$ with $\mathcal{K} \Subset \mathcal{B}_R(0)$.

Neumann Problem

Next, we discuss the exterior Neumann problem (ENP). We start with

Definition 3.69. $G^{(\mathrm{ENP})}(\Delta; \cdot, \cdot)$ is called *Green's function for $\mathbb{R}^3 \backslash \overline{\mathcal{B}_R(0)}$ with respect to the exterior Neumann problem* (ENP), if the following conditions are valid:

(i) for every $x \in \mathbb{R}^3 \backslash \overline{\mathcal{B}_R(0)}$,

$$y \mapsto \Phi(x,y) \;=\; G^{(\mathrm{ENP})}(\Delta; x, y) - G(\Delta; |x-y|) \tag{3.258}$$

is continuously differentiable on $\mathbb{R}^3 \backslash \mathcal{B}_R(0)$, twice continuously differentiable in $\mathbb{R}^3 \backslash \overline{\mathcal{B}_R(0)}$, harmonic in $\mathbb{R}^3 \backslash \overline{\mathcal{B}_R(0)}$, and regular at infinity,

(ii) for every $x \in \mathbb{R}^3 \backslash \overline{\mathcal{B}_R(0)}$,

$$\frac{\partial}{\partial \nu(y)} G^{(\mathrm{ENP})}(\Delta; x, y) = 0 \tag{3.259}$$

i.e.,

$$\frac{\partial}{\partial \nu(y)} \Phi(x, y) = -\frac{\partial}{\partial \nu(y)} G(\Delta; |x - y|) \tag{3.260}$$

on the sphere Ω_R, where ν is the inner (unit) normal field to Ω_R.

For every $x \in \mathbb{R}^3 \backslash \overline{\mathcal{B}_R(0)}$, Green's function for $\mathbb{R}^3 \backslash \overline{\mathcal{B}_R(0)}$ with respect to the exterior Neumann problem (EDP) is uniquely determined. Suppose that a function $U : \mathbb{R}^3 \backslash \mathcal{B}_R(0) \to \mathbb{R}$ is continuously differentiable on $\mathbb{R}^3 \backslash \mathcal{B}_R(0)$, twice continuously differentiable in $\mathbb{R}^3 \backslash \overline{\mathcal{B}_R(0)}$, harmonic in $\mathbb{R}^3 \backslash \overline{\mathcal{B}_R(0)}$, and regular at infinity. Then, from analogous arguments as in the Dirichlet case, we obtain for every $x \in \mathbb{R}^3 \backslash \overline{\mathcal{B}_R(0)}$

$$U(x) = \int_{\Omega_R} \left(G^{(\mathrm{ENP})}(\Delta; x, y) \frac{\partial U}{\partial \nu}(y) - U(y) \underbrace{\frac{\partial}{\partial \nu(y)} G^{(\mathrm{ENP})}(\Delta; x, y)}_{=0} \right) d\omega(y). \tag{3.261}$$

This leads us to the following statement: Let $U : \mathbb{R}^3 \backslash \mathcal{B}_R(0) \to \mathbb{R}$ satisfy the aforementioned assumptions with $\frac{\partial U}{\partial \nu}\big|_{\Omega_R} = F, F \in \mathrm{C}^{(0)}(\Omega_R)$. Then U allows the integral representation

$$U(x) = \int_{\Omega_R} F(y)\ G^{(\mathrm{ENP})}(\Delta; x, y)\, d\omega(y), \quad x \in \mathbb{R}^3 \backslash \mathcal{B}_R(0). \tag{3.262}$$

An elementary calculation shows that

$$\Phi(x, y) = \frac{R}{4\pi |x| \left| y - \frac{R^2}{|x|^2} x \right|} + \frac{1}{4\pi R} \ln \left(\frac{|x||y| + |x| \left| y - \frac{R^2}{|x|^2} x \right| - R^2}{|x||y| + |x| \left| y - \frac{R^2}{|x|^2} x \right| + R^2} \right) \tag{3.263}$$

satisfies the conditions (i) and (ii) of Definition 3.69. This leads us to the following representation of the solution for the exterior Neumann problem (ENP).

Theorem 3.70 (Exterior Neumann Problem). *Let F be of class $\mathrm{C}^{(0)}(\Omega_R)$. Then the function $U : \mathbb{R}^3 \backslash \mathcal{B}_R(0) \to \mathbb{R}$ given by*

$$U(x) = \frac{R}{4\pi} \int_{\Omega_R} F(y)\ N(x, y) d\omega(\eta), \quad x \in \mathbb{R}^3 \backslash \mathcal{B}_R(0), \tag{3.264}$$

with the Neumann kernel function

$$N(x, y) = \frac{2R}{|x - y|} + \ln \left(\frac{|x| + |x - y| - R}{|x| + |x - y| + R} \right) \tag{3.265}$$

is the unique solution of the exterior Neumann boundary-value problem:

(i) *U is continuously differentiable in $\mathbb{R}^3 \backslash \mathcal{B}_R(0)$, twice continuously differentiable in $\mathbb{R}^3 \backslash \overline{\mathcal{B}_R(0)}$, harmonic in $\mathbb{R}^3 \backslash \overline{\mathcal{B}_R(0)}$, i.e., $\Delta U = 0$ in $\mathbb{R}^3 \backslash \overline{\mathcal{B}_R(0)}$, and regular at infinity,*

(ii) $\left. \frac{\partial U^+}{\partial \nu} \right|_{\Omega_R} = F.$

Furthermore, U can be represented by a Fourier series expansion in terms of outer harmonics

$$U = \sum_{n=0}^{\infty} \sum_{k=1}^{2n+1} \frac{R}{n+1} F^{\wedge_{\mathrm{L}^2(\Omega_R)}}(n,k) H^R_{-n-1,k}, \tag{3.266}$$

where the Fourier coefficients are given by (3.251), and the series expansion is absolutely and uniformly convergent on each set $\overline{\mathcal{K}}$ with $\mathcal{K} \Subset \mathbb{R}^3 \setminus \overline{\mathcal{B}_R(0)}$.

It should be noted that the inner Neumann problem (INP) does not play a significant role in a geomathematically reflected potential theory. Nevertheless, the introduction of its Green function is of interest from mathematical point of view. To explain this approach in more detail, we again start from the representation formula

$$U(x) = \int_{\Omega_R} \left(G^{(\mathrm{INP})}(\Delta; x, y) \frac{\partial U}{\partial \nu}(y) - U(y) \frac{\partial}{\partial \nu(y)} G^{(\mathrm{INP})}(\Delta; x, y) \right) d\omega(y) \tag{3.267}$$

for every $x \in \mathcal{B}_R(0)$, where ν is the outward directed (unit) normal field to Ω_R. If we assume that $\frac{\partial}{\partial \nu(y)} G^{(\mathrm{INP})}(\Delta; x, y) = 0$ for all $y \in \Omega_R$, then we would get

$$\int_{\Omega_R} \frac{\partial}{\partial \nu(y)} \Phi(x,y) \, d\omega(y) \;=\; -\int_{\Omega_R} \frac{\partial}{\partial \nu(y)} G(\Delta; |x-y|) \, d\omega(y) \;=\; 1 \tag{3.268}$$

for every $x \in \mathcal{B}_R(0)$. This contradicts the necessary condition

$$\int_{\Omega_R} \frac{\partial}{\partial \nu(y)} \Phi(x,y) \, d\omega(y) \;=\; 0 \tag{3.269}$$

for every $x \in \mathcal{B}_R(0)$. In other words, we may not expect to represent U in the form

$$U(x) = \int_{\Omega_R} \frac{\partial U}{\partial \nu}(y) \; G^{(\mathrm{INP})}(\Delta; x, y) \, d\omega(y). \tag{3.270}$$

Consequently, we have to modify the usual procedure. Instead of $\frac{\partial}{\partial \nu(y)} \Phi(x,y) = -\frac{\partial}{\partial \nu(y)} G(\Delta; x, y) = 0$, for all $y \in \Omega_R$, we require

$$\frac{\partial}{\partial \nu(y)} \Phi(x,y) \;=\; -\frac{\partial}{\partial \nu(y)} G(\Delta; |x-y|) - \frac{1}{4\pi R^2}, \; y \in \Omega_R, \;\; x \in \mathcal{B}_R(0). \tag{3.271}$$

In this case, the necessary condition is fulfilled, and we are led to

Definition 3.71. $G^{(\mathrm{INP})}(\Delta;\cdot,\cdot)$ is called *Green's function for $\mathcal{B}_R(0)$ with respect to the interior Neumann problem* (INP), if the following conditions are valid:

(i) for every $x \in \mathbb{R}^3 \backslash \overline{\mathcal{B}_R(0)}$,

$$y \mapsto \Phi(x,y) = G^{(\mathrm{INP})}(\Delta;x,y) - G(\Delta;|x-y|) \tag{3.272}$$

is continuously differentiable on $\overline{\mathcal{B}_R(0)}$, twice continuously differentiable in $\mathcal{B}_R(0)$, and harmonic in $\mathcal{B}_R(0)$,

(ii) for every $x \in \mathcal{B}_R(0)$,

$$\frac{\partial}{\partial\nu(y)} G^{(\mathrm{INP})}(\Delta;x,y) = -\frac{1}{4\pi R^2}, \quad y \in \Omega_R, \tag{3.273}$$

where ν is the outward directed (unit) normal field to Ω_R.

Remark 3.72. The representation formula for a solution to (INP) reads

$$U(x) = \int_{\Omega_R} \frac{\partial U}{\partial \nu}(y)\, G^{(\mathrm{INP})}(\Delta;x,y)\, d\omega(y) - \underbrace{\frac{1}{4\pi R^2}\int_{\Omega_R} U(y)\, d\omega(y)}_{=C}, \tag{3.274}$$

for $x \in \mathcal{B}_R(0)$. Therefore, U is determined only up to an additive constant C. We already know that this is characteristic of the inner Neumann problem (INP). Finally, it should be remarked that $G^{(\mathrm{INP})}(\Delta;x,y)$ is explicitly given by

$$\begin{aligned} G^{(\mathrm{INP})}(\Delta;x,y) &= \frac{1}{4\pi}\left(\frac{1}{|x-y|} + \frac{R}{|x|\left|y-\frac{R^2}{|x|^2}x\right|}\right. \\ &\quad \left. + \frac{1}{R}\ln\left(\frac{2R^2}{R^2 - x\cdot y + |x|\left|y-\frac{R^2}{|x|^2}x\right|}\right)\right). \end{aligned} \tag{3.275}$$

From a geomathematical point of view the solutions of the exterior Dirichlet and Neumann problem, respectively, can be used in the so-called *Meissl scheme* (cf. P.A. Meissl [1971], R. Rummel [1997], H. Nutz [2002], W. Freeden, M. Schreiner [2009]) to characterize upward continuation of gravitational data (in a Meissl scheme (*frequency scheme*) based on the potential coefficients; see Table 3.1).

Ω_S-level:	$U^{\wedge_{L^2(\Omega_S)}}(n,k)$	$\overset{-\frac{n+1}{S}}{\longrightarrow}$	$\left(\frac{\partial U}{\partial \nu}\right)^{\wedge_{L^2(\Omega_S)}}(n,k)$
	$\uparrow \left(\frac{R}{S}\right)^n$		$\uparrow \left(\frac{R}{S}\right)^{n+1}$
Ω_R-level:	$U^{\wedge_{L^2(\Omega_R)}}(n,k)$	$\overset{-\frac{n+1}{R}}{\longrightarrow}$	$\left(\frac{\partial U}{\partial \nu}\right)^{\wedge_{L^2(\Omega_R)}}(n,k)$

TABLE 3.1
Meissl scheme for the upward continuation from the (terrestrial) height R to the (spaceborne) height S (involving outer harmonics).

Obviously, the vertical arrows, characterizing the *upward continuation*, amount to an attenuation by the factor $\left(\frac{R}{S}\right)^n$. The opposite directions, characterizing the *downward continuation*, amount to an amplification by the factor $\left(\frac{S}{R}\right)^n$.

3.4.3 Harnack's Inequality

From the Abel–Poisson representation (3.252) we obtain an interesting inequality.

Lemma 3.73 (Harnack's Inequality). *If U is harmonic and non-negative in $\mathcal{B}_R(0)$ as well as continuous in $\overline{\mathcal{B}_R(0)}$, then*

$$R\frac{R-|x|}{(R+|x|)^2}U(0) \le U(x) \le R\frac{R+|x|}{(R-|x|)^2}U(0) \tag{3.276}$$

for all $x \in \mathcal{B}_R(0)$.

Proof. First, the triangle inequality provides $R-|x| \le |x-y| \le R+|x|$ for the denominator of the Abel–Poisson kernel (3.249). From the Mean Value Property of harmonic functions, we obtain

$$\int_{\Omega_R} U(y)\, d\omega(y) = 4\pi R^2\, U(0). \tag{3.277}$$

By combination, we get

$$\frac{R^2-|x|^2}{4\pi R}\frac{4\pi R^2}{(R+|x|)^3}U(0) \;\le\; U(x) \;\le\; \frac{R^2-|x|^2}{4\pi R}\frac{4\pi R^2}{(R-|x|)^3}U(0). \tag{3.278}$$

This yields the desired result. □

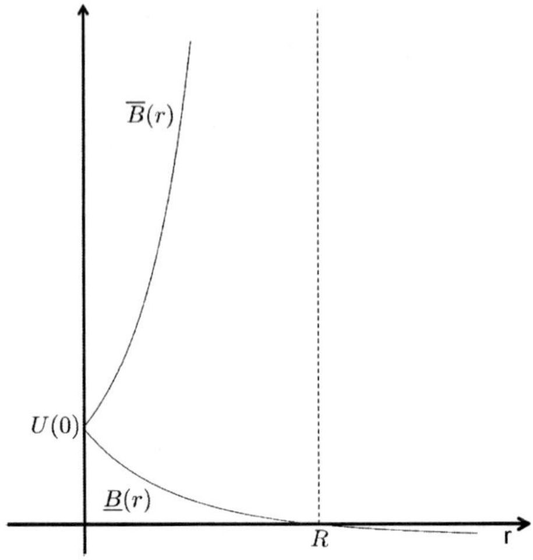

FIGURE 3.3
Graphical illustration of the bounds (3.279) in Harnack's inequality.

Harnack's inequality (Lemma 3.73) allows some interesting consequences by considering the bounds

$$\underline{B}(r) = \frac{1 - \frac{r}{R}}{(1 + \frac{r}{R})^2} U(0), \qquad \overline{B}(r) = \frac{1 + \frac{r}{R}}{(1 - \frac{r}{R})^2} U(0) \tag{3.279}$$

illustrated in Figure 3.3. Note that $\underline{B}$ possesses a zero at $r = R$ and a non-zero derivative everywhere. A simple shift of the coordinate system shows that the Harnack inequality also holds true for balls $\mathcal{B}_R(y)$, $y \neq 0$, and adapted bounds $\underline{B}$ and $\overline{B}$.

Lemma 3.74. *Let $\mathcal{G} \subset \mathbb{R}^3$ be a regular region. Suppose that $U \in \mathrm{C}^{(1)}(\overline{\mathcal{G}})$ is harmonic in $\mathcal{G}$ such that U reaches its minimum (maximum) at the point $\overline{x} \in \partial\mathcal{G}$. If there exists an open ball $\mathcal{B} \subset \mathcal{G}$ with $\overline{x} \in \partial\mathcal{B}$ such that $U(x) > U(\overline{x})$ $(U(x) < U(\overline{x}))$ for all $x \in \mathcal{B}$, then*

$$\frac{\partial U}{\partial \lambda}(\overline{x}) > 0 \qquad \left(\frac{\partial U}{\partial \lambda}(\overline{x}) < 0\right)$$

holds for every (unit) vector $\lambda(\overline{x})$ pointing from $\overline{x}$ toward $\mathcal{B}$ (see Figure 3.4).

Proof. Without loss of generality, we assume $U(\overline{x}) = 0$ to be the minimum. Moreover, the center of $\mathcal{B}$ is supposed to be the origin and the radius is R. Our intention is to replace U in $\mathcal{B}$ by its lower bound $\underline{B}$ and to estimate the directional derivative of U by the directional derivative of $\underline{B}$. The point of departure is the inequality

$$\frac{U(\overline{x} + t\lambda) - U(\overline{x})}{t} \geq \frac{\underline{B}(|\overline{x} + t\lambda|) - \underline{B}(|\overline{x}|)}{t}, \quad t > 0. \tag{3.280}$$

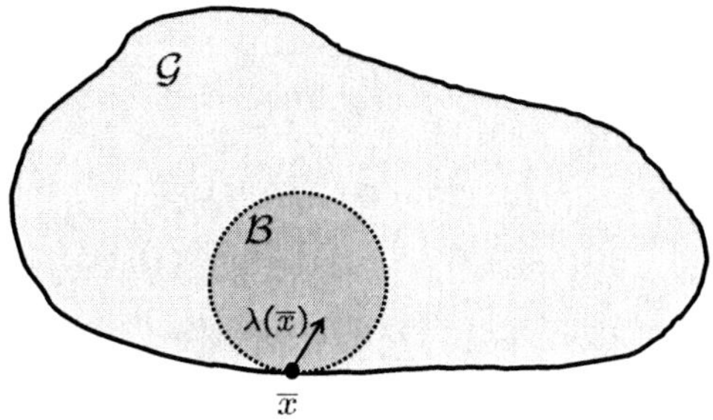

FIGURE 3.4
Illustration of the content for Lemma 3.74.

Passing to the limit $t \to 0+$ we therefore get

$$\frac{\partial U}{\partial \lambda}(\overline{x}) \geq \frac{\partial \underline{B}}{\partial \lambda}(\overline{x}). \tag{3.281}$$

Now it follows that

$$\frac{\partial \underline{B}}{\partial \lambda}(\overline{x}) = \lambda(\overline{x}) \cdot \nabla \underline{B}(|\overline{x}|) = \lambda(\overline{x}) \cdot \frac{\overline{x}}{R}\, \underline{B}'(|\overline{x}|), \tag{3.282}$$

where $\underline{B}'(|\overline{x}|) = \underline{B}'(R) < 0$ and $\lambda(\overline{x}) \cdot \overline{x} < 0$ (where $\underline{B}'$ denotes the one-dimensional derivative of $\underline{B}$). This proves Lemma 3.74. □

Another consequence of Harnack's inequality is an analogue to Liouville's Theorem known from univariate complex analysis.

Lemma 3.75. *Let U be harmonic in the entire space $\mathbb{R}^3$. If U is non-negative, then U is constant.*

Proof. Keep x fixed in Harnack's inequality and take the limit $R \to \infty$. Then we get $U(x) = U(0)$. □

We summarize our results: If U is harmonic in $\mathbb{R}^3$ and non-negative, then U is constant. Obviously, any bounded function can be shifted such that it is non-negative, and we obtain

Lemma 3.76. *Any bounded and harmonic function in $\mathbb{R}^3$ is constant.*

3.4.4 Integral Equation Method

Let $\mathcal{G}$ be a regular region. $\mathrm{Pot}(\mathcal{G})$ denotes the space of all functions $U \in \mathrm{C}^{(2)}(\mathcal{G})$ that are harmonic in $\mathcal{G}$, while $\mathrm{Pot}(\mathcal{G}^c)$ denotes the space of all functions $U \in \mathrm{C}^{(2)}(\mathcal{G}^c)$ that are harmonic in $\mathcal{G}^c$ and regular at infinity.

For $k = 0, 1, \ldots$ we denote by $\mathrm{Pot}^{(k)}(\overline{\mathcal{G}})$ the space of all $U \in \mathrm{C}^{(k)}(\overline{\mathcal{G}})$ such that U is of class $\mathrm{Pot}(\mathcal{G})$. Analogously, $\mathrm{Pot}^{(k)}(\overline{\mathcal{G}^c})$ denotes the space of all $U \in \mathrm{C}^{(k)}(\overline{\mathcal{G}^c})$ such that U is of class $\mathrm{Pot}(\mathcal{G}^c)$. In shorthand notation,

$$\mathrm{Pot}^{(k)}(\overline{\mathcal{G}}) = \mathrm{Pot}(\mathcal{G}) \cap \mathrm{C}^{(k)}(\overline{\mathcal{G}}), \tag{3.283}$$

$$\mathrm{Pot}^{(k)}(\overline{\mathcal{G}^c}) = \mathrm{Pot}(\mathcal{G}^c) \cap \mathrm{C}^{(k)}(\overline{\mathcal{G}^c}). \tag{3.284}$$

In the nomenclature of these function spaces, the boundary-value problems can be reformulated briefly as follows:

Interior Dirichlet Problem (IDP): Given a function F of class $\mathrm{C}^{(0)}(\partial\mathcal{G})$, find a function $V \in \mathrm{Pot}^{(0)}(\overline{\mathcal{G}})$ such that

$$V^-(x) = \lim_{\tau \to 0+} V(x - \tau\nu(x)) = F(x), \quad x \in \partial\mathcal{G}. \tag{3.285}$$

Interior Neumann Problem (INP): Given a function G of class $\mathrm{C}^{(0)}(\partial\mathcal{G})$ satisfying the property $\int_{\partial\mathcal{G}} G(x)\, d\omega(x) = 0$, find a function $V \in \mathrm{Pot}^{(1)}(\overline{\mathcal{G}})$ such that

$$\frac{\partial V^-}{\partial\nu}(x) = \lim_{\tau \to 0+} \nu(x) \cdot (\nabla V)(x - \tau\nu(x)) = G(x), \quad x \in \partial\mathcal{G}. \tag{3.286}$$

Exterior Dirichlet Problem (EDP): Given a function F of class $\mathrm{C}^{(0)}(\partial\mathcal{G})$, find a function $V \in \mathrm{Pot}^{(0)}(\overline{\mathcal{G}^c})$ such that

$$V^+(x) = \lim_{\tau \to 0+} V(x + \tau\nu(x)) = F(x), \quad x \in \partial\mathcal{G}. \tag{3.287}$$

Exterior Neumann Problem (ENP): Given a function G of class $\mathrm{C}^{(0)}(\partial\mathcal{G})$, find $V \in \mathrm{Pot}^{(1)}(\overline{\mathcal{G}^c})$ such that

$$\frac{\partial V^+}{\partial\nu}(x) = \lim_{\tau \to 0+} \nu(x) \cdot (\nabla V)(x + \tau\nu(x)) = G(x), \quad x \in \partial\mathcal{G}. \tag{3.288}$$

Following an idea of F. Neumann [1887] and I. Fredholm [1900], we use the double-layer potential

$$V(x) = \int_{\partial\mathcal{G}} S(y) \frac{\partial}{\partial\nu(y)} G(\Delta; |x - y|)\, d\omega(y) \tag{3.289}$$

to solve the Dirichlet problems (IDP) and (EDP), respectively. More explicitly, we search for a function $S \in \mathrm{C}^{(0)}(\partial\mathcal{G})$ such that $V^-|_{\partial\mathcal{G}} = F$ in the case if (IDP) and for a function $S \in \mathrm{C}^{(0)}(\partial\mathcal{G})$ such that $V^+|_{\partial\mathcal{G}} = F$ in the case of (EDP), where F is assumed to be of class $\mathrm{C}^{(0)}(\partial\mathcal{G})$. In fact, for every $x \in \partial\mathcal{G}$, the limit relations for the double-layer potential tell us that

$$\begin{aligned} F(x) &= V^{\pm}(x) = \lim_{\tau \to 0+} \int_{\partial\mathcal{G}} S(y) \frac{\partial}{\partial\nu(x)} G(\Delta; |x \pm \tau\nu(x) - y|)\, d\omega(y) \\ &= \pm\frac{1}{2} S(x) + \int_{\partial\mathcal{G}} S(y) \frac{\partial}{\partial\nu(y)} G(\Delta; |x - y|)\, d\omega(y). \end{aligned} \tag{3.290}$$

In an analogous way, we are interested in solving the Neumann problems (INP) and (ENP), respectively, by use of a single-layer-potential

$$U(x) = \int_{\partial \mathcal{G}} S(y)\, G(\Delta; |x-y|)\, d\omega(y). \tag{3.291}$$

Now, we look for a function $Q \in \mathrm{C}^{(0)}(\partial\mathcal{G})$ such that $\frac{\partial U^-}{\partial \nu}|_{\partial\mathcal{G}} = G$ in the case of (INP) and $\frac{\partial U^+}{\partial \nu}|_{\partial\mathcal{G}} = G$ in the case of (ENP), where G is assumed to be of class $\mathrm{C}^{(0)}(\partial\mathcal{G})$. For $x \in \partial\mathcal{G}$, the pointwise limit relations lead to the integral equations

$$\begin{aligned} G(x) &= \frac{\partial U^\pm}{\partial \nu}(x) = \lim_{\tau\to 0+} \int_{\partial\mathcal{G}} Q(y) \frac{\partial}{\partial \nu(x)} G(\Delta; |x \pm \tau\nu(x) - y|)\, d\omega(y) \\ &= \mp\frac{1}{2} Q(x) + \int_{\partial\mathcal{G}} Q(y) \frac{\partial}{\partial\nu(x)} G(\Delta; |x-y|)\, d\omega(y). \end{aligned} \tag{3.292}$$

Summarizing our results, we are confronted with the following integral equations:

$$\begin{aligned} &(\mathrm{ID}) : \frac{1}{2}S(x) - \int_{\partial\mathcal{G}} S(y)\frac{\partial}{\partial\nu(y)} G(\Delta;|x-y|)\, d\omega(y) = -F(x), \quad x\in\partial\mathcal{G}, \\ &(\mathrm{ED}) : \frac{1}{2}S(x) + \int_{\partial\mathcal{G}} S(y)\frac{\partial}{\partial\nu(y)} G(\Delta;|x-y|)\, d\omega(y) = F(x), \quad x\in\partial\mathcal{G}, \\ &(\mathrm{IN}) : \frac{1}{2}Q(x) + \int_{\partial\mathcal{G}} Q(y)\frac{\partial}{\partial\nu(x)} G(\Delta;|x-y|)\, d\omega(y) = G(x), \quad x\in\partial\mathcal{G}, \\ &(\mathrm{EN}) : \frac{1}{2}Q(x) - \int_{\partial\mathcal{G}} Q(y)\frac{\partial}{\partial\nu(x)} G(\Delta;|x-y|)\, d\omega(y) = -G(x), \quad x\in\partial\mathcal{G}. \end{aligned}$$

At this stage, it is useful to remember the potential operators $P(\tau,0)$, $P_{|1}(\tau,0)$, and $P_{|2}(\tau,0)$ from (3.204), (3.205), and (3.206), respectively. It is important to note that these linear operators map $\mathrm{C}^{(0)}(\partial\mathcal{G})$ into itself and that they are bounded with respect to $\|\cdot\|_{\mathrm{C}^{(0)}(\partial\mathcal{G})}$. Additionally, we know from (3.207) that $P_{|2}(0,0)^* = P_{|1}(0,0)$ and $P_{|1}(0,0)^* = P_{|2}(0,0)$. Thus, we are able to reformulate the previous integral equations in operator form as follows:

$$(\mathrm{ID}) : \frac{1}{2}S - P_{|2}(0,0)[S] = -F, \qquad S,F \in \mathrm{C}^{(0)}(\partial\mathcal{G}), \tag{3.293}$$

$$(\mathrm{ED}) : \frac{1}{2}S + P_{|2}(0,0)[S] = F, \qquad S,F \in \mathrm{C}^{(0)}(\partial\mathcal{G}), \tag{3.294}$$

$$(\mathrm{IN}) : \frac{1}{2}Q + P_{|2}(0,0)^*[Q] = G, \qquad Q,G \in \mathrm{C}^{(0)}(\partial\mathcal{G}), \tag{3.295}$$

$$(\mathrm{EN}) : \frac{1}{2}Q - P_{|2}(0,0)^*[Q] = -G \qquad Q,G \in \mathrm{C}^{(0)}(\partial\mathcal{G}). \tag{3.296}$$

All in all, we get two pairs of adjoint integral equations, viz. (ID),(EN) and

(ED), (IN), where the second integral equation is adjoint to the first integral equation. This is characteristic of so-called Fredholm integral equations. Indeed, there is a huge literature on Fredholm integral equations leading to the so-called Fredholm Alternative (see, for example, S.G. Michlin [1975]). As a matter of fact, in the middle of the last century, the specification of the Fredholm Alternative was a decisive step in creating a particular discipline of mathematics, namely functional analysis. In the following, we do not provide all functional analytic facets, but only provide the information (without proofs) required throughout this book. We adopt the Fredholm Alternative from functional analysis in a formulation (see, e.g., S.G. Michlin [1975]) that is particularly suited for our potential theoretic purposes.

Theorem 3.77 (Fredholm Alternative). *Let* $T : \mathrm{C}^{(0)}(\partial\mathcal{G}) \to \mathrm{C}^{(0)}(\partial\mathcal{G})$ *(or* $T : \mathrm{L}^2(\partial\mathcal{G}) \to \mathrm{L}^2(\partial\mathcal{G})$*) be a* linear, completely continuous operator, *i.e., for any sequence* $\{Q_n\}_{n\in\mathbb{N}} \subset \mathrm{C}^{(0)}(\partial\mathcal{G})$ *with* $\|Q_n\|_{\mathrm{C}^{(0)}(\partial\mathcal{G})} \leq 1$, $n \in \mathbb{N}$, *the sequences* $\{T[Q_n]\}_{n\in\mathbb{N}} \subset \mathrm{C}^{(0)}(\partial\mathcal{G})$ *and* $\{T^*[Q_n]\}_{n\in\mathbb{N}} \subset \mathrm{C}^{(0)}(\partial\mathcal{G})$ *contain convergent subsequences (the same should hold true if we substitute* $\mathrm{C}^{(0)}(\partial\mathcal{G})$ *by* $\mathrm{L}^2(\partial\mathcal{G})$*). Suppose that* $\lambda \neq 0$ *is a given real number. Then the following alternative holds true:*

(a) The homogeneous equation

$$T[S] + \lambda S = 0 \tag{3.297}$$

permits only the trivial solution. Then

$$T^*[Q] + \lambda Q = 0 \tag{3.298}$$

also allows only the trivial solution. Moreover,

$$T[S] + \lambda S = F, \qquad T^*[Q] + \lambda Q = G \tag{3.299}$$

possess a unique solution for every $F, G \in \mathrm{C}^{(0)}(\partial\mathcal{G})$ *(or* $F, G \in \mathrm{L}^2(\partial\mathcal{G})$*).*

(b) The homogeneous equation

$$T[S] + \lambda S = 0 \tag{3.300}$$

has non-trivial solutions. Then (3.300) permits only a finite number of linearly independent solutions $S_1, \ldots, S_p$. *In addition, the number of linearly independent solutions of*

$$T^*[Q] + \lambda Q = 0 \tag{3.301}$$

is also $p:$ $Q_1, \ldots, Q_p$. *In this case the inhomogeneous equation*

$$T[S] + \lambda S = F \tag{3.302}$$

is solvable for $F \in \mathrm{C}^{(0)}(\partial\mathcal{G})$ *(or* $F \in \mathrm{L}^2(\partial\mathcal{G})$*) if and only if*

$$(F, Q_i)_{\mathrm{L}^2(\partial\mathcal{G})} \quad = \quad \int_{\partial\mathcal{G}} F(y)\ Q_i(y)\ d\omega(y) \quad = \quad 0,\ i = 1, \ldots, p. \tag{3.303}$$

If S_{part} *is a particular solution of (3.302), then any solution of (3.302) can be represented in the form*

$$S_{\mathrm{part}} \ + \ \sum_{k=1}^{p} c_k S_k \tag{3.304}$$

with coefficients $c_k \in \mathbb{R}$*. Correspondingly, the inhomogeneous equation*

$$T^*[Q] + \lambda Q \ = \ G \tag{3.305}$$

is solvable for $G \in \mathrm{C}^{(0)}(\partial\mathcal{G})$ *(or* $G \in \mathrm{L}^2(\partial\mathcal{G})$*) if and only if*

$$(G, S_i)_{\mathrm{L}^2(\partial\mathcal{G})} \quad = \quad \int_{\partial\mathcal{G}} G(y)\ S_i(y)\ d\omega(y) \quad = \quad 0,\ i = 1, \ldots, p. \tag{3.306}$$

If Q_{part} *is a particular solution of (3.305), then any solution of (3.305) can be represented in the form*

$$Q_{\mathrm{part}} \ + \ \sum_{k=1}^{p} d_k Q_k \tag{3.307}$$

with coefficients $d_k \in \mathbb{R}$*.*

To be more specific, suppose that T denotes one of the operators $\pm P_{|2}(0,0)$. We know that T as well as T^* map $\mathrm{C}^{(0)}(\partial\mathcal{G})$ into itself. Furthermore, the famous Theorem of Arzéla-Ascoli (see, e.g., S.G. Michlin [1975]), in connection with (3.178), implies that both operators are completely continuous. Under this condition, the Fredholm Alternative is applicable for our potential operators.

We are now prepared to discuss the solvability of the aforementioned pairs (ID), (EN) and (IN), (ED) of integral equations. Already known potential theoretic results serve as essential tools for the discussion. We begin with the pair (ID), (EN) of adjoint equations (remember $P_{|1}(0,0) = P_{|2}(0,0)^*$, $P_{|1}(0,0)^* = P_{|2}(0,0)$):

$$\text{(ID)} \ : \ \frac{1}{2}S - P_{|2}(0,0)[S] \ = -F, \qquad S, F \in \mathrm{C}^{(0)}(\partial\mathcal{G}), \tag{3.308}$$

$$\text{(EN)} \ : \ \frac{1}{2}Q - P_{|2}(0,0)^*[Q] = -G, \qquad Q, G \in \mathrm{C}^{(0)}(\partial\mathcal{G}). \tag{3.309}$$

Theorem 3.78. *The homogeneous integral equation*

$$\frac{1}{2}S - P_{|2}(0,0)[S] = 0, \quad S \in \mathrm{C}^{(0)}(\partial\mathcal{G}), \tag{3.310}$$

has only a trivial solution in $\mathrm{C}^{(0)}(\partial\mathcal{G})$*.*

Proof. Let S be a solution of the integral equation (3.310). Suppose that U is the corresponding double-layer potential

$$U(x) = \int_{\partial\mathcal{G}} S(y) \frac{\partial}{\partial\nu(y)} G(\Delta; |x-y|)\, d\omega(y), \ x \in \mathcal{G}. \tag{3.311}$$

It follows (see (3.310)) that

$$\frac{1}{2}S(x) - \int_{\partial\mathcal{G}} S(y) \frac{\partial}{\partial\nu(y)} G(\Delta; |x-y|)\, d\omega(y) = 0, \ x \in \partial\mathcal{G}. \tag{3.312}$$

Since we are concerned with the inner Dirichlet boundary-value problem (IDP), we are able to conclude that U is of class $\mathrm{Pot}^{(0)}(\overline{\mathcal{G}})$ and

$$U(x) = \frac{1}{2}S(x) - \int_{\partial\mathcal{G}} S(y) \frac{\partial}{\partial\nu(y)} G(\Delta; |x-y|)\, d\omega(y) = 0, \quad x \in \partial\mathcal{G}. \tag{3.313}$$

From the Maximum/Minimum Principle it therefore follows that $U = 0$ in $\overline{\mathcal{G}}$. Thus, we have

$$\lim_{\tau\to 0+} \frac{\partial U}{\partial\nu}(x - \tau\nu(x)) = 0, \quad x \in \partial\mathcal{G}. \tag{3.314}$$

By the jump relation for the normal derivative of the double-layer potential we then get

$$\lim_{\tau\to 0+} \frac{\partial U}{\partial\nu}(x + \tau\nu(x)) = 0, \quad x \in \partial\mathcal{G}. \tag{3.315}$$

Due to the regularity at infinity of U and the unique solvability of (ENP), this yields $U(x) = 0$, $x \in \mathcal{G}^c$. Consequently, the jump relation for the double-layer potential implies

$$\lim_{\tau\to 0+} (\underbrace{U(x+\tau\nu(x))}_{=0} - \underbrace{U(x-\tau\nu(x))}_{=0}) = S(x), \quad x \in \partial\mathcal{G}. \tag{3.316}$$

Hence, $S = 0$ on $\partial\mathcal{G}$. This is the required result. □

Summarizing our considerations for the pair (ID), (EN), we are able to formulate the following statements:

Theorem 3.79. *For $F \in \mathrm{C}^{(0)}(\partial\mathcal{G})$, the inner Dirichlet problem* (IDP) *is uniquely solvable, and the solution can be represented in the form*

$$U(x) = \int_{\partial\mathcal{G}} S(y)\, \frac{\partial}{\partial\nu(y)} G(\Delta; |x-y|)\, d\omega(y), \ x \in \mathcal{G}, \tag{3.317}$$

where $S \in \mathrm{C}^{(0)}(\partial\mathcal{G})$ satisfies the integral equation (ID)

$$\frac{1}{2}S(x) - \int_{\partial\mathcal{G}} S(y)\, \frac{\partial}{\partial\nu(y)} G(\Delta; |x-y|)\, d\omega(y) = -F(x), \quad x \in \partial\mathcal{G}. \tag{3.318}$$

For $G \in \mathrm{C}^{(0)}(\partial\mathcal{G})$, the exterior Neumann problem (ENP) *is uniquely solvable, and the solution can be represented in the form*

$$U(x) = \int_{\partial\mathcal{G}} Q(y)\; G(\Delta; |x-y|)\; d\omega(y), \;\; x \in \mathcal{G}^c, \tag{3.319}$$

where $Q \in \mathrm{C}^{(0)}(\partial\mathcal{G})$ satisfies the integral equation (EN)

$$\frac{1}{2}Q(x) - \int_{\partial\mathcal{G}} Q(y)\; \frac{\partial}{\partial\nu(x)}G(\Delta; |x-y|)\; d\omega(y) = -G(x), \qquad x \in \partial\mathcal{G}. \tag{3.320}$$

Next we turn to the second pair (ED), (IN) of integral equations

$$\text{(ED)} \;\;:\;\; \frac{1}{2}S + P_{|2}(0,0)[S] \;= F, \qquad S, F \in \mathrm{C}^{(0)}(\partial\mathcal{G}), \tag{3.321}$$

$$\text{(IN)} \;\;:\;\; \frac{1}{2}Q + P_{|2}(0,0)^*[Q] = G, \qquad Q, G \in \mathrm{C}^{(0)}(\partial\mathcal{G}). \tag{3.322}$$

The homogeneous equation associated with (ED) is given by

$$\frac{1}{2}S(x) + \int_{\partial\mathcal{G}} S(y)\; \frac{\partial}{\partial\nu(y)}G(\Delta; |x-y|)\, d\omega(y) = 0 \tag{3.323}$$

For $x \in \partial\mathcal{G}$, we know from Lemma 3.5 that

$$\int_{\partial\mathcal{G}} \frac{\partial}{\partial\nu(y)}G(\Delta; |x-y|)\, d\omega(y) = -\frac{1}{2}. \tag{3.324}$$

Thus, the homogeneous equation (ED) has a non-trivial solution, viz. $\tilde{S} = 1$. Consequently, following the Fredholm Alternative, the homogeneous integral equation $\frac{1}{2}Q + P^*_{|2}(0,0)[Q] = 0$ associated with (IN) has at least one non-trivial solution, too. We denote this non-trivial solution by $\tilde{Q}$.

Lemma 3.80. *The homogeneous integral equations*

$$\text{(ED)} \;\;:\;\; \frac{1}{2}S + P_{|2}(0,0)[S] \;= 0, \qquad S \in \mathrm{C}^{(0)}(\partial\mathcal{G}), \tag{3.325}$$

$$\text{(IN)} \;\;:\;\; \frac{1}{2}Q + P_{|2}(0,0)^*[Q] = 0, \qquad Q \in \mathrm{C}^{(0)}(\partial\mathcal{G}) \tag{3.326}$$

permit no non-trivial solutions that are linearly independent of $\tilde{S}$ and $\tilde{Q}$, respectively.

Proof. In accordance with the Fredholm Alternative, it is sufficient to show this property for the second Equation (3.326). We start with the single-layer potential

$$\tilde{U}(x) = \int_{\partial\mathcal{G}} \tilde{Q}(y)\; G(\Delta; |x-y|)\, d\omega(y), \;\; x \in \mathcal{G}. \tag{3.327}$$

For $x \in \partial\mathcal{G}$, we have $\frac{\partial \tilde{U}}{\partial \nu}(x) = \frac{1}{2}\tilde{Q}(x) + P_{|2}(0,0)^*[\tilde{Q}](x) = 0$. Since $\tilde{U}$ is harmonic in $\mathcal{G}$ we can conclude by the uniqueness of the problem (INP) that $\tilde{U}(x) = C_1 = \text{const.}$ for $x \in \overline{\mathcal{G}}$, where $C_1 \neq 0$ (because otherwise $\tilde{Q} = 0$). Now, let us assume that $\frac{1}{2}Q + P_{|2}(0,0)^*[Q] = 0$ has a second non-trivial solution $\tilde{\tilde{Q}}$, for which we consider

$$\tilde{\tilde{U}}(x) = \int_{\partial\mathcal{G}} \tilde{\tilde{Q}}(y) G(\Delta; |x-y|)\, d\omega(y). \tag{3.328}$$

By the same arguments it follows that $\tilde{\tilde{U}} = C_2 = \text{const.} \neq 0$ in $\overline{\mathcal{G}}$. Finally, we consider $\overline{Q}$ given by $\overline{Q} = C_2\tilde{Q} - C_1\tilde{\tilde{Q}}$. Then $\overline{Q}$ solves $\frac{1}{2}\overline{Q} + P_{|2}(0,0)^*[\overline{Q}] = 0$. If $\bar{U}(x) = \int_{\partial\mathcal{G}} \overline{Q}(y) G(\Delta; |x-y|)\, d\omega(y)$, then we have $\bar{U} = C_2\tilde{U} - C_1\tilde{\tilde{U}}$. In other words, $\bar{U}(x) = C_2\tilde{U}(x) - C_1\tilde{\tilde{U}}(x) = C_2C_1 - C_1C_2 = 0$, $x \in \overline{\mathcal{G}}$. But this means that $\overline{Q} = 0$ on $\partial\mathcal{G}$; hence, $\tilde{\tilde{Q}} = \frac{C_2}{C_1}\tilde{Q}$. This is the desired result. □

Now we are in a position to deal with the solvability of (IN): The integral equation $\frac{1}{2}Q + P_{|2}(0,0)^*[Q] = G$ is solvable if and only if $G \in \mathrm{C}^{(0)}(\partial\mathcal{G})$ is orthogonal to all linearly independent solutions of the homogeneous adjoint equation $\frac{1}{2}S + P_{|2}(0,0)[S] = 0$. As we have seen before, this equation allows only one non-trivial linearly independent solution, namely $\tilde{S} = 1$. Consequently, for the solvability of (INP), it is necessary and sufficient to have the condition

$$\int_{\partial\mathcal{G}} G(y)\, \underbrace{\tilde{S}(y)}_{=1}\, d\omega(y) = \int_{\partial\mathcal{G}} G(y)\, d\omega(y) = 0. \tag{3.329}$$

In this case, the solution of (INP) can be represented as a single-layer potential.

Next, we come to the integral equation (ED): Considering the equation $\frac{1}{2}S + P_{|2}(0,0)[S] = F$, we are confronted with the solvability condition

$$(F, \tilde{Q})_{\mathrm{L}^2(\partial\mathcal{G})} = \int_{\partial\mathcal{G}} F(y)\tilde{Q}(y)\, d\omega(y) = 0. \tag{3.330}$$

If this condition is satisfied, then the integral equation is solvable. In this case, there exists a solution of the boundary-value problem (EDP) that is representable as a double-layer potential. However, we know that the double-layer potential is of order $\mathcal{O}\left(|x|^{-2}\right)$ for $|x| \to \infty$. The term $\mathcal{O}\left(|x|^{-2}\right)$ guarantees the regularity at infinity, but it states a faster decay than necessary. If the orthogonality condition $(F, \tilde{Q})_{\mathrm{L}^2(\partial\mathcal{G})} = 0$ is violated, then there exists no solution of the boundary-value problem (EDP) that is representable as a double-layer potential. But this does not mean that (EDP) is unsolvable. In order to find a solution of the boundary-value problem (EDP), which behaves at infinity like $\mathcal{O}\left(|x|^{-1}\right)$, we make a modified ansatz

$$U(x) = \int_{\partial\mathcal{G}} S(y) \frac{\partial}{\partial \nu(y)} G(\Delta; |x-y|)\, d\omega(y) + G(\Delta; |x|) \int_{\partial\mathcal{G}} S(y)\, d\omega(y). \tag{3.331}$$

Our goal is to determine $S \in C^{(0)}(\partial\mathcal{G})$ such that $U^+|_{\partial\mathcal{G}} = F$. In fact, we get the following (modified) integral equation (MED) for the boundary-value problem (EDP)

$$(\text{MED}) : \frac{1}{2}S(x) + \int_{\partial\mathcal{G}} S(y) \left(\frac{\partial}{\partial\nu(y)} G(\Delta; |x-y|) + G(\Delta; |x|) \right) d\omega(y) = F(x) \tag{3.332}$$

for $x \in \partial\mathcal{G}$. The Fredholm theory can also be applied to the operator T defined by

$$T[S](x) = \int_{\partial\mathcal{G}} S(y) \left(\frac{\partial}{\partial\nu(y)} G(\Delta; |x-y|) + G(\Delta; |x|) \right) d\omega(y), \quad x \in \partial\mathcal{G} \tag{3.333}$$

(note that the operator T is completely continuous). For the solvability we have to consider the homogeneous equation $(\frac{1}{2}I + A + P_{|2}(0,0))[S] = 0$, where the operator A is given by

$$A[S](x) = G(\Delta; |x|) \int_{\partial\mathcal{G}} S(y)\, d\omega(y), \quad x \in \partial\mathcal{G}, \tag{3.334}$$

such that $T = A + P_{|2}(0,0)$. Let $\tilde{S}$ be a solution of the homogeneous equation (MED). For $\tilde{U}$ given by

$$\tilde{U}(x) = \int_{\partial\mathcal{G}} \tilde{S}(y) \left(\frac{\partial}{\partial\nu(y)} G(\Delta; |x-y|) + G(\Delta; |x|) \right) d\omega(y), \quad x \in \mathbb{R}^3 \setminus \partial\mathcal{G}, \tag{3.335}$$

we find

$$\tilde{U}^+|_{\partial\mathcal{G}} = \left(\frac{1}{2}I + A + P_{|2}(0,0) \right) [\tilde{S}] = 0. \tag{3.336}$$

Since $\tilde{U}$ is regular at infinity, we get from the uniqueness theorem for the exterior Dirichlet problem that $\tilde{U} = 0$ in $\overline{\mathcal{G}^c}$. Furthermore, the double-layer potential is of the order $\mathcal{O}(|x|^{-2})$ at infinity, so that $|x|\tilde{U}(x) = 0$, $x \in \mathcal{G}^c$, implies $\int_{\partial\mathcal{G}} \tilde{S}(y) d\omega(y) = 0$. Therefore, for $F = 0$, Equation (3.332) reduces to

$$\frac{1}{2}\tilde{S}(x) + \int_{\partial\mathcal{G}} \tilde{S}(y) \frac{\partial}{\partial\nu(y)} G(\Delta; |x-y|)\, d\omega(y) = 0, \quad x \in \partial\mathcal{G}. \tag{3.337}$$

From Lemma 3.80 we know that the general solution of (3.337) is $\tilde{S} = \text{const.}$, so that $\int_{\partial\mathcal{G}} \tilde{S}(y) d\omega(y) = 0$ implies $\tilde{S}(x) = 0$, $x \in \partial\mathcal{G}$, as required. Summarizing our considerations we are therefore led to the following conclusion:

Theorem 3.81. *For $F \in C^{(0)}(\partial\mathcal{G})$, the exterior Dirichlet problem* (EDP) *is uniquely solvable, and the solution can be represented in the form*

$$U(x) = \int_{\partial\mathcal{G}} S(y) \left(\frac{\partial}{\partial\nu(y)} G(\Delta; |x-y|) + G(\Delta; |x|) \right) d\omega(y), \quad x \in \mathcal{G}^c, \tag{3.338}$$

where $S \in \mathrm{C}^{(0)}(\partial\mathcal{G})$ *satisfies the integral equation* (MED)

$$\frac{1}{2}S(x) + \int_{\partial\mathcal{G}} S(y) \left(\frac{\partial}{\partial\nu(y)} G(\Delta; |x-y|) + G(\Delta; |x|) \right) d\omega(y) = F(x), \quad x \in \partial\mathcal{G}. \tag{3.339}$$

Finally, we summarize the role of layer potentials for the classical Dirichlet and Neumann boundary-value problems. We restrict ourselves to the geoscientifically more relevant exterior problems.

(EDP): Let D^+ denote the *boundary-value space* of the Dirichlet type:

$$D^+ = \left\{ V^+|_{\partial\mathcal{G}} : V \in \mathrm{Pot}^{(0)}\left(\overline{\mathcal{G}^c}\right) \right\}. \tag{3.340}$$

The exterior Dirichlet problem (EDP) is always uniquely determined, hence,

$$D^+ = \mathrm{C}^{(0)}(\partial\mathcal{G}). \tag{3.341}$$

The solution of the problem $V \in \mathrm{Pot}^{(0)}\left(\overline{\mathcal{G}^c}\right), V^+|_{\partial\mathcal{G}} = F, F \in \mathrm{C}^{(0)}(\partial\mathcal{G})$, can be formulated in terms of a potential of the form

$$V(x) = \int_{\partial\mathcal{G}} S(y) \frac{\partial}{\partial\nu(y)} \left(G(\Delta; |x-y|) + G(\Delta; |x|) \right) d\omega(y), \quad S \in \mathrm{C}^{(0)}(\partial\mathcal{G}), \tag{3.342}$$

where S satisfies the integral equation

$$F = V^+|_{\partial\mathcal{G}} = \left(\frac{1}{2}I + A + P_{|2}(0,0) \right)[S], \quad F \in \mathrm{C}^{(0)}(\partial\mathcal{G}). \tag{3.343}$$

Furthermore, we have

$$\mathrm{kern}\left(\left(\frac{1}{2}I + A + P_{|2}(0,0) \right)^* \right) = \{0\}, \tag{3.344}$$

$$\left(\frac{1}{2}I + A + P_{|2}(0,0) \right)\left[\mathrm{C}^{(0)}(\partial\mathcal{G}) \right] = D^+. \tag{3.345}$$

Finally, by completion, we obtain

$$\mathrm{L}^2(\partial\mathcal{G}) = \overline{D^+}^{\|\cdot\|_{\mathrm{L}^2(\partial\mathcal{G})}} = \overline{\mathrm{C}^{(0)}(\partial\mathcal{G})}^{\|\cdot\|_{\mathrm{L}^2(\partial\mathcal{G})}}. \tag{3.346}$$

(ENP): Let N^+ denote the *boundary-value space* of the Neumann type:

$$N^+ = \left\{ \left.\frac{\partial V^+}{\partial\nu}\right|_{\partial\mathcal{G}} : V \in \mathrm{Pot}^{(1)}(\overline{\mathcal{G}^c}) \right\}. \tag{3.347}$$

The problem (ENP) is always uniquely determined, hence,

$$N^+ = \mathrm{C}^{(0)}(\partial\mathcal{G}). \tag{3.348}$$

The solution of the problem $V \in \mathrm{Pot}^{(1)}\left(\overline{\mathcal{G}^c}\right), \frac{\partial V^+}{\partial \nu}|_{\partial\mathcal{G}} = F, F \in \mathrm{C}^{(0)}(\partial\mathcal{G})$, can be formulated in terms of a single-layer potential

$$V(x) = \int_{\partial\mathcal{G}} Q(y)\; G(\Delta; |x-y|)\; d\omega(y), \quad Q \in \mathrm{C}^{(0)}(\partial\mathcal{G}), \tag{3.349}$$

where Q satisfies the integral equations

$$F = \frac{\partial V^+}{\partial \nu}\Big|_{\partial\mathcal{G}} = \left(-\frac{1}{2}I + P_{|1}(0,0)\right) Q. \tag{3.350}$$

We have

$$\begin{aligned} \mathrm{kern}\left(\left(-\frac{1}{2}I + P_{|1}(0,0)\right)^*\right) &= \{0\}, && (3.351)\\ \left(-\frac{1}{2}I + P_{|1}(0,0)\right)\left[\mathrm{C}^{(0)}(\partial\mathcal{G})\right] &= N^+. && (3.352)\end{aligned}$$

Again, by completion, we get

$$\mathrm{L}^2(\partial\mathcal{G}) = \overline{N^+}^{\|\cdot\|_{\mathrm{L}^2(\partial\mathcal{G})}} = \overline{\mathrm{C}^{(0)}(\partial\mathcal{G})}^{\|\cdot\|_{\mathrm{L}^2(\partial\mathcal{G})}}. \tag{3.353}$$

Similar arguments, of course, hold true for the inner boundary-value problems. The details are left to the reader.

3.4.5 Regularity Theorems

Let U be of class $\mathrm{Pot}^{(0)}(\overline{\mathcal{G}^c})$. Then the Maximum/Minimum Principle for the outer space $\mathcal{G}^c$ states

$$\sup_{x\in\overline{\mathcal{G}^c}} |U(x)| \leq \sup_{x\in\partial\mathcal{G}} |U(x)|. \tag{3.354}$$

From the theory of integral equations, we are able to verify the *continuous dependence of Neumann boundary data.*

Lemma 3.82. *Let U be of class $\mathrm{Pot}^{(1)}(\overline{\mathcal{G}^c})$. Then there exists a constant C (dependent on $\partial\mathcal{G}$) such that*

$$\sup_{x\in\overline{\mathcal{G}^c}} |U(x)| \leq C \sup_{x\in\partial\mathcal{G}} \left|\frac{\partial U^+}{\partial \nu}(x)\right|. \tag{3.355}$$

Proof. Let U be given as a single-layer potential (3.349) with density function $Q \in \mathrm{C}^{(0)}(\partial\mathcal{G})$ such that $\frac{\partial U^+}{\partial\nu}|_{\partial\mathcal{G}} = F$ with $F \in \mathrm{C}^{(0)}(\partial\mathcal{G})$. From (3.351) it follows that the operator $T = \frac{1}{2}I + P_{|1}(0,0)$ is invertible on $\mathrm{C}^{(0)}(\partial\mathcal{G})$ such that $Q = T^{-1}F$.

We have to prove (cf. S.G. Michlin [1975]) that T^{-1} is bounded with respect to $\|\cdot\|_{\mathrm{C}^{(0)}(\partial\mathcal{G})}$. Suppose the contrary. Then there exists a sequence $\{F_n\}_{n\in\mathbb{N}} \subset \mathrm{C}^{(0)}(\partial\mathcal{G})$ such that $\|Q_n\|_{\mathrm{C}^{(0)}(\partial\mathcal{G})} \geq n\, \|F_n\|_{\mathrm{C}^{(0)}(\partial\mathcal{G})}$, $Q_n = T^{-1}[F_n]$.

Put $\tilde{F}_n = \|Q_n\|^{-1}_{C^{(0)}(\partial\mathcal{G})} F_n$, $\tilde{Q}_n = \|Q_n\|^{-1}_{C^{(0)}(\partial\mathcal{G})} Q_n$. Then $\tilde{F}_n = T[\tilde{Q}_n]$, $\|\tilde{Q}_n\|_{C^{(0)}(\partial\mathcal{G})} = 1$, $\|\tilde{F}_n\|_{C^{(0)}(\partial\mathcal{G})} \to 0$. We know that $P_{|1}(0,0)$ is a completely continuous operator in $C^{(0)}(\partial\mathcal{G})$. The sequence $\{\tilde{Q}_n\}_{n\in\mathbb{N}} \subset C^{(0)}(\partial\mathcal{G})$ is bounded (note that $\|\tilde{Q}_n\|_{C^{(0)}(\partial\mathcal{G})} = 1$). We are able to pick out a subsequence $(\tilde{Q}_{n_k})_{k\in\mathbb{N}} \subset C^{(0)}(\partial\mathcal{G})$ such that $\lim_{k\to\infty} P_{|1}(0,0)[\tilde{Q}_{n_k}] = \tilde{Q}$ exists. In addition, we have $\tilde{F}_{n_k} = T[\tilde{Q}_{n_k}] \to T[\tilde{Q}] = 0$, i.e., $\frac{1}{2}\tilde{Q} + P_{|1}(0,0)[\tilde{Q}] = 0$. We know that the last equation has a unique solution, namely $\tilde{Q} = 0$. However, $0 = \|\tilde{Q}\|_{C^{(0)}(\partial\mathcal{G})} = \lim_{k\to\infty} \|\tilde{Q}_{n_k}\|_{C^{(0)}(\partial\mathcal{G})} = 1$. Thus we are confronted with a contradiction, which shows that T^{-1} is bounded with respect to $\|\cdot\|_{C^{(0)}(\partial\mathcal{G})}$.

In other words, there exists a constant $\tilde{C}$ such that $\|Q\|_{C^{(0)}(\partial\mathcal{G})} \leq \tilde{C}\|F\|_{C^{(0)}(\partial\mathcal{G})}$, where Q is the solution of $T[Q] = F$. From the representation of U as a single-layer potential we are able to deduce that

$$|U(x)| \leq \sup_{z\in\partial\mathcal{G}} |Q(z)| \int_{\partial\mathcal{G}} G(\Delta; |x-y|)\, d\omega(y), \quad x \in \partial\mathcal{G}. \tag{3.356}$$

The integral in (3.356) is the potential of a single layer with density equal to one. Therefore, this integral is continuous over $\mathbb{R}^3$, harmonic in $\mathcal{G}^c$, and regular at infinity. Hence, it follows that it is bounded by a constant K. Consequently,

$$|U(x)| \leq K \left(\sup_{y\in\partial\mathcal{G}} |Q(y)| \right) \leq \underbrace{\tilde{C}K}_{=C} \left(\sup_{y\in\partial\mathcal{G}} |F(y)| \right) = C \sup_{y\in\partial\mathcal{G}} \left| \frac{\partial U^+}{\partial\nu}(y) \right|, \; x \in \mathcal{G}^c. \tag{3.357}$$

This is the desired result. □

All in all, the exterior Neumann problem (ENP) is well-posed in the sense that its solution exists, is unique, and depends continuously on the boundary data.

In the following, we want to verify analogous *stability theorems* for the Dirichlet as well as Neumann problem in the $L^2(\partial\mathcal{G})$-context.

Theorem 3.83. *Let U be of class* $\mathrm{Pot}^{(0)}\left(\overline{\mathcal{G}^c}\right)$. *Then there exists a constant C (dependent on k, $\mathcal{K}$, and $\partial\mathcal{G}$) such that*

$$\sup_{x\in\overline{\mathcal{K}}} \left| \left(\nabla^{(k)} U\right)(x) \right| \leq C \left(\int_{\partial\mathcal{G}} \left|U^+(x)\right|^2 \, d\omega(x) \right)^{\frac{1}{2}} \tag{3.358}$$

for all $\mathcal{K} \Subset \mathcal{G}^c$ and all $k \in \mathbb{N}_0$.

Proof. First we recall (3.342), (3.343). The operator $T = \frac{1}{2}I + A - P_{|2}(0,0)$ as defined by (3.343) and its adjoint operator T^* with respect to $(\cdot,\cdot)_{L^2(\partial\mathcal{G})}$ are bijective with respect to the Banach space $(C^{(0)}(\partial\mathcal{G}), \|\cdot\|_{C^{(0)}(\partial\mathcal{G})})$. By virtue of the open mapping theorem of functional analysis, the operators T, T^{-1}: $C^{(0)}(\partial\mathcal{G}) \to C^{(0)}(\partial\mathcal{G})$ are linear and bounded with respect to $\|\cdot\|_{C^{(0)}(\partial\mathcal{G})}$. Furthermore, $(T^*)^{-1} = (T^{-1})^*$. Therefore, following the technique of P.D. Lax

[1954] (see also the proof of Theorem 3.56), the operators T, T^{-1}: $\mathrm{L}^2(\partial\mathcal{G}) \to \mathrm{L}^2(\partial\mathcal{G})$ are bounded with respect to $\|\cdot\|_{\mathrm{L}^2(\partial\mathcal{G})}$.

Now, for all points $x \in \overline{\mathcal{K}}$, with $\mathcal{K} \Subset \mathcal{G}^c$, the Cauchy–Schwarz inequality gives

$$\begin{aligned} &\left| \left(\nabla^{(k)} U\right)(x) \right| \qquad (3.359)\\ &= \left| \int_{\partial\mathcal{G}} Q(y) \nabla_x^{(k)} \left(\frac{\partial}{\partial\nu(y)} G(\Delta; |x-y|) + G(\Delta; |x|) \right) \, d\omega(y) \right| \\ &\leq \left(\int_{\partial\mathcal{G}} \left| \nabla_x^{(k)} \left(\frac{\partial}{\partial\nu(y)} G(\Delta; |x-y|) + G(\Delta; |x|) \right) \right|^2 \, d\omega(y) \right)^{\frac{1}{2}} \\ &\quad \times \left(\int_{\partial\mathcal{G}} |Q(y)|^2 \, d\omega(y) \right)^{\frac{1}{2}}. \end{aligned}$$

This shows us that

$$\sup_{x \in \overline{\mathcal{K}}} \left| \left(\nabla^{(k)} U\right)(x) \right| \leq D \left(\int_{\partial\mathcal{G}} |Q(y)|^2 \, d\omega(y) \right)^{\frac{1}{2}}, \tag{3.360}$$

where we have used the abbreviation

$$D = \sup_{x \in \overline{\mathcal{K}}} \left(\int_{\partial\mathcal{G}} \left| \nabla_x^{(k)} \left(\frac{\partial}{\partial\nu(y)} G(\Delta; |x-y|) + G(\Delta; |x|) \right) \right|^2 \, d\omega(y) \right)^{\frac{1}{2}}. \tag{3.361}$$

Assuming $U^+|_{\partial\mathcal{G}} = F$, $F \in \mathrm{C}^{(0)}(\partial\mathcal{G})$, we obtain as an immediate consequence

$$\sup_{x \in \overline{\mathcal{K}}} \left| \left(\nabla^{(k)} U\right)(x) \right| \leq D \left(\int_{\partial\mathcal{G}} \left| T^{-1}[F](y) \right|^2 \, d\omega(y) \right)^{\frac{1}{2}}. \tag{3.362}$$

Because of the boundedness of T^{-1} with respect to $\|\cdot\|_{\mathrm{L}^2(\partial\mathcal{G})}$, we see with $C = D\|T^{-1}\|_{\mathrm{L}^2(\partial\mathcal{G})\to\mathrm{L}^2(\partial\mathcal{G})}$ that the inequality

$$\sup_{x \in \overline{\mathcal{K}}} \left| \left(\nabla^{(k)} U\right)(x) \right| \leq C \left(\int_{\partial\mathcal{G}} |F(y)|^2 \, d\omega(y) \right)^{\frac{1}{2}} \tag{3.363}$$

holds true. Hence, the statement (3.358) is verified. □

An analogous argument yields the following theorem.

Theorem 3.84. *Let U be of class $\mathrm{Pot}^{(1)}(\overline{\mathcal{G}^c})$. Then there exists a constant C (dependent on k, $\mathcal{K}$, and $\partial\mathcal{G}$) such that*

$$\sup_{x \in \overline{\mathcal{K}}} \left| \left(\nabla^{(k)} U\right)(x) \right| \leq C \left(\int_{\partial\mathcal{G}} \left| \frac{\partial U^+}{\partial\nu}(x) \right|^2 \, d\omega(x) \right)^{\frac{1}{2}} \tag{3.364}$$

for all $\mathcal{K} \Subset \mathcal{G}^c$ and all $k \in \mathbb{N}_0$.

3.5 Locally and Globally Uniform Approximation

In boundary-value problems of potential theory, a result first motivated by C. Runge [1885] in one-dimensional complex analysis and later generalized by J.L. Walsh [1929] to potential theory is of basic interest. For our geoscientifically relevant purpose, it may be formulated as follows: Any harmonic function in $\mathcal{G}^c$ that is regular at infinity can be approximated by a function that is harmonic outside an arbitrarily given so-called Runge sphere contained in $\overline{\mathcal{G}}$ in the sense that, for any given $\varepsilon > 0$, the absolute error between the two functions is smaller than ε for all points outside and on any closed surface completely surrounding $\partial\mathcal{G}$ in its outer space. The value ε may be arbitrarily small, and the surrounding surface may be arbitrarily close to the surface $\partial\mathcal{G}$.

Obviously, the *Runge–Walsh theorem* in its preceding formulation is a pure existence theorem. It guarantees only the existence of an approximating function and does not provide a method to find it. Nothing is said about the approximation procedure and the structure of the approximation. The theorem describes merely the theoretical background for the approximation of a potential by another potential defined on a larger harmonicity domain.

The situation, however, is completely different in a spherical model. Assuming that the boundary $\partial\mathcal{G}$ is a sphere around the origin, a constructive approximation of a potential in the outer space is available, e.g., by (Abel–Poisson) integral representations leading to outer harmonic expansions. More concretely, in a spherical context, a constructive version of the Runge–Walsh theorem can be established by finite truncations of Fourier expansions in terms of outer harmonics. The only unknown information left when using an outer harmonic expansion is the a priori choice of the right truncation parameter.

From a superficial point of view, one could suggest that approximation by truncated series expansions in terms of outer harmonics is closely related to spherical boundaries. The purpose of our next work, however, is to show that the essential steps involved in the Fourier expansion method can be generalized to any regular region $\mathcal{G}$. The main techniques for establishing these results are the limit and jump relations and their formulations in the Hilbert space nomenclature of $(\mathrm{L}^2(\partial\mathcal{G}), \|\cdot\|_{\mathrm{L}^2(\partial\mathcal{G})})$.

We restrict ourselves to the geoscientifically relevant exterior cases. The interior cases follow by obvious arguments.

3.5.1 Closure in L^2-Topology

We begin our considerations with the proof of the following result.

Lemma 3.85. *Let $\mathcal{G} \subset \mathbb{R}^2$ be a regular surface such that $\underline{R} < \inf_{x\in\partial\mathcal{G}} |x|$ (cf. (1.74)). Then the following statements are valid:*

(a) $\left\{ H^{\underline{R}}_{-n-1,j}\Big|_{\partial\mathcal{G}} \right\}_{\substack{n=0,1,\ldots \\ j=1,\ldots,2n+1}}$ *is linearly independent,*

(b) $\left\{ \frac{\partial}{\partial\nu} H^{\underline{R}}_{-n-1,j}\Big|_{\partial\mathcal{G}} \right\}_{\substack{n=0,1,\ldots \\ j=1,\ldots,2n+1}}$ *is linearly independent.*

Proof. In order to verify statement (a) we have to derive that for any linear combination H of the form

$$H = \sum_{n=0}^{m} \sum_{j=1}^{2n+1} a_{n,j} H^{\underline{R}}_{-n-1,j}, \tag{3.365}$$

the condition $H|\partial\mathcal{G} = 0$ implies $a_{0,1} = \cdots = a_{m,1} = \cdots = a_{m,2m+1} = 0$. From the uniqueness theorem of the exterior Dirichlet problem we know that $H|\partial\mathcal{G} = 0$ yields $H|_{\overline{\mathcal{G}^c}} = 0$. Therefore, for every sphere $\Omega_{\overline{R}}$ with radius $\overline{R} > \sup_{x\in\partial\mathcal{G}} |x|$, it follows that

$$\int_{\Omega_{\overline{R}}} H^{\underline{R}}_{-n-1,j}(x) H(x)\, d\omega(x) = 0 \tag{3.366}$$

for $n = 0, \ldots, m, j = 1, \ldots, 2n+1$. Inserting (3.365) into (3.366) gives us, in connection with the completeness property of the spherical harmonics, that $a_{n,j} = 0$ for $n = 0, ..., m$, $j = 1, ..., 2n+1$, as required for statement (a).

For the proof of statement (b) we start from the homogeneous boundary condition

$$\frac{\partial H}{\partial\nu} = \sum_{n=0}^{m} \sum_{j=1}^{2n+1} a_{n,j} \frac{\partial H^{\underline{R}}_{-n-1,j}}{\partial\nu} = 0 \tag{3.367}$$

on $\partial\mathcal{G}$. The uniqueness theorem of the exterior Neumann problem then yields $H|_{\overline{\mathcal{G}^c}} = 0$. This gives us $a_{n,j} = 0$ for $n = 0, \ldots, m$, $j = 1, \ldots, 2n+1$, as required for statement (b). □

Next, our goal is to prove the completeness and closure theorems (see W. Freeden [1980]).

Theorem 3.86. *Let $\mathcal{G} \subset \mathbb{R}^3$ be a regular region such that $\underline{R} < \inf_{x\in\partial\mathcal{G}} |x|$. Then the following statements are valid:*

(a) $\left\{ H^{\underline{R}}_{-n-1,j}\Big|_{\partial\mathcal{G}} \right\}_{\substack{n=0,1,\ldots \\ j=1,\ldots,2n+1}}$ *is complete, i.e., dense in* $\mathrm{L}^2(\partial\mathcal{G}) = \overline{D^+}^{\|\cdot\|_{\mathrm{L}^2(\partial\mathcal{G})}}$,

(b) $\left\{ \frac{\partial}{\partial\nu} H^{\underline{R}}_{-n-1,j}\Big|_{\partial\mathcal{G}} \right\}_{\substack{n=0,1,\ldots \\ j=1,\ldots,2n+1}}$ *is complete, i.e., dense in* $\mathrm{L}^2(\partial\mathcal{G}) = \overline{N^+}^{\|\cdot\|_{\mathrm{L}^2(\partial\mathcal{G})}}$.

Proof. We restrict our attention to statement (a). Suppose that $F \in \mathrm{L}^2(\partial\mathcal{G})$ satisfies

$$(F, H^{\underline{R}}_{-n-1,j}|\partial\mathcal{G})_{\mathrm{L}^2(\partial\mathcal{G})} = \int_{\partial\mathcal{G}} F(y) H^{\underline{R}}_{-n-1,j}(y)\, d\omega(y) = 0 \tag{3.368}$$

for all $n \in \mathbb{N}_0$, $j = 1, ..., 2n+1$. We have to show that $F = 0$ in $\mathrm{L}^2(\partial\mathcal{G})$. We remember that the series expansion

$$G(\Delta; |x-y|) = \sum_{n=0}^{\infty} \frac{1}{2n+1} \frac{|x|^n}{|y|^{n+1}} \sum_{j=1}^{2n+1} Y_{n,j}(\xi) Y_{n,j}(\eta), \tag{3.369}$$

$x = |x|\xi$, $y = |y|\eta$, is analytic in the variable x on the ball $\mathcal{B}_{\underline{R}}(0)$, if $y \in \mathbb{R}^3 \setminus \mathcal{B}_{\underline{R}}(0)$. For all $x \in \mathcal{B}_{\underline{R}}(0)$, we thus find by virtue of (3.368)

$$\begin{aligned} U(x) &= \int_{\partial\mathcal{G}} F(y) G(\Delta; |x-y|)\, d\omega(y) \\ &= \sum_{n=0}^{\infty} \frac{R}{2n+1} \sum_{j=1}^{2n+1} H_{n,j}^{\underline{R}}(x) \int_{\partial\mathcal{G}} F(y) H_{-n-1,j}^{\underline{R}}(y)\, d\omega(y) \\ &= 0. \end{aligned} \tag{3.370}$$

Analytic continuation shows that the single-layer potential U vanishes at each point $x \in \mathcal{G}$. In other words, the equations

$$U(x - \tau\nu(x)) = 0, \tag{3.371}$$

$$\frac{\partial U}{\partial \nu}(x - \tau\nu(x)) = 0 \tag{3.372}$$

hold true for all $x \in \partial\mathcal{G}$ and all sufficiently small $\tau > 0$. Therefore, using the relations of Theorem 3.56, we obtain

$$\lim_{\tau \to 0+} \int_{\partial\mathcal{G}} \left| U(x + \tau\nu(x)) \right|^2 d\omega(x) = 0, \tag{3.373}$$

$$\lim_{\tau \to 0+} \int_{\partial\mathcal{G}} \left| \frac{\partial U}{\partial \nu}(x + \tau\nu(x)) + F(x) \right|^2 d\omega(x) = 0, \tag{3.374}$$

and

$$\lim_{\tau \to 0+} \int_{\partial\mathcal{G}} \left| \frac{\partial U}{\partial \nu}(x) + \frac{1}{2} F(x) \right|^2 d\omega(x) = 0. \tag{3.375}$$

Observing that the limit in the last equation can be omitted, (3.375) can also be understood as

$$-\int_{\partial\mathcal{G}} F(y) \frac{\partial}{\partial \nu(x)} G(\Delta; |x-y|)\, d\omega(y) = -\frac{1}{2} F(x), \tag{3.376}$$

in the sense of $\mathrm{L}^2(\partial\mathcal{G})$. The left-hand side of (3.376) constitutes a continuous function. Thus, the function F is continuous itself. For continuous functions, however, the classical limit and jump relations are valid:

$$\lim_{\tau \to 0+} U(x + \tau\nu(x)) = 0, \quad x \in \partial\mathcal{G}, \tag{3.377}$$

$$\lim_{\tau \to 0+} \frac{\partial U}{\partial \nu}(x + \tau\nu(x)) = -F(x), \quad x \in \partial\mathcal{G}. \tag{3.378}$$

Consequently, the uniqueness theorem of the exterior Dirichlet problem shows us that $U(x) = 0$ for all $x \in \mathbb{R}^3 \backslash \overline{\mathcal{G}^c}$. But this means that $F = 0$ on the surface $\partial\mathcal{G}$, as required. The remaining statement (b) follows by analogous arguments. □

From Constructive Approximation (see, e.g., P.J. Davis [1963]) we know that the properties of completeness and closure are equivalent in a Hilbert space such as $\mathrm{L}^2(\partial\mathcal{G})$. This leads us to the following statement.

Corollary 3.87. *Under the assumptions of Theorem 3.86 the following statements are valid:*

(a) $\{H^R_{-n-1,j}|_{\partial\mathcal{G}}\}_{\substack{n=0,1,\ldots,\\ j=1,\ldots,2n+1}}$ *is closed in* $\mathrm{L}^2(\partial\mathcal{G}) = \overline{D^+}^{\|\cdot\|_{\mathrm{L}^2(\partial\mathcal{G})}}$, *i.e., for any given* $F \in \mathrm{L}^2(\partial\mathcal{G})$ *and arbitrary* $\varepsilon > 0$ *there exists a linear combination*

$$H_m = \sum_{n=0}^{m} \sum_{j=1}^{2n+1} a_{n,j} H^R_{-n-1,j}\Big|_{\partial\mathcal{G}} \tag{3.379}$$

such that

$$\|F - H_m\|_{\mathrm{L}^2(\partial\mathcal{G})} \leq \varepsilon \ . \tag{3.380}$$

(b) $\{\frac{\partial}{\partial\nu} H^R_{-n-1,j}|_{\partial\mathcal{G}}\}_{\substack{n=0,1,\ldots,\\ j=1,\ldots,2n+1}}$ *is closed in* $\mathrm{L}^2(\partial\mathcal{G}) = \overline{N^+}^{\|\cdot\|_{\mathrm{L}^2(\partial\mathcal{G})}}$, *i.e., for any given* $F \in \mathrm{L}^2(\partial\mathcal{G})$ *and arbitrary* $\varepsilon > 0$ *there exists a linear combination*

$$S_m = \sum_{n=0}^{m} \sum_{j=1}^{2n+1} a_{n,j} \frac{\partial}{\partial\nu} H^R_{-n-1,j}\Big|_{\partial\mathcal{G}} \tag{3.381}$$

such that

$$\|F - S_m\|_{\mathrm{L}^2(\partial\mathcal{G})} \leq \varepsilon \ . \tag{3.382}$$

3.5.2 Fundamental Systems

Based on our results on outer harmonics, i.e., multipole expansions, a large number of polynomially-based countable systems of potentials can be shown to possess the L^2-closure property on $\partial\mathcal{G}$. Probably best known are *mass-pole representations* (i.e., fundamental solutions of the Laplace operator). Their $\mathrm{L}^2(\partial\mathcal{G})$-closure is adequately described by using the concept of *fundamental systems*, which should be recapitulated briefly (see W. Freeden [1980], W. Freeden, V. Michel [2004]).

Definition 3.88. Let $\mathcal{G} \subset \mathbb{R}^3$ be a (not necessarily bounded) region. A point set $\mathcal{Y} = \{y_n\}_{n=0,1,\ldots} \subset \mathcal{G}$ (with $y_n \neq y_l$ for $n \neq l$) is called a *fundamental system in* $\mathcal{G}$, if the following properties are satisfied:

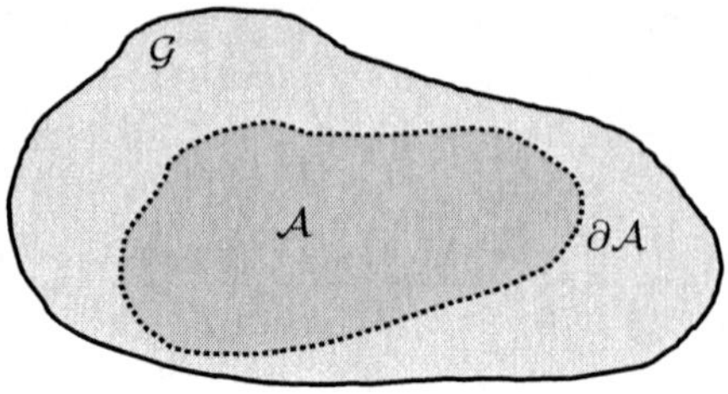

FIGURE 3.5
Illustration of the positioning of a fundamental system on $\partial\mathcal{A}$ in $\mathcal{G}$.

(i) $\mathrm{dist}(\mathcal{Y}, \partial\mathcal{G}) > 0$,

(ii) for each $U \in \mathrm{Pot}(\mathcal{G})$ the condition $U(y_n) = 0$, for $n = 0, 1, \ldots$, implies $U = 0$ in $\mathcal{G}$.

Some examples of fundamental systems should be listed for a regular region $\mathcal{G}$ (note that analogous arguments hold for fundamental systems in $\mathcal{G}^c$): $\mathcal{Y} = \{y_n\}_{n=0,1,\ldots}$ is a fundamental system in $\mathcal{G}$ if it is a dense set of points of one of the following subsets of $\mathcal{G}$: (1) regular region $\mathcal{A}$ with $\mathcal{A} \Subset \mathcal{G}$, (2) boundary $\partial\mathcal{A}$ of a regular region $\mathcal{A}$ with $\mathcal{A} \Subset \mathcal{G}$ (cf. Figure 3.5).

Theorem 3.89. *Let $\mathcal{G}$ be a regular region. Then the following statements are valid:*

(a) For every fundamental system $\mathcal{Y} = \{y_n\}_{n=0,1,\ldots}$ in $\mathcal{G}$, the system

$$\left\{ G(\Delta; |\cdot - y_n|)\Big|_{\partial\mathcal{G}} \right\}_{n=0,1,\ldots} \tag{3.383}$$

is closed in $\mathrm{L}^2(\partial\mathcal{G}) = \overline{D^+}^{\|\cdot\|_{\mathrm{L}^2(\partial\mathcal{G})}}$.

(b) For every fundamental system $\mathcal{Y} = \{y_n\}_{n=0,1,\ldots}$ in $\mathcal{G}$, the system

$$\left\{ \frac{\partial}{\partial\nu} G(\Delta; |\cdot - y_n|)\Big|_{\partial\mathcal{G}} \right\}_{n=0,1,\ldots} \tag{3.384}$$

is closed in $\mathrm{L}^2(\partial\mathcal{G}) = \overline{N^+}^{\|\cdot\|_{\mathrm{L}^2(\partial\mathcal{G})}}$.

Proof. We restrict ourselves to the proof of the statement (a). Since $y_n \neq y_m$ for all $n \neq m$, it immediately follows that the system $\{G(\Delta; |\cdot - y_n|)\}_{n=0,1,\ldots}$ is linearly independent.

Our purpose is to verify the completeness of the system (3.383) in $\mathrm{L}^2(\partial\mathcal{G})$. To this end, we consider a function $F \in \mathrm{L}^2(\partial\mathcal{G})$ with

$$\int_{\partial\mathcal{G}} F(x) G(\Delta; |x - y_n|) \, d\omega(x) = 0, \quad n \in \mathbb{N}_0. \tag{3.385}$$

We have to prove that $F = 0$ in $\mathrm{L}^2(\partial\mathcal{G})$. We consider the single-layer potential U given by

$$U(y) = \int_{\partial\mathcal{G}} F(x) G(\Delta; |x - y|)\, d\omega(x). \tag{3.386}$$

Since U is harmonic in $\mathcal{G}$, the properties of the fundamental system $\{y_n\}_{n=0,1,\ldots}$ in $\mathcal{G}$ imply that $U(y) = 0$ for all $y \in \mathcal{G}$. Then, the same arguments as given in the proof of Theorem 3.86 guarantee that $F = 0$ in the sense of $\mathrm{L}^2(\partial\mathcal{G})$, as required. The statement (b) follows by analogous arguments (cf. Exercise 3.14). □

Besides the outer harmonics, i.e., multipoles (see Corollary 3.87), and the mass (single-)poles (see Theorem 3.89), there exist a variety of countable systems of potentials showing the properties of completeness and closure in $\mathrm{L}^2(\partial\mathcal{G})$. Many systems, however, are much more difficult to handle numerically (for instance, the ellipsoidal systems of Lamé or Mathieu functions). Although they are very interesting in physical geodesy, they will not be discussed here (for more details see, e.g., E.W. Grafarend et al. [2010] and the references therein). Instead, we study some further kernel systems generated by superposition (i.e., infinite clustering) of outer harmonics, such that a certain amount of space localization can be expected on inner Runge spheres (as described by W. Freeden, M. Schreiner [2009]). In addition, if they are explicitly available as elementary functions, these systems turn out to be particularly suitable for numerical purposes.

Theorem 3.90. *Let $\mathcal{G} \subset \mathbb{R}^3$ be a regular region such that $\underline{R} < \inf_{x\in\partial\mathcal{G}} |x|$. We consider the kernel function $K(\cdot,\cdot) : \mathbb{R}^3\backslash\mathcal{B}_{\underline{R}}(0) \times \mathcal{B}_{\underline{R}}(0) \to \mathbb{R}$ given by*

$$\begin{aligned} K(x,y) &= \sum_{k=0}^{\infty}\sum_{l=1}^{2k+1} K^\wedge(k) H^{\underline{R}}_{-k-1,l}(x) H^{\underline{R}}_{k,l}(y) \\ &= \frac{\underline{R}}{|x|}\sum_{k=0}^{\infty}\frac{2k+1}{4\pi\underline{R}^2} K^\wedge(k)\left(\frac{|y|}{|x|}\right)^k P_k\left(\frac{x}{|x|}\cdot\frac{y}{|y|}\right) \end{aligned} \tag{3.387}$$

for $x \in \mathbb{R}^3\backslash\mathcal{B}_{\underline{R}}(0), y \in \mathcal{B}_{\underline{R}}(0)$. Let $\mathcal{Y} = \{y_n\}_{n=0,1,\ldots}$ be a fundamental system in $\mathcal{B}_{\underline{R}_0}(0)$ with $\underline{R}_0 < \underline{R} < \inf_{x\in\partial\mathcal{G}} |x|$. Suppose that

$$\sum_{k=0}^{\infty}(2k+1)\left|K^\wedge(k)\right|\left(\frac{\underline{R}_0}{\underline{R}}\right)^k < \infty \tag{3.388}$$

with $K^\wedge(k) \neq 0$ for $k \in \mathbb{N}_0$. Then the following statements are valid:

(a) The system

$$\left\{ K(\cdot, y_n)\Big|_{\partial\mathcal{G}} \right\}_{n=0,1,\ldots}$$

is closed in $\mathrm{L}^2(\partial\mathcal{G}) = \overline{D^+}^{\|\cdot\|_{\mathrm{L}^2(\partial\mathcal{G})}}$.

(b) The system

$$\left\{ \frac{\partial}{\partial \nu} K(\cdot, y_n)\Big|_{\partial \mathcal{G}} \right\}_{n=0,1,\ldots}$$

is closed in $\mathrm{L}^2(\partial\mathcal{G}) = \overline{N^+}^{\|\cdot\|_{\mathrm{L}^2(\partial\mathcal{G})}}$.

Proof. We only verify statement (a). Let F be of class $\mathrm{L}^2(\partial\mathcal{G})$. The function U given by

$$U(y) = \int_{\partial\mathcal{G}} F(x) K(x,y)\, d\omega(x), \tag{3.389}$$

is analytic in $\mathcal{B}_{\underline{R}_0}(0)$. Indeed, for all $y \in \mathbb{R}^3$ with $|y| < \underline{R}_0$, it follows from (3.389) that

$$U(y) = \sum_{k=0}^{\infty} K^\wedge(k) \sum_{j=1}^{2k+1} H_{k,j}^{\underline{R}}(y) \int_{\partial\mathcal{G}} F(x) H_{-k-1,j}^{\underline{R}}(x)\, d\omega(x). \tag{3.390}$$

Assume that $U(y_n) = 0$ for $n = 0, 1, \ldots$. Since $\mathcal{Y} = \{y_n\}_{n=0,1,\ldots}$ is a fundamental system in $\mathcal{B}_{\underline{R}_0}(0)$, the function U vanishes in $\mathcal{B}_{\underline{R}_0}(0)$. This implies that

$$\left(F, H_{-k-1,j}^{\underline{R}}\right)_{\mathrm{L}^2(\partial\mathcal{G})} = \int_{\partial\mathcal{G}} F(x) H_{-k-1,j}^{\underline{R}}(x)\, d\omega(x) = 0, \tag{3.391}$$

for $k \in \mathbb{N}_0$, $j = 1, \ldots, 2k+1$. Hence, by virtue of the completeness of the system of outer harmonics (Theorem 3.86 (i)), we get $F = 0$ in the topology of $\mathrm{L}^2(\partial\mathcal{G})$, as required. □

Examples of kernel representations (3.387) are easily obtainable from known series expansions in terms of Legendre polynomials (see, e.g., the representations in W. Freeden [1987], W. Freeden, V. Michel [2004] based on identities in W. Magnus et al. [1966]). Applying the Kelvin transform with respect to the sphere $\Omega_{\underline{R}}$, we are led to systems $\{\breve{K}(\cdot, \breve{y}_n)\}_{n=0,1,\ldots}$ with

$$\begin{aligned} \breve{K}(x, \breve{y}_n) &= \sum_{k=0}^{\infty} \sum_{l=1}^{2k+1} K^\wedge(k) H_{-k-1,l}^{\underline{R}}(x) H_{-k-1,l}^{\underline{R}}(\breve{y}_n) \\ &= \sum_{k=0}^{\infty} \frac{2k+1}{4\pi \underline{R}^2} K^\wedge(k) \left(\frac{\underline{R}^2}{|x||\breve{y}_n|}\right)^{k+1} P_k\left(\frac{x}{|x|} \cdot \frac{\breve{y}_n}{|\breve{y}_n|}\right), \end{aligned} \tag{3.392}$$

where $\breve{\mathcal{Y}} = \{\breve{y}_n\}_{n=0,1,\ldots}$ is the point system generated by application of the Kelvin transform to $\mathcal{Y}$, i.e., by letting $\breve{y}_n = \frac{\underline{R}^2}{|y_n|^2} y_n, n = 0, 1, \ldots$ (assuming that $0 \notin \mathcal{Y}$).

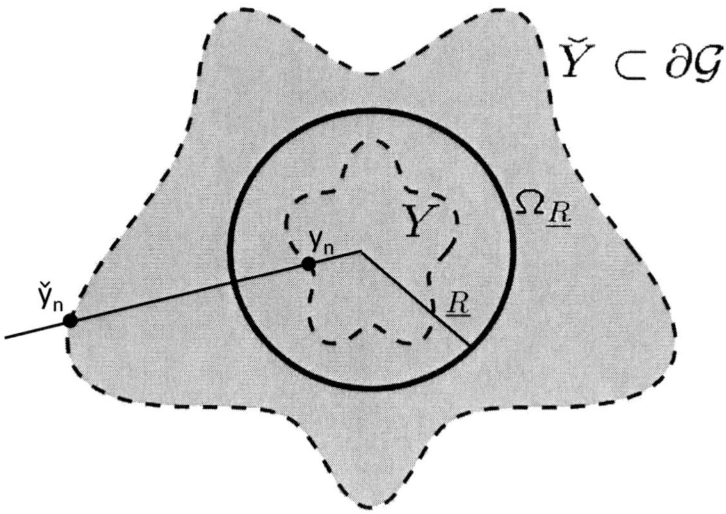

FIGURE 3.6
Fundamental system $\mathcal{Y}$ and the Kelvin transformed system $\check{\mathcal{Y}} \subset \partial\mathcal{G}$.

Theorem 3.91. *Suppose that* $\check{\mathcal{Y}} = \{\check{y}_n\}_{n=0,1,\ldots}$ *is given as described above. Then the following properties hold true:*

(a) The system

$$\left\{ \check{K}(\cdot, \check{y}_n)\Big|_{\partial\mathcal{G}} \right\}_{n=0,1,\ldots} \tag{3.393}$$

is closed in $\mathrm{L}^2(\partial\mathcal{G}) = \overline{D^+}^{\|\cdot\|_{\mathrm{L}^2(\partial\mathcal{G})}}$.

(b) The system

$$\left\{ \frac{\partial}{\partial\nu} \check{K}\left(\cdot, \check{y}_n\right) \Big|_{\partial\mathcal{G}} \right\}_{n=0,1,\ldots} \tag{3.394}$$

is closed in $\mathrm{L}^2(\partial\mathcal{G}) = \overline{N^+}^{\|\cdot\|_{\mathrm{L}^2(\partial\mathcal{G})}}$.

Remark 3.92. Particularly helpful in geosciences is a fundamental system $\mathcal{Y} = \{y_n\}_{n=0,1,\ldots}$ in $\mathcal{B}_{\underline{R}_0}(0)$ that yields $\check{\mathcal{Y}} = \{\check{y}_n\}_{n=0,1,\ldots} \subset \partial\mathcal{G}$ (cf. Figure 3.6). In other words, the closure property is related to points lying on the (Earth's) surface $\partial\mathcal{G}$.

The kernels $\check{K}(\cdot, \check{y}_n), n \in \mathbb{N}_0$, as given by (3.392), play a central role in the Sobolev space theory of harmonic functions (cf. W. Freeden [1999]). Typical examples are the following:

(a) *Abel–Poisson kernel:* $\breve{K}^\wedge(k) = 1$, $k \in \mathbb{N}_0$. The explicit representation of the kernel reads as follows:

$$\breve{K}(x, \breve{y}_n) = |x||\breve{y}_n| \frac{\underline{R}^2}{4\pi} \frac{|x|^2|\breve{y}_n|^2 - \underline{R}^2}{(L(x, \breve{y}_n))^{\frac{3}{2}}}, \qquad x \in \mathbb{R}^3 \backslash \mathcal{B}_{\underline{R}}(0), \tag{3.395}$$

where we have introduced the abbreviation

$$L(x, \breve{y}_n) = |x|^2|\breve{y}_n|^2 - 2\underline{R}^2 x \cdot \breve{y}_n + \underline{R}^4, \tag{3.396}$$

(b) *Singularity kernel:* $K^\wedge(k) = \frac{2}{2k+1}$, $k \in \mathbb{N}_0$. The kernel is now given by

$$\breve{K}(x, \breve{y}_n) = \frac{\underline{R}^2}{2\pi} \frac{1}{(L(x, \breve{y}_n))^{\frac{1}{2}}}, \quad x \in \mathbb{R}^3 \backslash \mathcal{B}_{\underline{R}}(0), \tag{3.397}$$

(c) *Logarithmic kernel:* $K^\wedge(k) = \frac{1}{(k+1)(2k+1)}$, $k \in \mathbb{N}_0$. Now we have

$$\breve{K}(x, \breve{y}_n) = \frac{1}{2\pi} \ln \left(1 + \frac{2\underline{R}^2}{(L(x, \breve{y}_n))^{\frac{1}{2}} + |x||\breve{y}_n| - \underline{R}^2} \right), \quad x \in \mathbb{R}^3 \backslash \mathcal{B}_{\underline{R}}(0). \tag{3.398}$$

Combining the $\mathrm{L}^2(\partial\mathcal{G})$-closure (Theorem 3.86) and the regularity theorems (Theorem 3.83 and Theorem 3.84), we obtain the following results for the system of outer harmonics.

Theorem 3.93. *Let $\mathcal{G} \subset \mathbb{R}^3$ be a regular region such that $\underline{R} < \inf_{x \in \partial\mathcal{G}} |x|$.*

(EDP): *For given $F \in \mathrm{C}^{(0)}(\partial\mathcal{G})$, let U be the potential of class $\mathrm{Pot}^{(0)}(\overline{\mathcal{G}^c})$ with $U^+|_{\partial\mathcal{G}} = F$. Then, for any given $\varepsilon > 0$ and $\mathcal{K} \Subset \mathcal{G}^c$, there exist an integer m (dependent on ε) and a set of coefficients $a_{0,1}, \ldots, a_{m,1}, \ldots, a_{m,2m+1}$ such that*

$$\left(\int_{\partial\mathcal{G}} \left| F(x) - \sum_{n=0}^{m} \sum_{j=1}^{2n+1} a_{n,j} H_{-n-1,j}^{\underline{R}}(x) \right|^2 d\omega(x) \right)^{\frac{1}{2}} \leq \varepsilon \tag{3.399}$$

and

$$\sup_{x \in \overline{\mathcal{K}}} \left| \left(\nabla^{(k)} U\right)(x) - \sum_{n=0}^{m} \sum_{j=1}^{2n+1} a_{n,j} \left(\nabla^{(k)} H_{-n-1,j}^{\underline{R}}\right)(x) \right| \leq C\varepsilon \tag{3.400}$$

hold for all $k \in \mathbb{N}_0$.

(ENP): *For given* $F \in \mathrm{C}^{(0)}(\partial\mathcal{G})$, *let* U *satisfy* $U \in \mathrm{Pot}^{(1)}(\overline{\mathcal{G}^c})$, $\frac{\partial U}{\partial\nu}\big|_{\partial\mathcal{G}} = F$. *Then, for any given* $\varepsilon > 0$ *and* $\mathcal{K} \Subset \mathcal{G}^c$, *there exist an integer* m *(dependent on* ε*) and a set of coefficients* $a_{0,1}, \dots, a_{m,1}, \dots, a_{m,2m+1}$ *such that*

$$\left(\int_{\partial\mathcal{G}} \left| F(x) - \sum_{n=0}^{m} \sum_{j=1}^{2n+1} a_{n,j} \frac{\partial H^{\underline{R}}_{-n-1,j}}{\partial\nu}(x) \right|^2 d\omega(x) \right)^{\frac{1}{2}} \leq \varepsilon \tag{3.401}$$

and

$$\sup_{x\in\overline{\mathcal{K}}} \left| \left(\nabla^{(k)} U\right)(x) - \sum_{n=0}^{m} \sum_{j=1}^{2n+1} a_{n,j} \left(\nabla^{(k)} H^{\underline{R}}_{-n-1,j}\right)(x) \right| \leq C\varepsilon \tag{3.402}$$

hold for all $k \in \mathbb{N}_0$.

In other words, the L^2-approximation in terms of outer harmonics on $\partial\mathcal{G}$ implies the uniform approximation (in the ordinary sense) on each subset $\overline{\mathcal{K}}$ with positive distance to $\partial\mathcal{G}$.

Unfortunately, the theorems developed until now are non-constructive since further information about the choice of m and the coefficients of the approximating linear combination is needed. In order to derive a constructive approximation theorem, the system of potential values and normal derivatives, respectively, can be orthonormalized on $\partial\mathcal{G}$. As a result, we obtain a *(generalized) Fourier expansion* (orthogonal Fourier expansion) that shows locally uniform approximation.

Theorem 3.94. *Let* $\mathcal{G} \subset \mathbb{R}^3$ *be a regular region such that* $\underline{R} < \inf_{x\in\partial\mathcal{G}} |x|$.

(EDP): *For given* $F \in \mathrm{C}^{(0)}(\partial\mathcal{G})$, *let* U *be the solution of the Dirichlet problem* $U \in \mathrm{Pot}^{(0)}(\overline{\mathcal{G}^c})$, $U^+|_{\partial\mathcal{G}} = F$. *Corresponding to the sequence* $\{H^{\underline{R}}_{-n-1,j}\}_{n\in\mathbb{N}_0}$, *there exists a system* $\{H_{-n-1,j}(\partial\mathcal{G};\cdot)\}_{n\in\mathbb{N}_0, j=1,\dots,2n+1} \subset \mathrm{Pot}^{(0)}(\mathbb{R}^3\backslash\mathcal{B}_{\underline{R}}(0))$ *such that* $\{H_{-n-1,j}(\partial\mathcal{G};\cdot)|_{\partial\mathcal{G}}\}_{n\in\mathbb{N}_0, j=,\dots,2n+1}$ *is orthonormal in the sense that*

$$\int_{\partial\mathcal{G}} H_{-n-1,j}(\partial\mathcal{G};y) H_{-l-1,k}(\partial\mathcal{G};y)\, d\omega(y) = \delta_{n,l}\, \delta_{j,k}. \tag{3.403}$$

Consequently, U *is representable in the form*

$$U(x) = \sum_{n=0}^{\infty} \sum_{j=1}^{2n+1} \left(\int_{\partial\mathcal{G}} F(y) H_{-n-1,j}(\partial\mathcal{G};y)\, d\omega(y) \right) H_{-n-1,j}(\partial\mathcal{G};x) \tag{3.404}$$

for all points $x \in \mathcal{K} \Subset \mathcal{G}^c$. *Moreover, for each* $U^{(m)}$ *given by*

$$U^{(m)}(x) = \sum_{n=0}^{m} \sum_{j=1}^{2n+1} \left(\int_{\partial\mathcal{G}} F(y) H_{-n-1,j}(\partial\mathcal{G};y)\, d\omega(y) \right) H_{-n-1,j}(\partial\mathcal{G};x) \tag{3.405}$$

we have the estimate

$$\sup_{x\in\overline{\mathcal{K}}}\left|\left(\nabla^{(k)}U\right)(x)-\left(\nabla^{(k)}U^{(m)}\right)(x)\right| \tag{3.406}$$

$$\leq C\left(\int_{\partial\mathcal{G}}|F(y)|^2\,d\omega(y)-\sum_{n=0}^{m}\sum_{j=1}^{2n+1}\left|\int_{\partial\mathcal{G}}F(y)H_{-n-1,j}(\partial\mathcal{G};y)\,d\omega(y)\right|^2\right)^{\frac{1}{2}}.$$

(ENP): *For given F of class $\mathrm{C}^{(0)}(\partial\mathcal{G})$, let U be the solution of the Neumann problem $U\in\mathrm{Pot}^{(1)}(\overline{\mathcal{G}^c})$, $\frac{\partial U^+}{\partial\nu}\big|_{\partial\mathcal{G}}=F$. Corresponding to the sequence of outer harmonics $\{H^{\underline{R}}_{-n-1,j}\}_{n\in\mathbb{N}_0,j=1,\ldots,2n+1}$, there exists a system $\{H_{-n-1,j}(\partial\mathcal{G};\cdot)\}_{n\in\mathbb{N}_0,j=1,\ldots,2n+1}\subset\mathrm{Pot}^{(0)}(\mathbb{R}^3\backslash\mathcal{B}_{\underline{R}}(0))$ such that $\{\frac{\partial}{\partial\nu}H_{-n-1,j}(\partial\mathcal{G};\cdot)\}_{n\in\mathbb{N}_0,j=1,\ldots,2n+1}$ is orthonormal in the sense that*

$$\int_{\partial\mathcal{G}}\frac{\partial}{\partial\nu}H_{-n-1,j}(\partial\mathcal{G};y)\frac{\partial}{\partial\nu}H_{-l-1,k}(\partial\mathcal{G};y)\,d\omega(y)=\delta_{n,l}\,\delta_{j,k}. \tag{3.407}$$

Consequently, U is representable in the form

$$U(x)=\sum_{n=0}^{\infty}\sum_{j=1}^{2n+1}\left(\int_{\partial\mathcal{G}}F(y)\frac{\partial}{\partial\nu}H_{-n-1,j}(\partial\mathcal{G};y)\,d\omega(y)\right)H_{-n-1,j}(\partial\mathcal{G};x) \tag{3.408}$$

for all points $x\in\mathcal{K}\Subset\mathcal{G}^c$. Moreover, for each $U^{(m)}$ given by

$$U^{(m)}(x)=\sum_{n=0}^{m}\sum_{j=1}^{2n+1}\left(\int_{\partial\mathcal{G}}F(y)\frac{\partial}{\partial\nu}H_{-n-1,j}(\partial\mathcal{G};y)\,d\omega(y)\right)H_{-n-1,j}(\partial\mathcal{G};x) \tag{3.409}$$

we have the estimate

$$\sup_{x\in\overline{\mathcal{K}}}\left|\left(\nabla^{(k)}U\right)(x)-\left(\nabla^{(k)}U^{(m)}\right)(x)\right| \tag{3.410}$$

$$\leq C\left(\int_{\partial\mathcal{G}}|F(y)|^2\,d\omega(y)-\sum_{n=0}^{m}\sum_{j=1}^{2n+1}\left|\int_{\partial\mathcal{G}}F(y)\frac{\partial}{\partial\nu}H_{-n-1,j}(\partial\mathcal{G};y)\,d\omega(y)\right|^2\right)^{\frac{1}{2}}.$$

Note that the orthonormalization procedure can be performed (e.g., by the well-known Gram–Schmidt orthonormalization process) once and for all when the boundary surface $\partial\mathcal{G}$ of a regular region $\mathcal{G}$ is specified.

In the same way, the inner boundary-value problems can be formulated by generalized Fourier expansions (orthogonal expansions) in terms of inner harmonics. Furthermore, locally uniform approximation by generalized Fourier expansions can be obtained not only for (the multipole system of inner/outer) harmonics, but also for mass point and related kernel representations. The details are omitted.

Finally, we rewrite our generalized Fourier approach in a more abstract form. For that purpose we introduce the concept of Dirichlet and Neumann bases.

FIGURE 3.7
The geometric situation of an $\mathrm{L}^2(\partial\mathcal{G})$-Dirichlet/Neumann Runge basis (with $\mathcal{A}$ an arbitrary regular region such that $\mathcal{A} \Subset \mathcal{G}$ (left) and $\mathcal{A}$ an inner Runge ball (right)).

Definition 3.95 (Dirichlet/Neumann Runge Basis). Let $\mathcal{A}, \mathcal{G} \subset \mathbb{R}^3$ be regular regions such that $\mathcal{A} \Subset \mathcal{G}$ holds true (cf. Figure 3.7). A linearly independent system $\{D_n\}_{n=0,1,\ldots} \subset \mathrm{Pot}(\mathcal{A}^c)$ is called a ($\mathrm{Pot}(\mathcal{A}^c)$-generated) $\mathrm{L}^2(\partial\mathcal{G})$-*Dirichlet Runge basis* if

$$\overline{\operatorname*{span}_{n\in\mathbb{N}_0}\{D_n|_{\partial\mathcal{G}}\}}^{\|\cdot\|_{\mathrm{L}^2(\partial\mathcal{G})}} = \mathrm{L}^2(\partial\mathcal{G}). \tag{3.411}$$

A linearly independent system $\{N_n\}_{n=0,1,\ldots} \subset \mathrm{Pot}(\mathcal{A}^c)$ is called a ($\mathrm{Pot}(\mathcal{A}^c)$-generated) $\mathrm{L}^2(\partial\mathcal{G})$-*Neumann Runge basis* if

$$\overline{\operatorname*{span}_{n\in\mathbb{N}_0}\left\{ \left.\frac{\partial N_n}{\partial\nu}\right|_{\partial\mathcal{G}} \right\}}^{\|\cdot\|_{\mathrm{L}^2(\partial\mathcal{G})}} = \mathrm{L}^2(\partial\mathcal{G}). \tag{3.412}$$

Corollary 3.96. *Let $\mathcal{A}, \mathcal{G} \subset \mathbb{R}^3$ be regular regions such that $\mathcal{A} \Subset \mathcal{G}$ holds true.*

(EDP): *Let $\{D_n^*\}_{n=0,1,\ldots} \subset \mathrm{Pot}(\mathcal{A}^c)$, be a function system generated by (Gram–Schmidt) orthonormalization of an $\mathrm{L}^2(\partial\mathcal{G})$-Dirichlet Runge basis $\{D_n\}_{n=0,1,\ldots} \subset \mathrm{Pot}(\mathcal{A}^c)$, such that*

$$(D_n^*, D_m^*)_{\mathrm{L}^2(\partial\mathcal{G})} = \int_{\partial\mathcal{G}} D_n^*(x) D_m^*(x)\, d\omega(x) \;=\; \delta_{n,m}. \tag{3.413}$$

If $F \in \mathrm{C}^{(0)}(\partial\mathcal{G})$, then

$$\lim_{m\to\infty} \left(\int_{\partial\mathcal{G}} \left| F(x) - F^{(m)}(x) \right|^2 \, d\omega(x) \right)^{\frac{1}{2}} = 0, \tag{3.414}$$

where

$$F^{(m)} = \sum_{n=0}^{m} (F, D_n^*)_{\mathrm{L}^2(\partial\mathcal{G})} \, D_n^*|_{\partial\mathcal{G}}. \tag{3.415}$$

The potential $U \in \mathrm{Pot}^{(0)}(\overline{\mathcal{G}^c})$ *satisfying* $U^+\big|_{\partial\mathcal{G}} = F$ *can be represented in the form*

$$\lim_{m\to\infty} \sup_{x\in\overline{\mathcal{K}}} \left| U(x) - U^{(m)}(x) \right| = 0, \tag{3.416}$$

for every $\mathcal{K} \Subset \mathcal{G}^c$, *where*

$$U^{(m)} = \sum_{n=0}^{m} (F, D_n^*)_{\mathrm{L}^2(\partial\mathcal{G})}\ D_n^*. \tag{3.417}$$

(ENP): *Let* $\{N_n^*\}_{n=0,1,\ldots} \subset \mathrm{Pot}(\mathcal{A}^c)$, *be a function system generated by (Gram–Schmidt) orthonormalization of an* $\mathrm{L}^2(\partial\mathcal{G})$*-Neumann Runge basis* $\{N_n\}_{n=0,1,\ldots} \subset \mathrm{Pot}(\mathcal{A}^c)$ *such that*

$$\left(\frac{\partial N_n^*}{\partial\nu}, \frac{\partial N_m^*}{\partial\nu} \right)_{\mathrm{L}^2(\partial\mathcal{G})} = \int_{\partial\mathcal{G}} \frac{\partial N_n^*}{\partial\nu}(x) \frac{\partial N_m^*}{\partial\nu}(x)\ d\omega(x) = \delta_{n,m}. \tag{3.418}$$

If $F \in \mathrm{C}^{(0)}(\partial\mathcal{G})$, *then*

$$\lim_{m\to\infty} \left(\int_{\partial\mathcal{G}} \left| F(x) - F^{(m)}(x) \right|^2\ d\omega(x) \right)^{\frac{1}{2}} = 0, \tag{3.419}$$

where

$$F^{(m)} = \sum_{n=0}^{m} \left(F, \frac{\partial N_n^*}{\partial\nu} \right)_{\mathrm{L}^2(\partial\mathcal{G})} \frac{\partial N_n^*}{\partial\nu}\Big|_{\partial\mathcal{G}}. \tag{3.420}$$

The potential $U \in \mathrm{Pot}^{(1)}(\overline{\mathcal{G}^c})$ *satisfying* $\frac{\partial U^+}{\partial\nu}\Big|_{\partial\mathcal{G}} = F$ *can be represented in the form*

$$\lim_{m\to\infty} \sup_{x\in\overline{\mathcal{K}}} \left| U(x) - U^{(m)}(x) \right| = 0, \tag{3.421}$$

for every $\mathcal{K} \Subset \mathcal{G}^c$, *where*

$$U^{(m)} = \sum_{n=0}^{m} \left(F, \frac{\partial N_n^*}{\partial\nu} \right)_{\mathrm{L}^2(\partial\mathcal{G})} N_n^*. \tag{3.422}$$

The *(generalized) Fourier expansions* (3.415) and (3.420) are indeed constructed to have the permanence property: The transition from $F^{(m)}$ to $F^{(m+1)}$, and therefore from $U^{(m)}$ to $U^{(m+1)}$, necessitates merely the addition of one more term; all the other terms obtained formerly remaining unchanged. This is characteristic of orthogonal expansions (generalized Fourier series).

In connection with the $\mathrm{L}^2(\partial\mathcal{G})$-regularity theorems, we find the following estimates:

(EDP): For given $F \in \mathrm{C}^{(0)}(\partial\mathcal{G})$, let U satisfy $U \in \mathrm{Pot}^{(0)}(\overline{\mathcal{G}^c})$, $U^+|_{\partial\mathcal{G}} = F$. Then

$$\sup_{x\in\overline{\mathcal{K}}} \left| \left(\nabla^{(k)} U\right)(x) - \sum_{n=0}^{m} (F, D_n^*)_{\mathrm{L}^2(\partial\mathcal{G})} \left(\nabla^{(k)} D_n^*\right)(x) \right| \leq C \left(\|F\|^2_{\mathrm{L}^2(\partial\mathcal{G})} - \sum_{n=0}^{m} (F, D_n^*)^2_{\mathrm{L}^2(\partial\mathcal{G})} \right)^{\frac{1}{2}} \tag{3.423}$$

holds for all $k \in \mathbb{N}_0$ and all subsets $\mathcal{K} \Subset \mathcal{G}^c$.

(ENP): For given $F \in \mathrm{C}^{(0)}(\partial\mathcal{G})$, let U satisfy $U \in \mathrm{Pot}^{(1)}(\overline{\mathcal{G}^c})$, $\frac{\partial U^+}{\partial\nu}|_{\partial\mathcal{G}} = F$. Then

$$\sup_{x\in\overline{\mathcal{K}}} \left| \left(\nabla^{(k)} U\right)(x) - \sum_{n=0}^{m} \left(F, \frac{\partial N_n^*}{\partial\nu}\right)_{\mathrm{L}^2(\partial\mathcal{G})} \left(\nabla^{(k)} N_n^*\right)(x) \right| \leq C \left(\|F\|^2_{\mathrm{L}^2(\partial\mathcal{G})} - \sum_{n=0}^{m} \left(F, \frac{\partial N_n^*}{\partial\nu}\right)^2_{\mathrm{L}^2(\partial\mathcal{G})} \right)^{\frac{1}{2}} \tag{3.424}$$

holds for all $k \in \mathbb{N}_0$ and all subsets $\mathcal{K} \Subset \mathcal{G}^c$.

In addition, Corollary 3.96 indicates that $F - F^{(m)}$ is $\mathrm{L}^2(\partial\mathcal{G})$-orthogonal to all members of the $\mathrm{L}^2(\partial\mathcal{G})$-orthonormal Runge basis up to the index m. This observation is valid for the Dirichlet as well as the Neumann case. It leads us to the following result.

Corollary 3.97. *Let $\mathcal{A}, \mathcal{G} \subset \mathbb{R}^3$ be regular regions such that $\mathcal{A} \Subset \mathcal{G}$ holds true.*

(EDP): *Let $\{D_n\}_{n=0,1,\ldots} \subset \mathrm{Pot}(\mathcal{A}^c)$, be an $\mathrm{L}^2(\partial\mathcal{G})$-Dirichlet Runge. If $F \in \mathrm{C}^{(0)}(\partial\mathcal{G})$, then*

$$\lim_{m\to\infty} \left(\int_{\partial\mathcal{G}} \left| F(x) - F^{(m)}(x) \right|^2 \, d\omega(x) \right)^{\frac{1}{2}} = 0, \tag{3.425}$$

where the coefficients $a_0^m, \ldots, a_m^m$ of the function

$$F^{(m)} = \sum_{n=0}^{m} a_n^m D_n\big|_{\partial\mathcal{G}} \tag{3.426}$$

satisfy the "normal equations"

$$\sum_{n=0}^{m} a_n^m (D_k, D_n)_{\mathrm{L}^2(\partial\mathcal{G})} = (D_k, F)_{\mathrm{L}^2(\partial\mathcal{G})}, \quad k = 0, \ldots, m. \tag{3.427}$$

The potential $U \in \mathrm{Pot}^{(0)}(\overline{\mathcal{G}^c})$ satisfying $U^+\Big|_{\partial\mathcal{G}} = F$, can be represented in the form

$$\lim_{m\to\infty} \sup_{x\in\overline{\mathcal{K}}} \left| U(x) - U^{(m)}(x) \right| = 0, \tag{3.428}$$

where

$$U^{(m)} = \sum_{n=0}^{m} a_n^m D_n \tag{3.429}$$

for every $\mathcal{K} \Subset \mathcal{G}^c$.

(ENP): *Let* $\{N_n\}_{n=0,1,\ldots}$, $N_n^* \in \mathrm{Pot}(\mathcal{A}), n = 0, 1, \ldots$ *be an* $\mathrm{L}^2(\partial\mathcal{G})$-*Neumann Runge basis. If* $F \in \mathrm{C}^{(0)}(\partial\mathcal{G})$, *then*

$$\lim_{m\to\infty} \left(\int_{\partial\mathcal{G}} \left| F(x) - F^{(m)}(x) \right|^2 \, d\omega(x) \right)^{\frac{1}{2}} = 0, \tag{3.430}$$

where the coefficients $a_0^m, \ldots, a_m^m$ *of the function*

$$F^{(m)} = \sum_{n=0}^{m} a_n^m \left. \frac{\partial N_n}{\partial \nu} \right|_{\partial\mathcal{G}} \tag{3.431}$$

satisfy the "normal equations"

$$\sum_{n=0}^{m} a_n^m \left(\frac{\partial N_k}{\partial \nu}, \frac{\partial N_n}{\partial \nu} \right)_{\mathrm{L}^2(\partial\mathcal{G})} = \left(\frac{\partial N_k}{\partial \nu}, F \right)_{\mathrm{L}^2(\partial\mathcal{G})}, \quad k = 0, \ldots, m. \tag{3.432}$$

The potential $U \in \mathrm{Pot}^{(1)}(\overline{\mathcal{G}^c})$ *satisfying* $\left. \frac{\partial U^+}{\partial \nu} \right|_{\partial\mathcal{G}} = F$, *can be represented in the form*

$$\lim_{m\to\infty} \sup_{x\in\overline{\mathcal{K}}} \left| U(x) - U^{(m)}(x) \right| = 0 \tag{3.433}$$

where

$$U^{(m)} = \sum_{n=0}^{m} a_n^m N_n \tag{3.434}$$

for every $\mathcal{K} \Subset \mathcal{G}^c$.

The approximation of boundary values and potential by the method of generalized Fourier expansion in terms of outer harmonics is achieved by superposition of functions with *oscillating character*. The oscillations grow in number, but they decrease in size with increasing truncation order. The oscillating character of the generalized Fourier expansions remains true (cf. W. Freeden [1983]) if other trial bases are used (for example, mass (single)poles, and certain kernel function representations such as Abel–Poisson and singularity kernels). Thus, generalized Fourier expansions provide least squares approximation by successive oscillations, which become larger and larger in number, but smaller and smaller in amplitude. It is therefore not (as A. Sommerfeld [1978] already pointed out) a technique of *osculating character* (as, e.g., interpolation or smoothing in reproducing Hilbert spaces by harmonic splines as proposed in Chapter 4 (see also W. Freeden [1981, 1987], L. Shure et al. [1982]).

Since C.F. Gauss [1838], there is evidence – at least in the spherical context using multipoles, i.e., outer harmonics – that a Fourier expansion provides an excellent *trend approximation* of a harmonic function such as the Earth's gravitational and magnetic potential. The ideal *frequency localization* of outer harmonics – each of them referring to a certain degree of oscillation – has proved to be extraordinarily advantageous due to to the physical interpretability and the immediate comparability of the Fourier coefficients for observables. From a numerical point of view, however, trial functions that show ideal frequency as well as space localization on the reference sphere would be desirable. The uncertainty principle (see, e.g., W. Freeden, O. Glockner, M. Schreiner [1999] and the references therein) teaches us that both properties are mutually exclusive (except in the trivial case). This explains some problems in the Fourier technique of approximation, at least by means of outer harmonics. Fourier expansions in terms of outer harmonics are well suited to resolve low-frequency ingredients in an observable, while their application is critical to obtain high-resolution phenomena.

As we know, e.g., from W. Freeden, M. Schreiner [2009], a suitable superposition of outer harmonics leads to so-called kernel functions (such as the Abel–Poisson kernel, the singularity kernel, etc.) showing a reduced frequency but increased space localization on the reference sphere $\Omega_{\underline{R}}$. The series conglomerates of outer harmonics, i.e., the kernel functions, are constructed to cover various spectral bands. Hence, they show certain intermediate stages of frequency and space localization. A particular kernel in potential theory is the mass (single)pole kernel, i.e., the fundamental solution to the Laplacian (being the Kelvin transformed counterpart of the singularity kernel). This kernel interrelates the length of its spectral bands to the distance of the mass point from the reference sphere. The fundamental solution (singularity kernel) is more and more space localized and simultaneously less frequency localized the closer the pole is to the sphere. All in all, seen from a methodological point of view, Fourier approaches using a sequence of kernel functions corresponding to an inner fundamental system can be realized in equivalent manner to outer harmonics expansions for completely recovering a gravitational potential. In fact, a sequence of kernel functions is even conceptually easier to implement than outer harmonic expansions, as long as the kernels are available in closed form as elementary functions. Consequently, kernel function approximations have a long history. Early attempts to make the *method of fundamental solutions* become reality date back to the first decades of the last century (cf. W. Ritz [1909], E. Trefftz [1926]). Further ideas are, e.g., due to I.N. Vekua [1953], V. Kupradze, M. Aleksidze [1964], E. Kita, N. Kamiya [1995], M. Golberg [1995]. The line to the Fourier approach as presented in this work follows J.L. Walsh [1929], I.N. Vekua [1953], W. Freeden [1980, 1983], M. Golberg, C. Chen [1998], W. Freeden, C. Mayer [2003],W. Freeden et al. [2003], and W. Freeden, V. Michel [2004]. In the meantime generalized Fourier expansions are theoretically established and practically applied not only to the Laplace equation, but also to more general elliptic partial differential equations, e.g., the reduced (Helmholtz) wave equation, the Cauchy-Navier equation, (reduced) Maxwell equations, the (linear) Stokes equations, etc. (see, e.g., W. Freeden, R. Reuter [1989], W. Freeden, V. Michel [2004], W. Freeden, C. Mayer [2007], C. Mayer [2007], W. Freeden, M. Schreiner [2009] for a more detailed list of references). The drawback of the numerical realization is the need for an adequate selection of a finite number of points out of the infinite inner fundamental system. An optimal strategy for positioning the finite system in a computationally efficient and physically relevant way remains a great challenge for future work.

3.5.3 Closure in $\mathrm{C}^{(0)}$-Topology

From our considerations leading to locally uniform approximation, we know for a regular region $\mathcal{G}$ with $\underline{R} < \inf_{x\in\partial\mathcal{G}} |x|$ that

$$\overline{D^+}^{\|\cdot\|_{\mathrm{L}^2(\partial\mathcal{G})}} = \overline{\operatorname*{span}_{\substack{n=0,1,\ldots\\ j=1,\ldots,2n+1}} \{H^{\underline{R}}_{-n-1,j}|_{\partial\mathcal{G}}\}}^{\|\cdot\|_{\mathrm{L}^2(\partial\mathcal{G})}} = \mathrm{L}^2(\partial\mathcal{G}), \tag{3.435}$$

$$\overline{N^+}^{\|\cdot\|_{\mathrm{L}^2(\partial\mathcal{G})}} = \overline{\operatorname*{span}_{\substack{n=0,1,\ldots\\ j=1,\ldots,2n+1}} \left\{ \frac{\partial}{\partial\nu} H^{\underline{R}}_{-n-1,j}\bigg|_{\partial\mathcal{G}} \right\}}^{\|\cdot\|_{\mathrm{L}^2(\partial\mathcal{G})}} = \mathrm{L}^2(\partial\mathcal{G}). \tag{3.436}$$

The same results remain valid when the regular surface $\partial\mathcal{G}$ is replaced by any inner parallel surface $\partial\mathcal{G}(-\tau)$ of distance $|\tau|$ to $\partial\mathcal{G}$ (where $|\tau|$ is chosen sufficiently small (cf. (3.105)). This fact will be exploited to verify the following closure properties (see W. Freeden [1980]).

Theorem 3.98. *Let $\partial\mathcal{G}$ be a regular region with $\underline{R} < \inf_{x\in\partial\mathcal{G}} |x|$. Then the following statements are true:*

(a) $\{H^{\underline{R}}_{-n-1,j}|_{\partial\mathcal{G}}\}_{n\in\mathbb{N}_0, j=1,\ldots,2n+1}$ *is closed in* $\mathrm{C}^{(0)}(\partial\mathcal{G})$:

$$\overline{\operatorname*{span}_{\substack{n=0,1,\ldots\\ j=1,\ldots,2n+1}} \{H^{\underline{R}}_{-n-1,j}|_{\partial\mathcal{G}}\}}^{\|\cdot\|_{\mathrm{C}^{(0)}(\partial\mathcal{G})}} = D^+ = \mathrm{C}^{(0)}(\partial\mathcal{G}), \tag{3.437}$$

(b) $\left\{\frac{\partial}{\partial\nu} H^{\underline{R}}_{-n-1,j}\big|_{\partial\mathcal{G}}\right\}_{n\in\mathbb{N}_0, j=1,\ldots,2n+1}$ *is closed in* $\mathrm{C}^{(0)}(\partial\mathcal{G})$:

$$\overline{\operatorname*{span}_{\substack{n=0,1,\ldots\\ j=1,\ldots,2n+1}} \left\{ \frac{\partial}{\partial\nu} H^{\underline{R}}_{-n-1,j}\bigg|_{\partial\mathcal{G}} \right\}}^{\|\cdot\|_{\mathrm{C}^{(0)}(\partial\mathcal{G})}} = N^+ = \mathrm{C}^{(0)}(\partial\mathcal{G}). \tag{3.438}$$

Proof. We restrict ourselves to statement (a). Let F be an element of $\mathrm{C}^{(0)}(\partial\mathcal{G})$. Then the operator equation between F and the function Q of a layer potential of type (3.343) is given by

$$F = \left(\frac{1}{2}I + A + P_{|2}(0,0)\right)[Q]. \tag{3.439}$$

Since we know that $P_{|2}(0,0) = P_{|1}(0,0)^*$, this equation is equivalent to

$$-F = \left(-\frac{1}{2}I - A - P_{|1}(0,0)^*\right)[Q]. \tag{3.440}$$

According to the limit formulas for the adjoint operators it follows that

$$\lim_{\tau\to 0+} \left\| P_{|1}(-\tau,0)^*[Q] - P_{|1}(0,0)^*[Q] - \frac{1}{2}Q \right\|_{C^{(0)}(\partial\mathcal{G})} = 0. \tag{3.441}$$

In connection with our operator equation (3.440), this means that

$$\lim_{\tau\to 0+} ||P_{|1}(-\tau,0)^*[Q] - F + A[Q]\,||_{C^{(0)}(\partial\mathcal{G})} = 0. \tag{3.442}$$

Next, we show that the integral extended over the surface $\partial\mathcal{G}$

$$P_{|1}(-\tau,0)^*[Q](x) = -\frac{1}{4\pi}\int_{\partial\mathcal{G}} Q(y)\frac{\nu(y)\cdot(y-\tau\nu(y)-x)}{|y-\tau\nu(y)-x|^3}\,d\omega(y) \tag{3.443}$$

can be expressed as an integral over the parallel surface $\partial\mathcal{G}(-\tau)$. To this end, we borrow from differential geometry (see, e.g., C. Müller [1969]) that, for sufficiently small $|\tau|$, the surface element $d\omega_{-\tau}$ of $\partial\mathcal{G}(-\tau)$ may be written in the form

$$d\omega_{-\tau} = (1 + 2H\tau + L\tau^2)\, d\omega, \tag{3.444}$$

where H is the mean curvature and L is the Gaussian curvature of $\partial\mathcal{G}$. Since the normals of the parallel surfaces $\partial\mathcal{G}(-\tau)$ coincide with the normals on $\partial\mathcal{G}$, a simple transformation gives

$$\begin{aligned} P_{|1}(-\tau,0)^*[Q](x) &= -\frac{1}{4\pi}\int_{\partial\mathcal{G}(-\tau)} Q_\tau(y)\frac{\nu(y)\cdot(y-x)}{|y-x|^3}\,d\omega_{-\tau}(y) \qquad (3.445)\\ &= \frac{1}{4\pi}\int_{\partial\mathcal{G}(-\tau)} Q_\tau(y)\frac{\partial}{\partial\nu(y)}\frac{1}{|x-y|}\,d\omega_{-\tau}(y), \end{aligned}$$

where

$$Q_\tau(x) = \frac{Q(x+\tau\nu(x))}{1+2H(x+\tau\nu(x))\tau + L(x+\tau\nu(x))\tau^2}. \tag{3.446}$$

$P_{|1}(-\tau,0)^*$ can be regarded as the double-layer potential operator with the density Q_τ on the (inner) parallel surface $\partial\mathcal{G}(-\tau)$. Furthermore, according to (3.442), $(P_{|1}(-\tau,0)^* + A)[Q] \to F$ in the norm $||\cdot||_{C^{(0)}(\partial\mathcal{G})}$ as $\partial\mathcal{G}(-\tau)\to\partial\mathcal{G}$. Therefore, for any given $\varepsilon > 0$, we can find a surface $\partial\mathcal{G}(-\tau_\varepsilon)$ such that

$$||P_{|1}(-\tau_\varepsilon,0)^*[Q] + A[Q] - F||_{C^{(0)}(\partial\mathcal{G})} \leq \frac{\varepsilon}{2}\,. \tag{3.447}$$

By $F_{-\tau_\varepsilon}$ we denote the restriction of the potential

$$U_{-\tau_\varepsilon}(x) = \frac{1}{4\pi}\int_{\partial\mathcal{G}(-\tau_\varepsilon)} Q_{\tau_\varepsilon}(y)\left(\frac{\partial}{\partial\nu(y)}\frac{1}{|x-y|} - \frac{1}{|x|}\right) d\omega_{\tau_\varepsilon}(y) \tag{3.448}$$

to the surface $\partial\mathcal{G}(-\tau_\varepsilon)$, i.e., $F_{-\tau_\varepsilon} = U_{-\tau_\varepsilon}|_{\partial\mathcal{G}(-\tau_\varepsilon)}$. The function $F_{-\tau_\varepsilon}$ is continuous on $\partial\mathcal{G}(-\tau_\varepsilon)$ and the potential $U_{-\tau_\varepsilon}$ represents the solution of Dirichlet's exterior problem corresponding to the boundary $\partial\mathcal{G}(-\tau_\varepsilon)$ and the boundary value $F_{-\tau_\varepsilon}$. According to our approach, $\{H^{\underline{R}}_{-n-1,j}|_{\partial\mathcal{G}(-\tau_\varepsilon)}\}_{n\in\mathbb{N}_0, j=1,\ldots,2n+1}$ is closed in $\mathrm{L}^2(\partial\mathcal{G}(-\tau_\varepsilon))$. Consequently, from the L^2-closure we are able to deduce that there exist an integer $m(=m(\varepsilon))$ and real numbers $a_{n,j}$ such that the inequalities

$$\left(\int_{\partial\mathcal{G}(-\tau_\varepsilon)} \left(F_{-\tau_\varepsilon}(y) - \sum_{n=0}^{m}\sum_{j=1}^{2n+1} a_{n,j}\; H^{\underline{R}}_{-n-1,j}(y)\right)^2 \, d\omega_{-\tau_\varepsilon}(y)\right)^{\frac{1}{2}} \leq \frac{\varepsilon}{2\mathrm{C}} \tag{3.449}$$

and

$$\sup_{x\in\overline{\mathcal{K}}} \left| U_{-\tau_\varepsilon}(x) - \sum_{n=0}^{m}\sum_{j=1}^{2n+1} a_{n,j}\; H^{\underline{R}}_{-n-1,j}(x)\right| \tag{3.450}$$

$$\leq \mathrm{C}\left(\int_{\partial\mathcal{G}(-\tau_\varepsilon)} \left|F_{-\tau_\varepsilon}(y) - \sum_{n=0}^{m}\sum_{j=1}^{2n+1} a_{n,j} H^{\underline{R}}_{-n-1,j}(y)\right|^2 \, d\omega_{-\tau_\varepsilon}(y)\right)^{\frac{1}{2}} \leq \frac{\varepsilon}{2}$$

hold true for each compact subset $\mathcal{K} \Subset \mathbb{R}^3\backslash\overline{\mathcal{G}(-\tau_\varepsilon)}$. In particular, for a compact set $\mathcal{K} \Subset \mathbb{R}^3 \setminus \overline{\mathcal{G}(-\tau_\varepsilon)}$ with $\partial\mathcal{G} \subset \mathcal{K}$ we get

$$\sup_{x\in\partial\mathcal{G}} \left| U_{-\tau_\varepsilon}(x) - \sum_{n=0}^{m}\sum_{j=1}^{2n+1} a_{n,j} H^{\underline{R}}_{-n-1,j}(x)\right| \leq \frac{\varepsilon}{2}. \tag{3.451}$$

From the relations (3.447) and (3.451) we are able to show via the triangle inequality that

$$\sup_{x\in\partial\mathcal{G}} \left| F(x) - \sum_{n=0}^{m}\sum_{j=1}^{2n+1} a_{n,j} H^{\underline{R}}_{-n-1,j}(x)\right| \tag{3.452}$$

$$\leq \sup_{x\in\partial\mathcal{G}} |F(x) - U_{-\tau_\varepsilon}(x)| + \sup_{x\in\partial\mathcal{G}} \left| U_{-\tau_\varepsilon}(x) - \sum_{n=0}^{m}\sum_{j=1}^{2n+1} a_{n,j} H^{\underline{R}}_{-n-1,j}(x)\right| \leq \varepsilon.$$

This proves Theorem 3.98 (a). Part (b) follows analogously. □

Remark 3.99. The same arguments leading to the $\mathrm{C}^{(0)}(\partial\mathcal{G})$-closure of outer harmonics on $\partial\mathcal{G}$ apply to all other systems for which the $\mathrm{L}^2(\partial\mathcal{G})$-closure is known.

Combining our results obtained by Theorem 3.98, we easily arrive at the following statement.

Theorem 3.100. *Let $\mathcal{G} \subset \mathbb{R}^3$ be a regular region with $\underline{R} < \inf_{x \in \partial\mathcal{G}} |x|$. Then the following statements are valid:*

(EDP): *For a given function $F \in \mathrm{C}^{(0)}(\partial\mathcal{G})$, let $U \in \mathrm{Pot}^{(0)}(\overline{\mathcal{G}^c})$ satisfy $U^+|_{\partial\mathcal{G}} = F$. Then, for every $\varepsilon > 0$, there exist an integer m (depending on ε) and a finite set of real numbers $a_{n,j}$ such that*

$$\begin{aligned} \sup_{x \in \overline{\mathcal{G}^c}} & \left| U(x) - \sum_{n=0}^{m} \sum_{j=1}^{2n+1} a_{n,j} H^{\underline{R}}_{-n-1,j}(x) \right| \\ \leq & \sup_{x \in \partial\mathcal{G}} \left| F(x) - \sum_{n=0}^{m} \sum_{j=1}^{2n+1} a_{n,j} H^{\underline{R}}_{-n-1,j}(x) \right| \\ \leq & \ \varepsilon. \end{aligned}$$

(ENP): *For a given function $F \in \mathrm{C}^{(0)}(\partial\mathcal{G})$, let $U \in \mathrm{Pot}^{(1)}(\overline{\mathcal{G}^c})$ satisfy $\frac{\partial U^+}{\partial \nu}\big|_{\partial\mathcal{G}} = F$. Then, for every $\varepsilon > 0$, there exist an integer m (depending on ε) and a finite set of real numbers $a_{n,j}$ such that*

$$\begin{aligned} \sup_{x \in \overline{\mathcal{G}^c}} & \left| U(x) - \sum_{n=0}^{m} \sum_{j=1}^{2n+1} a_{n,j} H^{\underline{R}}_{-n-1,j}(x) \right| \\ \leq & \ C \sup_{x \in \partial\mathcal{G}} \left| F(x) - \sum_{n=0}^{m} \sum_{j=1}^{2n+1} a_{n,j} \frac{\partial}{\partial \nu} H^{\underline{R}}_{-n-1,j}(x) \right| \\ \leq & \ C\varepsilon. \end{aligned}$$

Unfortunately, a constructive procedure of determining best approximate coefficients $a_{n,j}$ in the $\mathrm{C}^{(0)}(\partial\mathcal{G})$-topology seems to be unknown. Therefore, in Chapter 4, harmonic splines will be introduced in reproducing Hilbert subspaces of $\mathrm{Pot}^{(0)}(\mathbb{R}^3 \backslash \mathcal{B}_{\underline{R}}(0))$ (characterized by variational principles), so that the spline method can be regarded as an immediate extension of the method of generalized Fourier expansions to reproducing kernel subspaces of $\mathrm{Pot}^{(0)}(\mathbb{R}^3 \backslash \mathcal{B}_{\underline{R}}(0))$, hence, providing coefficients that are optimal in a different (Sobolev like) norm.

3.6 Exercises

Exercise 3.1. Verify that

$$\int_{\Omega_r(x)} |G(\Delta; |x-y|)| \ d\omega(y) \ = \ \mathcal{O}(r), \ r \to 0, \tag{3.453}$$

and

$$\int_{\Omega_r(x)} \left| \frac{\partial}{\partial x_i} G(\Delta; |x-y|) \right| \, d\omega(y) \;=\; \mathcal{O}(1), \; r \to 0, \; i = 1, 2, 3. \tag{3.454}$$

Exercise 3.2. Let F be of class $\mathrm{C}^{(0)}(\overline{\mathcal{B}_R(x)})$, $R > 0, x \in \mathbb{R}^3$. Show that the function $r \mapsto U(r)$, $r \in (0, R)$, given by

$$U(r) = \int_{\mathcal{B}_r(x)} F(y) \, dV(y) \tag{3.455}$$

is continuously differentiable, and that

$$U'(r) = r^2 \int_\Omega F(x + r\eta) \, d\omega(\eta), \tag{3.456}$$

where U' denotes the one-dimensional derivative of U.

Exercise 3.3. Let F be of class $\mathrm{C}^{(0)}(\overline{\mathcal{B}_R(x)})$, $R > 0$, $x \in \mathbb{R}^3$. Prove that

$$\int_\Omega F(x + r\eta) \, d\omega(\eta) = \int_\Omega F(x + r\mathbf{t}\eta) \, d\omega(\eta) \tag{3.457}$$

holds true for all $r \in (0, R)$ and all orthogonal transformations $\mathbf{t} \in \mathbb{R}^{3\times3}$ (remember $\Omega = \partial\mathcal{B}_1(0)$).

Exercise 3.4. Let F be of class $\mathrm{C}^{(1)}(\overline{\mathcal{B}_r(x)})$, $R > 0$, $x \in \mathbb{R}^3$. Verify that

$$P(r) = \int_\Omega F(x + r\eta) \, d\omega(\eta), \; r \in (0, R), \tag{3.458}$$

is continuously differentiable such that

$$P'(r) = \int_\Omega \eta \cdot \nabla F(x + r\eta) \, d\omega(\eta), \; r \in (0, R). \tag{3.459}$$

Exercise 3.5. Let $\mathcal{G} \subset \mathbb{R}^3$ be a regular region. Prove that the following statements are equivalent:

(a) $U \in \mathrm{C}^{(0)}(\mathcal{G})$ satisfies the Mean Value Property

(b) U is of class $\mathrm{C}^{(2)}(\mathcal{G})$ with $\Delta U = 0$ in $\mathcal{G}$.

Exercise 3.6. Let $\mathcal{G} \subset \mathbb{R}^3$ be a regular region. Assume that $U \in \mathrm{C}^{(2)}(\mathcal{G}) \cap \mathrm{C}^{(0)}(\overline{\mathcal{G}})$ is non-constant in $\mathcal{G}$. Verify that

(a) $\Delta U \geq 0$ in $\mathcal{G}$ implies $U(x) < \sup_{y \in \partial\mathcal{G}} U(y)$, $x \in \mathcal{G}$.

(b) $\Delta U \leq 0$ in $\mathcal{G}$ implies $U(x) > \inf_{y \in \partial\mathcal{G}} U(y)$, $x \in \mathcal{G}$

(c) $\Delta U = 0$ in $\mathcal{G}$ implies $|U(x)| < \sup_{y \in \partial\mathcal{G}} |U(y)|$, $x \in \mathcal{G}$.

Exercise 3.7. Let $\mathcal{G} \subset \mathbb{R}^3$ be a regular region. Show by use of Harnack's inequality that for every $\mathcal{K} \Subset \mathcal{G}$, there exists a constant C depending only on $\mathcal{G}$ such that

$$\max_{x \in \overline{\mathcal{K}}} U(x) \le \mathrm{C} \min_{x \in \overline{\mathcal{K}}} U(x), \tag{3.460}$$

provided that U is a non-negative harmonic function in $\mathcal{G}$.

Exercise 3.8. Let $\mathcal{G} \subset \mathbb{R}^3$ be a regular region. Suppose that U is of class $\mathrm{C}^{(2)}(\mathcal{G}) \cap \mathrm{C}^{(0)}(\overline{\mathcal{G}})$ and satisfies $U|_{\partial\mathcal{G}} = 0$ and

$$\int_{\mathcal{G}} |\Delta U(x)| \, dV(x) < \infty. \tag{3.461}$$

In addition, let $W \in \mathrm{C}^{(2)}(\mathcal{G}) \cap \mathrm{C}^{(0)}(\overline{\mathcal{G}})$ satisfy the same properties. Show that under this assumption, the following three integrals exist and are equal:

$$\begin{aligned} \int_{\mathcal{G}} \nabla U(x) \cdot \nabla W(x) \, dV(x) &= -\int_{\mathcal{G}} U(x) \Delta W(x) \, dV(x), \qquad (3.462) \\ &= -\int_{\mathcal{G}} W(x) \Delta U(x) \, dV(x). \end{aligned}$$

Exercise 3.9. Prove that, for $x, y \in \mathbb{R}^3 \backslash \{0\}$,

$$\left| \frac{x}{|x|} - y|x| \right| = \left| \frac{y}{|y|} - x|y| \right|. \tag{3.463}$$

Exercise 3.10. Let ${}^l Z^\rho(r)$, $r \in (0, \infty)$, be given by

$${}^l Z^\rho(r) = \begin{cases} \frac{1}{8\pi\rho^3}(5 - \frac{3}{\rho^2} r^3), & r \le \rho, \\ \frac{1}{4\pi r^3}, & r > \rho. \end{cases} \tag{3.464}$$

Show that

$$\nabla_x \cdot (x \, {}^l Z^\rho(|x|)) = \frac{15}{8\pi} \left(\frac{1}{\rho^3} - \frac{|x|^2}{\rho^5} \right), \qquad |x| \le \rho. \tag{3.465}$$

Exercise 3.11. Suppose that $|\tau| < \frac{1}{4M}$. Show that

$$\begin{aligned} & \left| |x - \tau\nu(x) - y|^k - |x - \tau\nu(x) - y|^k \right| \qquad (3.466) \\ \le \; & 2^{\frac{4-k}{2}} 3^{\frac{k-1}{2}} M k |\tau| |x - y|^2 (\tau^2 + |x - y|^2)^{\frac{k-2}{2}} \end{aligned}$$

holds true for $k \in \mathbb{N}$ and all $x \in \partial\mathcal{G}$, $y \in \partial\mathcal{G} \cap \mathcal{B}_\delta(x)$ (with δ, M specified by Definition 1.7).

Exercise 3.12. Show that the following exterior third boundary-value problem is uniquely solvable:

(ETBP): Let $F \in \mathrm{C}^{(0)}(\partial\mathcal{G})$ and $H \in \mathrm{C}^{(0)}(\partial\mathcal{G})$ be given functions. Furthermore, assume that H is non-zero and that $H(x) \ge 0$ for all $x \in \partial\mathcal{G}$. Find a function $V \in \mathrm{Pot}^{(1)}(\overline{\mathcal{G}})$ satisfying

$$\left. \frac{\partial V^+}{\partial \nu} \right|_{\partial\mathcal{G}} + H \left. V^+ \right|_{\partial\mathcal{G}} = F. \tag{3.467}$$

Exercise 3.13. Prove the following statement: Let U be of class $\mathrm{Pot}^{(1)}(\overline{\mathcal{G}^c})$. Then there exists a constant C (depending on $k, \mathcal{K}$, and $\partial\mathcal{G}$) such that

$$\sup_{x\in\overline{\mathcal{K}}} \left| \left(\nabla^{(k)} U\right)(x)\right| \leq C \left(\int_{\partial\mathcal{G}} \left| \frac{\partial U^+}{\partial\nu}(y)\right|^2 \, d\omega(y)\right)^{\frac{1}{2}} \tag{3.468}$$

for all $\mathcal{K} \Subset \mathcal{G}^c$ and all $k \in \mathbb{N}_0$.

Exercise 3.14. Prove that the system

$$\left\{ \left. \frac{\partial H^R_{-n-1,j}}{\partial\nu} \right|_{\partial\mathcal{G}} \right\}_{\substack{n=0,1,\ldots \\ j=1,\ldots,2n+1}} \tag{3.469}$$

is complete in $\mathrm{L}^2(\partial\mathcal{G})$.

Exercise 3.15. Prove that for every fundamental system $\mathcal{Y} = \{y_n\}_{n=0,1,\ldots}$ in $\mathcal{G}$, the system

$$\left\{ \left. \frac{\partial}{\partial\nu} G(\Delta; |\cdot - y_n|) \right|_{\partial\mathcal{G}} \right\}_{n=0,1,\ldots} \tag{3.470}$$

is complete in $\mathrm{L}^2(\partial\mathcal{G})$.

4

Gravitation

CONTENTS

If the Earth had a perfectly spherical shape and if the mass inside the Earth were distributed homogeneously or rotationally symmetric, then the line along which a test mass fell would be a straight line, directed radially and going exactly through the Earth's center of mass. The gravitational field obtained in this way would be spherically symmetric (see Figure 4.1, left illustration).

In reality, however, the situation is much more difficult (see Figure 4.1, right illustration). Gravity as observed on the Earth's surface is the combined effect of the gravitational mass attraction and the centrifugal force due to the Earth's rotation. The force of gravity provides a directional structure to the space above the Earth. It is tangential to the vertical plumb lines and perpendicular to all level surfaces. Any water surface at rest is part of a level surface.

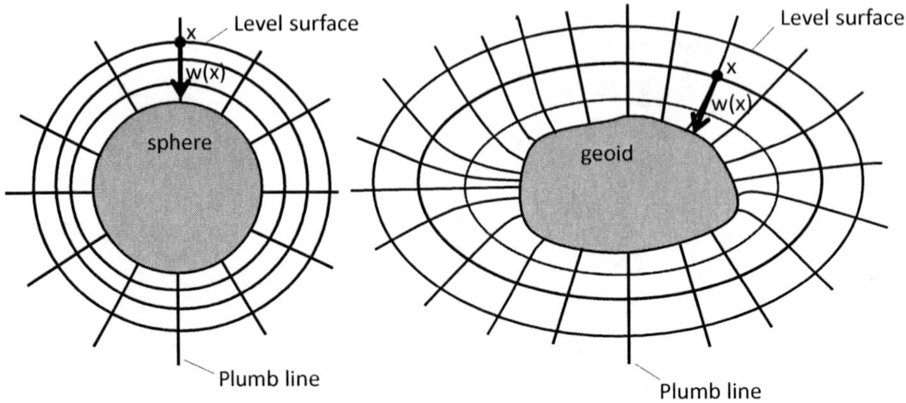

FIGURE 4.1
Level surfaces and plumb lines for a homogeneous ball (left) and an Earth-like body (right).

The *gravity acceleration (gravity)* w is the resultant of the *gravitation* v and the *centrifugal acceleration* c such that

$$w = v + c. \tag{4.1}$$

The centrifugal force c arises as a result of the rotation of the Earth about its axis. In this work we assume a rotation of constant angular velocity ω_0. The centrifugal acceleration acting on a unit mass is directed outward perpendicularly to the spin axis (see Figure 4.2). If the ε^3-axis of an Earth-fixed coordinate system with center of mass at the origin coincides with the axis of rotation, then we have $c(x) = -\omega_0^2 \varepsilon^3 \wedge (\varepsilon^3 \wedge x)$. Introducing the so-called *centrifugal potential* C, such that $c = \nabla C$, the function C turns out to be non-harmonic (cf. Exercise 4.1). The direction of the gravity w is known as the direction of the *plumb line*, the quantity $|w|$ is called the *gravity intensity* (often also just called *gravity*). Altogether, the *gravity potential of the Earth* can be expressed in the form

$$W = V + C, \tag{4.2}$$

and the gravity acceleration w is given by

$$w = \nabla W = \nabla V + \nabla C. \tag{4.3}$$

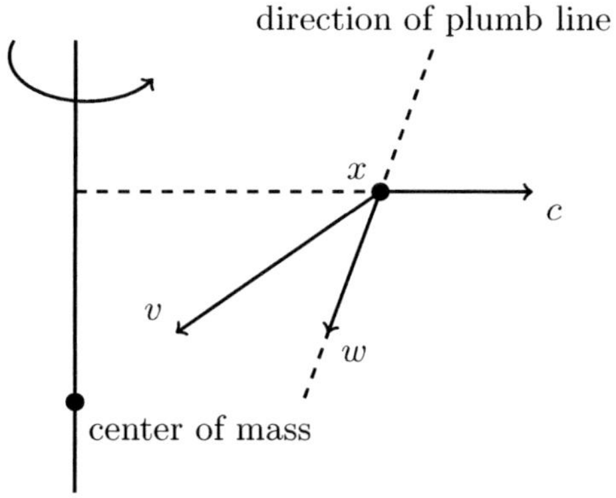

FIGURE 4.2
Gravitation v, centrifugal acceleration c, gravity acceleration w.

The surfaces of constant gravity potential, i.e., $W = \text{const.}$, are designated as *equipotential (level,* or *geopotential) surfaces of gravity* (for more details, the reader is referred to monographs in physical geodesy, e.g., W.A. Heiskanen, H. Moritz [1967], E. Groten [1979], W. Torge [1991], H. Moritz [2010]). The determination of equipotential surfaces of the potential W is strongly related to the knowledge of the potential V. The gravity vector w given by $w = \nabla W$ is normal to the equipotential surface passing through the same point. Thus, equipotential surfaces intuitively express the notion of tangential surfaces, as they are normal to the plumb lines given by the direction of the gravity vector. The equipotential surface at sea level is called *geoid.*

Seen from a mathematical point of view, the Earth's gravitational field v in the exterior of the Earth is a gradient field $v = \nabla V$, where the gravitational potential V is an infinitely often differentiable harmonic scalar field. In other words, the Earth's gravitational field v is an infinitely often differentiable vector field in the exterior of the Earth satisfying $\nabla \cdot v = 0$, $\nabla \wedge v = 0$. The gradient of the Earth's gravitational field, i.e., $\mathbf{v} = \nabla v = \nabla^{(2)} V$, obeys similar properties: $\mathbf{v}$ is an infinitely often differentiable tensor field in the exterior of the Earth obeying the equations $\nabla \cdot \mathbf{v} = 0$, $\nabla \wedge \mathbf{v} = 0$. We see that $\mathbf{v}$ represents the *Hesse tensor* of the scalar field V, i.e., $\mathbf{v} = \nabla^{(2)} V = (\nabla \otimes \nabla) V$.

This chapter deals with three essential potential theoretic aspects in Earth's gravitational field determination, namely *EODP (exterior oblique derivative problem), SST (satellite-to-satellite tracking), SGG (satellite-gravity-gradiometry),* and *GP (gravimetry problem)*: Some preparatory remarks are helpful to understand the proposed mathematical solution techniques:

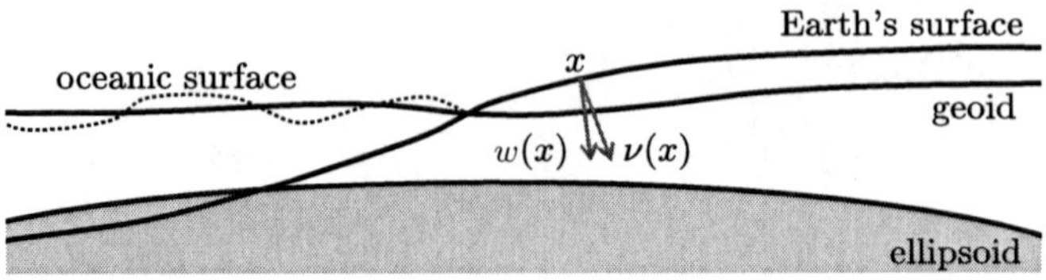

FIGURE 4.3
Earth's surface, geoid, ellipsoid (w = oblique gravity vector, normal to the geoid, ν = normal vector to the Earth's surface).

(EODP): The actual Earth's surface (globally available from modern satellite techniques such as DOPPLER, GPS (global positioning system), LASER, VLBI (very long baseline interferometry), etc.) does not coincide with a level surface (i.e., an equipotential surface). The force of gravity is generally not perpendicular to the actual Earth's surface (see Figure 4.3). Instead we are confronted with gravity as an oblique derivative (see Figure 4.3). The gravity vector is an oblique vector at any point on the Earth's surface and generally not the normal vector. Consequently, the boundary-value problem of determining the gravitational potential from terrestrial gravity vectors leads to a so-called *exterior oblique derivative problem*, or more precisely, an exterior oblique Neumann boundary-value problem. In Section 4.1, the oblique derivative problem is discussed as a discrete problem in a twofold way, viz. by volume as well as surface-based potentials resulting in a (Runge-type) reproducing Hilbert space framework. The reproducing kernel structure does not require any approximate integration, as long as the reproducing kernel is available in closed form by elementary function. This is a great computational advantage in Earth's gravitational field determination.

(SST/SGG): The terrestrial gravity data coverage now and in the foreseeable future is far from being satisfactory. This is the reason why spaceborne measurements have to come into play in modern Earth's gravitational field determination, at least for modeling low-frequency signatures. The three satellite concepts presently in operation by NASA (National Aeronautic and Space Administration), ESA (European Space Agency), GFZ (German Research Center for Geosciences), etc. are satellite-to-satellite tracking in the high-low mode (SST hi-lo), satellite-to-satellite tracking in the low-low mode (SST lo-lo), and satellite-gravity-gradiometry (SGG). Common to all three concepts is that the determination of the Earth's gravity field is based on the measurement of the relative motion of test masses (cf. Figure 4.4). In short, seen from a geomathematical viewpoint, the following gravitational data are obtainable at a satellite's altitude (for more details see, e.g., ESA [1996, 1998, 1999]):

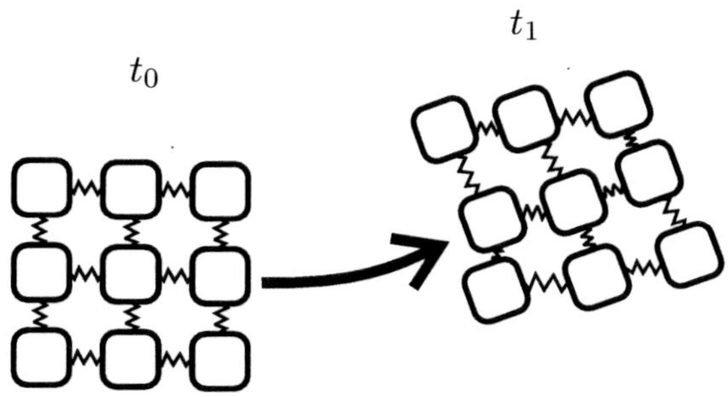

FIGURE 4.4
Test masses connected with springs positioned in the gravity space at different time steps: The principle of a gradiometer.

SST hi-lo:	3–D acceleration	$\simeq$	gravitational gradient,
SST lo-lo:	3-D acceleration difference	$\simeq$	difference in gravitat. gradient,
SGG:	acceleration differential	$\simeq$	gravitational tensor.

The satellite data obtained in this way therefore provide a transition from the first derivative of the gravitational potential via a difference in the first derivative to the second derivative. The guiding parameter that determines the sensitivity with respect to the spatial scales of the Earth's gravitational potential is the distance between the test masses, being almost infinity for SST hi-lo, finite for SST lo-lo, and almost zero for gradiometry. In a geomathematical sense, the *satellite problems* require the determination of the gravitational potential from vectorial gravitational gradients (SST) and/or gravitational tensors (SGG) at satellite's height. These problems amount to a new quality in potential theoretic research (cf. Section 4.2), which represent addenda to our boundary-value procedures. More concretely, the Meissl scheme (cf. Table 3.1) informs us about an exponential decay of the Fourier coefficients for the geopotential in height. This is a serious feature that must be observed. As long as the spaceborne data set is (ideally) assumed to constitute a fundamental system in the Earth's exterior, the Earth's exterior gravitational potential is uniquely recoverable even down to the Earth's surface. However, modeling Earth's gravitation at satellite altitude is an exponentially ill-posed problem (for the classification see, e.g., H. Engl et al. [1997]) if only finite data material is in use. In the case of finite data, potential theoretic methods do not suffice. Instead, appropriate techniques from the theory of inverse problems have to come into play (see, e.g., W. Freeden [1999] and the references therein). The price to be paid is that inverse problems do not enable us to recover the Earth's gravitational field at the Earth's surface without any attenuation (of high-frequency phenomena) of the gravity signals.

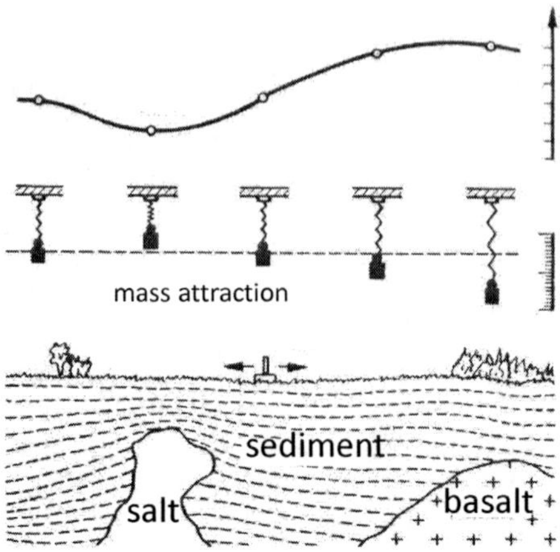

FIGURE 4.5
The principle of gravimetry (due to F. Jacobs, H. Meyer [1992]).

(GP): A central problem in Earth sciences is *gravimetry,* i.e., the determination of the Earth's mass density distribution from measurements of the gravitational potential or related quantities (cf. Figure 4.5 for an illustration of the observational situation).

The gravimetry problem amounts to the inversion of a Fredholm integral equation of the first kind, involving Newton's law of gravitation. Even if we restrict ourselves to a spherical Earth model (i.e., $\mathcal{G} = \mathcal{B}_R(0)$ with mean Earth's radius R) the gravimetry problem is severely ill-posed, as the inversion is not continuous (for classical inversion procedures see, e.g., E.W. Grafarend [1982], E. Groten [1979], W.A. Heiskanen, H. Moritz [1967], W. Torge [1991], B. Hofmann–Wellenhof, H. Moritz [2005], H. Moritz [2010], and the references therein). However, this is not the only contribution to the ill-posedness of the solution of the gravimetry problem. Within Hadamard's classification (existence, stability, uniqueness) we are confronted with the following situation:

Existence: It is well known that the gravitational potential is harmonic outside the Earth. Therefore, the Fredholm integral equation is unsolvable if the right-hand side is non-harmonic. Moreover, there is no solution for a certain set of harmonic right-hand sides. However, following suitable potential theoretic concepts, we are led to give a necessary and sufficient condition for the existence of a solution. Furthermore, the image of the corresponding Fredholm integral operator is dense in the space of harmonic functions with respect to the L^2-topology. A perturbed potential outside the image can still be treated in such a way that approximations to the exact solution of the unperturbed problem can be found in an appropriate way.

Stability: The inversion of the operator, i.e., the determination of a density distribution that corresponds to a given potential, is not continuous. This means that unavoidable errors in the measurements of the potential may lead to a completely different density function. In other words, as an addendum to potential theoretic concepts, regularization procedures as known from the theory of inverse problems become necessary (see, e.g., V. Michel [1999], W. Freeden, V. Michel [2004], and the references therein).

Uniqueness: The most serious difficulty is the non-uniqueness of the solution. Essential parts of the density distribution cannot be reconstructed from the gravitational potential. For every arbitrary density distribution, there exists an infinite-dimensional set of different density distributions that generate exactly the same potential. Within this context, it should be noted that a square-integrable function on a sphere can be approximated arbitrarily well by a harmonic function. However, this does not hold true for square-integrable functions defined on a ball. Another type of functions, i.e., anharmonic functions play an essential role (see, e.g., N. Weck [1972], L. Ballani et al. [1993], V. Michel, A.S. Fokas [2008], and the references therein). All in all, we are confronted with the problem that the determination of the density distribution exclusively by harmonic functions is insufficient. The observation of the anharmonic contribution is indispensable.

4.1 Oblique Derivative Problem

As already alluded, terrestrial observations of the gravity field on the real (known) Earth's surface do not generally provide normal derivatives (cf. Figure 4.3). Instead, oblique derivatives are measured, since the actual Earth's surface does not coincide with the equipotential surface of the geoid (at least not for large parts over continents). In the following, we are interested in discussing three solution methods for the oblique derivative problem, viz. a locally uniform approximation implied by generalized L^2-Fourier expansions with respect to certain trial functions (such as outer harmonics (multipoles), mass (single)poles (fundamental solutions), and kernel functions like the Abel–Poisson kernel (3.395) or the singularity kernel (3.397)), uniform approximation by (Newton-type) volume potential kernels, and uniform approximation by (Runge-type) surface layer-based kernels.

Classically, a solution procedure for the oblique derivative problem is undertaken by virtue of integral equations using the potential of a single-layer. These results were essentially worked out by A.V. Bitzadse [1968] and C. Miranda [1970]. In accordance with this work, K.R. Koch, A.J. Pope [1972] applied the integral equation procedure to the so-called *geodetic boundary-value problem using the known surface of the Earth.* However, the strong nature of the singularities demanding Cauchy's principal integral values turned out to be a serious obstacle. For numerical computation, alternative techniques must be taken into account.

The integral equation method also represents the point of departure for some subsequent work by W. Freeden, H. Kersten [1980, 1981]. They provide a new concept of approximation, viz. generalized Fourier expansions, thereby transferring strongly singular integrals into regular ones. As for the classical Dirichlet and Neumann boundary-value problems, the generalized Fourier approach yields Fourier coefficients of the boundary values within the $\mathrm{L}^2(\partial\mathcal{G})$-framework, and it simultaneously implies locally uniform approximation of the solution for subsets totally contained in the outer space. Even more, W. Freeden, H. Kersten [1980] and W. Freeden, V. Michel [2004] successfully provide the basis for closure theorems in uniform as well as Hölder topologies. But it should be noted that these results still lack a conversion into geoscientific numerics. Only if the approximation of the boundary values is implemented as a generalized Fourier expansion in the L^2-context or the method of generalized Fourier expansions is transferred to a reproducing kernel Hilbert $\mathrm{C}^{(0)}$-substructure, constructive realizations of the oblique derivative problem have been implemented successfully to real oblique data sets (see M. Gutting [2007] and the references therein).

4.1.1 Solution by Generalized Fourier Series

We start with some more notational background concerning the Hölder continuity. For $\mu \in (0,1)$, we let

$$\mathrm{Pot}^{(k,\mu)}(\overline{\mathcal{G}}) = \mathrm{Pot}(\mathcal{G}) \cap \mathrm{C}^{(k,\mu)}(\overline{\mathcal{G}}), \tag{4.4}$$

$$\mathrm{Pot}^{(k,\mu)}(\overline{\mathcal{G}^c}) = \mathrm{Pot}(\mathcal{G}^c) \cap \mathrm{C}^{(k,\mu)}(\overline{\mathcal{G}^c}). \tag{4.5}$$

Of particular importance for our later considerations on oblique derivative problems in gravitational theory is the space $\mathrm{C}^{(0,\mu)}(\partial\mathcal{G})$ of all μ-Hölder continuous functions on $\partial\mathcal{G}$. This is the reason why we list some properties of $\mathrm{C}^{(0,\mu)}(\partial\mathcal{G})$ in more detail (see Exercise 4.2). For $\mu' \leq \mu$ we have $\mathrm{C}^{(0,\mu)}(\partial\mathcal{G}) \subset \mathrm{C}^{(0,\mu')}(\partial\mathcal{G})$. $\mathrm{C}^{(0,\mu)}(\partial\mathcal{G})$ is a non-complete normed space with respect to

$$\|F\|_{\mathrm{C}^{(0)}(\partial\mathcal{G})} = \sup_{x\in\partial\mathcal{G}} \left|F(x)\right|, \tag{4.6}$$

and a Banach space under the norm

$$\|F\|_{\mathrm{C}^{(0,\mu)}(\partial\mathcal{G})} = \sup_{x\in\partial\mathcal{G}} |F(x)| + \sup_{\substack{x\in\partial\mathcal{G}\\ x\neq y}} \frac{|F(x)-F(y)|}{|x-y|^{\mu}}. \tag{4.7}$$

For $\mu' \leq \mu$ and $F \in \mathrm{C}^{(0,\mu)}(\partial\mathcal{G})$ we have, with a positive constant C (dependent on μ and μ'),

$$\|F\|_{\mathrm{C}^{(0,\mu')}(\partial\mathcal{G})} \leq C\, \|F\|_{\mathrm{C}^{(0,\mu)}(\partial\mathcal{G})}. \tag{4.8}$$

$\mathrm{C}^{(0,\mu)}(\partial\mathcal{G})$ is a non-complete normed space with respect to $\|\cdot\|_{\mathrm{C}^{(0,\mu')}(\partial\mathcal{G})}$ whenever $\mu' < \mu$. For $F, H \in \mathrm{C}^{(0,\mu)}(\partial\mathcal{G})$ it is not hard to verify that

$$\|FH\|_{\mathrm{C}^{(0)}(\partial\mathcal{G})} \leq \|F\|_{\mathrm{C}^{(0)}(\partial\mathcal{G})}\|H\|_{\mathrm{C}^{(0)}(\partial\mathcal{G})} \tag{4.9}$$

and

$$\begin{aligned}\|FH\|_{C^{(0,\mu)}(\partial\mathcal{G})} &\leq \|F\|_{C^{(0,\mu)}(\partial\mathcal{G})}\|H\|_{C^{(0)}(\partial\mathcal{G})} + \|F\|_{C^{(0)}(\partial\mathcal{G})}\|H\|_{C^{(0,\mu)}(\partial\mathcal{G})} \\ &\leq 2\|F\|_{C^{(0,\mu)}(\partial\mathcal{G})}\|H\|_{C^{(0,\mu)}(\partial\mathcal{G})}. \end{aligned} \tag{4.10}$$

In the space $C^{(0,\mu)}(\partial\mathcal{G})$, we are able to impose the inner product

$$(F,H)_{L^2(\partial\mathcal{G})} = \int_{\partial\mathcal{G}} F(x)H(x)\, d\omega(x). \tag{4.11}$$

The inner product $(\cdot,\cdot)_{L^2(\partial\mathcal{G})}$ implies the norm $\|F\|_{L^2(\partial\mathcal{G})} = (F,F)^{\frac{1}{2}}_{L^2(\partial\mathcal{G})}$. The space $(C^{(0,\mu)}(\partial\mathcal{G}), (\cdot,\cdot)_{L^2(\partial\mathcal{G})})$ is a pre-Hilbert space. For every $F \in C^{(0,\mu)}(\partial\mathcal{G})$ we have the norm estimate

$$\|F\|_{L^2(\partial\mathcal{G})} \leq \sqrt{\|\partial\mathcal{G}\|}\,\|F\|_{C^{(0)}(\partial\mathcal{G})} \leq \sqrt{\|\partial\mathcal{G}\|}\,\|F\|_{C^{(0,\mu)}(\partial\mathcal{G})}. \tag{4.12}$$

Obviously, $L^2(\partial\mathcal{G})$ is the completion of $C^{(0)}(\partial\mathcal{G})$ as well as $C^{(0,\mu)}(\partial\mathcal{G})$ with respect to the norm $\|\cdot\|_{L^2(\partial\mathcal{G})}$.

Formulation of the Problem

The *exterior oblique derivative problem* (EODP) can be formulated briefly as follows:

(EODP): Let $\mathcal{G} \subset \mathbb{R}^3$ be a regular region (in the sense of Definition 1.7). Given a function F of class $C^{(0,\mu)}(\partial\mathcal{G})$, $0 < \mu < 1$, find a function U of class $\mathrm{Pot}^{(1,\mu)}(\overline{\mathcal{G}^c})$ satisfying the boundary condition

$$\frac{\partial U^+}{\partial \lambda}(x) = \lim_{\tau\to 0+} \lambda(x)\cdot(\nabla U)(x+\tau\lambda(x)) = F(x), \quad x \in \partial\mathcal{G}, \tag{4.13}$$

where λ is a $c^{(1,\mu)}(\partial\mathcal{G})$-(unit) vector field satisfying

$$\inf_{x\in\partial\mathcal{G}} (\lambda(x)\cdot\nu(x)) > 0, \tag{4.14}$$

and ν is the (unit) normal field on $\partial\mathcal{G}$ directed outward into $\mathcal{G}^c$.

Remark 4.1. If the field λ coincides with the normal field ν on $\partial\mathcal{G}$, Eq. (4.13) becomes the boundary condition of the ordinary exterior Neumann boundary-value problem. In this case, we already know that the smoothness conditions imposed on the boundary values F may be weakened.

Our first task is to discuss the well-posedness of the exterior oblique derivative problem (EODP). We follow the standard solution theory by use of the potential of a single layer. Existence and uniqueness are recapitulated briefly in accordance with the work of A.V. Bitzadse [1968] and C. Miranda [1970]. Moreover, as in classical theory, we are interested in a regularity theorem providing the solution in locally uniform topology in the outer space from the L^2-approximation of the boundary values on $\partial\mathcal{G}$ (cf. W. Freeden, H. Kersten [1981]).

Well-Posedness of the Problem

For preparation we need some further tools concerning the geometry of a boundary surface $\partial\mathcal{G}$ of a regular region $\mathcal{G} \subset \mathbb{R}^3$. As already known, $\partial\mathcal{G}$ is representable, locally at $x \in \partial\mathcal{G}$, by a function with first derivatives vanishing at x. This implies the existence of a positive constant M such that

$$|\nu(x) \cdot (x-y)| \leq M\ |(x-y)_{\tan}|^2, \qquad y \in \partial\mathcal{G} \cap \mathcal{B}_\delta(x), \tag{4.15}$$

where the suffix "tan" denotes the tangential component of $(x-y)$ with respect to the tangent plane at x, i.e., $(x-y)_{\tan} = (x-y) - ((x-y)\cdot\nu(x))\ \nu(x)$. We know that the constants M and δ depend neither on x nor on y. By virtue of

$$|(x-y)_{\tan}|^2 \leq |(x-y)_{\tan}|^2 + |\nu(x)\cdot(x-y)|^2 = |x-y|^2 \tag{4.16}$$

we obtain, for $y \in \partial\mathcal{G} \cap \mathcal{B}_\delta(x)$, the estimate

$$|(x-y)_{\tan}| \leq |x-y| \leq \sqrt{|(x-y)_{\tan}|^2 + M^2|(x-y)_{\tan}|^4} = \mathcal{O}(|(x-y)_{\tan}|), \tag{4.17}$$

as $|x-y| \to 0$, i.e., $\delta \to 0$.

Lemma 4.2. *Let λ be a continuous unit vector field on $\partial\mathcal{G}$ forming at any point on $\partial\mathcal{G}$ an angle with the outside normal ν such that*

$$\inf_{x\in\partial\mathcal{G}}\ (\lambda(x)\cdot\nu(x)) > 0. \tag{4.18}$$

Then there exist constants $\delta > 0$, $\beta \in (0,1)$, such that

$$|\lambda(x)\cdot(x-y)| \leq \beta\ |x-y|, \quad y \in \partial\mathcal{G} \cap \mathcal{B}_\delta(x). \tag{4.19}$$

Proof. By definition, we let $C = \inf_{x\in\partial\mathcal{G}}(\lambda(x)\cdot\nu(x))^2 > 0$. Moreover, we set

$$\delta = \frac{1}{2M}(1-\sqrt{1-C}), \quad \beta = \frac{1}{2}(1+\sqrt{1-C}) < 1, \tag{4.20}$$

where M satisfies the estimate (4.15). Then, for $y \in \partial\mathcal{G} \cap \mathcal{B}_\delta(x)$, we get

$$\begin{aligned} |\lambda(x)\cdot(x-y)| &= |\lambda(x)_{\tan}\cdot(x-y)_{\tan} + (\lambda(x)\cdot\nu(x))(\nu(x)\cdot(x-y))| \\ &\leq |\lambda(x)_{\tan}|\ |(x-y)_{\tan}| + |\lambda(x)\cdot\nu(x)|\ |\nu(x)\cdot(x-y)| \\ &\leq \sqrt{1-(\lambda(x)\cdot\nu(x))^2}|x-y| + M|x-y|^2 \\ &\leq \left(\sqrt{1-C} + M|x-y|\right)|x-y| \\ &\leq \left(\sqrt{1-C} + \frac{1}{2}\left(1-\sqrt{1-C}\right)\right)|x-y| = \beta\ |x-y|. \end{aligned} \tag{4.21}$$

This is the desired result. □

We continue our considerations with the following estimates.

Lemma 4.3. *Let δ, β be as in Lemma 4.2. For $|\tau| \le \frac{1}{2}\delta$*

$$|\tau| = \inf_{x,y\in\partial\mathcal{G}} |x \pm \tau\lambda(x) - y| \ge \sqrt{1-\beta}\ |\tau|. \tag{4.22}$$

Proof. We observe that $|x \pm \tau\lambda(x) - x| = |\tau|$. For $r = |x-y| \le \delta$ we obtain

$$\begin{aligned} |x \pm \tau\lambda(x) - y|^2 &= r^2 + \tau^2 \pm 2\tau(\lambda(x)\cdot(x-y)) \ge r^2 + \tau^2 - 2\beta r|\tau| \\ &\ge r^2 + \tau^2 - \beta(r^2+\tau^2) \ge (1-\beta)\ \tau^2, \end{aligned} \tag{4.23}$$

and for $r \ge \delta$

$$|x \pm \tau\lambda(x) - y| \ge r - |\tau| \ge \delta - |\tau| \ge \frac{1}{2}\delta \ge |\tau|, \tag{4.24}$$

which guarantees the assertion of Lemma 4.3. □

Furthermore, we are able to verify that, for $x, y \in \partial\mathcal{G},\ \tau > 0,\ r = |x-y|$,

$$|y+\tau\lambda(y)-x|^2 \le r^2+\tau^2+2\tau|\lambda(y)\cdot(y-x)| \le r^2+\tau^2+2\tau r \le 2(r^2+\tau^2) \tag{4.25}$$

(note that $2\tau r \le r^2 + \tau^2$) and

$$|y+\tau\lambda(y)-x|^2 \ge r^2+\tau^2 - 2\tau|\lambda(y)\cdot(y-x)| \ge r^2+\tau^2-2\beta\tau r. \tag{4.26}$$

Then we get, for $r \le \delta$,

$$|y+\tau\lambda(y)-x|^2 \ge r^2+\tau^2-\beta(r^2+\tau^2) \ge (1-\beta)\ (r^2+\tau^2). \tag{4.27}$$

For $r \ge \delta,\ \tau \le \frac{1}{4}\delta \le \frac{1}{4}r$, we have

$$|y+\tau\lambda(y)-x|^2 \ge r^2+\tau^2-2\tau r \ge r^2+\tau^2-\frac{1}{2}r^2 \ge \frac{1}{2}(r^2+\tau^2). \tag{4.28}$$

Consequently, for sufficiently small τ, there exist positive constants C and $\tilde{C}$ such that

$$C\sqrt{|x-y|^2+\tau^2}\ \le\ |y-(x-\tau\lambda(y))|\ \le\ \tilde{C}\sqrt{|x-y|^2+\tau^2}. \tag{4.29}$$

By similar arguments we obtain

$$C\sqrt{|x-y|^2+\tau^2}\ \le\ |y-(x-\tau\lambda(x))|\ \le\ \tilde{C}\sqrt{|x-y|^2+\tau^2}. \tag{4.30}$$

From (4.29) and (4.30) it immediately follows that

$$\frac{C}{\tilde{C}}\ |y+\tau\lambda(y)-x|\ \le\ |y-(x-\tau\lambda(x))|\ \le\ \frac{\tilde{C}}{C}\ |y+\tau\lambda(y)-x| \tag{4.31}$$

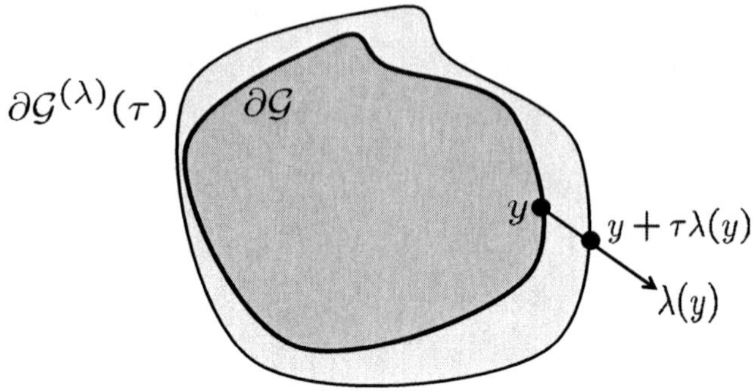

FIGURE 4.6
Illustration of an oblique neighboring surface $\partial\mathcal{G}^{(\lambda)}(\tau)$ to $\partial\mathcal{G}$.

and

$$\frac{C}{\tilde{C}}\ |y-(x-\tau\lambda(x))|\ \leq\ |y+\tau\lambda(y)-x|\ \leq\ \frac{\tilde{C}}{C}\ |y-(x-\tau\lambda(x))|. \tag{4.32}$$

The aforementioned estimates together with well-known integral estimates (known from Section 1.2) help to develop limit and jump relations involving *oblique neighboring surfaces* (cf. Figure 4.6)

$$\partial\mathcal{G}^{(\lambda)}(\tau)=\{x\in\mathbb{R}^3\ :\ x=y+\tau\lambda(y),\ y\in\partial\mathcal{G}\}, \tag{4.33}$$

where λ is a $\mathrm{c}^{(1,\mu)}(\partial\mathcal{G})$-(unit) vector field satisfying condition (4.14) (remember $\partial\mathcal{G}^{(\lambda)}(\tau)$ coincides with $\partial\mathcal{G}(\tau)$ provided that $\lambda=\nu$).

The formulation and the proof of the limit and jump relations in the $\mathrm{C}^{(0)}(\partial\mathcal{G})$-topology (see also W. Freeden, H. Kersten [1980, 1981]) are similar to the Dirichlet and Neumann case. Suppose that λ is a $\mathrm{c}^{(1,\mu)}$-(unit) vector field on $\partial\mathcal{G}$ satisfying (4.14). We consider the *potential operators* $P^{(\lambda)}(\tau,\sigma)$ defined by

$$P^{(\lambda)}(\tau,\sigma)[F](x)=\int_{\partial\mathcal{G}}F(y)\ G(\Delta;|x+\tau\lambda(x)-(y-\sigma\lambda(y))|)\ d\omega(y), \tag{4.34}$$

which coincide with the potential operators (3.180) for the case $\lambda=\nu$. For $F\in\mathrm{C}^{(0,\mu)}(\partial\mathcal{G})$, $0<\mu<1$, we canonically introduce the following operators:

$$P^{(\lambda)}(\tau,0)[F](x)=\int_{\partial\mathcal{G}}F(y)\ G(\Delta;|x+\tau\lambda(x)-y|)\ d\omega(y) \tag{4.35}$$

denotes the *operator of the single-layer potential on $\partial\mathcal{G}$ for values on $\partial\mathcal{G}^{(\lambda)}(\tau)$*. By

$$P^{(\lambda)}_{|1}(\tau,0)[F](x)=\frac{\partial}{\partial\tau}P^{(\lambda)}(\tau,0)[F](x) \tag{4.36}$$

we mean the *operator of the directional derivative of the single-layer potential on $\partial\mathcal{G}$ for values on $\partial\mathcal{G}^{(\lambda)}(\tau)$*, and

$$P_{|2}^{(\lambda)}(\tau,0)[F](x) = \int_{\partial\mathcal{G}} F(y) \frac{\partial}{\partial\lambda(y)} G(\Delta;|x+\tau\lambda(x)-y|)\, d\omega(y) \tag{4.37}$$

denotes the *operator of the double-layer potential on $\partial\mathcal{G}$ for values on the surface $\partial\mathcal{G}^{(\lambda)}(\tau)$*.

Analogously, $P_{|1}^{(\lambda)}(0,0)$ and $P_{|2}^{(\lambda)}(0,0)$ are introduced formally for functions $F \in \mathrm{C}^{(0,\mu)}(\partial\mathcal{G})$ as (strongly) singular integrals understood in the sense of Cauchy's principal value (see, e.g., A.V. Bitzadse [1968] and C. Miranda [1970]).

Suppose that, for sufficiently small $\tau > 0$, and $F \in \mathrm{C}^{(0,\mu)}(\partial\mathcal{G})$, $0 < \mu < 1$, the operators $(L_i^{\pm})^{(\lambda)}(\tau)$, $(J_i)^{(\lambda)}(\tau)$, $i = 1,2,3$, are defined as follows:

$$\left((L_1^{\pm})^{(\lambda)}(\tau)\right)[F](x) = \left(P^{(\lambda)}(\pm\tau,0) - P(0,0)\right)[F](x), \tag{4.38}$$

$$\left((L_2^{\pm})^{(\lambda)}(\tau)\right)[F](x) = \left(P_{|1}^{(\lambda)}(\pm\tau,0) - P_{|1}^{(\lambda)}(0,0) \pm \frac{1}{2}(\lambda(x)\cdot\nu(x))\right)[F](x), \tag{4.39}$$

$$\left((L_3^{\pm})^{(\lambda)}(\tau)\right)[F](x) = \left(P_{|2}^{(\lambda)}(\pm\tau,0) - P_{|2}^{(\lambda)}(0,0) \mp \frac{1}{2}(\lambda(x)\cdot\nu(x))\right)[F](x), \tag{4.40}$$

$$\left((J_1)^{(\lambda)}(\tau)\right)[F](x) = \left(P^{(\lambda)}(\tau,0) - P^{(\lambda)}(-\tau,0)\right)[F](x), \tag{4.41}$$

$$\left((J_2)^{(\lambda)}(\tau)\right)[F](x) = \left(P_{|1}^{(\lambda)}(\tau,0) - P_{|1}^{(\lambda)}(-\tau,0) + (\lambda(x)\cdot\nu(x))\right)[F](x), \tag{4.42}$$

$$\left((J_3)^{(\lambda)}(\tau)\right)[F](x) = \left(P_{|2}^{(\lambda)}(\tau,0) - P_{|2}^{(\lambda)}(-\tau,0) - (\lambda(x)\cdot\nu(x))\right)[F](x) \tag{4.43}$$

for $x \in \partial\mathcal{G}$ (observe that $(L_i^{\pm})^{(\nu)}(\tau) = L_i^{\pm}(\tau)$, $(J_i)^{(\nu)}(\tau) = J_i(\tau), i = 1,2,3$). W. Freeden, H. Kersten [1980] show that the limit and jump relations (cf. Subsection 3.3.2) hold true in a Hölder space context.

Theorem 4.4. *Let $\mathcal{G} \subset \mathbb{R}^3$ be a regular region. Then*

$$\lim_{\tau\to 0+} \left\| \left(L_i^{\pm}\right)^{(\lambda)}(\tau)[F] \right\|_{\mathrm{C}^{(0,\mu')}(\partial\mathcal{G})} = 0, \tag{4.44}$$

$$\lim_{\tau\to 0+} \left\| (J_i)^{(\lambda)}(\tau)[F] \right\|_{\mathrm{C}^{(0,\mu')}(\partial\mathcal{G})} = 0, \tag{4.45}$$

hold true for all $F \in \mathrm{C}^{(0,\mu)}(\partial\mathcal{G})$ and all positive values μ' with $\mu' < \mu < 1$, whereas with $\lambda = \nu$

$$\lim_{\tau\to 0+} \left\| \left(L_i^{\pm}\right)^{(\nu)}(\tau)[F] \right\|_{\mathrm{C}^{(0,\mu')}(\partial\mathcal{G})} = 0, \tag{4.46}$$

$$\lim_{\tau\to 0+} \left\| (J_i)^{(\nu)}(\tau)[F] \right\|_{\mathrm{C}^{(0,\mu')}(\partial\mathcal{G})} = 0, \tag{4.47}$$

are valid for all $F \in C^{(0,\mu)}(\partial\mathcal{G})$ and all positive values μ' with $\mu' \le \mu < 1$ (note that, in the case of (4.44) and (4.45), W. Freeden, H. Kersten [1980] assume $C^{(2,\mu)}$-smoothness locally on $\partial\mathcal{G}$, i.e., for each point $x \in \partial\mathcal{G}$ there exists a three-dimensional neighborhood $\mathcal{B}_\delta(x)$ such that $\partial\mathcal{G} \cap \mathcal{B}_\delta(x)$ can be mapped bijectively onto some open domain $\mathcal{U} \subset \mathbb{R}^2$ and that the mapping is twice μ-Hölder-continuously differentiable, $0 < \mu < 1$).

Evidently, the norm estimate (4.12) implies the limit relations

$$\lim_{\tau\to 0+} \left\| \left(L_i^{\pm}\right)^{(\lambda)} (\tau)[F] \right\|_{C^{(0)}(\partial\mathcal{G})} = 0, \tag{4.48}$$

$$\lim_{\tau\to 0+} \left\| (J_i)^{(\lambda)} (\tau)[F] \right\|_{C^{(0)}(\partial\mathcal{G})} = 0, \tag{4.49}$$

for all $F \in C^{(0,\mu)}(\partial\mathcal{G})$.

The *uniqueness* of the exterior oblique derivative problem (EODP) can be based on the extremum principle of Zaremba and Giraud (cf. A.V. Bitzadse [1968]) in connection with the regularity condition imposed on U at infinity (similar to results as provided by Lemma 3.74).

In order to prove the *existence of the solution*, we use a single-layer potential

$$U(x) = \int_{\partial\mathcal{G}} Q(y)\, G(\Delta; |x-y|)\, d\omega(y), \qquad Q \in C^{(0,\mu)}(\partial\mathcal{G}). \tag{4.50}$$

Observing the discontinuity of the directional derivative of the single-layer potential (see (4.39) and (4.44)), we obtain for each $Q \in C^{(0,\mu)}(\partial\mathcal{G})$ and all points $x \in \partial\mathcal{G}$

$$\frac{1}{2} Q(x)(\lambda(x)\cdot\nu(x)) + \int^*_{\partial\mathcal{G}} Q(y) \frac{\partial}{\partial\lambda(x)} G(\Delta; |x-y|)\, d\omega(y) = \frac{\partial U^+}{\partial\lambda}(x) = F(x), \tag{4.51}$$

where the integral $\int^*_{\partial\mathcal{G}} \ldots d\omega(y)$ exists only in the sense of Cauchy's principal value. The resulting integral equation

$$T[Q] = F, \quad Q \in C^{(0,\mu)}(\partial\mathcal{G}), \tag{4.52}$$

with

$$T[Q](x) = \frac{1}{2}(\lambda(x)\cdot\nu(x))Q(x) + \int^*_{\partial\mathcal{G}} Q(y) \frac{\partial}{\partial\lambda(x)} G(\Delta; |x-y|)\, d\omega(y) \tag{4.53}$$

is of (strongly) singular type. However, due to the work of A.V. Bitzadse [1968] and C. Miranda [1970], with λ of class $c^{(1,\mu)}$ satisfying (4.14), all standard Fredholm theorems are still valid. The homogeneous integral equation corresponding to (4.52) has no solution other than $Q = 0$. Thus, the solution of the scalar exterior oblique derivative problem exists and can be represented by a single-layer potential of the form (4.50). The operator T and its adjoint operator T^* (with respect to the $\mathrm{L}^2(\partial\mathcal{G})$-scalar product on $\mathrm{C}^{(0,\mu)}(\partial\mathcal{G})$) form mappings from $\mathrm{C}^{(0,\mu)}(\partial\mathcal{G})$ into $\mathrm{C}^{(0,\mu)}(\partial\mathcal{G})$, which are linear and bounded with respect to the norm $\|\cdot\|_{\mathrm{C}^{(0,\mu)}(\partial\mathcal{G})}$ (see also J. Schauder [1931]). The operators T, T^* in $\mathrm{C}^{(0,\mu)}(\partial\mathcal{G})$ are injective and by Fredholm's alternative bijective in the Banach space $\mathrm{C}^{(0,\mu)}(\partial\mathcal{G})$ (cf. A.V. Bitzadse [1968], C. Miranda [1970]). Consequently, by virtue of the open mapping theorem, the operators T^{-1} and $(T^*)^{-1} = (T^{-1})^*$ are linear and bounded with respect to $\|\cdot\|_{\mathrm{C}^{(0,\mu)}(\partial\mathcal{G})}$. Therefore, in accordance with a technique due to P.D. Lax [1954] (for more details, see also the approach developed by C. Müller [1969] or compare the proof of Theorem 3.56), both T^{-1} and $(T^*)^{-1}$ are bounded with respect to the norm $\|\cdot\|_{\mathrm{L}^2(\partial\mathcal{G})}$ in $\mathrm{C}^{(0,\mu)}(\partial\mathcal{G})$.

Next we deal with a *regularity theorem* involving the $\mathrm{L}^2(\partial\mathcal{G})$-topology.

Theorem 4.5. *Let* $U \in \mathrm{Pot}^{(1,\mu)}(\overline{\mathcal{G}^c})$, $\mu \in (0,1)$, *be the uniquely determined solution of the exterior oblique derivative problem* (EODP) *corresponding to the boundary values (4.13). Then there exists a constant* C *(depending on* $k, \mathcal{K}$, *and* $\partial\mathcal{G}$*) such that*

$$\sup_{x\in\overline{\mathcal{K}}} \left| \left(\nabla^{(k)} U\right)(x) \right| \leq C \left(\int_{\partial\mathcal{G}} \left(\frac{\partial U^+}{\partial\lambda}(x) \right)^2 d\omega(x) \right)^{\frac{1}{2}} \tag{4.54}$$

holds for all $\mathcal{K} \Subset \mathcal{G}^c$ *and all* $k \in \mathbb{N}_0$.

Proof. U allows a representation as a single-layer potential (4.50). For each subset $\mathcal{K} \Subset \mathcal{G}^c$, the estimate

$$\begin{aligned} \sup_{x\in\overline{\mathcal{K}}} \left| \left(\nabla^{(k)} U\right)(x) \right| &\leq \sup_{x\in\overline{\mathcal{K}}} \left(\int_{\partial\mathcal{G}} \left(\nabla_x^{(k)} \frac{1}{|x-y|} \right)^2 d\omega(y) \right)^{\frac{1}{2}} \|Q\|_{\mathrm{L}^2(\partial\mathcal{G})} \\ &\leq \tilde{C} \left\| T^{-1} \left(\frac{\partial U^+}{\partial\lambda} \right) \right\|_{\mathrm{L}^2(\partial\mathcal{G})} \end{aligned} \tag{4.55}$$

holds true (with positive $\tilde{C}$ dependent on $k, \mathcal{K}$, and $\partial\mathcal{G}$). By virtue of the boundedness of T^{-1} with respect to $\|\cdot\|_{\mathrm{L}^2(\partial\mathcal{G})}$, it follows that there exists a constant C $(= C(k; \mathcal{K}, \partial\mathcal{G}))$ such that

$$\sup_{x\in\overline{\mathcal{K}}} \left| \left(\nabla^{(k)} U\right)(x) \right| \leq C \left\| \frac{\partial U^+}{\partial\lambda} \right\|_{\mathrm{L}^2(\partial\mathcal{G})}, \tag{4.56}$$

as required. □

Closure Theorem

In the sequel, we consider the pre-Hilbert space $(\mathrm{C}^{(0,\mu)}(\partial\mathcal{G}), \|\cdot\|_{\mathrm{L}^2(\partial\mathcal{G})})$. Our purpose is to prove a closure theorem by use of a Hahn–Banach argument (cf. W. Freeden, H. Kersten [1981]).

Theorem 4.6. *Let $\mathcal{A}, \mathcal{G} \subset \mathbb{R}^3$ be regular regions such that $\mathcal{A} \Subset \mathcal{G}$ holds true. Assume that λ is a $\mathrm{c}^{(1,\mu)}$-(unit) vector field on $\partial\mathcal{G}$ satisfying (4.14). Then the linear space*

$$\frac{\partial \mathrm{Pot}(\mathcal{A}^c)}{\partial \lambda} = \left\{ \left. \frac{\partial U^+}{\partial \lambda} \right|_{\partial\mathcal{G}} \; : \; U \in \mathrm{Pot}(\mathcal{A}^c) \right\} \tag{4.57}$$

is a dense subspace of the pre-Hilbert space $(\mathrm{C}^{(0,\mu)}(\partial\mathcal{G}), \|\cdot\|_{\mathrm{L}^2(\partial\mathcal{G})})$.

Proof. Since $U \in \mathrm{Pot}(\mathcal{A}^c)$ has derivatives of arbitrary order in any neighborhood of $\partial\mathcal{G}$, both U and ∇U are μ-Hölder continuous on $\partial\mathcal{G}$. Consequently, the μ–Hölder continuity of λ shows us that $\frac{\partial \mathrm{Pot}(\mathcal{A}^c)}{\partial\lambda}$ is a subspace of $\mathrm{C}^{(0,\mu)}(\partial\mathcal{G})$.

Let Z be a continuous linear functional on $(\mathrm{C}^{(0,\mu)}(\partial\mathcal{G}), \|\cdot\|_{\mathrm{L}^2(\partial\mathcal{G})})$ fulfilling

$$Z\Big|_{\frac{\partial \mathrm{Pot}(\mathcal{A}^c)}{\partial\lambda}} = 0 \; . \tag{4.58}$$

We have to prove that Z is the zero functional. For this purpose we notice that, for each $x \in \mathcal{A}$, the function

$$y \mapsto F_x(y) = \frac{\partial}{\partial\lambda(y)} G(\Delta; |x-y|), \quad y \in \partial\mathcal{G}, \tag{4.59}$$

belongs to $\frac{\partial \mathrm{Pot}(\mathcal{A}^c)}{\partial\lambda}$. In other words, $Z[F_x] = 0$ for each $x \in \mathcal{A}$. Now it can be easily seen that the function $x \mapsto Z[F_x]$, $x \in \mathcal{G}$, is a solution of the Laplace equation in $\mathcal{G}$. Consequently, observing $\mathcal{A} \Subset \mathcal{G}$, we obtain by analytic continuation $Z\left[F_x\right] = 0$, $x \in \mathcal{G}$. We specialize the last relation to inner oblique neighboring surfaces, i.e., to points $x = y - \tau\lambda(y)$, $y \in \partial\mathcal{G}$, with $\tau > 0$ sufficiently small. Then we multiply by an arbitrary function $Q \in \mathrm{C}^{(0,\mu)}(\partial\mathcal{G})$ and integrate over $\partial\mathcal{G}$. As a result we find

$$\int_{\partial\mathcal{G}} G(y) \; Z\left[F_{y-\tau\lambda(y)}\right] \; d\omega(y) = 0, \quad \tau > 0 \; . \tag{4.60}$$

The mapping $A_\tau : \partial\mathcal{G} \to \mathrm{C}^{(0,\mu)}(\partial\mathcal{G})$ defined by

$$y \mapsto A_\tau(y) = F_{y-\tau\lambda(y)}, \; y \in \partial\mathcal{G}, \tag{4.61}$$

is continuous (note that $\|F_{y-\tau\lambda(y)} - F_{y_0-\tau\lambda(y_0)}\|_{\mathrm{L}^2(\partial\mathcal{G})} \to 0$ for $y \to y_0$ on $\partial\mathcal{G}$ with $\tau > 0$ fixed). Consequently, $A_\tau(y)$ is integrable. Thus, we obtain

$$Z\left[\int_{\partial\mathcal{G}} Q(y) \; F_{y-\tau\lambda(y)} \; d\omega(y)\right] = \int_{\partial\mathcal{G}} Q(y) \; Z\left[F_{y-\tau\lambda(y)}\right] \; d\omega(y) = 0 \; . \tag{4.62}$$

For $Q \in \mathrm{C}^{(0,\mu)}(\partial\mathcal{G})$ and $x \in \overline{\mathcal{G}^c}$, we consider the potentials

$$\begin{aligned} U(x) &= P^{(\lambda)}(0,0)[Q](x) = \int_{\partial\mathcal{G}} Q(y) \; G(\Delta; |x-y|) \; d\omega(y), & (4.63) \\ U_\tau(x) &= P^{(\lambda)}(0,\tau)[Q](x) = \int_{\partial\mathcal{G}} Q(y) G(\Delta; |x-(y-\tau\lambda(y))|) \; d\omega(y). & (4.64) \end{aligned}$$

In fact, we already know (see also Lemma 3.51 for the special case $\lambda = \nu$ and W. Freeden, H. Kersten [1981] for general λ) that the limit relation

$$\frac{\partial U_\tau}{\partial \lambda} \to \frac{\partial U^+}{\partial \lambda}, \quad \tau \to 0+, \tag{4.65}$$

holds true with respect to the norm $\|\cdot\|_{\mathrm{L}^2(\partial\mathcal{G})}$. By virtue of (4.65) and (4.62), it follows that

$$Z\left[\frac{\partial U^+}{\partial \lambda}\right] = \lim_{\tau\to 0+} Z\left[\frac{\partial U_\tau}{\partial \lambda}\right] = \lim_{\tau\to 0+} Z\left[\int_{\partial\mathcal{G}} Q(y)\, F_{y-\tau\lambda(y)}\, d\omega(y)\right] = 0 \tag{4.66}$$

for every single-layer potential U with $Q \in \mathrm{C}^{(0,\mu)}(\partial\mathcal{G})$. Since the operator T as defined by (4.53) is bijective, the space of boundary values $\frac{\partial U^+}{\partial\lambda}$ of such potentials is exactly equal to $\mathrm{C}^{(0,\mu)}(\partial\mathcal{G})$. Thus it follows that $Z = 0$ on the pre-Hilbert space $(\mathrm{C}^{(0,\mu)}(\partial\mathcal{G}), \|\cdot\|_{\mathrm{L}^2(\partial\mathcal{G})})$, as desired. □

Method of Generalized Fourier Expansion

The point of departure for our considerations concerning $\mathrm{L}^2(\partial\mathcal{G})$-approximation is the following result.

Theorem 4.7. *Let $\mathcal{A}, \mathcal{G} \subset \mathbb{R}^3$ be regular regions such that $\mathcal{A} \Subset \mathcal{G}$ holds true. If $\{D_n\}_{n=0,1,\ldots}$, is an $\mathrm{L}^2(\partial\mathcal{A})$-Dirichlet Runge basis (in the sense of Definition 3.95), then*

$$\operatorname*{span}_{n=0,1,\ldots} \left\{ \left.\frac{\partial D_n^+}{\partial \lambda}\right|_{\partial\mathcal{G}} \right\} \tag{4.67}$$

is dense in $(\mathrm{C}^{(0,\mu)}(\partial\mathcal{G}), \|\cdot\|_{\mathrm{L}^2(\partial\mathcal{G})})$.

Proof. On the one hand, given $\varepsilon > 0$ and $F \in \mathrm{C}^{(0,\mu)}(\partial\mathcal{G})$, there exists (by Theorem 4.6) a function $U \in \mathrm{Pot}(\mathcal{A}^c)$ such that

$$\left(\int_{\partial\mathcal{G}} \left(F(y) - \frac{\partial U^+}{\partial \lambda}(y)\right)^2 d\omega(y)\right)^{\frac{1}{2}} \le \frac{\varepsilon}{2}\,. \tag{4.68}$$

On the other hand, by arguments as for Corollary 3.96, there exists a member V within the set $\mathrm{span}_{n=0,1,\ldots}\{D_n\}$ satisfying

$$\sup_{x\in\partial\mathcal{G}} |(\nabla U)(x) - (\nabla V)(x)| \le \frac{\varepsilon}{2\|\partial\mathcal{G}\|}, \tag{4.69}$$

such that we get

$$\left(\int_{\partial\mathcal{G}} \left(\frac{\partial U^+}{\partial \lambda}(y) - \frac{\partial V^+}{\partial \lambda}(y)\right)^2 d\omega(y)\right)^{\frac{1}{2}} \le \frac{\varepsilon}{2}\,.$$

Combining our results by application of the triangle inequality we therefore obtain the estimate

$$\left(\int_{\partial \mathcal{G}} \left(F(y) - \frac{\partial V^+}{\partial \lambda}(y) \right)^2 \, d\omega(y) \right)^{\frac{1}{2}} \leq \varepsilon, \tag{4.70}$$

as wanted. □

For numerical purposes we orthonormalize the members of an $L^2(\partial\mathcal{A})$-Dirichlet Runge basis $\{D_n\}_{n=0,1,\ldots}$ (e.g., certain systems of single poles (fundamental solutions), outer harmonics (multipoles), and/or appropriate kernel functions such as Abel–Poisson kernel (3.395), singularity kernel (3.397), etc.) and obtain a system $\{D_n^*\}_{n=0,1,\ldots,}$, $D_n^* \in \mathrm{Pot}(\mathcal{A}^c)$, satisfying the orthonormality condition

$$\left(\frac{\partial D_n^*}{\partial \lambda}, \frac{\partial D_m^*}{\partial \lambda} \right)_{L^2(\partial \mathcal{G})} = \int_{\partial \mathcal{G}} \frac{\partial D_n^*}{\partial \lambda}(x) \frac{\partial D_m^*}{\partial \lambda}(x) \, d\omega(x) = \delta_{n,m} \ . \tag{4.71}$$

We are able to derive the following limit relation: If $F \in \mathrm{C}^{(0,\mu)}(\partial\mathcal{G})$, then

$$\lim_{m\to\infty} \left(\int_{\partial \mathcal{G}} \left| F(x) - \sum_{n=0}^{m} \left(F, \frac{\partial D_n^*}{\partial \lambda} \right)_{L^2(\partial \mathcal{G})} \frac{\partial D_n^*}{\partial \lambda}(x) \right|^2 \, d\omega(x) \right)^{\frac{1}{2}} = 0. \tag{4.72}$$

Consequently, the uniquely determined $U \in \mathrm{Pot}^{(1,\mu)}(\overline{\mathcal{G}^c})$, $\frac{\partial U^+}{\partial \lambda}\big|_{\partial\mathcal{G}} = F$, can be represented in the form

$$\lim_{m\to\infty} \sup_{x \in \overline{\mathcal{K}}} \left| U(x) - U^{(m)}(x) \right| = 0, \tag{4.73}$$

where

$$U^{(m)} = \sum_{n=0}^{m} \left(F, \frac{\partial D_n^*}{\partial \lambda} \right)_{L^2(\partial \mathcal{G})} D_n^* \tag{4.74}$$

for every totally contained subset $\mathcal{K}$ of $\mathcal{G}^c$ (compare Theorem 4.5).

Equivalently, $U^{(m)}$ can be obtained by use of the $L^2(\partial\mathcal{A})$-Dirichlet Runge basis $\{D_n\}_{n=0,1,\ldots,}$ $D_n \in \mathrm{Pot}(\mathcal{A}^c)$, in the form

$$U^{(m)} = \sum_{n=0}^{m} a_n \, D_n, \tag{4.75}$$

where the coefficients $a_0, \ldots, a_m$ satisfy the "normal equations"

$$\sum_{n=0}^{m} a_n \left(\frac{\partial D_n}{\partial \lambda}, \frac{\partial D_k}{\partial \lambda} \right)_{L^2(\partial \mathcal{G})} = \left(F, \frac{\partial D_k}{\partial \lambda} \right)_{L^2(\partial \mathcal{G})}, \quad k = 0, \ldots, m. \tag{4.76}$$

Remark 4.8. Our work leads to the surprising result that the method of generalized Fourier expansions in the $L^2(\partial\mathcal{G})$-norm as formulated in this Subsection 4.1 for kernel functions (such as the Abel–Poisson kernel, singularity kernel, etc.) canonically allows the transfer to minimum norm (spline) interpolation in the metric of the (Runge type) reproducing kernel Hilbert spaces (as presented in Subsections 4.1.2 and 4.1.3). Consequently, the generalized Fourier expansion as well as spline interpolation can be understood as (Runge) manifestations of the same approximation method, however, corresponding to different Hilbert reference space topologies.

4.1.2 Solution by Volume-Based Reprostructure

Contrary to $L^2(\partial\mathcal{G})$, the class $L^2(\mathcal{G})$ of square-integrable functions on a regular region $\mathcal{G}$ is not obtainable only by the L^2-completion of a harmonic function system. In addition, we have to take into account a so-called *anharmonic function system* (see, e.g., N. Weck [1972], L. Ballani et al. [1993], V. Michel [1999], W. Freeden, V. Michel [2004]). This observation should be studied in more detail.

Anharmonics

Let $\mathcal{G} \subset \mathbb{R}^3$ be a regular region. By $D(\mathcal{G})$ we denote the space of all infinitely often differentiable functions F in $\mathbb{R}^3$ possessing a compact support supp(F) in $\mathcal{G}$. We equip $D(\mathcal{G})$ with the following topology: a sequence $\{\phi_n\} \subset D(\mathcal{G})$ is called convergent to zero if and only if (1) there exists a bounded $\mathcal{B} \subset \mathbb{R}^3$ such that ϕ_n vanishes outside $\mathcal{B}$, (2) for every differential operator ∇^α, $\alpha \in \mathbb{N}_0^3$, the sequence $\{\nabla^\alpha \phi_n\}$ is convergent to zero with respect to the norm $\|\cdot\|_{C^{(0)}(\mathcal{G})}$. Members of $D(\mathcal{G})$ are called *test functions.* Elements of the dual space $D^*(\mathcal{G})$, i.e., continuous linear functionals $\mathcal{F} : D(\mathcal{G}) \to \mathbb{R}$, are called *distributions* (or generalized functions). Clearly, multiplication (by a scalar) and addition are defined canonically for linear functionals of the class $D^*(\mathcal{G})$; hence, they are in use for distributions in the same way, too. More details can be found in any textbook on distributions, e.g., L. Jantscher [1971], W. Walter [1994].

Let $\mathcal{F} \in D^*(\mathcal{G})$ be a given distribution. Assume that there exists a function $F : \mathcal{G} \to \mathbb{R}$ that is locally integrable, i.e., F is integrable on every compact subset of $\mathcal{G}$, such that $\mathcal{F}(\phi) = \int_{\mathcal{G}} F(x)\phi(x)\, dV(x)$ holds for all test functions $\phi \in D(\mathcal{G})$. Then $\mathcal{F}$ is called a *regular distribution.* If $\mathcal{F} \in D^*(\mathcal{G})$ is a regular distribution, then the associated function F is uniquely determined (except on a set of Lebesgue measure zero).

Remark 4.9. A well-known distribution that is not regular (see, e.g., W. Walter [1994]) is the *Delta distribution* δ given by $\delta(\phi) = \phi(0)$.

A sequence $\{\mathcal{F}_n\} \subset D^*(\mathcal{G})$ is called convergent to $\mathcal{F} \in D^*(\mathcal{G})$ if and only if $\lim_{n\to\infty} \mathcal{F}_n(\phi) = \mathcal{F}(\phi)$ for all $\phi \in D(\mathcal{G})$. This definition helps us to introduce derivatives of distributions: If, for a given distribution $\mathcal{F} \in D^*(\mathcal{G})$, there exists a distribution $\tilde{\mathcal{F}} \in \mathrm{D}^*(\mathcal{G})$ such that $\tilde{\mathcal{F}}(\phi) = (-1)^{[\alpha]}\mathcal{F}(\nabla^\alpha \phi)$, $\alpha \in \mathbb{N}_0^3$, $[\alpha] = \alpha_1 + \alpha_2 + \alpha_3$, for every $\phi \in \mathrm{D}(\mathcal{G})$, then we set $\tilde{\mathcal{F}} = \nabla^\alpha \mathcal{F}$. In our potential theoretic approach we are particularly interested in Laplace derivatives: A functional $\mathcal{F} \in \mathrm{D}^*(\mathcal{G})$ is called *distributionally harmonic* if and only if $\Delta \mathcal{F} = 0$. The set of all regular harmonic $\mathrm{L}^2(\mathcal{G})$-distributions in $\mathrm{D}^*(\mathcal{G})$ is denoted by $\mathrm{DistPot}(\mathcal{G})$. The space $\mathrm{DistPot}(\mathcal{G})$ apparently represents a generalization of the set $\mathrm{Pot}(\mathcal{G})$ of harmonic functions in $\mathcal{G}$. Indeed, the following characterization is valid (see, e.g., P.J. Davis [1963]):

Lemma 4.10. *The set* $\mathrm{DistPot}(\mathcal{G})$ *of all regular harmonic* $\mathrm{L}^2(\mathcal{G})$*-distributions is a closed subspace of* $\mathrm{L}^2(\mathcal{G})$.

Proof. The class $\mathrm{DistPot}(\mathcal{G})$ is a linear space. Suppose that $\{\mathcal{F}_n\}_{n=0,1,\ldots} \subset \mathrm{DistPot}(\mathcal{G})$ is a given sequence of regular distributions with $\lim_{n\to\infty} \mathcal{F}_n = \mathcal{F}$. Let $F_n, F \in \mathrm{L}^2(\mathcal{G})$ be the functions associated with the regular distributions $\mathcal{F}_n$ and $\mathcal{F}$, respectively. Then it follows that

$$0 = \int_{\mathcal{G}} F_n(x)\Delta\phi(x)\, dV(x), \qquad n = 0, 1, \ldots, \tag{4.77}$$

such that

$$0 = \lim_{n\to\infty} \int_{\mathcal{G}} F_n(x)\Delta\phi(x)\, dV(x) = \int_{\mathcal{G}} \lim_{n\to\infty} F_n(x)\Delta\phi(x)\, dV(x). \tag{4.78}$$

Therefore, the regular distribution $\mathcal{F}$, associated with the square-integrable function F, is a member of $\mathrm{DistPot}(\mathcal{G})$. □

It is known from the theory of distributions (see, e.g., W. Walter [1994]) that $\mathrm{Pot}(\mathcal{G}) \subset \mathrm{DistPot}(\mathcal{G})$. Moreover, the so-called Weyl's Lemma tells us that $\int_{\mathcal{G}} F(x)\Delta\phi(x)\, dV(x) = 0$ for all $\phi \in \mathrm{D}(\mathcal{G})$ implies $F \in \mathrm{Pot}(\mathcal{G})$, i.e., $\mathrm{DistPot}(\mathcal{G}) \subset \mathrm{Pot}(\mathcal{G})$. Consequently, we are led to the following remarkable result.

Theorem 4.11. *Let* $\mathcal{G} \subset \mathbb{R}^3$ *be a regular region. Then we have*

$$\mathrm{DistPot}(\mathcal{G}) = \mathrm{Pot}(\mathcal{G}). \tag{4.79}$$

Lemma 4.10, in connection with Theorem 4.11, tells us that $\mathrm{Pot}(\mathcal{G})$ is a closed linear subspace of $\mathrm{L}^2(\mathcal{G})$. We are therefore able, by a well-known result of functional analysis (see, e.g., P.J. Davis [1963]), to formulate the following decomposition of $\mathrm{L}^2(\mathcal{G})$.

Theorem 4.12. *Let* $\mathcal{G} \in \mathbb{R}^3$ *be a regular region. Then*

$$\mathrm{L}^2(\mathcal{G}) = \overline{\mathrm{Pot}(\mathcal{G})}^{\|\cdot\|_{\mathrm{L}^2(\mathcal{G})}} \oplus \left(\overline{\mathrm{Pot}(\mathcal{G})}^{\|\cdot\|_{\mathrm{L}^2(\mathcal{G})}}\right)^{\perp}. \tag{4.80}$$

The following convention is useful (see, e.g., L. Ballani et al. [1993], V. Michel [1999]).

Definition 4.13. The members of the class $\mathrm{AnPot}(\mathcal{G})$ defined by

$$\mathrm{AnPot}(\mathcal{G}) = \left(\overline{\mathrm{Pot}(\mathcal{G})}^{\|\cdot\|_{\mathrm{L}^2(\mathcal{G})}}\right)^{\perp} \tag{4.81}$$

are called *anharmonic functions* in $\mathcal{G}$.

Summarizing our results, we finally arrive at the following decomposition theorem.

Corollary 4.14. *Let $\mathcal{G} \subset \mathbb{R}^3$ be a regular region. Then*

$$\mathrm{L}^2(\mathcal{G}) = \mathrm{Pot}(\mathcal{G}) \oplus \mathrm{AnPot}(\mathcal{G}). \tag{4.82}$$

Reproducing Hilbert Space Structure

Newton's Law of Gravitation as formulated in (0.2) links the gravitational potential H to the mass density distribution F. More precisely, let us assume here that F is a member of class $\mathrm{L}^2(\mathcal{G})$. Then $T : \mathrm{L}^2(\mathcal{G}) \to \mathrm{H}(\mathcal{G}^c)$ given by

$$H = T[F] = \int_{\mathcal{G}} F(y)\, G(\Delta; |\cdot - y|)\, dV(y)\Big|_{\mathcal{G}^c}, \quad F \in L^2(\mathcal{G}), \tag{4.83}$$

defines a linear operator such that $T[F]$ is harmonic in $\mathcal{G}^c$ and regular at infinity. $\mathrm{H}(\mathcal{G}^c)$ simply denotes the image space of T. The operator T as defined by (4.83) is surjective, but it is not injective. Indeed, the null space (kernel) of T

$$\ker(T) = \{F \in \mathrm{L}^2(\mathcal{G}) : T[F] = 0\} \tag{4.84}$$

consists of all functions in $\mathrm{L}^2(\mathcal{G})$ that are orthogonal to harmonic functions in $\mathcal{G}$. By virtue of Corollary 4.14, $\ker(T)$ is the space of anharmonic functions in $\mathcal{G}$.

Theorem 4.15. *Let $\mathcal{G} \subset \mathbb{R}^3$ be a regular region. Then*

$$\ker(T) = \mathrm{AnPot}(\mathcal{G}) = \left(\overline{\mathrm{Pot}(\mathcal{G})}^{\|\cdot\|_{\mathrm{L}^2(\mathcal{G})}}\right)^{\perp}, \tag{4.85}$$

hence,

$$\mathrm{L}^2(\mathcal{G}) = \mathrm{Pot}(\mathcal{G}) \oplus \ker(T). \tag{4.86}$$

Let $\mathrm{Proj}_{\mathrm{Pot}(\mathcal{G})}$ and $\mathrm{Proj}_{\mathrm{AnPot}(\mathcal{G})}$ be the orthogonal projections of $\mathrm{L}^2(\mathcal{G})$ to $\mathrm{Pot}(\mathcal{G})$ and $\ker(T) = \mathrm{AnPot}(\mathcal{G})$, respectively. Then, every function F of the Hilbert space $\mathrm{L}^2(\mathcal{G})$ can be uniquely decomposed in the form

$$F = \mathrm{Proj}_{\mathrm{Pot}(\mathcal{G})}[F] + \mathrm{Proj}_{\mathrm{AnPot}(\mathcal{G})}[F] \tag{4.87}$$

such that

$$T[F] = T\left[\mathrm{Proj}_{\mathrm{Pot}(\mathcal{G})}[F]\right] + \underbrace{T\left[\mathrm{Proj}_{\mathrm{AnPot}(\mathcal{G})}[F]\right]}_{=0} = T\left[\mathrm{Proj}_{\mathrm{Pot}(\mathcal{G})}[F]\right]. \tag{4.88}$$

Furthermore, it is clear that

$$\|F\|^2_{\mathrm{L}^2(\mathcal{G})} = \left\|\mathrm{Proj}_{\mathrm{Pot}(\mathcal{G})}[F]\right\|^2_{\mathrm{L}^2(\mathcal{G})} + \left\|\mathrm{Proj}_{\mathrm{AnPot}(\mathcal{G})}[F]\right\|^2_{\mathrm{L}^2(\mathcal{G})}. \tag{4.89}$$

In conclusion, $T[\mathrm{Proj}_{\mathrm{Pot}(\mathcal{G})}[F]]$ is that function of class $\mathrm{L}^2(\mathcal{G})$ that has the smallest $\mathrm{L}^2(\mathcal{G})$-norm among all (density) functions F in $\mathrm{L}^2(\mathcal{G})$ generating the same potential in the space $\mathrm{H}(\mathcal{G}^c)$. Consequently, to every $H \in \mathrm{H}(\mathcal{G}^c)$, there corresponds a unique $F \in \mathrm{Pot}(\mathcal{G})$ such that $T[F] = T[\mathrm{Proj}_{\mathrm{Pot}(\mathcal{G})}[F]] = H$.

Lemma 4.16. *The restriction $T|_{\mathrm{Pot}(\mathcal{G})}$ is a linear bijective operator, i.e., for every $H \in \mathrm{H}(\mathcal{G}^c)$ there exists a unique $F \in \mathrm{Pot}(\mathcal{G})$ such that $T|_{\mathrm{Pot}(\mathcal{G})}[F] = H$.*

On the space $\mathrm{H}(\mathcal{G}^c)$ we are able to impose an inner product $(\cdot,\cdot)_{\mathrm{H}(\mathcal{G}^c)}$ by letting

$$\left(T|_{\mathrm{Pot}(\mathcal{G})}[F], T|_{\mathrm{Pot}(\mathcal{G})}[G]\right)_{\mathrm{H}(\mathcal{G}^c)} = (F,G)_{\mathrm{L}^2(\mathcal{G})}, \tag{4.90}$$

where $F, G \in \mathrm{L}^2(\mathcal{G})$. $\mathrm{H}(\mathcal{G}^c)$ equipped with the inner product $(\cdot,\cdot)_{\mathrm{H}(\mathcal{G}^c)}$ is a Hilbert space. $T|_{\mathrm{Pot}(\mathcal{G})}$ is an isometric operator relating $\mathrm{L}^2(\mathcal{G})$ to $\mathrm{H}(\mathcal{G}^c)$. Our goal is to show that $(\mathrm{H}(\mathcal{G}^c), (\cdot,\cdot)_{\mathrm{H}(\mathcal{G}^c)})$ is a reproducing kernel Hilbert space, i.e., a Hilbert space equipped with a reproducing kernel $K_{\mathrm{H}(\mathcal{G}^c)}(\cdot,\cdot)$. By a *reproducing kernel* $K_{\mathrm{H}(\mathcal{G}^c)}(\cdot,\cdot)$ of the Hilbert space $\mathrm{H}(\mathcal{G}^c)$ we mean a function $K_{\mathrm{H}(\mathcal{G}^c)}(\cdot,\cdot) : \mathcal{G}^c \times \mathcal{G}^c \to \mathbb{R}$ satisfying the conditions:

(i) $K_{\mathrm{H}(\mathcal{G}^c)}(x,\cdot) \in \mathrm{H}(\mathcal{G}^c)$ for every $x \in \mathcal{G}^c$, and $K_{\mathrm{H}(\mathcal{G}^c)}(\cdot,y) \in \mathrm{H}(\mathcal{G}^c)$ for every $y \in \mathcal{G}^c$,

(ii) $\left(K_{\mathrm{H}(\mathcal{G}^c)}(x,\cdot), F\right)_{\mathrm{H}(\mathcal{G}^c)} = F(x)$, for every $x \in \mathcal{G}^c$ and $F \in \mathrm{H}(\mathcal{G}^c)$.

In order to determine the reproducing kernel, we especially consider the fundamental solution of the Laplace operator: $G(\Delta; |\cdot - \cdot|) : \mathcal{G}^c \times \mathcal{G} \to \mathbb{R}$, with

$$G(\Delta;|x-y|) = \frac{1}{4\pi}\frac{1}{|x-y|}, \quad x \in \mathcal{G}^c,\ y \in \mathcal{G}. \tag{4.91}$$

It is clear that, for a fixed $x \in \mathcal{G}^c$, $G(\Delta;|x-\cdot|)$ is an element of $\mathrm{Pot}(\mathcal{G})$. Thus, it follows that any given potential $H \in \mathrm{H}(\mathcal{G}^c)$ can be represented in the form

$$H(x) = T|_{\mathrm{Pot}(\mathcal{G})}[F](x) = (F,\ G(\Delta;|x-\cdot|))_{\mathrm{L}^2(\mathcal{G})}, \quad x \in \mathcal{G}^c,\ F \in \mathrm{Pot}(\mathcal{G}). \tag{4.92}$$

For $x \in \mathcal{G}^c$, the evaluation functional $\mathcal{E}_x[H] = H(x)$ is a bounded functional on $\mathrm{H}(\mathcal{G}^c)$. Indeed, from the Cauchy–Schwarz inequality applied to (4.92) we get

$$|\mathcal{E}_x[H]| = |H(x)| \leq ||F||_{\mathrm{L}^2(\mathcal{G})}||G(\Delta;|x-\cdot|)||_{\mathrm{L}^2(\mathcal{G})}. \tag{4.93}$$

Consequently, we have

$$|\mathcal{E}_x[H]| = |H(x)| \leq C\ \|H\|_{\mathrm{H}(\mathcal{G})}, \qquad H \in \mathrm{H}(\mathcal{G}^c),\ x \in \mathcal{G}^c. \tag{4.94}$$

Thus, a necessary and sufficient condition for the Hilbert space $\mathrm{H}(\mathcal{G}^c)$ to possess a reproducing kernel (see, e.g., N. Aronszjain [1950]) is fulfilled. Even more, we are are able to find the explicit expression of the reproducing kernel $K_{\mathrm{H}(\mathcal{G}^c)}(\cdot,\cdot) : \mathcal{G}^c \times \mathcal{G}^c \to \mathbb{R}$ for the Hilbert space $\mathrm{H}(\mathcal{G}^c)$ such that, for every $H \in \mathrm{H}(\mathcal{G}^c)$, the reproducing property

$$H(x) = \left(H, K_{\mathrm{H}(\mathcal{G}^c)}(x,\cdot)\right)_{\mathrm{H}(\mathcal{G}^c)}, \quad x \in \mathcal{G}^c, \tag{4.95}$$

is valid. For $x \in \mathcal{G}^c$ and $F \in \mathrm{Pot}(\mathcal{G})$ such that $T[F] = H$, we obtain

$$\begin{aligned} H(x) = (F, G(\Delta; |x-\cdot|))_{\mathrm{L}^2(\mathcal{G})} &= (T[F], T[G(\Delta; |x-\cdot|)])_{\mathrm{H}(\mathcal{G}^c)} \\ &= (H, T[G(\Delta; |x-\cdot|)])_{\mathrm{H}(\mathcal{G}^c)}. \end{aligned} \tag{4.96}$$

Hence, $K_{\mathrm{H}(\mathcal{G}^c)}(x,\cdot) = T[G(\Delta; |x-\cdot|)]$, i.e., for $x, y \in \mathcal{G}^c$, the integral

$$K_{\mathrm{H}(\mathcal{G}^c)}(x,y) = (G(\Delta; |x-\cdot|), G(\Delta; |y-\cdot|))_{\mathrm{L}^2(\mathcal{G})} = \frac{1}{(4\pi)^2}\int_{\mathcal{G}} \frac{dV(z)}{|x-z||y-z|}, \tag{4.97}$$

represents the (unique) reproducing kernel of $\mathrm{H}(\mathcal{G}^c)$. Clearly, for geoscientifically relevant geometries $\mathcal{G}$ showing a more complicated boundary surface $\partial\mathcal{G}$, such as a geoid and real Earth, the integral (4.97) has to be determined by approximate integration rules (e.g., Euler summation formulas as presented in W. Freeden [2011]).

Theorem 4.17. $\left(\mathrm{H}(\mathcal{G}^c), (\cdot,\cdot)_{\mathrm{H}(\mathcal{G}^c)}\right)$ *as defined by (4.83) is a reproducing kernel Hilbert space possessing the reproducing kernel (4.97).*

Remark 4.18. All aforementioned arguments remain valid if the closure $\overline{\mathcal{G}^c}$ is used instead of $\mathcal{G}^c$. Thus, $\left(\mathrm{H}(\mathcal{G}^c), (\cdot,\cdot)_{\mathrm{H}(\mathcal{G}^c)}\right)$ is a suitable reference space for solving the discrete exterior Dirichlet problem.

The reproducing structure of $\mathrm{H}(\mathcal{G}^c)$ is of great importance in the discussion of the discrete exterior oblique derivative problem (DEODP), hence, it should be characterized in more detail: $K_{\mathrm{H}(\mathcal{G}^c)}(x,\cdot) = T[G(\Delta; |x-\cdot|)]$ states that, for a fixed $x \in \mathcal{G}^c$, the reproducing kernel $K_{\mathrm{H}(\mathcal{G}^c)}(x,\cdot)$ is the Newtonian potential corresponding to the harmonic density function $G(\Delta; |x-\cdot|)$.

Discrete Formulation of the Oblique Derivative Problem

Contrary to the classical boundary-value problems, where the solution process is based on continuous knowledge of the boundary function as a whole, the boundary information is given only in a set of discrete points in the case of a discrete boundary-value problem.

Discrete Exterior Oblique Derivative Problem (DEODP): Let λ be a unit $\mathrm{c}^{(1,\mu)}$-vector field on $\partial\mathcal{G}$, $0 < \mu < 1$, such that (4.14) is valid. Let $\mathcal{X}_N = \{x_1, ..., x_N\}$ be a discrete set of N mutually distinct points on $\partial\mathcal{G}$. Suppose that the values $\alpha_i = \frac{\partial U}{\partial \lambda}(x_i)$, $x_i \in \partial\mathcal{G}$, $i = 1, ..., N$, constitute a given data set associated to a potential $U \in \mathrm{Pot}^{(1,\mu)}(\overline{\mathcal{G}^c})$. Find an approximation S_N^U to the potential $U \in \mathrm{Pot}^{(1,\mu)}(\overline{\mathcal{G}^c})$ such that

$$\frac{\partial S_N^U}{\partial \lambda}(x_i) = \frac{\partial U}{\partial \lambda}(x_i) = \alpha_i, \quad i = 1, ..., N. \tag{4.98}$$

An oblique derivative value at a point $x \in \partial\mathcal{G}$ can be identified with a linear functional

$$\mathcal{D}_x : H \mapsto \mathcal{D}_x[H] = \frac{\partial H}{\partial \lambda}(x), \qquad H \in \mathrm{H}(\mathcal{G}^c). \tag{4.99}$$

The boundedness of $\mathcal{D}_x$ is an important aspect in our approach. This is the reason why, instead of considering integrals over $\mathcal{G}$, we consider (a Runge concept of) integrals extended over $\mathcal{A}$, with $\mathcal{A}$ totally contained in $\mathcal{G}$ (see Figure 3.7). In doing so, we start from the Hilbert space $\mathrm{H}(\mathcal{A}^c) = T[\mathrm{L}^2(\mathcal{A})]$ of potentials $T[F]$ defined by

$$T[F] = \int_{\mathcal{A}} F(y)\; G(\Delta; |\cdot - y|) dV(y)\bigg|_{\mathcal{A}^c}, \qquad F \in \mathrm{L}^2(\mathcal{A}). \tag{4.100}$$

Remembering the decomposition of $\mathrm{L}^2(\mathcal{A})$, we restrict the potential operator T to the closed subspace $\mathrm{Pot}(\mathcal{A})$ of harmonic density functions in $\mathrm{L}^2(\mathcal{A})$. The isometric operator $T|_{\mathrm{Pot}(\mathcal{A})}$ mapping $\mathrm{L}^2(\mathcal{A})$ to $\mathrm{H}(\mathcal{A}^c)$ imposes a Hilbert space structure on $\mathrm{H}(\mathcal{A}^c)$. In turn, the scalar product in $\mathrm{H}(\mathcal{A}^c)$ is defined by

$$(T|_{\mathrm{Pot}(\mathcal{A})}[F], T|_{\mathrm{Pot}(\mathcal{A})}[G])_{\mathrm{H}(\mathcal{A}^c)} = (F, G)_{\mathrm{L}^2(\mathcal{A})}, \quad F, G \in \mathrm{L}^2(\mathcal{A}). \tag{4.101}$$

By already known arguments it follows that the space $\mathrm{H}(\mathcal{A}^c)$ possesses a uniquely determined reproducing kernel $K_{\mathrm{H}(\mathcal{A}^c)}(\cdot,\cdot)$ relative to $\mathcal{A}$ given by

$$K_{\mathrm{H}(\mathcal{A}^c)}(x, y) = \int_{\mathcal{A}} G(\Delta; |x - z|)\; G(\Delta; |y - z|)\; dV(z), \quad x, y \in \mathcal{A}^c. \tag{4.102}$$

Now, we are in a position to guarantee the boundedness of $\mathcal{D}_x$ on $\mathrm{H}(\mathcal{A}^c)$ for $x \in \partial\mathcal{G}$. In fact, for $\varepsilon > 0$, it can be readily seen that

$$\frac{1}{\varepsilon^2}\|K_{\mathrm{H}(\mathcal{A}^c)}(x,\cdot) - K_{\mathrm{H}(\mathcal{A}^c)}(x + \varepsilon\lambda(x), \cdot)\|^2_{\mathrm{H}(\mathcal{A}^c)} = \mathcal{O}\left(\int_{\mathcal{A}} \frac{dV(z)}{|x - z|^4}\right). \tag{4.103}$$

The integral in (4.103) exists for $x \in \partial\mathcal{G}$; hence, for $H \in \mathrm{H}(\mathcal{A}^c)$, we obtain

$$\begin{aligned} &\frac{1}{\varepsilon}|H(x) - H(x + \varepsilon\lambda(x))| \\ &\leq \frac{1}{\varepsilon}\|H\|_{\mathrm{H}(\mathcal{A}^c)}\|K_{\mathrm{H}(\mathcal{A}^c)}(x,\cdot) - K_{\mathrm{H}(\mathcal{A}^c)}(x + \varepsilon\lambda(x), \cdot)\|_{\mathrm{H}(\mathcal{A}^c)} \\ &\leq C\|H\|_{\mathrm{H}(\mathcal{A}^c)} \end{aligned} \tag{4.104}$$

for some $C > 0$ depending on the choice of $\mathcal{A}$. Letting ε tend to 0 we are able to deduce that the functional $\mathcal{D}_x$ of the oblique derivative at $x \in \partial\mathcal{G}$ is bounded with respect to the $\mathrm{H}(\mathcal{A}^c)$-topology.

Lemma 4.19. *Let $\mathcal{A}, \mathcal{G} \subset \mathbb{R}^3$ be regular regions such that $\mathcal{A} \Subset \mathcal{G}$. For each $x \in \partial\mathcal{G}$,*

$$y \mapsto \frac{\partial}{\partial\lambda(x)} K_{\mathrm{H}(\mathcal{A}^c)}(x,y), \; y \in \mathcal{A}^c, \tag{4.105}$$

is the representer of the bounded linear functional

$$\mathcal{D}_x : H \mapsto \mathcal{D}_x[H] = \frac{\partial H}{\partial\lambda}(x), \;\; H \in \mathrm{H}(\mathcal{A}^c). \tag{4.106}$$

More precisely,

$$\mathcal{D}_x[H] = \left(H, \frac{\partial}{\partial\lambda(x)} K_{\mathrm{H}(\mathcal{A}^c)}(x,\cdot) \right)_{\mathrm{H}(\mathcal{A}^c)}, \qquad H \in \mathrm{H}(\mathcal{A}^c). \tag{4.107}$$

Minimum Norm Interpolation

The potential U for solving the discrete exterior oblique derivative problem (DEODP) is considered as an element of the Hilbert space $\mathrm{H}(\mathcal{A}^c)$ possessing the reproducing kernel $K_{\mathrm{H}(\mathcal{A}^c)}(\cdot,\cdot)$, while the observed values at the points $x_1, ..., x_N \in \partial\mathcal{G}$ are assumed to be associated with linearly independent bounded functionals $\mathcal{D}_{x_1}, ..., \mathcal{D}_{x_N}$. In doing so, we are able to find a minimum norm solution $S_N^U \in \mathrm{H}(\mathcal{A}^c)$ as a linear combination of the representers $\mathcal{D}_{x_i}[K_{\mathrm{H}(\mathcal{A}^c)}(\cdot,\cdot)]$ to the functionals $\mathcal{D}_{x_i}$, i.e., S_N^U is exactly the projection of U to the N-dimensional linear subspace spanned by the linearly independent representers $\mathcal{D}_{x_i}[K_{\mathrm{H}(\mathcal{A}^c)}(\cdot,\cdot)]$, $i = 1, ..., N$ (see, e.g., P.J. Davis [1963]). In other words, the solution of (DEODP) is sought in the reproducing Hilbert space $\mathrm{H}(\mathcal{A}^c)$ under the assumption that $\{\alpha_1, ..., \alpha_N\}$ with $\alpha_i = \mathcal{D}_{x_i}[U]$, $i = 1, \ldots, N$, forms the (observed) given data set for the unknown potential U, corresponding to the discrete set $\mathcal{X}_N = \{x_1, ..., x_N\}$ of points on $\partial\mathcal{G}$, such that (4.98) holds true. All in all, the aim of minimum norm interpolation in $\mathrm{H}(\mathcal{A}^c)$ is to find the "smoothest" $S_N^U \in \mathrm{H}(\mathcal{A}^c)$ within the set of all $\mathrm{H}(\mathcal{A}^c)$-interpolants, where the norm is minimized in the metric of $\mathrm{H}(\mathcal{A}^c)$. Equivalently, the problem is to find a function S_N^U within the set

$$\mathcal{I}^U_{\mathcal{D}_{x_1},...,\mathcal{D}_{x_N}} = \{F \in \mathrm{H}(\mathcal{A}^c) : \; \mathcal{D}_{x_i}[F] = \mathcal{D}_{x_i}[U], \quad i = 1, ..., N\}, \tag{4.108}$$

such that

$$\|S_N^U\|_{\mathrm{H}(\mathcal{A}^c)} = \inf_{F \in \mathcal{I}^U_{\mathcal{D}_{x_1},...,\mathcal{D}_{x_N}}} \|F\|_{\mathrm{H}(\mathcal{A}^c)} . \tag{4.109}$$

Remembering the integral representation of the reproducing kernel $K_{\mathrm{H}(\mathcal{A}^c)}(\cdot,\cdot)$ in $\mathrm{H}(\mathcal{A}^c)$, we are able to rewrite the functional of the oblique derivative at $x \in \partial\mathcal{G}$ in the form

$$\mathcal{D}_x \left[K_{\mathrm{H}(\mathcal{A}^c)}(\cdot,y) \right] = \left(\frac{1}{4\pi} \right)^2 \int_{\mathcal{A}} \frac{\partial}{\partial\lambda(x)} \frac{1}{|x-z||y-z|} dV(z). \tag{4.110}$$

For any *$\mathcal{D}$-unisolvent system* $\mathcal{X}_N = \{x_1, ..., x_N\} \subset \partial\mathcal{G}$, i.e., for any system $\mathcal{X}_N = \{x_1, \ldots, x_N\}$ such that $\{\mathcal{D}_{x_1}, ..., \mathcal{D}_{x_N}\}$ forms a set of N linearly independent bounded linear functionals on $\mathrm{H}(\mathcal{A}^c)$, we introduce $\mathrm{H}(\mathcal{A}^c)$-splines relative to $\{\mathcal{D}_{x_1}, ..., \mathcal{D}_{x_N}\}$ in the following way:

Definition 4.20. Let $\mathcal{X}_N = \{x_1, ..., x_N\} \subset \partial\mathcal{G}$ be a $\mathcal{D}$-unisolvent system on $\partial\mathcal{G}$. Then, any function $S \in \mathrm{H}(\mathcal{A}^c)$ given by

$$S(x) = \sum_{i=1}^{N} a_i \mathcal{D}_{x_i}[K_{\mathrm{H}(\mathcal{A}^c)}(x, \cdot)] = \sum_{i=1}^{N} a_i \left(\frac{1}{4\pi}\right)^2 \int_{\mathcal{A}} \frac{-\lambda(x_i) \cdot (x_i - z)}{|x_i - z|^3 |x - z|} \, dV(z) \tag{4.111}$$

with arbitrarily given (real) coefficients $a_1, ..., a_N$ is called an $\mathrm{H}(\mathcal{A}^c)$-*spline relative to* $\{\mathcal{D}_{x_1}, ..., \mathcal{D}_{x_N}\}$. The space of all $\mathrm{H}(\mathcal{A}^c)$-splines relative to $\{\mathcal{D}_{x_1}, ..., \mathcal{D}_{x_N}\}$ is denoted by $\mathrm{Spline}_{\mathrm{H}(\mathcal{A}^c)}(\mathcal{D}_{x_1}, ..., \mathcal{D}_{x_N})$.

Clearly, $\mathrm{Spline}_{\mathrm{H}(\mathcal{A}^c)}(\mathcal{D}_{x_1}, ..., \mathcal{D}_{x_N})$ is an N-dimensional subspace of $\mathrm{H}(\mathcal{A}^c)$. Moreover, by virtue of the reproducing property in $\mathrm{H}(\mathcal{A}^c)$, we immediately obtain the so-called $\mathrm{H}(\mathcal{A}^c)$-*spline formula.*

Lemma 4.21. *Let S be a function of class* $\mathrm{Spline}_{\mathrm{H}(\mathcal{A}^c)}(\mathcal{D}_{x_1}, ..., \mathcal{D}_{x_N})$. *Then, for each $F \in \mathrm{H}(\mathcal{A}^c)$,*

$$(S, F)_{\mathrm{H}(\mathcal{A}^c)} = \sum_{i=1}^{N} a_i \mathcal{D}_{x_i}[F]. \tag{4.112}$$

By virtue of the $\mathcal{D}$-unisolvence of the system $\mathcal{X}_N = \{x_1, \ldots, x_N\} \subset \partial\mathcal{G}$ it is not difficult to verify the *uniqueness of interpolation.*

Lemma 4.22. *For a given potential $U \in \mathrm{H}(\mathcal{A}^c)$, there exists a unique element S_N^U in* $\mathrm{Spline}_{\mathrm{H}(\mathcal{A}^c)}(\mathcal{D}_{x_1}, ..., \mathcal{D}_{x_N}) \cap \mathcal{I}^U_{\mathcal{D}_{x_1}, ..., \mathcal{D}_{x_N}}$.

Proof. The application of the N bounded linear functionals $\mathcal{D}_{x_1}, ..., \mathcal{D}_{x_N}$ on $\mathrm{H}(\mathcal{A}^c)$ to the $\mathrm{H}(\mathcal{A}^c)$-spline of the form (4.111) yields N linear equations in the unknowns $a_1^N, ..., a_N^N$, i.e.,

$$\sum_{j=1}^{N} a_j^N \mathcal{D}_{x_i} \mathcal{D}_{x_j}[K_{\mathrm{H}(\mathcal{A}^c)}(\cdot, \cdot)] = \mathcal{D}_{x_i}[U] \;=\; \alpha_i, \quad i = 1, ..., N, \tag{4.113}$$

where the coefficient matrix $\left(\mathcal{D}_{x_i} \mathcal{D}_{x_j}[K_{\mathrm{H}(\mathcal{A}^c)}(\cdot, \cdot)]\right)_{i,j=1,\ldots,N}$ is given by

$$\mathcal{D}_{x_i} \mathcal{D}_{x_j}[K_{\mathrm{H}(\mathcal{A}^c)}(\cdot, \cdot)] = \left(\frac{1}{4\pi}\right)^2 \int_{\mathcal{A}} \frac{\lambda(x_i) \cdot (x_i - z) \;\; \lambda(x_j) \cdot (x_j - z)}{|x_i - z|^3 |x_j - z|^3} \, dV(z). \tag{4.114}$$

From multivariate interpolation theory (see, e.g., P.J. Davis [1963]) we know that (4.114) constitutes a Gram matrix of N linearly independent functions $\mathcal{D}_{x_1}[K_{\mathrm{H}(\mathcal{A}^c)}(\cdot, \cdot)], ..., \mathcal{D}_{x_N}[K_{\mathrm{H}(\mathcal{A}^c)}(\cdot, \cdot)]$. Hence, it is non-singular such that the linear system (4.113) is uniquely solvable. The coefficients $a_1^N, ..., a_N^N$ form the unique interpolating spline S_N^U for which we are looking. □

The following minimum norm properties for the interpolating spline are easily derivable.

Lemma 4.23 (First Minimum Property). *If* $F \in \mathcal{I}^U_{\mathcal{D}_{x_1},...,\mathcal{D}_{x_N}}$, *then*

$$||F||^2_{\mathrm{H}(\mathcal{A}^c)} = ||S^U_N||^2_{\mathrm{H}(\mathcal{A}^c)} + ||S^U_N - F||^2_{\mathrm{H}(\mathcal{A}^c)}. \tag{4.115}$$

Lemma 4.24 (Second Minimum Property). *If* $S \in \mathrm{Spline}_{\mathrm{H}(\mathcal{A}^c)}(\mathcal{D}_{x_1}, ..., \mathcal{D}_{x_N})$ *and* $F \in \mathcal{I}^U_{\mathcal{D}_{x_1},...,\mathcal{D}_{x_N}}$, *then*

$$||S - F||^2_{\mathrm{H}(\mathcal{A}^c)} = ||S^U_N - F||^2_{\mathrm{H}(\mathcal{A}^c)} + ||S - S^U_N||^2_{\mathrm{H}(\mathcal{A}^c)}. \tag{4.116}$$

Summarizing our main results on $\mathrm{H}(\mathcal{A}^c)$-spline interpolation of a finite set of oblique derivatives we obtain

Theorem 4.25. *The minimum norm interpolation problem*

$$||S^U_N||_{\mathrm{H}(\mathcal{A}^c)} = \inf_{F \in \mathcal{I}^U_{\mathcal{D}_{x_1},...,\mathcal{D}_{x_N}}} ||F||_{\mathrm{H}(\mathcal{A}^c)} \tag{4.117}$$

with

$$\mathcal{I}^U_{\mathcal{D}_{x_1},...,\mathcal{D}_{x_N}} = \{F \in \mathrm{H}(\mathcal{A}^c) : \mathcal{D}_{x_i}[F] = \mathcal{D}_{x_i}[U] = \alpha_i,\ i = 1, \ldots, N\} \tag{4.118}$$

is well posed in the sense that its solution exists, is unique, and depends continuously on the data α_i, $i = 1, \ldots, N$. *The uniquely determined solution* S^U_N *is given in the explicit form*

$$S^U_N(x) = \sum_{i=1}^N a^N_i \left(\frac{1}{4\pi}\right)^2 \int_{\mathcal{A}} \frac{-\lambda(x_i)\cdot(x_i - z)}{|x_i - z|^3|x - z|}\, dV(z), \quad x \in \mathcal{A}^c, \tag{4.119}$$

where the coefficients $a^N_1, ..., a^N_N$ *satisfy the linear equations*

$$\sum_{i=1}^N a^N_i \left(\frac{1}{4\pi}\right)^2 \int_{\mathcal{A}} \frac{(\lambda(x_i)\cdot(x_i - z))\ (\lambda(x_j)\cdot(x_j - z))}{|x_i - z|^3|x_j - z|^3}\, dV(z) = \alpha_j, \quad j = 1, ..., N. \tag{4.120}$$

Stability of the Discrete Solution

Next we are concerned with the *stability of the solution.* As already known, for every $\mathcal{D}$-unisolvent system $\mathcal{X}_N = \{x_1, ..., x_N\} \subset \partial\mathcal{G}$ and for every function $U \in \mathrm{H}(\mathcal{A}^c)$ there exists a unique element $S^N_U \in \mathrm{H}(\mathcal{A}^c)$ that satisfies the conditions $\mathcal{D}_{x_i}[U] = \mathcal{D}_{x_i}[S^N_U]$, $i = 1, \ldots, N$.

Let $\Theta_{\mathcal{X}_N}$ denote the $\mathcal{X}_N$*-width on* $\partial\mathcal{G}$, i.e., the maximal distance for any point of $\partial\mathcal{G}$ to the system $\mathcal{X}_N$, i.e.,

$$\Theta_{\mathcal{X}_N} = \max_{x \in \partial\mathcal{G}} \left(\min_{y \in \mathcal{X}_N} |x - y| \right). \tag{4.121}$$

Lemma 4.26. *Let U be a member of class $\mathrm{H}(\mathcal{A}^c)$. Suppose that $\mathcal{X}_N \subset \partial\mathcal{G}$ is a $\mathcal{D}$-unisolvent system. Then there exists a constant $C > 0$ (dependent on $\partial\mathcal{G}$ and $\mathcal{A}$) such that*

$$\sup_{x\in\partial\mathcal{G}} \left|\mathcal{D}_x[U] - \mathcal{D}_x[S_N^U]\right| \leq C\, \Theta_{\mathcal{X}_N}\, \|U\|_{\mathcal{H}(\mathcal{A}^c)}. \tag{4.122}$$

Proof. For $x \in \partial\mathcal{G}$, there exists a point $y \in \mathcal{X}_N$ with $|x-y| \leq \Theta_{\mathcal{X}_N}$. Observing the interpolation property $\mathcal{D}_y[U] = \mathcal{D}_y[S_N^U], y \in \mathcal{X}_N$, we see that

$$\mathcal{D}_x[U] - \mathcal{D}_x[S_N^U] = (\mathcal{D}_x[U] - \mathcal{D}_y[U]) - \left(\mathcal{D}_x[S_N^U] - \mathcal{D}_y[S_N^U]\right). \tag{4.123}$$

The reproducing kernel structure of $\mathrm{H}(\mathcal{A}^c)$ enables us to derive the estimates

$$|\mathcal{D}_x[U] - \mathcal{D}_y[U]| \leq (\kappa(x,y))^{\frac{1}{2}}\, \|U\|_{\mathrm{H}(\mathcal{A}^c)}, \tag{4.124}$$

$$\left|\mathcal{D}_x[S_N^U] - \mathcal{D}_y[S_N^U]\right| \leq (\kappa(x,y))^{\frac{1}{2}}\, \|S_N^U\|_{\mathrm{H}(\mathcal{A}^c)}, \tag{4.125}$$

where

$$\kappa(x,y) = (\mathcal{D}_x\mathcal{D}_x - 2\mathcal{D}_x\mathcal{D}_y + \mathcal{D}_y\mathcal{D}_y)\,[K_{\mathrm{H}(\mathcal{A}^c)}(\cdot,\cdot)]. \tag{4.126}$$

S_N^U is the smoothest $\mathrm{H}(\mathcal{A}^c)$-interpolant, i.e., $\|S_N^U\|_{\mathrm{H}(\mathcal{A}^c)} \leq \|U\|_{\mathrm{H}(\mathcal{A}^c)}$. From (4.123), (4.124) and (4.125) we therefore obtain

$$\left|\mathcal{D}_x[U] - \mathcal{D}_x[S_N^U]\right| \leq 2\, (\kappa(x,y))^{\frac{1}{2}}\, \|U\|_{\mathrm{H}(\mathcal{A}^c)}. \tag{4.127}$$

An elementary calculation shows that

$$\begin{aligned}&(\mathcal{D}_x\mathcal{D}_x - 2\mathcal{D}_x\mathcal{D}_y + \mathcal{D}_y\mathcal{D}_y)\left[K_{\mathrm{H}(\mathcal{A}^c)}(\cdot,\cdot)\right] \\ &= \int_{\mathcal{A}} \left(\frac{\lambda(x)\cdot(x-z)}{|x-z|^3} - \frac{\lambda(y)\cdot(y-z)}{|y-z|^3}\right)^2 dV(z)\end{aligned} \tag{4.128}$$

for $x, y \in \partial\mathcal{G}$. Furthermore,

$$\begin{aligned}&\frac{\lambda(x)\cdot(x-z)}{|x-z|^3} - \frac{\lambda(y)\cdot(y-z)}{|y-z|^3} \\ &= \frac{\lambda(x)\cdot(x-z)|y-z|^3 - \lambda(y)\cdot(y-z)|x-z|^3}{|x-z|^3\;|y-z|^3},\end{aligned} \tag{4.129}$$

where the numerator of (4.129) can be rewritten in the form

$$\begin{aligned}&\lambda(x)\cdot(x-z)|y-z|^3 - \lambda(y)\cdot(y-z)|x-z|^3 \\ &= (\lambda(x) - \lambda(y))\cdot(x-z)|y-z|^3 \\ &\quad +\lambda(y)\cdot(x-y)|y-z|^3 + \lambda(y)\cdot(y-z)\left(|y-z|^3 - |x-z|^3\right).\end{aligned} \tag{4.130}$$

Clearly,

$$|y-z|^3 - |x-z|^3 = (|y-z| - |x-z|)\sum_{k=0}^{2} |y-z|^k |x-z|^{2-k}. \tag{4.131}$$

As a $c^{(1,\mu)}$-vector field, λ is Lipschitz continuous; hence, there exists a constant $L > 0$ such that $|\lambda(x) - \lambda(y)| \leq L\,|x - y|$ for $x, y \in \partial\mathcal{G}$. Moreover, the triangle inequality tells us that $||y - z| - |x - z|| \leq |x - y|$ for $x, y \in \partial\mathcal{G}$. Observing the assumption $\operatorname{dist}(\overline{\mathcal{A}}, \partial\mathcal{G}) > 0$, we finally get $|\kappa(x, y)| \leq \frac{C}{4}|x - y|^2$ with some constant $C > 0$ (depending on $\mathcal{A}$ and $\partial\mathcal{G}$). This proves Lemma 4.26. □

From Theorem 4.5 we know that there exists a constant $C > 0$ (dependent on $\partial\mathcal{G}$) such that

$$\sup_{x \in \overline{\mathcal{G}^c}} |U(x)| \leq C \sup_{x \in \partial\mathcal{G}} |\mathcal{D}_x[U]| \,. \tag{4.132}$$

Summarizing our results about (DEODP), we therefore obtain

Theorem 4.27. *Let $\mathcal{A}, \mathcal{G} \subset \mathbb{R}^3$ be regular regions with $\mathcal{A} \Subset \mathcal{G}$. Suppose that U is of class $\mathrm{H}(\mathcal{A}^c)$ and that $\mathcal{X}_N = \{x_1, \ldots, x_N\}$ is a $\mathcal{D}$-unisolvent system on $\partial\mathcal{G}$. Moreover, let S_N^U denote the uniquely determined solution of the spline interpolation problem*

$$\|S_N^U\|_{\mathrm{H}(\mathcal{A}^c)} = \inf_{F \in \mathcal{I}^U_{\mathcal{D}_{x_1}, \ldots, \mathcal{D}_{x_N}}} \|F\|_{\mathrm{H}(\mathcal{A}^c)}.$$

Then there exists a constant $C > 0$ (depending on $\mathcal{A}$ and $\mathcal{G}$) such that

$$\sup_{x \in \overline{\mathcal{G}^c}} \left| U(x) - S_N^U(x) \right| \leq C\, \Theta_{\mathcal{X}_N} \|U\|_{\mathrm{H}(\mathcal{A}^c)}.$$

From Theorem 4.27 we are canonically led to a sequence $\{\mathcal{X}_N\}_{N=0,1,\ldots}$ of $\mathcal{D}$-unisolvent systems $\mathcal{X}_N$ such that $\Theta_{\mathcal{X}_N} \to 0$ as $N \to \infty$. In fact, the solution of the exterior boundary-value problem

$$U \in \mathrm{H}(\mathcal{A}^c)|_{\overline{\mathcal{G}^c}}, \qquad \mathcal{D}|_{\partial\mathcal{G}}[U] \in \mathcal{D}|_{\partial\mathcal{G}}[\mathrm{H}(\mathcal{A}^c)], \tag{4.133}$$

with

$$\mathcal{D}|_{\partial\mathcal{G}}[U] : x \mapsto \mathcal{D}_x[U], \quad x \in \partial\mathcal{G}, \tag{4.134}$$

$$\mathcal{D}|_{\partial\mathcal{G}}[\mathrm{H}(\mathcal{A}^c)] = \{\mathcal{D}|_{\partial\mathcal{G}}[H] : H \in \mathrm{H}(\mathcal{A}^c)\}\,, \tag{4.135}$$

can be approximated in the sense that, for every $\varepsilon > 0$, there exists an integer $N (= N(\varepsilon))$ and a linear combination

$$S_N^U(x) = \sum_{i=1}^{N} a_i^N \mathcal{D}_{x_i}\left[K_{\mathrm{H}(\mathcal{A}^c)}(\cdot, x)\right], \quad x \in \overline{\mathcal{G}^c}, \tag{4.136}$$

uniquely determined by

$$\mathcal{D}_{x_j}[S_N^U] = \mathcal{D}_{x_j}[U], \quad j = 1, \ldots N,$$

such that

$$\sup_{x \in \overline{\mathcal{G}^c}} \left| U(x) - S_N^U(x) \right| \leq \varepsilon. \tag{4.137}$$

Thus, we are able to solve the exterior oblique boundary-value problem (EODP) in a constructive (Runge) way by spline interpolation in the $\mathcal{H}(\mathcal{A}^c)$-framework, provided that the $\mathcal{X}_N$-widths $\Theta_{\mathcal{X}_N}$ tend to zero as $N \to \infty$ and the boundary values are members of the class $\mathcal{D}|_{\partial\mathcal{G}}[\mathrm{H}(\mathcal{A}^c)]$. Furthermore, it is clear from our construction that $\mathcal{D}|_{\partial\mathcal{G}}[\mathrm{H}(\mathcal{A}^c)] \subset \mathrm{C}^{(0,\mu)}(\partial\mathcal{G})$. Thus, for any function $F \in \mathrm{C}^{(0,\mu)}(\partial\mathcal{G})$, there exists an element $\mathcal{D}|_{\partial\mathcal{G}}[U]$ in the space $\mathcal{D}|_{\partial\mathcal{G}}[\mathrm{H}(\mathcal{A}^c)]$ in an ε-neighborhood of F (with respect to the uniform topology on $\partial\mathcal{G}$).

Theorem 4.28. *Suppose that $\{\mathcal{X}_N\}_{N=0,1,\ldots}$ is a sequence of $\mathcal{D}$-unisolvent systems $\mathcal{X}_N = \{x_1, \ldots, x_N\} \subset \partial\mathcal{G}$ such that $\Theta_{\mathcal{X}_N} \to 0$ as $N \to \infty$. Then the solution V of the exterior oblique derivative problem* (EODP)

$$V \in \mathrm{Pot}^{(1,\mu)}(\overline{\mathcal{G}^c}), \qquad \left.\frac{\partial V^+}{\partial \lambda}\right|_{\partial\mathcal{G}} = F \tag{4.138}$$

corresponding to $F \in \mathrm{C}^{(0,\mu)}(\partial\mathcal{G})$, $0 < \mu < 1$, can be approximated in the sense that, for every $\varepsilon > 0$, there exists an integer $N(= N(\varepsilon))$ and a linear combination S_N of type (4.136) such that

$$\sup_{x \in \overline{\mathcal{G}_c}} |V(x) - S_N(x)| \leq \varepsilon. \tag{4.139}$$

The estimate (4.139) is of great significance from theoretical as well as practical point of view. It informs us that

$$\mathrm{L}^2(\partial\mathcal{G}) = \overline{\operatorname*{span}_{x \in \mathcal{X}} \{\mathcal{D}|_{\partial\mathcal{G}}[K_{\mathrm{H}(\mathcal{A}^c)}(x, \cdot)]\}}^{\|\cdot\|_{\mathrm{L}^2(\partial\mathcal{G})}} \tag{4.140}$$

provided that $\mathcal{X}$ is the union of a sequence $\{\mathcal{X}_N\}_{N=0,1,\ldots}$ of $\mathcal{D}$-unisolvent systems $\mathcal{X}_N \subset \partial\mathcal{G}$ with $\Theta_{\mathcal{X}_N} \to 0$ as $N \to \infty$.

Remark 4.29. The closure property (4.140) additionally implies that the method of generalized Fourier expansions in the $\mathrm{L}^2(\partial\mathcal{G})$-topology is also applicable for kernels $\mathcal{D}|_{\partial\mathcal{G}}[K_{\mathrm{H}(\mathcal{A}^c)}(x, \cdot)]$, $x \in \mathcal{X}$.

Even better, Theorem 4.28 can be exploited more generally to analyze the role of spline interpolation in the (EODP)-formulation (4.138). For that purpose we use, in addition to $\mathcal{D}|_{\partial\mathcal{G}}[\mathrm{H}(\mathcal{A}^c)] \subset \mathrm{C}^{(0,\mu)}(\partial\mathcal{G})$, an extended version of Helly's Theorem due to H. Yamabe [1950]. It tells us that, for any prescribed set $\mathcal{X}_M$ of M points $x_1, \ldots, x_M$ on $\partial\mathcal{G}$ and for any function $F \in \mathrm{C}^{(0,\mu)}(\partial\mathcal{G})$ there exists an element $\mathcal{D}|_{\partial\mathcal{G}}[V]$ of the space $\mathcal{D}|_{\partial\mathcal{G}}[\mathrm{H}(\mathcal{A}^c)]$ in an ε-neighborhood of F (with respect to the uniform topology on $\partial\mathcal{G}$) such that $F(x_i) = \mathcal{D}|_{\partial\mathcal{G}}[V](x_i)$, $i = 1, \ldots, M$. Observing these results we now obtain

Theorem 4.30. *Let $\tilde{\mathcal{X}}_M$ be a system consisting of points $\tilde{x}_1, \ldots, \tilde{x}_M$ on $\partial\mathcal{G}$. Suppose that $\{\mathcal{X}_N\}_{N=0,1,\ldots}$ is a sequence of $\mathcal{D}$-unisolvent systems $\mathcal{X}_N = \{x_1, \ldots, x_N\} \subset \partial\mathcal{G}$ such that $\tilde{\mathcal{X}}_M \subset \mathcal{X}_N$ for all N, and assume $\Theta_{\mathcal{X}_N} \to 0$*

as $N \to \infty$. Then the solution V of the exterior oblique derivative problem (EODP)

$$V \in \mathrm{Pot}^{(1,\mu)}(\overline{\mathcal{G}^c}), \qquad \left.\frac{\partial V^+}{\partial \lambda}\right|_{\partial\mathcal{G}} = F \tag{4.141}$$

corresponding to $F \in \mathrm{C}^{(0,\mu)}(\partial\mathcal{G})$, $0 < \mu < 1$, can be approximated in the sense that, for every $\varepsilon > 0$, there exists an integer $N(= N(\varepsilon))$ and a linear combination S_N of type (4.136) such that

$$\mathcal{D}_{\tilde{x}_j}[S_N] = F(\tilde{x}_j), \quad j = 1, \ldots, M, \tag{4.142}$$

and

$$\sup_{x \in \overline{\mathcal{G}_c}} |V(x) - S_N(x)| \leq \varepsilon. \tag{4.143}$$

Specialization to an Inner "Runge Ball"

We particularly investigate the reproducing kernel $K_{\mathrm{H}(\mathcal{A}^c)}(\cdot,\cdot)$ under the special situation that the regular region $\mathcal{A}$ is a ball of radius $\underline{R}$ around the origin, i.e., $\mathcal{A} = \mathcal{B}_{\underline{R}}(0)$ with $\underline{R} < \inf_{x\in\partial\mathcal{G}}|x|$. In this case we have, for all $x, y \in \mathbb{R}^3 \backslash \overline{\mathcal{B}_{\underline{R}}(0)}$,

$$K_{\mathrm{H}\left(\mathbb{R}^3\backslash\overline{\mathcal{B}_{\underline{R}}(0)}\right)}(x,y) = \left(\frac{1}{4\pi}\right)^2 \int_{\mathcal{B}_{\underline{R}}(0)} \frac{1}{|x-z||y-z|}\, dV(z). \tag{4.144}$$

Using the known spherical harmonic expansions for the fundamental solution, we easily get

$$\begin{aligned} & K_{\mathrm{H}\left(\mathbb{R}^3\backslash\overline{\mathcal{B}_{\underline{R}}(0)}\right)}(x,y) \\ &= \left(\frac{1}{4\pi}\right)^2 \int_{\mathcal{B}_{\underline{R}}(0)} \sum_{n=0}^{\infty} \frac{|z|^n}{|x|^{n+1}} P_n\left(\frac{x}{|x|}\cdot\frac{z}{|z|}\right) \sum_{m=0}^{\infty} \frac{|z|^m}{|y|^{m+1}} P_m\left(\frac{y}{|y|}\cdot\frac{z}{|z|}\right) dV(z). \end{aligned} \tag{4.145}$$

Observing the orthonormality and the addition theorem of spherical harmonics (cf. E. Kotevska [2011]), the above integral can be expressed as a Legendre series expansion

$$K_{\mathrm{H}\left(\mathbb{R}^3\backslash\overline{\mathcal{B}_{\underline{R}}(0)}\right)}(x,y) = \frac{\underline{R}}{4\pi} \sum_{n=0}^{\infty} \frac{1}{(2n+1)(2n+3)} \left(\frac{\underline{R}^2}{|x||y|}\right)^{n+1} P_n\left(\frac{x}{|x|}\cdot\frac{y}{|y|}\right). \tag{4.146}$$

4.1.3 Solution by Surface-Based Reprostructure

Let $\mathcal{A}, \mathcal{G} \subset \mathbb{R}^3$ be regular regions such that $\mathcal{A}$ is totally contained in $\mathcal{G}$ (see Figure 3.7). Suppose that $\{D_n^*\}_{n=0,1,\ldots}$ is an $\mathrm{L}^2(\partial\mathcal{A})$-Dirichlet Runge basis (in the sense of Definition 3.95) obeying the orthonormality condition

$$(D_n^*, D_m^*)_{\mathrm{L}^2(\partial\mathcal{A})} = \int_{\partial\mathcal{A}} D_n^*(x) D_m^*(x)\, d\omega(x) \;=\; \delta_{n,m}. \tag{4.147}$$

Reproducing Hilbert Space Structure

Assume that the system $\{E_n\}_{n=0,1,\ldots}$ is given by

$$E_n = \sigma_n D_n^*, \quad n = 0, 1, \ldots, \tag{4.148}$$

with $\sigma_n \in \mathbb{R}\backslash\{0\}$ such that

$$\sum_{n=0}^{\infty} \sigma_n^2 < \infty. \tag{4.149}$$

For applications, it turns out that the σ_n typically relate to the symbols of pseudodifferential operators. Then, for each $k \in \mathbb{N}_0$ and $\tilde{F} \in \mathrm{L}^2(\partial\mathcal{A})$, we define functions

$$F = \sum_{n=0}^{\infty} (\tilde{F}, D_n^*)_{\mathrm{L}^2(\partial\mathcal{A})} E_n, \tag{4.150}$$

which formally satisfy the estimate

$$\sup_{x\in\overline{\mathcal{K}}} \left|\nabla^{(k)} F(x)\right| \leq \left(\sum_{n=0}^{\infty} \left(\tilde{F}, D_n^*\right)^2_{\mathrm{L}^2(\partial\mathcal{A})}\right)^{\frac{1}{2}} \sup_{x\in\overline{\mathcal{K}}} \left(\sum_{n=0}^{\infty} \left|\nabla^{(k)} E_n(x)\right|^2\right)^{\frac{1}{2}}, \tag{4.151}$$

where $k \in \mathbb{N}_0$ and $\mathcal{K} \Subset \mathcal{A}^c$.

Remark 4.31. The function F in (4.150) can be regarded as the image of $\tilde{F}$ under a linear operator defined by the symbol $\{\sigma_n\}$ (see Example 4.38 for the symbol of the operator T relating to the volume-based approach of the previous subsection in the special case of a ball).

The regularity condition (3.358) implies the existence of a constant C (dependent on k, $\mathcal{A}, \mathcal{K}$) such that, for all $n \in \mathbb{N}_0$,

$$\sup_{x\in\overline{\mathcal{K}}} \left|\nabla^{(k)} D_n^*(x)\right|^2 \leq C^2 \underbrace{\int_{\partial\mathcal{A}} (D_n^*(x))^2 \, d\omega(x)}_{=1}. \tag{4.152}$$

Even more, the mean value theorem of multivariate analysis shows us that there exists a positive constant $\tilde{C}$ such that, for all $n \in \mathbb{N}_0$,

$$\left|\nabla^{(k)} D_n^*(x) - \nabla^{(k)} D_n^*(y)\right| \leq \tilde{C}|x-y| \tag{4.153}$$

is valid for $x, y \in \overline{\mathcal{K}}$. In addition, it follows from (4.148) and (4.152) that

$$\sup_{x\in\overline{\mathcal{K}}} \left(\sum_{n=0}^{\infty} \left|\nabla^{(k)} E_n(x)\right|^2\right)^{\frac{1}{2}} \leq C \left(\sum_{n=0}^{\infty} \sigma_n^2\right)^{\frac{1}{2}}. \tag{4.154}$$

Consequently, the expansion on the right-hand side of (4.150) exists, such that F is harmonic in $\mathcal{A}^c$ and regular at infinity. Moreover,

$$\sup_{x\in\overline{\mathcal{K}}} \left|\nabla^{(k)} F(x)\right| \leq C \left(\sum_{n=0}^{\infty} \sigma_n^2\right)^{\frac{1}{2}} \left(\sum_{n=0}^{\infty} \left(\tilde{F}, D_n^*\right)^2_{\mathrm{L}^2(\partial\mathcal{A})}\right)^{\frac{1}{2}}. \tag{4.155}$$

It is remarkable that all functions F defined by series of type (4.150) form a linear space $\mathrm{H}(\mathcal{A}^c)$ on which we are able to impose the structure of a separable Hilbert space (cf. W. Freeden [1981, 1982]). More precisely,

$$\mathrm{H}(\mathcal{A}^c) = \left\{ F = \sum_{n=0}^{\infty} (\tilde{F}, D_n^*)_{\mathrm{L}^2(\partial\mathcal{A})} E_n : \tilde{F} \in \mathrm{L}^2(\partial\mathcal{A}) \right\}. \tag{4.156}$$

For members $F, G \in \mathrm{H}(\mathcal{A}^c)$ associated to $\tilde{F}$, $\tilde{G} \in \mathrm{L}^2(\partial\mathcal{A})$, respectively, an inner product is given by

$$\begin{aligned} (F,G)_{\mathrm{H}(\mathcal{A}^c)} &= (\tilde{F},\tilde{G})_{\mathrm{L}^2(\partial\mathcal{A})} \\ &= \sum_{n=0}^{\infty} \left(\tilde{F}, D_n^*\right)_{\mathrm{L}^2(\partial\mathcal{A})} \left(\tilde{G}, D_n^*\right)_{\mathrm{L}^2(\partial\mathcal{A})} \\ &= \sum_{n=0}^{\infty} \frac{1}{\sigma_n^2} (F, D_n^*)_{\mathrm{L}^2(\partial\mathcal{A})} (G, D_n^*)_{\mathrm{L}^2(\partial\mathcal{A})} . \end{aligned} \tag{4.157}$$

Theorem 4.32. $(\mathrm{H}(\mathcal{A}^c), (\cdot,\cdot)_{\mathrm{H}(\mathcal{A}^c)})$ *is a separable Hilbert space possessing the (uniquely determined) reproducing kernel*

$$\begin{aligned} K_{\mathrm{H}(\mathcal{A}^c)}(x,y) &= \sum_{n=0}^{\infty} E_n(x)E_n(y) \\ &= \sum_{n=0}^{\infty} \sigma_n^2 D_n^*(x) D_n^*(y), \end{aligned} \tag{4.158}$$

for all $x, y \in \mathcal{A}^c$. *Furthermore, the Parseval identity in* $\mathrm{H}(\mathcal{A}^c)$,

$$\begin{aligned} (F,G)_{\mathrm{H}(\mathcal{A}^c)} &= \sum_{n=0}^{\infty} (F, E_n)_{\mathrm{H}(\mathcal{A}^c)} (G, E_n)_{\mathrm{H}(\mathcal{A}^c)} \\ &= \sum_{n=0}^{\infty} \left(\tilde{F}, D_n^*\right)_{\mathrm{L}^2(\partial\mathcal{A})} \left(\tilde{G}, D_n^*\right)_{\mathrm{L}^2(\partial\mathcal{A})} \\ &= \sum_{n=0}^{\infty} \frac{1}{\sigma_n^2} (F, D_n^*)_{\mathrm{L}^2(\partial\mathcal{A})} (G, D_n^*)_{\mathrm{L}^2(\partial\mathcal{A})} \end{aligned} \tag{4.159}$$

holds true for all $F, G \in \mathrm{H}(\mathcal{A}^c)$.

Minimum Norm Interpolation

The discrete exterior oblique boundary-value problem (DEODP) demands study of the boundedness of the functional $\mathcal{D}_x$ of the oblique derivative at a point $x \in \partial\mathcal{G}$ with respect to the $\mathrm{H}(\mathcal{A}^c)$-topology (as introduced by (4.157)), where $\mathcal{A}$ is given such that $\mathcal{A} \Subset \mathcal{G}$. Indeed, for $x \in \partial\mathcal{G}$ and arbitrary $\varepsilon > 0$, we get from (4.153)

$$\begin{aligned} &\frac{1}{\varepsilon}|F(x) - F(x + \varepsilon\lambda(x))| \\ &\leq \frac{1}{\varepsilon}\|F\|_{\mathrm{H}(\mathcal{A}^c)}\|K_{\mathrm{H}(\mathcal{A}^c)}(x,\cdot) - K_{\mathrm{H}(\mathcal{A}^c)}(x + \varepsilon\lambda(x),\cdot)\|_{\mathrm{H}(\mathcal{A}^c)} \\ &\leq C\ \|F\|_{\mathrm{H}(\mathcal{A}^c)} \end{aligned} \tag{4.160}$$

for some $C > 0$ (depending on $\mathcal{A}^c$ and $\partial\mathcal{G}$), provided that F is of class $\mathrm{H}(\mathcal{A}^c)$. Consequently, $\mathcal{D}_x$ is bounded and by the same minimum norm procedure as for the volume-based approach we obtain (in a self-explaining adaptation of the nomenclature)

Theorem 4.33. *The minimum norm interpolation problem*

$$\|S_N^U\|_{\mathrm{H}(\mathcal{A}^c)} = \inf_{F \in \mathcal{I}^U_{\mathcal{D}_{x_1},\ldots,\mathcal{D}_{x_n}}} \|F\|_{\mathrm{H}(\mathcal{A}^c)} \tag{4.161}$$

is well posed in the sense that its solution exists, is unique, and depends continuously on the data $\frac{\partial U}{\partial\lambda}(x_i) = \alpha_i$, $i = 1,\ldots,N$. *The uniquely determined solution is given in the form*

$$S_N^U(x) = \sum_{i=1}^{N} a_i^N \mathcal{D}_{x_i}[K_{\mathrm{H}(\mathcal{A}^c)}(\cdot,x)], \quad x \in \mathcal{A}^c, \tag{4.162}$$

where the coefficients $a_1^N,\ldots,a_N^N$ *satisfy the linear equations*

$$\sum_{i=1}^{N} a_i^N \mathcal{D}_{x_i}\mathcal{D}_{x_j}[K_{\mathrm{H}(\mathcal{A}^c)}(\cdot,\cdot)] = \alpha_j, \quad j = 1,\ldots,N. \tag{4.163}$$

Stability

Again, the *stability* should be investigated. Analogously to Lemma 4.26 we get

$$\sup_{x \in \partial\mathcal{G}} |\mathcal{D}_x[S_N^U] - \mathcal{D}_x[U]| \leq 2\ (\kappa(x,y))^{\frac{1}{2}}\ \|U\|_{\mathrm{H}(\mathcal{A}^c)} \tag{4.164}$$

where

$$\kappa(x,y) = (\mathcal{D}_x\mathcal{D}_x - 2\mathcal{D}_x\mathcal{D}_y + \mathcal{D}_y\mathcal{D}_y)[K_{\mathrm{H}(\mathcal{A}^c)}(\cdot,\cdot)] \tag{4.165}$$

and $x \in \partial\mathcal{G}$ and $y \in \mathcal{X}_N = \{x_1,\ldots,x_N\} \subset \partial\mathcal{G}$. More explicitly, we have

$$\kappa(x,y) = \sum_{n=0}^{\infty} \sigma_n^2 \left(\frac{\partial D_n^*}{\partial\lambda}(x) - \frac{\partial D_n^*}{\partial\lambda}(y)\right)^2. \tag{4.166}$$

By use of (4.153), we find

$$\left|\frac{\partial D_n^*}{\partial\lambda}(x) - \frac{\partial D_n^*}{\partial\lambda}\right| \leq C\ |x-y|, \tag{4.167}$$

where $C > 0$ is a constant (depending on $\mathcal{A}^c$ and $\partial\mathcal{G}$). Thus, we finally arrive at an analogous result to Theorem 4.27.

Theorem 4.34. *Suppose that U is of class $\mathrm{H}(\mathcal{A}^c)$. Let $\mathcal{X}_N = \{x_1, \ldots, x_N\}$ be a $\mathcal{D}$-unisolvent system on $\partial\mathcal{G}$. Let S_N^U denote that uniquely determined solution of the spline interpolation problem (4.161). Then there exists a constant C (dependent on $\mathcal{A}$ and $\mathcal{G}$) such that*

$$\sup_{x \in \overline{\mathcal{G}^c}} |S_N^U(x) - U(x)| \le C\Theta_{\mathcal{X}_N} \; \|U\|_{\mathrm{H}(\mathcal{A}^c)}. \tag{4.168}$$

Obviously,

$$\mathrm{L}^2(\partial\mathcal{G}) = \overline{\operatorname*{span}_{x\in\mathcal{X}} \left\{\mathcal{D}|_{\partial\mathcal{G}}[K_{\mathrm{H}(\mathcal{A}^c)}(x,\cdot)]\right\}}^{\|\cdot\|_{\mathrm{L}^2(\partial\mathcal{G})}}, \tag{4.169}$$

where $\mathcal{X}$ is the union of a sequence $\{\mathcal{X}_N\}_{N=0,1,\ldots}$ of $\mathcal{D}$-unisolvent systems $\mathcal{X}_N$ on $\partial\mathcal{G}$ with $\Theta_{\mathcal{X}_N} \to 0$ as $N \to \infty$.

Specialization to an Inner Runge Sphere

For computational reasons, reproducing kernels that have closed expressions in terms of elementary functions are welcome. In practical applications, an outer harmonics $\mathrm{L}^2(\Omega_{\underline{R}})$-Dirichlet Runge basis (with $\underline{R} < \inf_{x\in\partial\mathcal{G}}|x|$) is of frequent use (see, e.g., C.C. Tscherning, R.H. Rapp [1974], T. Krarup [1969], H. Moritz, H. Sünkel [1978], H. Moritz [1980]). In fact, for $x, y \in \mathbb{R}^3\backslash\overline{\mathcal{B}_{\underline{R}}(0)}$, a large number of representations can be derived from series expansions in terms of Legendre polynomials:

$$K_{\mathrm{H}\left(\mathbb{R}^3\backslash\overline{\mathcal{B}_{\underline{R}}(0)}\right)}(x,y) = \sum_{n=0}^{\infty} \sigma_n^2 \, \frac{2n+1}{4\pi \underline{R}^2} \left(\frac{\underline{R}^2}{|x||y|}\right)^{n+1} P_n\left(\frac{x}{|x|}\cdot\frac{y}{|y|}\right). \tag{4.170}$$

Following W. Freeden [1981, 1987], we restrict ourselves to some important cases.

Example 4.35. The choice $\sigma_n^2 = \left(\frac{\underline{R}_0^2}{\underline{R}^2}\right)^n$, $\underline{R}_0 < \underline{R}$, $n = 0, 1, \ldots$, leads to the *Abel–Poisson kernel (cf.* (3.395)*):*

$$K_{\mathrm{H}\left(\mathbb{R}^3\backslash\overline{\mathcal{B}_{\underline{R}}(0)}\right)}(x,y) = |x||y|\frac{\underline{R}^2}{4\pi}\,\frac{|x|^2|y|^2 - \underline{R}_0^4}{(L(x,y))^{\frac{3}{2}}} \tag{4.171}$$

with

$$L(x,y) = |x|^2|y|^2 - 2x\cdot y\underline{R}_0^2 + \underline{R}_0^4 \tag{4.172}$$

and

$$
\begin{aligned}
&\mathcal{D}_x\mathcal{D}_y[K_{\mathrm{H}\left(\mathbb{R}^3\setminus\overline{\mathcal{B}_{\underline{R}}(0)}\right)}(\cdot,\cdot)] \qquad (4.173)\\
&= \frac{\underline{R}^2}{4\pi\underline{R}_0^2}\left(\frac{9|x|^2|y|^2-\underline{R}_0^4}{(L(x,y))^{\frac{3}{2}}}\frac{(\lambda(x)\cdot x)(\lambda(y)\cdot y)}{|x|\ |y|}\right)\\
&\quad+\frac{3\underline{R}^2}{4\pi\underline{R}_0{}^2}\frac{3|x|^2|y|^2-\underline{R}_0{}^4}{(L(x,y))^{\frac{5}{2}}}\left(\frac{(\lambda(x)\cdot x)|y|}{|x|}(\underline{R}_0{}^2(\lambda(y)\cdot x)-(\lambda(y)\cdot y)|x|^2)\right)\\
&\quad+\frac{3\underline{R}^2}{4\pi\underline{R}_0{}^2}\frac{3|x|^2|y|^2-\underline{R}_0{}^4}{(L(x,y))^{\frac{5}{2}}}\left(\frac{(\lambda(y)\cdot y)|x|}{|y|}(\underline{R}_0{}^2(\lambda(x)\cdot y)-(\lambda(x)\cdot x)|y|^2)\right)\\
&\quad+\frac{3\underline{R}^2}{4\pi\underline{R}_0{}^2}\frac{|x|^2|y|^2-\underline{R}_0{}^4}{(L(x,y))^{\frac{5}{2}}}|x|\ |y|\left(\underline{R}_0{}^2(\lambda(x)\cdot\lambda(y))-2(\lambda(y)\cdot y)(\lambda(x)\cdot x)\right)\\
&\quad+\frac{15\underline{R}^2}{4\pi\underline{R}_0{}^2}\frac{|x|^2|y|^2-\underline{R}_0{}^4}{(L(x,y))^{7/2}}|x||y|(\underline{R}_0{}^2(\lambda(x)\cdot y)-(\lambda(x)\cdot x)|y|^2)\\
&\quad\times(\underline{R}_0{}^2(\lambda(y)\cdot y)-(\lambda(y)\cdot y)|x|^2).
\end{aligned}
$$

Example 4.36. For $\sigma_n^2=\frac{2}{2n+1}\left(\frac{\underline{R}_0^2}{\underline{R}^2}\right)^n$, $\underline{R}_0<\underline{R}$, $n=0,1,\ldots$, we get the *singularity kernel (cf.* (3.397)*):*

$$
K_{\mathrm{H}\left(\mathbb{R}^3\setminus\overline{\mathcal{B}_{\underline{R}}(0)}\right)}(x,y)=\frac{\underline{R}^2}{2\pi}\frac{1}{(L(x,y))^{\frac{1}{2}}} \qquad (4.174)
$$

and

$$
\begin{aligned}
&\mathcal{D}_x\mathcal{D}_y[K_{\mathrm{H}\left(\mathbb{R}^3\setminus\overline{\mathcal{B}_{\underline{R}}(0)}\right)}(\cdot,\cdot)] \qquad (4.175)\\
&= \frac{\underline{R}^2}{2\pi}\frac{1}{(L(x,y))^{\frac{3}{2}}}(\underline{R}_0{}^2(\lambda(x)\cdot\lambda(y))-2(\lambda(x)\cdot x)(\lambda(y)\cdot y))\\
&\quad+\frac{3\underline{R}^2}{2\pi}\frac{1}{(L(x,y))^{\frac{3}{2}}}(\underline{R}_0{}^2(\lambda(x)\cdot y)-(\lambda(x)\cdot x)|y|^2)\\
&\quad\times(\underline{R}_0{}^2(\lambda(y)\cdot x)-(\lambda(y)\cdot y)|x|^2).
\end{aligned}
$$

Example 4.37. For $\sigma_n^2=\frac{1}{(2n+1)(n+1)}\left(\frac{\underline{R}_0^2}{\underline{R}^2}\right)^n$, $\underline{R}_0<\underline{R}$, $n=0,1,\ldots$, we get the *logarithmic kernel (cf.* (3.398)*):*

$$
K_{\mathrm{H}\left(\mathbb{R}^3\setminus\overline{\mathcal{B}_{\underline{R}}(0)}\right)}(x,y)=\frac{\underline{R}^2}{2\pi\underline{R}_0^2}\ln\left(1+\frac{2\underline{R}_0{}^2}{M(x,y)}\right) \qquad (4.176)
$$

with

$$
M(x,y)=(L(x,y))^{\frac{1}{2}}+|x|\ |y|-\underline{R}_0{}^2 \qquad (4.177)
$$

and

$$\mathcal{D}_x\mathcal{D}_y[K_{\mathrm{H}\left(\mathbb{R}^3\setminus\overline{\mathcal{B}_{\underline{R}}(0)}\right)}(\cdot,\cdot)] \tag{4.178}$$

$$\begin{aligned}
= \quad & \frac{\underline{R}^2}{2\pi \underline{R}_0^2} \frac{1}{(M(x,y))^2 + 2\underline{R}_0{}^2 M(x,y)} \\
& \times \Big[(L(x,y))^{-\frac{3}{2}} (\underline{R}_0{}^2 (\lambda(y)\cdot x) - |x|^2 (\lambda(y)\cdot y)) \\
& \quad \times (\underline{R}_0{}^2 (n(x)\cdot y) - |y|^2 (\lambda(x)\cdot x)) \Big] \\
& + \frac{\underline{R}^2}{2\pi \underline{R}_0^2} \frac{1}{(M(x,y))^2 + 2\underline{R}_0{}^2 M(x,y)} \\
& \times \Big[(L(x,y)^{-\frac{1}{2}} (\underline{R}_0{}^2 (\lambda(x)\cdot \lambda(y)) - 2(\lambda(x)\cdot x)(\lambda(y)\cdot y))) \\
& \quad - \frac{(\lambda(x)\cdot x)(\lambda(y)\cdot y)}{|x||y|} \Big] \\
& + \frac{\underline{R}^2}{\pi \underline{R}_0^2} \frac{M(x,y) + \underline{R}_0{}^2}{((M(x,y))^2 + 2\underline{R}_0{}^2 M(x,y))^2} \\
& \times \left[(L(x,y))^{-\frac{1}{2}} (|x|^2 (\lambda(y)\cdot y) - \underline{R}_0{}^2 (\lambda(y)\cdot x)) \frac{|x|}{|y|} (\lambda(y)\cdot y) \right] \\
& \times \left[(L(x,y))^{-\frac{1}{2}} (|y|^2 (\lambda(x)\cdot x) - \underline{R}_0{}^2 (\lambda(x)\cdot y)) + \frac{|y|}{|x|} (\lambda(x)\cdot x) \right].
\end{aligned}$$

Example 4.38. Finally, the choice $\sigma_n^2 = \frac{1}{\underline{R}_0^3} \frac{1}{(2n+1)^2(2n+3)} \left(\frac{\underline{R}_0^2}{\underline{R}^2}\right)^n$, $\underline{R}_0 < \underline{R}$ and $n = 0, 1, \ldots$, leads to the *Newton kernel (cf.* (4.146)*):*

$$K_{\mathrm{H}\left(\mathbb{R}^3\setminus\overline{\mathcal{B}_{\underline{R}}(0)}\right)}(x,y) = \left(\frac{1}{4\pi}\right)^2 \int_{\mathcal{B}_{\underline{R}_0}(0)} \frac{1}{|x-z||y-z|}\, dV(z). \tag{4.179}$$

and

$$\mathcal{D}_x\mathcal{D}_y[K_{\mathrm{H}\left(\mathbb{R}^3\setminus\overline{\mathcal{B}_{\underline{R}}(0)}\right)}(\cdot,\cdot)] = \left(\frac{1}{4\pi}\right)^2 \int_{\mathcal{B}_{\underline{R}_0}(0)} \frac{\lambda(x)\cdot(x-z)\ \ \lambda(y)\cdot(y-z)}{|x-z|^3|y-z|^3}\, dV(z). \tag{4.180}$$

In other words, the Newton kernel leads back to the volume-based reproducing kernel structure presented in Subsection 4.1.2.

Finally, it should be noted that the advantage of a sphere-based reproducing kernel space $(\mathrm{H}(\mathcal{A}^c), (\cdot,\cdot)_{\mathrm{H}(\mathcal{A}^c)})$ is twofold: First, the reproducing kernel contains outer harmonics contributions of any degree like the Earth's gravitational potential itself. Second, the geometry of the regular region $\mathcal{G}$ may be arbitrary so that all geophysically relevant surfaces $\partial\mathcal{G}$ can be easily handled in numerical computations, thereby taking profit from the fact that there is no need for numerical integration. The coefficient matrix of the occurring linear (spline) systems are symmetric and positive definite; hence, they are solvable by standard methods of linear algebra. Even better, multipole (far and near field) methods in combination with suitable domain decomposition procedures (see W. Freeden, O. Glockner, M. Schreiner [1999], M. Gutting [2007, 2012], and the references therein) make spline interpolation (and/or smoothing in the case of error-affected data) efficient as well as economical for numerical application. Nevertheless, it should be mentioned that the choice of the reproducing kernel, i.e., the appropriate topology of $\mathrm{H}(\mathcal{A}^c)$ is a difficulty in minimum norm (spline) interpolation as proposed here. In principle, seen from a theoretical point of view, all topologies are equivalent. In practice, however, the reproducing kernel structure should be in adaptation to the characteristics of the available data set (if possible).

4.2 Satellite Problems

In the following, the most important techniques realized by the satellite technology today for gravitational field determination, namely *satellite-to-satellite tracking* (SST) *and satellite gravity gradiometry* (SGG) are characterized from potential theoretical point of view (for a more detailed geodetic/geoengineering information the reader is referred, e.g., to ESA [1996, 1998, 1999], R. Rummel et al. [2002], H. Laur, V. Liebig [2010], R. Rummel [2010]). Uniqueness results are formulated for the satellite problems (cf. W. Freeden [1999], W. Freeden et al. [2002], W. Freeden, V. Michel [2004], W. Freeden, M. Schreiner [2010]). Moreover, the mathematical justification is given for approximating the external gravitational field by finite linear combinations of certain types of gradient fields (for example, gradient fields of single poles, multipoles, and certain kernel functions) consistent to a given set of SST and/or SGG data.

4.2.1 Formulation of the Problems

In order to translate the satellite problems into a mathematical formulation (see W. Freeden [1999], W. Freeden et al. [2002], W. Freeden, V. Michel [2004] and for spherical harmonic approaches ESA [1996, 1998, 1999] and the references therein), we start from the following geometrical situation: Let the Earth's surface $\partial\mathcal{G}$ of the (regular Earth's) interior $\mathcal{G}$ and the orbits $\mathcal{S}$ of the low Earth orbiter (LEO) be known from geopositioning methods (e.g., DOPPLER, GPS, LASER, VLBI, etc.) in such a way that $\mathcal{S}$ is a strict subset of the Earth's exterior $\mathcal{G}^c$ satisfying (cf. Figure 4.7)

$$\underline{R} < \inf_{x\in\partial\mathcal{G}} |x| \leq \sup_{x\in\partial\mathcal{G}} |x| < \overline{R} \ll S \leq \inf_{x\in\mathcal{S}} |x| \ . \tag{4.181}$$

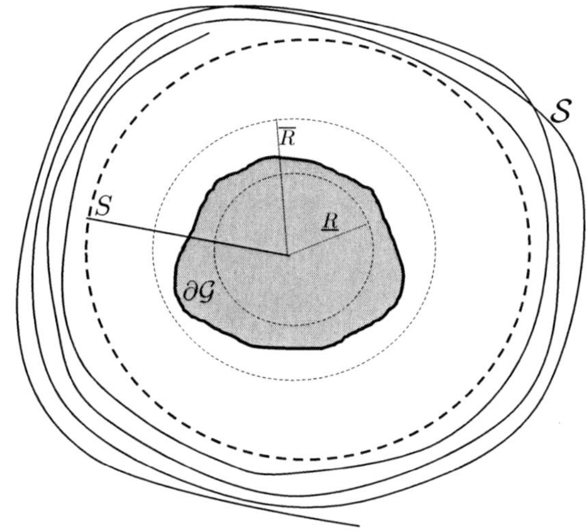

FIGURE 4.7
Satellite orbit $\mathcal{S}$ of a low Earth orbiter (LEO).

Mathematically spoken, the formulation of the *satellite problems* (after separating all non-gravitational influences) reads as follows:

(SST, high-low): Suppose that the (Earth's) surface $\partial\mathcal{G}$ and the (satellite) orbit $\mathcal{S}$ (i.e., the set of positional points of the low Earth orbiting satellite) are given such that (4.181) holds true. Furthermore, we assume that gradient vectors $v(x) = (\nabla V)(x), x \in \mathcal{X}$, for a subset $\mathcal{X} \subset \mathcal{S}$ of the orbit are known. Find an approximation u of the geopotential field v on $\overline{\mathcal{G}^c}$, i.e., on and outside the Earth's surface, such that v and its approximation u are in ε-accuracy (with respect to the uniform topology in $\overline{\mathcal{G}^c}$) and $v(x) = u(x)$ for all $x \in \mathcal{X}$.

(SST, low-low): Suppose that the (Earth's) surface $\partial\mathcal{G}$ and the (satellite) orbit $\mathcal{S}$ are given such that (4.181) is valid. Let the vectors $v(x) = (\nabla V)(x)$ and $v(x+h(x)) = (\nabla V)(x+h(x))$, $x \in \mathcal{X}$, be known for a subset $\mathcal{X} \subset \mathcal{S}$. The function h denotes the distance vector between the two low Earth orbiting satellites at the time of observation (GRACE concept). Find an approximation u of v in $\overline{\mathcal{G}^c}$, such that v and u are in ε-accuracy (with respect to the uniform topology in $\overline{\mathcal{G}^c}$) and $v(x) - v(x+h(x)) = u(x) - u(x+h(x))$ for all $x \in \mathcal{X}$.

(SGG): Suppose that the (Earth's) surface $\partial\mathcal{G}$ and the (satellite) orbit $\mathcal{S}$ are given such that (4.181) is valid. Let the tensors $\mathbf{v}(x) = (\nabla v)(x)$, $x \in \mathcal{X}$, be known for a subset $\mathcal{X}$ of the orbit $\mathcal{S}$. Find an approximation u of v in $\overline{\mathcal{G}^c}$, such that v and u are in ε-accuracy (with respect to the uniform topology in $\overline{\mathcal{G}^c}$) and $(\nabla v)(x) = \mathbf{v}(x) = (\nabla u)(x)$ for all $x \in \mathcal{X}$.

Our considerations start with the study of the uniqueness corresponding to the *model situation of an infinite (fundamental) system* $\mathcal{X} \subset \mathcal{S}$.

4.2.2 Uniqueness of the SST Problem

For a regular region $\mathcal{G} \subset \mathbb{R}^3$, let $\mathrm{pot}(\mathcal{G}^c)$ denote the space of vector fields $u : \mathcal{G}^c \to \mathbb{R}^3$ satisfying the properties:

(i) u is of class $\mathrm{c}^{(1)}(\mathcal{G}^c)$,

(ii) $\nabla \cdot u = 0$, $\nabla \wedge u = 0$ in $\mathcal{G}^c$,

(iii) $|u(x)| = \mathcal{O}(|x|^{-2})$, $|x| \to \infty$.

Analogous to the scalar case we set

$$\mathrm{pot}^{(k)}(\overline{\mathcal{G}^c}) = \mathrm{pot}(\mathcal{G}^c) \cap \mathrm{c}^{(k)}(\overline{\mathcal{G}^c}), \quad k \in \mathbb{N}_0. \tag{4.182}$$

We are now prepared to develop the following uniqueness result.

Theorem 4.39. *Assume that (4.181) is valid. Suppose that $\mathcal{X} \subset \mathcal{S}$ (i.e., the subset of observational points on the satellite orbit $\mathcal{S}$) is a fundamental system in $\mathcal{G}^c$. If v is of class $\mathrm{pot}^{(0)}\left(\overline{\mathcal{G}^c}\right)$ with $v(x) = 0$, $x \in \mathcal{X}$, then $v = 0$ in $\overline{\mathcal{G}^c}$.*

Proof. Any field $v \in \mathrm{pot}^{(0)}(\overline{\mathcal{G}^c})$ can be expressed in the form ∇V, $V \in \mathrm{Pot}^{(1)}(\overline{\mathcal{G}^c})$. Hence, the coordinate functions $v \cdot \varepsilon^i$, $i = 1,2,3$, satisfy $\Delta\left(v \cdot \varepsilon^i\right) = \Delta\left(\left(\varepsilon^i \cdot \nabla\right) V\right) = \left(\varepsilon^i \cdot \nabla\right) \Delta V = 0$ in $\mathcal{G}^c$, since the harmonic function V is arbitrarily often differentiable in $\mathcal{G}^c$. Moreover, according to our assumption, $(\varepsilon^i \cdot \nabla) V(x) = 0$ for all points x of the fundamental system $\mathcal{X}$ in $\mathcal{G}^c$. By virtue of the analyticity and the continuity up to $\partial\mathcal{G}$, this finally implies $v \cdot \varepsilon^i = 0$ in $\overline{\mathcal{G}^c}$, $i = 1,2,3$, as required. □

In addition, we are able to verify the following result.

Theorem 4.40. *Suppose that $\mathcal{X} \subset \mathcal{S}$ is a fundamental system in $\mathcal{G}^c$ such that condition (4.181) is satisfied. If v is of class $\mathrm{pot}^{(0)}(\overline{\mathcal{G}^c})$ with $x \cdot v(x) = 0$, $x \in \mathcal{X}$, then $v = 0$ in $\overline{\mathcal{G}^c}$.*

Proof. We base our arguments on the identity $v = \nabla V$ in $\mathcal{G}^c$. From our assumption (4.181) it is clear that $\mathbb{R}^3 \backslash \mathcal{B}_S(0)$, $S \leq \inf_{x \in \mathcal{S}} |x|$, is totally contained in $\mathcal{G}^c$. The potential $V \in \mathrm{Pot}^{(\infty)}(\mathbb{R}^3 \backslash \mathcal{B}_S(0))$ may be expanded by means of outer harmonics

$$V(x) = \sum_{n=0}^{\infty} \sum_{k=1}^{2n+1} V^{\wedge_{\mathrm{L}^2(\Omega_S)}}(n,k)\, H^S_{-n-1,k}(x), \quad x \in \mathbb{R}^3 \backslash \mathcal{B}_S(0), \tag{4.183}$$

where $V^{\wedge_{\mathrm{L}^2(\Omega_S)}}(n,k)$, $n \in \mathbb{N}_0$, $k = 1, \ldots, 2n+1$, are the expansion coefficients given by

$$V^{\wedge_{\mathrm{L}^2(\Omega_S)}}(n,k) = \int_{\Omega_S} V(x) H^S_{-n-1,k}(x)\, d\omega(x). \tag{4.184}$$

The series expansion in (4.183), together with all derivatives, is absolutely and uniformly convergent in $\mathbb{R}^3 \backslash \mathcal{B}_S(0)$. It is not hard to see by a straightforward calculation that

$$-\frac{x}{|x|} \cdot (\nabla V)(x) = \sum_{n=0}^{\infty} \sum_{k=1}^{2n+1} \frac{n+1}{|x|} V^{\wedge_{\mathrm{L}^2(\Omega_S)}}(n,k) H^S_{-n-1,k}(x), \quad x \in \mathbb{R}^3 \backslash \mathcal{B}_S(0). \tag{4.185}$$

Therefore, it is clear that $x \mapsto -x \cdot (\nabla V)(x)$, $x \in \mathbb{R}^3 \backslash \mathcal{B}_S(0)$, is a function of class $\mathrm{Pot}^{(\infty)}(\mathbb{R}^3 \backslash \mathcal{B}_S(0))$, for which we know that $-x \cdot (\nabla V)(x) = 0$, $x \in \mathcal{X}$. Since $\mathcal{X}$ is a fundamental system in $\mathbb{R}^3 \backslash \overline{\mathcal{B}_S(0)}$, we obtain $-x \cdot (\nabla V)(x) = 0$, $x \in \mathbb{R}^3 \backslash \mathcal{B}_S(0)$. Thus, the completeness property of spherical harmonics tells us that

$$(n+1) V^{\wedge_{\mathrm{L}^2(\Omega_S)}}(n,k) = 0. \tag{4.186}$$

Consequently, $V^{\wedge_{\mathrm{L}^2(\Omega_S)}}(n,k) = 0$ for all $n \in \mathbb{N}_0$, $k = 1, \ldots, 2n+1$. This yields $V = 0$ in $\mathbb{R}^3 \backslash \mathcal{B}_S(0)$. By analytical continuation and the continuity up to $\partial\mathcal{G}$, we get $V = 0$ in $\mathcal{G}^c$, and hence, $v = 0$ in $\overline{\mathcal{G}^c}$. This is the desired result. □

In other words, the Earth's external gravitational field is uniquely determinable on and outside the Earth's surface $\partial\mathcal{G}$ from SST data corresponding to a system of gradient vectors given on a fundamental system $\mathcal{X}$ on the satellite orbit $\mathcal{S}$. Furthermore, our results concerning SST have shown that the problem of developing the gravitational potential outside the Earth from given gradients in point systems on a spherical orbit $\mathcal{S}$ is overdetermined. In fact, it suffices to prescribe the radial component (cf. Theorem 4.40) on $\mathcal{S}$.

From potential theory it is clear that analogous uniqueness theorems cannot be deduced for the actual hi–lo SST problem of finding the external gravitational field of the Earth from a *finite* subsystem $\mathcal{X}$ on the satellite orbit $\mathcal{S}$. However, as we shall show, given the SST data for a finite subset $\mathcal{X} \subset \mathcal{S}$, we are able to find, for every value $\varepsilon > 0$, an approximation u of the external gravitational field v of the Earth in ε–accuracy so that u is consistent with the SST data on the finite subsystem $\mathcal{X}$.

4.2.3 Uniqueness of the SGG Problem

Let $\mathcal{G} \subset \mathbb{R}^3$ be a regular region. By $\mathbf{pot}\,(\mathcal{G}^c)$ we denote the space of tensor fields $\mathbf{u} : \mathcal{G}^c \to \mathbb{R}^{3\times 3}$ with the following properties:

(i) $\mathbf{u}$ is of class $\mathbf{c}^{(1)}(\mathcal{G}^c)$,

(ii) $\nabla \cdot \mathbf{u} = 0$, $\nabla \wedge \mathbf{u} = 0$ in $\mathcal{G}^c$,

(iii) $\mathbf{u}(x) = \mathcal{O}(|x|^{-3})$, $|x| \to \infty$.

Furthermore, we set

$$\mathbf{pot}^{(k)}\left(\overline{\mathcal{G}^c}\right) = \mathbf{pot}\,(\mathcal{G}^c) \cap \mathbf{c}^{(k)}(\overline{\mathcal{G}^c}), \quad k \in \mathbb{N}_0. \tag{4.187}$$

Theorem 4.41. *Assume that condition (4.181) is valid. Suppose that $\mathcal{X} \subset \mathcal{S}$ is a fundamental system in $\mathcal{G}^c$. If $\mathbf{v}$ is of class $\mathbf{pot}^{(0)}\left(\overline{\mathcal{G}^c}\right)$ with $\mathbf{v}(x) = 0, x \in \mathcal{X}$, then the associated field $v \in \mathrm{pot}^{(1)}\left(\overline{\mathcal{G}^c}\right)$ with $\mathbf{v} = \nabla v$ in $\mathcal{G}^c$ satisfies $v = 0$ in $\overline{\mathcal{G}^c}$.*

Proof. Any field $\mathbf{v}$ of the class $\mathbf{pot}^{(0)}\left(\overline{\mathcal{G}^c}\right)$ can be expressed in the form $\nabla^{(2)}V = (\nabla \otimes \nabla)V$, $V \in \mathrm{Pot}^{(2)}\left(\overline{\mathcal{G}^c}\right)$ (see, e.g., M.E. Gurtin [1972]). Furthermore, the coordinate functions $\mathbf{v}_{ij} = \varepsilon^i \cdot \mathbf{v}\varepsilon^j$, $i,j \in \{1,2,3\}$, satisfy $\Delta \mathbf{v}_{ij} = 0$ in $\mathcal{G}^c$. This implies $\mathbf{v}_{ij} = 0$ in $\mathcal{G}^c$, $i,j \in \{1,2,3\}$, because of the definition of a fundamental system. From $\mathbf{v} = \nabla^{(2)}V = (\nabla \otimes \nabla)V = 0$ and the regularity of V at infinity we finally get $V = 0$ in $\mathcal{G}^c$, and thus, $v = \nabla V = 0$ in $\overline{\mathcal{G}^c}$, as required. □

Consequently, the Earth's external gravitational field v is uniquely detectable on and outside the Earth's surface $\partial\mathcal{G}$ if SGG data (i.e., second-order derivatives of the Earth's gravitational potential V) are given on a fundamental system $\mathcal{X}$ (on the satellite orbit $\mathcal{S}$). Furthermore, we are able to verify the following result.

Theorem 4.42. *Suppose that $\mathcal{X}$ is a fundamental system in $\mathcal{G}^c$. If $\mathbf{v}$ is a field of class $\mathbf{pot}^{(0)}\left(\overline{\mathcal{G}^c}\right)$ with $x \cdot (\mathbf{v}(x)x) = 0, x \in \mathcal{X}$, then $v = 0$ in $\overline{\mathcal{G}^c}$, for the associated field $v \in \mathrm{pot}^{(1)}(\overline{\mathcal{G}^c})$ with $\mathbf{v} = \nabla v$ in $\mathcal{G}^c$.*

Proof. We start from the identity $\mathbf{v}(x) = \nabla v(x) = \nabla^{(2)}V(x), x \in \mathcal{G}^c$. It is clear that $\mathbb{R}^3\backslash\mathcal{B}_S(0)$ is totally contained in $\mathcal{G}^c$. Furthermore, the potential $V \in \mathrm{Pot}^{(\infty)}(\mathbb{R}^3\backslash\mathcal{B}_S(0))$ may be expanded in terms of outer harmonics as indicated by (4.183) and (4.184). By elementary calculations we get

$$\begin{aligned} \frac{x}{|x|} \cdot \left(\left(\nabla_x^{(2)}V\right)(x)\frac{x}{|x|}\right) &= \left(\frac{x}{|x|}\cdot\nabla_x\right)\left(\frac{x}{|x|}\cdot\nabla_x\right)V(x) \qquad (4.188) \\ &= \sum_{n=0}^{\infty}\sum_{k=1}^{2n+1} \frac{(n+1)(n+2)}{|x|^2} V^{\wedge_{\mathrm{L}^2(\Omega_S)}}(n,k)\; H^S_{-n-1,k}(x), \quad x \in \mathbb{R}^3\backslash\mathcal{B}_S(0). \end{aligned}$$

Hence, $x \mapsto x \cdot \left(\left(\nabla^{(2)}V\right)(x)\;x\right)$, $x \in \mathbb{R}^3\backslash\overline{\mathcal{B}_S(0)}$, is a harmonic function. In accordance with our above assumption $x \cdot \left((\nabla^{(2)}V)(x)\;x\right) = 0$, $x \in \mathcal{X}$, we thus obtain

$$\sum_{n=0}^{\infty}\sum_{k=1}^{2n+1} (n+1)(n+2)V^{\wedge_{\mathrm{L}^2(\Omega_S)}}(n,k)\; H^S_{-n-1,k}(x) = 0, \quad x \in \mathcal{X}. \qquad (4.189)$$

Since $\mathcal{X}$ is a fundamental system in $\mathbb{R}^3\backslash\overline{\mathcal{B}_S(0)}$, the identity (4.189) holds true in $\mathbb{R}^3\backslash\mathcal{B}_S(0)$. The theory of spherical harmonics tells us that

$$(n+1)(n+2)V^{\wedge_{\mathrm{L}^2(\Omega_S)}}(n,k) = 0, \qquad (4.190)$$

hence, $V^{\wedge_{\mathrm{L}^2(\Omega_S)}}(n,k) = 0$ for $n \in \mathbb{N}_0$, $k = 1,\ldots,2n+1$. This yields $V = 0$ in $\mathbb{R}^3\backslash\mathcal{B}_S(0)$. By analytical continuation we have $V = 0$ in $\mathcal{G}^c$, and by the continuity up to $\partial\mathcal{G}$ we find $v = \nabla V = 0$ in $\overline{\mathcal{G}^c}$. □

Theorem 4.42 means that the Earth's external gravitational field is uniquely recoverable from second-order radial derivatives corresponding to a fundamental system $\mathcal{X} \subset \mathcal{S}$. Once more, from potential theory it is clear that analogous uniqueness theorems cannot be deduced for the actual SGG problem of finding the external gravitational field v of the Earth from a *finite* subsystem $\mathcal{X}$ on the satellite orbit $\mathcal{S}$. In the following, however, we would like to show that, given the SGG data for a finite subset $\mathcal{X} \subset \mathcal{S}$, we are able to find for every value $\varepsilon > 0$ an approximation u of the external gravitational field v of the Earth in ε-accuracy so that u additionally is consistent with the SGG data on the finite subsystem $\mathcal{X}$.

4.2.4 Vectorial/Tensorial Basis Systems

Assume that $\mathcal{G}$ is a regular region such that $\mathcal{K}$ is totally contained in $\mathcal{G}^c$. Consider a potential $U \in \mathrm{Pot}^{(0)}(\overline{\mathcal{G}^c})$ with $U^+|_{\partial\mathcal{G}} = F$. In order to assure the basis property of vectorial/tensorial basis systems, we remember the regularity result

$$\sup_{x\in\overline{\mathcal{K}}} \left| \left(\nabla^{(k)} U\right)(x) \right| \leq C \left(\int_{\partial\mathcal{G}} |F(y)|^2 \, d\omega(y) \right)^{\frac{1}{2}}, \quad k \in \mathbb{N}_0, \tag{4.191}$$

where C is dependent on $\partial\mathcal{G}$ and $\mathcal{K}$.

Lemma 4.43 (Vectorial/Tensorial Basis Systems). *Let $\mathcal{G}$ be a regular region. Assume that $\mathcal{K}$ satisfies $\mathcal{K} \Subset \mathcal{G}^c$.*

(a) Each (scalar) $\mathrm{L}^2(\partial\mathcal{G})$-Dirichlet Runge basis $\{D_n\}_{n=0,1,\ldots}$ (in the sense of Definition 3.95) implies a vectorial basis system *in the following sense: For $v \in \mathrm{pot}(\overline{\mathcal{G}^c})$, there exists an approximation by a finite linear combination of vector fields $\{\nabla D_n\}_{n=0,1,\ldots}$, uniformly on $\overline{\mathcal{K}}$. More precisely, for every $\varepsilon > 0$, there exist $N(= N(\varepsilon, \mathcal{K}))$ and coefficients a_i, $i = 1, \ldots, N$, such that*

$$\sup_{x\in\overline{\mathcal{K}}} \left| v(x) - \sum_{n=0}^{N} a_n \nabla D_n(x) \right| \leq \varepsilon. \tag{4.192}$$

(b) Each (scalar) $\mathrm{L}^2(\partial\mathcal{G})$-Dirichlet Runge basis $\{D_n\}_{n=0,1,\ldots}$ (in the sense of Definition 3.95) implies a tensorial basis system *in the following sense: For $\mathbf{v} \in \mathbf{pot}(\overline{\mathcal{G}^c})$, there exists an approximation by a finite linear combination of tensor fields $\{\nabla^{(2)} D_n\}_{n=0,1,\ldots}$, uniformly on $\overline{\mathcal{K}}$. More precisely, for every $\varepsilon > 0$, there exist $N(= N(\varepsilon, \mathcal{K}))$ and coefficients a_i, $i = 1, \ldots, N$, such that*

$$\sup_{x\in\overline{\mathcal{K}}} \left| \mathbf{v}(x) - \sum_{n=0}^{N} a_n \nabla^{(2)} D_n(x) \right| \leq \varepsilon. \tag{4.193}$$

Proof. We restrict ourselves to assertion (b); part (a) can be proved analogously. Suppose that $\mathbf{v}$ is of class $\mathbf{pot}(\overline{\mathcal{G}^c})$ and $\overline{\mathcal{K}}$ is a subset of $\mathcal{G}^c$ with positive distance to $\partial\mathcal{G}$. Then, there exists a scalar potential $V \in \mathrm{Pot}(\overline{\mathcal{G}^c})$ such that $\mathbf{v}|_{\overline{\mathcal{K}}} = (\nabla^{(2)}V)|_{\overline{\mathcal{K}}}$. Now, for arbitrary $\varepsilon > 0$, we have an integer $N(= N(\varepsilon))$ and coefficients $a_0, \ldots, a_N$ such that

$$\left(\int_{\partial\mathcal{G}} \left| V(x) - \sum_{n=0}^{N} a_n D_n(x) \right|^2 \, d\omega(x) \right)^{\frac{1}{2}} \leq \varepsilon. \tag{4.194}$$

In connection with the regularity condition (4.191) we obtain

$$\sup_{x \in \overline{\mathcal{K}}} \left| \mathbf{v}(x) - \sum_{n=0}^{N} a_n \nabla^{(2)} D_n(x) \right| \leq C \; \varepsilon \; . \tag{4.195}$$

This is the desired result. □

Our interest now is to verify a closure theorem (of the Runge type) that can be applied in the potential theoretic description of satellite problems. For this purpose we assume that $\mathcal{A}, \mathcal{G}$ are regular regions satisfying $\mathcal{A} \Subset \mathcal{G}$ (cf. Figure 3.7). We start with the discussion of the relation between the spaces $\mathrm{pot}(\mathcal{A}^c)|_{\overline{\mathcal{G}^c}}$ and $\mathrm{pot}^{(0)}\left(\overline{\mathcal{G}^c}\right)$. Of course, we have

$$\mathrm{pot}(\mathcal{A}^c)\left|_{\overline{\mathcal{G}^c}} \subset \mathrm{pot}^{(0)}\left(\overline{\mathcal{G}^c}\right) . \right. \tag{4.196}$$

The inclusion is, in fact, strict: if $y \in \mathcal{A}^c \backslash \overline{\mathcal{G}^c}$, then the field

$$x \;\mapsto\; \nabla_x G(\Delta; |x-y|), \qquad x \neq y, \tag{4.197}$$

is an element of class $\mathrm{pot}^{(0)}\left(\overline{\mathcal{G}^c}\right)$, but it is not an element of $\mathrm{pot}(\mathcal{A}^c)\left|_{\overline{\mathcal{G}^c}}\right.$. Hence, it is clear that

$$\mathrm{pot}(\mathcal{A}^c)\left|_{\overline{\mathcal{G}^c}} \neq \mathrm{pot}^{(0)}\left(\overline{\mathcal{G}^c}\right) . \right. \tag{4.198}$$

Nevertheless, we are able to prove the following *closure theorem.*

Theorem 4.44. $\mathrm{pot}(\mathcal{A}^c)\left|_{\overline{\mathcal{G}^c}}\right.$ *is a dense subset of the space* $\mathrm{pot}^{(0)}\left(\overline{\mathcal{G}^c}\right)$ *with respect to* $\|\cdot\|_{\mathrm{c}^{(0)}(\overline{\mathcal{G}^c})}$, *i.e., for any given* $\varepsilon > 0$ *and any* $v \in \mathrm{pot}^{(0)}(\overline{\mathcal{G}^c})$ *there exists a field* $u \in \mathrm{pot}(\mathcal{A}^c)\left|_{\overline{\mathcal{G}^c}}\right.$ *such that*

$$\sup_{x \in \overline{\mathcal{G}^c}} |v(x) - u(x)| \leq \varepsilon. \tag{4.199}$$

Proof. The main tool is a Hahn–Banach argument known from functional analysis (see, e.g., L.W. Kantorowitsch, G. Akilow [1964]). Let Z be a linear functional on the space $\mathrm{pot}^{(0)}(\overline{\mathcal{G}^c})$, that is continuous with respect to $\|\cdot\|_{\mathrm{c}^{(0)}(\overline{\mathcal{G}^c})}$. Assume that Z is zero on the set $\mathrm{pot}(\mathcal{A}^c)\left|_{\overline{\mathcal{G}^c}}\right.$. Then we have to prove that Z is zero on the set $\mathrm{pot}^{(0)}(\mathcal{G}^c)$.

We know that every vector field $v \in \mathrm{pot}^{(0)}(\overline{\mathcal{G}^c})$ is representable as gradient field $v = \nabla V$, $V \in \mathrm{Pot}^{(1)}(\overline{\mathcal{G}^c})$, and vice versa. Thus, the scalar (exterior) Neumann problem is equivalent to the vectorial (exterior) normal derivative problem: If F is a given function of class $\mathrm{C}^{(0)}(\partial\mathcal{G})$, then we are able to find a field $v \in \mathrm{c}^{(0)}\left(\overline{\mathcal{G}^c}\right)$ such that $\nu \cdot v = F$ on $\partial\mathcal{G}$. As already known, the solution v is representable in the form

$$v(x) = \nabla_x \int_{\partial\mathcal{G}} Q(y) G(\Delta; |x-y|)\, d\omega(y), \quad Q \in \mathrm{C}^{(0)}(\partial\mathcal{G}). \tag{4.200}$$

In vectorial notation, Theorem 3.84 can be rewritten in the form

$$\sup_{x\in\overline{\mathcal{K}}} |v(x)| \leq C \left(\int_{\partial\mathcal{G}} |\nu(x) \cdot v(x)|^2 \, d\omega(x) \right)^{\frac{1}{2}} \tag{4.201}$$

for all $\mathcal{K} \Subset \mathcal{G}$, with constants $C > 0$ depending on $\mathcal{K}$ and $\partial\mathcal{G}$. Moreover, we have the estimate

$$\|v\|_{\mathrm{c}^{(0)}\left(\overline{\mathcal{G}^c}\right)} \leq \tilde{C}\, \|\nu \cdot v\|_{\mathrm{C}^{(0)}(\partial\mathcal{G})}, \tag{4.202}$$

for a constant $\tilde{C} > 0$ depending on $\partial\mathcal{G}$. On the one hand, for any point $y \in \mathcal{G}$, the vector field

$$x \mapsto a_y(x) = \nabla_y G(\Delta; |x-y|), \quad x \in \overline{\mathcal{G}^c}, \tag{4.203}$$

is of class $\mathrm{pot}^{(0)}\left(\overline{\mathcal{G}^c}\right)$. Thus, by setting

$$A(y) = Z\left[a_y|_{\overline{\mathcal{G}^c}} \right] \tag{4.204}$$

we get a function A defined on $\mathcal{G}$. The expressions

$$\left(\varepsilon^i \cdot \nabla_y\right) a_y(x) = \frac{\partial}{\partial y_i} a_y(x) = \frac{\partial}{\partial y_i} \nabla_y G(\Delta; |x-y|), \quad x \in \overline{\mathcal{G}^c},\ i \in \{1,2,3\}, \tag{4.205}$$

define vector fields $\varepsilon^i \cdot \nabla_y(a_y(\cdot))$ of class $\mathrm{pot}^{(0)}\left(\overline{\mathcal{G}^c}\right)$. Furthermore, it is not hard to see that A is a continuously differentiable function on $\mathcal{G}$ such that

$$\frac{\partial}{\partial y_i} A(y) = Z\left[\left. \frac{\partial}{\partial y_i} a_y \right|_{\overline{\mathcal{G}^c}} \right]. \tag{4.206}$$

By virtue of (4.206), we find that $\nabla_y \wedge \nabla_y A(y) = 0$, $\nabla_y \cdot \nabla_y A(y) = 0$, $y \in \mathcal{G}$. This means that A is analytic in $\mathcal{G}$. On the other hand, for $y \in \mathcal{A}$, we have $a_y \in \mathrm{pot}(\mathcal{A}^c)$ and $a_y|_{\overline{\mathcal{G}^c}} \in \mathrm{pot}(\mathcal{A}^c)|_{\overline{\mathcal{G}^c}}$. Hence, for $y \in \mathcal{A}$, $Z[a_y] = A(y) = 0$. Therefore, analytic continuation gives $A(y) = 0$ for all points $y \in \mathcal{G}$. In particular, this yields

$$0 = Q(x) A(x - s\nu(x)) = Z\left[Q(x) a_{x-s\nu(x)}|_{\overline{\mathcal{G}^c}} \right], \quad x \in \partial\mathcal{G},\ Q \in \mathrm{C}^{(0)}(\partial\mathcal{G}), \tag{4.207}$$

provided that $s > 0$ is sufficiently small. The mapping $b_s : \partial\mathcal{G} \to \mathrm{pot}^{(0)}\left(\overline{\mathcal{G}^c}\right)$ given by $b_s(x) = Q(x)a_{x-s\nu(x)}\big|_{\overline{\mathcal{G}^c}}$ may be investigated in parallel to arguments due to W. Freeden, H. Kersten [1980]. Indeed, b_s is continuous with respect to $\|\cdot\|_{\mathrm{c}^{(0)}(\partial\mathcal{G})}$, and thus, integrable over $\partial\mathcal{G}$. In other words,

$$0 = \int_{\partial\mathcal{G}} Q(x)A(x - s\nu(x))\, d\omega(x) = \int_{\partial\mathcal{G}} Z\left[Q(x)a_{x-s\nu(x)}|_{\overline{\mathcal{G}^c}}\right]\, d\omega(x). \quad (4.208)$$

By virtue of the continuity of Z and the integrability of b_s over $\partial\mathcal{G}$, the linear functional and the integral may be interchanged

$$0 = Z\left[\int_{\partial\mathcal{G}} Q(x)a_{x-s\nu(x)} d\omega(x)\bigg|_{\overline{\mathcal{G}^c}}\right], \quad s > 0. \quad (4.209)$$

Using the abbreviation

$$v_s(y) = \int_{\partial\mathcal{G}} Q(x)\nabla_z G(\Delta; |z-y|)\bigg|_{z=x-s\nu(x)} d\omega(x) \quad (4.210)$$

we notice that (4.209) reduces to $0 = Z[v_s]$, $s > 0$. The vector field v_s is of class $\mathrm{pot}^{(0)}\left(\overline{\mathcal{G}^c}\right)$ and we know that $v_s\cdot\nu \to v\cdot\nu$, as $s \to 0$, with respect to $\|\cdot\|_{\mathrm{c}^{(0)}(\partial\mathcal{G})}$. Therefore, by virtue of (4.202), we have $v_s \to v$, as $s \to 0$, with respect to $\|\cdot\|_{\mathrm{c}^{(0)}\left(\overline{\mathcal{G}^c}\right)}$. Consequently, the continuity of Z shows $Z[v] = \lim_{s\to 0+} Z[v_s] = 0$, as desired. □

Theorem 4.44 enables us to formulate the following results:

Theorem 4.45. *Assume that $\mathcal{A}, \mathcal{G}$ are regular regions satisfying $\mathcal{A} \Subset \mathcal{G}$. Let $\{D_n\}_{n=0,1,\ldots}$ be an $\mathrm{L}^2(\partial\mathcal{A})$-Dirichlet Runge basis (in the sense of Definition 3.95). Then, every field $v \in \mathrm{pot}^{(0)}(\overline{\mathcal{G}^c})$ can be approximated (with respect to the norm $\|\cdot\|_{\mathrm{c}^{(0)}(\overline{\mathcal{G}^c})}$) by a finite linear combination of gradient fields $\{\nabla D_n\}_{n=0,1,\ldots}$. More precisely, for given $\varepsilon > 0$ and $v \in \mathrm{pot}^{(0)}(\overline{\mathcal{G}^c})$, there exist an integer $N(= N(\varepsilon))$ and coefficients $a_0, \ldots, a_N$ such that*

$$\sup_{x\in\overline{\mathcal{G}^c}} \left| v(x) - \sum_{n=0}^{N} a_n \left(\nabla D_n\right)(x) \right| \leq \varepsilon\ . \quad (4.211)$$

Proof. In comparison to Theorem 4.44 it remains to be proven that any continuous linear functional Z on $\mathrm{pot}^{(0)}\left(\overline{\mathcal{G}^c}\right)$ satisfying $Z[\nabla D_n|_{\overline{\mathcal{G}^c}}] = 0$ for $n = 0, 1, \ldots$, is zero on the class $\mathrm{pot}\left(\mathcal{A}^c\right)\big|_{\overline{\mathcal{G}^c}}$.

Let u be a vector field of class $\mathrm{pot}(\mathcal{A}^c)$. Then there exists a function $U \in \mathrm{Pot}(\mathcal{A}^c)$ with $u = \nabla U$. Since $\{D_n\}_{n=0,1,\ldots}$ is assumed to be an $\mathrm{L}^2(\partial\mathcal{A})$-Dirichlet Runge basis, U can be approximated by finite linear combinations U_N of the system $\{D_n\}_{n=0,1,\ldots}$. More explicitly, $U_N \to U$ on each $\overline{\mathcal{K}}$ with $\mathcal{K} \Subset \mathcal{A}^c$. Even more, from Theorem 3.83 we are able to conclude that $\nabla U_N \to \nabla U$ on each $\overline{\mathcal{K}}$ with $\mathcal{K} \Subset \mathcal{A}^c$. Consequently, $\nabla U_N \to \nabla U$ in the sense of $\|\cdot\|_{\mathrm{c}^{(0)}(\overline{\mathcal{G}^c})}$. In accordance with the assumption $Z[\nabla U_N|_{\overline{\mathcal{G}^c}}] = 0$ and the continuity of Z we are allowed to conclude that

$$Z\left[u\left|_{\overline{\mathcal{G}^c}}\right.\right] = Z\left[\nabla U\left|_{\overline{\mathcal{G}^c}}\right.\right] = \lim_{N\to\infty} Z\left[\nabla U_N\left|_{\overline{\mathcal{G}^c}}\right.\right] = 0, \tag{4.212}$$

as desired. □

Remark 4.46. The results of Theorem 4.45 can be extended to the Hölder norm $\mathrm{c}^{(0,\mu)}\left(\overline{\mathcal{G}^c}\right)$ (cf. W. Freeden, H. Kersten [1980]). The details are omitted.

Example 4.47. Assume that $\mathcal{A} = \mathcal{B}_{\underline{R}}(0)$, $\underline{R} < \inf_{x\in\partial\mathcal{G}}|x|$ (cf. (4.181). Then, every field $v \in \mathrm{pot}^{(0)}(\overline{\mathcal{G}^c})$ can be approximated (with respect to the norm $\|\cdot\|_{\mathrm{c}^{(0)}(\overline{\mathcal{G}^c})}$) by a finite linear combination of vector outer harmonics $h^{(1);\underline{R}}_{-n-1,k}$, i.e., the gradient fields $\nabla H^{\underline{R}}_{-n-1,k}$ with respect to (scalar) outer harmonics, in such a way that, for given $\varepsilon > 0$ and $v \in \mathrm{pot}^{(0)}(\overline{\mathcal{G}^c})$, there exist an integer $N(= N(\varepsilon))$ and coefficients $a_{n,k}$ with

$$\left\| v - \sum_{n=0}^{N}\sum_{k=1}^{2n+1} a_{n,k} h^{(1);\underline{R}}_{-n-1,k}(x)\right\|_{\mathrm{c}^{(0)}(\overline{\mathcal{G}^c})} \leq \varepsilon\ . \tag{4.213}$$

Example 4.48. Under the assumptions of Example 4.47, every field $\mathbf{v} \in \mathbf{pot}^{(0)}(\overline{\mathcal{G}^c})$ can be approximated (with respect to the norm $\|\cdot\|_{\mathbf{c}^{(0)}(\overline{\mathcal{G}^c})}$) by a finite linear combination of tensor outer harmonics $\mathbf{h}^{(1);\underline{R}}_{-n-1,k}$, i.e., the tensor fields $\nabla^{(2)} H^{\underline{R}}_{-n-1;k}$ with respect to (scalar) outer harmonics, in such a way that, for given $\varepsilon > 0$ and $\mathbf{v} \in \mathbf{pot}^{(0)}(\overline{\mathcal{G}^c})$, there exist an integer $N(= N(\varepsilon))$ and coefficients $a_{n,k}$ with

$$\left\| \mathbf{v} - \sum_{n=0}^{N}\sum_{k=1}^{2n+1} a_{n,k} \mathbf{h}^{(1);\underline{R}}_{-n-1,k}(x)\right\|_{\mathbf{c}^{(0)}(\overline{\mathcal{G}^c})} \leq \varepsilon\ . \tag{4.214}$$

From an extended version of Helly's Theorem (in the formulation of H. Yamabe [1950]) we are able to derive the following corollaries, which are of importance in SST for determining the Earth's gravitational field from a finite set of SST data.

Corollary 4.49 (SST hi-lo). *Let the assumptions of Theorem 4.45 be fulfilled. Furthermore, let $\mathcal{X}$ be a finite subset of $\mathcal{S} \subset \mathcal{G}^c$ satisfying (4.181). Then, for any given $\varepsilon > 0$ and $v \in \mathrm{pot}^{(0)}\left(\overline{\mathcal{G}^c}\right)$, there exist an integer $N(= N(\varepsilon))$ and coefficients $a_0, \ldots, a_N$ such that*

$$\sup_{x\in\overline{\mathcal{G}^c}} \left| v(x) - \sum_{n=0}^{N} a_n \left(\nabla D_n\right)(x)\right| \leq \varepsilon \tag{4.215}$$

and

$$v(x) = \sum_{n=0}^{N} a_n \left(\nabla D_n\right)(x), \quad x \in \mathcal{X}. \tag{4.216}$$

Corollary 4.50 (SST lo-lo). *Under the assumptions of Corollary 4.49: For any given $\varepsilon > 0$ and $v \in \mathrm{pot}^{(0)}\left(\overline{\mathcal{G}^c}\right)$, there exist an integer $N(= N(\varepsilon))$ and coefficients $a_0, \ldots, a_N$ such that*

$$\sup_{x \in \overline{\mathcal{G}^c}} \left| v(x) - \sum_{n=0}^{N} a_n \left(\nabla D_n\right)(x) \right| \leq \varepsilon \tag{4.217}$$

and

$$h(x) \cdot (v(x) - v(x + h(x))) = \sum_{n=0}^{N} a_n h(x) \cdot \left(\left(\nabla D_n\right)(x) - \left(\nabla D_n\right)(x + h(x))\right) \tag{4.218}$$

for $x \in \mathcal{X}$, where h is the intersatellite distance.

Corollary 4.51 (SGG). *Under the assumptions of Corollary 4.49: For any given $\varepsilon > 0$ and $v \in \mathrm{pot}^{(0)}\left(\overline{\mathcal{G}^c}\right)$, there exist an integer $N(= N(\varepsilon))$ and coefficients $a_0, \ldots, a_N$ such that*

$$\sup_{x \in \overline{\mathcal{G}^c}} \left| v(x) - \sum_{n=0}^{N} a_n \left(\nabla D_n\right)(x) \right| \leq \varepsilon \tag{4.219}$$

and

$$x \cdot (\nabla v(x))\, x = \sum_{n=0}^{N} a_n \; (x \cdot \nabla_x)(x \cdot \nabla_x)\, D_n(x), \quad x \in \mathcal{X} \; . \tag{4.220}$$

Finally, it should be mentioned that the gravitational field in $\overline{\mathcal{G}^c}$ cannot only be approximated in ε-accuracy by a linear combination of gradient fields ∇D_n such that it is consistent to a finite set $\mathcal{X}$ of SST hi-lo, SST lo-lo or SGG data but also such that it is consistent to all three data types simultaneously.

Corollary 4.52 (Combined SST/SGG). *Under the assumptions of Corollary 4.49: For any given $\varepsilon > 0$ and $v \in \mathrm{pot}^{(0)}\left(\overline{\mathcal{G}^c}\right)$, there exist an integer $N(= N(\varepsilon))$ and coefficients $a_0, \ldots, a_N$ such that*

$$\sup_{x \in \overline{\mathcal{G}^c}} \left| v(x) - \sum_{n=0}^{N} a_n \left(\nabla D_n\right)(x) \right| \leq \varepsilon \tag{4.221}$$

with

$$-x \cdot v(x) = \sum_{n=0}^{N} a_n (-x) \cdot \left(\nabla D_n\right)(x), \tag{4.222}$$

for $x \in \mathcal{X}_1$,

$$h(x) \cdot (v(x) - v(x + h(x))) = \sum_{n=0}^{N} a_n h(x) \cdot \left(\left(\nabla D_n\right)(x) - \left(\nabla D_n\right)(x + h(x))\right), \tag{4.223}$$

for $x \in \mathcal{X}_2$*, and*

$$x \cdot (\nabla v(x))\, x = \sum_{n=0}^{N} a_n \,(x \cdot \nabla_x)\,(x \cdot \nabla_x)\, D_n(x), \tag{4.224}$$

for $x \in \mathcal{X}_3$*, where h is the intersatellite distance and*

$$\mathcal{X} = \mathcal{X}_1 \cup \mathcal{X}_2 \cup \mathcal{X}_3 \subset \mathcal{S}. \tag{4.225}$$

Of course, there remain several problems for practical realization, which are closely interrelated, namely the choice of the basis system $\{D_n\}_{n=0,1,\ldots}$ and the appropriate strategy of determining the coefficients in the linear combination consistent with the satellite data.

Concerning the *choice of the basis system*, a particular role is played by the system of outer harmonics. Their polynomial nature has tremendous advantages. In fact, outer harmonics are classical means for modeling the long-wavelength features of the Earth's gravitational field. But, according to the uncertainty principle (see W. Freeden, T. Gervens, M. Schreiner [1998], W. Freeden, M. Schreiner [2009]), the ideal frequency localization prohibits any space localization. Outer harmonics as non-space-localizing structures need a uniformly dense coverage of data everywhere. Local changes are not treatable locally. An advantage of satellite data is that they are available everywhere on a dense orbital set. However, the critical point is that information of possibly small data width must be handled by a trial system of non-space-localizing functions. Therefore, the numerical use of outer harmonics is limited for modeling satellite data containing medium- to short-wavelength features. As a matter of fact, the uncertainty principle tells us that there exists a hierarchy of the scalar basis functions, starting from the *polynomials* and ending up with *Dirac kernels* as ideal systems (for more details see, e.g., W. Freeden, M. Schreiner [2009] and the references therein). Our potential theoretic arguments given above show that all these functions can be used equitably as basis functions in satellite problems. Nevertheless, it should be pointed out that *downward continuation* based on a finite set of satellite data needs much more than potential theory-based superposition of harmonic trial functions.

What is really needed for the future satellite scenario are regularization methods by more and more space localizing basis systems with a canonical start from polynomials (outer harmonics) for global trend modeling and an efficient transition to model medium- to short-wavelength features of the Earth's gravitational potential via kernels establishing successively a *zooming-in* procedure for SST hi-lo, SST lo-lo, SGG data and finally, airborne and terrestrial data (Essential mathematical tools in the line of this zooming-in process are presented in W. Freeden, F. Schneider [1998], W. Freeden [1999], W. Freeden, V. Michel [2004], W. Freeden, M. Schreiner [2009], W. Freeden, H. Nutz [2011]).

4.3 Gravimetry Problem

The inversion of Newton's Law of Gravitation (0.2) i.e., the determination of the internal *density function* from information of the external gravitational potential including the boundary is known as the *gravimetry problem.* To be more concrete, we are interested in the problem of determining the density function $F \in L^2(\mathcal{G})$ from (information on) the gravitational potential H in $\mathbb{R}^3 \backslash \mathcal{G}$) in accordance with the integral equation

$$H(x) = T[F](x) = \int_{\mathcal{G}} F(y)\; G(\Delta; |x-y|)\; dV(y). \tag{4.226}$$

Once again, in the *classification due to Hadamard*, the gravimetry problem violates all criteria, viz. uniqueness, existence, and stability: (a) The potential H is harmonic outside $\mathcal{G}$. In accordance with the so-called Picard condition (see, e.g., A.N. Tykhonov [1963]), a solution exists only if H belongs to (an appropriate subset in) the space of harmonic functions. However, it should be pointed out that this observation does not cause a numerical problem since in practice the information on H is only finite dimensional. In particular, an approximation by an appropriate harmonic function is a natural ingredient of any practical method. (b) The most serious problem is the non-uniqueness of the solution: the associated Fredholm integral operator T is of the first kind and has a kernel (null space) that is already known to coincide with the $L^2(\mathcal{G})$-orthogonal space of the closed linear subspace of all harmonic functions on $\mathcal{G}$. As we know, the orthogonal complement, i.e., the class of anharmonic functions, is infinite dimensional. (c) Restricting the operator to harmonic densities leads to an injective mapping that has a discontinuous inverse, implying an unstable solution.

Concerning the historical background, the problem of non-uniqueness has been discussed extensively in literature, starting with a paper by G.G. Stokes [1867]. This problem can be bypassed by imposing some reasonable additional conditions on the density. A suitable condition, suggested by the mathematical structure of the Newton potential operator T, is to require that the density is harmonic. The approximate calculation of the harmonic density has already been implemented and covered in several papers, whereas the problem of determining the anharmonic part seems to remain a great challenge. Due to the lack of an appropriate physical interpretation of the harmonic part of the density, various alternative variants have been discussed in the literature. In general, gravitational data yield significant information only about the uppermost part of the Earth's interior, which is not laterally homogeneous.

4.3.1 Spectral Relation Between Potential and Density

In the following, the gravimetry problem is discussed in a model context starting from a spherical formulation, i.e., the underlying regular region $\mathcal{G}$ is simply understood to be the unit ball $\mathcal{B}_1(0)$ around the origin. In doing so, the classical spherical harmonic machinery becomes applicable. First, we are interested in a relation between the spherical harmonic coefficients of the gravitational potential and the inner harmonic coefficients of the density function. In doing so, our work closely follows the survey paper V. Michel, A.S. Fokas [2008].

Theorem 4.53. *Let the mass density function $F \in \mathrm{L}^2(\mathcal{B}_1(0))$ be given in such a way that*

$$F(x) = \sum_{n=0}^{\infty} \sum_{j=1}^{2n+1} F_{n,j}(|x|) Y_{n,j}\left(\frac{x}{|x|}\right) \tag{4.227}$$

converges in the $\mathrm{L}^2(\mathcal{B}_1(0))$-sense. Then the potential (4.226) admits a pointwise representation in $\mathbb{R}^3 \backslash \overline{\mathcal{B}_1(0)}$ (and, for its restriction $H|_\Omega$ to Ω, in the sense of $\|\cdot\|_{\mathrm{L}^2(\Omega)}$) by the series expansion

$$H(y) = \sum_{n=0}^{\infty} \sum_{j=1}^{2n+1} \left(\int_0^1 r^{n+2} F_{n,j}(r)\, dr\right) \frac{1}{2n+1} \frac{1}{|y|^{n+1}} Y_{n,j}\left(\frac{y}{|y|}\right). \tag{4.228}$$

Proof. Using the well-known formula for a fundamental solution and assuming y to be a point of $\mathbb{R}^3 \backslash \overline{\mathcal{B}_1(0)}$, we get, by introducing spherical coordinates,

$$\begin{aligned} &\int_{\mathcal{B}_1(0)} \frac{F(x)}{|x-y|}\, dV(x) \qquad\qquad (4.229)\\ &= \int_0^1 r^2 \int_\Omega F(r\xi) \frac{1}{|r\xi - y|} d\omega(\xi)\, dr \\ &= \sum_{m=0}^{\infty} \sum_{n=0}^{\infty} \sum_{j=1}^{2n+1} \frac{1}{|y|^{m+1}} \int_0^1 r^2 F_{n,j}(r) r^m\, dr \int_\Omega P_m\left(\xi \cdot \frac{y}{|y|}\right) Y_{n,j}(\xi)\, d\omega(\xi). \end{aligned}$$

Observing the addition theorem for scalar spherical harmonics we find that

$$\int_{\mathcal{B}_1(0)} \frac{F(x)}{|x-y|}\, dV(x) = \sum_{n=0}^{\infty} \sum_{j=1}^{2n+1} \int_0^1 r^{n+2} F_{n,j}(r)\, dr \frac{4\pi}{2n+1} \frac{1}{|y|^{n+1}} Y_{n,j}\left(\frac{y}{|y|}\right). \tag{4.230}$$

This series defined on $\mathbb{R}^3 \backslash \overline{\mathcal{B}_1(0)}$ can be formally extended to $\Omega = \partial\mathcal{B}_1(0)$ so that $H|_\Omega$ is of class $\mathrm{L}^2(\Omega)$. Indeed, it can be easily seen that

$$\begin{aligned} \|H\|^2_{\mathrm{L}^2(\Omega)} &= \sum_{n=0}^{\infty} \sum_{j=1}^{2n+1} \left(\int_0^1 r^{n+2} F_{n,j}(r)\, dr\right)^2 \left(\frac{1}{2n+1}\right)^2 \qquad (4.231)\\ &\leq \sum_{n=0}^{\infty} \sum_{j=1}^{2n+1} \left(\int_0^1 r^{2n+2}\, dr\right) \left(\int_0^1 r^2 (F_{n,j}(r))^2\, dr\right) \left(\frac{1}{2n+1}\right)^2 \\ &\leq \sum_{n=0}^{\infty} \sum_{j=1}^{2n+1} \int_0^1 r^2 (F_{n,j}(r))^2\, dr \leq \|F\|_{\mathrm{L}^2(\mathcal{B}_1(0))}. \end{aligned}$$

This guarantees our assertion. □

At this stage we already see that the solution of the inverse problem of determining F from the knowledge of H does not allow a unique solution, since the equations

$$\frac{1}{2n+1}\int_0^1 r^{n+2}F_{n,j}(r)\ dr = \int_\Omega H(\eta)Y_{n,j}(\eta)\ d\omega(\eta) = H^\wedge(n,j), \qquad (4.232)$$

for $n \in \mathbb{N}_0$ and $j = 1,\ldots,2n+1$, allow an infinite number of choices for the coefficients $F_{n,j}$. The coefficient equations (4.232) are well-known (see, for example, H. Moritz [1990], P. Pizzetti [1910a], C.C. Tscherning [1974], D.P. Rubincam [1979]). As a matter of fact, G.G. Stokes [1867] has already pointed out this fact, which was quantified more precisely by a large number of other scientists, later on. Clearly, the non-uniqueness of the solution can also be inferred in many other ways.

4.3.2 Characterization of a Basis for the Null Space

There are several ways of characterizing the null space $\ker(T) = \mathrm{AnPot}\,(\mathcal{B}_1(0))$ of the operator $T : \mathrm{L}^2\,(\mathcal{B}_1(0)) \to \mathrm{H}(\mathbb{R}^3 \setminus \overline{\mathcal{B}_1(0)})$ given by

$$T[F](y) = \int_{\mathcal{B}_1(0)} F(x)\ G(\Delta; |x-y|)\ dV(x), \quad y \in \mathbb{R}^3 \setminus \overline{\mathcal{B}_1(0)}. \qquad (4.233)$$

In order to make this observation more concrete, we use the orthonormal basis for $\mathrm{L}^2(\mathcal{B}_1(0))$ that is due to L. Ballani et al. [1993], H.M. Dufour [1977], V. Michel [1999], namely

$$B_{n,j,m}(x) = \gamma_{n,m} P_m^{(0,n+\frac{1}{2})}\left(2|x|^2-1\right)|x|^n\, Y_{n,j}\left(\frac{x}{|x|}\right), \qquad (4.234)$$

$m, n \in \mathbb{N}_0$, $j = 1, ..., 2n+1$, where $\{P_m^{(\alpha,\beta)}\}_{m\in\mathbb{N}_0}$ $(\alpha, \beta > -1)$ are the Jacobi polynomials (see, e.g., W. Magnus et al. [1966], G. Szegö [1939]) characterized uniquely by the following properties:

(i) $P_m^{(\alpha,\beta)}$ is a univariate polynomial of degree m,

(ii) $\int_{-1}^1 (1-t)^\alpha(1+t)^\beta P_n^{(\alpha,\beta)}(t)P_m^{(\alpha,\beta)}(t)\,\mathrm{d}t = 0$, $n \neq m$,

(iii) $P_m^{(\alpha,\beta)}(1) = \binom{n+\alpha}{n}$.

The normalization constants $\gamma_{n,m}$ are not of importance in our context (see, for example, V. Michel [2005]). The application of the basis (4.234) implies the following representation of the coefficients $F_{n,j}$ in (4.227):

$$F_{n,j}(r) = r^n \sum_{m=0}^{\infty} f_{n,j,m}\gamma_{n,m}P_m^{(0,n+\frac{1}{2})}\left(2r^2-1\right) \qquad (4.235)$$

in the sense of the $\mathrm{L}^2([0,1])$-topology. We insert (4.235) into the left-hand side of (4.232). Using the substitution $r=\sqrt{(t+1)/2}$, we get

$$\int_0^1 r^{n+2+n} \sum_{m=0}^{\infty} f_{n,j,m}\gamma_{n,m}P_m^{(0,n+\frac{1}{2})}\left(2r^2-1\right)\, dr \tag{4.236}$$

$$= \frac{1}{4}\int_{-1}^{1}\left[\frac{1}{2}(t+1)\right]^{n+1}\sum_{m=0}^{\infty} f_{n,j,m}\gamma_{n,m}P_m^{(0,n+\frac{1}{2})}(t)\left(\frac{1}{2}(t+1)\right)^{-\frac{1}{2}} dt$$

$$= \frac{1}{2^{n+\frac{5}{2}}}\int_{-1}^{1}(t+1)^{n+\frac{1}{2}}\left(\sum_{m=0}^{\infty} f_{n,j,m}\gamma_{n,m}P_m^{(0,n+\frac{1}{2})}(t)\right)P_0^{(0,n+\frac{1}{2})}(t)\, dt$$

$$= \frac{1}{2n+3}f_{n,j,0}\gamma_{n,0}.$$

In other words, $B_{n,j,m}$ is in the null space of T if and only if m is a positive integer. The functions $B_{n,j,0}$ constitute the system of inner harmonics, which form a basis for the harmonic functions on a ball. Consequently, all anharmonic functions, i.e., all functions orthogonal to all harmonic functions in the $\mathrm{L}^2(\mathcal{B}_1(0))$-sense, turn out to be precisely the density functions that produce a vanishing potential outside $\overline{\mathcal{B}_1(0)}$. This result has been shown by P. Pizzetti [1909, 1910a] and G. Lauricella [1912]. The same concept was later used by P. Novikoff [1938]. Needless to say, there exist various ways of establishing this result (for more details see also L. Ballani, D. Stromeyer [1990, 1993], W. Freeden, V. Michel [2004], V. Michel [1999], M. Thalhammer et al. [1996]).

4.3.3 Minimum Norm Solution

One way of treating the gravimetry problem is to look for a harmonic density function that assumes a minimum L^2-norm solution (see 4.89).

Corollary 4.54 (Harmonic Solution). *Let $H:\mathbb{R}^3\setminus\mathcal{B}_1(0)\to\mathbb{R}$ satisfy the conditions:*

(i) $H|_\Omega\in\mathrm{L}^2(\Omega)$ *and* $H\in\mathrm{C}^{(2)}(\mathbb{R}^3\setminus\overline{\mathcal{B}_1(0)})$,

(ii) $\sum_{n=0}^{\infty}\sum_{j=1}^{2n+1} H^\wedge(n,j)n^3<\infty$ *with* $H^\wedge(n,j)=\int_\Omega H(\eta)Y_{n,j}(\eta)\,d\omega(\eta)$,

(iii) $\Delta H=0$ *in* $\mathbb{R}^3\setminus\overline{\mathcal{B}_1(0)}$,

(iv) H is regular at infinity.

Then the unique harmonic function $F\in\mathrm{Pot}^{(0)}(\overline{\mathcal{B}_1(0)})$ with

$$H=T[F]=\frac{1}{4\pi}\int_{\mathcal{B}_1(0)}\frac{F(x)}{|x-\cdot|}\,dV(x) \tag{4.237}$$

is given by

$$F(x) = \sum_{n=0}^{\infty}(2n+1)(2n+3)|x|^n \sum_{j=1}^{2n+1} H^\wedge(n,j)\, Y_{n,j}\left(\frac{x}{|x|}\right), \tag{4.238}$$

provided that the series (4.238) converges in the $\mathrm{L}^2(\mathcal{B}_1(0))$*-sense.*

Proof. We observe that

$$B_{n,j,0}(x) = \sqrt{2n+3}\,|x|^n Y_{n,j}\left(\frac{x}{|x|}\right), \quad n \in \mathbb{N}_0, j = 1, \ldots, 2n+1, \tag{4.239}$$

forms an orthonormal inner harmonic basis of $(\mathrm{Pot}^{(0)}(\mathcal{B}_1(0)), (\cdot,\cdot)_{\mathrm{L}^2(\overline{\mathcal{B}_1(0)})})$, i.e., $\gamma_{n,0} = \sqrt{2n+3}$, $n \in \mathbb{N}_0$ (see, e.g., V. Michel [1999], F. Sansò et al. [1986]). It follows that the identity

$$F(x) = \sum_{n=0}^{\infty} \sum_{j=1}^{2n+1} f_{n,j}\,\sqrt{2n+3}\,|x|^n\, Y_{n,j}\left(\frac{x}{|x|}\right) \tag{4.240}$$

holds in the sense of $\mathrm{L}^2(\mathcal{B}_1(0))$. Hence, (4.232) together with $F_{n,j}(r) = f_{n,j}\sqrt{2n+3}\, r^n$ leads to

$$\frac{1}{2n+1}\int_0^1 r^{2n+2} dr\, f_{n,j}\sqrt{2n+3} = H^\wedge(n,j), \tag{4.241}$$

which is equivalent to

$$f_{n,j} = \sqrt{2n+3}\,(2n+1) H^\wedge(n,j), \tag{4.242}$$

leading to (4.238). □

The harmonic solution has been discussed in numerous publications such as G. Anger [1990], M.J. Fengler et al. [2006], W. Freeden, V. Michel [2004], G. Hein et al. [1989], V. Michel [1999], V. Michel, K. Wolf [2008], F. Sansò et al. [1986], M. Thalhammer et al. [1996], C.C. Tscherning, H. Sünkel [1981]. Nevertheless, this solution lacks a convincing physical interpretation, (see also M. Thalhammer et al. [1996]). Furthermore, an additional drawback is the maximum principle, according to which the harmonic density is maximal (and minimal) at the surface (which is in contrast to the real density). Nevertheless, the harmonic solution still offers certain advantages. The search for high-frequency density anomalies is supported by exactly this maximum principle, since such structures primarily occur at the uppermost Earth layer. Moreover, they can be derived in a remarkable qualitative resolution from gravitational data, see V. Michel [1999], V. Michel, K. Wolf [2008], M. Thalhammer et al. [1996]. In addition, V. Michel, A.S. Fokas [2008] point out that the behavior of r^n on the interval $[0,1]$ for different n is consistent with the following observation: the higher the frequency degree (band) of a density phenomenon, the more it is concentrated toward the surface.

Next, we present a straightforward derivation of a formula for the minimal $\mathrm{L}^2(\mathcal{B}_1(0))$-norm solution. It is a direct consequence of Corollary 4.54 and (4.89).

Theorem 4.55 (Minimum Norm Solution). *Let $H : \mathbb{R}^3 \backslash \mathcal{B}_1(0) \to \mathbb{R}$ be a function satisfying the conditions:*

(i) $H|_{\Omega}$ *is a member of class* $\mathrm{L}^2(\Omega)$,

(ii) $\sum\limits_{n=0}^{\infty} \sum\limits_{j=1}^{2n+1} H^{\wedge}(n,j)\, n^3 < \infty$,

(iii) $\Delta H = 0$ *in* $\mathbb{R}^3 \setminus \overline{\mathcal{B}_1(0)}$,

(iv) H *is regular at infinity.*

Then, among all functions $F \in \mathrm{L}^2(\mathcal{B}_1(0))$ *obeying*

$$H = T[F] = \frac{1}{4\pi} \int_{\mathcal{B}_1(0)} \frac{F(x)}{|x - \cdot|}\, dV(x), \tag{4.243}$$

there exists a unique minimizer of the functional

$$\int_{\mathcal{B}_1(0)} |F(x)|^2\, dV(x), \tag{4.244}$$

given within $\mathrm{L}^2(\mathcal{B}_1(0))$ *by*

$$F(x) = \sum_{n=0}^{\infty} (2n+1)(2n+3)|x|^n \sum_{j=1}^{2n+1} H^{\wedge}(n,j)\, Y_{n,j}\left(\frac{x}{|x|}\right), \tag{4.245}$$

provided that (4.245) converges in the $\mathrm{L}^2(\mathcal{B}_1(0))$*-sense.*

In the following, we are concerned with a condition for the convergence of the series (4.245) (which actually turns out to be identical with (4.238)).

Theorem 4.56. *The series in (4.245) converges in* $\mathrm{L}^2(\mathcal{B}_1(0))$ *if and only if*

$$\sum_{n=0}^{\infty} n^3 \sum_{j=1}^{2n+1} (H^{\wedge}(n,j))^2 < \infty\ . \tag{4.246}$$

Proof. In fact, the $\mathrm{L}^2(\mathcal{B}_1(0))$-norm of F can be calculated as follows

$$\begin{aligned} \|F\|^2_{\mathrm{L}^2(\mathcal{B}_1(0))} &= \int_0^1 r^2 \int_{\Omega} (F(r\xi))^2\, d\omega(\xi)\, dr \\ &= \sum_{n=0}^{\infty} \left(\frac{2n+1}{4\pi}(2n+3)\right)^2 \int_0^1 r^{2n+2}\, dr \sum_{j=1}^{2n+1} (H^{\wedge}(n,j))^2 \\ &= \sum_{n=0}^{\infty} \left(\frac{2n+1}{4\pi}\right)^2 (2n+3) \sum_{j=1}^{2n+1} (H^{\wedge}(n,j))^2\ . \end{aligned} \tag{4.247}$$

This guarantees Theorem 4.56. □

In conclusion, the condition for the convergence of the series (4.245) is precisely the condition for the solvability of the inverse problem. This result is not surprising, as it reflects the principle of the Picard condition for inverse problems (see, e.g., H. Engl et al. [1997]).

Remark 4.57. The (actual) gravitational potential of the Earth certainly satisfies the condition (4.246). Note that the *(empirical) Kaula's rule* (used in physical geodesy) states that

$$\sum_{j=1}^{2n+1} (H^\wedge(n,j))^2 = \mathcal{O}\left(\frac{\underline{R}^{n+1}}{n^3}\right), \quad n \to \infty, \tag{4.248}$$

for some constant $\underline{R} \in (0,1)$ specifying a Runge (in the jargon of physical geodesy usually called Bjerhammar) sphere (for more details concerning (4.248) the reader is referred to, e.g., F. Sansò, R. Rummel [1997] and the references therein).

4.3.4 Quasi-Harmonic Solution

An interesting modification of the *harmonicity* is *quasi-harmonicity*. In this case, the density F is assumed to satisfy an equation of the type

$$\Delta_x \left(\frac{F(x)}{G(|x|)}\right) = 0, \tag{4.249}$$

where G is a prescribed function showing no zeros. Such quasi-harmonic solutions are proposed by G. Hein et al. [1989], C.C. Tscherning [1992, 1995], C.C. Tscherning, G. Strykowski [1987], C.C. Tscherning, H. Sünkel [1981]. In C.C. Tscherning, G. Strykowski [1987], the case of a monomial function H is analyzed in detail and an associated spectral relation is derived. However, C.C. Tscherning [1992] reports rather disappointing numerical results.

In the sequel, we discuss a slight generalization of this approach following V. Michel, A.S. Fokas [2008].

Theorem 4.58 (Quasi-Harmonic Solution). *Let H satisfy the conditions of Theorem 4.55. Then the unique solution $F \in \mathrm{C}^{(2)}(\overline{\mathcal{B}_1(0)})$ of the equations*

$$H(y) = T[F](y) = \frac{1}{4\pi}\int_{\mathcal{B}_1(0)} F(x)\frac{1}{|x-y|}\, dV(x), \quad y \in \mathbb{R}^3 \backslash \mathcal{B}_1(0), \tag{4.250}$$

and

$$\Delta_x \left(F(x)|x|^{-p}\right) = 0, \quad x \in \mathcal{B}_1(0), \tag{4.251}$$

for a fixed $p > 0$, is given by

$$F(x) = \sum_{n=0}^{\infty} (2n+p+3)(2n+1)|x|^{n+p} \sum_{j=1}^{2n+1} H^\wedge(n,j)\, Y_{n,j}\left(\frac{x}{|x|}\right), \tag{4.252}$$

where the series (4.252) converges in the $\mathrm{L}^2(\mathcal{B}_1(0))$-sense.

Proof. The functions $x \mapsto |x|^{n+p}\ Y_{n,j}\left(\frac{x}{|x|}\right)$ provide an orthogonal basis for such quasi-harmonic functions with respect to the $\mathrm{L}^2(\mathcal{B}_1(0))$-inner product. Indeed, the square of the normalization constant can be calculated in a straightforward way

$$\int_{\mathcal{B}_1(0)} \left(|x|^{n+p} Y_{n,j}\left(\frac{x}{|x|}\right)\right)^2 dV(x) = \int_0^1 r^{2n+2p+2}\, dr \int_\Omega Y_{n,j}(\xi)^2\, d\omega(\xi) = \frac{1}{2n+2p+3}\ . \tag{4.253}$$

Hence, letting $F_{n,j}(r) = f_{n,j}\sqrt{2n+2p+3}\, r^{n+p}$ in (4.232), we obtain

$$(2n+1)H^\wedge(n,j) = \int_0^1 r^{2n+2+p}\, dr\, f_{n,j}\sqrt{2n+2p+3} = \frac{f_{n,j}}{\sqrt{2n+2p+3}}. \tag{4.254}$$

Thus we have

$$F(x) = \sum_{n=0}^{\infty}\sum_{j=1}^{2n+1} f_{n,j}\sqrt{2n+2p+3}\, |x|^{n+p}\, Y_{n,j}\left(\frac{x}{|x|}\right) = \sum_{n=0}^{\infty}(2n+p+3)(2n+1)|x|^{n+p}\sum_{j=1}^{2n+1} H^\wedge(n,j)\ Y_{n,j}\left(\frac{x}{|x|}\right) \tag{4.255}$$

in the topology of $\mathrm{L}^2(\mathcal{B}_1(0))$. The convergence of the series is guaranteed by the conditions imposed on H, since the Fourier coefficients of F obey the asymptotic expansion $f_{n,j} = \mathcal{O}\left(H^\wedge(n,j)\ n^{\frac{3}{2}}\right)$, $n \to \infty$. This explains Theorem 4.58. □

Of course, the case $p = 0$ in Theorem 4.58 leads back to the harmonic solution.

4.3.5 Biharmonic Solution

A biharmonic constraint imposed on the density, i.e., $\Delta\Delta F = 0$ in $\mathcal{B}_1(0)$, is discussed, e.g., in M. Skorvanek [1981], C.C. Tscherning, G. Strykowski [1987]. Clearly, the knowledge of $H|_\Omega$ is insufficient for obtaining a unique solution, since the harmonic solution represents only a particular case of a biharmonic solution. This is the reason why it is assumed in many geoscientific approaches that F is known on Ω. According to M. Skorvanek [1981], the minimization of the $\mathrm{L}^2(\mathcal{B}_1(0))$-norm of ∇F, provided that F and H are given on Ω, implies the biharmonicity of the solution F. A spectral relation for the spherical harmonics coefficients is given in M. Skorvanek [1981]. However, these results have to be slightly modified, since the knowledge of H and F on Ω is not sufficient for a unique solution. In fact, V. Michel, A.S. Fokas [2008] show the following result.

Theorem 4.59 (Biharmonic Solution). *All solutions $F \in \mathrm{C}^{(4)}(\overline{\mathcal{B}_1(0)})$ of the problem*

$$\Delta\Delta F(x) = 0, \quad x \in \mathcal{B}_1(0), \tag{4.256}$$

and

$$H(x) = T[F](x) = \frac{1}{4\pi}\int_{\mathcal{B}_1(0)} \frac{F(y)}{|x-y|}\, dV(y), \quad x \in \Omega,\ F \in \mathrm{L}^2(\Omega), \tag{4.257}$$

can be represented in the form

$$\begin{aligned} F(x) &= \sum_{n=0}^{1}\sum_{j=1}^{2n+1} \left(a_{n,j}|x|^n + b_{n,j}|x|^{n+2} + c_{n,j}|x|^{-n+1}\right) Y_{n,j}\left(\frac{x}{|x|}\right) \qquad (4.258)\\ &\quad + \frac{1}{2}\sum_{n=2}^{\infty}(2n+1)(2n+3)(2n+5)\left(|x|^n - |x|^{n+2}\right)\\ &\qquad \times \sum_{j=1}^{2n+1} H^\wedge(n,j)\, Y_{n,j}\left(\frac{x}{|x|}\right)\\ &\quad + \frac{1}{2}\sum_{n=2}^{\infty}\left((2n+5)|x|^{n+2} - (2n+3)|x|^n\right)\\ &\qquad \times \sum_{j=1}^{2n+1} F^\wedge(n,j)\, Y_{n,j}\left(\frac{x}{|x|}\right), \end{aligned}$$

provided that these series converge at least in the $\mathrm{L}^2(\mathcal{B}_1(0))$–sense, where the coefficients $a_{0,1}, \ldots, c_{1,3}$ are given by the underdetermined system of linear equations

$$\begin{array}{ccccccc} \frac{1}{2n+3}\, a_{n,j} & + & \frac{1}{2n+5}\, b_{n,j} & + & \frac{1}{4}\, c_{n,j} & = & (2n+1)H^\wedge(n,j) \\ a_{n,j} & + & b_{n,j} & + & c_{n,j} & = & F^\wedge(n,j) \end{array} \tag{4.259}$$

for $n = 0, 1$, $j = 1, \ldots, 2n+1$.

Proof. First, our goal is to derive a biharmonic basis for F. It is easy to verify that $\Delta_x\Delta_x(r^\alpha Y_{n,j}(\xi)) = 0$, $r = |x|$, $\xi = \frac{x}{|x|}$, has exactly four solutions for fixed $n \in \mathbb{N}_0$, namely $\alpha \in \{-n-1, -n+1, n, n+2\}$, where $-n-1$ and n correspond to the harmonic case and the non-negative solutions n and $n+2$ yield bounded functions. However, for $n = 0$ and $n = 1$, the choice $\alpha = -n+1$ gives another bounded basis function. Hence, we are able to use

$$F_{n,j}(r) = a_{n,j}r^n + b_{n,j}r^{n+2} \tag{4.260}$$

for $n \geq 2$ and

$$F_{n,j}(r) = a_{n,j}r^n + b_{n,j}r^{n+2} + c_{n,j}r^{-n+1} \tag{4.261}$$

for $n \leq 1$ in (4.232). The case $n \geq 2$ yields

$$\begin{aligned}(2n+1)H^{\wedge}(n,j) &= a_{n,j}\int_0^1 r^{2n+2}\,dr + b_{n,j}\int_0^1 r^{2n+4}\,dr \\ &= \frac{a_{n,j}}{2n+3} + \frac{b_{n,j}}{2n+5}\end{aligned} \tag{4.262}$$

and

$$F^{\wedge}(n,j) = a_{n,j} + b_{n,j}. \tag{4.263}$$

This is uniquely solvable by

$$\begin{aligned}a_{n,j} &= \frac{1}{2}(2n+1)(2n+3)(2n+5)\,H^{\wedge}(n,j) \\ &\quad -\frac{1}{2}(2n+3)F^{\wedge}(n,j),\end{aligned} \tag{4.264}$$

$$\begin{aligned}b_{n,j} &= -\frac{1}{2}(2n+1)(2n+3)(2n+5)\,H^{\wedge}(n,j) \\ &\quad +\frac{1}{2}(2n+5)F^{\wedge}(n,j)\ .\end{aligned} \tag{4.265}$$

If $n \leq 1$, the two equations (4.262) and (4.263) take the particular form

$$\frac{a_{n,j}}{2n+3} + \frac{b_{n,j}}{2n+5} + \frac{c_{n,j}}{4} = (2n+1)\,H^{\wedge}(n,j), \tag{4.266}$$

$$a_{n,j} + b_{n,j} + c_{n,j} = F^{\wedge}(n,j), \tag{4.267}$$

which leaves exactly one degree of freedom for every pair (n,j). □

The formula presented in M. Skorvanek [1981] corresponds to the choice $c_{n,j} = 0$. The case of a harmonic solution is consistent with this result, since harmonicity implies $c_{n,j} = 0$ and $b_{n,j} = 0$, which is equivalent to

$$F^{\wedge}(n,j) = \frac{1}{4\pi}(2n+1)(2n+3)H^{\wedge}(n,j). \tag{4.268}$$

This is the result of Corollary 4.54.

Finally, it should be noted that the availability of F on Ω cannot be assumed on a sufficiently dense point grid in reality.

4.3.6 Discussion of the Radial Mean

In the following, we discuss a particular case of a non-radially dependent density.

Theorem 4.60 (Spectral Relation for a Layer Density). *Let a spherical shell be given by*

$$\mathcal{B}_{R,R+\varepsilon}(0) = \{x \in \mathbb{R}^3 : R < |x| < R+\varepsilon\}, \qquad \varepsilon, R > 0, R+\varepsilon \leq 1. \tag{4.269}$$

Furthermore, let F^{L} be a square-integrable density function which admits the spherical harmonic expansion

$$F^{\mathrm{L}}(x) = \sum_{n=0}^{\infty} \sum_{j=1}^{2n+1} (F^L)^\wedge(n,j) Y_{n,j}\left(\frac{x}{|x|}\right), \quad x \in \mathcal{B}_{R,R+\varepsilon}(0)$$

in the shell $\mathcal{B}_{R,R+\varepsilon}(0)$ and that vanishes outside the shell. Suppose that $H = T[F^L]$ is the corresponding gravitational potential. Then

$$(F^L)^\wedge(n,j) = \frac{(2n+1)(n+3)}{(R+\varepsilon)^{n+3} - R^{n+3}} H^\wedge(n,j) \ .$$

Proof. Clearly,

$$\begin{aligned} H(y) &= \frac{1}{4\pi} \sum_{n=0}^{\infty} \int_R^{R+\varepsilon} r^2 \int_\Omega F^{\mathrm{L}}(r\xi) \frac{r^n}{|y|^{n+1}} P_n\left(\xi \cdot \frac{y}{|y|}\right) d\omega(\xi)\, dr \quad (4.270) \\ &= \sum_{n=0}^{\infty} \int_R^{R+\varepsilon} r^{n+2}\, dr\, |y|^{-n-1} \frac{1}{2n+1} \sum_{j=1}^{2n+1} (F^L)^\wedge(n,j)\, Y_{n,j}\left(\frac{y}{|y|}\right) \\ &= \sum_{n=0}^{\infty} \frac{(R+\varepsilon)^{n+3} - R^{n+3}}{(2n+1)(n+3)} \frac{1}{|y|^{n+1}} \sum_{j=1}^{2n+1} (F^L)^\wedge(n,j) Y_{n,j}\left(\frac{y}{|y|}\right). \end{aligned}$$

The expansion (4.270) holds pointwise for all $y \in \mathbb{R}^3 \backslash \overline{\mathcal{B}_1(0)}$. Due to the square–integrability of F^{L} the convergence domain can be extended to Ω such that we get a function of class $\mathrm{L}^2(\Omega)$ obeying

$$\begin{aligned} &\sum_{n=0}^{\infty} \left(\frac{(R+\varepsilon)^{n+3} - R^{n+3}}{(2n+1)(n+3)}\right)^2 \sum_{j=1}^{2n+1} \left((F^L)^\wedge(n,j)\right)^2 \qquad (4.271) \\ &\leq \sum_{n=0}^{\infty} \sum_{j=1}^{2n+1} \left((F^L)^\wedge(n,j)\right)^2 . \end{aligned}$$

This yields the desired relation. □

Regarding the inverse gravimetric problem there are several other approaches discussed in the literature. For example, a functional of type $\int_{\mathcal{B}_1(0)} (|x| F(x))^2\, dV(x)$ is minimized by M. Skorvanek [1981], and a formula for the spherical harmonics expansion of F is given as a canonical consequence. A variant that postulates a fluid mantle is due to E.W. Schwiderski [1967]. In Z. Martinec, K. Pec [1989], in addition to the constraint of harmonicity, it is postulated that the lateral density variations in the interior of the Earth's mantle and the density contrast on the undulated core mantle boundary do not influence the gravitational field in the core. This approach leads to a system of linear equations from which the minimal energy solution is chosen. A measure-theoretic approach is discussed, e.g., in G. Anger [1981, 1990], B.-W. Schulze, G. Wildenhain [1977]. Note that G. Anger [1981] also contains a list of references on historical keystones in research concerning the inverse gravimetric problem. The condition of positive solutions of linear inverse problems is discussed by M. Bertero, P. Brianzi, E.R. Pike, L. Rebolia [1988]. A more detailed overview of recent activities can be found in V. Michel, A.S. Fokas [2008]. Initiated by concepts on the sphere as described by W. Freeden, T. Gervens, M. Schreiner [1998] and W. Freeden, V. Michel [2004], the note by V. Michel, A.S. Fokas [2008] also presents spectral multiscale methods. Finally, it should be noted that the approach presented here does not consider the problem where the determination of the shape of the gravitating body is combined with the calculation of the density function. For such problems or similar ones, the reader is refereed to, e.g., R. Barzaghi, F. Sansò [1986], P. Novikoff [1938], D.P. Zidarov [1974, 1980, 1986], D.P. Zidarov [1990].

4.3.7 Approximate Solution by Haar Kernels

In constructive approximation, locally supported functions are nothing new, having been discussed already by A. Haar [1910]. The primary importance of locally supported Haar kernels in the classical one-dimensional Euclidean space is the birth of an entire basis family by means of two operations, viz. dilations and translations. In other words, an entire set of approximants is available from the single locally supported *Haar mother kernel*, and this basis family provides useful building block functions that enable the multiscale modeling and the decorrelation of data.

In the following, we make the attempt to apply the Haar philosophy to an approximate determination of the mass density distribution inside a body. The regularization procedure of the Newton potential as proposed in Section 3.2 is the essential tool.

Definition 4.61. Let $\mathcal{G} \subset \mathbb{R}^3$ be a regular region, and $\{H_\rho\}_{\rho>0}$ be the family of Haar kernels $H_\rho : r \mapsto H_\rho(r)$, $r > 0$ given by

$$H_\rho(r) = \begin{cases} 0, & r > \rho, \\ \frac{3}{4\pi\rho^3}, & r \leq \rho \end{cases} \tag{4.272}$$

(note that $\|\mathcal{B}_\rho(0)\| = \frac{4}{3}\pi\rho^3$). Then $\{I_\rho\}_{\rho>0}$, defined by

$$I_\rho[F](x) = \int_{\mathcal{G}} H_\rho(|x-z|)F(z)\, dV(z), \quad x \in \overline{\mathcal{G}}, F \in \mathrm{C}^{(0)}(\overline{\mathcal{G}}), \tag{4.273}$$

is called the *Haar singular integral* on $\overline{\mathcal{G}}$.

From the mean value theorem of multidimensional analysis we immediately obtain

Lemma 4.62. *If $\{I_\rho\}_{\rho>0}$ is the Haar singular integral on $\overline{\mathcal{G}}$, then*

$$\lim_{\rho\to 0+} I_\rho[F](x) = \frac{\alpha(x)}{4\pi} F(x) = \begin{cases} F(x) & , \quad x \in \mathcal{G}, \\ \frac{1}{2}F(x) & , \quad x \in \partial\mathcal{G}, \end{cases}$$

where α is the solid angle subtended by the boundary $\partial\mathcal{G}$ at the point $x \in \overline{\mathcal{G}}$ (cf Lemma 3.5) and F is of class $\mathrm{C}^{(0)}(\overline{\mathcal{G}})$.

Let $\{\rho_j\}_{j\in\mathbb{N}_0}$ be a monotonically decreasing sequence of positive values ρ_j such that $\lim_{j\to\infty}\rho_j = 0$ (for example, $\rho_j = 2^{-j}, j \in \mathbb{N}_0$). Then, Lemma 4.62 enables us to specify a sufficiently large integer J such that

$$\frac{\alpha(x)}{4\pi} F(x) \simeq I_{\rho_J}[F](x) = \int_{\mathcal{G}} H_{\rho_J}(|x-z|)F(z)\, dV(z), \tag{4.274}$$

(as always, "$\simeq$ "means that the error is negligible). An elementary calculation shows that the linear regularization ${}^lG^\rho(\Delta;\cdot)$ of the fundamental solution $G(\Delta;\cdot)$ as defined by (3.78), i.e.,

$${}^lG^{\rho_J}(\Delta;r) = \begin{cases} \frac{1}{8\pi\rho_J}\left(3-\frac{r^2}{\rho_j^2}\right), & r \le \rho_J, \\ \frac{1}{4\pi r}, & r > \rho_J, \end{cases} \tag{4.275}$$

satisfies the differential equation

$$\Delta_x\, {}^lG^{\rho_J}(\Delta;|x-z|) = -\, H_{\rho_J}(|x-z|) \tag{4.276}$$

for all $x, z \in \mathbb{R}^3$. From (4.274) it therefore follows that

$$\Delta_x \int_{\mathcal{G}} {}^lG^{\rho_J}(\Delta;|x-z|)F(z)\, dV(z) \simeq -\, F(x), \quad x \in \mathcal{G}. \tag{4.277}$$

In doing so, we are aware of the fact (cf. Theorem 3.30) that

$$U(x) \simeq \int_{\mathcal{G}} {}^lG^{\rho_J}(\Delta;|x-y|)F(y)\, dV(y), \quad x \in \overline{G}, \tag{4.278}$$

provides an approximation of the Newton integral (3.75) with negligible error. In order to realize a fully discrete approximation of F, we have to apply approximate integration formulas over $\overline{\mathcal{G}}$ leading to

$$U(x) \simeq \sum_{i=1}^{N_J} {}^lG^{\rho_J}(\Delta;|x-y_i^{N_J}|)w_i^{N_J}F(y_i^{N_J}), \tag{4.279}$$

where $w_i^{N_J}, y_i^{N_J}, i = 1,\ldots,N_J$, are the known weights and knots, respectively.

For the numerical realization of mass density modeling by means of Haar kernels, we assume that all coefficients $a_i^{N_J} = w_i^{N_J}F(y_i^{N_J})$, $i = 1,\ldots,N_J$, are unknown. Then we have to solve a linear system, namely

$$U(x_k^{M_J}) = \sum_{i=1}^{N_J} {}^lG^{\rho_J}(\Delta;|x_k^{M_J} - y_i^{N_J}|)a_i^{N_J}, \quad k = 1,\ldots,M_J, \tag{4.280}$$

to determine $a_i^{N_J}, i = 1, \ldots, N_J$, from known gravitational values $U(x_k^{M_J})$ at knots $x_k^{M_J} \in \overline{\mathcal{G}}$, $k = 1, \ldots, M_J$.

Remark 4.63. The linear system (4.280) can be efficiently and economically attacked by, e.g., use of domain decomposition techniques in connection with fast multipole methods (see, e.g., W. Freeden, O. Glockner, M. Schreiner [1999], K. Hesse [2002], M. Gutting [2007, 2012] and the references therein).

Once all density values $F(y_i^{N_J})$, $i = 1, \ldots, N_J$, are available (note that the integration weights $w_i^{N_J}, i = 1, \ldots, N_J$, are known), the density distribution F can be obtained from the formula

$$\frac{\alpha(x)}{4\pi} F(x) \simeq I_{\rho_J}[F](x) = F_{\rho_J}(x) = \sum_{i=1}^{N_J} H_{\rho_J}(|x - y_i^{N_J}|) w_i^{N_J} F(y_i^{N_J}), \quad x \in \overline{\mathcal{G}}. \tag{4.281}$$

In addition, fully discrete Haar filtered versions of F at lower scales can be derived in accordance with the approximate integration rules

$$\int_{\mathcal{G}} H_{\rho_j}(|y - z|) F(z)\, dV(z) \simeq \sum_{i=1}^{N_j} H_{\rho_j}(|y - y_i^{N_j}|) w_i^{N_j} F(y_i^{N_j}) \tag{4.282}$$

for $j = J_0, \ldots, J$, where $w_i^{N_j}$, $y_i^{N_j}$, $i = 1, \ldots, N_j$, are known weights and knots, respectively, such that $\{y_1^{N_j}, \ldots, y_{N_j}^{N_j}\} \subset \{y_1^{N_J}, \ldots, y_{N_J}^{N_J}\} \subset \overline{\mathcal{G}}$, i.e., the sequence of knots $\{y_1^{N_j}, \ldots, y_{N_j}^{N_j}\} \subset \overline{\mathcal{G}}$ shows a hierarchical positioning.

Altogether, our approach yields Haar filtered versions (4.282) establishing a fully discrete (space-based) multiscale decomposition $F_{\rho_J}, \ldots, F_{\rho_{J_0}}$ of the density distribution F, such that an entire set of approximations is available from a single locally supported mother function, i.e., the Haar kernel function (4.272), and this set provides useful building block functions, which enable suitable storage and fast decorrelation of density data.

4.4 Exercises

Exercise 4.1. Prove that the scalar function $C : \mathbb{R}^3 \to \mathbb{R}$ given by

$$C(x) \;=\; \frac{1}{2}\left(|\omega|^2 |x|^2 - (\omega \cdot x)^2\right), \qquad x \in \mathbb{R}^3, \tag{4.283}$$

with $\omega \in \mathbb{R}^3$ fixed, is the centrifugal potential of $c : \mathbb{R}^3 \to \mathbb{R}^3$, i.e., ∇C is the centrifugal force c given by

$$c(x) \;=\; -\,\omega \wedge (\omega \wedge x), \qquad x \in \mathbb{R}^3. \tag{4.284}$$

Exercise 4.2. Let $\mathcal{G} \subset \mathbb{R}^3$ be a regular region. Show that

$$\mathrm{C}^{(1)}(\partial\mathcal{G}) \subset \mathrm{C}^{(0,\mu)}(\partial\mathcal{G}) \subset \mathrm{C}^{(0,\mu')}(\partial\mathcal{G}) \subset \mathrm{C}^{(0)}(\partial\mathcal{G}) \tag{4.285}$$

is valid for all μ', μ with $0 < \mu' \leq \mu \leq 1$.

Exercise 4.3. Calculate the Newton potential

$$\int_{\mathcal{B}_R(0)} F(y) G(\Delta; |x-y|)\, dV(y) \tag{4.286}$$

for the density distribution

$$F(y) = \begin{cases} 2, & 0 \leq |y| \leq \frac{R}{2}, \\ 1, & \frac{R}{2} \leq |y| \leq R, \end{cases} \tag{4.287}$$

in $x = (0,0,2R)^T$.

Exercise 4.4. Smoothing is a method for the determination of a spline in $\mathrm{H}(\mathcal{A}^c)$ such that the quantity

$$\rho_{\beta_1,\ldots,\beta_n,\delta}(F) = \sum_{i=1}^{N} \left(\frac{\mathcal{D}_{x_i}[F] - \alpha_i}{\beta_i} \right)^2 + \delta(F,F)_{\mathcal{H}(\mathcal{A}^c)} \tag{4.288}$$

is minimal in the reproducing kernel Hilbert space $(\mathrm{H}(\mathcal{A}^c, (\cdot,\cdot)_{\mathrm{H}(\mathcal{A}^c)}))$, where $\beta_1^2, \ldots, \beta_N^2$ are positive weights adapted to the standard deviation of the measured values $\alpha_1, \ldots, \alpha_N$. Prove the following results:

(a) There exists a unique spline function $S \in \mathrm{Spline}_{\mathrm{H}(\mathcal{A}^c)}(\mathcal{D}_{x_1}, \ldots, \mathcal{D}_{x_N})$ such that the inequality

$$\rho_{\beta_1,\ldots,\beta_n,\delta}(S) \leq \rho_{\beta_1,\ldots,\beta_n,\delta}(F) \tag{4.289}$$

holds for all $F \in \mathrm{H}(\mathcal{A}^c)$.

(b) If S is expressed in the form

$$S(x) = \sum_{i=1}^{N} a_i \mathcal{D}_{x_i}[K_{\mathrm{H}(\mathcal{A}^c)}(x,\cdot)], \quad x \in \mathcal{A}^c, \tag{4.290}$$

then it is uniquely determined by the equations

$$\mathcal{D}_{x_i}[S] + \delta\beta_i^2 a_i = \alpha_i, \quad i = 1, \ldots, N. \tag{4.291}$$

Exercise 4.5. Consider an approximation of $\mathcal{D}_x$, $x \in \partial\mathcal{G}$, by a linear combination $I = \sum_{i=1}^{N} a_i \mathcal{D}_{x_i}$, $x_i \in \partial\mathcal{G}$. Show that, for all $F \in \mathcal{H}(\mathcal{A}^c)$, the following a priori estimate holds true:

$$\begin{aligned} &\left| \mathcal{D}_x[F] - \sum_{i=1}^{N} a_i \mathcal{D}_{x_i}[F] \right| \\ &\leq \sqrt{\left(\mathcal{D}_x\mathcal{D}_x - 2D_x \sum_{i=1}^{N} a_i \mathcal{D}_{x_i} + \sum_{i=1}^{N}\sum_{j=1}^{N} a_i a_j \mathcal{D}_{x_i}\mathcal{D}_{x_j} \right) [K_{\mathcal{H}(\mathcal{A}^c)}(\cdot,\cdot)]} \sqrt{(F,F)_{\mathcal{H}(\mathcal{A}^c)}}. \end{aligned} \tag{4.292}$$

Exercise 4.6. Let $\mathcal{X}_N = \{x_1, \ldots, x_N\} \subset \partial\mathcal{G}$ be $\mathcal{D}$-unisolvent. Prove the following equivalences:

(a) Let the function $F \in \mathrm{H}(\mathcal{A}^c)$ be given, and let S_N^F denote the unique spline of class $\mathrm{Spline}_{\mathrm{H}(\mathcal{A}^c)}(\mathcal{D}_{x_i}, \ldots, \mathcal{D}_{x_N})$ satisfying $\mathcal{D}_{x_i}(F) = \mathcal{D}_{x_i}(S_N^F)$, $i = 1, \ldots, N$. Then

$$\mathcal{J} = \sum_{i=1}^{N} a_i \mathcal{D}_{x_i} \tag{4.293}$$

is uniquely determined by the property that $\mathcal{J}[F] = \mathcal{D}_x[S_N^F]$, $x \in \partial\mathcal{G}$, for every $F \in \mathrm{H}(\mathcal{A}^c)$,

(b) $\mathcal{J}$ given by (4.293) is uniquely determined by the property $\mathcal{D}_x[S] = \mathcal{J}[S]$ whenever $S \in \mathrm{Spline}_{\mathrm{H}(\mathcal{A}^c)}(\mathcal{D}_{x_1}, \ldots, \mathcal{D}_{x_N})$.

Exercise 4.7. Let $\{D_n\}_{n=0,1,\ldots}$ be an $\mathrm{L}^2(\partial\mathcal{G})$-Dirichlet Runge basis.

(a) Derive (from the structure of the Gram–Schmidt orthonormalizing process) that the associated $\mathrm{L}^2(\partial\mathcal{G})$-orthonormalized system $\{D_n^*\}_{n=0,1,\ldots}$ is related to $\{D_n\}_{n=0,1,\ldots}$ by the following linear equations

$$D_n^* = \sum_{j=0}^{N} a_{n,j} D_j, \tag{4.294}$$

where the matrix

$$\mathbf{a} = (a_{n,j})_{n,j=0,\ldots,N} \tag{4.295}$$

is a lower triangular matrix.

(b) Show that, for $F \in \mathrm{L}^2(\partial\mathcal{G})$,

$$\sum_{n=0}^{N} (F, D_n^*)_{\mathrm{L}^2(\partial\mathcal{G})} D_n^* \tag{4.296}$$

can be equivalently represented by

$$\sum_{n=0}^{N} a_n D_n, \tag{4.297}$$

where the coefficients $a = (a_0, \ldots, a_N)^T$ satisfy the *normal equations* $\mathbf{g}a = b$ given by

$$\mathbf{g} = \begin{pmatrix} (D_0, D_0)_{\mathrm{L}^2(\partial\mathcal{G})} & \cdots & (D_0, D_N)_{\mathrm{L}^2(\partial\mathcal{G})} \\ \vdots & \ddots & \vdots \\ (D_N, D_0)_{\mathrm{L}^2(\partial\mathcal{G})} & \cdots & (D_N, D_N)_{\mathrm{L}^2(\partial\mathcal{G})} \end{pmatrix} \tag{4.298}$$

and

$$b = \begin{pmatrix} (F, D_0)_{\mathrm{L}^2(\partial\mathcal{G})} \\ \vdots \\ (F, D_N)_{\mathrm{L}^2(\partial\mathcal{G})} \end{pmatrix}. \tag{4.299}$$

(c) Show that

$$\mathbf{g} = \mathbf{a}^{-1}(\mathbf{a}^{-1})^T. \tag{4.300}$$

Exercise 4.8.

(a) Verify the Meissl scheme as illustrated in Figure 4.8.

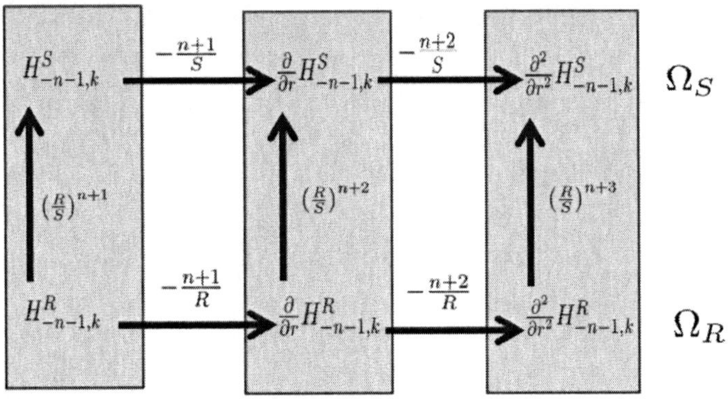

FIGURE 4.8
The classical Meissl scheme involving upward continuation.

(b) Verify the inverse Meissl scheme as illustrated in Figure 4.9.

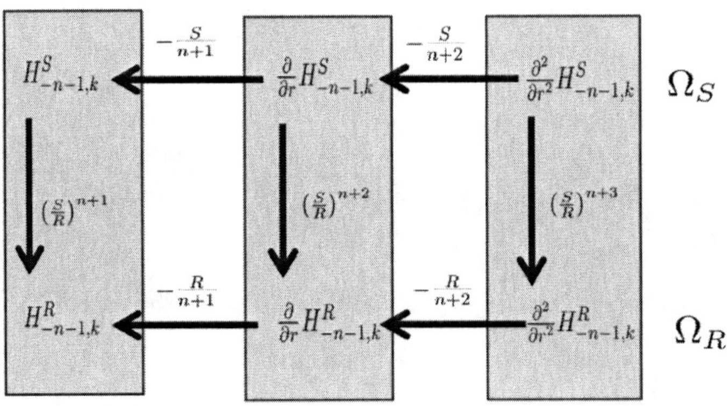

FIGURE 4.9
The inverse Meissl scheme involving downward continuation.

Exercise 4.9. The radial SST integral operator $T_{SST} : \mathrm{L}^2(\Omega_R) \to \mathrm{L}^2(\Omega_S)$ is given by

$$T_{SST}[F](x) = \int_{\Omega_R} K_{SST}(x,y)F(y)\, d\omega(y), \quad x \in \Omega_S, \tag{4.301}$$

where $R < S$, $F \in \mathrm{L}^2(\Omega_R)$, and the SST kernel $K_{SST}(\cdot,\cdot)$ reads as follows

$$K_{SST}(S\xi, R\eta) = -\frac{\partial}{\partial r}\left(\frac{1}{4\pi R} \frac{r^2 - R^2}{(r^2 + R^2 - 2Rr(\xi\cdot\eta))^{\frac{3}{2}}} \right)\Bigg|_{r=S}. \tag{4.302}$$

Show that

$$T_{SST}[F](S\xi) = \frac{1}{S}\sum_{n=0}^{\infty}\sum_{k=1}^{2n+1}\left(\frac{R}{S}\right)^n \frac{n+1}{S} F^{\wedge_{\mathrm{L}^2(\Omega_R)}}(n,k)\; Y_{n,k}(\xi),\ \xi \in \Omega. \tag{4.303}$$

From functional analysis it follows that $T_{SST} : \mathrm{L}^2(\Omega_R) \to \mathrm{L}^2(\Omega_S)$ is a compact operator with range $\mathrm{im}(T_{SST})$ such that

$$\overline{\mathrm{im}(T_{SST})}^{\|\cdot\|_{\mathrm{L}^2(\Omega_S)}} = \mathrm{L}^2(\Omega_S). \tag{4.304}$$

The triple $\{\sigma_n, Y^R_{n,k}, Y^S_{n,k}\}_{n\in\mathbb{N}_0, k=1,\ldots,2n+1}$ satisfying

$$\sigma_n = \frac{n+1}{S}\left(\frac{R}{S}\right)^n, \tag{4.305}$$

$$T_{SST}[Y^R_{n,k}] = \sigma_n Y^S_{n,k} = \frac{\sigma_n}{S} Y_{n,k}, \tag{4.306}$$

$$T^*_{SST}[Y^S_{n,k}] = \sigma_n Y^R_{n,k} = \frac{\sigma_n}{R} Y_{n,k} \tag{4.307}$$

is called the *singular system* for the integral operator T_{SST} (see page 273, Section 5.5, for a more general setting).

Exercise 4.10. The radial SGG integral operator $T_{SGG} : \mathrm{L}^2(\Omega_R) \to \mathrm{L}^2(\Omega_S)$ is given by

$$T_{SGG}[F](x) = \int_{\Omega_R} K_{SGG}(x,y)F(y)\, d\omega(y), \quad x \in \Omega_S, \tag{4.308}$$

where $R < S$, $F \in \mathrm{L}^2(\Omega_R)$, and the SGG-kernel $K_{SGG}(\cdot,\cdot)$ reads as follows

$$K_{SGG}(S\xi, R\eta) = \frac{\partial^2}{\partial r^2}\left(\frac{1}{4\pi R} \frac{r^2 - R^2}{(r^2 + R^2 - 2Rr(\xi\cdot\eta))^{\frac{3}{2}}} \right)\Bigg|_{r=S}. \tag{4.309}$$

Verify that $\{\sigma_n, Y^R_{n,k}, Y^S_{n,k}\}_{n\in\mathbb{N}_0, k=1,\ldots,2n+1}$, with

$$\sigma_n = \frac{(n+1)(n+2)}{S^2}\left(\frac{R}{S}\right)^n, \tag{4.310}$$

forms the singular system for the integral operator T_{SGG}.

Exercise 4.11. Let T be one of the integral operators (4.301), (4.308). Consider the integral equation

$$T[F] = G, \ F \in \mathrm{L}^2(\Omega_R), \ G \in \mathrm{L}^2(\Omega_S), \tag{4.311}$$

with an (undisturbed) right-hand side $G \in \mathrm{im}(T)$ of the form

$$G = \sum_{n=0}^{\infty} \sum_{k=1}^{2n+1} G^{\wedge_{\mathrm{L}^2(\Omega_S)}}(n,k) Y_{n,k}^S, \tag{4.312}$$

where $Y_{n,k}^S = \frac{1}{S} Y_{n,k}$ and the orthogonal coefficients $G^{\wedge_{\mathrm{L}^2(\Omega_S)}}(n,k)$ are given by

$$G^{\wedge_{\mathrm{L}^2(\Omega_S)}}(n,k) = \int_{\Omega_S} G(y) Y_{n,k}^S(y) \, d\omega(y), \ n \in \mathbb{N}_0, \ k = 1, \ldots, 2n+1. \tag{4.313}$$

Suppose that $\{F_J(\sigma_n)\}_{J\in\mathbb{Z}, n\in\mathbb{N}_0} \subset (0,\infty)$, is a sequence of positive, real numbers that satisfy

(i) $\sup_{n\in\mathbb{N}_0} |F_J(\sigma_n)| < \infty$, for every $J \in \mathbb{Z}$,

(ii) $\lim_{J\to\infty} F_J(\sigma_n) = \sigma_n^{-1}$, for every $n \in \mathbb{N}_0$,

(iii) $\sup_{n\in\mathbb{N}_0, J\in\mathbb{Z}} |F_J(\sigma_n)\sigma_n| \leq C$, for some fixed constant $C > 0$.

In the (inverse problem) context, the notation F_J typically describes a *filter*. Furthermore, let $\{R_J\}_{J\in\mathbb{Z}}$ be a family of operators $R_J : \mathrm{L}^2(\Omega_R) \to \mathrm{L}^2(\Omega_S)$ with

$$R_J[Y_{n,k}^S] = F_J(\sigma_n) \, Y_{n,k}^R, \quad J \in \mathbb{Z}, \ n \in \mathbb{N}_0, \ k = 1, \ldots, 2n+1. \tag{4.314}$$

Prove that the family $\{R_J\}_{J\in\mathbb{Z}}$ is a regularization of T^{-1} in the sense that each $R_J : \mathrm{L}^2(\Omega_S) \to \mathrm{L}^2(\Omega_R)$ is a bounded operator and that

$$\lim_{J\to\infty} \|R_J[G] - F\|_{\mathrm{L}^2(\Omega_R)} = \lim_{J\to\infty} \left\|R_J[G] - T^{-1}[G]\right\|_{\mathrm{L}^2(\Omega_R)} = 0, \quad G \in \mathrm{im}(T). \tag{4.315}$$

More details concerning the regularization technique in our examples of gravitational field determination from satellite data can be found in W. Freeden [1999], W. Freeden, V. Michel [2004].

In W. Freeden, F. Schneider [1998] the compact operator equation is dealt with using a two step (decomposition and reconstruction) regularization method, based on filtering techniques of T^*T. The general setup of this regularization method in form of a multiresolution analysis using reconstruction and decomposition scaling functions also applies to the inversion of the spherical Biot–Savart operator for ionospheric current field determination in geomagnetics (see Section 5.5).

5

Geomagnetism

CONTENTS

5.1 Geomagnetic Background

We start by giving a brief overview on the Earth's magnetic field, also called the *geomagnetic field*. It characterizes the magnetic field generated by all sources inside and outside the solid Earth up to the magnetopause. The magnetopause forms the transition layer between the geomagnetic field and the interplanetary magnetic field (IMF) originating from solar processes. Restricting ourselves to the Earth's magnetic field, we are led to a subdivision into the following major source regions (see Figure 5.1):

Core: Convection in the Earth's liquid outer core drives dynamo processes that generate by far the largest part of the geomagnetic field (with a field strength varying between 30,000 nT and 60,000 nT at the Earth's surface). Thus, the core field is sometimes also called the *main field*. It has a dominating dipole component and is of rather large scale, concerning its spatial as well as temporal variation.

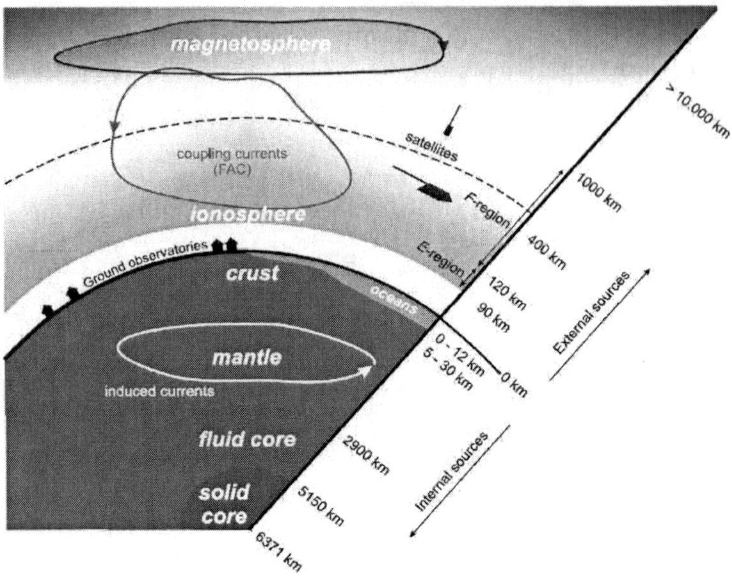

FIGURE 5.1
Schematic description of the contributions to the Earth's magnetic field. (Courtesy of N. Olsen et al. [2010b])

Crust/Lithosphere: Ferromagnetic minerals in the Earth's crust and mantle produce a further contribution to the magnetic field. Magnetization of such minerals can only take place at temperatures below the specific Curie temperature, so that the crustal field has its sources in depths no more than a few tens of kilometers below the Earth's surface. The magnetization can be remanent (i.e., it has taken place in the past) or induced by an ambient magnetic field (in first place, the core field). The magnetic signature of such magnetized minerals reveals a strong spatial variation and can reach from a few nanoteslas to more than 1,000 nT (locally). Together with the core field, the crustal field forms the so-called *internal field.*

Ionosphere: The ionosphere denotes approximately the region between 50 km and 1,000 km above the Earth's surface, where solar heating leads to higher conductivity on the dayside than on the nightside and drives different electric current systems. An example is the eastward directed equatorial electrojet (EEJ), which is due to an enhanced conductivity along the dip equator, while an enhanced conductivity in the polar regions drives the auroral electrojet (AEJ). These current systems produce additional magnetic fields contributing to the geomagnetic field. A further permanently present magnetic field is the so-called solar-quiet (Sq) variation of about 20–50 nT (which can be significantly stronger during magnetically disturbed times).

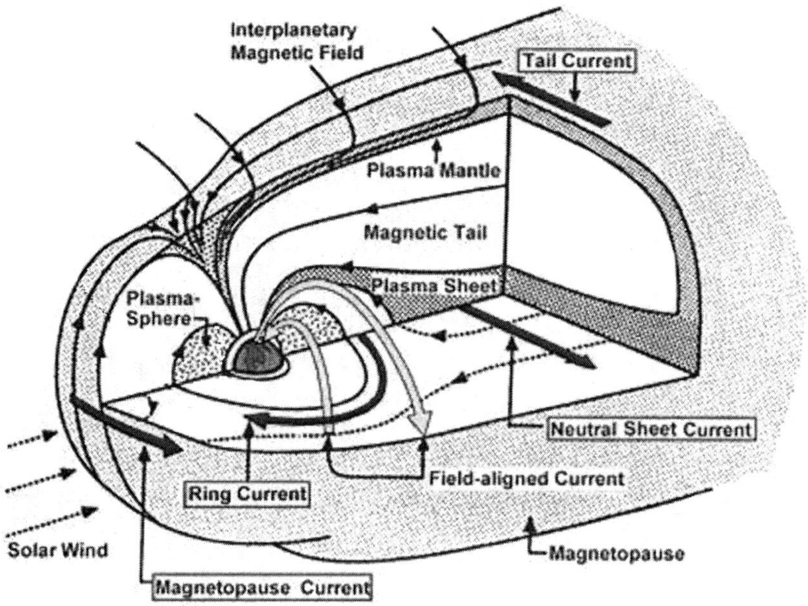

FIGURE 5.2
Illustration of current systems in the magnetosphere and their coupling with the ionosphere.

The ionospheric field in general shows strong spatial as well as temporal variations.

Magnetosphere: The magnetosphere extends beyond the ionosphere up to the magnetopause. Due to deformation by the solar wind, it has an extension of a few tens of Earth's radii on the dayside, but reaches up to several hundred Earth's radii on the nightside. Major current systems are the magnetopause current, the magnetotail current, and the equatorial ring current. Their magnetic effect is spatially of rather large scale but can vary significantly in time. A coupling with the ionosphere is established by field-aligned currents (FACs) that flow along the core field lines and can be detected at satellite altitude as non-potential magnetic fields. Together with the ionospheric contribution, the magnetospheric magnetic field forms the *external field* (see Figure 5.2).

The description above gives a first impression of the complexity of the Earth's magnetic field. The crustal field has hardly any noticeable time dependence, and the core field only shows long-term variations in time. The secular variation, i.e., the time change of the core field, is of the order of 50 nT per year. Occasional changes at shorter timescales with an increased rate of secular variation are so-called geomagnetic jerks. At larger timescales of several thousand or million years, a complete reversal of the direction of the main field becomes possible. Paleomagnetic records show that this has happened many times during the Earth's life span. Changes of the external field take place on significantly smaller timescales. There are the daily variations due to the different solar influence on the day- and nightside, variations due to changes in the interplanetary magnetic field, as well as disturbances by magnetic storms or substorms due to increased solar activity. The initial phase of a magnetic storm usually lasts only a few minutes, while the main phase can extend up to several hours, and the subsequent recovery phase up to a few days. Amplitudes of such storms are generally about a few tens of nanoteslas but can reach more than 1,000 nT. In the year 1989, a particularly strong magnetic storm caused the shutdown of parts of the Canadian electrical power grid.

Concerning the spatial extension, the core field and magnetospheric field change at significantly larger scales (of several thousands of kilometers) than the ionospheric and the crustal field. Spatial scales for the crustal field variability are often below 500 km. A typical and well-known example is the Bangui anomaly in central Africa. Actually, the crustal field can reveal variations on severely smaller spatial scales. Concentrations of magnetic ores can produce magnetic fields of up to several thousand nanoteslas in an area of tens of meters. The detection of such local variations, however, requires ground or aeromagnetic measurements.

With all these contributions, it is a tremendous effort to achieve a good description of the geomagnetic field from the available data (satellite, aeromagnetic and ground measurements). The different spatial and temporal scales make it difficult to find appropriate modeling approaches. Yet, these differences allow the separation of the different contributions at least to a certain degree. For those interested in the internal field, e.g., the selection of satellite measurements taken at nighttime and during magnetically quiet periods minimizes the disturbances by the external field. A study by F.J. Lowes [1974] of the power spectrum of the internal contributions shows a significant change around spherical harmonic degree 15, which is interpreted as the transition from the large-scale core field-dominated part to the small-scale crustal field-dominated part of the magnetic field. This is still considered to be the (empirical) criterion concerning where to truncate models of the main field. Also the choice of the coordinate system can improve the understanding of the different contributions. While the internal field is modeled in an Earth-fixed coordinate system using geographic longitude and colatitude, the external field is usually modeled in a sun-fixed coordinate system, e.g., using local time and dipole latitude (thus, taking into account the dominating solar influence for the external field).

The introduction presented here gives only a very brief insight into the Earth's magnetic field. Recent extensive treatments are gathered in D. Gubbins, E. Herrero-Bervera [2007] and M. Kono [2009]. A comprehensive description with focus on the lithospheric field and satellite data processing is given in R.A. Langel, W.J. Hinze [1998]. Some shorter overviews can be found, e.g., in G. Hulot et al. [2010] and N. Olsen et al. [2010b].

In this book (more precisely, in Chapters 5 and 8), we focus on mathematical tools in geomagnetic modeling that have a strong relation to potential theoretic concepts. In particular, we study boundary-value problems for the magnetic field in source-free as well as source regions, and we derive various representations (Fourier expansions and integral representations with explicitly known kernels) for different contributions to the magnetic field and corresponding current systems in the ionosphere. For the integral representations, we introduce spatial regularization techniques and indicate how to obtain multiscale approximations for spatially localizing features or irregularly distributed data. Applications are presented for crustal field modeling and the reconstruction of ionospheric current systems from satellite measurements.

5.1.1 Maxwell's Equations

The fundamental relations of any electrodynamic process are described by *Maxwell's equations*:

$$\nabla \wedge e = -\frac{\partial}{\partial t} b, \tag{5.1}$$

$$\nabla \cdot d = \rho, \tag{5.2}$$

$$\nabla \wedge h = j + \frac{\partial}{\partial t} d, \tag{5.3}$$

$$\nabla \cdot b = 0, \tag{5.4}$$

with b denoting the magnetic field, e the electric field, j the density of free currents and ρ the density of free charges. The magnetic displacement h, and the electric displacement d are given by

$$d = \varepsilon_0 e + p, \tag{5.5}$$

$$h = \frac{1}{\mu_0} b - m, \tag{5.6}$$

with m being the magnetization, p the polarization, μ_0 the vacuum permeability, and ε_0 the vacuum permittivity.

For today's satellite data (at each time step the satellite can measure the magnetic field only at its current position), this set of equations is too detailed. To circumvent this problem, we observe that many of the geomagnetic phenomena we are interested in change at length scales L of about 100–1000 km and timescales T of hours to days, or even longer (e.g., for the core and crustal field), such that the quotient L/T nearly vanishes in comparison to the speed of light. Under this assumption, the displacement current $\frac{\partial}{\partial t} d$ can be neglected in (5.3). Furthermore, the magnetization m vanishes in the Earth's atmosphere, so that Equations (5.3), (5.4), and (5.6) reduce to the so-called *pre-Maxwell equations*

$$\nabla \wedge b = \mu_0 j, \tag{5.7}$$
$$\nabla \cdot b = 0. \tag{5.8}$$

They form the foundation for our further modeling approaches (for the sake of simplicity, we usually set the vacuum permeability μ_0 to be equal to one for the theoretical considerations, but for numerical examples we use its exact value). In the neutral atmosphere near the Earth's surface, one can additionally assume that the current density j vanishes, which is a necessary condition for the Gauss Representation as discussed in Section 5.3. At the altitude of low Earth orbiting satellites (approximately 300–600 km), this does not hold true, such that the magnetic field cannot be assumed to be a potential field in these regions. A more detailed description of our previous considerations can be found, e.g., in G.E. Backus et al. [1996].

An important integral representation of the magnetic field b in dependence on the current density j (i.e., a solution of the pre-Maxwell equations (5.7), (5.8)) is given by the *Law of Biot–Savart*. This relation is made mathematically more precise by use of the Helmholtz decomposition in the upcoming Section 5.2.

5.2 Mie and Helmholtz Decompositions

The Mie and the Helmholtz decompositions are two decompositions of particular importance in electromagnetism. It is the aim of this section to explain some of their basic properties.

5.2.1 Helmholtz Decomposition

We begin with the Helmholtz decomposition, splitting a vector field f into a divergence-free contribution $\nabla \wedge v$ and a curl-free contribution ∇U, such that

$$f = \nabla U + \nabla \wedge v. \tag{5.9}$$

This representation can be found in most textbooks related to electromagnetism (e.g., G.E. Backus et al. [1996] and R.J. Blakely [1996]). Mathematically, there exist formulations for several function spaces and underlying domains. In our approach, we formulate the Helmholtz decomposition under the classical condition of continuously differentiable functions on regular regions or the entire space $\mathbb{R}^3$. Nevertheless, it should be pointed out that the assumptions can be weakened in various ways (see, e.g., Y.F. Gui, W.B. Dou [2007] and W. Sprössig [2010] for brief overviews).

Theorem 5.1 (Helmholtz Decomposition).

(a) *Let f be of class $\mathrm{c}^{(1)}(\mathbb{R}^3)$ satisfying $|f(x)| = \mathcal{O}(|x|^{-(2+\varepsilon)})$, $\varepsilon > 0$, for $|x| \to \infty$. Then there exist functions v of class $\mathrm{c}^{(1)}(\mathbb{R}^3)$ and U of class $\mathrm{C}^{(1)}(\mathbb{R}^3)$ such that*

$$f(x) = \nabla U(x) + \nabla \wedge v(x), \quad x \in \mathbb{R}^3. \tag{5.10}$$

If additionally $|\nabla \otimes f(x)| = \mathcal{O}(|x|^{-(2+\varepsilon)})$, $\varepsilon > 0$, for $|x| \to \infty$, then the functions U, v can be represented by

$$\begin{aligned} U(x) &= -\frac{1}{4\pi}\int_{\mathbb{R}^3} \frac{\nabla_y \cdot f(y)}{|x-y|} dV(y) \qquad (5.11)\\ &= -\frac{1}{4\pi}\int_{\mathbb{R}^3} f(y) \cdot \frac{x-y}{|x-y|^3} dV(y), \\ v(x) &= \frac{1}{4\pi}\int_{\mathbb{R}^3} \frac{\nabla_y \wedge f(y)}{|x-y|} dV(y) \qquad (5.12)\\ &= \frac{1}{4\pi}\int_{\mathbb{R}^3} f(y) \wedge \frac{x-y}{|x-y|^3} dV(y), \end{aligned}$$

for $x \in \mathbb{R}^3$.

(b) *Let $\mathcal{G} \subset \mathbb{R}^3$ be a regular region, and assume that f is of class $\mathrm{c}^{(1)}(\mathcal{G})$. Then there exist functions v of class $\mathrm{c}^{(1)}(\mathcal{G})$ and U of class $\mathrm{C}^{(1)}(\mathcal{G})$ such that*

$$f(x) = \nabla U(x) + \nabla \wedge v(x), \quad x \in \mathcal{G}. \tag{5.13}$$

If f is of class $\mathrm{c}^{(1)}(\overline{\mathcal{G}})$, then a representation for U, v is given by

$$\begin{aligned} U(x) &= -\frac{1}{4\pi}\int_{\mathcal{G}} \frac{\nabla_y \cdot f(y)}{|x-y|} dV(y) + \int_{\partial\mathcal{G}} \frac{\nu(y)\cdot f(y)}{|x-y|} d\omega(y) \qquad (5.14)\\ &= -\frac{1}{4\pi}\int_{\mathcal{G}} f(y) \cdot \frac{x-y}{|x-y|^3} dV(y), \\ v(x) &= \frac{1}{4\pi}\int_{\mathcal{G}} \frac{\nabla_y \wedge f(y)}{|x-y|} dV(y) - \int_{\partial\mathcal{G}} \frac{\nu(y)\wedge f(y)}{|x-y|} d\omega(y) \qquad (5.15)\\ &= \frac{1}{4\pi}\int_{\mathcal{G}} f(y) \wedge \frac{x-y}{|x-y|^3} dV(y), \end{aligned}$$

for $x \in \mathcal{G}$.

Proof. The proofs of (a) and (b) can be essentially based on the same arguments. We only treat part (a) in more detail. Similar to Theorem 3.30, it follows that a function $g \in \mathrm{c}^{(2)}(\mathbb{R}^3)$ exists that satisfies

$$\Delta g(x) = f(x), \quad x \in \mathbb{R}^3. \tag{5.16}$$

The assumption $|f(x)| = \mathcal{O}(|x|^{-(2+\varepsilon)})$, $|x| \to \infty$, is necessary to guarantee the existence of g as a Newton potential, i.e.,

$$g(x) = -\frac{1}{4\pi} \int_{\mathbb{R}^3} \frac{f(y)}{|x-y|} dV(y), \quad x \in \mathbb{R}^3. \tag{5.17}$$

Thus, (5.17) represents a solution to (5.16). Observing Exercise 1.2, this implies

$$f(x) = \Delta g(x) = \nabla(\nabla \cdot g(x)) - \nabla \wedge (\nabla \wedge g(x)), \quad x \in \mathbb{R}^3, \tag{5.18}$$

such that $U = \nabla \cdot g$ and $v = -\nabla \wedge g$ satisfy the desired properties. The integral representations of U, v immediately follow from (5.17) in connection with Green's formulas from Subsection 1.2.3. The additional decay condition on $\nabla \otimes f$ guarantees the existence of the integral expressions for U, v. □

Remark 5.2. In general, the functions U, v are not uniquely determined since any function $\tilde{v} = v + \nabla V$, with V twice continuously differentiable, admits a representation

$$f = \nabla U + \nabla \wedge \tilde{v} \tag{5.19}$$

as well. Not even the quantities ∇U and $\nabla \wedge v$ are uniquely determined: For any harmonic function W, there exists a vector field w such that $\nabla W = -\nabla \wedge w$. Consequently,

$$f = \nabla \tilde{U} + \nabla \wedge \tilde{v}, \tag{5.20}$$

for $\tilde{U} = U + W$ and $\tilde{v} = v + w$.

Uniqueness can be achieved by adequate decay conditions at infinity or boundary conditions on $\partial\mathcal{G}$. A specific formulation is given in the following lemma.

Lemma 5.3 (Uniqueness of the Helmholtz Decomposition).

(a) Let f be of class $\mathrm{c}^{(1)}(\mathbb{R}^3)$ with $|f(x)| = \mathcal{O}(|x|^{-(2+\varepsilon)})$, $\varepsilon > 0$, for $|x| \to \infty$. Then the following statements hold true:

(α) The quantities ∇U and $\nabla \wedge v$ occurring in Theorem 5.1 are uniquely determined if U is regular at infinity.

(β) The functions U, v occurring in Theorem 5.1 are uniquely determined if U is regular at infinity and, additionally,

$$\nabla \cdot v(x) = 0, \quad x \in \mathbb{R}^3, \tag{5.21}$$

with $|v(x)| = \mathcal{O}(|x|^{-2})$, for $|x| \to \infty$.

(b) *Let $\mathcal{G} \subset \mathbb{R}^3$ be a regular region, and assume that f is of class $\mathrm{c}^{(1)}(\overline{\mathcal{G}})$. Then the following statements are valid:*

 (α) *The quantities ∇U and $\nabla \wedge v$ occurring in Theorem 5.1 are uniquely determined if $U(x) = 0$, $x \in \partial\mathcal{G}$.*

 (β) *The functions U, v from Theorem 5.1 are uniquely determined if $U(x) = 0$, $x \in \partial\mathcal{G}$, and, additionally,*

$$\nabla \cdot v(x) = 0, \quad x \in \mathcal{G}, \tag{5.22}$$

 with $\nu(x) \cdot v(x) = 0$, $x \in \partial\mathcal{G}$.

Proof. We start with part (a). It is clear that U satisfies

$$\nabla \cdot f(x) = \Delta U(x), \quad x \in \mathbb{R}^3. \tag{5.23}$$

From the regularity of U at infinity, we know that the solution of the Poisson equation (5.23) is unique. Thus, ∇U and $\nabla \wedge v = f - \nabla U$ are uniquely determined, and (α) is proven. Now, we assume that a second vector field $\tilde{v}$ exists such that $\nabla \wedge v(x) = \nabla \wedge \tilde{v}(x) = f - \nabla U(x)$ and $\nabla \cdot v(x) = \nabla \cdot \tilde{v}(x) = 0$, $x \in \mathbb{R}^3$. For the difference $w = v - \tilde{v}$ it follows that

$$\begin{aligned} \nabla \wedge w(x) &= 0, \quad x \in \mathbb{R}^3, && (5.24)\\ \nabla \cdot w(x) &= 0, \quad x \in \mathbb{R}^3. && (5.25) \end{aligned}$$

The assumptions $|v(x)| = \mathcal{O}(|x|^{-2})$ and $|\tilde{v}(x)| = \mathcal{O}(|x|^{-2})$, $|x| \to \infty$, imply $|w(x)| = \mathcal{O}(|x|^{-2})$, $|x| \to \infty$. Thus, Lemma 5.5 yields $w(x) = 0$, $x \in \mathbb{R}^3$. This is the desired result for (β).

Taking a look at part (b), we are led to

$$\begin{aligned} \Delta U(x) &= \nabla \cdot f(x), \quad x \in \mathcal{G}, && (5.26)\\ U(x) &= 0, \quad x \in \partial\mathcal{G}. && (5.27) \end{aligned}$$

Since U is determined uniquely by the Dirichlet problem (5.26), (5.27), this implies the assertion of (α). Analogous arguments as in part (a) result in the problem

$$\begin{aligned} \nabla \wedge w(x) &= 0, \quad x \in \mathcal{G}, && (5.28)\\ \nabla \cdot w(x) &= 0, \quad x \in \mathcal{G}, && (5.29)\\ \nu(x) \cdot w(x) &= 0, \quad x \in \partial\mathcal{G}. && (5.30) \end{aligned}$$

This means that we have verified the assertion of (β) because Lemma 5.5 again provides $w(x) = 0$, $x \in \mathcal{G}$. □

Remark 5.4. The special choice of boundary values $U(x) = 0$, $\nu(x) \cdot v(x) = 0$, $x \in \partial\mathcal{G}$, as given in part (b) of Theorem 5.3 has been made for convenience only. Any other choice of sufficiently smooth boundary functions F_U, F_v, such that $U(x) = F_U(x)$ and $\nu(x) \cdot v(x) = F_v(x)$, $x \in \partial\mathcal{G}$, leads to uniqueness as well.

Concerning a solution of the pre-Maxwell equations, we are not so much interested in determining unique functions U, v for the Helmholtz decomposition, but rather in uniquely specifying the magnetic field by its curl and its divergence.

Lemma 5.5. *The following assertions on the uniqueness of the pre-Maxwell equations hold true:*

(a) Let f be of class $c^{(1)}(\mathbb{R}^3)$ with $|f(x)| = \mathcal{O}(|x|^{-2})$, $|x| \to \infty$. If f additionally satisfies

$$\nabla \wedge f(x) = 0, \quad x \in \mathbb{R}^3, \tag{5.31}$$

$$\nabla \cdot f(x) = 0, \quad x \in \mathbb{R}^3, \tag{5.32}$$

then $f(x) = 0$, $x \in \mathbb{R}^3$.

(b) Let $\mathcal{G} \subset \mathbb{R}^3$ be a regular region, and suppose that f is a vector field of class $c^{(1)}(\mathcal{G}) \cap c^{(0)}(\overline{\mathcal{G}})$. If f additionally satisfies

$$\nabla \wedge f(x) = 0, \quad x \in \mathcal{G}, \tag{5.33}$$

$$\nabla \cdot f(x) = 0, \quad x \in \mathcal{G}, \tag{5.34}$$

$$\nu(x) \cdot f(x) = 0, \quad x \in \partial\mathcal{G}, \tag{5.35}$$

then $f(x) = 0$, $x \in \mathcal{G}$.

Proof. We begin with part (a). Let γ_x be a regular curve of finite length connecting the origin with $x \in \mathbb{R}^3$ (by a regular curve, we mean an injective mapping $\gamma_x : [0,1] \to \mathbb{R}^3$ with $\gamma_x(0) = 0$, $\gamma_x(1) = x$, that is at least of class $c^{(2)}([0,1])$ and satisfies $\frac{\partial}{\partial t}\gamma_x(t) \neq 0$ for all $t \in [0,1]$). Due to (5.31), the function

$$U(x) = \int_{\gamma_x} \tau(y) \cdot f(y) d\sigma(y), \quad x \in \mathbb{R}^3, \tag{5.36}$$

where τ denotes the unit tangential vector field of the curve γ_x, is well-defined and path-independent. Furthermore, standard arguments from vector analysis imply

$$\nabla U(x) = f(x), \quad x \in \mathbb{R}^3, \tag{5.37}$$

and, by virtue of (5.32), U is harmonic, i.e., $\Delta U(x) = 0$, $x \in \mathbb{R}^3$. Since any function U satisfying (5.37) is only unique up to an additive constant, the decay condition on f implies that U can be chosen such that $|U(x)| = \mathcal{O}(|x|^{-1})$, $|x| \to \infty$. Observing

$$\begin{aligned} |f(x)|^2 = |\nabla U(x)|^2 &= \nabla \cdot (U(x)\nabla U(x)) - U(x)\Delta U(x) \\ &= \nabla \cdot (U(x) f(x)), \end{aligned} \tag{5.38}$$

for $x \in \mathbb{R}^3$, we obtain from the Theorem of Gauss (cf. Subsection 1.2.3) that

$$\int_{\mathcal{B}_R(0)} |f(x)|^2 dV(x) = \int_{\Omega_R} \nu(x) \cdot f(x) U(x) d\omega(x). \tag{5.39}$$

Since $|U(x)f(x)| = \mathcal{O}(|x|^{-3})$, $|x| \to \infty$, and $\|\Omega_R\| = 4\pi R^2$, Equation (5.39) leads to the estimate

$$\int_{\mathcal{B}_R(0)} |f(x)|^2 dV(x) = \mathcal{O}\left(\frac{1}{R}\right), \quad R \to \infty. \tag{5.40}$$

Thus, letting $R \to \infty$, we end up with $\int_{\mathbb{R}^3} |f(x)|^2 dV(x) = 0$, or in other words, $f(x) = 0$, $x \in \mathbb{R}^3$.

Analogous arguments hold for part (b): We choose the function U similar as before such that

$$\nabla U(x) = f(x), \quad x \in \mathcal{G}, \tag{5.41}$$

and

$$\int_{\mathcal{G}} |f(x)|^2 dV(x) = \int_{\partial\mathcal{G}} \nu(x) \cdot f(x) U(x) d\omega(x). \tag{5.42}$$

Now it is clear from (5.35) that $\int_{\mathcal{G}} |f(x)|^2 dV(x) = 0$. Thus, $f(x) = 0$, $x \in \mathcal{G}$. □

Finally, the uniqueness known from Lemma 5.5 and the representations from Theorem 5.1 lead us to the already announced Law of Biot–Savart for a magnetic field b and a current density j.

Theorem 5.6 (Law of Biot–Savart). *Let b be of class $\mathrm{c}^{(2)}(\mathbb{R}^3)$ with $|b(x)| = \mathcal{O}(|x|^{-2})$, $|x| \to \infty$. Furthermore, suppose that j is of class $\mathrm{c}^{(2)}(\mathbb{R}^3)$ with $|j(x)| = \mathcal{O}(|x|^{-(2+\varepsilon)})$ and $|\nabla \otimes j(x)| = \mathcal{O}(|x|^{-(2+\varepsilon)})$, $\varepsilon > 0$, for $|x| \to \infty$. If the pre-Maxwell equations*

$$\nabla \wedge b(x) = j(x), \quad x \in \mathbb{R}^3, \tag{5.43}$$

$$\nabla \cdot b(x) = 0, \quad x \in \mathbb{R}^3, \tag{5.44}$$

are satisfied, then b admits an integral representation of the form

$$b(x) = \frac{1}{4\pi} \int_{\mathbb{R}^3} j(y) \wedge \frac{x-y}{|x-y|^3} dV(y), \quad x \in \mathbb{R}^3. \tag{5.45}$$

A local version of the Law of Biot–Savart on regular regions $\mathcal{G} \subset \mathbb{R}^3$ can be formulated as well and is part of the exercises. In this case, adequate boundary values on $\partial\mathcal{G}$ are required to obtain uniqueness of b.

Remark 5.7. The decay conditions mentioned in Theorem 5.6 as well as in parts (a) of Theorem 5.1 and Lemma 5.3 and 5.5 are rather strict. It is a well-known fact (see, e.g., O. Blumenthal [1905] for the Helmholtz decomposition) that the uniform decay to zero at infinity is sufficient for many assertions. This observation is made more precise in Exercise 5.2. In our approach, we have chosen the stricter conditions to guarantee the existence of the integral representations. Furthermore, most geoscientifically relevant magnetic fields satisfy these decay conditions (e.g., a dipole field behaves like $|x|^{-3}$ at infinity), and the current systems j are typically restricted to bounded regions like the ionosphere.

5.2.2 Mie Decomposition

Next, we turn to the second important decomposition in geomagnetic modeling, namely, the Mie decomposition.

Definition 5.8. Let $\mathcal{G} \subset \mathbb{R}^3$ be a region. A vector field f of class $\mathrm{c}^{(1)}(\mathcal{G})$ is said to be *solenoidal* if

$$\int_\Sigma \nu(y) \cdot f(y) d\omega(y) = 0 \tag{5.46}$$

for every closed, locally $\mathrm{C}^{(2)}$-smooth surface $\Sigma \subset \mathcal{G}$. It is called *toroidal* if a scalar field Q of class $\mathrm{C}^{(1)}(\mathcal{G})$ exists such that

$$f(x) = \mathrm{L}Q(x), \quad x \in \mathcal{G}. \tag{5.47}$$

If a scalar field P of class $\mathrm{C}^{(2)}(\mathcal{G})$ exists such that

$$f(x) = \nabla \wedge \mathrm{L}Q(x), \quad x \in \mathcal{G}, \tag{5.48}$$

then f is called *poloidal.* The fields P and Q are also known as *Mie scalars.*

Remark 5.9. As a consequence of the Theorem of Gauss, any solenoidal vector field is divergence-free (i.e., $\nabla \cdot f = 0$). Concerning the entire space $\mathbb{R}^3$, the converse holds true as well. Thus, functions satisfying the pre-Maxwell equations everywhere are solenoidal. In subregions $\mathcal{G} \subset \mathbb{R}^3$, however, a divergence-free function is not necessarily solenoidal (a counterexample is given in Exercise 5.3).

Theorem 5.10 (Mie Decomposition)**.** *Let $f : \mathcal{B}_{R_0,R_1}(0) \to \mathbb{R}^3$ be a solenoidal vector field. Then there exist scalar fields P, Q of class $\mathrm{C}^{(1)}(\mathcal{B}_{R_0,R_1}(0))$ such that*

$$f(x) = \nabla \wedge LP(x) + LQ(x), \quad x \in \mathcal{B}_{R_0,R_1}(0). \tag{5.49}$$

P, Q are uniquely determined by the additional conditions

$$\frac{1}{4\pi r^2} \int_{\Omega_r} P(y) d\omega(y) = \frac{1}{4\pi r^2} \int_{\Omega_r} Q(y) d\omega(y) = 0, \tag{5.50}$$

for every $r \in (R_0, R_1)$.

The proof requires techniques that are intrinsic to the sphere, as treated in the third part of this book. Thus, it is deferred to Section 8.1, where the Mie decomposition is discussed from a spherical perspective. We see that the Mie scalars P, Q are only determined up to a constant (due to the assumption of vanishing integral mean values). However, the poloidal part $p = \nabla \wedge \mathrm{L}P$ and the toroidal part $q = \mathrm{L}Q$ are determined uniquely without further assumptions.

One of the most significant properties of the Mie decomposition is its capability to reduce the vectorial pre-Maxwell equations (5.7), (5.8) to a set of simple scalar equations. More precisely, let P_b, Q_b be the Mie scalars of the magnetic field b, and P_j, Q_j the Mie scalars of the current density j. Then

$$\begin{aligned} \nabla \wedge b &= \nabla \wedge (\nabla \wedge \mathrm{L}P_b + \mathrm{L}Q_b) \\ &= \nabla(\nabla \cdot \mathrm{L}P_b) - \Delta \mathrm{L}P_b + \nabla \wedge \mathrm{L}Q_b \\ &= -\mathrm{L}\Delta P_b + \nabla \wedge \mathrm{L}Q_b, \end{aligned} \tag{5.51}$$

where we have used $\nabla \cdot \mathrm{L} = 0$ and $\Delta \mathrm{L} = \mathrm{L}\Delta$ in the last step. Since $\nabla \wedge b = j$ and $j = \nabla \wedge \mathrm{L}P_j + \mathrm{L}Q_j$, the uniqueness of the Mie scalars (under the assumption of vanishing integral mean values) yields

$$Q_b = P_j, \tag{5.52}$$

$$\Delta P_b = -Q_j. \tag{5.53}$$

Thus, the original pre-Maxwell equations have been reduced to scalar equations involving the Laplace operator, which can be handled by potential-theoretic tools as introduced in Chapter 3 (note that the vacuum permeability μ_0 is chosen to be equal to 1).

Remark 5.11. Equations (5.52) and (5.53) show that toroidal magnetic fields are solely generated by poloidal currents, and poloidal magnetic fields are solely generated by toroidal currents.

Because of its significance in geomagnetism, the previous result is summarized in the next lemma.

Lemma 5.12. *Let $b \in \mathrm{c}^{(2)}(\mathbb{R}^3)$ and $j \in \mathrm{c}^{(1)}(\mathbb{R}^3)$ satisfy the pre-Maxwell equations*

$$\nabla \wedge b(x) = j(x), \quad x \in \mathbb{R}^3, \tag{5.54}$$

$$\nabla \cdot b(x) = 0, \quad x \in \mathbb{R}^3. \tag{5.55}$$

Furthermore, let P_b, Q_b, P_j, Q_j be the uniquely determined Mie scalars of b and j, respectively, as introduced in Theorem 5.10. Then

$$Q_b(x) = P_j(x), \quad x \in \mathbb{R}^3, \tag{5.56}$$

$$\Delta P_b(x) = -Q_j(x), \quad x \in \mathbb{R}^3. \tag{5.57}$$

We conclude this section with some properties of poloidal and toroidal fields. The simple proofs are left to the reader (however, note that they require the spherical Helmholtz decomposition and further tools from Section 8.1).

Lemma 5.13. *Let $f : \mathcal{B}_{R_0,R_1}(0) \to \mathbb{R}^3$, $0 \leq R_0 < R_1$, be a sufficiently often differentiable vector field. Then the following statements are valid:*

(a) f is toroidal if and only if f is solenoidal and tangential.

(b) f is poloidal if and only if f is solenoidal and $\nabla \wedge f$ is tangential.

(c) If f is toroidal, then $\nabla \wedge f$ is poloidal. And vice versa, if f is poloidal, then $\nabla \wedge f$ is toroidal.

5.3 Gauss Representation and Uniqueness

In source-free regions with vanishing current density j, we have

$$\nabla \wedge b = 0, \tag{5.58}$$

$$\nabla \cdot b = 0, \tag{5.59}$$

and the magnetic field b can be represented as a potential field, i.e.,

$$b = \nabla U, \tag{5.60}$$

where U is a harmonic function (cf. Subsection 5.2.1). For the geomagnetic field, the assumption of a vanishing current density holds true only in the neutral atmosphere, i.e., the region between the Earth's surface and the ionosphere. In spherical approximation of the Earth, this amounts to the validity of (5.60) in the spherical shell $\mathcal{B}_{R_0,R_1}(0) = \{x \in \mathbb{R}^3 : R_0 < |x| < R_1\}$, with R_0 typically denoting the mean Earth radius and R_1 the radius up to the lower bound of the ionosphere. Since U is harmonic, we can use methods derived in Chapter 3 for establishing an approximation of the magnetic field.

5.3.1 Uniqueness from Vectorial and Radial Boundary Data

Different from gravitation, we are not dealing with an exterior boundary problem but a combination of an interior (with respect to $\mathcal{B}_{R_1}(0)$) and an exterior (with respect to $\mathbb{R}^3 \setminus \overline{\mathcal{B}_{R_0}(0)}$) problem. Thus, the potential U has to be expanded simultaneously in terms of inner and outer harmonics, and boundary information is required on the sphere Ω_{R_0} as well as on the sphere Ω_{R_1}. Since $f = \nabla U$ is the observed quantity (in many cases, we more generally write f instead of b since the upcoming considerations are not restricted to magnetic fields but hold for general potential fields), Neumann boundary values $\frac{\partial}{\partial\nu} U = \nu \cdot \nabla U = \nu \cdot f$ can be assumed to be known. Therefore, we have to solve the Neumann boundary-value problem

$$\Delta U(x) = 0, \quad x \in \mathcal{B}_{R_0,R_1}(0), \tag{5.61}$$

$$\frac{\partial}{\partial\nu} U(x) = \nu \cdot f(x), \quad x \in \Omega_{R_0} \cup \Omega_{R_1}. \tag{5.62}$$

The solution U is unique up to a constant, so that $f = \nabla U$ is uniquely determined.

Theorem 5.14. *Let f be of class $\mathrm{c}^{(1)}(\overline{\mathcal{B}_{R_0,R_1}(0)})$, $0 < R_0 < R_1$, satisfying*

$$\nabla \wedge f(x) = 0, \quad x \in \mathcal{B}_{R_0,R_1}(0), \tag{5.63}$$

$$\nabla \cdot f(x) = 0, \quad x \in \mathcal{B}_{R_0,R_1}(0). \tag{5.64}$$

Then there exists a harmonic function $U : \mathcal{B}_{R_0,R_1}(0) \to \mathbb{R}$ such that

$$f(x) = \nabla U(x), \quad x \in \mathcal{B}_{R_0,R_1}(0), \tag{5.65}$$

and

$$U(x) = \sum_{n=0}^{\infty} \sum_{k=1}^{2n+1} \alpha_{n,k} H_{n,k}^{R_1}(x) + \beta_{n,k} H_{-n-1,k}^{R_0}(x), \quad x \in \mathcal{B}_{R_0,R_1}(0). \tag{5.66}$$

The coefficients $\alpha_{n,k}$, $\beta_{n,k}$ are uniquely determined by the linear equations

$$\frac{n}{R_1}\alpha_{n,k} - \frac{n+1}{R_1}\left(\frac{R_0}{R_1}\right)^n \beta_{n,k} = \left(\nu \cdot f, Y_{n,k}^{R_1}\right)_{L^2(\Omega_{R_1})}, \tag{5.67}$$

$$\frac{n}{R_1}\left(\frac{R_0}{R_1}\right)^{n-1} \alpha_{n,k} - \frac{n+1}{R_0}\beta_{n,k} = \left(\nu \cdot f, Y_{n,k}^{R_0}\right)_{L^2(\Omega_{R_0})}, \tag{5.68}$$

for $n \in \mathbb{N}$, $k = 1, \ldots, 2n+1$. The coefficient $\beta_{0,1}$ is given by

$$\beta_{0,1} = -R_1 \left(\nu \cdot f, Y_{0,1}^{R_1}\right)_{L^2(\Omega_{R_1})}, \tag{5.69}$$

while $\alpha_{0,1} \in \mathbb{R}$ can be chosen arbitrarily.

Proof. From the Helmholtz decomposition, the existence of a harmonic function U with $f(x) = \nabla U(x)$, $x \in \mathcal{B}_{R_0,R_1}(0)$, is clear. A representation of U as stated in (5.66) follows from considerations in Chapter 3. It remains to investigate the boundary conditions. The regularity of f on the boundary implies that $\nu \cdot f$ can be expanded uniformly on Ω_{R_0} and Ω_{R_1} in terms of spherical harmonics. Representation (5.66) leads to

$$\begin{aligned}
\left(\nu \cdot f, Y_{n,k}^{R_1}\right)_{L^2(\Omega_{R_1})} &= \left(\frac{\partial}{\partial \nu} U, Y_{n,k}^{R_1}\right)_{L^2(\Omega_{R_1})} \qquad (5.70)\\
&= \sum_{m=0}^{\infty} \sum_{j=1}^{2n+1} \left(m\alpha_{m,j} \frac{|\cdot|^{m-1}}{{R_1}^m} Y_{m,j}^{R_1} - (m+1)\beta_{m,j} \frac{{R_0}^{m+1}}{|\cdot|^{m+2}} Y_{m,j}^{R_0}, Y_{n,k}^{R_1}\right)_{L^2(\Omega_{R_1})}\\
&= \alpha_{n,k}\frac{n}{R_1}\left(Y_{n,k}^{R_1}, Y_{n,k}^{R_1}\right)_{L^2(\Omega_{R_1})} - \beta_{n,k}\frac{n+1}{R_1}\left(\frac{R_0}{R_1}\right)^n \left(Y_{n,k}^{R_1}, Y_{n,k}^{R_1}\right)_{L^2(\Omega_{R_1})}\\
&= \alpha_{n,k}\frac{n}{R_1} - \beta_{n,k}\frac{n+1}{R_1}\left(\frac{R_0}{R_1}\right)^n,
\end{aligned}$$

for $n \in \mathbb{N}$, $k = 1, \ldots, 2n+1$. Analogous calculations for $(\nu \cdot f, Y^{R_0}_{n,k})_{L^2(\Omega_{R_0})}$ yield the second set of equations stated in (5.68). Setting $n = 0$, (5.67) and (5.68) turn into

$$-\frac{1}{R_1}\beta_{0,1} = \left(\nu \cdot f, Y^{R_1}_{0,1}\right)_{L^2(\Omega_{R_1})}, \tag{5.71}$$

$$-\frac{1}{R_0}\beta_{0,1} = \left(\nu \cdot f, Y^{R_0}_{0,1}\right)_{L^2(\Omega_{R_0})}. \tag{5.72}$$

Since all terms involving $\alpha_{0,1}$ vanish, this coefficient can be chosen arbitrarily. Observing that f is divergence-free, an application of the Theorem of Gauss yields

$$R_1\left(\nu \cdot f, Y^{R_1}_{0,1}\right)_{L^2(\Omega_{R_1})} = \int_{\Omega_{R_1}} \nu(y) \cdot f(y) Y_{0,1}\left(\frac{y}{|y|}\right) d\omega(y) \tag{5.73}$$

$$= \int_{\mathcal{B}_{R_0,R_1}(0)} \nabla_y \cdot \left(f(y) Y_{0,1}\left(\frac{y}{|y|}\right)\right) dV(y) + \int_{\Omega_{R_0}} \nu(y) \cdot f(y) Y_{0,1}\left(\frac{y}{|y|}\right) d\omega(y)$$

$$= \int_{\Omega_{R_0}} \nu(y) \cdot f(y) Y_{0,1}\left(\frac{y}{|y|}\right) d\omega(y)$$

$$= R_0\left(\nu \cdot f, Y^{R_0}_{0,1}\right)_{L^2(\Omega_{R_0})}.$$

Thus, (5.71) and (5.72) generate the same coefficient $\beta_{0,1}$ (in the previous equations one has to be aware of the orientation of the normal ν. In accordance with the typical geomagnetic convention, against our standard nomenclature that assumes the normal to be directed into the exterior of $\mathcal{B}_{R_0,R_1}(0)$, we have chosen the unit normal field on Ω_{R_0} and Ω_{R_1} to point into the exterior spaces $\mathbb{R}^3 \setminus \overline{\mathcal{B}_{R_0}(0)}$ and $\mathbb{R}^3 \setminus \overline{\mathcal{B}_{R_1}(0)}$, respectively). □

Remark 5.15. The geomagnetic field b is divergence-free everywhere in $\mathbb{R}^3$ and not only in the spherical shell $\mathcal{B}_{R_0,R_1}(0)$. Thus, b is solenoidal and the coefficient $\beta_{0,1}$ as defined in (5.69) is zero. In other words, this fact states the non-existence of magnetic monopoles. The coefficient $\alpha_{0,1}$ is typically set to zero by choice.

The above representation requires the knowledge of the normal component $\nu \cdot f$ on both spheres Ω_{R_0} and Ω_{R_1}. Since modern magnetic field measurements generally supply the entire vectorial function f and not just its radial contribution, one may ask the question if it suffices to know f only on one sphere $\Omega_R \subset \mathcal{B}_{R_0,R_1}(0)$. Indeed, the answer is positive.

Theorem 5.16. *Let f be of class $\mathrm{c}^{(1)}(\overline{\mathcal{B}_{R_0,R_1}(0)})$, $0 < R_0 < R_1$, satisfying*

$$\nabla \wedge f(x) = 0, \quad x \in \mathcal{B}_{R_0,R_1}(0), \tag{5.74}$$

$$\nabla \cdot f(x) = 0, \quad x \in \mathcal{B}_{R_0,R_1}(0). \tag{5.75}$$

Then there exists a harmonic function $U : \mathcal{B}_{R_0,R_1}(0) \to \mathbb{R}$ *such that*

$$f(x) = \nabla U(x), \quad x \in \mathcal{B}_{R_0,R_1}(0), \tag{5.76}$$

and

$$U(x) = \sum_{n=0}^{\infty} \sum_{k=1}^{2n+1} \alpha_{n,k} H_{n,k}^{R_1}(x) + \beta_{n,k} H_{-n-1,k}^{R_0}(x), \quad x \in \mathcal{B}_{R_0,R_1}(0). \tag{5.77}$$

Let $R \in (R_0, R_1)$ *be fixed. Then the coefficients* $\alpha_{n,k}$, $\beta_{n,k}$ *are uniquely determined by the linear equations*

$$\frac{n}{R_1}\left(\frac{R}{R_1}\right)^n \alpha_{n,k} - \frac{n+1}{R_0}\left(\frac{R_0}{R}\right)^{n+1} \beta_{n,k} = \left(\nu \cdot f, Y_{n,k}^R\right)_{L^2(\Omega_R)}, \tag{5.78}$$

$$\frac{1}{R_1}\left(\frac{R}{R_1}\right)^n \alpha_{n,k} + \frac{1}{R_0}\left(\frac{R_0}{R}\right)^{n+1} \beta_{n,k} = \frac{1}{\sqrt{n(n+1)}}\left(f, y_{n,k}^{(2),R}\right)_{l^2(\Omega_R)} \tag{5.79}$$

for $n \in \mathbb{N}$, $k = 1, \ldots, 2n+1$. *The coefficient* $\beta_{0,1}$ *is given by*

$$\beta_{0,1} = -R\left(\nu \cdot f, Y_{0,1}^R\right)_{L^2(\Omega_R)}, \tag{5.80}$$

while $\alpha_{0,1} \in \mathbb{R}$ *can be chosen arbitrarily.*

Proof. The existence of a harmonic function U with $f(x) = \nabla U(x)$, $x \in \mathcal{B}_{R_0,R_1}(0)$, and its representation (5.77) follow by the same arguments as in Theorem 5.14. It remains to calculate the scalars $\alpha_{n,k}$ and $\beta_{n,k}$. Equations (5.78) and (5.80) can be obtained in the same manner as in Theorem 5.14 by use of the normal component $\nu \cdot f$. The equations for the second boundary surface are now substituted by equations for the tangential part $f_{\tan}$. More precisely, $\nabla_x = \xi \frac{\partial}{\partial r} + \frac{1}{r}\nabla_\xi^*$, with $r = |x|$ and $\xi = \frac{x}{|x|}$, implies $f_{\tan}(x) = \frac{1}{r}\nabla_\xi^* U(r\xi)$. Thus,

$$\begin{aligned}
\left(f, y_{n,k}^{(2),R}\right)_{l^2(\Omega_R)} &= \left(f_{\tan}, y_{n,k}^{(2),R}\right)_{l^2(\Omega_R)} = \frac{1}{R}\left(\nabla^* U, y_{n,k}^{(2),R}\right)_{l^2(\Omega_R)} \qquad (5.81)\\
&= \frac{1}{R}\sum_{m=1}^{\infty}\sum_{j=1}^{2m+1}\left(\alpha_{m,j}\frac{|\cdot|^m}{R_1^{m+1}}\nabla^* Y_{m,j} + \beta_{m,j}\frac{R_0^m}{|\cdot|^{m+1}}\nabla^* Y_{m,j}, y_{n,k}^{(2),R}\right)_{l^2(\Omega_R)}\\
&= \alpha_{n,k}\frac{\sqrt{n(n+1)}R^n}{R_1^{n+1}}\left(y_{n,k}^{(2),R}, y_{n,k}^{(2),R}\right)_{l^2(\Omega_R)}\\
&\quad +\beta_{n,k}\frac{\sqrt{n(n+1)}R_0^n}{R^{n+1}}\left(y_{n,k}^{(2),R}, y_{n,k}^{(2),R}\right)_{l^2(\Omega_R)}\\
&= \frac{\sqrt{n(n+1)}}{R_1}\left(\frac{R}{R_1}\right)^n \alpha_{n,k} + \frac{\sqrt{n(n+1)}}{R_0}\left(\frac{R_0}{R}\right)^{n+1}\beta_{n,k},
\end{aligned}$$

for $n \in \mathbb{N}$, $k = 1, \ldots, 2n+1$. This completes the proof. □

Remark 5.17. In geomagnetism, the representation derived in Theorem 5.16 is known as the *Gauss Representation* because it was first used by Gauss in connection with the Earth's magnetic field (cf. C.F. Gauss [1838]). In spite of the small amount of magnetic observatories at that time, Gauss was able to approximate the coefficients $\alpha_{n,k}$, $\beta_{n,k}$ up to the spherical harmonic degree $n = 4$.

Remark 5.18. An expansion of the potential U in terms of (scalar) inner and outer harmonics implies an expansion of the vector field f in terms of vector spherical harmonics. Applying the gradient $\nabla_x = \xi\frac{\partial}{\partial r} + \frac{1}{r}\nabla^*_\xi$ to the occurring (scalar) inner and outer harmonics leads to vector spherical harmonics of type 1 and 2, i.e., $y^{(1)}_{n,k}$, $y^{(2)}_{n,k}$ or $\tilde{y}^{(1)}_{n,k}$, $\tilde{y}^{(2)}_{n,k}$, respectively. The first set of vector spherical harmonics mixes up the contributions of the inner and outer harmonics, while due to Lemma 2.54, the second set maintains the separation (this is the reason why it is of interest in Sections 5.4 and 8.3 for the separation of the magnetic field with respect to its sources). The use of $\tilde{y}^{(1)}_{n,k}$, $\tilde{y}^{(2)}_{n,k}$ also simplifies the calculation of the coefficients $\alpha_{n,k}$, $\beta_{n,k}$. Alternatively to (5.78) and (5.79), we obtain

$$\alpha_{n,k} = \frac{1}{R_1\sqrt{n(2n+1)}}\left(\frac{R_1}{R}\right)^n \left(f, \tilde{y}^{(2);R}_{n,k}\right)_{l^2(\Omega_R)}, \tag{5.82}$$

$$\beta_{n,k} = \frac{1}{R_0\sqrt{(n+1)(2n+1)}}\left(\frac{R}{R_0}\right)^{n+1} \left(f, \tilde{y}^{(1);R}_{n,k}\right)_{l^2(\Omega_R)}. \tag{5.83}$$

A similar representation in the more general framework of the Mie Representation, as presented in Section 5.4, can be found Exercise 5.5.

5.3.2 Uniqueness from Intensity Data

Up to now, we have assumed to know the magnetic field b or its normal component $\nu\cdot b$ on a sphere Ω_R, which is a valid assumption since most modern instruments provide this vectorial quantity. However, satellite missions prior to MAGSAT (1979/80) could only measure the magnetic field intensity $|b|$. Today, still the intensity is measurable with a higher accuracy than the actual vectorial magnetic field b. Thus, it is worth asking if intensity measurements on a sphere Ω_R are sufficient to reconstruct the magnetic field in a source-free region. In general, the answer is negative, as can be seen in the following example (for simplicity, we restrict our upcoming investigation to magnetic fields that are source-free in the exterior of a sphere, i.e., in a set $\mathbb{R}^3 \setminus \overline{\mathcal{B}_R(0)}$).

Example 5.19. We begin with a vector field f_0 that consists of a finite linear combination of vector outer harmonics of type 1 (cf. Section 2.5):

$$f_0(x) = \sum_{n=1}^{N}\sum_{k=1}^{2n+1} \alpha_{n,k} h^{R;(1)}_{-n-1,k}. \tag{5.84}$$

Clearly, this vector field satisfies

$$\begin{aligned} \nabla \wedge f_0(x) &= 0, \quad x \in \mathbb{R}^3 \setminus \overline{\mathcal{B}_R(0)}, & (5.85) \\ \nabla \cdot f_0(x) &= 0, \quad x \in \mathbb{R}^3 \setminus \overline{\mathcal{B}_R(0)}, & (5.86) \end{aligned}$$

i.e., f_0 is source-free in $\mathbb{R}^3 \setminus \overline{\mathcal{B}_R(0)}$. Our goal is to find another non-zero vector field f_1 that obeys the corresponding pre-Maxwell equations (5.85), (5.85), and that additionally satisfies

$$f_0(x) \cdot f_1(x) = 0, \quad x \in \Omega_R. \tag{5.87}$$

Under these assumptions, we can introduce

$$f = f_1 + f_0, \qquad \tilde{f} = f_1 - f_0. \tag{5.88}$$

These functions again obey (5.85), (5.86) and have (due to the orthogonality condition (5.87)) the same intensities on the boundary Ω_R, i.e.,

$$|f(x)| = |\tilde{f}(x)|, \quad x \in \Omega_R. \tag{5.89}$$

By construction, both cases $f = \tilde{f}$ and $f = -\tilde{f}$ are excluded. Thus, we have found an example that illustrates the non-uniqueness from intensity-only measurements.

It only remains to construct a function f_1 with the desired properties. Starting with an axial dipole field $f_0 = h^{R;(1)}_{-2,2}$, this has first been worked out in G.E. Backus [1970] (thus, this non-uniqueness phenomenon in geomagnetic field models is called the *Backus effect*). For the series

$$f_1(x) = \sum_{n=1}^{\infty} \sum_{k=1}^{2n+1} \beta_{n,k} h^{R;(1)}_{-n-1,k}, \tag{5.90}$$

a recursion formula can be derived that determines the coefficients $\beta_{n,k}$ (this is left as a task to the interested reader). More recently, P. Alberto et al. [2004] presented a general construction principle that can also deal with initial functions f_0 forming multipole fields of higher order.

However, under certain additional assumptions, uniqueness (up to the sign) of the magnetic field in a source-free region $\mathbb{R}^3 \setminus \mathcal{B}_R(0)$ under prescribed intensity data on the boundary Ω_R can be realized. One such condition is that the magnetic field has a finite expansion in terms of vector outer harmonics (see, e.g., G.E. Backus [1968]). As a consequence, we directly see that the two functions f_0 and f_1 occurring in Example 5.19 cannot both have finite expansions if they form a counterexample for uniqueness.

Theorem 5.20. *Let a function $f \in \mathrm{c}^{(1)}(\mathbb{R}^3 \setminus \mathcal{B}_R(0))$, $R > 0$, satisfy $|f(x)| = \mathcal{O}(|x|^{-(2+\varepsilon)})$, $\varepsilon > 0$, for $|x| \to \infty$, and*

$$\begin{aligned} \nabla \wedge f(x) &= 0, \quad x \in \mathbb{R}^3 \setminus \overline{\mathcal{B}_R(0)}, & (5.91) \\ \nabla \cdot f(x) &= 0, \quad x \in \mathbb{R}^3 \setminus \overline{\mathcal{B}_R(0)}. & (5.92) \end{aligned}$$

We assume that the intensity $|f|$ is known on Ω_R and that f can be expressed by a finite linear combination of vector outer harmonics (of type 1), i.e.,

$$f = \sum_{n=0}^{N} \sum_{k=1}^{2n+1} \alpha_{n,k} h^{(1);R}_{-n-1,k}, \tag{5.93}$$

for some $N \in \mathbb{N}_0$ and coefficients $\alpha_{n,k} \in \mathbb{R}$. Then f is uniquely determined (up to the sign).

Proof. We suppose that there exist two functions $f, \tilde{f}$ of class $c^{(1)}(\mathbb{R}^3 \setminus \mathcal{B}_R(0))$ that satisfy the conditions of Theorem 5.20, and that have the same intensities on Ω_R, i.e.,

$$|f(x)| = |\tilde{f}(x)|, \quad x \in \Omega_R. \tag{5.94}$$

We know from (5.91), (5.92), and the Helmholtz decomposition that harmonic functions $U, \tilde{U} \in C^{(2)}(\mathbb{R}^3 \setminus \overline{\mathcal{B}_R(0)}) \cap C^{(1)}(\mathbb{R}^3 \setminus \mathcal{B}_R(0))$ exist such that

$$f(x) = \nabla U(x), \quad \tilde{f}(x) = \nabla \tilde{U}(x), \quad x \in \mathbb{R}^3 \setminus \overline{\mathcal{B}_R(0)}, \tag{5.95}$$

where U, $\tilde{U}$ are regular at infinity in the sense of Definition 3.23. Since $f, \tilde{f}$ are assumed to have a finite expansion in terms of vector outer harmonics, we find that

$$U = \sum_{n=0}^{N} \sum_{k=1}^{2n+1} \alpha_{n,k} H^{R}_{-n-1,k}, \tag{5.96}$$

$$\tilde{U} = \sum_{n=0}^{\tilde{N}} \sum_{k=1}^{2n+1} \tilde{\alpha}_{n,k} H^{R}_{-n-1,k}, \tag{5.97}$$

for $N, \tilde{N} \in \mathbb{N}_0$ and coefficients $\alpha_{n,k}, \tilde{\alpha}_{n,k} \in \mathbb{R}$. Identity (5.94) implies that the functions $V = U + \tilde{U}$ and $W = U - \tilde{U}$ satisfy

$$\nabla V(x) \cdot \nabla W(x) = |f(x)|^2 - |\tilde{f}(x)|^2 = 0, \quad x \in \Omega_R. \tag{5.98}$$

Applying the Kelvin transform to V and W, we obtain

$$\breve{V} = \mathcal{K}^R[V] = \sum_{n=0}^{L} \sum_{k=1}^{2n+1} \beta_{n,k} H^{R}_{n,k}, \tag{5.99}$$

$$\breve{W} = \mathcal{K}^R[W] = \sum_{n=0}^{M} \sum_{k=1}^{2n+1} \gamma_{n,k} H^{R}_{n,k}, \tag{5.100}$$

for degrees $L, M \in \mathbb{N}_0$ and coefficients $\beta_{n,k}, \gamma_{n,k} \in \mathbb{R}$ depending on the representations (5.96) and (5.97). In other words, $\breve{V}$ and $\breve{W}$ are harmonic polynomials defined on $\mathbb{R}^3$. For these Kelvin transformed functions, the relation (5.98) becomes

$$\breve{V}(x)\breve{W}(x) + r\frac{\partial}{\partial r}\left(\breve{V}(x)\breve{W}(x)\right) + \nabla\breve{V}(x) \cdot \nabla\breve{W}(x) = 0, \; x \in \Omega_R, \tag{5.101}$$

where $x = r\xi$, with $r = |x|$, $\xi = \frac{x}{|x|}$. Identity (5.101) can also be expressed as

$$(L + M + 1)H_L(x)H_M(x) + R_{L+M-1}(x) = 0, \quad x \in \Omega_R, \tag{5.102}$$

where $H_L = \sum_{k=1}^{2L+1} \beta_{L,k} H_{L,k}^R$ and $H_M = \sum_{k=1}^{2M+1} \gamma_{M,k} H_{M,k}^R$ are homogeneous, harmonic polynomials of degree L and M, respectively, and R_{L+M-1} is the remaining polynomial of degree $L + M - 1$. Obviously, the product $H_L H_M$ forms a homogeneous (not necessarily harmonic) polynomial of degree $L+M$. A well-known decomposition states that any homogeneous polynomial K_n of degree n can be represented in the form

$$K_n(x) = \sum_{i=0}^{\lfloor \frac{n}{2} \rfloor} |x|^{2i} \tilde{H}_{n-2i}(x), \quad x \in \mathbb{R}^3, \tag{5.103}$$

where $\tilde{H}_{n-2i}$ are uniquely determined homogeneous, harmonic polynomials of degree $n-2i$ (see, e.g., W. Freeden, T. Gervens, M. Schreiner [1998] and earlier references therein). Thus, we can find homogeneous, harmonic polynomials H_{L+M} of degree $L + M$ and $\tilde{H}_{L+M-2i}$ of degree $L + M - 2i$ such that

$$H_L(x)H_M(x) = H_{L+M}(x) + \sum_{i=1}^{\lfloor \frac{L+M}{2} \rfloor} |x|^{2i} \tilde{H}_{L+M-2i}(x), \quad x \in \mathbb{R}^3. \tag{5.104}$$

Observing a similar decomposition for the remainder polynomial R_{L+M-1}, Equation (5.102) becomes

$$(L + M + 1)H_{L+M}(x) + \sum_{i=0}^{L+M-1} F_i(|x|^2)\tilde{\tilde{H}}_i(x) = 0, \quad x \in \Omega_R, \tag{5.105}$$

where $\tilde{\tilde{H}}_i(x)$ is a homogeneous, harmonic polynomials of degree i and F_i is a one-dimensional polynomial (not necessarily of degree i). Due to the orthogonality property of spherical harmonics, it is clear that $(H_{L+M}, \tilde{\tilde{H}}_i)_{\mathrm{L}^2(\Omega_R)} = 0$, $i = 0, \ldots, L + M - 1$. Thus, the identity (5.105) implies $H_{L+M} = 0$ everywhere in $\mathbb{R}^3$. The uniqueness of (5.104), and decompositions of type (5.103) in general, leads us to the conclusion that H_L or H_M vanishes everywhere in $\mathbb{R}^3$ (compare Exercise 5.4). Since H_L and H_M are chosen such that they represent the contributions to $\check{V}$ and $\check{W}$, respectively, with the highest polynomial degree, we find that $\check{V}$ or $\check{W}$ vanishes everywhere in $\mathbb{R}^3$. As a consequence, the original function V or W, respectively, is zero in $\mathbb{R}^3 \setminus \overline{\mathcal{B}_R(0)}$. Finally, we are led to $U = \tilde{U}$ or $U = -\tilde{U}$, i.e.,

$$f(x) = \nabla U(x) = \nabla \tilde{U}(x) = \tilde{f}(x), \tag{5.106}$$

for all $x \in \mathbb{R}^3 \setminus \overline{\mathcal{B}_R(0)}$, or

$$f(x) = \nabla U(x) = -\nabla \tilde{U}(x) = -\tilde{f}(x), \tag{5.107}$$

for all $x \in \mathbb{R}^3 \setminus \overline{\mathcal{B}_R(0)}$, which completes the proof. □

One might argue that Theorem 5.20 is suitable for many models of the Earth's magnetic field. Because the models are often given in terms of Fourier series expansions that are truncated at some maximal spherical harmonic degree, Theorem 5.20 claims uniqueness (up to the sign) of this representation if only intensity measurements are given on the boundary. Nevertheless, the true magnetic field has contributions at all degrees, which can lead to strongly varying magnetic field models if only intensity measurements are used (a practical demonstration is illustrated, e.g., in D.P. Stern [1976]).

The requirement of a finite series expansion can be dropped if one knows the location of the dip equator on Ω_R, i.e., the set of curves where the radial component of b vanishes. This guarantees uniqueness (up to the sign) of the magnetic field as well if only measurements of $|b|$ are given on the boundary Ω_R (cf. A. Khokhlov et al. [1997]).

Theorem 5.21. *Let a function* $f \in \mathrm{c}^{(1)}(\mathbb{R}^3 \setminus \mathcal{B}_R(0))$, $R > 0$, *satisfy* $|f(x)| = \mathcal{O}(|x|^{-(2+\varepsilon)})$, $\varepsilon > 0$, *for* $|x| \to \infty$, *and*

$$\nabla \wedge f(x) = 0, \quad x \in \mathbb{R}^3 \setminus \overline{\mathcal{B}_R(0)}, \tag{5.108}$$

$$\nabla \cdot f(x) = 0, \quad x \in \mathbb{R}^3 \setminus \overline{\mathcal{B}_R(0)}. \tag{5.109}$$

We assume that $|f|$ *and the dip equator* $\mathcal{N}_f = \{x \in \Omega_R : \frac{x}{|x|} \cdot f(x) = 0\}$ *are known on* Ω_R, *that* $\mathcal{N}_f$ *consists of a finite set of closed regular curves, and that* $\frac{x}{|x|} \cdot f(x)$ *changes its sign whenever* x *crosses* $\mathcal{N}_f$. *Then* f *is uniquely determined (up to the sign).*

To prove Theorem 5.21, we need a certain variant of the maximum/minimum principle providing properties for the gradient of a harmonic function in its extremal points. For more details concerning the next lemma, the reader is referred to L. Bers et al. [1964].

Lemma 5.22. *Let* $\mathcal{G} \subset \mathbb{R}^3$ *be a regular region. Suppose that the function* $U \in \mathrm{C}^{(2)}(\mathcal{G}) \cap \mathrm{C}^{(1)}(\overline{\mathcal{G}})$ *is harmonic in* $\mathcal{G}$ *and non-constant. If* U *takes its maximum or minimum at a point* $x_0 \in \partial\mathcal{G}$, *then* $\nabla U(x_0) = \alpha\, \nu(x_0)$ *for some* $\alpha \neq 0$.

Proof. Theorem 5.21. Let there exist two functions f, $\tilde{f}$ of class $\mathrm{c}^{(1)}(\mathbb{R}^3 \setminus \mathcal{B}_R(0))$ that satisfy the conditions of Theorem 5.21. Furthermore, we assume

$$|f(x)| = |\tilde{f}(x)|, \quad x \in \Omega_R, \tag{5.110}$$

as well as

$$\mathcal{N}_f = \mathcal{N}_{\tilde{f}}. \tag{5.111}$$

Just like in the proof of Theorem 5.20, there exist harmonic functions $U, \tilde{U} \in \mathrm{C}^{(2)}(\mathbb{R}^3 \setminus \overline{\mathcal{B}_R(0)}) \cap \mathrm{C}^{(1)}(\mathbb{R}^3 \setminus \mathcal{B}_R(0))$ such that

$$f(x) = \nabla U(x), \quad \tilde{f}(x) = \nabla \tilde{U}(x), \quad x \in \mathbb{R}^3 \setminus \overline{\mathcal{B}_R(0)}. \tag{5.112}$$

Additionally, U and $\tilde{U}$ are regular at infinity. Setting $V = U + \tilde{U}$ and $W = U - \tilde{U}$, we find

$$\nabla V(x) \cdot \nabla W(x) = |f(x)|^2 - |\tilde{f}(x)|^2 = 0, \quad x \in \Omega_R. \tag{5.113}$$

In other words, ∇V is orthogonal to ∇W on Ω_R. Since $\mathcal{N}_f$, $\mathcal{N}_{\tilde{f}}$ consist of finitely many closed regular curves, the sphere Ω_R is divided into regular regions $\Gamma_k \subset \Omega_R$, $k = 1, \ldots, N$, with $\bigcup_{k=1}^N \partial\Gamma_k = \mathcal{N}_f = \mathcal{N}_{\tilde{f}}$. The definition of $\mathcal{N}_f$ and $\mathcal{N}_{\tilde{f}}$ tells us that the radial contributions of f and $\tilde{f}$ (i.e., $\frac{x}{|x|} \cdot f(x)$ and $\frac{x}{|x|} \cdot \tilde{f}(x)$, respectively) do not change their sign on a fixed Γ_k. Furthermore, due to the assumption that the radial contributions change their sign whenever they cross the dip equator $\mathcal{N}_f$, we are in the situation that either

$$\operatorname{sgn}\left(\frac{x}{|x|} \cdot f(x)\right) = \operatorname{sgn}\left(\frac{x}{|x|} \cdot \tilde{f}(x)\right), \tag{5.114}$$

for all $x \in \Omega_R$, or

$$\operatorname{sgn}\left(\frac{x}{|x|} \cdot f(x)\right) = -\operatorname{sgn}\left(\frac{x}{|x|} \cdot \tilde{f}(x)\right), \tag{5.115}$$

for all $x \in \Omega_R$. First, we assume

$$\frac{x}{|x|} \cdot f(x) > 0, \quad \frac{x}{|x|} \cdot \tilde{f}(x) > 0, \quad x \in \Gamma_k,\ k = 1, \ldots, N. \tag{5.116}$$

This implies that $\frac{x}{|x|} \cdot \nabla V(x) = \frac{x}{|x|} \cdot f(x) + \frac{x}{|x|} \cdot \tilde{f}(x) = 0$ holds true only for $x \in \mathcal{N}_f$. In other words, $\nabla V(x)$ cannot be tangential to the sphere Ω_R at a point $x \in \Omega_R \setminus \mathcal{N}_f$. Thus, by use of (5.113), the gradient $\nabla W(x)$ is radial to the sphere Ω_R only for $x \in \mathcal{N}_f$.

As a consequence, assuming that $|W|$ is non-constant and reaches its maximum on Ω_R, Lemma 5.22 states that the maximum is attained in a point $x_0 \in \mathcal{N}_f$ with $\frac{x_0}{|x_0|} \cdot \nabla W(x_0) = \alpha \neq 0$. Simultaneously, we get $\frac{x}{|x|} \cdot \nabla W(x) = \frac{x}{|x|} \cdot f(x) - \frac{x}{|x|} \cdot \tilde{f}(x) = 0$, $x \in \mathcal{N}_f$, by definition of the dip equators $\mathcal{N}_f$, $\mathcal{N}_{\tilde{f}}$. This is a contradiction, so that, by use of the Maximum/Minimum Principle from Theorem 3.10, either $|W|$ is constant on all of $\mathbb{R}^3 \setminus \mathcal{B}_R(0)$ or $|W|$ reaches its maximum at infinity. Since W is regular at infinity, both cases imply $W(x) = 0$, for $x \in \mathbb{R}^3 \setminus \mathcal{B}_R(0)$. In other words, $U(x) = \tilde{U}(x)$ and $f(x) = \tilde{f}(x)$, for $x \in \mathbb{R}^3 \setminus \mathcal{B}_R(0)$.

The assumption

$$\frac{x}{|x|} \cdot f(x) > 0, \quad \frac{x}{|x|} \cdot \tilde{f}(x) < 0, \quad x \in \Gamma_k,\ k = 1, \ldots, N, \tag{5.117}$$

implies $f(x) = -\tilde{f}(x)$, $x \in \mathbb{R}^3 \setminus \mathcal{B}_R(0)$ by the same argumentation as before, simply interchanging the roles of V and W. Thus, the theorem is proven. □

Remark 5.23. In a source-free region, we know that the magnetic field b can be expressed as $b = \nabla U$, for some harmonic function $U : \mathbb{R}^3 \setminus \overline{\mathcal{B}_R(0)} \to \mathbb{R}$. Choosing $\lambda(x) = \frac{b(x)}{|b(x)|}$, it can easily be seen that

$$|b(x)| = \lambda(x) \cdot b(x) = \frac{\partial}{\partial \lambda} U(x), \quad x \in \Omega_R. \tag{5.118}$$

Thus, the problem of determining b in $\mathbb{R}^3 \setminus \overline{\mathcal{B}_R(0)}$ from the knowledge of its intensities on Ω_R, can also be expressed as an oblique derivative problem:

$$\Delta U(x) = 0, \quad x \in \mathbb{R}^3 \setminus \overline{\mathcal{B}_R(0)}, \tag{5.119}$$

$$\frac{\partial}{\partial \lambda} U(x) = |b(x)|, \quad x \in \Omega_R. \tag{5.120}$$

It is important to note that the oblique vector λ in the current example depends on the function U itself, which makes the problem nonlinear and significantly more difficult. The situation is essentially the same in gravitational modeling when only the gravity intensity is available. However, the oblique derivative problems treated mathematically in Section 4.1 assume that the oblique direction λ is a known quantity, thus making the problem linear. Another way in physical geodesy to use gravitational intensities is the transition from the gravitational potential to the disturbing potential, resulting in a process of linearization by the additional introduction of an (ellipsoidal) level surface associated with a so-called normal potential (see Chapter 7).

Besides the problem of intensity measurements, one might also ask if the magnetic field is uniquely determined only from directional measurements on a sphere Ω_R. This is a question of less significance, but when working with historical data sets of inclination and declination, e.g., it has a certain relevance. In general, measurements of the direction only do not yield uniqueness. An overview on the entire topic of uniqueness with respect to magnetic field measurements is given, e.g., in T.J. Sabaka et al. [2010], however, with no mathematical proofs but with several references.

5.4 Separation of Sources

In the introduction of this chapter we mentioned the difficulty of separating the different contributions to the Earth's magnetic field. It is, however, possible to some extent to separate the contributions due to sources in the exterior of a given sphere Ω_R (or a spherical shell $\mathcal{B}_{R_0,R_1}(0)$) from those due to sources in the interior. This is of interest, e.g., when modeling the Earth's interior magnetic field. As a consequence, exterior influences like magnetospheric currents can be filtered out using this separation.

When the magnetic field measurements are conducted in the source-free shell $\mathcal{B}_{R_0,R_1}(0)$, the Gauss Representation already provides such a separation (for an illustration, see Figure 5.3). More precisely, the magnetic field $b = \nabla U$ can be split into

$$b(x) = b^{\text{int}}(x) + b^{\text{ext}}(x), \quad x \in \mathcal{B}_{R_0,R_1}(0), \tag{5.121}$$

with

$$\begin{aligned} \nabla \wedge b^{\text{int}}(x) &= 0, \quad x \in \mathbb{R}^3 \setminus \overline{\mathcal{B}_{R_0}(0)}, &(5.122)\\ \nabla \cdot b^{\text{int}}(x) &= 0, \quad x \in \mathbb{R}^3 \setminus \overline{\mathcal{B}_{R_0}(0)}, &(5.123)\end{aligned}$$

and

$$\begin{aligned} \nabla \wedge b^{\text{ext}}(x) &= 0, \quad x \in \mathcal{B}_{R_1}(0), &(5.124)\\ \nabla \cdot b^{\text{ext}}(x) &= 0, \quad x \in \mathcal{B}_{R_1}(0). &(5.125)\end{aligned}$$

Thus, b^{int} denotes the part of the magnetic field due to sources inside the Earth $\mathcal{B}_{R_0}(0)$ and can be represented as a potential field $b^{\text{int}}(x) = \nabla U^{\text{int}}(x)$, $x \in \mathbb{R}^3 \setminus \overline{\mathcal{B}_{R_0}(0)}$, where

$$U^{\text{int}}(x) = \sum_{n=0}^{\infty} \sum_{k=1}^{2n+1} \beta_{n,k} H^{R_0}_{-n-1,k}(x), \quad x \in \mathbb{R}^3 \setminus \mathcal{B}_{R_0}(0). \tag{5.126}$$

Analogously, b^{ext} denotes the part of the magnetic field due to sources in the iono- and magnetosphere, i.e., in the exterior $\mathbb{R}^3 \setminus \overline{\mathcal{B}_{R_1}(0)}$, and can be represented as a potential field $b^{\text{ext}}(x) = \nabla U^{\text{ext}}(x)$, $x \in \mathcal{B}_{R_1}(0)$, where

$$U^{\text{ext}}(x) = \sum_{n=0}^{\infty} \sum_{k=1}^{2n+1} \alpha_{n,k} H^{R_1}_{n,k}(x), \quad x \in \mathcal{B}_{R_1}(0). \tag{5.127}$$

The coefficients $\alpha_{n,k}$ and $\beta_{n,k}$ are determined as indicated for the Gauss Representation in Theorems 5.14 and 5.16.

However, if the measurements are not conducted in a source-free region (i.e., we have a current density $j \neq 0$), as is the case for satellite missions orbiting in or above the ionosphere, the magnetic field b cannot be represented by a potential field ∇U as in the case of the Gauss Representation. As a remedy, the *Mie Representation* is used. If $p_b = \nabla \wedge \mathrm{L} P_b$ and $q_b = \mathrm{L} Q_b$ denote the poloidal and toroidal part of the magnetic field and $p_j = \nabla \wedge \mathrm{L} P_j$ and $q_j = \mathrm{L} Q_j$ the poloidal and toroidal part of the current density, then the magnetic field can be divided into

$$b(x) = p_b^{\text{int}}(x) + p_b^{\text{ext}}(x) + q_b(x), \quad x \in \Omega_R, \tag{5.128}$$

with

$$\nabla \wedge p_b^{\text{int}}(x) = \begin{cases} 0, & x \in \mathbb{R}^3 \setminus \overline{\mathcal{B}_R(0)}, \\ q_j(x), & x \in \mathcal{B}_R(0), \end{cases} \tag{5.129}$$

$$\nabla \cdot p_b^{\text{int}}(x) = 0, \quad x \in \mathbb{R}^3, \tag{5.130}$$

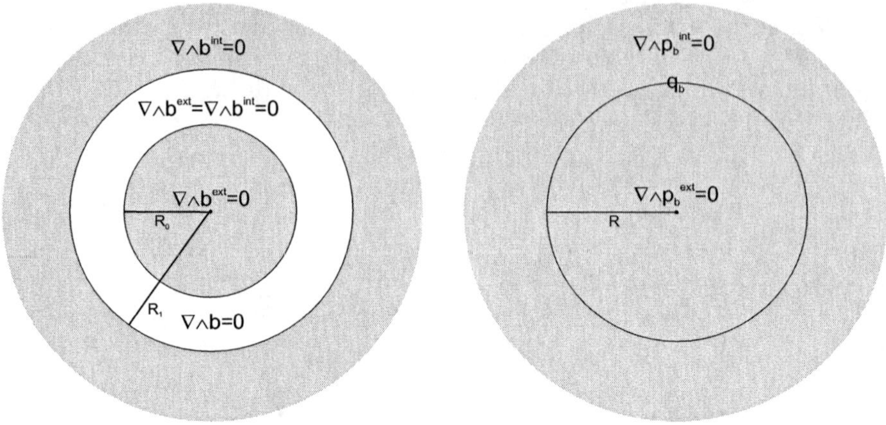

FIGURE 5.3
Separation into interior and exterior sources with respect to measurements conducted in a source-free shell (left) and measurements in a source-carrying region (right).

and

$$\nabla \wedge p_b^{\rm ext}(x) = \begin{cases} q_j(x), & x \in \mathbb{R}^3 \setminus \overline{\mathcal{B}_R(0)}, \\ 0, & x \in \mathcal{B}_R(0), \end{cases} \tag{5.131}$$

$$\nabla \cdot p_b^{\rm ext}(x) = 0, \quad x \in \mathbb{R}^3. \tag{5.132}$$

For a graphical illustration, we again refer to Figure 5.3. The toroidal magnetic field q_b is closely related to the radial current density on Ω_R (more details are given in Section 8.4). It represents the part of the magnetic field that is due to poloidal currents crossing the sphere. The poloidal part p_b can be split into a part $p_b^{\rm int}$ due to toroidal currents in the interior of the sphere, i.e., $\mathcal{B}_R(0)$, and a part $p_b^{\rm ext}$ due to toroidal currents in the exterior, i.e., $\mathbb{R}^3 \setminus \overline{\mathcal{B}_R(0)}$. Just as for the source-free case in (5.121)–(5.127), we find potentials $U^{\rm int}$, $U^{\rm ext}$ such that

$$p_b^{\rm int}(x) = \nabla U^{\rm int}(x), \quad x \in \mathbb{R}^3 \setminus \overline{\mathcal{B}_R(0)}, \tag{5.133}$$

$$p_b^{\rm ext}(x) = \nabla U^{\rm ext}(x), \quad x \in \mathcal{B}_R(0). \tag{5.134}$$

The representations of $U^{\rm int}$ and $U^{\rm ext}$ are analogous to (5.126) and (5.127), respectively, simply with the radii R_0, R_1 substituted by R. This allows us to expand $p_b^{\rm int}$ and $p_b^{\rm ext}$ (as well as $b^{\rm int}$ and $b^{\rm ext}$ in the source-free case) in terms of the vector inner/outer harmonics $h^{(1);R}_{-n-1,k}$ and $h^{(2);R}_{n,k}$, respectively. Exercise 5.5 indicates how the coefficients $\alpha_{n,k}$ and $\beta_{n,k}$ can be specified. Some more details and applications to real satellite data are given in Section 8.3.

5.4.1 Geophysical Motivation

It is the aim of this subsection to give a more physically oriented explanation of the separation (5.128)–(5.132) by means of the Law of Biot–Savart (see also G.E. Backus et al. [1996]). Assuming that all currents are located inside $\overline{\mathcal{G}}$, where $\mathcal{G} \subset \mathbb{R}^3$ is a regular region or its open complement, the Law of Biot–Savart motivates the representation

$$\begin{aligned} b(x) &= \frac{1}{4\pi}\int_{\mathcal{G}} j(y) \wedge \frac{x-y}{|x-y|^3}\, dV(y) \\ &= \frac{1}{4\pi}\int_{\mathcal{G}} \left(\nabla_x \frac{1}{|x-y|}\right) \wedge j(y)\, dV(y) \\ &= \nabla \wedge v(x), \end{aligned} \tag{5.135}$$

for $x \in \mathbb{R}^3$, where

$$v(x) = \frac{1}{4\pi}\int_{\mathcal{G}} \frac{j(y)}{|x-y|}\, dV(y). \tag{5.136}$$

It follows that

$$\nabla \wedge b(x) = \nabla \wedge (\nabla \wedge v(x)) = \nabla(\nabla \cdot v(x)) - \Delta v(x), \quad x \in \mathbb{R}^3. \tag{5.137}$$

From Theorems 3.28 and 3.30 it is known that

$$\Delta v(x) = \begin{cases} -j(x), & x \in \mathcal{G}, \\ 0, & x \in \mathbb{R}^3 \setminus \overline{\mathcal{G}}, \end{cases} \tag{5.138}$$

which leads to

$$\nabla \wedge b(x) = \begin{cases} j(x) + \nabla(\nabla \cdot v(x)), & x \in \mathcal{G}, \\ \nabla(\nabla \cdot v(x)), & x \in \mathbb{R}^3 \setminus \overline{\mathcal{G}}. \end{cases} \tag{5.139}$$

We see that the representation of b as provided in (5.135) includes additional sources $\nabla(\nabla \cdot v)$. The term $\nabla \cdot v$ can be rewritten in the form

$$\begin{aligned} \nabla_x \cdot v(x) &= -\frac{1}{4\pi}\int_{\mathcal{G}} \nabla_y \cdot \left(\frac{j(y)}{|x-y|}\right) dV(y) + \frac{1}{4\pi}\int_{\mathcal{G}} \frac{\nabla_y \cdot j(y)}{|x-y|}\, dV(y) \quad (5.140) \\ &= -\frac{1}{4\pi}\int_{\partial\mathcal{G}} \frac{\nu(y) \cdot j(y)}{|x-y|}\, d\omega(y), \end{aligned}$$

where $\nabla \cdot j = 0$ and the Theorem of Gauss have been used (since this section has only a motivating character, all appearing functions are supposed to be sufficiently often differentiable and have a sufficient decay at infinity). Thus, $\nabla(\nabla \cdot v)$ can be interpreted as an electric field generated by a surface charge density due to the normal current density $\nu \cdot j$ on $\partial\mathcal{G}$. Returning to our original problem concerning the separation of the magnetic field measured at the satellite's orbit Ω_R, we choose $\mathcal{G}$ to be $\mathcal{B}_R(0)$ or $\mathbb{R}^3 \setminus \overline{\mathcal{B}_R(0)}$. Then $\nu \cdot j$ denotes the radial current density, and in the case that j is tangential (in particular if it is toroidal) the undesired contributions $\nabla(\nabla \cdot v)$ vanish. Remembering that poloidal magnetic fields are exclusively generated by toroidal currents (cf. Remark 5.11), we are lead to introduce

$$p_b^{\mathrm{int}}(x) = \frac{1}{4\pi} \int_{\mathcal{B}_R(0)} q_j(y) \wedge \frac{x-y}{|x-y|^3} \, dV(y), \quad x \in \mathbb{R}^3, \tag{5.141}$$

$$p_b^{\mathrm{ext}}(x) = \frac{1}{4\pi} \int_{\mathbb{R}^3 \setminus \overline{\mathcal{B}_R(0)}} q_j(y) \wedge \frac{x-y}{|x-y|^3} \, dV(y), \quad x \in \mathbb{R}^3. \tag{5.142}$$

With these definitions at hand, we obtain the desired decomposition (5.128)–(5.132).

5.4.2 Connection of Gauss and Mie Representation

The previous considerations show that the Mie Representation implies a generalized version of the Gauss Representation. In source-free regions, both representations are valid and a precise relation between the representation $f = \nabla U$ and $f = \nabla \wedge \mathrm{L}P + \mathrm{L}Q$ can be realized.

Lemma 5.24. *Let $f : \mathcal{B}_{R_0,R_1}(0) \to \mathbb{R}^3$, $0 \leq R_0 < R_1$, be solenoidal with*

$$\nabla \wedge f(x) = 0, \quad x \in \mathcal{B}_{R_0,R_1}(0), \tag{5.143}$$

$$\nabla \cdot f(x) = 0, \quad x \in \mathcal{B}_{R_0,R_1}(0). \tag{5.144}$$

Then there exist uniquely determined harmonic functions U, P of class $\mathrm{C}^{(2)}(\mathcal{B}_{R_0,R_1}(0))$ such that

$$f(x) = \nabla U(x) = \nabla \wedge \mathrm{L}P(x), \quad x \in \mathcal{B}_{R_0,R_1}(0), \tag{5.145}$$

and

$$\frac{1}{4\pi r^2} \int_{\Omega_r} U(y) d\omega(y) = \frac{1}{4\pi r^2} \int_{\Omega_r} P(y) d\omega(y) = 0, \quad r \in (R_0, R_1). \tag{5.146}$$

The functions U and P are connected via

$$U(r\xi) = -r \frac{\partial}{\partial r} \left(rP(r\xi) \right), \tag{5.147}$$

$$\Delta_\xi^* P(r\xi) = r \frac{\partial}{\partial r} U(r\xi), \tag{5.148}$$

for $\xi \in \Omega$ and $r \in (R_0, R_1)$.

Proof. The existence of a harmonic function U satisfying $f(x) = \nabla U(x)$, $x \in \mathcal{B}_{R_0,R_1}(0)$, and (5.146) is known from Sections 5.2 and 5.3. From the Mie

Representation in Theorem 5.10, we know that uniquely determined P, Q of class $\mathrm{C}^{(1)}(\mathcal{B}_{R_0,R_1}(0))$ exist such that

$$f(x) = \nabla \wedge \mathrm{L}P(x) + \mathrm{L}Q(x), \quad x \in \mathcal{B}_{R_0,R_1}(0), \tag{5.149}$$

where P and Q have a vanishing integral mean value on Ω_r, $r \in (R_0, R_1)$. The Theorem of Stokes implies

$$\int_\Gamma \nabla_y \wedge f(y) d\omega(y) = \int_{\partial\Gamma} \tau(y) \cdot f(y) d\sigma(y) = 0 \tag{5.150}$$

for any regular region $\Gamma \subset \Omega_r$. Thus, by Lemma 6.61, there exists a function $V(r\cdot)$ of class $\mathrm{C}^{(1)}(\Omega)$ with $f_{\tan}(r\xi) = \nabla^*_\xi V(r\xi)$, for $\xi \in \Omega$, $r \in (R_0, R_1)$ (here and in the remainder of the proof, we use some auxiliary tools that are worked out in more detail in Chapters 6 and 8). In other words, f cannot have a toroidal contribution, and therefore

$$f(x) = \nabla \wedge \mathrm{L}P(x), \quad x \in \mathcal{B}_{R_0,R_1}(0). \tag{5.151}$$

From Lemma 8.6 and the representation $\nabla_x = \xi\frac{\partial}{\partial r} + \frac{1}{r}\nabla^*_\xi$, $r = |x|$, $\xi = \frac{x}{|x|}$, we obtain

$$\begin{aligned} f(r\xi) &= \xi\frac{\partial}{\partial r}U(r\xi) + \frac{1}{r}\nabla^*_\xi U(\xi), && (5.152)\\ f(r\xi) &= \xi\frac{1}{r}\Delta^*_\xi P(r\xi) - \frac{1}{r}\nabla^*_\xi \frac{\partial}{\partial r}\left(rP(r\xi)\right), && (5.153) \end{aligned}$$

for $r \in (R_0, R_1)$ and $\xi \in \Omega$. The uniqueness of the spherical Helmholtz decomposition in Theorem 8.1 then implies the representations (5.147) and (5.148). Equations (5.147) and (5.148) carry the differentiability of the harmonic function U over to P, so that P is also of class $\mathrm{C}^{(\infty)}(\mathcal{B}_{R_0,R_1}(0))$. Thus, we can apply the curl to (5.151) and obtain

$$\begin{aligned} 0 &= \nabla \wedge f(x) = \nabla \wedge \nabla \wedge \mathrm{L}P(x) && (5.154)\\ &= \nabla(\nabla \cdot \mathrm{L}P(x)) - \Delta \mathrm{L}P(x) && (5.155)\\ &= -\mathrm{L}\Delta P(x), && (5.156) \end{aligned}$$

for $x \in \mathcal{B}_{R_0,R_1}(0)$. Observing Lemma 6.59 and the subsequent remark, we see that ΔP is constant in $\mathcal{B}_{R_0,R_1}(0)$. Since ΔP has a vanishing integral mean value, the uniqueness of the Helmholtz scalars from Theorem 8.1 yields $\Delta P(x) = 0$, $x \in \mathcal{B}_{R_0,R_1}(0)$. As a consequence, we find that P is harmonic. □

5.5 Ionospheric Current Systems

We assume the magnetic field b to be generated by ionospheric currents j. As usual, the quantities are connected by the pre-Maxwell equations

$$\nabla \wedge b = j, \tag{5.157}$$
$$\nabla \cdot b = 0, \tag{5.158}$$

everywhere in $\mathbb{R}^3$. If j is known, the Law of Biot–Savart

$$b(x) = \frac{1}{4\pi} \int_{\mathbb{R}^3} j(y) \wedge \frac{x-y}{|x-y|^3} \, dV(y) \tag{5.159}$$

represents a solution of the pre-Maxwell equations. However, the data situation is such that the current density j is unknown. Typically, magnetic field data sets are only provided on a sphere Ω_R. Thus, we do not have sufficient information to solve the pre-Maxwell equations for b or j, respectively.

5.5.1 Spherical Biot–Savart Operator

To overcome the problem of insufficient data, we introduce a spherical restriction of the Biot–Savart operator and assume that all currents are located on a sphere Ω_{R_1} in the ionosphere. This concept is elaborated in the current subsection.

Definition 5.25. The operator $T_{R_1,R} : \mathrm{l}^2(\Omega_{R_1}) \to \mathrm{l}^2(\Omega_R)$, $R \neq R_1$, given by

$$T_{R_1,R}[g](x) = \frac{1}{4\pi} \int_{\Omega_{R_1}} g(y) \wedge \frac{x-y}{|x-y|^3} d\omega(y), \quad x \in \Omega_R, \tag{5.160}$$

for a vector field g of class $\mathrm{l}^2(\Omega_{R_1})$, is called the *spherical Biot–Savart operator.*

The spherical Biot–Savart operator relates the magnetic field b on Ω_R to its source currents j on Ω_{R_1}, i.e.,

$$T_{R_1,R}[j](x) = b(x), \quad x \in \Omega_R. \tag{5.161}$$

In other words, the inversion of $T_{R_1,R}$ allows the determination of the current density j from the measured magnetic field b (under the assumption that all currents are located on a sphere Ω_{R_1}). Unfortunately, this approach leads to an ill-posed problem.

Lemma 5.26. *The operator $T_{R_1,R} : \mathrm{l}^2(\Omega_{R_1}) \to \mathrm{l}^2(\Omega_R)$ is linear, bounded, and completely continuous.*

The complete continuity can be realized by standard arguments from functional analysis, similar to the proof of Lemma 6.25 (see also K. Yoshida [1980] and H.W. Alt [2006]). The ill-posedness requires a regularization of the inverse operator $T^{-1}_{R_1,R}$ to obtain a stable reconstruction of j. One possibility to obtain such a regularization is by use of a multiscale method. In this subsection, our goal is to derive a singular system that allows us to formally invert the spherical Biot–Savart operator. In Subsection 5.5.2 we then present some numerical results for real satellite data that are obtained following the regularization method as described in W. Freeden, F. Schneider [1998] (and applied to the spherical Biot–Savart operator in C. Mayer [2004]). In doing so, by a *singular system* we mean the set of non-zero singular values $\{\sigma_n\}_{n\in\mathbb{N}_0}$, where σ_n^2 are the non-zero eigenvalues of the self-adjoint operator $T^*_{R_1,R}T_{R_1,R}$, and two orthonormal function systems $\{f_n\}_{n\in\mathbb{N}_0} \subset \mathrm{l}^2(\Omega_R)$ and $\{g_n\}_{n\in\mathbb{N}_0} \subset \mathrm{l}^2(\Omega_{R_1})$ such that

$$T_{R_1,R}[g_n] = \sigma_n f_n, \qquad T_{R_1,R}[g] = \sum_{n=0}^{\infty} \sigma_n (g_n, g)_{\mathrm{l}^2(\Omega_{R_1})} f_n, \tag{5.162}$$

$$T^*_{R_1,R}[f_n] = \sigma_n g_n, \qquad T^*_{R_1,R}[f] = \sum_{n=0}^{\infty} \sigma_n (f_n, f)_{\mathrm{l}^2(\Omega_R)} g_n, \tag{5.163}$$

for f of class $\mathrm{l}^2(\Omega_R)$, g of class $\mathrm{l}^2(\Omega_{R_1})$, and $T^*_{R_1,R}$ the adjoint operator of $T_{R_1,R}$ (compare Exercises 4.9–4.11 for related examples). An easy calculation using Fubini's theorem shows that

$$T^*_{R_1,R} = T_{R,R_1}. \tag{5.164}$$

We start by calculating the image of the the set of vector spherical harmonics $\{y^{(i)}_{n,k}\}_{i=1,2,3,\, n\in\mathbb{N}_{0_i},k=1,\dots 2n+1}$ with respect to the spherical Biot–Savart operator (cf. C. Mayer [2004]).

Theorem 5.27. *Let $y^{(i);R_1}_{n,k}$ and $y^{(i);R}_{n,k}$, $i = 1,2,3$, $n \in \mathbb{N}_{0_i}$, $k = 1,\dots 2n+1$, be given as in Section 2.5 and assume $0 < R < R_1$. Then we have*

$$T_{R_1,R}\left[y^{(1);R_1}_{n,k}\right] = -\frac{\sqrt{n(n+1)}}{2n+1}\left(\frac{R}{R_1}\right)^{n+1} y^{(3);R}_{n,k}, \tag{5.165}$$

$$T_{R_1,R}\left[y^{(2);R_1}_{n,k}\right] = \frac{n}{2n+1}\left(\frac{R}{R_1}\right)^{n+1} y^{(3);R}_{n,k}, \tag{5.166}$$

$$T_{R_1,R}\left[y^{(3);R_1}_{n,k}\right] = -\frac{n+1}{2n+1}\left(\frac{R}{R_1}\right)^{n} \left(\sqrt{\frac{n}{n+1}}\, y^{(1);R}_{n,k} + y^{(2);R}_{n,k}\right). \tag{5.167}$$

Proof. First, we observe that the spherical Biot–Savart operator can also be expressed in the form

$$T_{R_1,R}[g](x) = \nabla_x \wedge P_{R_1,R}[g](x), \quad x \in \Omega_R, \tag{5.168}$$

where

$$P_{R_1,R}[g](x) = \frac{1}{4\pi}\int_{\Omega_{R_1}} \frac{g(y)}{|x-y|}\, d\omega(y), \quad x \in \Omega_R. \tag{5.169}$$

From (2.99), we know that

$$\frac{1}{|x-y|} = \frac{1}{R_1} \sum_{m=0}^{\infty} \left(\frac{r}{R_1}\right)^m P_m(\xi \cdot \eta), \tag{5.170}$$

for $\xi = \frac{x}{|x|}$, $\eta = \frac{y}{|y|}$, $r = |x|$, $R_1 = |y|$, and $r \leq R < R_1$. Consequently, we get

$$\begin{aligned} &P_{R_1,R}\left[y_{n,k}^{(1);R_1}\right](x) \qquad (5.171)\\ &= \frac{1}{4\pi R_1^2} \sum_{m=0}^{\infty} \int_{\Omega_{R_1}} \left(\frac{r}{R_1}\right)^m P_m\left(\frac{x}{|x|}\cdot\frac{y}{|y|}\right) o^{(1)}_{\frac{y}{|y|}} Y_{n,k}\left(\frac{y}{|y|}\right) d\omega(y)\\ &= \frac{1}{4\pi} \sum_{m=0}^{\infty} \int_{\Omega} \left(\frac{r}{R_1}\right)^m P_m(\xi\cdot\eta)\, o^{(1)}_{\eta} Y_{n,k}(\eta)\, d\omega(\eta), \end{aligned}$$

for $\xi = \frac{x}{|x|}$ and $r = |x|$. The vectorial Funk–Hecke formula (2.178) leads to

$$\begin{aligned} &P_{R_1,R}\left[y_{n,k}^{(1);R_1}\right](x) \qquad (5.172)\\ &= \frac{1}{4\pi} \sum_{m=0}^{\infty} \left(\frac{r}{R_1}\right)^m \frac{1}{2n+1}\Big(\big((n+1)P_m^{\wedge}(n+1) + nP_m^{\wedge}(n-1)\big)o^{(1)}_{\xi} Y_{n,k}(\xi)\\ &\qquad + \big(P_m^{\wedge}(n-1) - P_m^{\wedge}(n+1)\big)o^{(2)}_{\xi} Y_{n,k}(\xi)\Big). \end{aligned}$$

The Legendre coefficients of the Legendre polynomials are given by $P_m^{\wedge}(n) = 4\pi(2n+1)^{-1}\delta_{m,n}$. Thus, by virtue of Exercise 2.12, we find

$$\begin{aligned} &P_{R_1,R}\left[y_{n,k}^{(1);R_1}\right](x) \qquad (5.173)\\ &= \left(\frac{r}{R_1}\right)^{n+1} \frac{1}{(2n+1)(2n+3)} \Big((n+1)o^{(1)}_{\xi} Y_{n,k}(\xi) - o^{(2)}_{\xi} Y_{n,k}(\xi)\Big)\\ &\quad + \left(\frac{r}{R_1}\right)^{n-1} \frac{1}{(2n+1)(2n-1)} \Big(n o^{(1)}_{\xi} Y_{n,k}(\xi) + o^{(2)}_{\xi} Y_{n,k}(\xi)\Big)\\ &= \frac{1}{R_1^{n+1}} \frac{1}{(2n+1)(2n+3)} k_n^{(1)}(r^n Y_{n,k}(\xi))\\ &\quad + \frac{1}{R_1^{n-1}} \frac{1}{(2n+1)(2n-1)} k_n^{(2)}(r^n Y_{n,k}(\xi)). \end{aligned}$$

Applying the curl to (5.173) and observing Exercise 2.13, we get

$$\begin{aligned}\nabla_x \wedge P_{R_1,R}\left[y_{n,k}^{(1);R_1}\right](x) =& \frac{1}{R_1^{n+1}} \frac{1}{(2n+1)(2n+3)} \nabla_x \wedge k_n^{(1)}(r^n Y_{n,k}(\xi)) \\ =& - \frac{1}{R_1^{n+1}} \frac{2n+3}{(2n+1)(2n+3)} r^n k_n^{(3)} Y_{n,k}(\xi) \\ =& - \frac{1}{2n+1}\left(\frac{r}{R_1}\right)^{n+1} \frac{1}{r} o_\xi^{(3)} Y_{n,k}(\xi) \\ =& - \frac{(\mu_n^{(3)})^{\frac{1}{2}}}{2n+1}\left(\frac{r}{R_1}\right)^{n+1} y_{n,k}^{(3);r}(\xi),\end{aligned} \tag{5.174}$$

for $\xi = \frac{x}{|x|}$ and $r = |x|$. Restricting (5.174) to the sphere Ω_R and observing (5.168), we finally obtain the representation (5.165), i.e.,

$$T_{R_1,R}\left[y_{n,k}^{(1);R_1}\right] = - \frac{(\mu_n^{(3)})^{\frac{1}{2}}}{2n+1}\left(\frac{R}{R_1}\right)^{n+1} y_{n,k}^{(3);R}. \tag{5.175}$$

To verify the representation (5.166), we proceed in a similar manner as for (5.165). First, we get

$$P_{R_1,R}\left[y_{n,k}^{(2);R_1}\right](x) = \frac{1}{4\pi}\sum_{m=0}^{\infty}\int_\Omega \left(\frac{r}{R_1}\right)^m (\mu_n^{(2)})^{-\frac{1}{2}} P_m(\xi\cdot\eta)\, o_\eta^{(2)} Y_{n,k}(\eta)\, d\omega(\eta), \tag{5.176}$$

for $\xi = \frac{x}{|x|}$ and $r = |x|$. From the vectorial Funk–Hecke formula (2.179), we are able to deduce that

$$\begin{aligned}&P_{R_1,R}\left[y_{n,k}^{(2);R_1}\right](x) \\ &= \frac{1}{4\pi}\sum_{m=0}^{\infty}\left(\frac{r}{R_1}\right)^m \frac{(\mu_n^{(2)})^{-\frac{1}{2}}}{2n+1}\Big(\big(n(n+1)P_m^\wedge(n-1) - P_m^\wedge(n+1)\big) o_\xi^{(1)} Y_{n,k}(\xi) \\ &\qquad + \big(nP_m^\wedge(n+1) + (n+1)P_m^\wedge(n-1)\big) o_\xi^{(2)} Y_{n,k}(\xi)\Big).\end{aligned} \tag{5.177}$$

Exercise 2.12 together with $P_m^\wedge(n) = 4\pi(2n+1)^{-1}\delta_{m,n}$ implies

$$
\begin{aligned}
&P_{R_1,R}\left[y_{n,k}^{(2);R_1}\right](x) \qquad (5.178)\\
&= \left(\frac{r}{R_1}\right)^{n+1} \frac{(\mu_n^{(2)})^{-\frac{1}{2}}}{(2n+1)(2n+3)} \left(-n(n+1)o_\xi^{(1)}Y_{n,k}(\xi) + no_\xi^{(2)}Y_{n,k}(\xi)\right)\\
&\quad + \left(\frac{r}{R_1}\right)^{n-1} \frac{(\mu_n^{(2)})^{-\frac{1}{2}}}{(2n+1)(2n-1)} \left(n(n+1)o_\xi^{(1)}Y_{n,k}(\xi) + (n+1)o_\xi^{(2)}Y_{n,k}(\xi)\right)\\
&= -\frac{1}{R_1^{n+1}} \frac{n(\mu_n^{(2)})^{-\frac{1}{2}}}{(2n+1)(2n+3)} k_n^{(1)}(r^nY_{n,k}(\xi))\\
&\quad + \frac{1}{R_1^{n-1}} \frac{n(\mu_n^{(2)})^{-\frac{1}{2}}}{(2n+1)(2n-1)} k_n^{(2)}(r^nY_{n,k}(\xi)).
\end{aligned}
$$

Applying the curl to (5.178) and observing Exercise 2.13, we get

$$
\begin{aligned}
\nabla_x \wedge P_{R_1,R}\left[y_{n,k}^{(2);R_1}\right](x) &= -\frac{1}{R_1^{n+1}} \frac{n(\mu_n^{(2)})^{-\frac{1}{2}}}{(2n+1)(2n+3)} \nabla_x \wedge k_n^{(1)}(r^nY_{n,k}(\xi))\\
&= -\frac{1}{R_1^{n+1}} \frac{n(2n+3)(\mu_n^{(2)})^{-\frac{1}{2}}}{(2n+1)(2n+3)} r^n k_n^{(3)} Y_{n,k}(\xi)\\
&= \frac{n}{2n+1}\left(\frac{r}{R_1}\right)^{n+1} \frac{(\mu_n^{(3)})^{-\frac{1}{2}}}{r} o_\xi^{(3)}Y_{n,k}(\xi) \qquad (5.179)\\
&= \frac{n}{2n+1}\left(\frac{r}{R_1}\right)^{n+1} y_{n,k}^{(3);r}(\xi),
\end{aligned}
$$

for $\xi = \frac{x}{|x|}$ and $r = |x|$. In the third row, we have used $\mu_n^{(3)} = \mu_n^{(2)}$. Restricting (5.179) to the sphere Ω_R and observing (5.168), we obtain the desired representation (5.166).

At last, we turn to the representation (5.167). The vectorial Funk–Hecke formula (2.180), Exercise 2.12, and an analogous argumentation as for the previous two relations leads to

$$
\begin{aligned}
P_{R_1,R}\left[y_{n,k}^{(3);R_1}\right](x) &= \frac{1}{4\pi}\sum_{m=0}^{\infty} \int_\Omega \left(\frac{r}{R_1}\right)^m (\mu_n^{(3)})^{-\frac{1}{2}} P_m(\xi\cdot\eta)\, o_\eta^{(3)}Y_{n,k}(\eta)\, d\omega(\eta)\\
&= \left(\frac{r}{R_1}\right)^n \frac{(\mu_n^{(3)})^{-\frac{1}{2}}}{2n+1} o_\xi^{(3)}Y_{n,k}(\xi) \qquad (5.180)\\
&= \frac{1}{R_1^n}\frac{(\mu_n^{(3)})^{-\frac{1}{2}}}{2n+1} k_n^{(3)}(r^nY_{n,k}(\xi)),
\end{aligned}
$$

for $\xi = \frac{x}{|x|}$ and $r = |x|$. Applying the curl to (5.180) and observing Exercise

2.13, we get

$$
\begin{aligned}
&\nabla_x \wedge P_{R_1,R}\left[y_{n,k}^{(3);R_1}\right](x) && (5.181)\\
&=\frac{1}{R_1^n}\frac{(\mu_n^{(3)})^{-\frac{1}{2}}}{2n+1}\nabla_x \wedge k_n^{(3)}(r^n Y_{n,k}(\xi))\\
&=\frac{1}{R_1^n}\frac{(\mu_n^{(3)})^{-\frac{1}{2}}}{2n+1}\left(o_\xi^{(1)}\left(\frac{1}{r}\Delta_\xi^*(r^n Y_{n,k}(\xi))\right)+o_\xi^{(2)}\left(-\frac{1}{r}\frac{\partial}{\partial r}(r^{n+1}Y_{n,k}(\xi))\right)\right)\\
&=-(\mu_n^{(2)})^{-\frac{1}{2}}\frac{n+1}{2n+1}\left(\frac{r}{R_1}\right)^n\frac{1}{r}\left(n\, o_\xi^{(1)}Y_{n,k}(\xi)+o_\xi^{(2)}Y_{n,k}(\xi)\right)\\
&=-\frac{n+1}{2n+1}\left(\frac{r}{R_1}\right)^n\left(\sqrt{\frac{n}{n+1}}y_{n,k}^{(1);r}(\xi)+y_{n,k}^{(2);r}(\xi)\right)
\end{aligned}
$$

Restricting (5.181) to the sphere Ω_R and observing (5.168), we obtain the desired result. □

By an argument similar to the proof of Theorem 5.27 we are able to describe the behavior of $T_{R_1,R}$ for $R > R_1$.

Theorem 5.28. *Let $y_{n,k}^{(i);R_1}$ and $y_{n,k}^{(i);R}$, $i=1,2,3$, $n \in \mathbb{N}_{0_i}$, $k=1,\ldots 2n+1$, be given as in Section 2.5 and assume $0 < R_1 < R$. Then we have*

$$T_{R_1,R}\left[y_{n,k}^{(1);R_1}\right] = -\frac{\sqrt{n(n+1)}}{2n+1}\left(\frac{R_1}{R}\right)^n y_{n,k}^{(3);R}, \tag{5.182}$$

$$T_{R_1,R}\left[y_{n,k}^{(2);R_1}\right] = -\frac{n+1}{2n+1}\left(\frac{R_1}{R}\right)^n y_{n,k}^{(3);R}, \tag{5.183}$$

$$T_{R_1,R}\left[y_{n,k}^{(3);R_1}\right] = -\frac{n}{2n+1}\left(\frac{R_1}{R}\right)^{n+1}\left(\sqrt{\frac{n+1}{n}}y_{n,k}^{(1);R}-y_{n,k}^{(2);R}\right). \tag{5.184}$$

Observing the identities (2.218)–(2.220), the use of the vector spherical harmonics $\tilde{y}_{n,k}^{(i)}$, $i=1,2,3$, enables the following more concise characterization of $T_{R_1,R}$.

Corollary 5.29. *Let $\tilde{y}_{n,k}^{(i);R_1}$ and $\tilde{y}_{n,k}^{(i);R}$, $i=1,2,3$, $n \in \mathbb{N}_{0_i}$, $k=1,\ldots 2n+1$, be given as in Section 2.5. Then we have for $0 < R < R_1$,*

$$T_{R_1,R}\left[\tilde{y}_{n,k}^{(1);R_1}\right] = -\sqrt{\frac{n}{2n+1}}\left(\frac{R}{R_1}\right)^{n+1}\tilde{y}_{n,k}^{(3);R}, \tag{5.185}$$

$$T_{R_1,R}\left[\tilde{y}_{n,k}^{(2);R_1}\right] = 0, \tag{5.186}$$

$$T_{R_1,R}\left[\tilde{y}_{n,k}^{(3);R_1}\right] = -\sqrt{\frac{n+1}{2n+1}}\left(\frac{R}{R_1}\right)^n\tilde{y}_{n,k}^{(2);R}. \tag{5.187}$$

For $0 < R_1 < R$, *we get*

$$T_{R_1,R}\left[\tilde{y}_{n,k}^{(1);R_1}\right] = 0, \tag{5.188}$$

$$T_{R_1,R}\left[\tilde{y}_{n,k}^{(2);R_1}\right] = -\sqrt{\frac{n+1}{2n+1}}\left(\frac{R_1}{R}\right)^n \tilde{y}_{n,k}^{(3);R}, \tag{5.189}$$

$$T_{R_1,R}\left[\tilde{y}_{n,k}^{(3);R_1}\right] = -\sqrt{\frac{n}{2n+1}}\left(\frac{R_1}{R}\right)^{n+1} \tilde{y}_{n,k}^{(1);R}. \tag{5.190}$$

Since we know that $T^*_{R_1,R} = T_{R,R_1}$, the following corollary is a further direct consequence.

Corollary 5.30. *Let* $\tilde{y}_{n,k}^{(i);R_1}$ *and* $\tilde{y}_{n,k}^{(i);R}$, $i = 1,2,3$, $n \in \mathbb{N}_{0_i}$, $k = 1,\ldots 2n+1$, *be given as in Section 2.5. Then we have for* $0 < R < R_1$,

$$T^*_{R_1,R}T_{R_1,R}\left[\tilde{y}_{n,k}^{(1);R_1}\right] = \frac{n}{2n+1}\left(\frac{R}{R_1}\right)^{2n+2} \tilde{y}_{n,k}^{(1);R}, \tag{5.191}$$

$$T^*_{R_1,R}T_{R_1,R}\left[\tilde{y}_{n,k}^{(2);R_1}\right] = 0, \tag{5.192}$$

$$T^*_{R_1,R}T_{R_1,R}\left[\tilde{y}_{n,k}^{(3);R_1}\right] = \frac{n+1}{2n+1}\left(\frac{R}{R_1}\right)^{2n} \tilde{y}_{n,k}^{(3);R}. \tag{5.193}$$

For $0 < R_1 < R$, *we have*

$$T^*_{R_1,R}T_{R_1,R}\left[\tilde{y}_{n,k}^{(1);R_1}\right] = 0, \tag{5.194}$$

$$T^*_{R_1,R}T_{R_1,R}\left[\tilde{y}_{n,k}^{(2);R_1}\right] = \frac{n+1}{2n+1}\left(\frac{R_1}{R}\right)^{2n} \tilde{y}_{n,k}^{(2);R}, \tag{5.195}$$

$$T^*_{R_1,R}T_{R_1,R}\left[\tilde{y}_{n,k}^{(3);R_1}\right] = \frac{n}{2n+1}\left(\frac{R_1}{R}\right)^{2n+2} \tilde{y}_{n,k}^{(3);R}. \tag{5.196}$$

Corollaries 5.29 and 5.30 not only constitute the basis for the formulation of a two step (decomposition and reconstruction) singular system, allowing a representation of the inverse $T^{-1}_{R_1,R}$, but they provide a generalized mathematical formulation of the well-known fact that field-aligned currents together with their corresponding Pedersen currents do not produce any magnetic effect beneath the ionosphere (see, e.g., N. Fukushima [1976]). More precisely, Equation (5.186) tells us that spherical current densities composed of vector spherical harmonics $\tilde{y}_{n,k}^{(2);R_1}$ (i.e., those vector spherical harmonics obtained by application of the gradient to (scalar) inner harmonics) produce no magnetic effect inside $\mathcal{B}_{R_1}(0)$. The same holds for an exterior formulation. In fact, (5.188) states that spherical current densities composed of vector spherical harmonics $\tilde{y}_{n,k}^{(1);R_1}$ (i.e., those vector spherical harmonics obtained by application of the gradient to (scalar) outer harmonics) produce no magnetic effect in $\mathbb{R}^3 \setminus \overline{\mathcal{B}_{R_1}(0)}$.

Remark 5.31. From Corollaries 5.29 and 5.30, we see that the singular system for decomposition and reconstruction of $T_{R_1,R}$, $0 < R < R_1$, is given by the singular values

$$\left\{ \sqrt{\frac{n}{2n+1}} \left(\frac{R}{R_1}\right)^{n+1}, \sqrt{\frac{n+1}{2n+1}} \left(\frac{R}{R_1}\right)^{n} \right\}_{n\in\mathbb{N},\, k=1,\ldots,2n+1} \tag{5.197}$$

and the orthonormal function systems

$$\left\{ -\tilde{y}_{n,k}^{(1);R_1}, -\tilde{y}_{n,k}^{(3);R_1} \right\}_{n\in\mathbb{N},\, k=1,\ldots,2n+1} \subset \mathrm{l}^2(\Omega_{R_1}), \tag{5.198}$$

$$\left\{ \tilde{y}_{n,k}^{(3);R}, \tilde{y}_{n,k}^{(2);R} \right\}_{n\in\mathbb{N},\, k=1,\ldots,2n+1} \subset \mathrm{l}^2(\Omega_{R}). \tag{5.199}$$

Thus, the inverse of $T_{R_1,R}$, $0 < R < R_1$, is formally representable by

$$\begin{aligned} T_{R_1,R}^{-1} f \quad = \quad & -\sum_{n=1}^{\infty} \sum_{k=1}^{2n+1} \sqrt{\frac{2n+1}{n}} \left(\frac{R_1}{R}\right)^{n+1} \left(f, \tilde{y}_{n,k}^{(3);R}\right)_{\mathrm{l}^2(\Omega_R)} \tilde{y}_{n,k}^{(1);R_1} \\ & -\sum_{n=1}^{\infty} \sum_{k=1}^{2n+1} \sqrt{\frac{2n+1}{n+1}} \left(\frac{R_1}{R}\right)^{n} \left(f, \tilde{y}_{n,k}^{(2);R}\right)_{\mathrm{l}^2(\Omega_R)} \tilde{y}_{n,k}^{(3);R_1}, \end{aligned} \tag{5.200}$$

for f of class $\tilde{\mathrm{l}}^2_{(2)}(\Omega_R) \oplus \tilde{\mathrm{l}}^2_{(3)}(\Omega_R)$. Since $(2n+1)^{\frac{1}{2}} n^{-\frac{1}{2}} {R_1}^{n+1} R^{-(n+1)} \to \infty$ and $(2n+1)^{\frac{1}{2}} (n+1)^{-\frac{1}{2}} {R_1}^{n} R^{-n} \to \infty$, for $n \to \infty$, the operator $T_{R_1,R}^{-1}$ is unbounded, therefore, showing the ill-posedness of the original inverse problem. Similar considerations hold true for $T_{R_1,R}^{-1}$, $0 < R_1 < R$.

Remark 5.32. A possible regularization scheme (compare Exercise 4.11 for a similar linear concept) for the inverse $T_{R_1,R}^{-1}$, $0 < R < R_1$ can briefly be described as follows: Let a sequence $\left\{F_J^{(1)}(\sigma_n), F_J^{(3)}(\sigma_n)\right\}_{J,n\in\mathbb{N}}$ be given such that

(i) for all $n \in \mathbb{N}$,

$$\lim_{J\to\infty} F_J^{(1)}(\sigma_n) = \left(\frac{2n+1}{n}\right)^{\frac{1}{4}} \left(\frac{R_1}{R}\right)^{\frac{n+1}{2}}, \tag{5.201}$$

$$\lim_{J\to\infty} F_J^{(3)}(\sigma_n) = \left(\frac{2n+1}{n+1}\right)^{\frac{1}{4}} \left(\frac{R_1}{R}\right)^{\frac{n}{2}}, \tag{5.202}$$

(ii) for all $J \in \mathbb{N}$,

$$\left(F_J^{(1)}(\sigma_n)\right)^2 \le \left(F_{J+1}^{(1)}(\sigma_n)\right)^2, \quad \left(F_J^{(3)}(\sigma_n)\right)^2 \le \left(F_{J+1}^{(3)}(\sigma_n)\right)^2, \tag{5.203}$$

(iii) for all $\xi \in \Omega$ and $J \in \mathbb{N}$,

$$\sum_{n=1}^{\infty} \sum_{k=1}^{2n+1} \left| F_J^{(i)}(\sigma_n) \tilde{y}_{n,k}^{(j)}(\xi) \right|^2 < \infty, \quad i,j \in \{1,3\}. \tag{5.204}$$

If the above properties are satisfied, the sequence $\{F_J^{(1)}(\sigma_n), F_J^{(3)}(\sigma_n)\}_{J,n\in\mathbb{N}}$ is called admissible, and we are able to introduce so-called scaling kernels

$$k_d^{J;(1)}(x,\eta) = \sum_{n=1}^{\infty}\sum_{k=1}^{2n+1} F_J^{(1)}(\sigma_n)\tilde{y}_{n,k}^{(3);R}(x)Y_{n,k}(\eta), \quad x \in \Omega_R, \eta \in \Omega, \tag{5.205}$$

$$k_d^{J;(3)}(x,\eta) = \sum_{n=1}^{\infty}\sum_{k=1}^{2n+1} F_J^{(3)}(\sigma_n)\tilde{y}_{n,k}^{(2);R}(x)Y_{n,k}(\eta), \quad x \in \Omega_R, \eta \in \Omega, \tag{5.206}$$

$$k_r^{J;(1)}(x,\eta) = -\sum_{n=1}^{\infty}\sum_{k=1}^{2n+1} F_J^{(1)}(\sigma_n)\tilde{y}_{n,k}^{(1);R_1}(x)Y_{n,k}(\eta), \quad x \in \Omega_{R_1}, \eta \in \Omega, \tag{5.207}$$

$$k_r^{J;(3)}(x,\eta) = -\sum_{n=1}^{\infty}\sum_{k=1}^{2n+1} F_J^{(3)}(\sigma_n)\tilde{y}_{n,k}^{(3);R_1}(x)Y_{n,k}(\eta), \quad x \in \Omega_{R_1}, \eta \in \Omega \tag{5.208}$$

Note that the indices r and k simply stand for *reconstruction*, and *decomposition*. Furthermore, different from the linear case in Exercise 4.11, the quantities $F_J^{(i)}(\sigma_n)$ converge to the square root of the inverse singular value in the bilinear approach presented here, not to the inverse singular value itself. It is not difficult to show that

$$\begin{aligned}\lim_{J\to\infty} \Bigg\| T_{R_1,R}^{-1} f - \int_\Omega k_r^{J;(1)}(\cdot,\eta) \int_{\Omega_R} k_d^{J;(1)}(y,\eta)\cdot f(y)d\omega(y)d\omega(\eta) \\ - \int_\Omega k_r^{J;(3)}(\cdot,\eta)\int_{\Omega_R} k_d^{J;(3)}(y,\eta)\cdot f(y)d\omega(y)d\omega(\eta)\Bigg\|_{\mathrm{l}^2(\Omega_{R_1})} = 0,\end{aligned} \tag{5.209}$$

for f of class $\tilde{\mathrm{l}}^2_{(2)}(\Omega_R) \oplus \tilde{\mathrm{l}}^2_{(3)}(\Omega_R)$. The kernels $k_d^{J;(i)}(\cdot,\eta)$, $k_r^{J;(i)}(\cdot,\eta)$, $i = 1,3$, $\eta \in \Omega$, typically show a better spatial localization than spherical harmonics, depending on the choice of the coefficients $F_J^{(i)}(\sigma_n)$, $i = 1,3$. Beside the regularization of the ill-posed inverse problem, this is an advantageous feature for local reconstructions. For the numerical example in the upcoming subsection, we make the particular choice of a cubic polynomial regularization:

$$F_J^{(1)}(\sigma_n) = \begin{cases} \left(\frac{2n+1}{n}\right)^{\frac14}\left(\frac{R_1}{R}\right)^{\frac{n+1}{2}}\left(1-\frac{n}{2^J-1}\right)^2\left(1+\frac{2n}{2^J-1}\right), & n = 1,\ldots,2^J-1, \\ 0, & n = 2^J, 2^J+1,\ldots, \end{cases} \tag{5.210}$$

$$F_J^{(3)}(\sigma_n) = \begin{cases} \left(\frac{2n+1}{n+1}\right)^{\frac14}\left(\frac{R_1}{R}\right)^{\frac{n}{2}}\left(1-\frac{n}{2^J-1}\right)^2\left(1+\frac{2n}{2^J-1}\right), & n = 1,\ldots,2^J-1, \\ 0, & n = 2^J, 2^J+1,\ldots. \end{cases} \tag{5.211}$$

The sequence $\{F_J^{(1)}(\sigma_n), F_J^{(3)}(\sigma_n)\}_{n\in\mathbb{N}}$ is obviously bounded (for any fixed $J \in \mathbb{N}$). Consequently, (5.209) allows an approximation of $T_{R_1,R}^{-1}$ by bounded operators, which provides the desired stability of the regularization. For more details, the reader is referred to C. Mayer [2004]. The case $T_{R_1,R}^{-1}$, $0 < R_1 < R$, can be treated analogously.

5.5.2 Numerical Application: Tangential Currents

We conclude our investigations with the reconstruction of a toroidal current system on a sphere Ω_{R_1} in the ionosphere, with radius $R_1 = R_0 + 110$ km (where $R_0 = 6371.2$ km denotes the mean Earth radius). The used magnetic field dataset has been collected by the MAGSAT satellite (at an average altitude of 450km above the spherical Earth surface, i.e., $S = R_0 + 450$ km) during evening local time and has been pre-processed at DTU Space (i.e., after an adequate data selection, the magnetic field model GSFC(12/83) up to spherical harmonic degree 12 has been subtracted in order to correct for the internal magnetic field of the Earth).

Since we are only interested in the toroidal part q_j of the current density j, relation (5.190) is the one of importance to us in order to obtain $q_j = T^{-1}_{R_1,S}[p_b]$, with p_b denoting the known poloidal part of the measured magnetic field b. Remembering Remark 5.32, we need to evaluate the quantity

$$q_j(x) \approx \int_{\Omega} k_r^{J;(3)}(x,\eta) \int_{\Omega_S} k_d^{J;(3)}(y,\eta) \cdot p_b(y) d\omega(y) d\omega(\eta), \quad x \in \Omega_{R_1}, \quad (5.212)$$

for some sufficiently large $J \in \mathbb{N}$ (note that only the case $R_1 > S$ was treated in more detail in Remark 5.32, so that in the present case the kernels $k_r^{J;(3)}(\cdot,\cdot)$ and $k_d^{J;(3)}(\cdot,\cdot)$ need to be modified). For the numerical evaluation of the integrals in (5.212), we use the method proposed by J.R. Driscoll, R.M. Healy [1994], which uses an equiangular grid that is well-suited for satellite data.

The results for $J = 6$ are shown in Figure 5.4. One can clearly recognize strong polar current systems and the equatorial electrojet directed along the magnetic dip equator. The North-South effects appearing along the zero meridian are probably due to errors in the data-averaging process and have no geophysical interpretation. Furthermore, it has to be stressed that the reconstructed current system does not represent the true toroidal current system in the ionosphere. Due to the assumption that all currents are located on the sphere Ω_{R_1}, we have only reconstructed an equivalent current system (*equivalent* is meant in the sense that the reconstructed current system produces the same magnetic effect at satellite altitude as the true current system, but it can differ from the true three-dimensional currents). However, since the current carrying ionosphere is relatively thin in comparison to the entire iono-/magnetosphere structure, this still gives a pretty good insight into the true toroidal currents in the ionosphere.

Finally, it can be seen from Figure 5.4 that the toroidal current density has no sinks or sources. This is due to the fact that toroidal fields are surface divergence-free, i.e., $\nabla^* \cdot q_j = 0$. Consequently, toroidal currents are often called *surface divergence-free currents* and are denoted by $j_{\mathrm{df}} = \mathrm{L}^* J_3$ (where J_3 denotes the corresponding Helmholtz scalar of the vectorial current density j). The determination of the surface curl-free, tangential currents $j_{\mathrm{cf}} = \nabla^* J_2$ additionally requires the knowledge of the radial contributions J_1 (see, e.g., Subsection 8.4.2 for some more details).

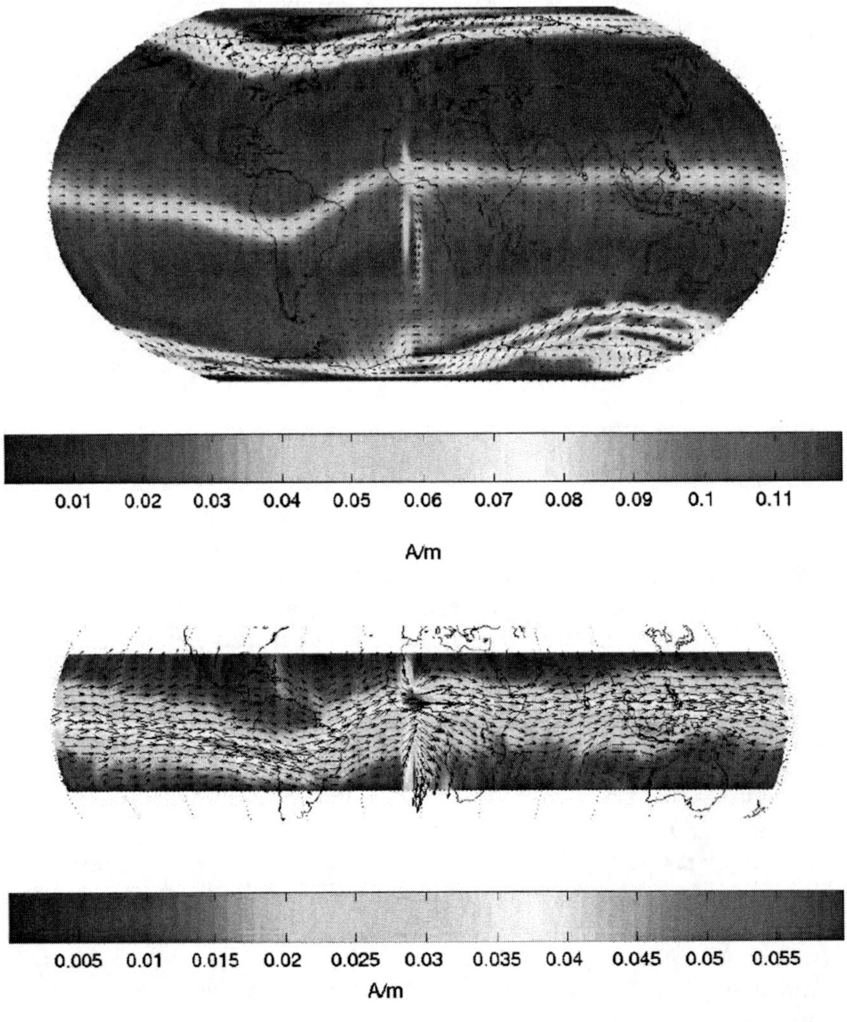

FIGURE 5.4
Global (top) and local (bottom) approximation of the toroidal current densities q_j at an altitude of 110 km.

5.6 Exercises

Exercise 5.1. Formulate the Law of Biot–Savart (cf. Theorem 5.6) for regular regions $\mathcal{G} \subset \mathbb{R}^3$. Specify (boundary) conditions that guarantee uniqueness of the magnetic field b.

Exercise 5.2. Let f be of class $c^{(1)}(\mathbb{R}^3)$ such that $\lim_{|x|\to\infty} |f(x)| = \lim_{|x|\to\infty} |\nabla \otimes f(x)| = 0$, uniformly with respect to all directions of x.

(a) Let $y_0 \in \mathbb{R}^3$ be fixed. Show that the integral

$$u(x) = \frac{1}{4\pi} \int_{\mathbb{R}^3} (\nabla_y \cdot f(y)) \left(\nabla_y \frac{1}{|x-y|} - \nabla_y \frac{1}{|y_0 - y|} \right) dV(y) \qquad (5.213)$$

exists for all $x \in \mathbb{R}^3$. Furthermore, prove that

$$\nabla \cdot u(x) = \nabla \cdot f(x), \quad x \in \mathbb{R}^3. \qquad (5.214)$$

(b) Use part (a) to verify that vector fields u, v of class $c^{(1)}(\mathbb{R}^3)$ exist, where u is curl-free, i.e., $\nabla \wedge u(x) = 0$, $x \in \mathbb{R}^3$, and v is divergence-free, i.e., $\nabla \cdot v(x) = 0$, $x \in \mathbb{R}^3$, such that

$$f(x) = u(x) + v(x), \quad x \in \mathbb{R}^3. \qquad (5.215)$$

Exercise 5.3 (Solenoidal and divergence-free vector fields)**.**

(a) Consider the function $f(x) = \frac{x}{|x|^2}$, for $x \in \mathcal{B}_{R_0,R_1}(0)$, $0 < R_0 < R_1$. Show that f is divergence-free in $\mathcal{B}_{R_0,R_1}(0)$ but not solenoidal.

(b) Let $\mathcal{G} \subset \mathbb{R}^3$ be a region, and suppose that f is of class $c^{(1)}(\mathcal{G})$. Show that, if f is divergence-free on $\mathcal{G}$ and additionally satisfies $\int_{\Sigma_0} \nu(y) \cdot f(y) d\omega(y) = 0$ for one fixed closed regular surface $\Sigma_0 \subset \mathcal{G}$, then f is solenoidal.

Exercise 5.4. Suppose that H_n and H_m are non-zero homogeneous, harmonic polynomials of degree n and m, respectively. It is known that there exist uniquely determined homogeneous, harmonic polynomials H_{n+m} of degree $n+m$ and $\tilde{H}_{n+m-2i}$ of degree $n+m-2i$, $i = 1, \ldots, \lfloor \frac{n+m}{2} \rfloor$, such that

$$H_n H_m = H_{n+m} + \sum_{i=1}^{\lfloor \frac{n+m}{2} \rfloor} |x|^{2i} \tilde{H}_{n+m-2i}. \qquad (5.216)$$

Show that the leading coefficient H_{n+m} has to be non-zero. (Hint: An irreducible polynomial P divides the product $Q_1 Q_2$ of two polynomials Q_1 and Q_2 only if P divides at least one of the factors Q_1 or Q_2.)

Exercise 5.5. Assume that f is of class $c^{(2)}(\mathbb{R}^3)$ and g of class $c^{(1)}(\mathbb{R}^3)$. Furthermore, suppose that f is regular at infinity and that the pre-Maxwell equations

$$\nabla \wedge f(x) = g(x), \quad x \in \mathbb{R}^3, \tag{5.217}$$

$$\nabla \cdot f(x) = 0, \quad x \in \mathbb{R}^3, \tag{5.218}$$

are satisfied. We already know from Section 5.4 that f can be decomposed into

$$f(x) = p_f^{\text{int}}(x) + p_f^{\text{ext}}(x) + q_f(x), \quad x \in \Omega_R, \tag{5.219}$$

for a fixed radius $R > 0$. Show that

$$p_f^{\text{int}}(x) = \sum_{n=0}^{\infty}\sum_{k=1}^{2n+1} \alpha_{n,k} h^{(1);R}_{-n-1,k}(x), \quad x \in \mathbb{R}^3 \setminus \overline{\mathcal{B}_R(0)}, \tag{5.220}$$

$$p_f^{\text{ext}}(x) = \sum_{n=1}^{\infty}\sum_{k=1}^{2n+1} \beta_{n,k} h^{(2);R}_{n,k}(x), \quad x \in \mathcal{B}_R(0), \tag{5.221}$$

$$q_f(x) = \sum_{n=1}^{\infty}\sum_{k=1}^{2n+1} \gamma_{n,k} \tilde{y}^{(3);R}_{n,k}(x), \quad x \in \Omega_R, \tag{5.222}$$

where the coefficients $\alpha_{n,k}$, $\beta_{n,k}$, and $\gamma_{n,k}$ are determined by

$$\alpha_{n,k} = \left(f, \tilde{y}^{(1);R}_{n,k}\right)_{l^2(\Omega_R)}, \tag{5.223}$$

$$\beta_{n,k} = \left(f, \tilde{y}^{(2);R}_{n,k}\right)_{l^2(\Omega_R)}, \tag{5.224}$$

$$\gamma_{n,k} = \left(f, \tilde{y}^{(3);R}_{n,k}\right)_{l^2(\Omega_R)}. \tag{5.225}$$

Observe that p_f^{int} and p_f^{ext} are continuous up to Ω_R.

Exercise 5.6. Prove Lemma 5.26, i.e., show that the operators $T_{R_0,R}$, $R \neq R_0$, are completely continuous.

Exercise 5.7. Prove Theorem 5.28: For $0 < R_0 < R$, we have

$$T_{R_0,R}\left[y^{(1);R_0}_{n,k}\right] = -\frac{\sqrt{n(n+1)}}{2n+1}\left(\frac{R_0}{R}\right)^n y^{(3);R}_{n,k}, \tag{5.226}$$

$$T_{R_0,R}\left[y^{(2);R_0}_{n,k}\right] = -\frac{n+1}{2n+1}\left(\frac{R_0}{R}\right)^n y^{(3);R}_{n,k}, \tag{5.227}$$

$$T_{R_0,R}\left[y^{(3);R_0}_{n,k}\right] = -\frac{n}{2n+1}\left(\frac{R_0}{R}\right)^{n+1}\left(\sqrt{\frac{n+1}{n}}\, y^{(1);R}_{n,k} - y^{(2);R}_{n,k}\right). \tag{5.228}$$

Part III

Potential Theory on the Unit Sphere Ω

6

Basic Concepts

CONTENTS

6.1 Background Material

In this section, we collect some basic material concerning potential theory on the sphere. The structure of the entire chapter is designed in parallel to Chapter 3 to point out the similarities as well as the differences in comparison to potential theory in the Euclidean space $\mathbb{R}^3$.

6.1.1 Fundamental Solution

We start by defining the fundamental solution for the Beltrami operator.

Definition 6.1. The function $G(\Delta^*;\cdot) : [-1,1) \to \mathbb{R}$ given by

$$G(\Delta^*;t) = \frac{1}{4\pi}\ln(1-t) - \frac{1}{4\pi}(1-\ln(2)), \quad t \in [-1,1), \tag{6.1}$$

is called the *fundamental solution for the Beltrami operator.*

$G(\Delta^*;\xi\cdot\eta)$, $1-\xi\cdot\eta > 0$, possesses the following interesting properties, which are listed here for later use:

Symmetry: $(\xi,\eta) \mapsto G(\Delta^*;\xi\cdot\eta)$, $1-\xi\cdot\eta>0$, is rotational invariant, i.e., it depends only on $t=\xi\cdot\eta$.

Differential Equation: For any fixed $\xi \in \Omega$, the function $\eta \mapsto G(\Delta^*;\xi\cdot\eta)$ is twice continuously differentiable on the set $\{\eta\in\Omega : 1-\xi\cdot\eta>0\}$, and

$$\Delta^*_\eta G(\Delta^*;\xi\cdot\eta) = -\frac{1}{4\pi} \tag{6.2}$$

for $\eta \in \Omega$ with $1-\xi\cdot\eta > 0$.

Regularity: For any fixed $\xi\in\Omega$, the function

$$\eta \mapsto G(\Delta^*;\xi\cdot\eta) - \frac{1}{4\pi}\ln(1-\xi\cdot\eta) \tag{6.3}$$

is continuously differentiable on Ω.

Normalization: For any fixed $\xi\in\Omega$,

$$\frac{1}{4\pi}\int_\Omega G(\Delta^*;\xi\cdot\eta)\,d\omega(\eta) = 0. \tag{6.4}$$

These properties actually determine the fundamental solution of the Beltrami operator uniquely (see Exercise 6.1). Using the Addition Theorem for spherical harmonics, we are able to verify the following bilinear series expansion of the fundamental solution:

$$\begin{aligned} G(\Delta^*;\xi\cdot\eta) &= \sum_{n=1}^{\infty}\sum_{k=1}^{2n+1} \frac{1}{-n(n+1)} Y_{n,k}(\xi)Y_{n,k}(\eta) \\ &= \frac{1}{4\pi}\sum_{n=1}^{\infty}\frac{2n+1}{-n(n+1)}P_n(\xi\cdot\eta), \end{aligned} \tag{6.5}$$

for $\xi,\eta\in\Omega$ with $1-\xi\cdot\eta>0$. Note that the series (6.5) contains no zero-order term. Thus, $\eta\mapsto\Delta^*_\eta G(\Delta^*;\xi\cdot\eta)$ differs from the Dirac distribution by the term $-\frac{1}{4\pi}$ (in contrast to the fundamental solution for the Laplace operator, where $y\mapsto\Delta_y G(\Delta;|x-y|)$ actually yields the Dirac distribution, all understood in a formal sense). More detailed information on $G(\Delta^*;\cdot)$ as well as fundamental solutions for the more general spherical Helmholtz operator $\Delta^* + \lambda$, $\lambda \in \mathbb{R}$, can be found, e.g., in W. Freeden, M. Schreiner [2009]. An overview on related topics involving sphericity and multiperiodicity in $\mathbb{R}^3$ (and general dimensions $\mathbb{R}^n$, $n \geq 2$) is given in W. Freeden [2011].

6.1.2 Third Green Formula

By use of the fundamental solution, we can formulate the Third Green Theorem for the Beltrami operator.

Theorem 6.2 (Third Green Theorem for Δ^*). *Let $\Gamma \subset \Omega$ be a regular region, and F be of class $\mathrm{C}^{(2)}(\overline{\Gamma})$. Then*

$$\begin{aligned}\frac{\alpha(\xi)}{2\pi}F(\xi) &= \frac{1}{4\pi}\int_\Gamma F(\eta)d\omega(\eta) + \int_\Gamma G(\Delta^*;\xi\cdot\eta)\Delta^*_\eta F(\eta)d\omega(\eta) \\ &\quad + \int_{\partial\Gamma} F(\eta)\frac{\partial}{\partial\nu(\eta)}G(\Delta^*;\xi\cdot\eta)d\sigma(\eta) - \int_{\partial\Gamma} G(\Delta^*;\xi\cdot\eta)\frac{\partial}{\partial\nu(\eta)}F(\eta)d\sigma(\eta),\end{aligned} \tag{6.6}$$

for $\xi \in \Omega$, where the solid angle α is given by

$$\alpha(\xi) = \begin{cases} 2\pi, & \xi \in \Gamma, \\ \pi, & \xi \in \partial\Gamma, \\ 0, & \xi \in \Gamma^c. \end{cases} \tag{6.7}$$

Proof. Letting $\rho \in (0,2)$ and $\xi \in \Omega$, we get from Green's formula (cf. Theorem 2.9) that

$$\begin{aligned}&\int_{\substack{\eta\in\Gamma\\ 1-\xi\cdot\eta\geq\rho}} F(\eta)\Delta^*_\eta G(\Delta^*;\xi\cdot\eta)d\omega(\eta) - \int_{\substack{\eta\in\Gamma\\ 1-\xi\cdot\eta\geq\rho}} G(\Delta^*;\xi\cdot\eta)\Delta^*_\eta F(\eta)d\omega(\eta) \\ &= \int_{\substack{\eta\in\Gamma\\ 1-\xi\cdot\eta=\rho}} F(\eta)\frac{\partial}{\partial\nu(\eta)}G(\Delta^*;\xi\cdot\eta)d\sigma(\eta) - \int_{\substack{\eta\in\Gamma\\ 1-\xi\cdot\eta=\rho}} G(\Delta^*;\xi\cdot\eta)\frac{\partial}{\partial\nu(\eta)}F(\eta)d\sigma(\eta) \\ &\quad + \int_{\partial\Gamma} F(\eta)\frac{\partial}{\partial\nu(\eta)}G(\Delta^*;\xi\cdot\eta)d\sigma(\eta) - \int_{\partial\Gamma} G(\Delta^*;\xi\cdot\eta)\frac{\partial}{\partial\nu(\eta)}F(\eta)d\sigma(\eta) \\ &= \frac{1}{4\pi}\int_{\substack{\eta\in\Gamma\\ 1-\xi\cdot\eta=\rho}} F(\eta)\frac{\partial}{\partial\nu(\eta)}\ln(1-\xi\cdot\eta)d\sigma(\eta) - \int_{\substack{\eta\in\Gamma\\ 1-\xi\cdot\eta=\rho}} G(\Delta^*;\xi\cdot\eta)\frac{\partial}{\partial\nu(\eta)}F(\eta)d\sigma(\eta) \\ &\quad + \int_{\partial\Gamma} F(\eta)\frac{\partial}{\partial\nu(\eta)}G(\Delta^*;\xi\cdot\eta)d\sigma(\eta) - \int_{\partial\Gamma} G(\Delta^*;\xi\cdot\eta)\frac{\partial}{\partial\nu(\eta)}F(\eta)d\sigma(\eta).\end{aligned} \tag{6.8}$$

The second term on the right-hand side of (6.8), i.e., the integral

$$\int_{\substack{\eta\in\Gamma\\ 1-\xi\cdot\eta=\rho}} G(\Delta^*;\xi\cdot\eta)\frac{\partial}{\partial\nu(\eta)}F(\eta)d\sigma(\eta) \tag{6.9}$$

vanishes uniformly with respect to $\xi \in \Omega$ as ρ tends to zero, since F is continuously differentiable on $\overline{\Gamma}$. Observing that the unit normal vector is given by $\nu(\eta) = (1-(\xi\cdot\eta)^2)^{-\frac{1}{2}}(\xi-(\xi\cdot\eta)\eta)$, we find for the first term on the right-hand side of (6.8)

$$
\begin{aligned}
& \frac{1}{4\pi} \int\limits_{\substack{\eta\in\Gamma \\ 1-\xi\cdot\eta=\rho}} F(\eta) \frac{\partial}{\partial\nu(\eta)} \ln(1-\xi\cdot\eta)\, d\sigma(\eta) \qquad (6.10) \\
= & -\frac{1}{4\pi} \int\limits_{\substack{\eta\in\Gamma \\ 1-\xi\cdot\eta=\rho}} F(\eta) \frac{\xi-(\xi\cdot\eta)\eta}{\sqrt{1-(\xi\cdot\eta)^2}} \cdot \frac{\xi-(\xi\cdot\eta)\eta}{1-\xi\cdot\eta}\, d\sigma(\eta) \\
= & -\frac{1}{4\pi} \int\limits_{\substack{\eta\in\Gamma \\ 1-\xi\cdot\eta=\rho}} F(\eta) \frac{\sqrt{1-(\xi\cdot\eta)^2}}{1-\xi\cdot\eta}\, d\sigma(\eta) \\
= & -\frac{1}{4\pi} \frac{\sqrt{1-(1-\rho)^2}}{\rho} \int\limits_{\substack{\eta\in\Gamma \\ 1-\xi\cdot\eta=\rho}} F(\eta)\, d\sigma(\eta) \\
= & -\frac{1}{4\pi} \frac{\sqrt{1-(1-\rho)^2}}{\rho} F(\eta_\rho) 2\pi\alpha_\rho(\xi) \sqrt{1-(1-\rho)^2} \\
= & -\frac{2-\rho}{2} \alpha_\rho(\xi) F(\eta_\rho),
\end{aligned}
$$

where $\eta_\rho \in \partial\Gamma_\rho(\xi) = \{\eta \in \Omega : 1-\xi\cdot\eta = \rho\}$ and α_ρ is determined by

$$
2\pi\alpha_\rho(\xi)\sqrt{1-(1-\rho)^2} = \int\limits_{\substack{\eta\in\Gamma \\ 1-\xi\cdot\eta=\rho}} d\sigma(\eta). \qquad (6.11)
$$

In other words, η_ρ tends to ξ as ρ tends to zero, and thus, $\frac{2-\rho}{2}F(\eta_\rho)$ tends to $F(\xi)$. Furthermore, $\alpha_\rho(\xi)$ tends to the solid angle $\alpha(\xi)$ as ρ tends to zero. Last, we observe that

$$
\int\limits_{\substack{\eta\in\Gamma \\ 1-\xi\cdot\eta\geq\rho}} F(\eta)\Delta^*_\eta G(\Delta^*;\xi\cdot\eta)\, d\omega(\eta) = -\frac{1}{4\pi} \int\limits_{\substack{\eta\in\Gamma \\ 1-\xi\cdot\eta\geq\rho}} F(\eta)\, d\omega(\eta). \qquad (6.12)
$$

Combining the previous results, reconsidering Equation (6.8), and taking the limit $\rho \to 0$, we obtain the desired Third Green Theorem (6.6). □

Remark 6.3. Choosing $F = 1$ in Theorem 6.2, we get

$$
\int_{\partial\Gamma} \frac{\partial}{\partial\nu(\eta)} G(\Delta^*;\xi\cdot\eta)\, d\sigma(\eta) = \begin{cases} 1 - \frac{\|\Gamma\|}{4\pi}, & \xi \in \Gamma, \\ \frac{1}{2} - \frac{\|\Gamma\|}{4\pi}, & \xi \in \partial\Gamma, \\ -\frac{\|\Gamma\|}{4\pi}, & \xi \in \Gamma^c. \end{cases} \qquad (6.13)
$$

Analogous to Theorem 6.2, we obtain the following identities for the surface gradient and the surface curl gradient.

Theorem 6.4 (Third Green Theorem for ∇^* and L^*). *Let $\Gamma \subset \Omega$ be a regular region, and suppose that F is of class $\mathrm{C}^{(1)}(\overline{\Gamma})$. Then*

$$\begin{aligned} \frac{\alpha(\xi)}{2\pi} F(\xi) &= \frac{1}{4\pi} \int_\Gamma F(\eta) d\omega(\eta) - \int_\Gamma \nabla^*_\eta G(\Delta^*; \xi \cdot \eta) \cdot \nabla^*_\eta F(\eta) d\omega(\eta) \quad (6.14) \\ &\quad + \int_{\partial\Gamma} F(\eta) \frac{\partial}{\partial\nu(\eta)} G(\Delta^*; \xi \cdot \eta) d\sigma(\eta), \\ &= \frac{1}{4\pi} \int_\Gamma F(\eta) d\omega(\eta) - \int_\Gamma \mathrm{L}^*_\eta G(\Delta^*; \xi \cdot \eta) \cdot \mathrm{L}^*_\eta F(\eta) d\omega(\eta) \\ &\quad + \int_{\partial\Gamma} F(\eta) \frac{\partial}{\partial\nu(\eta)} G(\Delta^*; \xi \cdot \eta) d\sigma(\eta), \end{aligned}$$

for $\xi \in \Omega$, where α is the solid angle as defined in Theorem 6.2.

Comparing Theorems 6.2 and 6.4 to Theorem 3.3 in the Euclidean framework of $\mathbb{R}^3$, we find the integral mean value $\frac{1}{4\pi} \int_\Gamma F(\eta) d\omega(\eta)$ as an additional term. Once more, this observation reflects the property that $\eta \mapsto \Delta^*_\eta G(\Delta^*; \xi \cdot \eta)$ formally describes the Dirac distribution only up to an additive constant $-\frac{1}{4\pi}$.

Remark 6.5. Concerning the entire sphere Ω, the Third Green Theorems are still valid, but the boundary integrals vanish and $\alpha(\xi) = 2\pi$ for all $\xi \in \Omega$. More precisely,

$$F(\xi) = \frac{1}{4\pi} \int_\Omega F(\eta) d\omega(\eta) + \int_\Omega G(\Delta^*; \xi \cdot \eta) \Delta^*_\eta F(\eta) d\omega(\eta), \quad (6.15)$$

for $\xi \in \Omega$ and $F \in \mathrm{C}^{(2)}(\Omega)$.

6.1.3 Mean Value Property

Next, we turn to the Mean Value Property of harmonic functions.

Definition 6.6. Let $\Gamma \subset \Omega$ be a regular region. A function $U : \Gamma \to \mathbb{R}$ is called *harmonic with respect to the Beltrami operator* on Γ if U is of class $\mathrm{C}^{(2)}(\Gamma)$ and satisfies

$$\Delta^* U(\xi) = 0, \quad \xi \in \Gamma. \quad (6.16)$$

Note that whenever it is obvious from the context that the harmonicity is meant with respect to the Beltrami operator, we just call U harmonic on Γ.

Theorem 6.7 (Mean Value Property). *Let $\Gamma \subset \Omega$ be a regular region. Then a continuous function $U : \Gamma \to \mathbb{R}$ is harmonic with respect to the Beltrami operator if and only if*

$$U(\xi) = \frac{1}{4\pi} \int_{\Gamma_\rho(\xi)} U(\eta) d\omega(\eta) + \frac{\sqrt{2-\rho}}{4\pi\sqrt{\rho}} \int_{\partial\Gamma_\rho(\xi)} U(\eta) d\sigma(\eta), \quad \xi \in \Gamma, \quad (6.17)$$

for any spherical cap $\Gamma_\rho(\xi) \Subset \Gamma$.

Proof. "$\Rightarrow$": We assume U to be harmonic with respect to the Beltrami operator in Γ. If $\xi \in \Gamma$ and $\rho \in (0,2)$ such that $\overline{\Gamma_\rho(\xi)} \subset \Gamma$, then Theorem 6.2 yields

$$\begin{aligned} U(\xi) &= \frac{1}{4\pi}\int_{\Gamma_\rho(\xi)} U(\eta)d\omega(\eta) + \int_{\partial\Gamma_\rho(\xi)} U(\eta)\frac{\partial}{\partial\nu(\eta)}G(\Delta^*;\xi\cdot\eta)d\sigma(\eta) \\ &\quad - \int_{\partial\Gamma_\rho(\xi)} G(\Delta^*;\xi\cdot\eta)\frac{\partial}{\partial\nu(\eta)}U(\eta)\,d\sigma(\eta). \end{aligned} \tag{6.18}$$

For $\eta \in \partial\Gamma_\rho(\xi)$, we have $G(\Delta^*;\xi\cdot\eta) = \frac{1}{4\pi}\ln(\rho) + \frac{1}{4\pi}(1-\ln(2))$. Using the spherical variant of the Gauss Theorem, this implies

$$\begin{aligned} &\int\limits_{\partial\Gamma_\rho(\xi)} G(\Delta^*;\xi\cdot\eta)\frac{\partial}{\partial\nu(\eta)}U(\eta)d\sigma(\eta) \\ &= \left(\frac{1}{4\pi}\ln(\rho) + \frac{1}{4\pi}(1-\ln(2))\right)\int\limits_{\partial\Gamma_\rho(\xi)} \frac{\partial}{\partial\nu(\eta)}U(\eta)d\sigma(\eta) \\ &= \left(\frac{1}{4\pi}\ln(\rho) + \frac{1}{4\pi}(1-\ln(2))\right)\int\limits_{\Gamma_\rho(\xi)} \Delta^*_\eta U(\eta)d\omega(\eta) \\ &= 0. \end{aligned} \tag{6.19}$$

Furthermore, an easy calculation shows that, for $\eta \in \partial\Gamma_\rho(\xi)$,

$$\begin{aligned} \frac{\partial}{\partial\nu(\eta)}G(\Delta^*;\xi\cdot\eta) &= -\frac{\xi-(\xi\cdot\eta)\eta}{\sqrt{1-(\xi\cdot\eta)^2}}\cdot\left(-\frac{1}{4\pi}\frac{\xi-(\xi\cdot\eta)\eta}{1-\xi\cdot\eta}\right) \\ &= \frac{1}{4\pi}\frac{\sqrt{1-(\xi\cdot\eta)^2}}{1-\xi\cdot\eta} = \frac{\sqrt{2-\rho}}{4\pi\sqrt{\rho}}. \end{aligned} \tag{6.20}$$

Inserting the equalities (6.19) and (6.20) into (6.18), we are led to (6.17).

"$\Leftarrow$": We defer this direction to Section 6.4, where we are able to use the theory of boundary-value problems (for the proof, the reader is referred to page 321). □

6.1.4 Maximum/Minimum Principle

Once the Mean Value Property is established, a maximum/minimum principle can be derived in the same manner as for the Laplace operator in Theorem 3.10 (see Exercise 6.3).

Theorem 6.8 (Maximum/Minimum Principle). *Let* $\Gamma \subset \Omega$ *be a regular region. Suppose that* $U : \Gamma \to \mathbb{R}$ *is harmonic with respect to the Beltrami operator. Then the following statements hold true:*

(a) *If U is non-constant, then there exists neither a minimum nor a maximum of U in Γ.*

(b) *If U is additionally of class $C^{(0)}(\overline{\Gamma})$, then U reaches its minimum and maximum in $\overline{\Gamma}$, and all extremal points are located on $\partial\Gamma$.*

Corollary 6.9. *Let $\Gamma \subset \Omega$ be a regular region, and assume that $U \in C^{(0)}(\overline{\Gamma})$ is harmonic with respect to the Beltrami operator in Γ. Then*

$$\sup_{\xi \in \Gamma} |U(\xi)| \leq \sup_{\xi \in \partial\Gamma} |U(\xi)|. \tag{6.21}$$

Harnack's theorem for harmonic functions with respect to the Beltrami operator can as well be formulated in analogy to the corresponding statement for the Laplace operator (cf. Theorem 3.9). Due to the close similarity to the Euclidean case, the precise elaboration of the proof is left to the reader.

Theorem 6.10. *Let $\Gamma \subset \Omega$ be a regular region, and suppose that the functions $U_n : \Gamma \to \mathbb{R}$, $n \in \mathbb{N}$, form a sequence of harmonic functions with respect to the Beltrami operator. If $\{U_n\}_{n \in \mathbb{N}_0}$ converges locally uniformly toward a function $U : \Gamma \to \mathbb{R}$, i.e.,*

$$\lim_{n \to \infty} \sup_{\xi \in \Sigma} |U(\xi) - U_n(\xi)| = 0, \tag{6.22}$$

for every $\Sigma \Subset \Omega$, then U is harmonic with respect to the Beltrami operator as well. Furthermore, $\{\nabla^* U_n\}_{n \in \mathbb{N}_0}$ converges locally uniformly toward $\nabla^* U$.

Remark 6.11. If the functions U_n appearing in Theorem 6.10 are additionally of class $C^{(0)}(\overline{\Gamma})$, then the Maximum/Minimum Principle tells us that uniform convergence of $\{U_n\}_{n \in \mathbb{N}_0}$ only on the boundary $\partial\Gamma$ is sufficient to guarantee the harmonicity of the limit function U (with respect to the Beltrami operator) in Γ.

6.2 Surface Potentials

Let $\Gamma \subset \Omega$ be a regular region and suppose that $F : \Gamma \to \mathbb{R}$ is sufficiently often differentiable. By

$$U(\xi) = \int_\Gamma G(\Delta^*; \xi \cdot \eta) F(\eta) d\omega(\eta), \quad \xi \in \Omega, \tag{6.23}$$

we denote the spherical analogon to the Newton potential. Although U is defined as a surface integral, it essentially takes over the role of volume potentials (and not surface potentials) known from the Euclidean context.

Theorem 6.12. *Let $\Gamma \subset \Omega$ be a regular region, and assume that $F : \Gamma \to \mathbb{R}$ is an integrable, bounded function. If U is given by (6.23), then U is of class $\mathrm{C}^{(2)}(\Gamma^c)$ and*

$$\Delta_\xi^* \int_\Gamma G(\Delta^*; \xi \cdot \eta) F(\eta) d\omega(\eta) = -\frac{1}{4\pi} \int_\Gamma F(\eta) d\omega(\eta), \quad \xi \in \Gamma^c. \tag{6.24}$$

Theorem 6.12 follows by obvious arguments from the properties of the fundamental solution $G(\Delta^*; \cdot)$ since the integrand of U is non-singular as long as ξ is chosen from $\Gamma^c = \Omega \setminus \overline{\Gamma}$. Observing the potential U on Γ, we are able to verify a spherical counterpart to the Poisson equation. In doing so, we follow the course of Section 3.2.

6.2.1 Differentiability

Theorem 6.13. *Let $\Gamma \subset \Omega$ be a regular region. Suppose that F is of class $\mathrm{C}^{(0)}(\overline{\Gamma})$. If U is given by (6.23), then U is of class $\mathrm{C}^{(1)}(\Omega)$ and*

$$\nabla_\xi^* \int_\Gamma G(\Delta^*; \xi \cdot \eta) F(\eta) d\omega(\eta) = \int_\Gamma F(\eta) \nabla_\xi^* G(\Delta^*; \xi \cdot \eta) d\omega(\eta), \quad \xi \in \Omega. \tag{6.25}$$

Proof. For $\rho \in (0, 2)$, we define the linearly regularized fundamental solution

$$G^\rho(\Delta^*; t) = \begin{cases} G(\Delta^*; t), & 1 - t > \rho, \\ \frac{1-t}{4\pi\rho} + \frac{1}{4\pi}(\ln \rho - \ln 2), & 1 - t \leq \rho. \end{cases} \tag{6.26}$$

This regularized function is of class $\mathrm{C}^{(1)}([-1, 1])$, and we get

$$\nabla_\xi^* G^\rho(\Delta^*; \xi \cdot \eta) = \begin{cases} -\frac{1}{4\pi(1-\xi\cdot\eta)}(\eta - (\xi \cdot \eta)\xi), & 1 - \xi \cdot \eta > \rho, \\ -\frac{1}{4\pi\rho}(\eta - (\xi \cdot \eta)\xi), & 1 - \xi \cdot \eta \leq \rho, \end{cases} \tag{6.27}$$

for $\xi, \eta \in \Omega$. Now, it follows that

$$\begin{aligned} & \left| \int_\Gamma \nabla_\xi^* G^\rho(\Delta^*; \xi \cdot \eta) F(\eta) d\omega(\eta) - \int_\Gamma \nabla_\xi^* G(\Delta^*; \xi \cdot \eta) F(\eta) d\omega(\eta) \right| \qquad (6.28) \\ = & \left| \int_{\substack{\eta \in \Gamma \\ 1-\xi\cdot\eta \leq \rho}} \nabla_\xi^* G^\rho(\Delta^*; \xi \cdot \eta) F(\eta) d\omega(\eta) - \int_{\substack{\eta \in \Gamma \\ 1-\xi\cdot\eta \leq \rho}} \nabla_\xi^* G(\Delta^*; \xi \cdot \eta) F(\eta) d\omega(\eta) \right| \\ \leq & \sup_{\eta \in \Gamma} |F(\eta)| \int_{\substack{\eta \in \Gamma \\ 1-\xi\cdot\eta \leq \rho}} \left| \nabla_\xi^* G^\rho(\Delta^*; \xi \cdot \eta) \right| d\omega(\eta) \\ & + \sup_{\eta \in \Gamma} |F(\eta)| \int_{\substack{\eta \in \Gamma \\ 1-\xi\cdot\eta \leq \rho}} \left| \nabla_\xi^* G(\Delta^*; \xi \cdot \eta) \right| d\omega(\eta) \end{aligned}$$

$$
\begin{aligned}
&= \sup_{\eta\in\Gamma}|F(\eta)| \int\limits_{\substack{\eta\in\Gamma\\ 1-\xi\cdot\eta\le\rho}} \left|\frac{\sqrt{1-(\xi\cdot\eta)^2}}{4\pi\rho}\right| d\omega(\eta) \\
&\quad + \sup_{\eta\in\Gamma}|F(\eta)| \int\limits_{\substack{\eta\in\Gamma\\ 1-\xi\cdot\eta\le\rho}} \left|\frac{\sqrt{1+\xi\cdot\eta}}{4\pi\sqrt{1-\xi\cdot\eta}}\right| d\omega(\eta) \\
&= \sup_{\eta\in\Gamma}|F(\eta)| \int\limits_{1-\rho}^{1} \frac{\sqrt{1-t^2}}{2\rho} dt + \sup_{\eta\in\Gamma}|F(\eta)| \int\limits_{1-\rho}^{1} \frac{\sqrt{1+t}}{2\sqrt{1-t}} dt.
\end{aligned}
$$

For the integrals in the last equation, it can easily be seen that they vanish uniformly with respect to $\xi \in \Omega$ as ρ tends to zero. By analogous arguments, we obtain the relation

$$
\lim_{\rho\to 0+} \sup_{\xi\in\Omega} \left| \int_\Gamma G^\rho(\Delta^*;\xi\cdot\eta)F(\eta)d\omega(\eta) - \int_\Gamma G(\Delta^*;\xi\cdot\eta)F(\eta)d\omega(\eta) \right| = 0. \tag{6.29}
$$

Since $\xi \mapsto \int_\Gamma G^\rho(\Delta^*;\xi\cdot\eta)F(\eta)d\omega(\eta)$ is of class $\mathrm{C}^{(1)}(\Omega)$ and

$$
\int_\Gamma \nabla^*_\xi G^\rho(\Delta^*;\xi\cdot\eta)F(\eta)d\omega(\eta) = \nabla^*_\xi \int_\Gamma G^\rho(\Delta^*;\xi\cdot\eta)F(\eta)d\omega(\eta), \quad \xi\in\Omega, \tag{6.30}
$$

the relations (6.28) and (6.29) imply the assertions stated in the theorem. □

6.2.2 Poisson Differential Equation

Our goal is to solve the (spherical) Poisson differential equation $\Delta^* U = F$ in Γ. The upcoming theorem presents an integral representation for a solution. Furthermore, it shows that a function U of the form (6.23) can only solve the Poisson equation if F has vanishing integral mean value on Γ, i.e., $\frac{1}{\|\Gamma\|}\int_\Gamma F(\eta)d\omega(\eta) = 0$ (nevertheless, solutions of a different form may exist if this condition on F is not satisfied, as we see later on).

Theorem 6.14. *Let $\Gamma \subset \Omega$ be a regular region, and suppose that F is of class $\mathrm{C}^{(1)}(\overline{\Gamma})$. If U is given by* (6.23)*, then U is of class $\mathrm{C}^{(2)}(\Gamma)$ and satisfies*

$$
\Delta^*_\xi \int_\Gamma G(\Delta^*;\xi\cdot\eta)F(\eta)d\omega(\eta) = F(\xi) - \frac{1}{4\pi}\int_\Gamma F(\eta)d\omega(\eta), \quad \xi\in\Gamma. \tag{6.31}
$$

Proof. Using Green's formulas (see Section 2.2) and Theorem 6.13, we obtain

$$
\begin{aligned}
\Delta^*_\xi U(\xi) &= \mathrm{L}^*_\xi \cdot \int_\Gamma \mathrm{L}^*_\xi G(\Delta^*;\xi\cdot\eta)F(\eta)d\omega(\eta) \qquad (6.32)\\
&= \mathrm{L}^*_\xi \cdot \int_\Gamma G(\Delta^*;\xi\cdot\eta)\mathrm{L}^*_\eta F(\eta)d\omega(\eta) \\
&\quad -\mathrm{L}^*_\xi \cdot \int_{\partial\Gamma} G(\Delta^*;\xi\cdot\eta)\tau(\eta)F(\eta)d\sigma(\eta)
\end{aligned}
$$

$$
\begin{aligned}
&= \int_{\Gamma} \mathrm{L}^*_{\xi} G(\Delta^*; \xi\cdot\eta) \cdot \mathrm{L}^*_{\eta} F(\eta) d\omega(\eta) \\
&\quad - \int_{\partial\Gamma} \underbrace{\tau(\eta)\cdot \mathrm{L}^*_{\xi} G(\Delta^*; \xi\cdot\eta)}_{=-\frac{\partial}{\partial\nu(\eta)} G(\Delta^*;\xi\cdot\eta)} F(\eta) d\sigma(\eta),
\end{aligned}
$$

for $\xi \in \Gamma$. A further application of Green's formulas requires the removal of a spherical cap with radius $\rho \in (0,2)$ around the singularity $\eta = \xi$. This yields

$$
\begin{aligned}
\Delta^*_{\xi} U(\xi) &= -\lim_{\rho\to 0+} \int_{\substack{\eta\in\Gamma \\ 1-\xi\cdot\eta\geq\rho}} \mathrm{L}^*_{\eta} G(\Delta^*; \xi\cdot\eta)\cdot \mathrm{L}^*_{\eta} F(\eta) d\omega(\eta) \qquad (6.33)\\
&\quad - \int_{\partial\Gamma} \frac{\partial}{\partial\nu(\eta)} G(\Delta^*; \xi\cdot\eta)\, F(\eta) d\sigma(\eta) \\
&= \lim_{\rho\to 0+} \int_{\substack{\eta\in\Gamma \\ 1-\xi\cdot\eta\geq\rho}} F(\eta)\, \Delta^*_{\eta} G(\Delta^*; \xi\cdot\eta) d\omega(\eta) \\
&\quad - \int_{\partial\Gamma} \frac{\partial}{\partial\nu(\eta)} G(\Delta^*; \xi\cdot\eta)\, F(\eta) d\sigma(\eta) \\
&\quad - \lim_{\rho\to 0+} \int_{\substack{\eta\in\Gamma \\ 1-\xi\cdot\eta=\rho}} \frac{\partial}{\partial\nu(\eta)} G(\Delta^*; \xi\cdot\eta)\, F(\eta) d\sigma(\eta) \\
&\quad + \int_{\partial\Gamma} \frac{\partial}{\partial\nu(\eta)} G(\Delta^*; \xi\cdot\eta)\, F(\eta) d\sigma(\eta) \\
&= F(\xi) - \frac{1}{4\pi} \int_{\Gamma} F(\eta) d\omega(\eta),
\end{aligned}
$$

for $\xi \in \Gamma$. The last equation follows from the properties of the fundamental solution $G(\Delta^*;\cdot)$ and the limit relation

$$
\lim_{\rho\to 0+} \int_{\substack{\eta\in\Gamma \\ 1-\xi\cdot\eta=\rho}} \frac{\partial}{\partial\nu(\eta)} G(\Delta^*; \xi\cdot\eta)\, F(\eta) d\sigma(\eta) = -F(\xi), \tag{6.34}
$$

for $\xi \in \Gamma$ (compare the proof of Theorem 6.2). An application of Green's formulas as in Equation (6.32) also implies $U \in \mathrm{C}^{(2)}(\Omega)$. □

Remark 6.15. Theorems 6.14 and 6.2 also hold true if Γ is chosen to be the entire sphere Ω. Clearly, the boundary integrals do not occur in this case. Moreover, the Poisson equation

$$
\Delta^* U(\xi) = F(\xi), \quad \xi \in \Omega, \tag{6.35}
$$

is solvable if and only if F has vanishing integral mean value on Ω (different than for strict subdomains $\Gamma \subset \Omega$, $\Gamma \neq \Omega$, this is a necessary condition). As a matter of fact, the solution is determined uniquely up to a constant and has the form

$$U(\xi) = \frac{1}{4\pi} \int_\Omega U(\eta) d\omega(\eta) + \int_\Omega G(\Delta^*; \xi \cdot \eta) F(\eta) d\omega(\eta), \quad \xi \in \Omega. \tag{6.36}$$

6.3 Curve Potentials

Let $\Gamma \subset \Omega$ be a regular region. The so-called *single-layer potential (with respect to the Beltrami operator)* of a function F of class $\mathrm{C}^{(0)}(\partial\Gamma)$ is defined as

$$\xi \mapsto \int_{\partial\Gamma} G(\Delta^*; \xi \cdot \eta) F(\eta) d\sigma(\eta), \quad \xi \in \Omega, \tag{6.37}$$

while the *double-layer potential (with respect to the Beltrami operator)* is given by

$$\xi \mapsto \int_{\partial\Gamma} \frac{\partial}{\partial\nu(\eta)} G(\Delta^*; \xi \cdot \eta) F(\eta) d\sigma(\eta), \quad \xi \in \Omega. \tag{6.38}$$

These potentials are actually defined as integrals along boundary curves. Nevertheless, they have a similar meaning as surface potentials in the Euclidean context of $\mathbb{R}^3$.

6.3.1 Preparatory Estimates

Next, we derive some estimates for the kernels of the single- and double-layer potentials. They are of particular importance for the verification of the upcoming limit and jump relations.

Lemma 6.16. *Let $\Gamma \subset \Omega$ be a regular region. Then,*

$$\lim_{\rho \to 0+} \int_{\substack{\eta \in \partial\Gamma \\ 1 - \tilde{\xi} \cdot \eta < \rho}} |G(\Delta^*; \xi \cdot \eta)| d\sigma(\eta) = 0 \tag{6.39}$$

holds uniformly with respect to $\xi, \tilde{\xi} \in \Omega$.

Proof. Since $|G(\Delta^*; \xi \cdot \eta)| \leq |\frac{1}{4\pi} \ln(1 - \xi \cdot \eta)| + \frac{1}{4\pi}(1 - \ln(2))$, we only need to estimate the integral over the logarithm. The integral over the constant obviously vanishes as $\rho \to 0$. Denoting by $\partial\Gamma^+$ the set $\{\eta \in \partial\Gamma : 1 - \xi \cdot \eta \geq 1\}$

and by $\partial\Gamma^-$ the set $\{\eta \in \partial\Gamma : 1 - \xi \cdot \eta < 1\}$, we obtain

$$\int_{\substack{\eta\in\partial\Gamma \\ 1-\tilde{\xi}\cdot\eta<\rho}} \left| \frac{1}{4\pi} \ln(1 - \xi \cdot \eta) \right| d\sigma(\eta) \tag{6.40}$$

$$\leq \int_{\substack{\eta\in\partial\Gamma^- \\ 1-\tilde{\xi}\cdot\eta<\rho}} \left| \frac{1}{4\pi} \ln(1 - \xi \cdot \eta) \right| d\sigma(\eta) + \int_{\substack{\eta\in\partial\Gamma^+ \\ 1-\tilde{\xi}\cdot\eta<\rho}} \left| \frac{1}{4\pi} \ln(2) \right| d\sigma(\eta).$$

Due to the constant integrand, the second integral vanishes as ρ tends to zero. To estimate the first integral, let $\bar{\xi}$ be the point on $\partial\Gamma$ with the least distance to $\tilde{\xi} \in \Omega$. For further calculations, we switch to the local (tangential-normal) coordinate system with center $\bar{\xi}$ (cf. Definition 2.1), and we assume ρ to be sufficiently small, such that (2.49) and (2.50) are satisfied. Additionally, I^- designates the subset of $(-\tilde{\rho}, \tilde{\rho})$ satisfying $(t, \gamma_1(t), \gamma_2(t)) \in \partial\Gamma^-$, for $t \in I^-$ (where $\gamma(t) = (\gamma_1(t), \gamma_2(t))^T$ denotes the parameterization from Definition 2.1). Then, for $\xi = (\xi_1, \xi_2, \xi_3)^T \in \Omega$, we get

$$\int_{\substack{\eta\in\partial\Gamma^- \\ 1-\tilde{\xi}\cdot\eta<\rho}} |\ln(1 - \xi \cdot \eta)| d\sigma(\eta) \leq M \int_{I^-} \left| \ln\left(\frac{1}{2}|\xi - (t, \gamma_1(t), \gamma_2(t))^T|^2\right) \right| dt \tag{6.41}$$

$$= M \int_{I^-} \left| \ln\left(\frac{1}{2}((t - \xi_1)^2 + (\xi_2 - \gamma_1(t))^2 + (\xi_3 - \gamma_2(t))^2)\right) \right| dt$$

$$\leq M \int_{-\tilde{\rho}}^{\tilde{\rho}} \left| \ln\left(\frac{1}{2}(t - \xi_1)^2\right) \right| dt = M \int_{-\tilde{\rho}-\xi_1}^{\tilde{\rho}-\xi_1} \left| \ln\left(\frac{t^2}{2}\right) \right| dt.$$

Since $t \mapsto |\ln(\frac{t^2}{2})|$ is integrable, the integral (6.41) vanishes uniformly with respect to $\xi, \tilde{\xi} \in \Omega$ as ρ (and thus $\tilde{\rho}$ as well) tends to zero, which proves the assertion of Lemma 6.16. □

Choosing $\tilde{\xi} = \xi$, a direct consequence of Lemma 6.16 is that the integral over the curve $\partial\Gamma$ has to be bounded. More precisely, there exists a constant $M > 0$ such that

$$\int_{\partial\Gamma} |G(\Delta^*; \xi \cdot \eta)| d\sigma(\eta) \leq M, \tag{6.42}$$

for arbitrary $\xi \in \Omega$. The corresponding estimate for the normal derivative requires some more effort.

Lemma 6.17. *Let $\Gamma \subset \Omega$ be a regular region. There exists a constant $M > 0$ such that*

$$\int_{\partial\Gamma} \left| \frac{\partial}{\partial\nu(\eta)} G(\Delta^*; \xi \cdot \eta) \right| d\sigma(\eta) \leq M, \tag{6.43}$$

for arbitrary $\xi \in \Omega$.

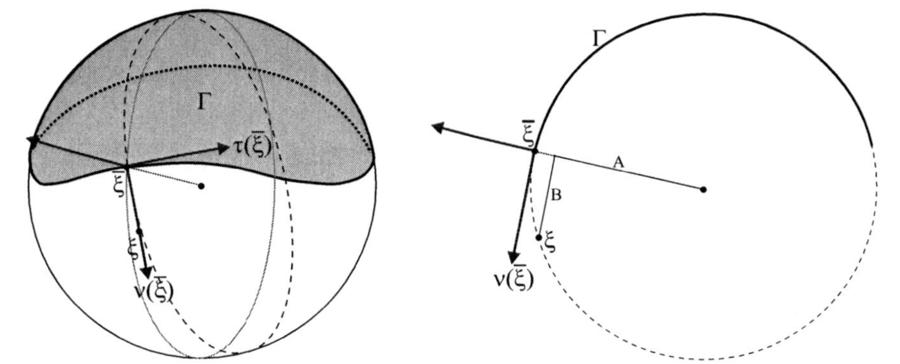

FIGURE 6.1
Illustration of the local (tangential-normal) coordinate system with center $\bar{\xi}$ (left), and a two-dimensional sketch of the choice of the point $\bar{\xi}$ based on ξ (right). Note that the right figure indicates the cut along the dashed circle in the left figure.

Proof. Let $\rho \in (0,2)$ be such that the conditions (2.49) and (2.50) are satisfied. If the point $\xi \in \Omega$ satisfies $1 - \xi \cdot \eta \geq \rho$ for every $\eta \in \partial\Gamma$, we have

$$\left| \frac{\partial}{\partial \nu(\eta)} G(\Delta^*; \xi \cdot \eta) \right| = \frac{1}{4\pi} \left| \frac{\xi \cdot \nu(\eta)}{1 - \xi \cdot \eta} \right| \leq \frac{1}{4\pi\rho}, \tag{6.44}$$

such that the integral (6.43) under consideration is bounded by $\frac{1}{4\pi\rho}\|\partial\Gamma\|$, independently of $\xi \in \Omega$. In the case that points $\eta \in \partial\Gamma$ exist with $1 - \xi \cdot \eta < \rho$, we are only able to estimate

$$\int\limits_{\substack{\eta \in \partial\Gamma \\ 1-\xi\cdot\eta \geq \rho}} \left| \frac{\partial}{\partial \nu(\eta)} G(\Delta^*; \xi \cdot \eta) \right| d\sigma(\eta) \leq \frac{1}{4\pi\rho} \|\partial\Gamma\|. \tag{6.45}$$

To handle the integral over $\{\eta \in \partial\Gamma : 1 - \xi \cdot \eta < \rho\}$, we choose $\bar{\xi}$ as the point on $\partial\Gamma$ with the least distance to $\xi \in \Omega$. Then there exist constants A, B with $\xi = A\bar{\xi} + B\nu(\bar{\xi})$ (where $B \to 0$ and $A \to 1$ as ξ tends to $\partial\Gamma$). Consequently, we get

$$\begin{aligned} \int\limits_{\substack{\eta \in \partial\Gamma \\ 1-\xi\cdot\eta < \rho}} \left| \frac{\partial}{\partial \nu(\eta)} G(\Delta^*; \xi \cdot \eta) \right| d\sigma(\eta) &= \frac{1}{4\pi} \int\limits_{\substack{\eta \in \partial\Gamma \\ 1-\xi\cdot\eta < \rho}} \left| \frac{\xi \cdot \nu(\eta)}{1 - \xi \cdot \eta} \right| d\sigma(\eta) \qquad (6.46) \\ &\leq \frac{1}{8\pi} \int\limits_{\substack{\eta \in \partial\Gamma \\ 1-\xi\cdot\eta < \rho}} \left| \frac{A\bar{\xi} \cdot \nu(\eta)}{|\xi - \eta|^2} \right| d\sigma(\eta) + \frac{1}{8\pi} \int\limits_{\substack{\eta \in \partial\Gamma \\ 1-\xi\cdot\eta < \rho}} \left| \frac{B\nu(\bar{\xi}) \cdot \nu(\eta)}{|\xi - \eta|^2} \right| d\sigma(\eta). \end{aligned}$$

For the estimate of the second summand in the last row of the inequality

(6.46), we switch to the local (tangential-normal) coordinate system with center $\bar{\xi}$. By $\gamma(t) = (\gamma_1(t), \gamma_2(t))^T$, we mean the parameterization from Definition 2.1. In particular, we have $\bar{\xi} = (0,0,0)^T$, $\xi = (0, B, A-1)^T$, and $\eta = (t, \gamma_1(t), \gamma_2(t))^T$ in this coordinate system (see Figure 6.1). In this framework, we find

$$\begin{aligned}
\int\limits_{\substack{\eta\in\partial\Gamma \\ 1-\xi\cdot\eta<\rho}} \left| \frac{B\nu(\bar{\xi})\cdot\nu(\eta)}{|\xi-\eta|^2} \right| d\sigma(\eta) &\leq M \int\limits_{-\tilde{\rho}}^{\tilde{\rho}} \frac{B}{t^2 + (B-\gamma_1(t))^2 + (A-1-\gamma_2(t))^2} dt \\
&\leq M \int\limits_{-\tilde{\rho}}^{\tilde{\rho}} \frac{B}{t^2 + B^2 + \gamma_1(t)^2 - 2B\gamma_1(t)} dt \\
&\leq M \int\limits_{-\tilde{\rho}}^{\tilde{\rho}} \frac{B}{t^2 + \frac{1}{2}B^2 - \gamma_1(t)^2} dt \qquad (6.47) \\
&\leq M \int\limits_{-\tilde{\rho}}^{\tilde{\rho}} \frac{B}{t^2 + \frac{1}{2}B^2 - M^2t^4} dt \\
&\leq M \int\limits_{-\tilde{\rho}}^{\tilde{\rho}} \frac{B}{\frac{1}{2}t^2 + \frac{1}{2}B^2} dt.
\end{aligned}$$

In the third row of (6.47), we have applied Young's inequality to deduce that

$$2|B\gamma_1(t)| \leq \frac{B^2}{2} + \frac{(2\gamma_1(t))^2}{2} = \frac{B^2}{2} + 2\gamma_1(t)^2. \qquad (6.48)$$

In the fourth row of (6.47), the estimate $|\gamma_i(t)| \leq Mt^2$, $i = 1, 2$, for the parameterization γ has been used (see (2.49)). Choosing $\rho > 0$ sufficiently small, such that the corresponding $\tilde{\rho}$ satisfies the relation

$$M^2t^2 \leq \frac{1}{2}, \quad t \in (-\tilde{\rho}, \tilde{\rho}), \qquad (6.49)$$

we achieve the last estimate in (6.47). Substituting $t = Bs$, we finally arrive at

$$\begin{aligned}
\int\limits_{\substack{\eta\in\partial\Gamma \\ 1-\xi\cdot\eta<\rho}} \left| \frac{B\nu(\bar{\xi})\cdot\nu(\eta)}{|\xi-\eta|^2} \right| d\sigma(\eta) &\leq M \int\limits_{-\infty}^{\infty} \frac{B}{\frac{1}{2}t^2 + \frac{1}{2}B^2} dt \qquad (6.50) \\
&\leq M \int\limits_{-\infty}^{\infty} \frac{1}{\frac{1}{2}s^2 + \frac{1}{2}} ds.
\end{aligned}$$

The last integral in (6.50) is bounded independently of $\xi \in \Omega$. To handle the first integral on the right-hand side of (6.46), we notice that $|\nu(\bar{\xi})\cdot\eta| \leq M|\bar{\xi}-\eta|^2$. Switching to the local (tangential-normal) coordinate system, we obtain similarly as before that

$$\begin{aligned}\int\limits_{\substack{\eta\in\partial\Gamma\\1-\xi\cdot\eta<\rho}} \left|\frac{A\bar{\xi}\cdot\nu(\eta)}{|\xi-\eta|^2}\right| d\sigma(\eta) &\leq MA \int\limits_{\substack{\eta\in\partial\Gamma\\1-\xi\cdot\eta<\rho}} \frac{|\bar{\xi}-\eta|^2}{|\xi-\eta|^2} d\sigma(\eta) \\ &\leq MA \int\limits_{-\tilde{\rho}}^{\tilde{\rho}} \frac{(t^2+\gamma_1(t)^2+\gamma_2(t)^2)}{\frac{1}{2}t^2+\frac{1}{2}B^2} dt \\ &\leq 2M(1+M^2)A \int\limits_{-\tilde{\rho}}^{\tilde{\rho}} \frac{t^2}{t^2+B^2} dt \\ &\leq 2M(1+M^2)A \int\limits_{-\tilde{\rho}}^{\tilde{\rho}} dt. \end{aligned} \tag{6.51}$$

The last expression is again integrable and therefore bounded independently of $\xi \in \Omega$. For the estimate in the second row of (6.51), we have again used $|\gamma(t)| \leq Mt^2$ in connection with (6.49). Altogether, this proves the assertion of Lemma 6.17. □

Remark 6.18. Taking a closer look at the integral relation (6.51), we find the following relation

$$\lim_{\rho\to 0+} \sup_{\xi\in\partial\Gamma} \int\limits_{\substack{\eta\in\partial\Gamma\\1-\xi\cdot\eta<\rho}} \frac{|\xi\cdot\nu(\eta)|}{1-\frac{\xi+\tau\nu(\xi)}{\sqrt{1+\tau^2}}\cdot\eta} d\sigma(\eta) = 0, \tag{6.52}$$

which holds uniformly with respect to $\tau \in \mathbb{R}$. However, it should be mentioned that the integral

$$\int\limits_{\substack{\eta\in\partial\Gamma\\1-\xi\cdot\eta<\rho}} \left|\frac{\partial}{\partial\nu(\eta)} G(\Delta^*;\xi\cdot\eta)\right| d\sigma(\eta) \tag{6.53}$$

does not vanish uniformly as ρ tends to zero.

For later use, it is not only advisable to estimate the integral over the normal derivative of the fundamental solution, but also to discuss the normal derivative of the integral over the fundamental solution. Using Proposition 2.3 and analogous steps as in the preceding proof, we obtain the following result.

Lemma 6.19. *Let $\Gamma \subset \Omega$ be a regular region. There exists a $M > 0$ such that*

$$\int_{\partial\Gamma} \left| \left.\frac{\partial}{\partial\nu(\zeta)} G(\Delta^*;\zeta\cdot\eta)\right|_{\zeta=\frac{\xi+\tau\nu(\xi)}{\sqrt{1+\tau^2}}} \right| d\sigma(\eta) \leq M \tag{6.54}$$

holds for arbitrary $\xi \in \partial\Gamma$ and $\tau \in \mathbb{R}$.

6.3.2 Limit and Jump Relations in $C^{(0)}$-Topology

To formulate the limit and jump relations for layer potentials and their normal derivatives, we turn to a more convenient representation in terms of *potential operators* for $\sigma, \tau \in \mathbb{R}$ (analogous to the potential operators defined in Subsection 3.3.3 within the Euclidean framework):

$$P(\tau,\sigma)[F](\xi) = \int_{\partial\Gamma} G\left(\Delta^*; \frac{\xi+\tau\nu(\xi)}{\sqrt{1+\tau^2}} \cdot \frac{\eta+\sigma\nu(\eta)}{\sqrt{1+\sigma^2}}\right) F(\eta)d\sigma(\eta), \tag{6.55}$$

$$P_{|1}(\tau,\sigma)[F](\xi) = \int_{\partial\Gamma} \frac{\partial}{\partial\tau} G\left(\Delta^*; \frac{\xi+\tau\nu(\xi)}{\sqrt{1+\tau^2}} \cdot \frac{\eta+\sigma\nu(\eta)}{\sqrt{1+\sigma^2}}\right) F(\eta)d\sigma(\eta), \tag{6.56}$$

$$P_{|2}(\tau,\sigma)[F](\xi) = \int_{\partial\Gamma} \frac{\partial}{\partial\sigma} G\left(\Delta^*; \frac{\xi+\tau\nu(\xi)}{\sqrt{1+\tau^2}} \cdot \frac{\eta+\sigma\nu(\eta)}{\sqrt{1+\sigma^2}}\right) F(\eta)d\sigma(\eta), \tag{6.57}$$

for $\xi \in \partial\Gamma$ and functions F of class $C^{(0)}(\partial\Gamma)$. In order to emphasize the case $\tau = \sigma = 0$, we formally write

$$P_{|1}(0,0)[F](\xi) = \int_{\partial\Gamma} \frac{\partial}{\partial\nu(\xi)} G\left(\Delta^*; \xi\cdot\eta\right) F(\eta)d\sigma(\eta), \tag{6.58}$$

$$P_{|2}(0,0)[F](\xi) = \int_{\partial\Gamma} \frac{\partial}{\partial\nu(\eta)} G\left(\Delta^*; \xi\cdot\eta\right) F(\eta)d\sigma(\eta). \tag{6.59}$$

These formal definitions should, however, not be misunderstood as the limit of the operators $P_{|1}(\tau,0)[F]$ and $P_{|2}(\tau,0)[F]$ for τ tending to zero. If we differentiate (6.55) twice with respect to τ and σ, the corresponding integrals do not exist anymore for the choice $\tau = \sigma = 0$ (at least if we take F of class $C^{(0)}(\partial\Gamma)$). But, for $\tau \neq 0$ and/or $\sigma \neq 0$, we can still write

$$P_{|1|2}(\tau,\sigma)[F](\xi) = \int_{\partial\Gamma} \frac{\partial}{\partial\tau}\frac{\partial}{\partial\sigma} G\left(\Delta^*; \frac{\xi+\tau\nu(\xi)}{\sqrt{1+\tau^2}} \cdot \frac{\eta+\sigma\nu(\eta)}{\sqrt{1+\sigma^2}}\right) F(\eta)d\sigma(\eta). \tag{6.60}$$

One can easily see the following relationships to the single- and double-layer potentials known from (6.37) and (6.38):

$$P(\tau,0)[F](\xi) = \int_{\partial\Gamma} G\left(\Delta^*; \frac{\xi+\tau\nu(\xi)}{\sqrt{1+\tau^2}} \cdot \eta\right) F(\eta)d\sigma(\eta), \tag{6.61}$$

$$P_{|2}(\tau,0)[F](\xi) = \int_{\partial\Gamma} \frac{\partial}{\partial\nu(\eta)} G\left(\Delta^*; \frac{\xi+\tau\nu(\xi)}{\sqrt{1+\tau^2}} \cdot \eta\right) F(\eta)d\sigma(\eta), \tag{6.62}$$

for $\xi \in \partial\Gamma$. Choosing $\tau \neq 0$, we additionally obtain

$$P_{|1}(\tau,0)[F](\xi) = \frac{1}{\sqrt{1+\tau^2}} \frac{\partial}{\partial\nu(\zeta)} \int_{\partial\Gamma} G\left(\Delta^*; \zeta\cdot\eta\right) F(\eta)d\sigma(\eta)\Bigg|_{\zeta=\frac{\xi+\tau\nu(\xi)}{\sqrt{1+\tau^2}}}, \tag{6.63}$$

$$P_{|1|2}(\tau,0)[F](\xi) = \frac{1}{\sqrt{1+\tau^2}} \frac{\partial}{\partial\nu(\zeta)} \int_{\partial\Gamma} \frac{\partial}{\partial\nu(\eta)} G\left(\Delta^*; \zeta\cdot\eta\right) F(\eta)d\sigma(\eta)\Bigg|_{\zeta=\frac{\xi+\tau\nu(\xi)}{\sqrt{1+\tau^2}}}. \tag{6.64}$$

The potential operators as defined here allow a more concise formulation of the limit and jump relations, but what we are typically interested in are the relations for the single- and double-layer potentials, and their normal derivatives. Indeed, relations (6.61)–(6.64) ensure that both formulations are equivalent (the pre-factors appearing in (6.63) and (6.64) are irrelevant as they tend to one if τ tends to zero).

Next, we deal with the desired limit and jump relations for functions F of class $C^{(0)}(\partial\Gamma)$. As expected, the layer potentials show the same behavior as in the Euclidean case (compare Subsections 3.3.2 and 3.3.3).

Theorem 6.20 (Limit Relations). *Let $\Gamma \subset \Omega$ be a regular region and suppose that F is of class $C^{(0)}(\partial\Gamma)$.*

(a) For the single-layer potential we have

$$\lim_{\tau\to 0+} \sup_{\xi\in\partial\Gamma} \left| P(\pm\tau,0)[F](\xi) - P(0,0)[F](\xi) \right| = 0, \tag{6.65}$$

$$\lim_{\tau\to 0+} \sup_{\xi\in\partial\Gamma} \left| P_{|1}(\pm\tau,0)[F](\xi) - P_{|1}(0,0)[F](\xi) \mp \frac{1}{2}F(\xi) \right| = 0. \tag{6.66}$$

(b) For the double-layer potential we have

$$\lim_{\tau\to 0+} \sup_{\xi\in\partial\Gamma} \left| P_{|2}(\pm\tau,0)[F](\xi) - P_{|2}(0,0)[F](\xi) \pm \frac{1}{2}F(\xi) \right| = 0. \tag{6.67}$$

Theorem 6.21 (Jump Relations). *Let $\Gamma \subset \Omega$ be a regular region and suppose that F is of class $C^{(0)}(\partial\Gamma)$.*

(a) For the single-layer potential, we have

$$\lim_{\tau\to 0+} \sup_{\xi\in\partial\Gamma} \left| P(\tau,0)[F](\xi) - P(-\tau,0)[F](\xi) \right| = 0, \tag{6.68}$$

$$\lim_{\tau\to 0+} \sup_{\xi\in\partial\Gamma} \left| P_{|1}(\tau,0)[F](\xi) - P_{|1}(-\tau,0)[F](\xi) - F(\xi) \right| = 0. \tag{6.69}$$

(b) For the double-layer potential, we have

$$\lim_{\tau\to 0+} \sup_{\xi\in\partial\Gamma} \left| P_{|2}(\tau,0)[F](\xi) - P_{|2}(-\tau,0)[F](\xi) + F(\xi) \right| = 0, \tag{6.70}$$

$$\lim_{\tau\to 0+} \sup_{\xi\in\partial\Gamma} \left| P_{|1|2}(\tau,0)[F](\xi) - P_{|1|2}(-\tau,0)[F](\xi) \right| = 0. \tag{6.71}$$

The jump relations for the layer potentials are a direct consequence of Theorem 6.20. The second relation under point (b) has no analogue in the case of the limit relations from Theorem 6.20 since $P_{|1|2}(0,0)[F]$ is not necessarily defined for functions F of class $C^{(0)}(\partial\Gamma)$. Note that, for functions F with additional differentiability properties, one can formulate a limit relation as well. But for our purposes, it is sufficient to consider only the jump relation, so that we restrict ourselves to that case.

Proof. *Theorem 6.20 (b).* Using the equality (6.13), we get

$$\begin{aligned}
&\left| P_{|2}(\tau,0)[F](\xi) - P_{|2}(0,0)[F](\xi) + \frac{1}{2}F(\xi)\right| \qquad (6.72)\\
&= \left| \int_{\partial\Gamma} F(\eta)\frac{\partial}{\partial\nu(\eta)} G\left(\Delta^*; \frac{\xi+\tau\nu(\xi)}{\sqrt{1+\tau^2}}\cdot\eta\right) d\sigma(\eta) \right.\\
&\quad - \int_{\partial\Gamma} F(\eta)\frac{\partial}{\partial\nu(\eta)} G\left(\Delta^*; \xi\cdot\eta\right) d\sigma(\eta)\\
&\quad \left. + \int_{\partial\Gamma} F(\xi)\frac{\partial}{\partial\nu(\eta)} G\left(\Delta^*; \xi\cdot\eta\right) d\sigma(\eta) + \frac{\|\Gamma\|}{4\pi}F(\xi)\right|\\
&\leq \left| \int_{\partial\Gamma} (F(\xi)-F(\eta))\frac{\partial}{\partial\nu(\eta)} G\left(\Delta^*; \xi\cdot\eta\right) d\sigma(\eta) \right.\\
&\quad \left. + \int_{\partial\Gamma} (F(\eta)-F(\xi))\frac{\partial}{\partial\nu(\eta)} G\left(\Delta^*; \frac{\xi+\tau\nu(\xi)}{\sqrt{1+\tau^2}}\cdot\eta\right) d\sigma(\eta)\right|\\
&\quad + \left| \underbrace{\int_{\partial\Gamma} F(\xi)\frac{\partial}{\partial\nu(\eta)} G\left(\Delta^*; \frac{\xi+\tau\nu(\xi)}{\sqrt{1+\tau^2}}\cdot\eta\right) d\sigma(\eta)}_{=-\frac{\|\Gamma\|}{4\pi}F(\xi),\ \text{since}\ \frac{\xi+\tau\nu(\xi)}{\sqrt{1+\tau^2}}\in\Gamma^c} + \frac{\|\Gamma\|}{4\pi}F(\xi)\right|,
\end{aligned}$$

for $\xi \in \partial\Gamma$. We immediately see that the last summand on the right-hand side of (6.72) vanishes. Let now $\mu(\cdot;F)$ be the continuity modulus of F. Since F is uniformly continuous on $\partial\Gamma$, $\mu(\cdot;F)$ does not depend on the variable at which F is evaluated. Therefore, we are led to

$$\begin{aligned}
&\left| P_{|2}(\tau,0)[F](\xi) - P_{|2}(0,0)[F](\xi) + \frac{1}{2}F(\xi)\right| \qquad (6.73)\\
&\leq \mu(\rho;F) \int\limits_{\substack{\eta\in\partial\Gamma\\ 1-\xi\cdot\eta<\rho}} \left|\frac{\partial}{\partial\nu(\eta)} G\left(\Delta^*; \xi\cdot\eta\right)\right| d\sigma(\eta)\\
&\quad + \mu(\rho;F) \int\limits_{\substack{\eta\in\partial\Gamma\\ 1-\xi\cdot\eta<\rho}} \left|\frac{\partial}{\partial\nu(\eta)} G\left(\Delta^*; \frac{\xi+\tau\nu(\xi)}{\sqrt{1+\tau^2}}\cdot\eta\right)\right| d\sigma(\eta)\\
&\quad + 2\Big(\sup_{\eta\in\partial\Gamma}|F(\eta)|\Big)\\
&\quad \times \int\limits_{\substack{\eta\in\partial\Gamma\\ 1-\xi\cdot\eta\geq\rho}} \left|\frac{\partial}{\partial\nu(\eta)} G\left(\Delta^*; \frac{\xi+\tau\nu(\xi)}{\sqrt{1+\tau^2}}\cdot\eta\right) - \frac{\partial}{\partial\nu(\eta)} G\left(\Delta^*; \xi\cdot\eta\right)\right| d\sigma(\eta).
\end{aligned}$$

In order to investigate the last integral in the above inequality, we start from $\frac{\partial}{\partial\nu(\eta)}G(\Delta^*;\xi\cdot\eta) = (1-\xi\cdot\eta)^{-1}\xi\cdot\nu(\eta)$. If $\tilde{\xi}$ is a point on Ω that is sufficiently close to ξ, such that $1-\tilde{\xi}\cdot\eta \geq \frac{\rho}{2}$ holds true for all $\eta\in\partial\Gamma$ satisfying $1-\xi\cdot\eta\geq\rho$, we are able to verify

$$\begin{aligned} & \left| \frac{1}{1-\xi\cdot\eta}\xi\cdot\nu(\eta) - \frac{1}{1-\tilde{\xi}\cdot\eta}\tilde{\xi}\cdot\nu(\eta)\right| \\ = & \left| \frac{(1-\tilde{\xi}\cdot\eta)\xi\cdot\nu(\eta) - (1-\xi\cdot\eta)\tilde{\xi}\cdot\nu(\eta)}{(1-\xi\cdot\eta)(1-\tilde{\xi}\cdot\eta)}\right| \\ \leq & \frac{2}{\rho^2}|(1-\xi\cdot\eta)(\tilde{\xi}-\xi)\cdot\nu(\eta) + (\tilde{\xi}-\xi)\cdot\eta(\xi\cdot\nu(\eta))| \\ \leq & \frac{6}{\rho^2}|\xi-\tilde{\xi}|. \end{aligned} \tag{6.74}$$

The choice $\tilde{\xi} = (\xi + \tau\nu(\xi))(1+\tau^2)^{-\frac{1}{2}}$ in (6.74) implies that the third summand on the right-hand side of the inequality (6.73) vanishes as τ tends to zero. More specifically,

$$\lim_{\tau\to 0+} \int\limits_{\substack{\eta\in\partial\Gamma \\ 1-\xi\cdot\eta\geq\rho}} \left| \frac{\partial}{\partial\nu(\eta)} G\left(\Delta^*; \frac{\xi+\tau\nu(\xi)}{\sqrt{1+\tau^2}}\cdot\eta\right) - \frac{\partial}{\partial\nu(\eta)} G\left(\Delta^*;\xi\cdot\eta\right)\right| d\sigma(\eta) = 0 \tag{6.75}$$

is valid uniformly with respect to $\xi \in \partial\Gamma$. Lemma 6.17 guarantees the uniform boundedness of the first two integrals of the right-hand side of the inequality (6.73) with respect to $\xi \in \partial\Gamma$. Thus,

$$\lim_{\tau\to 0+} \sup_{\xi\in\partial\Gamma} \left| P_{|2}(\tau,0)[F](\xi) - P_{|2}(0,0)[F](\xi) + \frac{1}{2}F(\xi)\right| \leq 2M\mu(\rho;F). \tag{6.76}$$

This justifies the desired assertion since $\rho \in (0,2)$ can be chosen arbitrarily. The case $P_{|2}(-\tau,0)$ follows analogously, considering that

$$\int_{\partial\Gamma} F(\xi) \frac{\partial}{\partial\nu(\eta)} G\left(\Delta^*; \frac{\xi-\tau\nu(\xi)}{\sqrt{1+\tau^2}}\cdot\eta\right) d\sigma(\eta) = \left(1 - \frac{\|\Gamma\|}{4\pi}\right) F(\xi), \tag{6.77}$$

since $\frac{\xi-\tau\nu(\xi)}{\sqrt{1+\tau^2}} \in \Gamma$. □

Proof. Theorem 6.20 (a). We begin with the single-layer. It is not hard to see that

$$\begin{aligned} & |P(\tau,0)[F](\xi) - P(0,0)[F](\xi)| \\ = & \frac{1}{4\pi}\left| \int\limits_{\partial\Gamma} F(\eta)\left(\ln\left(1 - \frac{\xi+\tau\nu(\xi)}{\sqrt{1+\tau^2}}\cdot\eta\right) - \ln(1-\xi\cdot\eta)\right) d\sigma(\eta)\right| \end{aligned}$$

$$
\begin{aligned}
&\leq \frac{1}{4\pi}\left(\sup_{\eta\in\partial\Gamma}|F(\eta)|\right) \int\limits_{\substack{\eta\in\partial\Gamma\\1-\xi\cdot\eta\geq\rho}} \left|\ln\left(\frac{1-\frac{\xi+\tau\nu(\xi)}{\sqrt{1+\tau^2}}\cdot\eta}{1-\xi\cdot\eta}\right)\right| d\sigma(\eta) \qquad (6.78)\\
&+\frac{1}{4\pi}\left(\sup_{\eta\in\partial\Gamma}|F(\eta)|\right) \int\limits_{\substack{\eta\in\partial\Gamma\\1-\xi\cdot\eta<\rho}} |G(\Delta^*;\xi\cdot\eta)|\, d\sigma(\eta)\\
&+\frac{1}{4\pi}\left(\sup_{\eta\in\partial\Gamma}|F(\eta)|\right) \int\limits_{\substack{\eta\in\partial\Gamma\\1-\xi\cdot\eta<\rho}} \left|G\left(\Delta^*;\frac{\xi+\tau\nu(\xi)}{\sqrt{1+\tau^2}}\cdot\eta\right)\right| d\sigma(\eta).
\end{aligned}
$$

According to Lemma 6.16, we can choose, for every $\varepsilon > 0$, a sufficiently small $\rho \in (0, 2)$ such that the second and third summand of the inequality (6.78) are smaller than ε. Furthermore, for $\xi, \tilde{\xi}, \eta \in \Omega$ with $1-\xi\cdot\eta \geq \rho$ and $1-\tilde{\xi}\cdot\eta \geq \frac{\rho}{2}$, it follows that

$$
\left|1-\frac{1-\tilde{\xi}\cdot\eta}{1-\xi\cdot\eta}\right| \leq \frac{|\tilde{\xi}-\xi|}{\rho}. \qquad (6.79)
$$

Clearly, (6.79) converges to zero, uniformly with respect to ξ, η, as $\tilde{\xi}$ tends to ξ. Therefore, we are able to deduce from the uniform continuity of $\ln(\frac{1-s}{1-t})$ for $1-t \geq \rho$, $1-s \geq \frac{\rho}{2}$, that

$$
\lim_{\tau\to 0+} \int\limits_{\substack{\eta\in\partial\Gamma\\1-\xi\cdot\eta\geq\rho}} \left|\ln\left(\frac{1-\frac{\xi+\tau\nu(\xi)}{\sqrt{1+\tau^2}}\cdot\eta}{1-\xi\cdot\eta}\right)\right| d\sigma(\eta) = 0, \qquad (6.80)
$$

uniformly with respect $\xi \in \partial\Gamma$. Combining (6.80) with the estimate (6.78), we obtain

$$
\lim_{\tau\to 0+} \sup_{\xi\in\partial\Gamma} |P(\pm\tau,0)[F](\xi) - P(0,0)[F](\xi)| \leq 2\varepsilon. \qquad (6.81)
$$

Since $\varepsilon > 0$ can be chosen arbitrarily, we have proven the continuity of the single-layer potential. The discussion of $P(-\tau, 0)$ follows analogously.

In order to guarantee the limit relation (6.66) for the normal derivative $P_{|1}(\tau, 0)$ of the single-layer potential, it is sufficient to show that

$$
\begin{aligned}
\lim_{\tau\to 0+} \sup_{\xi\in\partial\Gamma} \Big| & P_{|1}(\tau,0)[F](\xi) + P_{|2}(\tau,0)[F](\xi) \qquad (6.82)\\
& -\left(P_{|1}(0,0)[F](\xi) + P_{|2}(0,0)[F](\xi)\right)\Big| = 0.
\end{aligned}
$$

If the relation (6.82) has been established, the proof is complete by use of the already known limit relation for $P_{|2}(\tau, 0)$ from Theorem 6.20 (b). Therefore, we begin with the estimate

$$
\begin{aligned}
& \left| P_{|1}(\tau,0)[F](\xi) + P_{|2}(\tau,0)[F](\xi) - \left(P_{|1}(0,0)[F](\xi) + P_{|2}(0,0)[F](\xi)\right)\right| \\
= & \left| \frac{1}{4\pi} \int_{\partial\Gamma} \frac{1}{1-\frac{\xi+\tau\nu(\xi)}{\sqrt{1+\tau^2}}\cdot\eta} \left(\frac{\xi+\tau\nu(\xi)}{\sqrt{1+\tau^2}}\cdot\nu(\eta) + \left(\frac{\nu(\xi)}{\sqrt{1+\tau^2}} - \tau\frac{\xi+\tau\nu(\xi)}{(1+\tau^2)^{\frac{3}{2}}}\right)\cdot\eta\right)\right. \\
& \qquad \left. \times F(\eta)\, d\sigma(\eta) - \frac{1}{4\pi}\int_{\partial\Gamma} \frac{1}{1-\xi\cdot\eta} \left(\xi\cdot\nu(\eta) + \nu(\xi)\cdot\eta\right) F(\eta) d\sigma(\eta)\right| \\
\leq & \frac{1}{4\pi}\Big(\sup_{\eta\in\partial\Gamma}|F(\eta)|\Big)\int_{\partial\Gamma}\left|\frac{1}{1-\frac{\xi+\tau\nu(\xi)}{\sqrt{1+\tau^2}}\cdot\eta}\frac{\xi\cdot\nu(\eta)}{\sqrt{1+\tau^2}} - \frac{1}{1-\xi\cdot\eta}\xi\cdot\nu(\eta)\right| d\sigma(\eta) \\
& + \frac{1}{4\pi}\Big(\sup_{\eta\in\partial\Gamma}|F(\eta)|\Big)\int_{\partial\Gamma}\left|\frac{1}{1-\frac{\xi+\tau\nu(\xi)}{\sqrt{1+\tau^2}}\cdot\eta}\frac{\nu(\xi)\cdot\eta}{\sqrt{1+\tau^2}} - \frac{1}{1-\xi\cdot\eta}\nu(\xi)\cdot\eta\right| d\sigma(\eta) \\
& + \frac{1}{4\pi}\Big(\sup_{\eta\in\partial\Gamma}|F(\eta)|\Big)\int_{\partial\Gamma}\left|\frac{1}{1-\frac{\xi+\tau\nu(\xi)}{\sqrt{1+\tau^2}}\cdot\eta}\left(\frac{\tau\nu(\xi)\cdot\nu(\eta)}{\sqrt{1+\tau^2}} - \frac{\tau\xi\cdot\eta}{\sqrt{1+\tau^2}}\right)\right| d\sigma(\eta) \\
& + \frac{1}{4\pi}\Big(\sup_{\eta\in\partial\Gamma}|F(\eta)|\Big)\int_{\partial\Gamma}\left|\frac{1}{1-\frac{\xi+\tau\nu(\xi)}{\sqrt{1+\tau^2}}\cdot\eta}\left(\frac{\tau\xi\cdot\eta}{\sqrt{1+\tau^2}} - \frac{\tau\xi\cdot\eta}{(1+\tau^2)^{\frac{3}{2}}}\right)\right| d\sigma(\eta) \\
& + \frac{1}{4\pi}\Big(\sup_{\eta\in\partial\Gamma}|F(\eta)|\Big)\tau^2\int_{\partial\Gamma}\left|\frac{1}{1-\frac{\xi+\tau\nu(\xi)}{\sqrt{1+\tau^2}}\cdot\eta}\frac{\nu(\xi)\cdot\eta}{(1+\tau^2)^{\frac{3}{2}}}\right| d\sigma(\eta).
\end{aligned} \tag{6.83}
$$

The first integral on the right-hand side of (6.83) can be split up as follows:

$$
\begin{aligned}
& \int\limits_{\partial\Gamma}\left|\frac{1}{1-\frac{\xi+\tau\nu(\xi)}{\sqrt{1+\tau^2}}\cdot\eta}\frac{\xi\cdot\nu(\eta)}{\sqrt{1+\tau^2}} - \frac{1}{1-\xi\cdot\eta}\xi\cdot\nu(\eta)\right| d\sigma(\eta) \qquad (6.84) \\
\leq & \int\limits_{\substack{\eta\in\partial\Gamma \\ 1-\xi\cdot\eta<\rho}}\left|\frac{1}{1-\frac{\xi+\tau\nu(\xi)}{\sqrt{1+\tau^2}}\cdot\eta}\frac{\xi\cdot\nu(\eta)}{\sqrt{1+\tau^2}}\right| d\sigma(\eta) + \int\limits_{\substack{\eta\in\partial\Gamma \\ 1-\xi\cdot\eta<\rho}}\left|\frac{1}{1-\xi\cdot\eta}\xi\cdot\nu(\eta)\right| d\sigma(\eta) \\
& + \int\limits_{\substack{\eta\in\partial\Gamma \\ 1-\xi\cdot\eta\geq\rho}}\left|\frac{1}{1-\frac{\xi+\tau\nu(\xi)}{\sqrt{1+\tau^2}}\cdot\eta}\frac{\xi\cdot\nu(\eta)}{\sqrt{1+\tau^2}} - \frac{1}{1-\xi\cdot\eta}\xi\cdot\nu(\eta)\right| d\sigma(\eta).
\end{aligned}
$$

From (6.52) we see that, for any $\varepsilon > 0$, a constant $\rho \in (0,2)$ can be chosen such that the first and second integral on the right-hand side of (6.84) can be estimated by ε, independently of $\xi \in \partial\Gamma$ and $\tau > 0$. As a consequence of the estimate (6.74), the third integral in (6.84) vanishes uniformly with respect to $\xi \in \partial\Gamma$ as τ tends to zero. Thus, we arrive at

$$
\lim_{\tau\to 0+}\sup_{\xi\in\partial\Gamma}\int_{\partial\Gamma}\left|\frac{1}{1-\frac{\xi+\tau\nu(\xi)}{\sqrt{1+\tau^2}}\cdot\eta}\frac{\xi\cdot\nu(\eta)}{\sqrt{1+\tau^2}} - \frac{1}{1-\xi\cdot\eta}\xi\cdot\nu(\eta)\right| d\sigma(\eta) \leq 2\varepsilon. \tag{6.85}
$$

Since $\varepsilon > 0$ can be chosen arbitrarily, the left hand side of (6.84) converges to zero, uniformly with respect to $\xi \in \partial\Gamma$, as τ tends to zero. The same arguments also apply to the second integral on the right-hand side of (6.83). To handle the third integral in (6.83), we first observe

$$\begin{aligned} |\nu(\xi)\cdot\nu(\eta)-\xi\cdot\eta| &\leq |1-\xi\cdot\eta|+|1-\nu(\xi)\cdot\nu(\eta)| \\ &= \frac{1}{2}|\xi-\eta|^2+\frac{1}{2}|\nu(\xi)-\nu(\eta)|^2 \\ &\leq M|\xi-\eta|^2, \end{aligned} \tag{6.86}$$

for $\xi,\eta\in\partial\Gamma$, where the last estimate follows from (2.52). Consequently, using the estimate from Proposition 2.3, we end up with

$$\begin{aligned} &\int_{\partial\Gamma}\left|\frac{1}{1-\frac{\xi+\tau\nu(\xi)}{\sqrt{1+\tau^2}}\cdot\eta}\left(\frac{\tau\nu(\xi)\cdot\nu(\eta)}{\sqrt{1+\tau^2}}-\frac{\tau\xi\cdot\eta}{\sqrt{1+\tau^2}}\right)\right|d\sigma(\eta) \\ &\leq \tau\int_{\partial\Gamma}\frac{2M|\xi-\eta|^2}{\left|\frac{\xi+\tau\nu(\xi)}{\sqrt{1+\tau^2}}-\eta\right|^2}d\sigma(\eta) \\ &\leq \tau\int_{\partial\Gamma}4\sqrt{2}M\,d\sigma(\eta). \end{aligned} \tag{6.87}$$

The last term in (6.87) obviously converges to zero as τ tends to zero. What remains to complete the proof is the investigation of the fourth and fifth terms on the right-hand side of (6.83). Due to the similarity of the argumentation, we only treat the fourth term. It can be rewritten as follows:

$$\begin{aligned} &\int_{\partial\Gamma}\left|\frac{1}{1-\frac{\xi+\tau\nu(\xi)}{\sqrt{1+\tau^2}}\cdot\eta}\left(\frac{\tau\xi\cdot\eta}{\sqrt{1+\tau^2}}-\frac{\tau\xi\cdot\eta}{(1+\tau^2)^{\frac{3}{2}}}\right)\right|d\sigma(\eta) \\ &= \frac{\tau^3}{(1+\tau^2)^{\frac{3}{2}}}\int_{\partial\Gamma}\left|\frac{\xi\cdot\eta}{1-\frac{\xi+\tau\nu(\xi)}{\sqrt{1+\tau^2}}\cdot\eta}\right|d\sigma(\eta). \end{aligned} \tag{6.88}$$

The last integral in (6.88) is bounded uniformly with respect to $\xi\in\partial\Gamma$ and $\tau>0$. Thus, the integral (6.88) vanishes as τ tends to zero, and we have shown that (6.82) holds true. This concludes the limit relation for $P_{|1}(\tau,0)$. The limit relation for $P_{|1}(-\tau,0)$ follows in complete analogy. □

Proof. Theorem 6.21. It only remains to show the second assertion of (b). The other ones follow directly from Theorem 6.20, e.g., by splitting up the jump relation under consideration in the following way:

$$\begin{aligned} &P_{|1}(\tau,0)[F]-P_{|1}(-\tau,0)[F]-F \\ &= \left(P_{|1}(\tau,0)[F]-P_{|1}(0,0)[F]-\frac{1}{2}F\right)-\left(P_{|1}(-\tau,0)[F]-P_{|1}(0,0)[F]+\frac{1}{2}F\right). \end{aligned} \tag{6.89}$$

To obtain the second assertion of (b), we formally calculate the second-order normal derivatives of the fundamental solution for the Beltrami operator:

$$
\begin{aligned}
&\frac{\partial}{\partial\tau}\frac{\partial}{\partial\sigma}G\left(\Delta^*;\frac{\xi+\tau\nu(\xi)}{\sqrt{1+\tau^2}}\cdot\frac{\eta+\sigma\nu(\eta)}{\sqrt{1+\sigma^2}}\right) \qquad (6.90)\\
&= \frac{1}{4\pi}\frac{1}{\left(1-\frac{\xi+\tau\nu(\xi)}{\sqrt{1+\tau^2}}\cdot\frac{\eta+\sigma\nu(\eta)}{\sqrt{1+\sigma^2}}\right)^2}\left(\left(\frac{\nu(\xi)}{\sqrt{1+\tau^2}}-\tau\frac{\xi+\tau\nu(\xi)}{(1+\tau^2)^{\frac{3}{2}}}\right)\cdot\frac{\eta+\sigma\nu(\eta)}{\sqrt{1+\sigma^2}}\right)\\
&\qquad\times\left(\frac{\xi+\tau\nu(\xi)}{\sqrt{1+\tau^2}}\cdot\left(\frac{\nu(\eta)}{\sqrt{1+\sigma^2}}-\sigma\frac{\eta+\sigma\nu(\eta)}{(1+\sigma^2)^{\frac{3}{2}}}\right)\right)\\
&\quad-\frac{1}{4\pi}\frac{1}{1-\frac{\xi+\tau\nu(\xi)}{\sqrt{1+\tau^2}}\cdot\frac{\eta+\sigma\nu(\eta)}{\sqrt{1+\sigma^2}}}\left(\frac{\nu(\xi)}{\sqrt{1+\tau^2}}-\tau\frac{\xi+\tau\nu(\xi)}{(1+\tau^2)^{\frac{3}{2}}}\right)\\
&\qquad\times\left(\frac{\nu(\eta)}{\sqrt{1+\sigma^2}}-\sigma\frac{\eta+\tau\nu(\eta)}{(1+\sigma^2)^{\frac{3}{2}}}\right).
\end{aligned}
$$

Now, we are able to consider the difference $P_{|1|2}(\tau,0)[F]-P_{|1|2}(-\tau,0)[F]$. Using Equation (6.90), this difference is split up into portions that can be estimated by already available tools, so that we eventually end up with the convergence toward zero as τ tends to zero. More specifically, the following lengthy estimate has to be discussed in order to specify its limit behavior:

$$
\begin{aligned}
&\left|P_{|1|2}(\tau,0)[F](\xi)-P_{|1|2}(-\tau,0)[F](\xi)\right|\\
&= \left|\frac{1}{4\pi}\int_{\partial\Gamma}\frac{1}{\left(1-\frac{\xi+\tau\nu(\xi)}{\sqrt{1+\tau^2}}\cdot\eta\right)^2}\left(\left(\frac{\nu(\xi)}{\sqrt{1+\tau^2}}-\tau\frac{\xi+\tau\nu(\xi)}{(1+\tau^2)^{\frac{3}{2}}}\right)\cdot\eta\right)\right.\\
&\qquad\times\left(\frac{\xi+\tau\nu(\xi)}{\sqrt{1+\tau^2}}\cdot\nu(\eta)\right)d\sigma(\eta)\\
&\quad-\frac{1}{4\pi}\int_{\partial\Gamma}\frac{1}{1-\frac{\xi+\tau\nu(\xi)}{\sqrt{1+\tau^2}}\cdot\eta}\left(\frac{\nu(\xi)}{\sqrt{1+\tau^2}}-\tau\frac{\xi+\tau\nu(\xi)}{(1+\tau^2)^{\frac{3}{2}}}\right)\cdot\nu(\eta)d\sigma(\eta)\\
&\quad-\frac{1}{4\pi}\int_{\partial\Gamma}\frac{1}{\left(1-\frac{\xi-\tau\nu(\xi)}{\sqrt{1+\tau^2}}\cdot\eta\right)^2}\left(\left(\frac{\nu(\xi)}{\sqrt{1+\tau^2}}+\tau\frac{\xi-\tau\nu(\xi)}{(1+\tau^2)^{\frac{3}{2}}}\right)\cdot\eta\right)\\
&\qquad\times\left(\frac{\xi-\tau\nu(\xi)}{\sqrt{1+\tau^2}}\cdot\nu(\eta)\right)d\sigma(\eta)\\
&\quad\left.+\frac{1}{4\pi}\int_{\partial\Gamma}\frac{1}{1-\frac{\xi-\tau\nu(\xi)}{\sqrt{1+\tau^2}}\cdot\eta}\left(\frac{\nu(\xi)}{\sqrt{1+\tau^2}}+\tau\frac{\xi-\tau\nu(\xi)}{(1+\tau^2)^{\frac{3}{2}}}\right)\cdot\nu(\eta)d\sigma(\eta)\right|
\end{aligned}
$$

$$
\begin{aligned}
\leq\ & \left| \frac{1}{4\pi} \int_{\partial\Gamma} \left(\frac{1}{\left(1 - \frac{\xi + \tau\nu(\xi)}{\sqrt{1+\tau^2}} \cdot \eta\right)^2} - \frac{1}{\left(1 - \frac{\xi - \tau\nu(\xi)}{\sqrt{1+\tau^2}} \cdot \eta\right)^2} \right) \right. \\
& \qquad \left. \times \frac{\nu(\xi)\cdot\eta}{\sqrt{1+\tau^2}} \frac{\xi\cdot\nu(\eta)}{\sqrt{1+\tau^2}} d\sigma(\eta) \right| \\
& + \left| \frac{1}{4\pi} \int_{\partial\Gamma} \left(\frac{1}{\left(1 - \frac{\xi + \tau\nu(\xi)}{\sqrt{1+\tau^2}} \cdot \eta\right)^2} - \frac{1}{\left(1 - \frac{\xi - \tau\nu(\xi)}{\sqrt{1+\tau^2}} \cdot \eta\right)^2} \right) \right. \\
& \qquad \left. \times \frac{\tau\xi\cdot\eta}{\sqrt{1+\tau^2}} \frac{\tau\nu(\xi)\cdot\nu(\eta)}{\sqrt{1+\tau^2}} d\sigma(\eta) \right| \\
& + \left| \frac{1}{4\pi} \int_{\partial\Gamma} \left(\frac{1}{1 - \frac{\xi + \tau\nu(\xi)}{\sqrt{1+\tau^2}} \cdot \eta} - \frac{1}{1 - \frac{\xi - \tau\nu(\xi)}{\sqrt{1+\tau^2}} \cdot \eta} \right) \frac{\nu(\xi)\cdot\nu(\eta)}{\sqrt{1+\tau^2}} d\sigma(\eta) \right| \\
& + \left| \frac{1}{4\pi} \int_{\partial\Gamma} \frac{1}{\left(1 - \frac{\xi + \tau\nu(\xi)}{\sqrt{1+\tau^2}} \cdot \eta\right)^2} \right. \\
& \qquad \left. \times \left(\frac{\nu(\xi)\cdot\eta}{\sqrt{1+\tau^2}} \frac{\tau\nu(\xi)\cdot\nu(\eta)}{\sqrt{1+\tau^2}} - \frac{\tau\xi\cdot\eta}{(1+\tau^2)^{\frac{3}{2}}} \frac{\xi\cdot\nu(\eta)}{\sqrt{1+\tau^2}} \right) d\sigma(\eta) \right| \\
& + \left| \frac{1}{4\pi} \int_{\partial\Gamma} \frac{1}{\left(1 - \frac{\xi - \tau\nu(\xi)}{\sqrt{1+\tau^2}} \cdot \eta\right)^2} \right. \\
& \qquad \left. \times \left(\frac{\nu(\xi)\cdot\eta}{\sqrt{1+\tau^2}} \frac{\tau\nu(\xi)\cdot\nu(\eta)}{\sqrt{1+\tau^2}} - \frac{\tau\xi\cdot\eta}{(1+\tau^2)^{\frac{3}{2}}} \frac{\xi\cdot\nu(\eta)}{\sqrt{1+\tau^2}} \right) d\sigma(\eta) \right| \\
& + \tau^2 \left| \frac{1}{4\pi} \int_{\partial\Gamma} \frac{1}{\left(1 - \frac{\xi + \tau\nu(\xi)}{\sqrt{1+\tau^2}} \cdot \eta\right)^2} \right. \\
& \qquad \left. \times \left(\frac{\nu(\xi)\cdot\eta}{(1+\tau^2)^{\frac{3}{2}}} \frac{\xi\cdot\nu(\eta)}{\sqrt{1+\tau^2}} + \frac{\nu(\xi)\cdot\eta}{(1+\tau^2)^{\frac{3}{2}}} \frac{\tau\nu(\xi)\cdot\nu(\eta)}{\sqrt{1+\tau^2}} \right) d\sigma(\eta) \right| \\
& + \tau^2 \left| \frac{1}{4\pi} \int_{\partial\Gamma} \frac{1}{\left(1 - \frac{\xi - \tau\nu(\xi)}{\sqrt{1+\tau^2}} \cdot \eta\right)^2} \right. \\
& \qquad \left. \times \left(\frac{\nu(\xi)\cdot\eta}{(1+\tau^2)^{\frac{3}{2}}} \frac{\xi\cdot\nu(\eta)}{\sqrt{1+\tau^2}} - \frac{\nu(\xi)\cdot\eta}{(1+\tau^2)^{\frac{3}{2}}} \frac{\tau\nu(\xi)\cdot\nu(\eta)}{\sqrt{1+\tau^2}} \right) d\sigma(\eta) \right| \\
& + \tau \left| \frac{1}{4\pi} \int_{\partial\Gamma} \frac{1}{1 - \frac{\xi - \tau\nu(\xi)}{\sqrt{1+\tau^2}} \cdot \eta} \left(\frac{\xi\cdot\nu(\eta)}{(1+\tau^2)^{\frac{3}{2}}} - \frac{\tau\nu(\xi)\cdot\nu(\eta)}{(1+\tau^2)^{\frac{3}{2}}} \right) d\sigma(\eta) \right| \\
& + \tau \left| \frac{1}{4\pi} \int_{\partial\Gamma} \frac{1}{1 - \frac{\xi + \tau\nu(\xi)}{\sqrt{1+\tau^2}} \cdot \eta} \left(\frac{\xi\cdot\nu(\eta)}{(1+\tau^2)^{\frac{3}{2}}} + \frac{\tau\nu(\xi)\cdot\nu(\eta)}{(1+\tau^2)^{\frac{3}{2}}} \right) d\sigma(\eta) \right|.
\end{aligned} \tag{6.91}
$$

The last four terms on the right-hand side of (6.91) vanish as τ tends to zero, due to the uniform boundedness of the occurring integrals. Concerning the first three terms, we only treat the third one in more detail. The convergence to zero of the first two integrals follows by analogous argumentation. In doing so, we obtain

$$\left| \int_{\partial\Gamma} \left(\frac{1}{1 - \frac{\xi+\tau\nu(\xi)}{\sqrt{1+\tau^2}} \cdot \eta} - \frac{1}{1 - \frac{\xi-\tau\nu(\xi)}{\sqrt{1+\tau^2}} \cdot \eta} \right) \frac{\nu(\xi)\cdot\nu(\eta)}{\sqrt{1+\tau^2}} d\sigma(\eta) \right| \tag{6.92}$$

$$\leq \frac{1}{\sqrt{1+\tau^2}} \int\limits_{\substack{\eta\in\partial\Gamma \\ 1-\xi\cdot\eta\geq\rho}} \left| \frac{1}{1 - \frac{\xi+\tau\nu(\xi)}{\sqrt{1+\tau^2}} \cdot \eta} - \frac{1}{1 - \frac{\xi-\tau\nu(\xi)}{\sqrt{1+\tau^2}} \cdot \eta} \right| d\sigma(\eta)$$

$$+ \frac{1}{1+\tau^2} \int\limits_{\substack{\eta\in\partial\Gamma \\ 1-\xi\cdot\eta<\rho}} \frac{2\tau|\nu(\xi)\cdot\eta|}{\left(1 - \frac{\xi+\tau\nu(\xi)}{\sqrt{1+\tau^2}} \cdot \eta\right)\left(1 - \frac{\xi-\tau\nu(\xi)}{\sqrt{1+\tau^2}} \cdot \eta\right)} d\sigma(\eta).$$

By use of an estimate similar to (6.74), we see that the first term on the right-hand side of (6.92) vanishes uniformly with respect to $\xi \in \partial\Gamma$ as τ tends to zero. The second term can be estimated by the Cauchy–Schwarz inequality:

$$\int\limits_{\substack{\eta\in\partial\Gamma \\ 1-\xi\cdot\eta<\rho}} \frac{2\tau|\nu(\xi)\cdot\eta|}{\left(1 - \frac{\xi+\tau\nu(\xi)}{\sqrt{1+\tau^2}} \cdot \eta\right)\left(1 - \frac{\xi-\tau\nu(\xi)}{\sqrt{1+\tau^2}} \cdot \eta\right)} d\sigma(\eta) \tag{6.93}$$

$$\leq \left(\int\limits_{\substack{\eta\in\partial\Gamma \\ 1-\xi\cdot\eta<\rho}} \frac{|\nu(\xi)\cdot\eta|^2}{\left(1 - \frac{\xi+\tau\nu(\xi)}{\sqrt{1+\tau^2}} \cdot \eta\right)^2} d\sigma(\eta) \right)^{\frac{1}{2}} \left(\int\limits_{\substack{\eta\in\partial\Gamma \\ 1-\xi\cdot\eta<\rho}} \frac{4\tau^2}{\left(1 - \frac{\xi-\tau\nu(\xi)}{\sqrt{1+\tau^2}} \cdot \eta\right)^2} d\sigma(\eta) \right)^{\frac{1}{2}}.$$

Arguments as in (6.50) show that the second integral on the right-hand side of (6.93) is bounded uniformly with respect to τ, ρ, and $\xi \in \partial\Gamma$. The first integral converges to zero, uniformly with respect to τ and $\xi \in \partial\Gamma$, as ρ tends to zero. This is guaranteed by arguments as presented in (6.51) and (6.52). In other words, for every $\varepsilon > 0$, there exists a sufficiently small $\rho > 0$ such that (6.93) turns out to be smaller than ε. Thus, we end up with

$$\lim_{\tau\to 0+} \sup_{\xi\in\partial\Gamma} \left| \int_{\partial\Gamma} \left(\frac{1}{1 - \frac{\xi+\tau\nu(\xi)}{\sqrt{1+\tau^2}} \cdot \eta} - \frac{1}{1 - \frac{\xi-\tau\nu(\xi)}{\sqrt{1+\tau^2}} \cdot \eta} \right) \frac{\nu(\xi)\cdot\nu(\eta)}{\sqrt{1+\tau^2}} d\sigma(\eta) \right| < \varepsilon. \tag{6.94}$$

Since $\varepsilon > 0$ can be chosen arbitrarily small, (6.92) converges to zero, uniformly with respect to $\xi \in \partial\Gamma$, as τ tends to zero. At last, we have to take a closer look at the fourth and fifth term of the right-hand side of (6.91). We find

$$\begin{aligned} &\left| \int_{\partial\Gamma} \frac{1}{\left(1-\frac{\xi+\tau\nu(\xi)}{\sqrt{1+\tau^2}}\cdot\eta\right)^2} \left(\frac{\nu(\xi)\cdot\eta}{\sqrt{1+\tau^2}} \frac{\tau\nu(\xi)\cdot\nu(\eta)}{\sqrt{1+\tau^2}} - \frac{\tau\xi\cdot\eta}{(1+\tau^2)^{\frac{3}{2}}} \frac{\xi\cdot\nu(\eta)}{\sqrt{1+\tau^2}} \right) d\sigma(\eta) \right| \\ \leq\ & \left| \int_{\partial\Gamma} \frac{1}{\left(1-\frac{\xi+\tau\nu(\xi)}{\sqrt{1+\tau^2}}\cdot\eta\right)^2} \frac{\tau(\nu(\xi)\cdot\eta)(\nu(\xi)\cdot\nu(\eta)) - \tau(\xi\cdot\eta)(\xi\cdot\nu(\eta))}{1+\tau^2} d\sigma(\eta) \right| \\ &+ \left| \frac{1}{1+\tau^2} - \frac{1}{(1+\tau^2)^2} \right| \left| \int_{\partial\Gamma} \frac{1}{\left(1-\frac{\xi+\tau\nu(\xi)}{\sqrt{1+\tau^2}}\cdot\eta\right)^2} \tau(\xi\cdot\eta)(\xi\cdot\nu(\eta)) d\sigma(\eta) \right|. \end{aligned} \tag{6.95}$$

Since the integral in the second term of the right-hand side of (6.95) is uniformly bounded due to an estimate similar to (6.93), the whole term vanishes as τ tends to zero. The first term on the right-hand side of (6.95) can be rewritten as

$$\begin{aligned} &\left| \int_{\partial\Gamma} \frac{1}{\left(1-\frac{\xi+\tau\nu(\xi)}{\sqrt{1+\tau^2}}\cdot\eta\right)^2} \frac{\tau(\nu(\xi)\cdot\eta)(\nu(\xi)\cdot\nu(\eta)) - \tau(\xi\cdot\eta)(\xi\cdot\nu(\eta))}{1+\tau^2} d\sigma(\eta) \right| \\ \leq\ & \int_{\partial\Gamma} \frac{1}{\left(1-\frac{\xi+\tau\nu(\xi)}{\sqrt{1+\tau^2}}\cdot\eta\right)^2} \frac{\tau}{1+\tau^2} |(\nu(\xi)\cdot\eta - \xi\cdot\nu(\eta))\nu(\xi)\cdot\nu(\eta)| d\sigma(\eta) \\ &+ \frac{\tau}{1+\tau^2} \int_{\partial\Gamma} \frac{1}{\left(1-\frac{\xi+\tau\nu(\xi)}{\sqrt{1+\tau^2}}\cdot\eta\right)^2} |\xi\cdot\nu(\eta)(\nu(\xi)\cdot\nu(\eta) - \xi\cdot\eta)| d\sigma(\eta). \end{aligned} \tag{6.96}$$

Again, splitting both integrals on the right-hand side of (6.96) into integrals over $\{\eta \in \partial\Gamma : 1-\xi\cdot\eta \geq \rho\}$ and over $\{\eta \in \partial\Gamma : 1-\xi\cdot\eta < \rho\}$, respectively, we are able to assure that they vanish uniformly with respect to $\xi \in \partial\Gamma$ as τ tends to zero. The argumentation is similar as for (6.92) and uses the estimates (2.53) and (6.86). Thus, we are finally led to

$$\begin{aligned} \lim_{\tau\to 0+} \sup_{\xi\in\partial\Gamma} \Bigg| \int_{\partial\Gamma} & \frac{1}{\left(1-\frac{\xi+\tau\nu(\xi)}{\sqrt{1+\tau^2}}\cdot\eta\right)^2} \\ & \times \left(\frac{\nu(\xi)\cdot\eta}{\sqrt{1+\tau^2}} \frac{\tau\nu(\xi)\cdot\nu(\eta)}{\sqrt{1+\tau^2}} - \frac{\tau\xi\cdot\eta}{(1+\tau^2)^{\frac{3}{2}}} \frac{\xi\cdot\nu(\eta)}{\sqrt{1+\tau^2}} \right) d\sigma(\eta) \Bigg| = 0, \end{aligned} \tag{6.97}$$

which completes the proof. □

Remark 6.22. Although the previous proofs are rather lengthy and technical, the jumps of the double-layer potential for the special case $F = 1$ can already be observed in the identity (6.13), depending on the location of the evaluation point ξ (i.e., ξ in Γ, $\partial\Gamma$, or Γ^c).

Next, our goal is to formulate the limit relations for the *adjoint operators*. Indeed, it follows for the single-layer potential that the transfer to the adjoint operator amounts to an interchange of the parameters σ and τ. This can be easily seen from Fubini's theorem:

$$\begin{aligned} & (P(\tau,\sigma)F,G)_{\mathrm{L}^2(\partial\Gamma)} && (6.98) \\ = & \int_{\partial\Gamma}\int_{\partial\Gamma} G\left(\Delta^*; \frac{\xi+\tau\nu(\xi)}{\sqrt{1+\tau^2}}\cdot\frac{\eta+\sigma\nu(\eta)}{\sqrt{1+\sigma^2}}\right) F(\eta)\,d\sigma(\eta)\,G(\xi)\,d\sigma(\xi) \\ = & \int_{\partial\Gamma}\int_{\partial\Gamma} G\left(\Delta^*; \frac{\xi+\tau\nu(\xi)}{\sqrt{1+\tau^2}}\cdot\frac{\eta+\sigma\nu(\eta)}{\sqrt{1+\sigma^2}}\right) G(\xi)\,d\sigma(\xi)\,F(\eta)\,d\sigma(\eta) \\ = & (F,P(\sigma,\tau)G)_{\mathrm{L}^2(\partial\Gamma)}, \end{aligned}$$

for $\tau,\sigma\in\mathbb{R}$ and F,G of class $\mathrm{C}^{(0)}(\partial\Gamma)$. Similar manipulations are valid for $P_{|1}(\tau,\sigma)$, $P_{|2}(\tau,\sigma)$, and $P_{|1|2}(\tau,\sigma)$. Furthermore, for $\tau,\sigma\in\mathbb{R}$, the operators $P(\tau,\sigma)$, $P_{|1}(\tau,\sigma)$, and $P_{|2}(\tau,\sigma)$ are well-defined on $\mathrm{L}^2(\partial\Gamma)$. However, the operator $P_{|1|2}(\tau,\sigma)$ is well-defined only for $\tau\neq 0$ and/or $\sigma\neq 0$. It turns out that the adjoint operators in the L^2-topology are characterized in the form

$$P(\tau,\sigma)^* = P(\sigma,\tau), \tag{6.99}$$
$$P_{|1}(\tau,\sigma)^* = P_{|2}(\sigma,\tau), \tag{6.100}$$
$$P_{|2}(\tau,\sigma)^* = P_{|1}(\sigma,\tau), \tag{6.101}$$

for $\sigma,\tau\in\mathbb{R}$. Furthermore,

$$P_{|1|2}(\tau,\sigma)^* = P_{|1|2}(\sigma,\tau), \tag{6.102}$$

for $\tau\neq 0$ and/or $\sigma\neq 0$. The upcoming limit and jump relations are still restricted to functions F of class $\mathrm{C}^{(0)}(\partial\Gamma)$. An extension to $\mathrm{L}^2(\partial\Gamma)$ is given in the next subsection.

Theorem 6.23 (Limit Relations)**.** *Let $\Gamma\subset\Omega$ be a regular region, and suppose that F is of class $\mathrm{C}^{(0)}(\partial\Gamma)$.*

(a) For the adjoint operator of the single-layer potential, we have

$$\lim_{\tau\to 0+}\sup_{\xi\in\partial\Gamma}\left|P(\pm\tau,0)^*[F](\xi)-P(0,0)^*[F](\xi)\right| = 0, \tag{6.103}$$

$$\lim_{\tau\to 0+}\sup_{\xi\in\partial\Gamma}\left|P_{|1}(\pm\tau,0)^*[F](\xi)-P_{|1}(0,0)^*[F](\xi)\mp\frac{1}{2}F(\xi)\right| = 0. \tag{6.104}$$

(b) For the adjoint operator of the double-layer potential, we have

$$\lim_{\tau\to 0+}\sup_{\xi\in\partial\Gamma}\left|P_{|2}(\pm\tau,0)^*[F](\xi)-P_{|2}(0,0)^*[F](\xi)\pm\frac{1}{2}F(\xi)\right| = 0. \tag{6.105}$$

Theorem 6.24 (Jump Relations)**.** *Let $\Gamma\subset\Omega$ be a regular region, and assume that F is of class $\mathrm{C}^{(0)}(\partial\Gamma)$.*

(a) For the adjoint operator of the single-layer potential, we have

$$\lim_{\tau\to 0+}\sup_{\xi\in\partial\Gamma} |P(\tau,0)^*[F](\xi) - P(-\tau,0)^*[F](\xi)| = 0, \quad (6.106)$$

$$\lim_{\tau\to 0+}\sup_{\xi\in\partial\Gamma} \left|P_{|1}(\tau,0)^*[F](\xi) - P_{|1}(-\tau,0)^*[F](\xi) - F(\xi)\right| = 0. \quad (6.107)$$

(b) For the adjoint operator of the double-layer potential, we have

$$\lim_{\tau\to 0+}\sup_{\xi\in\partial\Gamma} \left|P_{|2}(\tau,0)^*[F](\xi) - P_{|2}(-\tau,0)^*[F](\xi) + F(\xi)\right| = 0, \quad (6.108)$$

$$\lim_{\tau\to 0+}\sup_{\xi\in\partial\Gamma} \left|P_{|1|2}(\tau,0)^*[F](\xi) - P_{|1|2}(-\tau,0)^*[F](\xi)\right| = 0. \quad (6.109)$$

Proof. Theorems 6.23 and 6.24. The proof of the limit relation for $P(\pm\tau,0)^*$ in Theorem 6.23 is analogous to that of $P(\pm\tau,0)$. For the case of the adjoint operator to the double-layer potential $P_{|2}(\pm\tau,0)^*$, we have to verify that

$$\begin{aligned}\lim_{\tau\to 0+}\sup_{\xi\in\partial\Gamma} \Big| & P_{|2}(\pm\tau,0)^*[F](\xi) - P_{|2}(\pm\tau,0)[F](\xi) \\ & -\big(P_{|2}(0,0)^*[F](\xi) - P_{|2}(0,0)[F](\xi)\big)\Big| = 0.\end{aligned} \quad (6.110)$$

Then, the desired limit relation for the adjoint operator follows from the already known limit relation for $P_{|2}(\pm\tau,0)$ (cf. Theorem 6.20). A detailed elaboration of the proof is left to the reader. The same idea works for $P_{|1}(\pm\tau,0)$ as well.

The first three jump relations in Theorem 6.24 are again a direct consequence of the limit relations. The relation for the adjoint operator $P_{|1|2}(\tau,0)^*$ follows similarly as for the original operator $P_{|1|2}(\tau,0)$ in Theorem 6.21. □

We conclude this subsection by showing that the occurring potential operators are uniformly bounded and completely continuous. Like in the Euclidean context, both properties are of importance for the existence of solutions to the linear integral equations arising from Dirichlet and Neumann boundary-value problems with respect to the Beltrami operator.

Lemma 6.25. *Let $\Gamma\subset\Omega$ be a regular region. The linear operators $P(\tau,0)$: $\mathrm{C}^{(0)}(\partial\Gamma)\to\mathrm{C}^{(0)}(\partial\Gamma)$, $P_{|1}(\tau,0):\mathrm{C}^{(0)}(\partial\Gamma)\to\mathrm{C}^{(0)}(\partial\Gamma)$, $P_{|2}(\tau,0):\mathrm{C}^{(0)}(\partial\Gamma)\to\mathrm{C}^{(0)}(\partial\Gamma)$ and their adjoint operators are completely continuous and uniformly bounded with respect to $\tau\in\mathbb{R}$.*

Proof. The uniform boundedness of the operators follows from (6.42) in connection with Lemma 6.17 and 6.19. If $\tau\neq 0$, the kernel of $P_{|2}(\tau,0)$ is uniformly continuous on $\partial\Gamma$, and a standard procedure can be applied to assure the complete continuity of these convolution operators. The set $\{P(\tau,0)[F] : F\in\mathrm{C}^{(0)}(\partial\Gamma), \|F\|_{\mathrm{C}^{(0)}(\partial\Gamma)}<1\}$ is equicontinuous and, by virtue of Lemma 6.17, uniformly bounded. The Theorem of Arzelà-Ascoli then leads to the complete continuity of the operator $P(\tau,0)$ (cf., e.g., H.W. Alt [2006] and K. Yoshida [1980]). To treat the case $\tau = 0$, we need a convergent sequence of completely continuous operators. For that purpose, let $G^\rho(\Delta^*;\cdot)$ be a linearly regularized Green function, as provided, e.g., in the proof of Theorem 6.13, such that

$$\frac{\partial}{\partial\nu(\eta)} G^\rho(\Delta^*;\xi\cdot\eta) = \begin{cases} -\frac{1}{4\pi}\frac{1}{1-\xi\cdot\eta}\xi\cdot\nu(\eta), & 1-\xi\cdot\eta \geq \rho, \\ -\frac{1}{4\pi\rho}\xi\cdot\nu(\eta), & 1-\xi\cdot\eta < \rho. \end{cases} \tag{6.111}$$

We introduce the following set of linear operators on $\mathrm{C}^{(0)}(\partial\Gamma)$:

$$T^\rho[F](\xi) = \int_{\partial\Gamma} F(\eta)\frac{\partial}{\partial\nu(\eta)} G^\rho(\Delta^*;\xi\cdot\eta)d\sigma(\eta), \quad \xi\in\partial\Gamma,\ \rho\in(0,2). \tag{6.112}$$

Based on the boundedness and uniform continuity of $\eta\mapsto \frac{\partial}{\partial\nu(\eta)}G^\rho(\Delta^*;\xi\cdot\eta)$, these operators can be seen to be completely continuous. Furthermore, by aid of (2.53), we get

$$\begin{aligned} &\left|P_{|2}(0,0)[F](\xi) - T^\rho[F](\xi)\right| \\ \leq\ & \frac{1}{4\pi}\left(\sup_{\eta\in\partial\Gamma}|F(\eta)|\right)\int\limits_{\substack{\eta\in\partial\Gamma\\1-\xi\cdot\eta<\rho}} \frac{1}{1-\xi\cdot\eta}|\xi\cdot\nu(\eta)|d\sigma(\eta) \\ & +\frac{1}{4\pi\rho}\left(\sup_{\eta\in\partial\Gamma}|F(\eta)|\right)\int\limits_{\substack{\eta\in\partial\Gamma\\1-\xi\cdot\eta<\rho}} |\xi\cdot\nu(\eta)|d\sigma(\eta) \\ \leq\ & \frac{1}{2\pi}\left(\sup_{\eta\in\partial\Gamma}|F(\eta)|\right)\int\limits_{\substack{\eta\in\partial\Gamma\\1-\xi\cdot\eta<\rho}} M d\sigma(\eta) \\ & +\frac{1}{4\pi\rho}\left(\sup_{\eta\in\partial\Gamma}|F(\eta)|\right)\int\limits_{\substack{\eta\in\partial\Gamma\\1-\xi\cdot\eta<\rho}} M|\xi-\eta|^2 d\sigma(\eta). \end{aligned} \tag{6.113}$$

The upper bound on the right-hand side of (6.113) converges to zero, uniformly with respect to $\xi\in\partial\Gamma$, as ρ tends to zero. This way, we find

$$\lim_{\rho\to 0}\|P_{|2}(0,0) - T^\rho\|_{\mathrm{C}^{(0)}(\partial\Gamma)\to\mathrm{C}^{(0)}(\partial\Gamma)} = 0, \tag{6.114}$$

where, as usual, $\|T\|_{X\to Y} = \sup_{x\in X}\frac{\|Tx\|_Y}{\|x\|_X}$ denotes the operator norm of a linear operator $T: X\to Y$, with Banach spaces X, Y. Thus, $P_{|2}(0,0)$ is completely continuous as the limit of a sequence of completely continuous operators. The complete continuity of $P(0,0)$ and $P_{|1}(0,0)$ follows by a similar argumentation, and therefore, the assertions for the corresponding adjoint operators are valid as well. □

6.3.3 Limit and Jump Relations in L^2-Topology

Once we have established the limit and jump relations in the $C^{(0)}$-norm, the same procedure as proposed in Theorem 3.56 leads us to the corresponding relations in the L^2-norm. For the sake of brevity, we introduce a set of *limit and jump operators* by

$$L_1^{\pm}(\tau)[F] = P(\pm\tau,0)[F] - P(0,0)[F], \tag{6.115}$$

$$L_2^{\pm}(\tau)[F] = P_{|1}(\pm\tau,0)[F] - P_{|1}(0,0)[F] \mp \frac{1}{2}F, \tag{6.116}$$

$$L_3^{\pm}(\tau)[F] = P_{|2}(\pm\tau,0)[F] - P_{|2}(0,0)[F] \pm \frac{1}{2}F, \tag{6.117}$$

and

$$J_1(\tau)[F] = P(\tau,0)[F] - P(-\tau,0)[F], \tag{6.118}$$

$$J_2(\tau)[F] = P_{|1}(\tau,0)[F] - P_{|1}(-\tau,0)[F] - F, \tag{6.119}$$

$$J_3(\tau)[F] = P_{|2}(\tau,0)[F] - P_{|2}(-\tau,0)[F] + F, \tag{6.120}$$

$$J_4(\tau)[F] = P_{|1|2}(\tau,0)[F] - P_{|1|2}(-\tau,0)[F], \tag{6.121}$$

where $\tau > 0$ and $F \in L^2(\partial\Gamma)$. Then, the limit and jump relations in an L^2-context read as follows:

Theorem 6.26 (Limit Relations). *Let $\Gamma \subset \Omega$ be a regular region, and suppose that F is of class $L^2(\partial\Gamma)$. Then*

$$\lim_{\tau\to 0+} \left\|L_i^{\pm}(\tau)[F]\right\|_{L^2(\partial\Gamma)} = 0, \quad i = 1,2,3, \tag{6.122}$$

$$\lim_{\tau\to 0+} \left\|L_i^{\pm}(\tau)^*[F]\right\|_{L^2(\partial\Gamma)} = 0, \quad i = 1,2,3. \tag{6.123}$$

Theorem 6.27 (Jump Relations). *Let $\Gamma \subset \Omega$ be a regular region, and assume that F is of class $L^2(\partial\Gamma)$. Then*

$$\lim_{\tau\to 0+} \|J_i(\tau)[F]\|_{L^2(\partial\Gamma)} = 0, \quad i = 1,2,3,4, \tag{6.124}$$

$$\lim_{\tau\to 0+} \|J_i(\tau)^*[F]\|_{L^2(\partial\Gamma)} = 0, \quad i = 1,2,3,4. \tag{6.125}$$

Later on, the limit and jump relations in the L^2-topology enable us to formulate generalized Fourier expansions as constructive solutions to boundary-value problems.

6.4 Boundary-Value Problems

Boundary-value problems for the Beltrami equation $\Delta^* U = H$ in a regular region Γ, corresponding to Neumann or Dirichlet boundary values on $\partial\Gamma$, appear in many applications where spherical decompositions are required. For purposes in geomagnetism, e.g., the case $H \neq 0$ is often of greater importance than the harmonic case $H = 0$. Furthermore, spherical differential equations with respect to the vectorial operators ∇^* and L^* frequently appear in gravitational as well as geomagnetic modeling. However, understood in a strict sense, those formulations are not real boundary-value problems on regular regions $\Gamma \subset \Omega$ since the solutions are already uniquely determined (up to a constant) from information inside Γ. Yet, they profit from some considerations for the Beltrami equation.

6.4.1 Formulation and Uniqueness

We begin with the formulation of the two classical boundary-value problems. Different from the Euclidean framework, we do not need to distinguish between inner and outer problems since the open complement of a regular region on the sphere again constitutes a regular region.

Dirichlet Problem (DP): Let F be of class $\mathrm{C}^{(0)}(\partial\Gamma)$. We are looking for a function U of class $\mathrm{C}^{(2)}(\Gamma) \cap \mathrm{C}^{(0)}(\overline{\Gamma})$ such that

(i) $\Delta^* U(\xi) = 0, \quad \xi \in \Gamma,$

(ii) $U(\xi) = F(\xi), \quad \xi \in \partial\Gamma.$

Neumann Problem (NP): Let G be of class $\mathrm{C}^{(0)}(\partial\Gamma)$. We are looking for a function U of class $\mathrm{C}^{(2)}(\Gamma) \cap \mathrm{C}^{(1)}(\overline{\Gamma})$ such that

(i) $\Delta^* U(\xi) = 0, \quad \xi \in \Gamma,$

(ii) $\frac{\partial}{\partial \nu} U(\xi) = G(\xi), \quad \xi \in \partial\Gamma.$

Once the problems (DP) and (NP) are treated, the case $\Delta^* U = H$, for $H \neq 0$, is a simple consequence of Theorem 6.14. Furthermore, the Maximum/Minimum Principle and Green's formulas imply uniqueness in exactly the same manner as in the Euclidean case (cf. Lemma 3.58 and 3.59).

Lemma 6.28. *A solution of* (DP) *is uniquely determined.*

Lemma 6.29. *A solution of* (NP) *is uniquely determined up to a constant.*

The Maximum/Minimum Principle additionally assures the continuous dependence on the boundary values for (DP).

Lemma 6.30. *Let U, $\tilde{U} : \Gamma \to \mathbb{R}$ satisfy* (DP) *for boundary values F, $\tilde{F}$ of class $\mathrm{C}^{(0)}(\partial\Gamma)$. Then*

$$\sup_{\xi \in \Gamma} |U(\xi) - \tilde{U}(\xi)| \leq \sup_{\xi \in \partial\Gamma} |F(\xi) - \tilde{F}(\xi)|. \tag{6.126}$$

6.4.2 Integral Equation Methods

The relation $U(\xi) = F(\xi)$, $\xi \in \partial\Gamma$, in the case of the Dirichlet problem, can be reformulated in the following sense:

$$U^-(\xi) = \lim_{\tau\to 0+} U\left(\frac{\xi - \tau\nu(\xi)}{\sqrt{1+\tau^2}}\right) = F(\xi), \quad \xi \in \partial\Gamma. \tag{6.127}$$

Approaching the boundary from the exterior, i.e., from Γ^c, we write

$$U^+(\xi) = \lim_{\tau\to 0+} U\left(\frac{\xi + \tau\nu(\xi)}{\sqrt{1+\tau^2}}\right), \quad \xi \in \partial\Gamma. \tag{6.128}$$

For Neumann boundary values, $\frac{\partial}{\partial\nu}U(\xi) = G(\xi)$, for $\xi \in \partial\Gamma$, is meant in the sense

$$\frac{\partial}{\partial\nu}U^-(\xi) = \lim_{\tau\to 0+} \frac{\partial}{\partial\nu(\zeta)}U(\zeta)\Big|_{\zeta=\frac{\xi-\tau\nu(\xi)}{\sqrt{1+\tau^2}}} = G(\xi), \quad \xi \in \partial\Gamma. \tag{6.129}$$

And the exterior limit reads

$$\frac{\partial}{\partial\nu}U^+(\xi) = \lim_{\tau\to 0+} \frac{\partial}{\partial\nu(\zeta)}U(\zeta)\Big|_{\zeta=\frac{\xi+\tau\nu(\xi)}{\sqrt{1+\tau^2}}}, \quad \xi \in \partial\Gamma. \tag{6.130}$$

In any case, one has to be aware of the orientation of the normal vector ν. Concerning an underlying regular region $\Gamma \subset \Omega$, we always choose ν to be directed into the exterior Γ^c.

Dirichlet Problem (DP): Making an ansatz with a double-layer potential

$$U(\xi) = \int_{\partial\Gamma} Q(\eta)\frac{\partial}{\partial\nu(\eta)}G(\Delta^*;\xi\cdot\eta)d\sigma(\eta), \quad \xi \in \Omega, \tag{6.131}$$

for some $Q \in \mathrm{C}^{(0)}(\partial\Gamma)$, we see that $\Delta^*_\xi U(\xi) = 0$, for $\xi \in \Gamma$. Taking the limit in the sense of (6.127), we see that the following integral equation has to hold true for (DP):

$$F(\xi) = P_{|2}(0,0)[Q](\xi) + \frac{1}{2}Q(\xi), \quad \xi \in \partial\Gamma. \tag{6.132}$$

We have used the limit relations from Theorem 6.20.

Neumann Problem (NP): Here, we make the ansatz of a single-layer potential

$$U(\xi) = \int_{\partial\Gamma} \tilde{Q}(\eta)G(\Delta^*;\xi\cdot\eta)d\sigma(\eta), \quad \xi \in \Omega, \tag{6.133}$$

for some $\tilde{Q} \in \mathrm{C}^{(0)}(\partial\Gamma)$. In order to guarantee the harmonicity of U with respect to the Beltrami operator, we need to require

$$\frac{1}{\|\partial\Gamma\|}\int_{\partial\Gamma} \tilde{Q}(\eta)d\sigma(\eta) = 0, \tag{6.134}$$

since $\Delta^* U(\xi) = -\frac{1}{4\pi}\int_{\partial\Gamma} \tilde{Q}(\eta) d\sigma(\eta)$, $\xi \in \Gamma$. For convenience, we let $\mathrm{C}_0^{(0)}(\partial\Gamma)$ denote the class of functions in $\mathrm{C}^{(0)}(\partial\Gamma)$ that satisfy the condition (6.134), i.e.,

$$\mathrm{C}_0^{(0)}(\partial\Gamma) = \left\{ F \in \mathrm{C}^{(0)}(\partial\Gamma) : \frac{1}{\|\partial\Gamma\|}\int_{\partial\Gamma} F(\eta) d\sigma(\eta) = 0 \right\}. \tag{6.135}$$

Taking the limit in the sense of (6.129), we see that in order to satisfy (NP), the following integral equation has to be valid:

$$G(\xi) = P_{|1}(0,0)[\tilde{Q}](\xi) - \frac{1}{2}\tilde{Q}(\xi), \quad \xi \in \partial\Gamma. \tag{6.136}$$

Again, we have used the limit relations from Theorem 6.20.

Summarizing, we have to consider the following two integral equations on $\partial\Gamma$:

$$(\mathrm{D}): \qquad F = P_{|2}(0,0)[Q] + \frac{1}{2}Q, \tag{6.137}$$

$$(\mathrm{N}): \qquad G = P_{|1}(0,0)[\tilde{Q}] - \frac{1}{2}\tilde{Q}, \tag{6.138}$$

relating to the Dirichlet problem (DP) and the Neumann problem (NP), respectively. Since $P_{|1}(0,0)$, $P_{|2}(0,0)$, and their adjoint operators are completely continuous linear operators, we can apply the Fredholm Theorem 3.77 to investigate the existence of solutions $Q \in \mathrm{C}^{(0)}(\partial\Gamma)$ for (D) and solutions $\tilde{Q} \in \mathrm{C}_0^{(0)}(\partial\Gamma)$ for (N). Numerical applications of these boundary integral equations can be found, e.g., in S. Gemmrich et al. [2008]. A more theoretical investigation is presented in C. Gerhards [2011a].

Dirichlet Problem

We begin with the investigation of the integral equation (D).

Lemma 6.31. *Let $\Gamma \subset \Omega$ be a regular region. The homogeneous integral equation*

$$P_{|2}(0,0)[Q] + \frac{1}{2}Q = 0 \tag{6.139}$$

admits only the trivial solution $Q = 0$ in $\mathrm{C}^{(0)}(\partial\Gamma)$.

Proof. Setting $U(\xi) = \int_{\partial\Gamma} Q(\eta)\frac{\partial}{\partial\nu(\eta)} G(\Delta^*;\xi\cdot\eta)\, d\sigma(\eta)$, for $\xi \in \Gamma$, we obtain, by observing the boundary values in connection with the properties of the operators $P_{|2}(\tau,\sigma)$, that

$$U^-(\xi) = \lim_{\tau\to 0+} U\left(\frac{\xi - \tau\nu(\xi)}{\sqrt{1+\tau^2}}\right) = \lim_{\tau\to 0+} P_{|2}(-\tau,0)[Q](\xi), \quad \xi \in \partial\Gamma. \tag{6.140}$$

The limit relations for the double-layer potential in combination with the integral equation (6.139) additionally imply

$$U^-(\xi) = P_{|2}(0,0)[Q](\xi) + \frac{1}{2}Q(\xi) = 0, \quad \xi \in \partial\Gamma. \tag{6.141}$$

The uniqueness of the Dirichlet problem $\Delta^* U(\xi) = 0$, $\xi \in \Gamma$, corresponding to boundary values $U(\xi) = 0$, $\xi \in \partial\Gamma$, shows us

$$U(\xi) = 0, \quad \xi \in \Gamma, \tag{6.142}$$

such that,

$$\frac{\partial}{\partial\nu} U^-(\xi) = 0, \quad \xi \in \Gamma. \tag{6.143}$$

The jump relations for $P_{|1|2}(-\tau, 0)$ (cf. Theorem 6.21) then lead to

$$\frac{\partial}{\partial\nu} U^+(\xi) = \lim_{\tau\to 0+} \frac{\partial}{\partial\nu(\zeta)} U(\zeta)\Big|_{\zeta = \frac{\xi+\tau\nu(\xi)}{\sqrt{1+\tau^2}}} = 0, \quad \xi \in \partial\Gamma. \tag{6.144}$$

In other words, U satisfies the (exterior) Neumann problem

$$\Delta^* U(\xi) = 0, \quad \xi \in \Gamma^c, \tag{6.145}$$

$$\frac{\partial}{\partial\nu} U^+(\xi) = 0, \quad \xi \in \partial\Gamma. \tag{6.146}$$

Thus, U has to be constant on Γ^c by the uniqueness result known from Lemma 6.28. More precisely, $U(\xi) = K$, for $\xi \in \Gamma^c$, and $U(\xi) = 0$, for $\xi \in \Gamma$. Therefore, by another application of the jump relation for the double-layer potential, we find

$$-Q(\xi) = \lim_{\tau\to 0+} U\left(\frac{\xi - \tau\nu(\xi)}{\sqrt{1+\tau^2}}\right) - \lim_{\tau\to 0+} U\left(\frac{\xi + \tau\nu(\xi)}{\sqrt{1+\tau^2}}\right) = K, \tag{6.147}$$

for $\xi \in \partial\Gamma$. Inserting this into our original integral equation (6.139), and observing the equality (6.13), we end up with

$$0 = P_{|2}(0,0)[Q](\xi) + \frac{1}{2}Q(\xi) = -\frac{1}{2}K + \frac{\|\Gamma\|}{4\pi}K - \frac{1}{2}K = \left(\frac{\|\Gamma\|}{4\pi} - 1\right)K, \tag{6.148}$$

for $\xi \in \partial\Gamma$. Since $\Gamma \neq \Omega$, this implies $K = 0$, which completes the proof. □

As an immediate consequence of Theorem 3.77, the integral equation (D), i.e., Equation (6.137), is solvable for any F of class $\mathrm{C}^{(0)}(\partial\Gamma)$. Thus, the Dirichlet boundary-value problem (DP) possesses a solution of the form (6.131).

Theorem 6.32. *The Dirichlet problem* (DP) *is uniquely solvable for any boundary value F of class* $\mathrm{C}^{(0)}(\partial\Gamma)$.

Theorem 6.33. *Let $\Gamma \subset \Omega$ be a regular region. Furthermore, we assume that F is of class $\mathrm{C}^{(0)}(\partial\Gamma)$ and H is of class $\mathrm{C}^{(1)}(\overline{\Gamma})$. Then there exists a unique solution U of class $\mathrm{C}^{(2)}(\Gamma) \cap \mathrm{C}^{(0)}(\overline{\Gamma})$ of the Dirichlet problem*

$$\begin{aligned} \Delta^* U(\xi) &= H(\xi), \quad \xi \in \Gamma, & (6.149)\\ U(\xi) &= F(\xi), \quad \xi \in \partial\Gamma. & (6.150) \end{aligned}$$

Proof. The function $\bar{H}(\xi) = H(\xi) - \frac{1}{\|\Gamma\|}\int_\Gamma H(\eta)d\omega(\eta)$, $\xi \in \Gamma$, obviously satisfies $\int_\Gamma \bar{H}(\eta)d\omega(\eta) = 0$. Introducing

$$\bar{U}(\xi) = \int_\Gamma G(\Delta^*; \xi \cdot \eta)\bar{H}(\eta)d\omega(\eta), \quad \xi \in \Gamma, \tag{6.151}$$

we obtain from Theorem 6.14 that $\Delta^*\bar{U}(\xi) = \bar{H}(\xi)$, $\xi \in \Gamma$. In other words, if we set $\tilde{U}(\xi) = \bar{U}(\xi) - \frac{1}{\|\Gamma\|}\ln(1 - \xi \cdot \tilde{\xi})\int_\Gamma H(\eta)d\omega(\eta)$, for $\xi \in \Gamma$ and a fixed $\tilde{\xi} \in \Gamma^c$, we end up with

$$\Delta^*\tilde{U}(\xi) = \bar{H}(\xi) + \frac{1}{\|\Gamma\|}\int_\Gamma H(\eta)d\omega(\eta) = H(\xi), \quad \xi \in \Gamma. \tag{6.152}$$

Furthermore, $\tilde{U}$ is of class $\mathrm{C}^{(2)}(\Gamma) \cap \mathrm{C}^{(1)}(\overline{\Gamma})$. Thus, the proof is complete if we find a solution to the Dirichlet boundary-value problem

$$\begin{aligned} \Delta^*\tilde{\tilde{U}}(\xi) &= 0, \quad \xi \in \Gamma, & (6.153)\\ \tilde{\tilde{U}}(\xi) &= F(\xi) - \tilde{U}(\xi), \quad \xi \in \partial\Gamma. & (6.154) \end{aligned}$$

This homogeneous problem is solvable due to Theorem 6.32, such that the desired solution to our original Dirichlet problem is given by $U = \tilde{U} + \tilde{\tilde{U}}$. Since the double-layer potential only guarantees $\tilde{\tilde{U}}$ to be a member of class $\mathrm{C}^{(2)}(\Gamma) \cap \mathrm{C}^{(0)}(\overline{\Gamma})$ for boundary values $F \in \mathrm{C}^{(0)}(\partial\Gamma)$, we may expect U to be only of class $\mathrm{C}^{(2)}(\Gamma) \cap \mathrm{C}^{(0)}(\overline{\Gamma})$. Uniqueness of the solution is guaranteed by Lemma 6.28. □

Before we proceed with the Neumann boundary-value problem, we fill the missing part of the proof of the Mean Value Property in Theorem 6.7.

Proof. *Theorem 6.7.* "$\Leftarrow$": We assume that the function U of class $\mathrm{C}^{(0)}(\Gamma)$ satisfies

$$U(\xi) = \frac{1}{4\pi}\int_{\Gamma_\rho(\xi)} U(\eta)d\omega(\eta) + \frac{\sqrt{2-\rho}}{4\pi\sqrt{\rho}}\int_{\partial\Gamma_\rho(\xi)} U(\eta)d\sigma(\eta), \quad \xi \in \Gamma, \tag{6.155}$$

for any spherical cap $\Gamma_\rho(\xi)$ with radius $\rho \in (0, 2)$ and center $\xi \in \Gamma$ such that

$\overline{\Gamma_\rho(\xi)} \subset \Gamma$. Now, let $\tilde{U}$ be a function of class $C^{(2)}(\Gamma_{\tilde{\rho}}(\tilde{\xi})) \cap C^{(0)}(\overline{\Gamma_{\tilde{\rho}}(\tilde{\xi})})$, for some fixed $\tilde{\xi} \in \Gamma$ and $\tilde{\rho} \in (0, 2)$, obeying

$$\Delta^* \tilde{U}(\xi) = 0, \quad \xi \in \Gamma_{\tilde{\rho}}(\tilde{\xi}), \tag{6.156}$$

$$\tilde{U}(\xi) = U(\xi), \quad \xi \in \partial\Gamma_{\tilde{\rho}}(\tilde{\xi}). \tag{6.157}$$

According to Theorem 6.33, we are able to guarantee the existence of such a function $\tilde{U}$. Moreover, the already known direction of Theorem 6.7 implies that a mean value representation similar to (6.155) also holds true for $\tilde{U}$. This way, we get

$$\begin{aligned} U(\xi) - \tilde{U}(\xi) &= \frac{1}{4\pi} \int_{\Gamma_\rho(\xi)} U(\eta) - \tilde{U}(\eta) d\omega(\eta) \\ &\quad + \frac{\sqrt{2-\rho}}{4\pi\sqrt{\rho}} \int_{\partial\Gamma_\rho(\xi)} U(\eta) - \tilde{U}(\eta) d\sigma(\eta), \end{aligned} \tag{6.158}$$

for $\xi \in \Gamma_{\tilde{\rho}}(\tilde{\xi})$ and any spherical cap $\Gamma_\rho(\xi)$ with radius $\rho \in (0, 2)$ and center $\xi \in \Gamma_{\tilde{\rho}}(\tilde{\xi})$ such that $\overline{\Gamma_\rho(\xi)} \subset \Gamma_{\tilde{\rho}}(\tilde{\xi})$. Remembering the fact that the Mean Value Property, not the harmonicity, is the essential feature in the proof of the Maximum/Minimum Principle, we see that the Maximum/Minimum Principle holds true for $U - \tilde{U}$. Consequently, by virtue of (6.157),

$$\sup_{\xi \in \Gamma_{\tilde{\rho}}(\tilde{\xi})} |U(\xi) - \tilde{U}(\xi)| \leq \sup_{\xi \in \partial\Gamma_{\tilde{\rho}}(\tilde{\xi})} |U(\xi) - \tilde{U}(\xi)| = 0. \tag{6.159}$$

In other words, $U = \tilde{U}$ in $\Gamma_{\tilde{\rho}}(\tilde{\xi})$. Thus, U is harmonic in $\Gamma_{\tilde{\rho}}(\tilde{\xi})$ by (6.156). Since this property holds true for any spherical cap $\Gamma_{\tilde{\rho}}(\tilde{\xi}) \subset \Gamma$, the function U is harmonic on Γ, which completes the proof of Theorem 6.7. □

Neumann Problem

Next, we turn to the Neumann boundary-value problem (NP). First, we notice that

$$\int_{\partial\Gamma} G(\eta) d\sigma(\eta) = 0 \tag{6.160}$$

is a necessary condition for its solvability. This can easily be derived from

$$0 = \int_\Gamma \Delta_\eta^* U(\eta) d\omega(\eta) = \int_{\partial\Gamma} \frac{\partial}{\partial\nu(\eta)} U(\eta) d\sigma(\eta) = \int_{\partial\Gamma} G(\eta) d\sigma(\eta). \tag{6.161}$$

Furthermore, the single-layer ansatz in (6.133) requires $\tilde{Q}$ to be of class $C_0^{(0)}(\partial\Gamma)$. We observe that $P_{|1}(0,0)$ maps $C_0^{(0)}(\partial\Gamma)$ into $C_0^{(0)}(\partial\Gamma)$, so that the Fredholm theory is applicable on the function space $C_0^{(0)}(\partial\Gamma)$. Since we have already treated the integral equation (D) for the Dirichlet case, it is advisable to investigate the adjoint equation to (N) for the Neumann case instead of the original integral equation (N) itself (remember $P_{|2}(0,0) = P_{|1}(0,0)^*$).

Lemma 6.34. *Let $\Gamma \subset \Omega$ be a regular region. The homogeneous integral equation*

$$P_{|2}(0,0)[\tilde{Q}] - \frac{1}{2}\tilde{Q} = P_{|1}(0,0)^*[\tilde{Q}] - \frac{1}{2}\tilde{Q} = 0 \tag{6.162}$$

has only the trivial solution $\tilde{Q} = 0$ in $C_0^{(0)}(\partial\Gamma)$.

Proof. The same arguments as provided in the proof of Lemma 6.31 tell us that the ansatz $U(\xi) = \int_{\partial\Gamma} \tilde{Q}(\eta) \frac{\partial}{\partial\nu(\eta)} G(\Delta^*; \xi \cdot \eta)\, d\sigma(\eta)$ leads to $\tilde{Q}(\xi) = -K$, $\xi \in \partial\Gamma$, for a constant $K \in \mathbb{R}$. Since $\tilde{Q}$ is of class $C_0^{(0)}(\partial\Gamma)$, this implies $\tilde{Q}(\xi) = 0$, $\xi \in \partial\Gamma$, as desired. □

Remark 6.35. Observing $\tilde{Q}(\xi) = -K$, $\xi \in \partial\Gamma$, in the previous proof, we also find $\tilde{Q}(\xi) = 0$, $\xi \in \partial\Gamma$, from the property

$$0 = P_{|2}(0,0)[\tilde{Q}](\xi) - \frac{1}{2}\tilde{Q}(\xi) = -\frac{1}{2}K + \frac{\|\Gamma\|}{4\pi}K + \frac{1}{2}K = \frac{\|\Gamma\|}{4\pi}K. \tag{6.163}$$

Thus, the integral equation

$$P_{|2}(0,0)[\tilde{Q}] - \frac{1}{2}\tilde{Q} = P_{|1}(0,0)^*[\tilde{Q}] - \frac{1}{2}\tilde{Q} = 0 \tag{6.164}$$

allows only the trivial solution $\tilde{Q} = 0$, even when applied to the space $\mathrm{C}^{(0)}(\partial\Gamma)$ instead of $\mathrm{C}_0^{(0)}(\partial\Gamma)$. This result is different from the Euclidean theory, where non-trivial solutions exist for the corresponding integral equation. For the solution of the Neumann problem (NP), this observation has no consequences since $G \in \mathrm{C}^{(0)}(\partial\Gamma)\backslash\mathrm{C}_0^{(0)}(\partial\Gamma)$ implies a solution $\tilde{Q} \in \mathrm{C}^{(0)}(\partial\Gamma)\backslash\mathrm{C}_0^{(0)}(\partial\Gamma)$ for the integral equation (N), i.e., Equation (6.138). This, however, would not be in accordance with the single-layer ansatz (6.133), which requires $\tilde{Q} \in \mathrm{C}_0^{(0)}(\partial\Gamma)$.

As a direct consequence, the existence of solutions to (NP) is guaranteed.

Theorem 6.36. *The Neumann problem* (NP) *is uniquely solvable (up to an additive constant) for any boundary value G of class $\mathrm{C}_0^{(0)}(\partial\Gamma)$.*

Remark 6.37. Concerning the uniqueness of (NP), it should be noted that the solution obtained by the single-layer ansatz (6.133) is always uniquely determined. Yet, any constant can be added, and we still have a solution to (NP) (which is in accordance with the already known fact that the Neumann problem is only unique up to a constant).

Theorem 6.38. *Let $\Gamma \subset \Omega$ be a regular region. Furthermore, we assume that G is of class $\mathrm{C}^{(0)}(\partial\Gamma)$ and H is of class $\mathrm{C}^{(1)}(\overline{\Gamma})$. Then there exists a unique (up to an additive constant) solution U of class $\mathrm{C}^{(2)}(\Gamma) \cap \mathrm{C}^{(0)}(\overline{\Gamma})$ to the Neumann problem*

$$\Delta^* U(\xi) = H(\xi), \quad \xi \in \Gamma, \tag{6.165}$$

$$\frac{\partial}{\partial\nu} U(\xi) = G(\xi), \quad \xi \in \partial\Gamma, \tag{6.166}$$

if and only if

$$\int_{\partial\Gamma} G(\eta)d\sigma(\eta) - \int_{\Gamma} H(\eta)d\omega(\eta) = 0. \tag{6.167}$$

Proof. In the same manner as in the proof of Theorem 6.33, the problem can be reduced to finding a solution of

$$\Delta^* \tilde{\tilde{U}}(\xi) = 0, \quad \xi \in \Gamma, \tag{6.168}$$

$$\frac{\partial}{\partial\nu}\tilde{\tilde{U}}(\xi) = G(\xi) - \frac{\partial}{\partial\nu(\xi)}\tilde{U}(\xi), \quad \xi \in \partial\Gamma. \tag{6.169}$$

From Theorem 6.36 it is known that a solution exists if and only if $G - \frac{\partial}{\partial\nu}\tilde{U}$ is of class $\mathrm{C}_0^{(0)}(\partial\Gamma)$. Application of the Gauss Theorem leads to the condition (6.167). Uniqueness up to a constant is guaranteed by Lemma 6.29. □

Remark 6.39. The solution given in Theorem 6.38 is generally not of class $\mathrm{C}^{(2)}(\Gamma)\cap\mathrm{C}^{(1)}(\overline{\Gamma})$ but only of class $\mathrm{C}^{(2)}(\Gamma)\cap\mathrm{C}^{(0)}(\overline{\Gamma})$. However, the constructed solution has a well-defined normal derivative $\frac{\partial}{\partial\nu}U$ in any point on $\partial\Gamma$.

6.4.3 Well-Posedness

The well-posedness of (DP) has already been provided by Lemma 6.30 with the help of the Maximum/Minimum Principle. We just recapitulate it here for the convenience of the reader.

Lemma 6.40. *Let $\Gamma \subset \Omega$ be a regular region. If U, $\tilde{U} : \Gamma \to \mathbb{R}$ satisfy* (DP) *corresponding to boundary values F, $\tilde{F}$ of class* $\mathrm{C}^{(0)}(\partial\Gamma)$, *respectively, then*

$$\sup_{\xi\in\Gamma} |U(\xi) - \tilde{U}(\xi)| \leq \sup_{\xi\in\partial\Gamma} |F(\xi) - \tilde{F}(\xi)|. \tag{6.170}$$

Now that we have the tools of boundary integral equations at hand, the same argumentation as in Lemma 3.82 leads us to the well-posedness of the Neumann problem (NP).

Lemma 6.41. *Let $\Gamma \subset \Omega$ be a regular region. If U, $\tilde{U} : \Gamma \to \mathbb{R}$ satisfy* (NP) *for boundary values G, $\tilde{G}$ of class* $\mathrm{C}^{(0)}(\partial\Gamma)$, *respectively, then*

$$\sup_{\xi\in\Gamma} |U(\xi) - \tilde{U}(\xi)| \leq \sup_{\xi\in\partial\Gamma} |G(\xi) - \tilde{G}(\xi)|. \tag{6.171}$$

Additionally, the same methods as discussed in Subsection 3.4.5 enable us to formulate regularity properties in the L^2-topology. Because of the similarity, we omit details and refer to the proofs within the Euclidean framework.

Lemma 6.42. *Let $\Gamma \subset \Omega$ be a regular region, and suppose that U of class $C^{(0)}(\overline{\Gamma})$ is harmonic in Γ. Then there exists a constant $K = K(k, \Sigma)$ such that*

$$\sup_{\xi \in \Sigma} |(\nabla^*)^{(k)} U(\xi)| \leq K \|U\|_{L^2(\partial\Gamma)}, \tag{6.172}$$

for $k \in \mathbb{N}_0$ and any subregion $\Sigma \Subset \Gamma$. (By $(\nabla^*)^{(k)}$, we denotes the k-fold application of ∇^*, especially $(\nabla^*)^{(0)}U = U$, $(\nabla^*)^{(1)}U = \nabla^* U$ and $(\nabla^*)^{(2)}U = \nabla^* \otimes \nabla^* U$.)

Lemma 6.43. *Let $\Gamma \subset \Omega$ be a regular region, and assume that U of class $C^{(1)}(\overline{\Gamma})$ is harmonic in Γ. Then there exists a constant $K = K(k, \Sigma)$ such that*

$$\sup_{\xi \in \Sigma} |(\nabla^*)^{(k)} U(\xi)| \leq K \left\| \frac{\partial}{\partial \nu} U \right\|_{L^2(\partial\Gamma)}, \tag{6.173}$$

for $k \in \mathbb{N}_0$ and any subregion $\Sigma \Subset \Gamma$.

6.4.4 Green's Functions

After the existence and well-posedness of a solution to (DP) and (IP) have been resolved, the next question is the representation of such a solution. A viable way is indicated in Theorem 6.2. However, this representation requires the simultaneous knowledge of U and $\frac{\partial}{\partial \nu} U$ on the boundary $\partial\Gamma$, which is not necessary (and generally not allowed since the two quantities are not independent from each other). As a remedy, we use Green's functions for Dirichlet and Neumann boundary values, respectively.

Definition 6.44. Let $\Gamma \subset \Omega$ be a regular region. A function $G^{(D)}(\Delta^*; \cdot, \cdot)$ that is well-defined on the set $\{(\xi, \eta) \in \Gamma \times \overline{\Gamma} : 1 - \xi \cdot \eta > 0\}$ is called a *Dirichlet Green function (with respect to the Beltrami operator)* if it can be decomposed in the form

$$G^{(D)}(\Delta^*; \xi, \eta) = G(\Delta^*; \xi \cdot \eta) - \Phi^{(D)}(\xi, \eta), \quad \eta \in \overline{\Gamma}, \xi \in \Gamma, 1 - \xi \cdot \eta > 0, \tag{6.174}$$

where $\Phi^{(D)}(\xi, \cdot)$ is of class $C^{(2)}(\Gamma) \cap C^{(1)}(\overline{\Gamma})$ and satisfies

$$\Delta^*_\eta \Phi^{(D)}(\xi, \eta) = -\frac{1}{4\pi}, \quad \eta \in \Gamma, \tag{6.175}$$

$$\Phi^{(D)}(\xi, \eta) = G(\Delta^*; \xi \cdot \eta), \quad \eta \in \partial\Gamma, \tag{6.176}$$

for every $\xi \in \Gamma$.

From Green's formulas it follows that the function $\Phi^{(D)}(\cdot, \cdot)$ in (6.174) satisfies the integral identity

$$\begin{aligned} 0 = & -\frac{1}{4\pi} \int_\Gamma U(\eta) d\omega(\eta) + \int_{\partial\Gamma} \Phi^{(D)}(\xi, \eta) \frac{\partial}{\partial \nu(\eta)} U(\eta) d\sigma(\eta) \\ & - \int_{\partial\Gamma} U(\eta) \frac{\partial}{\partial \nu(\eta)} \Phi^{(D)}(\xi, \eta) d\sigma(\eta) - \int_\Gamma \Phi^{(D)}(\xi, \eta) \Delta^*_\eta U(\eta) d\omega(\eta), \end{aligned} \tag{6.177}$$

for $\xi \in \Gamma$ and U of class $\mathrm{C}^{(2)}(\Gamma) \cap \mathrm{C}^{(1)}(\overline{\Gamma})$, with $\Delta^* U$ integrable on Γ. In combination with the integral formula of Theorem 6.2, we are therefore led to

$$U(\xi) = \int_{\Gamma} G^{(D)}(\Delta^*; \xi, \eta) \Delta^*_\eta U(\eta) d\omega(\eta) + \int_{\partial\Gamma} U(\eta) \frac{\partial}{\partial\nu(\eta)} G^{(D)}(\Delta^*; \xi, \eta) d\sigma(\eta), \tag{6.178}$$

for $\xi \in \Gamma$. Consequently, the following representation for a solution of the Dirichlet problem is obtainable.

Theorem 6.45. *Let $\Gamma \subset \Omega$ be a regular region. Furthermore, we assume that F is of class $\mathrm{C}^{(0)}(\partial\Gamma)$ and H is of class $\mathrm{C}^{(1)}(\overline{\Gamma})$. If $U \in \mathrm{C}^{(2)}(\Gamma) \cap \mathrm{C}^{(1)}(\overline{\Gamma})$ is the solution of the Dirichlet problem*

$$\Delta^* U(\xi) = H(\xi), \quad \xi \in \Gamma, \tag{6.179}$$

$$U(\xi) = F(\xi), \quad \xi \in \partial\Gamma, \tag{6.180}$$

and $G^{(D)}(\Delta^*; \cdot, \cdot)$ is a Dirichlet Green function, then U is of the form

$$U(\xi) = \int_{\Gamma} G^{(D)}(\Delta^*; \xi, \eta) H(\eta) d\omega(\eta) + \int_{\partial\Gamma} F(\eta) \frac{\partial}{\partial\nu(\eta)} G^{(D)}(\Delta^*; \xi, \eta) d\sigma(\eta), \tag{6.181}$$

for $\xi \in \Gamma$.

Remark 6.46. The representation (6.181) requires U to be a member of class $\mathrm{C}^{(2)}(\Gamma) \cap \mathrm{C}^{(1)}(\overline{\Gamma})$ (or at least of class $\mathrm{C}^{(2)}(\Gamma) \cap \mathrm{C}^{(0)}(\overline{\Gamma})$ with an existent normal derivative $\frac{\partial}{\partial\nu} U$ on $\partial\Gamma$) to guarantee the applicability of the Third Green Theorem, while Theorem 6.33 only yields $U \in \mathrm{C}^{(2)}(\Gamma) \cap \mathrm{C}^{(0)}(\overline{\Gamma})$ for the solution. Yet, if the boundary value F is sufficiently smooth (e.g., F of class $\mathrm{C}^{(2)}(\partial\Gamma)$), the condition $U \in \mathrm{C}^{(2)}(\Gamma) \cap \mathrm{C}^{(1)}(\overline{\Gamma})$ is fulfilled. Since $F(\eta) = G(\Delta^*; \xi \cdot \eta)$, $\xi \in \Gamma$ fixed, obviously satisfies this smoothness condition, the existence of the auxiliary function $\Phi^{(D)}(\cdot, \cdot)$ in Definition 6.44 is guaranteed. However, it is not our aim to discuss the boundary regularity in more detail. For the Euclidean theory with respect to the Laplace operator, the reader is referred to D. Gilbart, N.S Trudinger [1977] (by use of the stereographic projection, the results can be transferred to the sphere and the Beltrami operator).

Next, we deal with the Neumann boundary-value problem.

Definition 6.47. Let $\Gamma \subset \Omega$ be a regular region. A function $G^{(N)}(\Delta^*; \cdot, \cdot)$ that is well-defined on the set $\{(\xi, \eta) \in \Gamma \times \overline{\Gamma} : 1 - \xi \cdot \eta > 0\}$ is called a *Neumann Green function (with respect to the Beltrami operator)* if it can be decomposed in the form

$$G^{(N)}(\Delta^*; \xi, \eta) = G(\Delta^*; \xi \cdot \eta) - \Phi^{(N)}(\xi, \eta), \quad \eta \in \overline{\Gamma}, \xi \in \Gamma, 1 - \xi \cdot \eta > 0, \tag{6.182}$$

where $\Phi^{(N)}(\xi,\cdot)$ is of class $C^{(2)}(\Gamma)\cap C^{(1)}(\overline{\Gamma})$ and satisfies the conditions

$$\Delta^*_\eta \Phi^{(N)}(\xi,\eta) = \frac{1}{\|\Gamma\|} - \frac{1}{4\pi}, \quad \eta \in \Gamma, \tag{6.183}$$

$$\frac{\partial}{\partial\nu(\eta)}\Phi^{(N)}(\xi,\eta) = \frac{\partial}{\partial\nu(\eta)} G(\Delta^*;\xi\cdot\eta), \quad \eta \in \partial\Gamma, \tag{6.184}$$

for every $\xi \in \Gamma$.

Analogous to (6.177), we get

$$\begin{aligned} 0 &= \left(\frac{1}{\|\Gamma\|} - \frac{1}{4\pi}\right)\int_\Gamma U(\eta)d\omega(\eta) + \int_{\partial\Gamma} \Phi^{(N)}(\xi,\eta)\frac{\partial}{\partial\nu(\eta)}U(\eta)d\sigma(\eta) \\ &\quad - \int_{\partial\Gamma} U(\eta)\frac{\partial}{\partial\nu(\eta)}\Phi^{(N)}(\xi,\eta)d\sigma(\eta) - \int_\Gamma \Phi^{(N)}(\xi,\eta)\Delta^*_\eta U(\eta)d\omega(\eta), \end{aligned} \tag{6.185}$$

where $\xi \in \Gamma$ and $U \in C^{(2)}(\Gamma)\cap C^{(1)}(\overline{\Gamma})$, with $\Delta^* U$ being integrable on Γ. In combination with Theorem 6.2, this yields

$$\begin{aligned} U(\xi) &= \frac{1}{\|\Gamma\|}\int_\Gamma U(\eta)d\omega(\eta) + \int_\Gamma G^{(N)}(\Delta^*;\xi,\eta)\Delta^*_\eta U(\eta)d\omega(\eta) \\ &\quad - \int_{\partial\Gamma} G^{(N)}(\Delta^*;\xi,\eta)\frac{\partial}{\partial\nu(\eta)}U(\eta)d\sigma(\eta), \end{aligned} \tag{6.186}$$

for $\xi \in \Gamma$. Consequently, we are led to a solution of the Neumann problem.

Theorem 6.48. *Let $\Gamma \subset \Omega$ be a regular region. Furthermore, we assume that G is of class $C^{(0)}(\partial\Gamma)$ and H is of class $C^{(1)}(\overline{\Gamma})$. If $U \in C^{(2)}(\Gamma)\cap C^{(1)}(\overline{\Gamma})$ is a solution of the Neumann problem*

$$\Delta^* U(\xi) = H(\xi), \quad \xi \in \Gamma, \tag{6.187}$$

$$\frac{\partial}{\partial\nu}U(\xi) = G(\xi), \quad \xi \in \partial\Gamma, \tag{6.188}$$

and $G^{(N)}(\Delta^*;\cdot,\cdot)$ a Neumann Green function, then U is of the form

$$\begin{aligned} U(\xi) &= \frac{1}{\|\Gamma\|}\int_\Gamma U(\eta)d\omega(\eta) + \int_\Gamma G^{(N)}(\Delta^*;\xi,\eta)H(\eta)d\omega(\eta) \\ &\quad - \int_{\partial\Gamma} G^{(N)}(\Delta^*;\xi,\eta)G(\eta)d\sigma(\eta), \end{aligned} \tag{6.189}$$

for $\xi \in \Gamma$.

Remark 6.49. The uniqueness only up to a constant is reflected by the term $\frac{1}{\|\Gamma\|}\int_\Gamma U(\eta)d\omega(\eta)$ in (6.189). Theorem 6.38 again yields $U \in C^{(2)}(\Gamma)\cap C^{(0)}(\overline{\Gamma})$ for the solution of the Neumann problem (6.187), (6.188), but additionally with the guarantee that $\frac{\partial}{\partial\nu}U$ exists on $\partial\Gamma$. This condition is actually sufficient for the validity of (6.189).

6.4.5 Explicit Representations on Spherical Caps

We finish this section with the explicit construction of Green's functions for spherical caps.

Dirichlet Problem

Analogous to the Dirichlet case for the Laplace operator in the Euclidean context, the construction of the Dirichlet Green function (with respect to the Beltrami operator) on a spherical cap $\Gamma_\rho(\tilde{\xi})$ is straightforward once one has found a reflection point $\breve{\xi} \in \left(\Gamma_\rho(\tilde{\xi})\right)^c$ and a scaling factor $\breve{r} \in \mathbb{R}$ such that

$$1 - \xi \cdot \eta = \breve{r}(1 - \breve{\xi} \cdot \eta), \quad \eta \in \partial\Gamma_\rho(\tilde{\xi}), \, \xi \in \Gamma_\rho(\tilde{\xi}). \tag{6.190}$$

Indeed, under this assumption, it is clear that

$$\Phi^{(D)}(\xi, \eta) = \frac{1}{4\pi} \ln(\breve{r}(1 - \breve{\xi} \cdot \eta)) + \frac{1}{4\pi}(1 - \ln(2)), \tag{6.191}$$

satisfies the conditions required in Definition 6.44, for $\eta \in \overline{\Gamma_\rho(\tilde{\xi})}$, $\xi \in \Gamma_\rho(\tilde{\xi})$. It should be noted that $\breve{\xi}$ and $\breve{r}$ depend on ξ, $\tilde{\xi}$, and ρ. For brevity, we usually omit these dependencies in our notation.

Lemma 6.50. *Let $\Gamma_\rho(\tilde{\xi}) = \{\eta \in \Omega : 1 - \tilde{\xi} \cdot \eta < \rho\}$ be the spherical cap with center $\tilde{\xi} \in \Omega$ and radius $\rho \in (0, 2)$. Then, we have*

$$G^{(D)}(\Delta^*; \xi, \eta) = \frac{1}{4\pi} \ln(1 - \xi \cdot \eta) - \frac{1}{4\pi} \ln(\breve{r}(1 - \breve{\xi} \cdot \eta)), \tag{6.192}$$

for $\eta \in \overline{\Gamma_\rho(\tilde{\xi})}$, $\xi \in \Gamma_\rho(\tilde{\xi})$ with $1 - \xi \cdot \eta > 0$, where $\breve{\xi}$ and $\breve{r}$ are given by

$$\breve{\xi} = \frac{1}{\breve{r}}\xi - \frac{\breve{r} - 1}{\breve{r}(\rho - 1)}\tilde{\xi}, \tag{6.193}$$

$$\breve{r} = -\frac{1 + 2\xi \cdot \tilde{\xi}(\rho - 1) + (\rho - 1)^2}{\rho(\rho - 2)}. \tag{6.194}$$

Proof. We consider the cap $\Gamma_\rho(\varepsilon^3)$ with center ε^3. What remains is to show that $\breve{\xi}$ and $\breve{r}$, as defined in (6.193) and (6.194), actually yield $\breve{\xi} \in \left(\Gamma_\rho(\varepsilon^3)\right)^c$ if $\xi \in \Gamma_\rho(\varepsilon^3)$, and that (6.190) is satisfied for every $\eta \in \partial\Gamma_\rho(\varepsilon^3)$. This task is left to the reader (observe that $\eta = (\eta_1, \eta_2, \eta_3)^T \in \partial\Gamma_\rho(\varepsilon^3)$ implies $\eta_3 = 1 - \rho$, and that $\xi_3 > 1 - \rho$ gives $\xi = (\xi_1, \xi_2, \xi_3)^T \in \Gamma_\rho(\varepsilon^3)$). Applying (6.191) to Definition 6.44, we obtain the desired representation for $G^{(D)}(\Delta^*; \cdot, \cdot)$ on spherical caps $\Gamma_\rho(\varepsilon^3)$.

The case for spherical caps $\Gamma_\rho(\tilde{\xi})$ with arbitrary centers $\tilde{\xi} \in \Omega$ is a direct consequence of the substitution of $\Phi^{(D)}(\xi, \eta)$ by $\Phi^{(D)}(\mathbf{t}^T\xi, \mathbf{t}^T\eta)$, where $\mathbf{t} \in \mathbb{R}^{3\times 3}$ denotes an orthogonal transformation with $\mathbf{t}\varepsilon^3 = \tilde{\xi}$ and $\det(\mathbf{t}) = 1$. □

An explicit calculation of $\frac{\partial}{\partial\nu(\eta)} G^{(D)}(\Delta^*; \xi, \eta)$ in combination with Theorem 6.45 enables us to verify the next result.

Theorem 6.51. *Let $\Gamma_\rho(\tilde{\xi})$ be the spherical cap with center $\tilde{\xi} \in \Omega$ and radius $\rho \in (0,2)$. Furthermore, suppose that F is of class $\mathrm{C}^{(1)}(\partial\Gamma)$. Then the solution $U \in \mathrm{C}^{(2)}(\Gamma_\rho(\tilde{\xi})) \cap \mathrm{C}^{(0)}(\overline{\Gamma_\rho(\tilde{\xi})})$ of the Dirichlet problem*

$$\begin{aligned} \Delta^* U(\xi) &= 0, \quad \xi \in \Gamma_\rho(\tilde{\xi}), & (6.195)\\ U(\xi) &= F(\xi), \quad \xi \in \partial\Gamma_\rho(\tilde{\xi}), & (6.196) \end{aligned}$$

is given in the form

$$U(\xi) = \frac{1}{2\pi} \frac{\xi \cdot \tilde{\xi} + \rho - 1}{\sqrt{\rho(2-\rho)}} \int\limits_{\partial\Gamma_\rho(\tilde{\xi})} \frac{1}{1 - \xi \cdot \eta} F(\eta)\, d\sigma(\eta), \tag{6.197}$$

for $\xi \in \Gamma_\rho(\tilde{\xi})$.

Remark 6.52. For the special case of spherical caps, Theorem 6.51 does not require U to be of class $\mathrm{C}^{(2)}(\Gamma) \cap \mathrm{C}^{(1)}(\overline{\Gamma})$ as stated in Theorem 6.45 (for general regular regions $\Gamma \subset \Omega$). The kernel appearing in (6.197) is the spherical counterpart to the Abel–Poisson kernel known from the Euclidean theory and has similar properties when ξ approaches the boundary $\partial\Gamma_\rho(\tilde{\xi})$. The validation of Theorem 6.51 for U of class $\mathrm{C}^{(2)}(\Gamma_\rho(\tilde{\xi})) \cap \mathrm{C}^{(0)}(\overline{\Gamma_\rho(\tilde{\xi})})$ is left to the reader (cf. Exercise 6.8).

Furthermore, Theorem 6.51 allows an alternative version of the Mean Value Property as stated in Theorem 6.7. It is more related to the Euclidean case in the sense that it does not involve a surface integral over $\Gamma_\rho(\xi)$ anymore. This assertion can easily be seen from the expression (6.197) by choosing $\xi = \tilde{\xi}$.

Theorem 6.53 (Mean Value Property). *Let $\Gamma \subset \Omega$ be a regular region. Then a function U of class $\mathrm{C}^{(0)}(\Gamma)$ is harmonic with respect to the Beltrami operator if and only if*

$$U(\xi) = \frac{1}{2\pi\sqrt{\rho(2-\rho)}} \int_{\partial\Gamma_\rho(\xi)} U(\eta) d\sigma(\eta), \quad \xi \in \Gamma, \tag{6.198}$$

for any spherical cap $\Gamma_\rho(\xi) \Subset \Gamma$.

To get a better geometric understanding of the reflection point $\breve{\xi}$ in (6.193), we need the stereographic projection (with respect to $\tilde{\xi}$) as described in Subsection 2.4.3. Indeed, we use it to map a point $\xi \in \Gamma_\rho(\tilde{\xi})$ onto the plane $\mathbb{R}^2$. We notice that $p_{\mathrm{stereo}}(\Gamma_\rho(\tilde{\xi})) = \mathcal{B}^{(2)}_{2\rho^{\frac{1}{2}}(2-\rho)^{-\frac{1}{2}}}(0)$. In other words, the spherical cap is mapped onto a disc in the plane. In the projection plane, the Kelvin transform $\breve{x} = \frac{R}{|x|^2} x$, for $x \in \mathcal{B}^{(2)}_R(0)$, can then be applied in the same way as known from Section 1.1.3 for the three-dimensional Euclidean space $\mathbb{R}^3$. In our spherical framework, this leads to

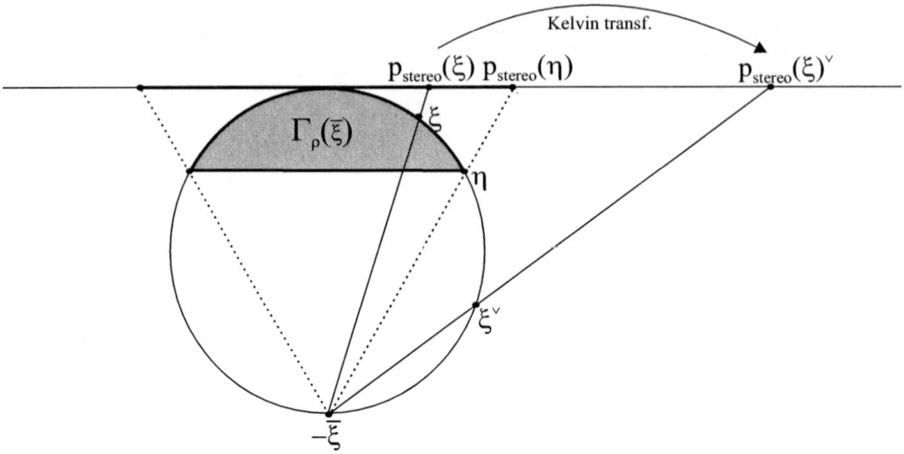

FIGURE 6.2
Schematic description of the construction of the reflection point $\breve{\xi}$.

$$p_{\text{stereo}}(\xi)\breve{} = \frac{\rho}{2-\rho}\left(\frac{2\xi\cdot(\mathbf{t}\varepsilon^1)}{1-\xi_3}, \frac{2\xi\cdot(\mathbf{t}\varepsilon^2)}{1-\xi\cdot\tilde{\xi}}\right)^T, \quad \xi\in\Gamma_\rho(\tilde{\xi}). \tag{6.199}$$

Projecting $p_{\text{stereo}}(\xi)\breve{}$ back onto the sphere, we obtain the same point $\breve{\xi}\in\left(\Gamma_\rho(\tilde{\xi})\right)^c$ as we have derived in (6.193). A graphical illustration is given in Figure 6.2. However, it should be noted that the scaling factor $\breve{r}$ has to be calculated separately.

Remark 6.54. Lemma 2.35 tells us that the stereographic projection of a harmonic function with respect to the Laplace operator is a harmonic function with respect to the Beltrami operator. This fact offers another possibility for obtaining the Dirichlet Green function $G^{(D)}(\Delta^*;\cdot,\cdot)$: One simply takes the stereographic projection of the known Dirichlet Green function with respect to the Laplace operator on a disc in $\mathbb{R}^2$ (see, e.g., E. DiBenedetto [1995] for the Green function on a disc). However, the Neumann Green function $G^{(N)}(\Delta^*;\cdot,\cdot)$ is not harmonic with respect to the Beltrami operator, so that the approach via stereographic projections does not work. Some more general construction principles for Green functions are presented, e.g., in E. Gutkin, K.P. Newton [2004].

Neumann Problem

In order to find a Neumann Green function on spherical caps, we again derive a function $\Phi^{(N)}(\cdot,\cdot)$ obeying the conditions of Definition 6.47. By $\Phi^{(D)}(\cdot,\cdot)$, we denote the known auxiliary function from (6.191) for the Dirichlet problem.

We observe, for $\xi \in \Gamma_\rho(\tilde{\xi})$ and $\eta \in \partial\Gamma_\rho(\tilde{\xi})$, that

$$\begin{aligned} &\frac{\partial}{\partial\nu(\eta)} G(\Delta^*;\xi\cdot\eta) + \frac{\partial}{\partial\nu(\eta)} \Phi^{(D)}(\xi,\eta) \qquad (6.200) \\ &= -\frac{1}{4\pi}\frac{1}{1-\xi\cdot\eta}\xi\cdot\nu(\eta) - \frac{1}{4\pi}\frac{1}{\check{r}(1-\check{\xi}\cdot\eta)}\check{r}\check{\xi}\cdot\nu(\eta) \\ &= -\frac{1}{4\pi}\frac{2}{1-\xi\cdot\eta}\xi\cdot\nu(\eta) + \frac{1}{4\pi}\frac{1}{1-\xi\cdot\eta}\frac{\check{r}-1}{\rho-1}\tilde{\xi}\cdot\nu(\eta) \\ &= \frac{1-\rho}{2\pi\sqrt{\rho(2-\rho)}}. \end{aligned}$$

In the last row we have used $\nu(\eta) = -(\rho(2-\rho))^{-\frac{1}{2}}\left(\tilde{\xi} - (1-\rho)\eta\right)$ in connection with the expressions (6.193), (6.194). This shows us that the normal derivative of $\eta \mapsto \Phi^{(D)}(\xi,\eta)$ just differs by an additive constant from the negative of the normal derivative of $\eta \mapsto G(\Delta^*;\xi\cdot\eta)$. It is our aim to find a function $\Phi^{(N)}(\cdot,\cdot)$ that satisfies the equations

$$\Delta^*_\eta \Phi^{(N)}(\xi,\eta) = \frac{1}{\|\Gamma_\rho(\tilde{\xi})\|} - \frac{1}{4\pi}, \quad \eta,\xi \in \Gamma_\rho(\tilde{\xi}), \qquad (6.201)$$

$$\frac{\partial}{\partial\nu(\eta)}\Phi^{(N)}(\xi,\eta) = \frac{\partial}{\partial\nu(\eta)} G(\Delta^*;\xi\cdot\eta), \quad \eta \in \partial\Gamma_\rho(\tilde{\xi}), \xi \in \Gamma_\rho(\tilde{\xi}). \qquad (6.202)$$

If we set $\Phi^{(N)}(\xi,\eta) = -\Phi^{(D)}(\xi,\eta) + \tilde{\Phi}(\eta)$ and use (6.200), the conditions (6.201) and (6.202) reduce to finding a function that satisfies

$$\Delta^*_\eta \tilde{\Phi}(\eta) = \frac{1}{\|\Gamma_\rho(\tilde{\xi})\|} - \frac{1}{2\pi}, \quad \eta \in \Gamma_\rho(\tilde{\xi}), \qquad (6.203)$$

$$\frac{\partial}{\partial\nu(\eta)}\tilde{\Phi}(\eta) = \frac{1-\rho}{2\pi\sqrt{\rho(2-\rho)}}, \quad \eta \in \partial\Gamma_\rho(\tilde{\xi}). \qquad (6.204)$$

Observing $\|\Gamma_\rho(\tilde{\xi})\| = 2\pi\rho$, we are led to the choice

$$\tilde{\Phi}(\eta) = -\frac{1-\rho}{2\pi\rho}\ln(1+\tilde{\xi}\cdot\eta), \quad \eta \in \overline{\Gamma_\rho(\tilde{\xi})}. \qquad (6.205)$$

Summarizing our results, we finally obtain the following lemma.

Lemma 6.55. *Let $\Gamma_\rho(\tilde{\xi})$ be the spherical cap with center $\tilde{\xi} \in \Omega$ and radius $\rho \in (0,2)$. Then, a Neumann Green function is given by*

$$G^{(N)}(\Delta^*;\xi,\eta) = \frac{1}{4\pi}\ln(1-\xi\cdot\eta) + \frac{1}{4\pi}\ln(\check{r}(1-\check{\xi}\cdot\eta)) + \frac{1-\rho}{2\pi\rho}\ln(1+\tilde{\xi}\cdot\eta),$$

for $\xi \in \Gamma_\rho(\tilde{\xi})$, $\eta \in \overline{\Gamma_\rho(\tilde{\xi})}$ with $1-\xi\cdot\eta > 0$, where $\check{\xi}$, $\check{r}$ are defined as indicated by (6.193) and (6.194).

Theorem 6.56. *Let $\Gamma_\rho(\tilde{\xi})$ be the spherical cap with center $\tilde{\xi} \in \Omega$ and radius $\rho \in (0,2)$. Furthermore, we assume that G is of class $C^{(0)}(\partial\Gamma)$. Then a solution $U \in \mathrm{C}^{(2)}(\Gamma_\rho(\tilde{\xi})) \cap \mathrm{C}^{(0)}(\overline{\Gamma_\rho(\tilde{\xi})})$ of the Neumann problem*

$$\Delta^* U(\xi) = 0, \quad \xi \in \Gamma_\rho(\tilde{\xi}), \tag{6.206}$$

$$\frac{\partial}{\partial \nu} U(\xi) = G(\xi), \quad \xi \in \partial\Gamma_\rho(\tilde{\xi}), \tag{6.207}$$

is given in the form

$$\begin{aligned} U(\xi) = & \frac{1}{2\pi\rho} \int\limits_{\Gamma_\rho(\tilde{\xi})} U(\eta) d\omega(\eta) \\ & - \int\limits_{\partial\Gamma_\rho(\tilde{\xi})} G(\eta) \left(\frac{1}{2\pi} \ln(1-\xi\cdot\eta) + \frac{1-\rho}{2\pi\rho} \ln(2-\rho) \right) d\sigma(\eta), \end{aligned} \tag{6.208}$$

for $\xi \in \Gamma_\rho(\tilde{\xi})$. Even though the solution U is generally only of class $\mathrm{C}^{(2)}(\Gamma_\rho(\tilde{\xi})) \cap \mathrm{C}^{(0)}(\overline{\Gamma_\rho(\tilde{\xi})})$, the normal derivative is well-defined on $\partial\Gamma_\rho(\tilde{\xi})$.

6.4.6 Harnack's Inequality

By virtue of our previous results on spherical caps, we are able to derive a spherical version of Harnack's inequality.

Lemma 6.57. *Let $U : \Gamma_\rho(\tilde{\xi}) \to \mathbb{R}$ be harmonic on the spherical cap $\Gamma_\rho(\tilde{\xi})$ with radius $\rho \in (0,2)$ and center $\tilde{\xi} \in \Omega$. Moreover, assume that U is non-negative. Then*

$$\frac{\sqrt{\rho} - \sqrt{1-\xi\cdot\tilde{\xi}}}{\sqrt{\rho} + \sqrt{1-\xi\cdot\tilde{\xi}}} U(\tilde{\xi}) \le U(\xi) \le \frac{\sqrt{\rho} + \sqrt{1-\xi\cdot\tilde{\xi}}}{\sqrt{\rho} - \sqrt{1-\xi\cdot\tilde{\xi}}} U(\tilde{\xi}), \tag{6.209}$$

holds true for $\xi \in \Gamma_\rho(\tilde{\xi})$.

Proof. First, we see that

$$\begin{aligned} 1-\xi\cdot\eta &= \frac{1}{2}|\xi-\eta|^2 = \frac{1}{2}|\xi-\tilde{\xi}+\tilde{\xi}-\eta|^2 \\ &\le \frac{1}{2}\left(|\xi-\tilde{\xi}| + |\eta-\tilde{\xi}|\right)^2 \\ &= \left(\sqrt{1-\xi\cdot\tilde{\xi}} + \sqrt{\rho}\right)^2, \end{aligned} \tag{6.210}$$

for $\xi \in \Gamma_\rho(\tilde{\xi})$ and $\eta \in \partial\Gamma_\rho(\tilde{\xi})$. Analogously, we find

$$1-\xi\cdot\eta \ge \left(\sqrt{\rho} - \sqrt{1-\xi\cdot\tilde{\xi}}\right)^2, \tag{6.211}$$

for $\xi \in \Gamma_\rho(\tilde{\xi})$ and $\eta \in \partial\Gamma_\rho(\tilde{\xi})$. Together with

$$\xi \cdot \tilde{\xi} + \rho - 1 = \left(\sqrt{\rho} - \sqrt{1 - \xi \cdot \tilde{\xi}}\right)\left(\sqrt{1 - \xi \cdot \tilde{\xi}} + \sqrt{\rho}\right), \tag{6.212}$$

Theorem 6.51 shows that

$$U(\xi) \leq \frac{1}{2\pi} \frac{1}{\sqrt{\rho(2-\rho)}} \frac{\sqrt{\rho} + \sqrt{1 - \xi \cdot \tilde{\xi}}}{\sqrt{\rho} - \sqrt{1 - \xi \cdot \tilde{\xi}}} \int_{\partial\Gamma_\rho(\tilde{\xi})} U(\eta)\, d\sigma(\eta), \tag{6.213}$$

$$U(\xi) \geq \frac{1}{2\pi} \frac{1}{\sqrt{\rho(2-\rho)}} \frac{\sqrt{\rho} - \sqrt{1 - \xi \cdot \tilde{\xi}}}{\sqrt{\rho} + \sqrt{1 - \xi \cdot \tilde{\xi}}} \int_{\partial\Gamma_\rho(\tilde{\xi})} U(\eta)\, d\sigma(\eta), \tag{6.214}$$

for $\xi \in \Gamma_\rho(\tilde{\xi})$. Now, the Mean Value Property (6.198) guarantees the desired estimate. □

Different from the Euclidean case for the Laplace operator, the fact that a harmonic function (with respect to the Beltrami operator) on the entire sphere is constant is not a consequence of Harnack's inequality. The reason is that the sphere is bounded and that the radius ρ can only be chosen from the interval $(0, 2)$. Nonetheless, the desired property for harmonic functions on the entire sphere follows directly from the Third Green Theorem (cf. Theorem 6.2).

Lemma 6.58. *If $U : \Omega \to \mathbb{R}$ is harmonic with respect to the Beltrami operator on the entire sphere Ω, then U is constant.*

6.5 Differential Equations for ∇^* and L^*

In this section, we take a brief look at the vectorial differential equations $\nabla^* U = f$ and $\mathrm{L}^* U = f$ on Γ. They are of special importance when we treat decompositions of spherical vector fields later on. More detailed information can be found, e.g., in G.E. Backus et al. [1996], T. Fehlinger et al. [2007], T. Fehlinger et al. [2008], T. Fehlinger [2009], and references therein.

6.5.1 Existence and Uniqueness

Different from the Beltrami differential equation, the solutions of the differential equations with respect to the surface gradient and the surface curl gradient are uniquely determined (up to a constant) on regular regions $\Gamma \subset \Omega$, without the necessity of boundary values.

Lemma 6.59. *Let $\Gamma \subset \Omega$ be a regular region, and suppose that U is of class $C^{(1)}(\Gamma)$. Then*

$$\nabla^* U(\xi) = 0, \quad \xi \in \Gamma, \tag{6.215}$$

if and only if U is constant on Γ.

Proof. We assume that $\nabla^* U(\xi) = 0$, for $\xi \in \Gamma$. If γ_ξ is a regular spherical curve connecting a fixed point $\tilde{\xi} \in \Gamma$ with $\xi \in \Gamma$, then standard arguments of vector analysis tell us that

$$U(\xi) - U(\tilde{\xi}) = \int_{\gamma_\xi} \tau(\eta) \cdot \nabla^*_\eta U(\eta) d\sigma(\eta) = 0, \quad \xi \in \Gamma. \tag{6.216}$$

In other words, $U(\xi) = U(\tilde{\xi})$, for all $\xi \in \Gamma$, so that U is constant on Γ. (By a regular spherical curve in Γ with starting point $\tilde{\xi}$ and endpoint ξ, we mean any injective mapping $\gamma_\xi : [0,1] \to \Gamma$ with $\gamma_\xi(0) = \tilde{\xi}$ and $\gamma_\xi(1) = \xi$ that is at least of class $C^{(2)}([0,1])$, of finite length, and satisfies $\frac{\partial}{\partial t}\gamma_\xi(t) \neq 0$ for all $t \in [0,1]$.)

Conversely, if U is constant, then obviously $\nabla^* U(\xi) = 0$, $\xi \in \Gamma$. □

Remark 6.60. Observing that $\xi \wedge L^*_\xi = -\nabla^*_\xi$, for $\xi \in \Omega$, we directly see that Lemma 6.59 also holds true for the operator L^* instead of ∇^*.

Lemma 6.61. *Let $\Gamma \subset \Omega$ be a regular region. Furthermore, suppose that f of class $c^{(0)}(\Gamma)$ is a tangential vector field, i.e., $\xi \cdot f(\xi) = 0$, for $\xi \in \Gamma$. If*

$$\int_\gamma \tau(\eta) \cdot f(\eta) d\sigma(\eta) = 0 \tag{6.217}$$

for every closed regular spherical curve γ contained in Γ (closed means that starting point and endpoint coincide, i.e., $\gamma(0) = \gamma(1)$), then there exists a function U of class $c^{(1)}(\Gamma)$ such that

$$f(\xi) = \nabla^* U(\xi), \quad \xi \in \Gamma. \tag{6.218}$$

The function U is uniquely determined up to a constant.

Proof. Let γ_ξ be a regular spherical curve connecting $\tilde{\xi} \in \Gamma$ to $\xi \in \Gamma$. Then we define the function

$$U(\xi) = U(\tilde{\xi}) + \int_{\gamma_\xi} \tau(\eta) \cdot f(\eta) d\sigma(\eta), \quad \xi \in \Gamma. \tag{6.219}$$

Because of (6.217), the function U is well-defined since it is independent of the choice of γ_ξ. Furthermore, U is of class $C^{(1)}(\Gamma)$ since γ_ξ is regular. Furthermore, we get from (6.216) that

$$U(\xi) = U(\tilde{\xi}) + \int_{\gamma_\xi} \tau(\eta) \cdot \nabla^*_\eta U(\eta) d\sigma(\eta), \quad \xi \in \Gamma. \tag{6.220}$$

Combining (6.219) and (6.220), we get

$$U(\xi) = \int_{\gamma_\xi} \tau(\eta) \cdot \big(f(\eta) - \nabla^*_\eta U(\eta)\big)\, d\sigma(\eta), \quad \xi \in \Gamma. \tag{6.221}$$

Since (6.221) is valid for any regular spherical curve γ_ξ, we obtain by contraction of the curve to $\xi = \tilde{\xi}$ that

$$f(\xi) = \nabla^* U(\xi), \quad \xi \in \Gamma, \tag{6.222}$$

which is the desired assertion. The uniqueness of U up to a constant follows from Lemma 6.59 and is reflected by the freedom of choice of the value for $U(\tilde{\xi})$. □

Since $\mathrm{L}^* \cdot f(\xi) = 0$, $\xi \in \Gamma$, implies the property (6.217) for tangential vector fields f by use of the Gauss Theorem, we are led to formulate the following existence theorems.

Theorem 6.62. *Let $\Gamma \subset \Omega$ be a regular region. Furthermore, let $f \in \mathrm{c}^{(1)}(\Gamma)$ be a tangential vector field satisfying*

$$\mathrm{L}^* \cdot f(\xi) = 0, \quad \xi \in \Gamma. \tag{6.223}$$

Then there exists a function U of class $\mathrm{C}^{(2)}(\Gamma)$, which is uniquely determined up to a constant, such that

$$f(\xi) = \nabla^* U(\xi), \quad \xi \in \Gamma. \tag{6.224}$$

Theorem 6.63. *Let $\Gamma \subset \Omega$ be a regular region. Furthermore, let $f \in \mathrm{c}^{(1)}(\Gamma)$ be a tangential vector field satisfying*

$$\nabla^* \cdot f(\xi) = 0, \quad \xi \in \Gamma. \tag{6.225}$$

Then there exists a function U of class $\mathrm{C}^{(2)}(\Gamma)$, which is uniquely determined up to a constant, such that

$$f(\xi) = \mathrm{L}^* U(\xi), \quad \xi \in \Gamma. \tag{6.226}$$

6.5.2 Integral Representations

From Theorem 6.4, we already know a possible expression of the solutions to the differential equations for the surface gradient and the surface curl gradient. However, this representation requires the knowledge of U on the boundary of $\partial\Gamma$, which is actually not necessary according to Lemma 6.61. Using a Neumann Green function, (6.185) and the identity in Theorem 6.4 directly imply the following result.

Theorem 6.64. *Let $\Gamma \subset \Omega$ be a regular region. Assume that f of class $c^{(1)}(\Gamma)$ is a tangential vector field satisfying $L^* \cdot f(\xi) = 0$, $\xi \in \Gamma$. If $G^{(N)}(\Delta^*;\cdot,\cdot)$ denotes a Neumann Green function as introduced in Definition 6.47, then the solution of*

$$f(\xi) = \nabla^* U(\xi), \quad \xi \in \Gamma, \tag{6.227}$$

is given in the form

$$U(\xi) = \frac{1}{\|\Gamma\|} \int_\Gamma U(\eta) d\omega(\eta) - \int_\Gamma \left(\nabla^*_\eta G^{(N)}(\Delta^*;\xi,\eta) \right) \cdot f(\eta) d\omega(\eta). \tag{6.228}$$

Theorem 6.65. *Let $\Gamma \subset \Omega$ be a regular region. Assume that f of class $c^{(1)}(\Gamma)$ is a tangential vector field satisfying $\nabla^* \cdot f(\xi) = 0$, $\xi \in \Gamma$. If $G^{(N)}(\Delta^*;\cdot,\cdot)$ denotes a Neumann Green function as introduced in Definition 6.47, then the solution of*

$$f(\xi) = L^* U(\xi), \quad \xi \in \Gamma, \tag{6.229}$$

is given in the form

$$U(\xi) = \frac{1}{\|\Gamma\|} \int_\Gamma U(\eta) d\omega(\eta) - \int_\Gamma \left(L^*_\eta G^{(N)}(\Delta^*;\xi,\eta) \right) \cdot f(\eta) d\omega(\eta). \tag{6.230}$$

Remark 6.66. If we deal with the entire sphere Ω instead of regular regions $\Gamma \subset \Omega$, the same results as in the preceding theorems hold true. For the integral representations, one simply has to substitute the Neumann Green function by the fundamental solution $G(\Delta^*;\cdot)$.

6.6 Locally and Globally Uniform Approximation

Similar to Section 3.5 for the Euclidean approach, we are next concerned with the approximation of harmonic functions (with respect to the Beltrami operator) on arbitrary regular regions $\Gamma \subset \Omega$.

6.6.1 Closure in L^2-Topology

Using the limit and jump relations for layer potentials, we first discuss the completeness property of function systems with respect to the fundamental solution for the Beltrami operator $G(\Delta^*;\cdot)$, i.e., the spherical counterpart to the Euclidean mass point representation.

Definition 6.67. Let $\Gamma \subset \Omega$ be a regular region and suppose that $\{\xi_k\}_{k\in\mathbb{N}} \subset \Gamma$ is a set of points satisfying

$$\mathrm{dist}(\{\xi_k\}_{k\in\mathbb{N}}, \partial\Gamma) > 0. \tag{6.231}$$

If $F(\xi_k) = 0$, $k \in \mathbb{N}$, implies $F(\xi) = 0$, $\xi \in \Gamma$, for every harmonic function F with respect to the Beltrami operator in Γ, then we call $\{\xi_k\}_{k\in\mathbb{N}}$ a *fundamental system (with respect to Γ)*.

If $\Gamma \subset \Omega$ is a regular region and Σ is a compactly contained regular subregion of Γ (e.g., a spherical cap), then a dense point set $\{\xi_k\}_{k\in\mathbb{N}} \subset \partial\Sigma$ constitutes an example for a fundamental system.

Theorem 6.68. *Let $\Gamma \subset \Omega$ be a regular region, and suppose that $\{\xi_k\}_{k\in\mathbb{N}} \subset \Gamma$ is a fundamental system with respect to Γ. Then the following statements hold true:*

(a) The function system $\{G_k\}_{k\in\mathbb{N}_0} \subset \mathrm{L}^2(\partial\Gamma)$ given by

$$G_k(\xi) = \frac{1}{4\pi}\ln(1-\xi_k\cdot\xi),\ k\in\mathbb{N}, \qquad G_0(\xi) = \frac{1}{4\pi}, \qquad \xi\in\partial\Gamma,$$

is complete, and hence closed in $\mathrm{L}^2(\partial\Gamma)$.

(b) The function system $\{\tilde{G}_k\}_{k\in\mathbb{N}_0} \subset \mathrm{L}^2(\partial\Gamma)$, given by

$$\tilde{G}_k(\xi) = \frac{1}{4\pi}\frac{\partial}{\partial\nu(\xi)}\ln(1-\xi_k\cdot\xi),\ k\in\mathbb{N}, \qquad \tilde{G}_0(\xi) = \frac{1}{4\pi}, \qquad \xi\in\partial\Gamma,$$

is complete, and hence closed in $\mathrm{L}^2(\partial\Gamma)$.

Proof. (a) To show the completeness, we assume that for $F \in \mathrm{L}^2_0(\partial\Gamma)$, where

$$\mathrm{L}^2_0(\partial\Gamma) = \left\{H\in\mathrm{L}^2(\partial\Gamma) : \int_{\partial\Gamma} H(\eta)d\sigma(\eta) = 0\right\}, \tag{6.232}$$

the following condition holds true:

$$(F, G_k)_{\mathrm{L}^2(\partial\Gamma)} = \int_{\partial\Gamma} F(\eta)\frac{1}{4\pi}\ln(1-\xi_k\cdot\eta)d\sigma(\eta) = 0, \tag{6.233}$$

for every $k \in \mathbb{N}$. Setting

$$U(\xi) = \int_{\partial\Gamma} F(\eta)\frac{1}{4\pi}\ln(1-\xi\cdot\eta)d\sigma(\eta), \quad \xi\in\Gamma, \tag{6.234}$$

we see that $U(\xi_k) = 0$, $k \in \mathbb{N}$, and that U is of class $\mathrm{C}^{(2)}(\Omega\setminus\partial\Gamma)$ with

$$\Delta^* U(\xi) = -\frac{1}{4\pi}\int_{\partial\Gamma} F(\eta)d\sigma(\eta) = 0, \quad \xi\in\Omega\setminus\partial\Gamma. \tag{6.235}$$

As $\{\xi_k\}_{k\in\mathbb{N}}$ constitutes a fundamental system, it follows that $U(\xi) = 0$, $\xi \in \Gamma$. In other words, for $\tau > 0$,

$$U\left(\frac{\xi-\tau\nu(\xi)}{\sqrt{1+\tau^2}}\right) = 0, \quad \xi\in\partial\Gamma. \tag{6.236}$$

This yields a vanishing derivative with respect to τ, i.e.,

$$P_{|1}(-\tau,0)[F](\xi) = \frac{\partial}{\partial\tau} U\left(\frac{\xi - \tau\nu(\xi)}{\sqrt{1+\tau^2}}\right) = 0, \quad \xi \in \partial\Gamma. \tag{6.237}$$

From the limit relations of Theorem 6.26, we then get

$$P_{|1}(0,0)[F](\xi) = \int_{\partial\Gamma} F(\eta) \frac{\partial}{\partial\nu(\xi)} G(\Delta^*; \xi\cdot\eta)\, d\sigma(\eta) = \frac{1}{2}F(\xi), \tag{6.238}$$

in the $L^2(\partial\Gamma)$-sense. Since the left-hand side of (6.238) is continuous on $\partial\Gamma$, it follows that F is of class $C^{(0)}(\partial\Gamma)$, so that (6.238) even holds uniformly for every $\xi \in \partial\Gamma$. Observing that $U\left(\frac{\cdot - \tau\nu}{\sqrt{1+\tau^2}}\right)$ only differs by an additive constant from $P(-\tau,0)[F]$, the jump relations from Theorem 6.21 imply

$$\lim_{\tau\to 0+} U\left(\frac{\xi + \tau\nu(\xi)}{\sqrt{1+\tau^2}}\right) = 0, \quad \xi \in \partial\Gamma, \tag{6.239}$$

$$\lim_{\tau\to 0+} P_{|1}(\tau,0)[F](\xi) = F(\xi), \quad \xi \in \partial\Gamma. \tag{6.240}$$

Equation (6.239) in combination with (6.235) tells us that

$$\Delta^* U(\xi) = 0, \quad \xi \in \Gamma^c, \tag{6.241}$$

$$U^+(\xi) = 0, \quad \xi \in \partial\Gamma. \tag{6.242}$$

The uniqueness of the solution to the Dirichlet problem yields $U(\xi) = 0$, for $\xi \in \Gamma^c$. As a consequence, $P_{|1}(\tau,0)[F](\xi) = 0$, for every $\xi \in \partial\Gamma$ and $\tau > 0$. Equation (6.240) finally implies $F(\xi) = 0$, $\xi \in \partial\Gamma$, which guarantees the completeness property of the function system $\{G_k\}_{k\in\mathbb{N}}$ in $L^2_0(\partial\Gamma)$. Adding the constant function G_0, we arrive at the desired completeness in $L^2(\partial\Gamma)$.

(b) The argumentation is very similar to that of part (a). We assume that, for a function F of class $L^2_0(\partial\Gamma)$,

$$(F, \tilde{G}_k)_{L^2(\partial\Gamma)} = \int_{\partial\Gamma} F(\eta) \frac{1}{4\pi} \frac{\partial}{\partial\nu(\eta)} \ln(1 - \xi_k\cdot\eta)\, d\sigma(\eta) = 0, \tag{6.243}$$

for every $k \in \mathbb{N}$. Moreover, we set

$$U(\xi) = \int_{\partial\Gamma} F(\eta) \frac{1}{4\pi} \frac{\partial}{\partial\nu(\eta)} \ln(1 - \xi\cdot\eta)\, d\sigma(\eta), \quad \xi \in \Gamma. \tag{6.244}$$

By the same steps as in part (a), this results in $P_{|2}(-\tau,0)[F](\xi) = 0$ and $P_{|1|2}(-\tau,0)[F](\xi) = 0$, $\xi \in \partial\Gamma$, which informs us, together with Theorem 6.26 and Theorem 6.21, that

$$\lim_{\tau\to 0+} \frac{\partial}{\partial\nu(\zeta)} U(\zeta)\Big|_{\zeta = \frac{\xi + \tau\nu(\xi)}{\sqrt{1+\tau^2}}} = 0, \quad \xi \in \partial\Gamma, \tag{6.245}$$

$$\lim_{\tau\to 0+} P_{|2}(\tau,0)[F](\xi) = -F(\xi), \quad \xi \in \partial\Gamma, \tag{6.246}$$

where we have observed (6.64) for the first relation. Thus, U satisfies the Neumann boundary-value problem

$$\Delta^* U(\xi) = 0, \quad \xi \in \Gamma^c, \tag{6.247}$$

$$\frac{\partial}{\partial \nu} U^+(\xi) = 0, \quad \xi \in \partial\Gamma, \tag{6.248}$$

such that $U(\xi) = K$, $\xi \in \Gamma^c$, for some constant $K \in \mathbb{R}$. Equation (6.246) now implies $F(\xi) = -K$, $\xi \in \partial\Gamma$, and therefore, $F(\xi) = 0$, $\xi \in \partial\Gamma$, since F is of class $\mathrm{L}_0^2(\partial\Gamma)$. In combination with the constant function $\tilde{G}_0$, this justifies the completeness property in $\mathrm{L}^2(\partial\Gamma)$. □

Remark 6.69. In order to approximate harmonic functions (with respect to the Beltrami operator), the functions G_k from Theorem 6.68 need to be modified since they only satisfy $\Delta^* G_k(\xi) = -\frac{1}{4\pi}$, for $\xi \in \Gamma$ and $k \in \mathbb{N}$. An auxiliary function G of class $\mathrm{C}^{(2)}(\overline{\Gamma})$ that satisfies $\Delta^* G(\xi) = \frac{1}{4\pi}$, $\xi \in \Gamma$, can be added to G_k without changing the completeness property. In other words,

$$G_k^{(\mathrm{mod})}(\xi) = G_k(\xi) + G(\xi), \quad \xi \in \Gamma, k \in \mathbb{N}, \qquad G_0^{(\mathrm{mod})}(\xi) = G_0(\xi), \quad \xi \in \Gamma,$$

is still a complete function system in $\mathrm{L}^2(\partial\Gamma)$ (when restricted to $\partial\Gamma$) but additionally satisfies $\Delta^* G_k^{(\mathrm{mod})}(\xi) = 0$, $\xi \in \Gamma$. If a spherical cap with radius $\rho \in (0,2)$ and center $\tilde{\xi} \in \Omega$ is chosen such that $\Gamma \Subset \Gamma_\rho(\tilde{\xi})$, then the auxiliary function G can be constructed by the tools from Subsection 6.4.5. Note that the functions $\tilde{G}_k$, $k \in \mathbb{N}_0$, are already harmonic.

Theorem 6.70. *Let $\Gamma \subset \Omega$ be a regular region. Assume that $\Gamma_\rho(\tilde{\xi})$ is a spherical cap with center $\tilde{\xi} \in \Gamma$ and radius $\rho \in (0,2)$ such that $\Gamma \Subset \Gamma_\rho(\tilde{\xi})$. Then, the following statements are valid:*

(a) The inner harmonics with respect to the spherical cap $\Gamma_\rho(\tilde{\xi})$, i.e.,

$$\{H_{0,1}^{\rho,\mathrm{stereo}}\} \cup \{H_{n,k}^{\rho,\mathrm{stereo}}\}_{n\in\mathbb{N},k=1,2}.$$

form a closed system in $\mathrm{L}^2(\partial\Gamma)$.

(b) The normal derivatives of the inner harmonics with respect to the spherical cap $\Gamma_\rho(\tilde{\xi})$ together with a constant function, i.e.,

$$\{H_{0,1}^{\rho,\mathrm{stereo}}\} \cup \left\{\frac{\partial}{\partial\nu} H_{n,k}^{\rho,\mathrm{stereo}}\right\}_{n\in\mathbb{N},k=1,2},$$

form a closed system in $\mathrm{L}^2(\partial\Gamma)$.

Proof. We restrict our attention to part (a). Suppose F is of class $\mathrm{L}^2(\partial\Gamma)$ with

$$(F, H_{n,k}^{\rho,\mathrm{stereo}})_{\mathrm{L}^2(\partial\Gamma)} = \int_{\partial\Gamma} F(\eta) H_{n,k}^{\rho,\mathrm{stereo}}(\eta) d\sigma(\eta) = 0, \tag{6.249}$$

for $n \in \mathbb{N}$, $k = 1, 2$, and $n = 0$, $k = 1$. Using Lemma 2.42, we find

$$
\begin{aligned}
U(\xi) &= \int_{\partial\Gamma} \left(\ln(1 - \xi\cdot\eta) - \ln(1 + \eta\cdot\tilde{\xi})\right) F(\eta) d\sigma(\eta) \qquad (6.250)\\
&= \int_{\partial\Gamma} \left(\ln(1 - \xi\cdot\eta) - \ln(1 + \eta\cdot\tilde{\xi}) + \ln(2) - \ln(1 - \xi\cdot\tilde{\xi})\right) F(\eta) d\sigma(\eta)\\
&= -\frac{4\pi\rho}{2-\rho} \sum_{n=1}^{\infty}\sum_{k=1}^{2} H^{\rho,\mathrm{stereo}}_{-n,k}(\xi) \int_{\partial\Gamma} H^{\rho,\mathrm{stereo}}_{n,k}(\eta) F(\eta) d\sigma(\eta)\\
&= 0,
\end{aligned}
$$

for $\xi \in \left(\Gamma_\rho(\tilde{\xi})\right)^c$. In the second row of (6.250), we have used

$$
\int_{\partial\Gamma} \left(\ln(2) - \ln(1 - \xi\cdot\tilde{\xi})\right) F(\eta) d\sigma(\eta) = 0, \qquad (6.251)
$$

since $(F, H^{\rho,\mathrm{stereo}}_{0,1})_{\mathrm{L}^2(\partial\Gamma)} = 0$. Analytic continuation yields $U(\xi) = 0$, for $\xi \in \Gamma^c$. From here on, the same argumentation as in the proof of Theorem 6.68 (a) can be used (the term $\ln(1 + \eta\cdot\tilde{\xi})$ does not influence the required limit and jump relations since it only becomes singular for $\eta = -\tilde{\xi} \notin \partial\Gamma$). □

Analogous to Section 3.5, closure and completeness can be deduced for a variety of other related function systems. We omit the considerations here for reasons of brevity and similarity. The same argument applies to the discussion of generalized Fourier series, so that we only mention it in its general form. It is a direct result of the well-posedness known from Subsection 6.4.3.

Theorem 6.71. *Let $\Gamma \subset \Omega$ be a regular region and Σ a compactly contained subregion of Γ.*

(a) (DP) *Let $\{D_k\}_{k\in\mathbb{N}_0} \subset \mathrm{C}^{(2)}(\overline{\Gamma})$ be a set of harmonic functions that are orthonormal with respect to the $\mathrm{L}^2(\partial\Gamma)$-inner product and complete in $\mathrm{L}^2(\partial\Gamma)$ (when restricted to $\partial\Gamma$). If F is of class $\mathrm{C}^{(0)}(\partial\Gamma)$, then*

$$
\lim_{M\to\infty} \left\| F - \sum_{k=0}^{M} (F, D_k)_{\mathrm{L}^2(\partial\Gamma)} D_k \right\|_{\mathrm{L}^2(\partial\Gamma)} = 0. \qquad (6.252)
$$

The solution of the Dirichlet boundary-value problem

$$
\begin{aligned}
\Delta^* U(\xi) &= 0, \quad \xi \in \Gamma, \qquad (6.253)\\
U(\xi) &= F(\xi), \quad \xi \in \partial\Gamma, \qquad (6.254)
\end{aligned}
$$

can be approximated by

$$
\lim_{M\to\infty} \sup_{\xi\in\Sigma} \left| U(\xi) - \sum_{k=0}^{M} (F, D_k)_{\mathrm{L}^2(\partial\Gamma)} D_k(\xi) \right| = 0. \qquad (6.255)
$$

(*b*) (NP) *Let* $\{N_k\}_{k\in\mathbb{N}_0} \subset \mathrm{C}^{(2)}(\overline{\Gamma})$ *be a set of harmonic functions such that the restriction of* $\left\{\frac{\partial}{\partial\nu}N_k\right\}_{k\in\mathbb{N}_0}$ *to* $\partial\Gamma$ *is orthonormal with respect to the* $\mathrm{L}^2(\partial\Gamma)$*-inner product and complete in* $\mathrm{L}^2(\partial\Gamma)$*. If* G *is of class* $\mathrm{C}^{(0)}(\partial\Gamma)$*, then*

$$\lim_{M\to\infty}\left\|G-\sum_{k=0}^{M}\left(G,\frac{\partial}{\partial\nu}N_k\right)_{\mathrm{L}^2(\partial\Gamma)}\frac{\partial}{\partial\nu}N_k\right\|_{\mathrm{L}^2(\partial\Gamma)}=0. \tag{6.256}$$

A solution of the Neumann boundary-value problem

$$\Delta^*U(\xi) = 0, \quad \xi\in\Gamma, \tag{6.257}$$

$$\frac{\partial}{\partial\nu}U(\xi) = G(\xi), \quad \xi\in\partial\Gamma, \tag{6.258}$$

can be approximated by

$$\lim_{M\to\infty}\sup_{\xi\in\Sigma}\left|U(\xi)-\sum_{k=0}^{M}\left(G,\frac{\partial}{\partial\nu}N_k\right)_{\mathrm{L}^2(\partial\Gamma)}N_k(\xi)\right|=0. \tag{6.259}$$

Possible choices for the bases D_k and N_k are, e.g., the inner harmonics with respect to spherical caps or an orthonormalized version of the functions $G_k^{(\mathrm{mod})}$ known from Remark 6.69.

6.6.2 Closure in $\mathrm{C}^{(0)}$-Topology

In this concluding subsection, we discuss the closure of the previously mentioned function systems in the space $\mathrm{C}^{(0)}(\partial\Gamma)$. Since the proofs are analogous to that of Theorems 3.98 and 3.100 in the Euclidean case, we just state the results.

Theorem 6.72. *Let* $\Gamma\subset\Omega$ *be a regular region. Assume that* $\{\Phi_k\}_{k\in\mathbb{N}_0}$ *denotes one of the function systems introduced in Remark 6.69 or Theorem 6.70. Then, for* F *of class* $\mathrm{C}^{(0)}(\partial\Gamma)$ *and* $\varepsilon>0$*, there exists a constant* $M(\varepsilon)\in\mathbb{N}$ *and coefficients* $a_k=a_k(\varepsilon)\in\mathbb{R}$ *such that*

$$\sup_{\xi\in\partial\Gamma}\left|F(\xi)-\sum_{k=0}^{M}a_k\Phi_k(\xi)\right|<\varepsilon. \tag{6.260}$$

Theorem 6.73. *Let* $\Gamma\subset\Omega$ *be a regular region. Assume that* $\{\Phi_k\}_{k\in\mathbb{N}_0}$ *denotes one of the function systems introduced in Remark 6.69 or Theorem 6.70.*

(*a*) (DP) *Let* F *be of class* $\mathrm{C}^{(0)}(\partial\Gamma)$*, and assume that* $U\in\mathrm{C}^{(2)}(\Gamma)\cap\mathrm{C}^{(0)}(\overline{\Gamma})$ *is a solution of the boundary-value problem*

$$\Delta^*U(\xi) = 0, \quad \xi\in\Gamma, \tag{6.261}$$

$$U(\xi) = F(\xi), \quad \xi\in\partial\Gamma. \tag{6.262}$$

Then, for any $\varepsilon > 0$, there exists a constant $M = M(\varepsilon) \in \mathbb{N}$ and coefficients $a_k = a_k(\varepsilon) \in \mathbb{R}$ such that

$$\begin{aligned} \sup_{\xi \in \overline{\Gamma}} \left| U(\xi) - \sum_{k=0}^{M} a_k \Phi_k(\xi) \right| & \\ \leq \sup_{\xi \in \partial\Gamma} \left| F(\xi) - \sum_{k=0}^{M} a_k \Phi_k(\xi) \right| & \\ < \varepsilon. & \end{aligned} \tag{6.263}$$

(b) (NP) *Let G be of class $\mathrm{C}^{(0)}(\partial\Gamma)$, and suppose that $U \in \mathrm{C}^{(2)}(\Gamma) \cap \mathrm{C}^{(1)}(\overline{\Gamma})$ is a solution of the boundary-value problem*

$$\Delta^* U(\xi) = 0, \quad \xi \in \Gamma, \tag{6.264}$$

$$\frac{\partial}{\partial \nu} U(\xi) = G(\xi), \quad \xi \in \partial\Gamma. \tag{6.265}$$

Then, for any $\varepsilon > 0$, there exists a constant $K \in \mathbb{R}$, a constant $M = M(\varepsilon) \in \mathbb{N}$, and coefficients $a_k = a_k(\varepsilon) \in \mathbb{R}$ such that

$$\begin{aligned} \sup_{\xi \in \overline{\Gamma}} \left| U(\xi) - \sum_{k=0}^{M} a_k \Phi_k(\xi) \right| & \\ \leq K \sup_{\xi \in \partial\Gamma} \left| G(\xi) - \sum_{k=0}^{M} a_k \frac{\partial}{\partial \nu} \Phi_k(\xi) \right| & \\ < K\varepsilon. & \end{aligned} \tag{6.266}$$

As already mentioned for the Euclidean case, a constructive procedure to determine the coefficients a_k for an approximation in the $\mathrm{C}^{(0)}$-topology is not known, yet. Nevertheless, they can be derived for (Sobolev-like) reproducing kernel subspaces in connection with spline variational methods (similar as presented in Chapter 4). We omit the details.

6.7 Exercises

Exercise 6.1. Prove that the fundamental solution $G(\Delta^*; \cdot)$ is uniquely determined by the properties of symmetry, differential equation, regularity, and normalization, as listed after Definition 6.1.

Exercise 6.2. Let $\Gamma \subset \Omega$ be a regular region. Show that every harmonic function $U : \Gamma \to \mathbb{R}$ is of class $\mathrm{C}^{(\infty)}(\Gamma)$.

Exercise 6.3. Prove the spherical Maximum/Minimum Principle from Theorem 6.8: For regular regions $\Gamma \subset \Omega$ and functions $U : \Gamma \to \mathbb{R}$ that are harmonic with respect to the Beltrami operator, the following statements hold true:

(a) If U is non-constant, then there exists neither a minimum nor a maximum of U in Γ.

(b) If U is additionally of class $\mathrm{C}^{(0)}(\overline{\Gamma})$, then U reaches its minimum and maximum in $\overline{\Gamma}$, and all extremal points are located on $\partial\Gamma$.

Exercise 6.4. Find a linear regularization of $\nabla^* G(\Delta^*;\cdot)$ similar to the one used for the scalar case $G(\Delta^*;\cdot)$ in the proof of Theorem 6.13. Use it to verify Theorem 6.14, i.e.,

$$\Delta_\xi^* \int_\Gamma G(\Delta^*;\xi\cdot\eta)F(\eta)d\omega(\eta) = F(\xi) - \frac{1}{4\pi}\int_\Gamma F(\eta)d\omega(\eta), \tag{6.267}$$

for $\xi \in \Gamma$ and $F \in \mathrm{C}^{(1)}(\overline{\Gamma})$, in a similar manner as has been done to establish Theorem 3.30 in the Euclidean framework. Can the constraints on the function F be reduced using this type of proof?

Exercise 6.5. Let $\Gamma \subset \Omega$ be a regular region. Find defining properties for a Green function $G^{(I)}(\Delta^*;\cdot,\cdot)$, similar to the Definitions 6.44 and 6.47, such that

$$\begin{aligned} U(\xi) &= \frac{1}{\|\Gamma\|}\int_\Gamma U(\eta)d\omega(\eta) + \int_\Gamma G^{(I)}(\Delta^*;\xi,\eta)\Delta_\eta^* U(\eta)d\omega(\eta) \qquad (6.268)\\ &\quad + \int_{\partial\Gamma} U(\eta)\frac{\partial}{\partial\nu(\eta)}G^{(I)}(\Delta^*;\xi,\eta)d\sigma(\eta), \end{aligned}$$

for $\xi \in \Gamma$ and U of class $\mathrm{C}^{(2)}(\Gamma)\cap\mathrm{C}^{(1)}(\overline{\Gamma})$, with $\Delta^* U$ integrable on Γ. Calculate an explicit representation of $G^{(I)}(\Delta^*;\cdot,\cdot)$ for spherical caps $\Gamma_\rho(\tilde{\xi})$ with center $\tilde{\xi} \in \Omega$ and radius $\rho \in (0,2)$.

Exercise 6.6. Let $\Gamma \subset \Omega$ be a regular region and $\mathcal{G} = p_{\mathrm{stereo}}(\Gamma)$ its stereographic image (with respect to some fixed $\tilde{\xi} \in \Omega$ that satisfies $-\tilde{\xi} \in \Gamma^c$) in the plane $\mathbb{R}^2$. Verify the following assertion: If $G^{(D)}(\Delta;\cdot,\cdot)$ is the Dirichlet Green function for the Laplace operator in $\mathbb{R}^2$, i.e.,

$$\Delta_y G^{(D)}(\Delta;x,y) = 0, \quad x,y \in \mathcal{G}, \tag{6.269}$$

$$G^{(D)}(\Delta;x,y) = 0, \quad x \in \mathcal{G}, y \in \partial\mathcal{G}, \tag{6.270}$$

then $G^D(\Delta^*;\xi,\eta) = G^{(D)}(\Delta;p_{\mathrm{stereo}}(\xi),p_{\mathrm{stereo}}(\eta))$, $\xi,\eta \in \Gamma$, denotes the Dirichlet Green function for the Beltrami operator as given in Definition 6.44.

Exercise 6.7. Let $\Gamma \subset \Omega$ be a regular region, and assume that H is of class $\mathrm{C}^{(2)}(\overline{\Gamma})$ and f is of class $\mathrm{c}^{(2)}(\overline{\Gamma})$. Furthermore, let u be of class $\mathrm{c}^{(2)}(\Gamma)\cap\mathrm{c}^{(1)}(\overline{\Gamma})$ and d of class $\mathrm{c}^{(0)}(\partial\Gamma)$, such that

$$
\begin{aligned}
\nabla^* \wedge u(\xi) &= h(\xi), \quad \xi \in \Gamma, && (6.271)\\
\nabla^* \cdot u(\xi) &= H(\xi), \quad \xi \in \Gamma, && (6.272)\\
u(\xi) &= f(\xi), \quad \xi \in \partial\Gamma. && (6.273)
\end{aligned}
$$

Show that u is given by

$$
\begin{aligned}
u(\xi) = & \int_\Gamma \Big(\eta G^{(D)}(\Delta^*;\xi,\eta) - \nabla^*_\eta G^{(D)}(\Delta^*;\xi,\eta)\Big) H(\eta)d\omega(\eta) \qquad (6.274)\\
& - \int_\Gamma \Big(\eta G^{(D)}(\Delta^*;\xi,\eta) - \nabla^*_\eta G^{(D)}(\Delta^*;\xi,\eta)\Big) \wedge h(\eta)d\omega(\eta)\\
& + \int_{\partial\Gamma} G^{(D)}(\Delta^*;\xi,\eta)\big(\nu(\eta)H(\eta) - \nu(\eta)\wedge h(\eta)\big)d\sigma(\eta)\\
& + \int_{\partial\Gamma} f(\eta)\frac{\partial}{\partial\nu(\eta)} G^{(D)}(\Delta^*;\xi,\eta)d\sigma(\eta),
\end{aligned}
$$

for $\xi \in \Gamma$. (Hint: Use Exercise 2.2 to reduce Equations (6.271)–(6.273) to a Dirichlet boundary-value problem for the Beltrami operator.)

Exercise 6.8. Let $\Gamma_\rho(\tilde{\xi})$ be the spherical cap with center $\tilde{\xi} \in \Omega$ and radius $\rho \in (0,2)$. Suppose that F is of class $\mathrm{C}^{(0)}(\partial\Gamma)$. Show that

$$
U(\xi) = \frac{1}{2\pi}\frac{\xi\cdot\tilde{\xi}+\rho-1}{\sqrt{\rho(2-\rho)}} \int\limits_{\partial\Gamma_\rho(\tilde{\xi})} \frac{1}{1-\xi\cdot\eta} F(\eta)\, d\sigma(\eta), \qquad (6.275)
$$

satisfies

$$
\begin{aligned}
\Delta^* U(\xi) &= 0, \quad \xi \in \Gamma_\rho(\tilde{\xi}), && (6.276)\\
U(\xi) &= F(\xi), \quad \xi \in \partial\Gamma_\rho(\tilde{\xi}). && (6.277)
\end{aligned}
$$

Exercise 6.9. Let $\Gamma \subset \Omega$ be a regular region and $\Gamma_\rho(\tilde{\xi})$, with $\tilde{\xi} \in \Gamma$, a spherical cap such that $\overline{\Gamma_\rho(\tilde{\xi})} \subset \Gamma$. Prove that the normal derivatives of the outer harmonics with respect to the spherical cap $\Gamma_\rho(\tilde{\xi})$ together with a constant function, i.e.,

$$
\{H_{0,1}^{\rho,\mathrm{stereo}}\} \cup \left\{\frac{\partial}{\partial\nu} H_{-n,k}^{\rho,\mathrm{stereo}}\right\}_{n\in\mathbb{N},k=1,2},
$$

form a complete system in $\mathrm{L}^2(\partial\Gamma)$.

Exercise 6.10. Prove Theorem 6.72: Let $\Gamma \subset \Omega$ be a regular region and $\{\Phi_k\}_{k\in\mathbb{N}_0}$ denote the function system from Theorem 6.68 (a). Then, for F

of class $C^{(0)}(\partial\Gamma)$ and $\varepsilon > 0$, there exists a constant $M = M(\varepsilon) \in \mathbb{N}$ and coefficients $a_k = a_k(\varepsilon) \in \mathbb{R}$ such that

$$\sup_{\xi \in \partial\Gamma} \left| F(\xi) - \sum_{k=0}^{M} a_k \Phi_k(\xi) \right| < \varepsilon. \tag{6.278}$$

7

Gravitation

CONTENTS

7.1 Disturbing Potential

The lines that intersect all equipotential surfaces orthogonally are not exactly straight but slightly curved (cf. Figure 4.1). They are called *lines of gravity force* or *plumb lines.* The gravity vector at any point is tangential to the plumb line. Hence, *direction of the gravity vector*, *vertical* and *direction of the plumb line* are synonymous. As the level surfaces are, so to speak, "horizontal", i.e., orthogonal to the plumb lines, they play an important part in our daily life (e.g., in civil engineering for the purpose of height determination). Equipotential surfaces of the Earth's gravity potential W allow, in general, no simple mathematical representation. This is the reason why physical geodesy and geophysics choose a suitable reference surface for modeling the geoid i.e., the equipotential surface at sea level. The reference surface is constructed as an equipotential surface of an artificial *normal gravity potential* U. Its gradient field, i.e., $u = \nabla U$, is called *normal gravity.* For reasons of simplicity, physical geodesy usually uses an ellipsoid of revolution in such a way that a good adaption to the Earth's surface is guaranteed. Closed representations of normal gravity potentials, in consideration of the centrifugal force, can be found extensively in the geodetic literature (cf. G.G. Stokes [1849], W.A. Heiskanen, H. Moritz [1967], P.A. Meissl [1971], E. Groten [1979], W. Torge [1991], B. Hofmann–Wellenhof, H. Moritz [2005], H. Moritz [2010], E.W. Grafarend et al. [2010]), and the references therein). The deviations of the gravity field of the Earth from the normal field of such an ellipsoid are small. The remaining parts of the gravity field are gathered in a so-called *disturbing gravity field* ∇T corresponding to the *disturbing potential* T.

Knowing the gravity potential, all equipotential surfaces, including the geoid, are given by an equation of the form $W(x) = \text{const}$. By introducing U as the normal gravity potential corresponding to the ellipsoidal field, the disturbing potential T is the difference of the gravity potential W and the normal gravity potential U, i.e., we are led to a decomposition of the gravity potential in the form $W = U+T$. According to the concept developed by G.G. Stokes [1849] and P. Pizzetti [1909, 1910a], we assume that (a) the center of the ellipsoid coincides with the center of the gravity of the Earth, and (b) the difference of the mass of the Earth and the mass of the ellipsoid is zero.

7.1.1 Gravity Disturbances and Anomalies, Deflections of the Vertical

A point x of the geoid can be projected onto its associated point y of the ellipsoid by means of the ellipsoidal normal. The distance $N(x)$ between x and y is called the *geoidal height* or *geoidal undulation* in x (cf. Figure 7.1). The *gravity anomaly vector* $a(x)$ at the point x of the geoid is defined as the difference between the gravity vector $w(x)$ and the normal gravity vector $u(y)$, i.e.,

$$a(x) = w(x) - u(y). \tag{7.1}$$

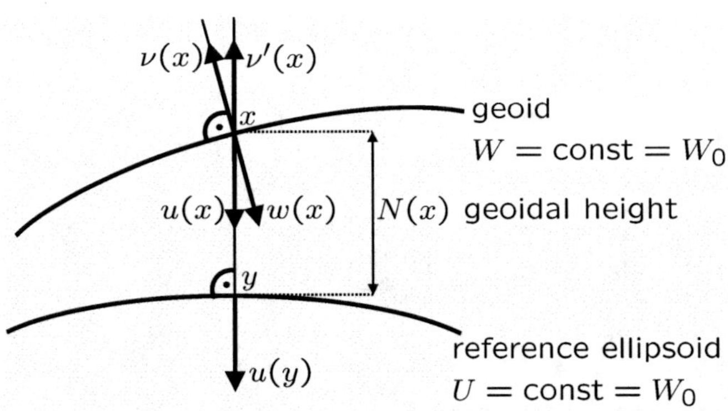

FIGURE 7.1
Illustration of the gravity anomaly vector $a(x) = w(x) - u(y)$, the gravity disturbance vector $d(x) = w(x) - u(x)$, and the geoidal height $N(x)$.

Another possibility is to form the difference between the vectors w and u at the same point x such that we get the *gravity disturbance vector* $d(x)$ to be defined by

$$d(x) = w(x) - u(x). \tag{7.2}$$

In physical geodesy, several basic mathematical relations between the scalar fields $|w|$ and $|u|$, as well as between the vector fields a and d are known. In the following we only describe the fundamental relations heuristically. The point of departure for our excursion into physical geodesy is the observation that the gravity disturbance vector $d(x)$ at the point x on the geoid can be written as follows

$$d(x) = w(x) - u(x) = \nabla\left(W(x) - U(x)\right) = \nabla T(x). \tag{7.3}$$

According to Taylor's formula, $U(y) + \frac{\partial U}{\partial \nu'}(y)N(x)$ is the linearization of $U(x)$, i.e., by expanding the potential U at the point x and truncating the Taylor series at the linear term we get

$$U(x) \simeq U(y) + \frac{\partial U}{\partial \nu'}(y)N(x), \tag{7.4}$$

where

$$\nu'(y) = -\frac{u(y)}{|u(y)|} \tag{7.5}$$

is the ellipsoidal normal at y and the geoidal undulation $N(x)$ is the aforementioned distance between x and y. The symbol "$\simeq$" means that the error between the left and the right-hand side may be assumed to be insignificantly small. Using the fact that $T(x) = W(x) - U(x)$ and observing the relations

$$|u(y)| = -\nu'(y) \cdot u(y) = -\nu'(y) \cdot \nabla U(y) = -\frac{\partial U}{\partial \nu'}(y), \tag{7.6}$$

we obtain from (7.4) that

$$N(x) = \frac{U(y) - U(x)}{|u(y)|} = \frac{T(x) - (W(x) - U(y))}{|u(y)|}. \tag{7.7}$$

Finally, considering $U(y) = W(x) = \text{const.} = W_0$, we end up with the so-called *Bruns' formula* (cf. E.H. Bruns [1878])

$$N(x) = \frac{T(x)}{|u(y)|}. \tag{7.8}$$

This formula relates the physical quantity $T(x)$ to the geometric quantity $N(x)$ for points x on the geoid. It is helpful to study the vector field $\nu(x)$ in more detail:

$$\nu(x) = -\frac{w(x)}{|w(x)|}. \tag{7.9}$$

Due to the definition of the normal vector field (7.9) we obtain the following identity

$$w(x) = \nabla W(x) = \; - \; |w(x)| \; \nu(x). \tag{7.10}$$

In an analogous way we obtain

$$u(x) = \nabla U(x) = \; - \; |u(x)| \; \nu'(x). \tag{7.11}$$

The *deflection of the vertical* $\Theta(x)$ at the point x on the geoid is understood to be the angular (i.e., tangential) difference between the directions $\nu(x)$ and $\nu'(x)$. More concretely, the deflection of the vertical is determined by the angle between the plumb line and the ellipsoidal normal through the same point:

$$\Theta(x) = \nu(x) - \nu'(x) - ((\nu(x) - \nu'(x)) \cdot \nu(x)) \, \nu(x). \tag{7.12}$$

According to its construction, the deflection of the vertical $\Theta(x)$ at x is orthogonal to the normal vector field $\nu(x)$, i.e., $\Theta(x) \cdot \nu(x) = 0$. Since the plumb lines are orthogonal to the level surfaces of the geoid and the ellipsoid, respectively, the deflection of the vertical gives a measure of the gradient of the level surfaces (cf. W.A. Heiskanen, H. Moritz [1967], W. Torge [1991]). From (7.10), in connection with (7.12), it follows that

$$\begin{aligned} w(x) &= \nabla W(x) = -|w(x)| \; \nu(x) \qquad (7.13) \\ &= -|w(x)| \left(\Theta(x) + \nu'(x) + ((\nu(x) - \nu'(x)) \cdot \nu(x)) \, \nu(x)\right). \end{aligned}$$

Therefore, with the aid of (7.11) and (7.13), we obtain for the gravity disturbing vector $d(x)$ at the point x

$$\begin{aligned} d(x) &= \nabla T(x) = w(x) - u(x) \qquad (7.14) \\ &= -|w(x)| \left(\Theta(x) + \nu'(x) + \left((\nu(x) - \nu'(x)) \cdot \nu(x)\right) \nu(x)\right) - \left(-|u(x)|\nu'(x)\right) \\ &= -|w(x)| \left(\Theta(x) + \left((\nu(x) - \nu'(x)) \cdot \nu(x)\right) \nu(x)\right) - \left(|w(x)| - |u(x)|\right) \nu'(x). \end{aligned}$$

The quantity

$$D(x) = |w(x)| - |u(x)| \tag{7.15}$$

is called the *gravity disturbance*, whereas

$$A(x) = |w(x)| - |u(y)| \tag{7.16}$$

is called the *gravity anomaly*. Splitting the gradient $\nabla T(x)$ of the disturbing potential T at x into a normal part (pointing into the direction of $\nu(x)$) and an angular (tangential) part (using the representation of the surface gradient ∇^*) we have

$$\nabla T(x) = \nu(x) \frac{\partial T}{\partial \nu}(x) + \frac{1}{|x|} \nabla^* T(x). \tag{7.17}$$

Since the gravity disturbances represent at most a factor 10^{-4} of the Earth's gravitational force (for more details see W.A. Heiskanen, H. Moritz [1967]), the error between $\nu(x)\frac{\partial T}{\partial \nu}(x)$ and $\nu'(x)\frac{\partial T}{\partial \nu'}(x)$ has no significance. Consequently, we may assume

$$d(x) \simeq \nu'(x)\frac{\partial T}{\partial \nu'}(x) + \frac{1}{|x|}\nabla^* T(x). \tag{7.18}$$

Moreover, the scalar product $(\nu(x) - \nu'(x)) \cdot \nu(x)$ can also be neglected. Thus, in connection with (7.15), we obtain

$$d(x) \simeq -|w(x)|\ \Theta(x) - (|w(x)| - |u(x)|)\,\nu'(x). \tag{7.19}$$

By comparison of (7.18) and (7.19), we therefore get

$$D(x) = |w(x)| - |u(x)| = -\frac{\partial T}{\partial \nu'}(x), \tag{7.20}$$

and the angular (tangential) differential equation

$$|w(x)|\Theta(x) = -\frac{1}{|x|}\nabla^* T(x). \tag{7.21}$$

In other words, the gravity disturbance $D(x)$, besides being the difference in magnitude of the actual and the normal gravity vector, is also the normal component of the gravity disturbance vector $d(x)$. In addition, we are led to the angular differential equation given in (7.21). Applying Bruns' formula we obtain for the gravity disturbance

$$D(x) = |w(x)| - |u(x)| = -|u(y)|\ \frac{\partial N}{\partial \nu'}(x). \tag{7.22}$$

For the deflections of the vertical we find

$$|w(x)|\ \Theta(x) = -\frac{1}{|x|}\ \nabla^* T(x) = -\frac{1}{|x|}|u(y)|\ \nabla^* N(x). \tag{7.23}$$

Since $|\Theta(x)|$ is a small quantity, it may be (without loss of precision) multiplied either by $|w(x)|$ or by $|u(x)|$, i.e., by $g(x)$ or by $\gamma(x)$.

Turning to the gravity anomalies, it follows from the identity (7.20) by linearization that

$$-\frac{\partial T}{\partial \nu'}(x) = D(x) \simeq A(x) - \frac{\partial |u(y)|}{\partial \nu'} N(x). \tag{7.24}$$

Using Bruns' formula (7.8) we obtain for the gravity anomalies that

$$A(x) = -\frac{\partial T}{\partial \nu'}(x) + \frac{1}{|u(y)|}\frac{\partial |u(y)|}{\partial \nu'} T(x). \tag{7.25}$$

Summing up our results, (7.20) and (7.25) we are led to the so-called *fundamental equations of physical geodesy* (see, for example, W.A. Heiskanen, H. Moritz [1967], P.A. Meissl [1971]):

$$D(x) = |w(x)| - |u(x)| = -\frac{\partial T}{\partial \nu'}(x), \tag{7.26}$$

$$A(x) = |w(x)| - |u(y)| = -\frac{\partial T}{\partial \nu'}(x) + \frac{1}{|u(y)|}\frac{\partial |u(y)|}{\partial \nu'}T(x). \tag{7.27}$$

The first equation shows a relation between the disturbing potential T and the gravity disturbance D on the geoid, whereas the second equation relates the disturbing potential T with the gravity anomaly A on the geoid. The Equations (7.26) and (7.27) constitute relations given on the geoid, and later they are used as boundary conditions in boundary-value problems.

Remark 7.1. The geoidal heights N, i.e., the deviations of the equipotential surface on the mean ocean level from the reference ellipsoid, are extremely small. Their order is of about 30–50 m and only at some places 100 m are achieved, which is only a factor 10^{-5} of the Earth's radius (see W.A. Heiskanen, H. Moritz [1967] for more details). Even more, the reference ellipsoid only differs from a sphere Ω_R with (mean Earth's) radius R in the order of the flattening of about $3 \cdot 10^{-3}$. Therefore, since the time of G.G. Stokes [1849], it is common use that, in theory, an ellipsoidal reference surface should be taken into account. However, in practice, the reference ellipsoid is treated as a sphere and Equations (7.22) and (7.23) are solved in spherical approximation. In doing so, a relative error of the order of the flattening of the Earth's body at the poles, i.e., a relative error of 10^{-3}, is accepted in all equations containing the disturbing potential. Considering appropriately performed reductions in numerical calculations, this error seems to be quite permissible (cf. W.A. Heiskanen, H. Moritz [1967] and the important remarks in A.A. Aardalan et al. [2006], E.W. Grafarend et al. [2010] for comparison with ellipsoidal approaches).

Remark 7.2. In addition, according to the Pizzetti assumptions (see P. Pizzetti [1909, 1910a,b]), it follows that the first moment integrals of the disturbing potential vanish, i.e.,

$$T^{\wedge_{\mathrm{L}^2(\Omega_R)}}(n,k) = \int_{\Omega_R} T(y) H^R_{-n-1,k}(y)\, d\omega(y) = 0 \tag{7.28}$$

for $n = 0, 1,\ k = 1, \ldots, 2n+1$. Indeed, if the Earth's center of gravity is the origin, there are no first-degree terms in the spherical harmonic expansion of T. If the mass of the spherical Earth and the mass of the normal ellipsoid is equal, there is no zero term. In this way, together with the indicated processes in gravity modeling, formulas and structures are obtained that are rigorously valid for the sphere, and, hence, suitable and applicable for computational purposes.

In the well-known spherical nomenclature, involving a sphere Ω_R (R being the mean Earth's radius) with a mass M distributed homogeneously in its interior, we are simply led to

$$U(y) = \frac{GM}{|y|}, \qquad u(y) = \nabla U(y) = -\frac{GM}{|y|^2}\frac{y}{|y|}. \tag{7.29}$$

Hence, we obtain

$$\begin{aligned} |u(y)| &= \frac{GM}{|y|^2}, & (7.30)\\ \frac{\partial |u(y)|}{\partial \nu'} &= -\frac{u(y)}{|u(y)|} \cdot \nabla |u(y)| = -2\frac{GM}{|y|^3}, & (7.31)\\ \frac{1}{|u(y)|}\frac{\partial |u(y)|}{\partial \nu'} &= -\frac{2}{|y|}. & (7.32) \end{aligned}$$

Furthermore, on the sphere Ω_R, the following relation holds true

$$-\frac{\partial T}{\partial \nu'}(x) = -\frac{x}{|x|} \cdot \nabla T(x), \qquad x \in \Omega_R. \tag{7.33}$$

Therefore, we end up with the formulation of the so-called *fundamental equations of physical geodesy*:

$$D(x) = -\frac{x}{|x|} \cdot \nabla T(x), \qquad x \in \Omega_R, \tag{7.34}$$

and

$$A(x) = -\frac{x}{|x|} \cdot \nabla T(x) - \frac{2}{|x|}T(x), \qquad x \in \Omega_R. \tag{7.35}$$

In addition, for the differential equation (7.21) we obtain in a vector spherical context (see also W. Freeden, M. Schreiner [2009]),

$$-\nabla^* T(x) = \frac{GM}{R}\Theta(x), \qquad x \in \Omega_R, \tag{7.36}$$

and, by virtue of Brun's formula, we finally find

$$-\frac{GM}{R^2}\nabla^* N(x) = \frac{GM}{R}\Theta(x), \qquad x \in \Omega_R, \tag{7.37}$$

i.e.,

$$-\nabla^* N(x) = R\Theta(x), \qquad x \in \Omega_R. \tag{7.38}$$

Remark 7.3. In physical geodesy (see, e.g., W.A. Heiskanen, H. Moritz [1967], B. Hofmann–Wellenhof, H. Moritz [2005]), a componentwise scalar determination of the deflections of the vertical is usually used. Our work prefers the vectorial framework, i.e., the vector equation (7.36). In doing so, we are concerned with an isotropic vector approach by means of the fundamental solution with respect to the Beltrami operator (see also W. Freeden, M. Schreiner [2009]) instead of the conventional anisotropic scalar decomposition into vector components due to F.A. Vening Meinesz [1928].

All in all, the determination of the disturbing potential and resulting geoid undulations can be based either on knowledge of the gravity anomalies or on knowledge of the gravity disturbances. Additionally, the disturbing potential can be derived from the deflections of the vertical.

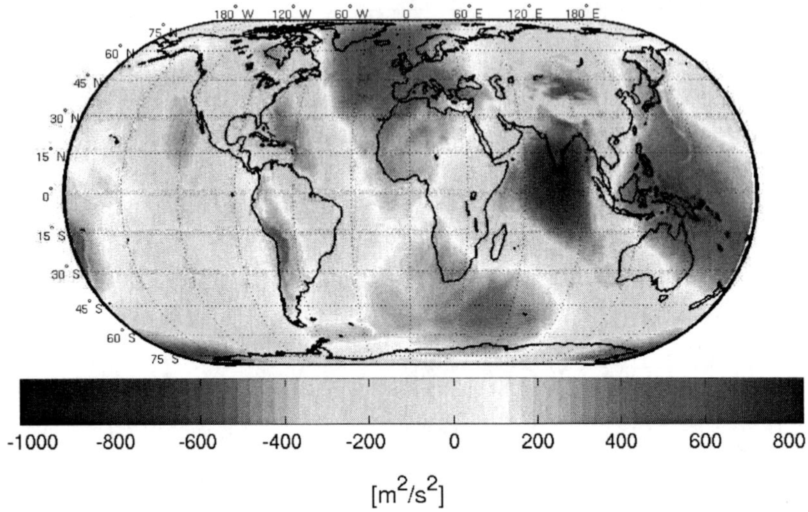

FIGURE 7.2
Illustration of the disturbing potential computed from EGM96 (Earth Gravitational Model 1996 due to F.G. Lemoine et al. [1998]) from degree 2 up to degree 360.

The disturbing potential enables us to make the following *geophysical interpretations* (for more details the reader is referred, e.g., R. Rummel [1997], D. Schubert et al. [2001], J. Kusche [2010], E.W. Grafarend et al. [2010], and the references therein):

Gravity disturbances and/or the gravity anomalies, which give a relation between the real Earth and an ellipsoidal Earth model, represent the expression of an imbalance of forces in the interior of the Earth. In accordance with Newton's Law of Gravitation they permit the conclusion of an irregular density distribution inside the Earth. Clearly, gravity anomalies and/or gravity disturbances do not uniquely determine the interior density distribution of the Earth. They may be interpreted (see Figures 7.3 and 7.4) as filtered signatures, which give major weight to the density contrasts close to the surface and simultaneously suppress the influence of deeper structures inside the Earth.

Geoid undulations provide a measure for the perturbations of the Earth from a hydrostatic equilibrium. They form the deviations of the equipotential surfaces at mean sea level from the reference ellipsoid. Geoid undulations show no essential correlation to the distributions of the continents (see Figure 7.2). They seem to be generated by density contrasts much deeper inside the Earth.

The deflections of the vertical (see Figure 7.5) represent a measure of how far the direction of the gravitational field is shifted by anomalies compared to the normal field.

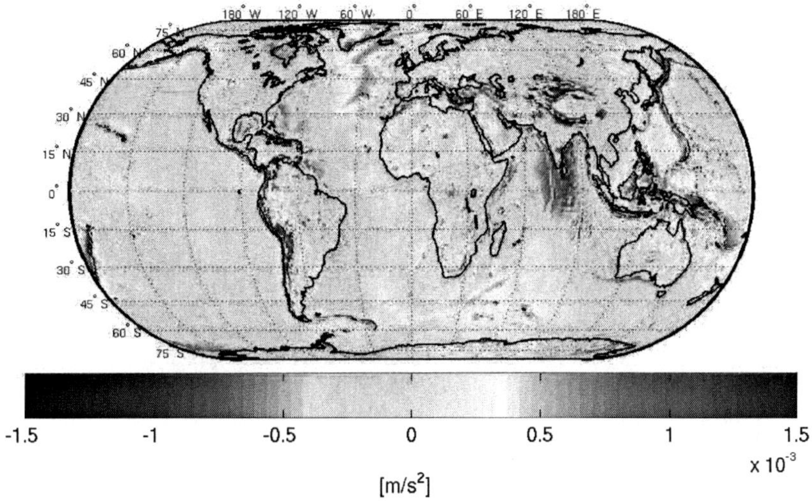

FIGURE 7.3
Illustration of the gravity disturbances computed from EGM96 from degree 2 up to degree 360.

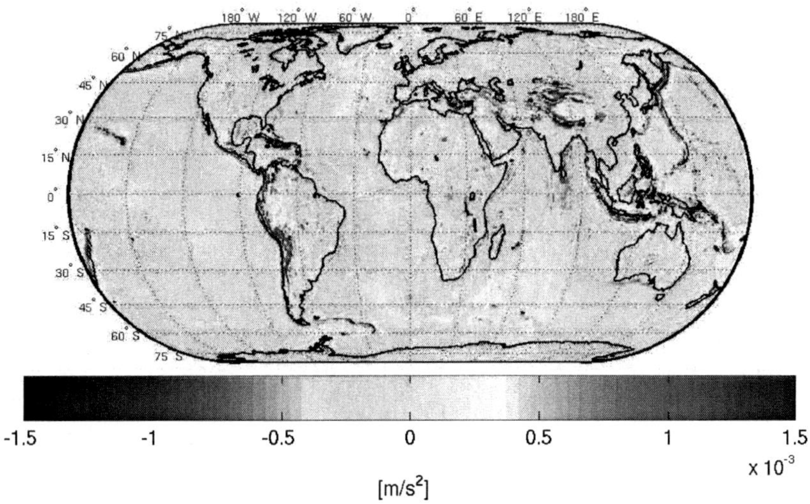

FIGURE 7.4
Illustration of gravity anomalies computed from EGM96 from degree 2 up to degree 360.

7.1.2 Gravity Disturbances as Scalar Boundary Data

As already explained, the task of determining the disturbing potential T (see Figure 7.2) from gravity disturbances or gravity anomalies, respectively, leads to boundary-value problems corresponding to a spherical boundary. Numerical realizations of such boundary-value problems have a long tradition, starting from G.G. Stokes [1849], F. Neumann [1887]. Nonetheless, our work presents some new aspects in their potential theoretic treatment by proposing appropriate space-regularization techniques applied to the resulting integral representations of their solutions. For both boundary-value problems, viz. the Neumann and the Stokes problem, we are interested in presenting two solution methods: On the one hand, the disturbing potential may be solved by a Fourier (orthogonal) expansion method in terms of spherical harmonics. On the other hand, it can be described by a singular integral representation over the boundary Ω_R.

Remark 7.4. So far, much more data on gravity anomalies $A(x) = |w(x)| - |u(y)|$ are available than on gravity disturbances $D(x) = |w(x)| - |u(x)|$. However, by modern GPS technology, the point x is determined rather than y. Therefore, in the future, it can be expected that D will become more important than A (as B. Hofmann–Wellenhof, H. Moritz [2005] point out in their monograph on physical geodesy). This is the reason why we continue to work with D. Nevertheless, the results of our (multiscale) approach applied to A are of significance. Therefore, the key ideas and concepts concerning A are treated in parallel as exercises in Section 7.4.

In the language of potential theory, the *exterior Neumann boundary-value problem of physical geodesy* (ENPPG) corresponding to known gravity disturbances D (compare (7.34)) reads as follows:

(ENPPG): We are given a continuous function D on Ω_R, i.e., $D \in \mathrm{C}^{(0)}(\Omega_R)$ with

$$D^{\wedge_{\mathrm{L}^2(\Omega_R)}}(n,k) = \int_{\Omega_R} D(y) H^R_{-n-1,k}(y)\, d\omega(y) = 0, \tag{7.39}$$

for $n = 0, 1,\ k = 1, \ldots, 2n+1$. Find a function $T \in \mathrm{Pot}^{(1)}\left(\mathbb{R}^3 \backslash \mathcal{B}_R(0)\right)$, $D = \frac{\partial T^+}{\partial \nu}\big|_{\Omega_R}$ such that T fulfills the conditions

$$T^{\wedge_{\mathrm{L}^2(\Omega_R)}}(n,k) = \int_{\Omega_R} T(y) H^R_{-n-1,k}(y)\, d\omega(y) = 0 \tag{7.40}$$

for $n = 0, 1,\ k = 1, \ldots, 2n+1$.

From the same arguments leading to (3.264) and (3.266), we are able to deduce that the solution of the boundary-value problem (ENPPG) can be represented in the form

$$T(x) = \frac{1}{4\pi R} \int_{\Omega_R} D(y)\ N(x,y)\ d\omega(y), \qquad x \in \mathbb{R}^3 \backslash \mathcal{B}_R(0), \tag{7.41}$$

where the *Neumann kernel* $N(\cdot,\cdot)$ in (7.41) possesses the spherical harmonic expansion

$$N(x,y) = \sum_{n=2}^{\infty} \left(\frac{R^2}{|x||y|}\right)^{n+1} \frac{2n+1}{n+1} P_n\left(\frac{x}{|x|}\cdot\frac{y}{|y|}\right). \tag{7.42}$$

By well-known manipulations, the series in terms of Legendre polynomials can be expressed as an elementary function leading to the integral representation

$$T(x) = \frac{1}{4\pi R}\int_{\Omega_R} D(y)\left(\frac{2R}{|x-y|} - \ln\left(\frac{|y| + \left|y - \frac{R^2}{|x|^2}x\right| + \frac{R^2}{|x|}}{|y| + \left|y - \frac{R^2}{|x|^2}x\right| - \frac{R^2}{|x|}}\right)\right) d\omega(y). \tag{7.43}$$

It is not difficult to see that the integral (7.43) is equivalent to

$$T(x) = \frac{1}{4\pi R}\int_{\Omega_R} D(y)\left(\frac{2R}{|x-y|} - \ln\left(\frac{|x| + |x-y| + R}{|x| + |x-y| - R}\right)\right) d\omega(y). \tag{7.44}$$

Note that this is actually the integral representation known from (3.264). Written out in the spherical nomenclature $x = R\frac{x}{|x|}$, $y = R\frac{y}{|y|}$, $x \neq y$ on Ω_R, we find

$$N\left(R\frac{x}{|x|}, R\frac{y}{|y|}\right) = \frac{2}{\left|\frac{x}{|x|} - \frac{y}{|y|}\right|} + \ln\left(\frac{R\left|\frac{x}{|x|} - \frac{y}{|y|}\right|}{2R + R\left|\frac{x}{|x|} - \frac{y}{|y|}\right|}\right). \tag{7.45}$$

If we replace $\left|\frac{x}{|x|} - \frac{y}{|y|}\right|$ by $\left(2 - 2\frac{x\cdot y}{|x|\,|y|}\right)^{\frac{1}{2}}$, then, for $x \neq y$, we are led to the identity

$$\begin{aligned} N\left(R\frac{x}{|x|}, R\frac{y}{|y|}\right) &= N\left(\frac{x}{|x|}, \frac{y}{|y|}\right) \\ &= \frac{\sqrt{2}}{\sqrt{1 - \frac{x}{|x|}\cdot\frac{y}{|y|}}} - \ln\left(1 + \frac{\sqrt{2}}{\sqrt{1 - \frac{x}{|x|}\cdot\frac{y}{|y|}}}\right). \end{aligned} \tag{7.46}$$

Consequently, for points $x \in \Omega_R$, we (formally) get the so-called *Neumann formula*, which constitutes an improper integral over Ω_R:

$$T\left(R\frac{x}{|x|}\right) = \frac{1}{4\pi R}\int_{\Omega_R} D\left(R\frac{y}{|y|}\right) N\left(\frac{x}{|x|}, \frac{y}{|y|}\right) d\omega(y), \tag{7.47}$$

where (7.42) takes the form (7.46), i.e., the Neumann kernel constitutes a zonal function on Ω.

In accordance with the conventional approach of physical geodesy, the Neumann formula (7.47) is valid under the following constraints (see also W.A. Heiskanen, H. Moritz [1967], W. Freeden, K. Wolf [2008], H. Moritz [2010]):

(i) the mass within the reference ellipsoid is equal to the mass of the Earth,

(ii) the center of the reference ellipsoid coincides with the center of the Earth,

(iii) the formulation is given in the spherical context to guarantee economical and efficient numerics.

An equivalent formulation of the improper integral (7.47) over the unit sphere Ω by use of the unit vectors $\xi = \frac{x}{|x|}$ and $\eta = \frac{y}{|y|}$ is given by

$$T(R\xi) = \frac{R}{4\pi} \int_{\Omega} D(R\eta) N(\xi \cdot \eta)\, d\omega(\eta), \tag{7.48}$$

where the Neumann kernel is the zonal function of the form

$$N(\xi \cdot \eta) = S(\xi \cdot \eta) - \ln\left(1 + S(\xi \cdot \eta)\right), \qquad 1 - \xi \cdot \eta > 0. \tag{7.49}$$

The single-layer kernel $S : [-1, 1) \to \mathbb{R}$ (cf. (2.213)) is given by

$$S(t) = \frac{\sqrt{2}}{\sqrt{1-t}}, \qquad t \in [-1, 1). \tag{7.50}$$

Note that we are able to set $N(R\xi, R\eta) = N(\xi \cdot \eta)$, $\xi, \eta \in \Omega$, which simplifies our notation. Once again, it should be noted that the Stokes problem of determining the disturbing potential from known gravity anomalies can be handled in a quite analogous way, providing the so-called Stokes integral associated to the Stokes kernel as an improper integral on Ω_R (cf. Section 7.4).

7.1.3 Deflections of the Vertical as Vectorial Input Data

Suppose that T fulfills the conditions (7.40). We consider the differential equation

$$\nabla^*_\xi T(R\xi) = -\frac{GM}{R} \Theta(R\xi), \qquad \xi \in \Omega, \tag{7.51}$$

where $T(R\cdot)$ represents the disturbing potential and $\Theta(R\cdot)$ denotes the deflections of the vertical (compare (7.36)). The differential equation (7.51) can be solved in a unique way by means of the fundamental solution with respect to the Beltrami operator

$$T(R\xi) = \frac{GM}{R} \int_{\Omega} \Theta(R\eta) \cdot \nabla^*_\eta G\left(\Delta^*; \xi \cdot \eta\right) d\omega(\eta), \qquad \xi \in \Omega. \tag{7.52}$$

The identity (7.52) immediately follows from the Third Green Theorem 6.4 for ∇^* on Ω in connection with (7.40). By virtue of the identity

$$\nabla^*_\eta G\left(\Delta^*; \xi \cdot \eta\right) = -\frac{\xi - (\xi \cdot \eta)\eta}{4\pi(1 - \xi \cdot \eta)}, \quad \xi \neq \eta, \tag{7.53}$$

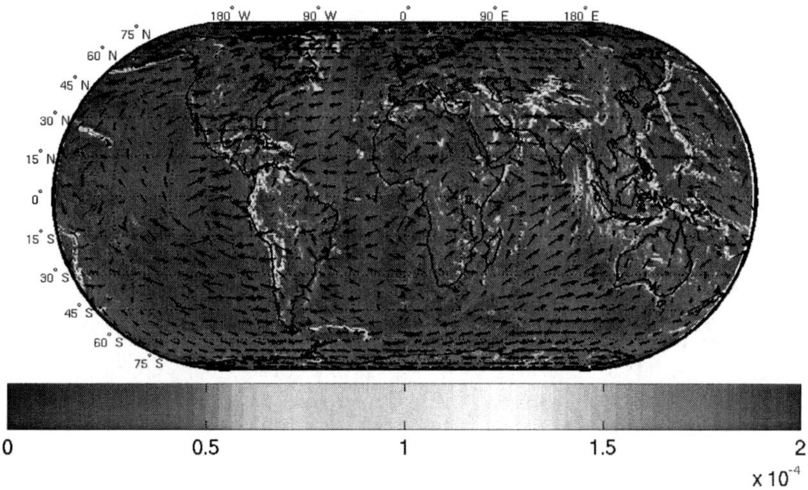

FIGURE 7.5
Illustration of the absolute values of the deflections of the vertical and their directions based on EGM96 from degree 2 up to degree 360.

the integral (7.52) can be written out in the form

$$T(R\xi) = \frac{R}{4\pi} \int_{\Omega} \Theta(R\eta) \cdot g\,(\Delta^*; \xi, \eta)\; d\omega(\eta), \tag{7.54}$$

where the vector kernel $g(\Delta^*; \xi, \eta),\ \xi \neq \eta$, is given by

$$g\,(\Delta^*; \xi, \eta) \quad = \quad -\frac{GM}{R^2} \frac{\xi - (\xi \cdot \eta)\eta}{1 - \xi \cdot \eta}. \tag{7.55}$$

Again we are confronted with a representation of the disturbing potential T as an improper integral over the sphere Ω_R.

7.2 Linear Regularization Method

All settings leading to the disturbing potential on the sphere Ω_R turn out to be improper integrals. By inspection it follows that they have either the singularity behavior of the single-layer kernel or the characteristic logarithmic singularity of the fundamental solution with respect to the Beltrami operator. Indeed, the fundamental solution and the single-layer kernel are interrelated by the identities

$$S(\xi \cdot \eta) = \frac{\sqrt{2}}{\sqrt{1 - \xi \cdot \eta}} = \sqrt{2}\; e^{-2\pi G(\Delta^*; \xi \cdot \eta) + \frac{1}{2}} \tag{7.56}$$

and

$$G(\Delta^*;\xi\cdot\eta) = -\frac{1}{2\pi}\ln(S(\xi\cdot\eta)) - \frac{1}{4\pi}(1-2\ln(2)). \tag{7.57}$$

Therefore, we are confronted with the remarkable situation that a regularization of the single-layer kernel implies a regularization of the fundamental solution, and vice versa.

In our work, we follow the space regularization methods following the ideas in W. Freeden, M. Schreiner [2006] for linear regularization of the fundamental solution, W. Freeden, K. Wolf [2008] for linear regularization of the single-layer kernel, and C. Gerhards [2011a] for higher-order regularization of both kernels. More specifically, for the multiscale modeling of gravitation as intended in this chapter, we restrict ourselves to linear regularization involving the single-layer kernel. In the next chapter concerned with geomagnetic modeling, we extend our techniques to higher-order regularization of the fundamental solution as well as the single-layer kernel.

7.2.1 Linear Regularization of the Single-Layer Kernel

The essential idea is to regularize the single-layer kernel function (2.213)

$$S(t) = \frac{\sqrt{2}}{\sqrt{1-t}} \tag{7.58}$$

by replacing it with a Taylor linearization. To this end, we notice that the first derivative of the kernel S reads as follows

$$S'(t) = \frac{1}{\sqrt{2}(1-t)^{\frac{3}{2}}}, \qquad t\in[-1,1). \tag{7.59}$$

Consequently, we obtain as the (Taylor) linearized approximation corresponding to the expansion point $1-\frac{\rho^2}{2R^2}$, $\rho\in(0,2R]$,

$$S(t) \simeq S\left(1-\frac{\rho^2}{2R^2}\right) + S'\left(1-\frac{\rho^2}{2R^2}\right)\left(t-(1-\frac{\rho^2}{2R^2})\right). \tag{7.60}$$

In more detail, the kernel S is replaced by its (Taylor) linearized approximation ${}^{l}S^{\rho}$ at the point $1-\frac{\rho^2}{2R^2}$, $\rho\in(0,2R]$, given by

$${}^{l}S^{\rho}(t) = \begin{cases} \frac{R}{\rho}\left(3-\frac{2R^2}{\rho^2}(1-t)\right), & 0\le 1-t\le \frac{\rho^2}{2R^2}, \\ \frac{\sqrt{2}}{\sqrt{1-t}}, & \frac{\rho^2}{2R^2} < 1-t \le 2. \end{cases} \tag{7.61}$$

Note that the expansion point $1-(\rho^2/2R^2)$, $\rho\in(0,2R]$, is chosen consistently with the notation of the initial paper W. Freeden, M. Schreiner [2006] and the subsequent papers by T. Fehlinger et al. [2007], T. Fehlinger et al. [2008], W. Freeden, K. Wolf [2008], T. Fehlinger [2009], K. Wolf [2009]. The upper

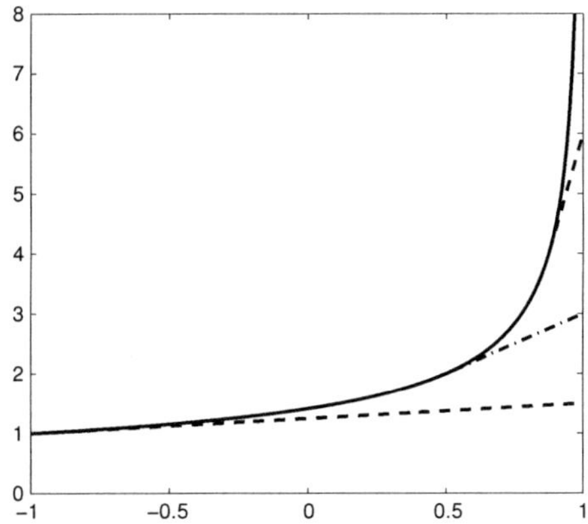

FIGURE 7.6
Single-layer kernel $S(t)$ (continuous black line) and its Taylor linearized regularization ${}^{l}S^{\rho}(t)$, for $\rho = \frac{1}{2}, 1, 2$ (dotted lines).

left index "l" indicates that (7.61) is the special case of a linear regularization related to the particular expansion point $1-(\rho^2/2R^2)$. A graphical illustration of the original kernel $S(t)$ and a ρ-scale dependent version of its linear space-regularized kernel ${}^{l}S^{\rho}(t)$ is shown in Figure 7.6.

7.2.2 Integral Relations for the Regularized Single-Layer Kernel

Clearly, the function ${}^{l}S^{\rho}$ is continuously differentiable on the interval $[-1,1]$, and we have

$$\left({}^{l}S^{\rho}\right)'(t) = \begin{cases} \frac{2R^3}{\rho^3}, & 0 \le 1-t \le \frac{\rho^2}{2R^2}, \\ \frac{1}{\sqrt{2}(1-t)^{\frac{3}{2}}}, & \frac{\rho^2}{2R^2} < 1-t \le 2. \end{cases} \tag{7.62}$$

Furthermore, the functions S and ${}^{l}S^{\rho}$ are monotonically increasing on the interval $[-1,1)$, such that $S(t) \ge {}^{l}S^{\rho}(t) \ge S(-1) = {}^{l}S^{\rho}(-1) = 1$ holds true on the interval $[-1,1)$. Considering the difference between the kernel S and its linearly regularized version ${}^{l}S^{\rho}$, we find

$$S(t) - {}^{l}S^{\rho}(t) = \begin{cases} \frac{\sqrt{2}}{\sqrt{1-t}} - \frac{R}{\rho}\left(3 - \frac{2R^2}{\rho^2}(1-t)\right), & 0 < 1-t \le \frac{\rho^2}{2R^2}, \\ 0, & \frac{\rho^2}{2R^2} < 1-t \le 2. \end{cases} \tag{7.63}$$

Based on the properties of (7.61), we readily get by elementary manipulations of one-dimensional analysis

$$\int_{-1}^{1} \left|S(t) - {}^{l}S^{\rho}(t)\right| \, dt = \int_{-1}^{1} \left|S(t) - {}^{l}S^{\rho}(t)\right| \, dt \;=\; O(\rho). \tag{7.64}$$

As a consequence, we obtain

Lemma 7.5. *For $F \in \mathrm{C}^{(0)}(\Omega)$ and ${}^l S^\rho$ as defined by (7.61),*

$$\lim_{\rho\to 0+} \sup_{\xi\in\Omega} \left| \int_\Omega S(\xi\cdot\eta)F(\eta)\, d\omega(\eta) - \int_\Omega {}^l S^\rho(t)(\xi\cdot\eta)F(\eta)\, d\omega(\eta) \right| = 0. \quad (7.65)$$

In a similar way, by some lengthier calculations, one can find the following relations that are also of some importance for the Stokes boundary-value problem, which is described in the exercises.

Lemma 7.6. *Let S be the singular kernel given by (7.58) and let ${}^l S^\rho, \rho \in (0, 2R]$, be the corresponding (Taylor) linearized regularized kernel as defined by (7.61). Then*

$$\lim_{\rho\to 0+} \int_{-1}^{1} \left|\ln\left(1+S(t)\right) - \ln\left(1+{}^l S^\rho(t)\right)\right| dt = 0, \quad (7.66)$$

$$\lim_{\rho\to 0+} \int_{-1}^{1} \left|\ln\left(\frac{1}{S(t)}+\frac{1}{(S(t))^2}\right) - \ln\left(\frac{1}{{}^l S^\rho(t)}+\frac{1}{\left({}^l S^\rho(t)\right)^2}\right)\right| dt = 0, \quad (7.67)$$

$$\lim_{\rho\to 0+} \int_{-1}^{1} \left((S(t))^2 - \left({}^l S^\rho(t)\right)^2\right)\sqrt{1-t^2}\, dt = 0. \quad (7.68)$$

For a context involving the surface gradient and the surface curl gradient, we have to pay some attention, since the vector fields $\eta \mapsto \nabla^*_\xi S(\xi\cdot\eta)$ as well as $\eta \mapsto \mathrm{L}^*_\xi S(\xi\cdot\eta)$ are not integrable on the unit sphere Ω. One option is to choose F to be of class $\mathrm{C}^{(1)}(\Omega)$. Letting $\mathbf{t}_\xi \in \mathbb{R}^{3\times 3}$ be the orthogonal matrix (with $\det(\mathbf{t}) = 1$) leaving ε^3 fixed such that $\mathbf{t}_\xi\xi = \varepsilon^3$, we get

$$\begin{aligned} \nabla^*_\xi \int_\Omega S(\xi\cdot\eta)F(\eta)d\omega(\eta) &= \nabla^*_\xi \int_\Omega S(\eta_3)F(\mathbf{t}_\xi^T\eta)d\omega(\eta) \\ &= \int_\Omega S(\eta_3)\nabla^*_\xi F(\mathbf{t}_\xi^T\eta)d\omega(\eta), \quad \xi\in\Omega, \end{aligned} \quad (7.69)$$

for $\xi \in \Omega$ and $\eta = (\eta_1, \eta_2, \eta_3)^T$. By regularizing the single-layer kernel, we obtain

$$\begin{aligned} \int_\Omega \nabla^{*l}_\xi S^\rho(\xi\cdot\eta)F(\eta)d\omega(\eta) &= \nabla^*_\xi \int_\Omega {}^l S^\rho(\xi\cdot\eta)F(\eta)d\omega(\eta) \\ &= \nabla^*_\xi \int_\Omega {}^l S^\rho(\eta_3)F(\mathbf{t}_\xi^T\eta)d\omega(\eta) \\ &= \int_\Omega {}^l S^\rho(\eta_3)\nabla^*_\xi F(\mathbf{t}_\xi^T\eta)d\omega(\eta), \end{aligned} \quad (7.70)$$

for $\xi \in \Omega$. The same argumentation holds true for the operator L^*. Therefore, Lemma 7.5 leads us to following limit relations.

Lemma 7.7. *Let F be of class $\mathrm{C}^{(1)}(\Omega)$. Suppose that S^ρ is given by (7.61). Then*

$$\lim_{\rho\to 0+} \sup_{\xi\in\Omega} \left| \int_\Omega \nabla_\xi^{*l} S^\rho(\xi\cdot\eta)F(\eta)d\omega(\eta) - \nabla_\xi^* \int_\Omega S(\xi\cdot\eta)F(\eta)d\omega(\eta)\right| = 0, \quad (7.71)$$

$$\lim_{\rho\to 0+} \sup_{\xi\in\Omega} \left| \int_\Omega \mathrm{L}_\xi^{*l} S^\rho(\xi\cdot\eta)F(\eta)d\omega(\eta) - \mathrm{L}_\xi^* \int_\Omega S(\xi\cdot\eta)F(\eta)d\omega(\eta)\right| = 0. \quad (7.72)$$

Using the kernel ${}^lG^\rho(\Delta^*;\cdot)$, given by

$${}^lG^\rho(\Delta^*;t) = -\frac{1}{2\pi}\ln({}^lS^\rho(t)) - \frac{1}{4\pi}(1-2\ln(2)), \ -1\le t\le 1, \quad (7.73)$$

as *single-layer kernel regularization of the fundamental solution* $G(\Delta^*;\cdot)$, we are led to the following integral relations.

Lemma 7.8. *For $F\in \mathrm{C}^{(0)}(\Omega)$ and ${}^lG^\rho(\Delta^*;\cdot)$ as defined by (7.73), we have*

$$\lim_{\rho\to 0+} \sup_{\xi\in\Omega} \left| \int_\Omega G(\Delta^*;\xi\cdot\eta)F(\eta)\ d\omega(\eta) - \int_\Omega {}^lG^\rho(\Delta^*;\xi\cdot\eta)F(\eta)\ d\omega(\eta)\right| = 0, \quad (7.74)$$

and

$$\lim_{\rho\to 0+} \sup_{\xi\in\Omega} \left| \nabla_\xi^* \int_\Omega G(\Delta^*;\xi\cdot\eta)F(\eta)\ d\omega(\eta) - \int_\Omega \nabla_\xi^{*l} G^\rho(\Delta^*;\xi\cdot\eta)F(\eta)\ d\omega(\eta)\right| = 0, \quad (7.75)$$

$$\lim_{\rho\to 0+} \sup_{\xi\in\Omega} \left| \int_\Omega \mathrm{L}_\xi^* G(\Delta^*;\xi\cdot\eta)F(\eta)\ d\omega(\eta) - \mathrm{L}_\xi^* \int_\Omega {}^lG^\rho(\Delta^*;\xi\cdot\eta)F(\eta)\ d\omega(\eta)\right| = 0. \quad (7.76)$$

Remark 7.9. Numerical implementations and computational aspects of the linear regularization techniques as presented here have been applied (even for subsets of Ω) to different fields of geoscientific research, namely physical geodesy (W. Freeden, M. Schreiner [2006, 2009], T. Fehlinger et al. [2008], W. Freeden, K. Wolf [2008], K. Wolf [2009], T. Fehlinger [2009]), W. Freeden et al. [2009], W. Freeden [2009], W. Freeden [2010]), geostrophic ocean circulation (W. Freeden, M. Schreiner [2009], T. Fehlinger et al. [2007], W. Freeden [2009], W. Freeden [2010]), and geomagnetic modeling (W. Freeden, C. Gerhards [2010], C. Gerhards [2011a], C. Gerhards [2011b], C. Gerhards [2012]).

7.3 Multiscale Solution

The linear space regularization techniques as proposed in Section 7.2 enable us to formulate multiscale solutions for the disturbing potential from gravity disturbances or deflections of the vertical (note that we need higher-order regularizations (as presented in Chapter 8 for geomagnetism) whenever gravitational observables containing second- or higher-order derivatives come into play. An example is gravity gradiometry, which will not be discussed here. The point of departure for our consideration is the special case study of the linear regularization of the single-layer kernel in the integral representation of the solution of the Neumann boundary-value problem (ENPPG).

7.3.1 Disturbing Potential from Gravity Disturbances

As we already know, the solution of the (Earth's) disturbing potential $T \in \mathrm{Pot}^{(1)}\left(\mathbb{R}^3 \backslash \mathcal{B}_R(0)\right)$ from known gravity disturbances $D = \frac{\partial T^+}{\partial \nu}\big|_{\Omega_R}$, satisfying the conditions (7.40) on the sphere Ω_R, can be formulated as an improper integral

$$T(R\xi) = \frac{R}{4\pi} \int_{\Omega} N(\xi \cdot \eta)\ D(R\eta)\, d\omega(\eta),\ \xi \in \Omega, \tag{7.77}$$

where the Neumann kernel (7.49) reads as follows:

$$N(\xi \cdot \eta) = S(\xi \cdot \eta) - \ln\left(1 + S(\xi \cdot \eta)\right),\ 1 - \xi \cdot \eta > 0. \tag{7.78}$$

Our interest is to formulate regularizations of the disturbing potential T by use of the (Taylor) linearized version ${}^{l}S^{\rho} : [-1, 1] \to \mathbb{R},\ \rho \in (0, 2R]$, introduced in (7.61). As a result, we obtain the regularized Neumann kernels

$$\begin{aligned} &{}^{l}N^{\rho}(\xi \cdot \eta) \qquad\qquad (7.79)\\ &= \begin{cases} {}^{l}S^{\rho}(\xi \cdot \eta) - \ln\left(1 + {}^{l}S^{\rho}(\xi \cdot \eta)\right), & 0 \le 1 - \xi \cdot \eta \le \frac{\rho^2}{2R^2}, \\ S(\xi \cdot \eta) - \ln(1 + S(\xi \cdot \eta)), & \frac{\rho^2}{2R^2} < 1 - \xi \cdot \eta \le 2, \end{cases} \\ &= \begin{cases} \frac{R}{\rho}\left(3 - \frac{2R^2}{\rho^2}(1 - \xi \cdot \eta)\right) \\ \quad - \ln\left(1 + \frac{R}{\rho}\left(3 - \frac{2R^2}{\rho^2}(1 - \xi \cdot \eta)\right)\right), & 0 \le 1 - \xi \cdot \eta \le \frac{\rho^2}{2R^2}, \\ \frac{\sqrt{2}}{\sqrt{1 - \xi \cdot \eta}} - \ln\left(1 + \frac{\sqrt{2}}{\sqrt{1 - \xi \cdot \eta}}\right), & \frac{\rho^2}{2R^2} < 1 - \xi \cdot \eta \le 2. \end{cases} \end{aligned}$$

In doing so, we are immediately led to the regularized representation of the disturbing potential T corresponding to the known gravity disturbances:

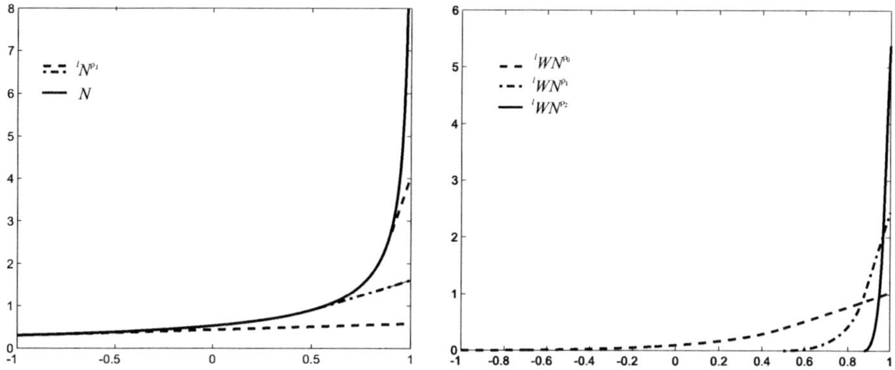

FIGURE 7.7
Illustration of the Neumann kernel $N(t)$ (left, continuous black line) and its Taylor linearized regularization ${}^lN^{\rho_J}(t)$, $J = 1, 2, 3$ (left, dotted lines). The corresponding Taylor linearized Neumann wavelets ${}^lWN^{\rho_J}(t)$ for scales $J = 0, 1, 2$, are shown on the right.

$$
\begin{aligned}
{}^lT^\rho(R\xi) = & \frac{R}{4\pi} \int\limits_{\substack{\eta\in\Omega \\ 1-\xi\cdot\eta > \frac{\rho^2}{2R^2}}} D(R\eta) \frac{\sqrt{2}}{\sqrt{1-\xi\cdot\eta}}\, d\omega(\eta) \\
& - \frac{R}{4\pi} \int\limits_{\substack{\eta\in\Omega \\ 1-\xi\cdot\eta > \frac{\rho^2}{2R^2}}} D(R\eta) \ln\left(1 + \frac{\sqrt{2}}{\sqrt{1-\xi\cdot\eta}}\right) d\omega(\eta) \\
& + \frac{R}{4\pi} \int\limits_{\substack{\eta\in\Omega \\ 1-\xi\cdot\eta \le \frac{\rho^2}{2R^2}}} D(R\eta) \frac{R}{\rho}\left(3 - \frac{2R^2}{\rho^2}(1-\xi\cdot\eta)\right) d\omega(\eta) \\
& - \frac{R}{4\pi} \int\limits_{\substack{\eta\in\Omega \\ 1-\xi\cdot\eta \le \frac{\rho^2}{2R^2}}} D(R\eta) \ln\left(1 + \frac{R}{\rho}\left(3 - \frac{2R^2}{\rho^2}(1-\xi\cdot\eta)\right)\right) d\omega(\eta).
\end{aligned}
\tag{7.80}
$$

The representation (7.80) is remarkable, since the integrands of T and ${}^lT^\rho$ only differ on the spherical cap $\Gamma_{\rho^2/2R^2}(\xi)$. By aid of Lemma 7.5 and 7.6, we obtain

Theorem 7.10. *Suppose that T is the solution of the Neumann boundary-value problem* (ENPPG) *of the form (7.77). Let ${}^lT^\rho$, $\rho \in (0, 2R]$, represent its regularization (7.80). Then*

$$
\lim_{\rho\to 0+} \sup_{\xi\in\Omega} \left| T(R\xi) - {}^lT^\rho(R\xi) \right| = 0. \tag{7.81}
$$

For numerical applications we have to go over to scale-discretized approximations of the solution to the boundary-value problem (ENPPG). For that purpose, we choose a monotonically decreasing sequence $\{\rho_j\}_{j\in\mathbb{N}_0}$, such that

$$\lim_{j\to\infty} \rho_j = 0, \ \ \rho_0 = 2R. \tag{7.82}$$

A particularly important example, which we use in our numerical implementations below, is the dyadic sequence with

$$\rho_j = 2^{1-j}R, \ \ j\in\mathbb{N} \qquad \rho_0 = 2R. \tag{7.83}$$

It is easily seen that $2\rho_{j+1} = \rho_j$, $j\in\mathbb{N}_0$, is the relation between two consecutive elements of the sequence. In correspondence to the sequence $\{\rho_j\}_{j\in\mathbb{N}_0}$, a sequence $\{{}^lN^{\rho_j}\}_{j\in\mathbb{N}_0}$ of discrete versions of the regularized Neumann kernels (7.79), so-called *Neumann scaling functions*, is available. Figure 7.7 shows a graphical illustration of the regularized Neumann kernels for different scales j.

The *regularized Neumann wavelets*, forming the sequence $\{{}^lWN^{\rho_j}\}_{j\in\mathbb{N}_0}$, are understood to be the difference of two consecutive regularized Neumann scaling functions, respectively,

$${}^lWN^{\rho_j} = {}^lN^{\rho_{j+1}} - {}^lN^{\rho_j}, \qquad j\in\mathbb{N}_0. \tag{7.84}$$

The Neumann wavelets are illustrated in Figure 7.7 as well. These wavelets possess the numerically nice property of a local support. More specifically, $\eta\mapsto {}^lWN^{\rho_j}(\xi\cdot\eta)$, $\eta\in\Omega$, vanishes everywhere outside the spherical cap $\Gamma_{\rho_j^2/2R^2}(\xi)$. Explicitly written, we have

$$\begin{aligned}
&{}^lWN^{\rho_j}(\xi\cdot\eta) \\
&= \ {}^lN^{\rho_{j+1}}(\xi\cdot\eta) \ - \ {}^lN^{\rho_j}(\xi\cdot\eta) \\
&= \begin{cases}
\frac{R}{\rho_{j+1}}\left(3-\frac{2R^2}{\rho_{j+1}^2}(1-\xi\cdot\eta)\right) & \\
-\ln\left(1+\frac{R}{\rho_{j+1}}\left(3-\frac{2R^2}{\rho_{j+1}^2}(1-\xi\cdot\eta)\right)\right) & \\
-\frac{R}{\rho_j}\left(3-\frac{2R^2}{\rho_j^2}(1-\xi\cdot\eta)\right) & \\
+\ln\left(1+\frac{R}{\rho_j}\left(3-\frac{2R^2}{\rho_j^2}(1-\xi\cdot\eta)\right)\right), & 0\le 1-\xi\cdot\eta\le\frac{\rho_{j+1}^2}{2R^2}, \\
\frac{\sqrt{2}}{\sqrt{1-\xi\cdot\eta}}-\ln\left(1+\frac{\sqrt{2}}{\sqrt{1-\xi\cdot\eta}}\right) & \\
-\frac{R}{\rho_j}\left(3-\frac{2R^2}{\rho_j^2}(1-\xi\cdot\eta)\right) & \\
+\ln\left(1+\frac{R}{\rho_j}\left(3-\frac{2R^2}{\rho_j^2}(1-\xi\cdot\eta)\right)\right), & \frac{\rho_{j+1}^2}{2R^2}<1-\xi\cdot\eta\le\frac{\rho_j^2}{2R^2}, \\
0, & \frac{\rho_j^2}{2R^2}<1-\xi\cdot\eta\le 2.
\end{cases}
\end{aligned} \tag{7.85}$$

Let $J\in\mathbb{N}_0$ be an arbitrary scale. Suppose that ${}^lN^{\rho_J}$ is the regularized

Neumann scaling function at scale J. Furthermore, let ${}^{l}WN^{\rho_j}$, $j = 0, \ldots, J$, be the regularized Neumann wavelets as given by (7.85). Then an easy manipulation shows that

$$ {}^{l}N^{\rho_J} = {}^{l}N^{\rho_0} + \sum_{j=0}^{J} {}^{l}WN^{\rho_j}. \tag{7.86} $$

The local support of the Neumann wavelets within the framework of (7.86) should be studied in more detail: Following the sequence (7.83), we start with a globally supported scaling kernel ${}^{l}N^{\rho_0} = {}^{l}N^{2R}$. Then we add more and more wavelet kernels ${}^{l}WN^{\rho_j}$, $j = 0, \ldots, J$, to achieve the required scaling kernel ${}^{l}N^{\rho_J}$. It is of particular importance that $\eta \mapsto {}^{l}WN^{\rho_j}(\xi \cdot \eta)$, $\xi \in \Omega$ fixed, are ξ-zonal functions and possess spherical caps as local supports. Figure 7.8 (see also K. Wolf [2009]) illustrates the computationally relevant regions for the different wavelet scales j. For a better understanding, the areas outside the caps are chosen to be uncolored. Clearly, the support of the wavelets ${}^{l}WN^{\rho_j}$ become more and more localized for increasing scales j. In conclusion, a calculation of an integral representation for the disturbing potential T starts with a global trend approximation using the scaling kernel at scale $j = 0$ (of course, this requires data on the whole sphere, but the data can be rather sparsely distributed since they only serve as a trend approximation). Step by step, we are able to refine this approximation by use of wavelets. The increasing spatial localization of the wavelets successively allows a better spatial resolution of the disturbing potential T. Additionally, the local supports of the wavelets have a computational advantage since the integration is reduced from the entire sphere to smaller and smaller spherical caps. Consequently, the presented numerical technique becomes capable of handling heterogeneously distributed data.

All in all, keeping the space-localizing properties of the regularized Neumann scaling and wavelet functions in mind, we are able to establish an approximation of the solution of the disturbing potential T from gravity disturbances D in the form of a *zooming-in* multiscale method. A low-pass filtered version of the disturbing potential T at the scale j in an integral representation over the unit sphere Ω is given by

$$ {}^{l}T^{\rho_j}(R\xi) = \frac{R}{4\pi} \int_{\Omega} D(R\eta)\, {}^{l}N^{\rho_j}(\xi \cdot \eta)\, d\omega(\eta), \qquad \xi \in \Omega, \tag{7.87} $$

while the j-scale band-pass filtered version of T leads to the integral representation

$$ {}^{l}WT^{\rho_j}(R\xi) = \frac{R}{4\pi} \int_{\Gamma_{\rho_j^2/2R^2}(\xi)} D(R\eta)\, {}^{l}WN^{\rho_j}(\xi \cdot \eta)\, d\omega(\eta), \qquad \xi \in \Omega. \tag{7.88} $$

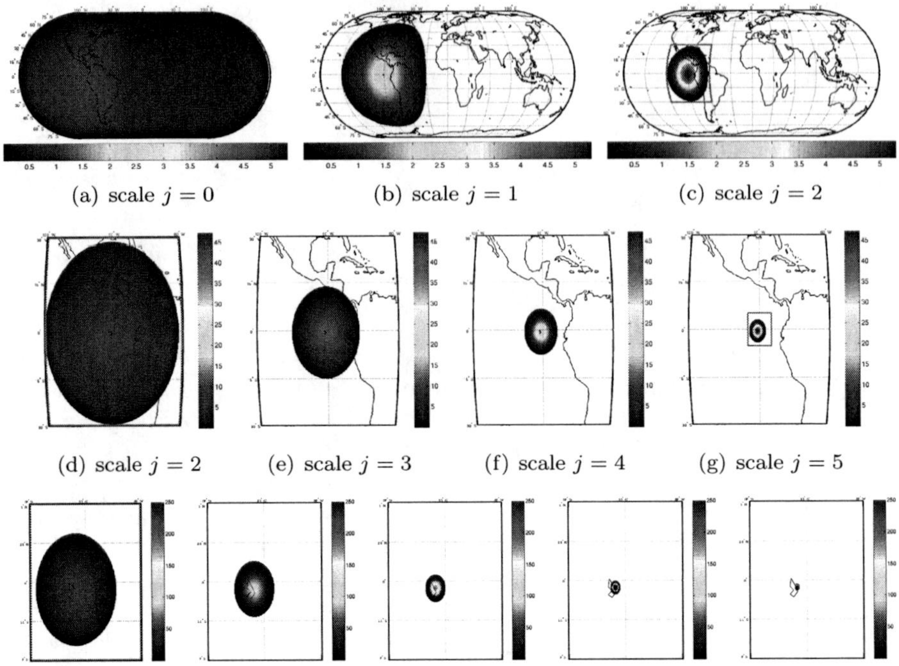

(a) scale $j = 0$ (b) scale $j = 1$ (c) scale $j = 2$

(d) scale $j = 2$ (e) scale $j = 3$ (f) scale $j = 4$ (g) scale $j = 5$

(h) scale $j = 5$ (i) scale $j = 6$ (j) scale $j = 7$ (k) scale $j = 8$ (l) scale $j = 9$

FIGURE 7.8
Illustration of the linearized Neumann wavelet kernel $\eta \mapsto {}^{l}WN^{\rho_j}(\xi \cdot \eta)$ for scales $j = 0, \ldots, 9$ to visualize the local supports $\Gamma_{\rho_j^2/2R^2}(\xi)$ for a fixed ξ .

Theorem 7.11. *Let ${}^{l}T^{\rho_{J_0}}$ be the regularized version of the disturbing potential at some arbitrary initial scale J_0 as given in (7.87), and let ${}^{l}WT^{\rho_j}$, $j = 0, 1, \ldots,$ be given by (7.88). Then, the following reconstruction formula holds true:*

$$\lim_{N\to\infty} \sup_{\xi\in\Omega} \left| T(R\xi) - \left({}^{l}T^{\rho_{J_0}}(R\xi) + \sum_{j=0}^{N} {}^{l}WT^{\rho_{J_0+j}}(R\xi) \right) \right| = 0.$$

The multiscale procedure (wavelet reconstruction) as developed here can be illustrated by the following scheme

$$\begin{array}{ccccccccccc} {}^{l}WT^{\rho_{J_0}} & & & & & {}^{l}WT^{\rho_{J_0+1}} & & & & & \\ & \searrow & & & & & \searrow & & & & \\ {}^{l}T^{\rho_{J_0}} & \longrightarrow & + & \longrightarrow & & {}^{l}T^{\rho_{J_0+1}} & \longrightarrow & + & \longrightarrow & {}^{l}T^{\rho_{J_0+2}} & \ldots . \end{array} \tag{7.89}$$

Consequently, a tree algorithm based on regularization in the space domain has been realized for determining the disturbing potential T from locally available data sets of gravity disturbances D. An example is shown in Figure 7.9.

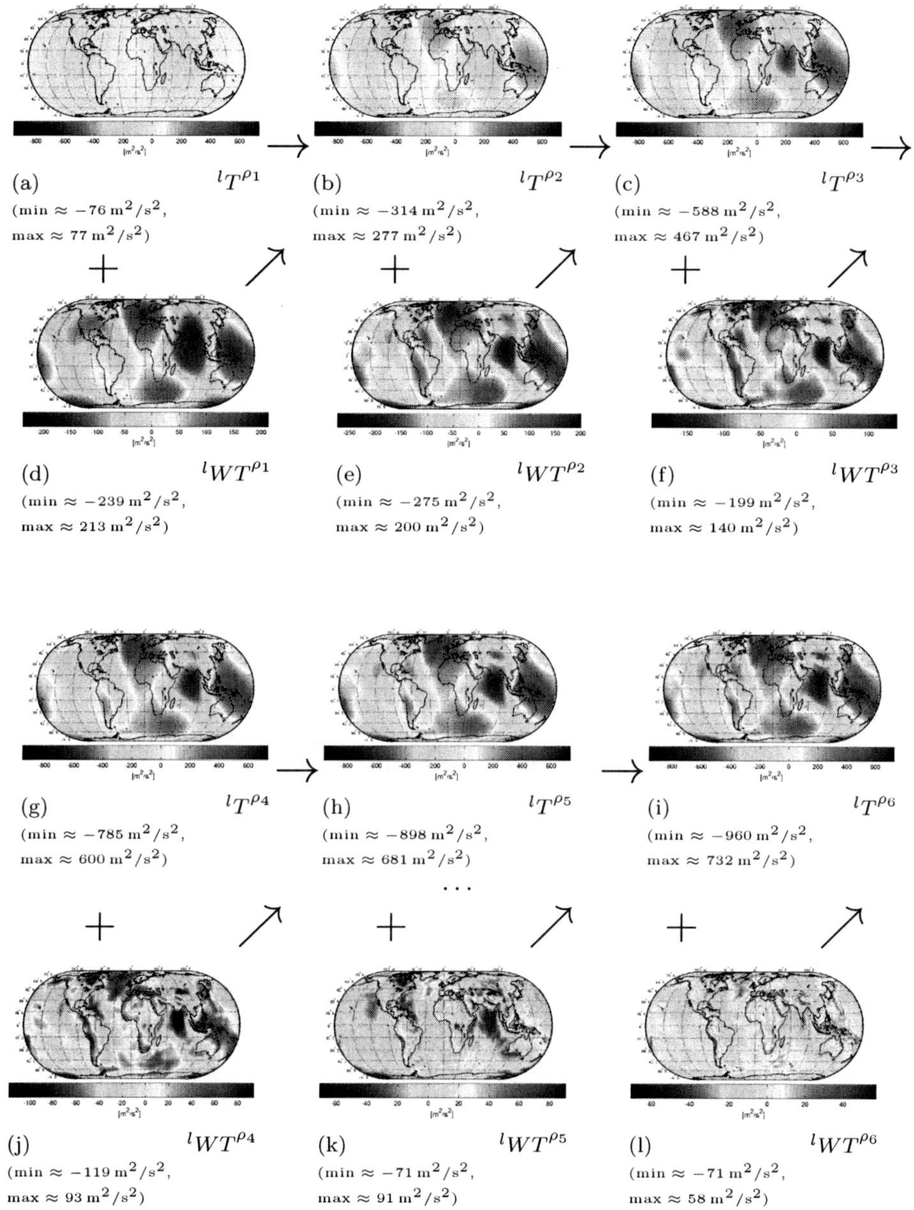

FIGURE 7.9
Illustration of a (global) multiscale approximation of the Earth's disturbing potential in $[\frac{m^2}{s^2}]$ from gravity disturbances, i.e., low-pass filtered versions $^{l}T^{\rho_J}$ and detail information (band-pass filtered versions) $^{l}WT^{\rho_j}$ for scales $j = 1, \ldots, 6$, by use of the linear Neumann scaling functions and wavelets computed from 4,000,000 data distributed over the whole sphere Ω_R.

In order to get a fully discretized solution of the Neumann boundary-value problem (ENPPG), approximate integration by use of appropriate cubature formulas is necessary (see, e.g., W. Freeden, T. Gervens, M. Schreiner [1998], K. Hesse et al. [2010] for more details about approximate integration on the (unit) sphere). The fully discretized multiscale approximations have the following representations

$$ {}^{l}T^{\rho_J}(R\xi) \simeq \frac{R}{4\pi} \sum_{k=1}^{N_J} w_k^{N_J} D\left(R\eta_k^{N_J}\right) \, {}^{l}N^{\rho_J}\left(\xi \cdot \eta_k^{N_J}\right), \quad \xi \in \Omega, \tag{7.90} $$

$$ {}^{l}WT^{\rho_j}(R\xi) \simeq \frac{R}{4\pi} \sum_{k=1}^{N_j} w_k^{N_j} D\left(R\eta_k^{N_j}\right) \, {}^{l}WN^{\rho_j}\left(\xi \cdot \eta_k^{N_j}\right), \quad \xi \in \Omega, \tag{7.91} $$

where $\eta_k^{N_j}$ are the integration knots and $w_k^{N_j}$ the integration weights. Whereas the sum in (7.90) has to be extended over the whole sphere Ω, the summation in (7.91) has only to be computed for the local supports of the wavelets (note that the symbol $\simeq$ means that the error between the right- and the left-hand side can be neglected).

Remark 7.12. In our examples, we have used the algorithm of J.R. Driscoll, R.M. Healy [1994] for the numerical integration in (7.90) and (7.90). Although the knots of this integration method do not generate an equidistribution in the sense of Weyl's Law of Equidistribution (see Weyl [1916], W. Freeden, O. Glockner, M. Schreiner [1999]), this rule combines two advantages, namely a simple structure and a certain degree of (odd) polynomial exactness.

Seen from the geodetic reality, Figures 7.10 to 7.15 are remarkable in the following sense: To get a better accuracy in numerical integration procedures providing the (global) solution of the boundary-value problem (ENPPG) as illustrated in Figure 7.10, we need denser, globally equidistributed data sets over the whole sphere Ω_R (most notably, in the sense of Weyl's Law of Equidistribution). However, in today's reality of gravitational field observation, we are confronted with the problem that terrestrial gravitational data (such as gravity disturbances, gravity anomalies) of sufficient width and quality are only available for certain parts of the Earth's surface (for more details concerning the observational aspects see, e.g., ESA [1996, 1998, 1999], R. Rummel et al. [2002]). As a matter of fact, there are large gaps, particularly at sea, where no data sets of sufficient quality are available at all. This is the reason why the observational situation implies the need for specific geodetically oriented modeling techniques reflecting heterogeneous, local data availability (usually related to latitude–longitude data grids). In this respect, our zooming-in realization based on single-layer space regularization is a suitable efficient and economic mathematical answer.

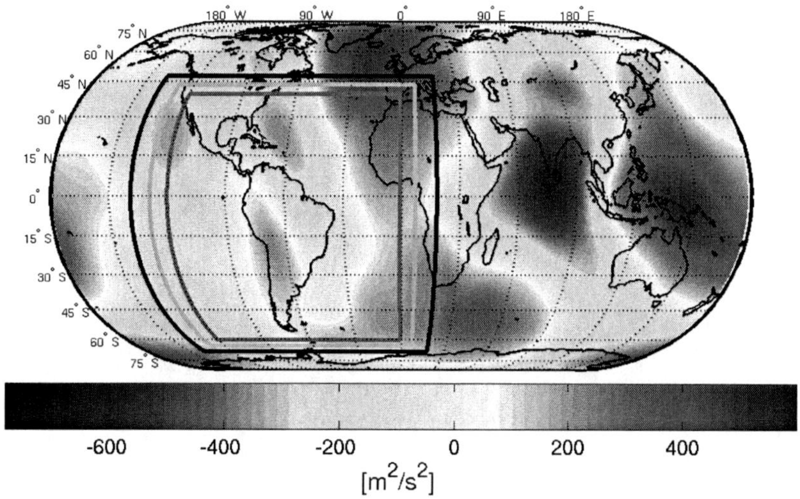

FIGURE 7.10
Low-pass filtered version ${}^{l}T^{\rho_4}$ of the disturbing potential in $[\frac{m^2}{s^2}]$ calculated from 490,000 data points distributed over the whole sphere Ω_R .

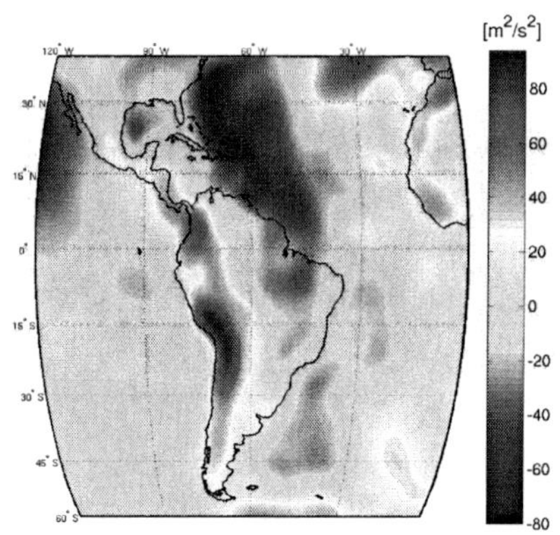

(a) details at scale 4 from 281 428 data points distributed within the outer-bordered subregion in Figure 7.10

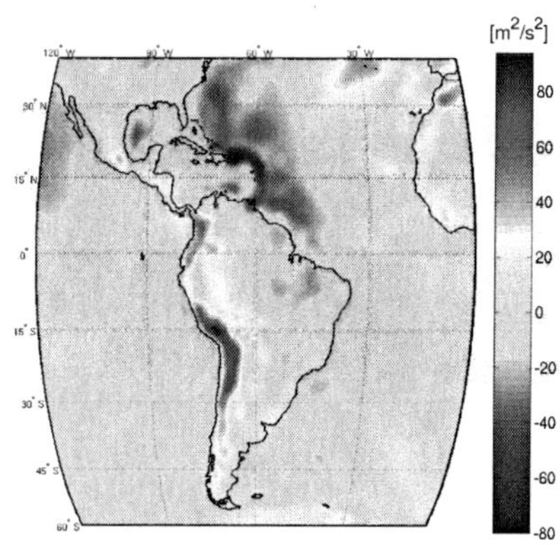

(b) details at scale 5 from 226 800 data points distributed within the mid-bordered subregion in Figure 7.10

FIGURE 7.11
Band-pass filtered versions ${}^{l}WT^{\rho_j}$ of the disturbing potential in $[\frac{m^2}{s^2}]$ for subregions in Figure 7.10 for scales $j = 4, 5$ calculated from different numbers of data points (note that both data sets would correspond to a data set with 1,000,000 data points distributed over the whole sphere Ω_R).

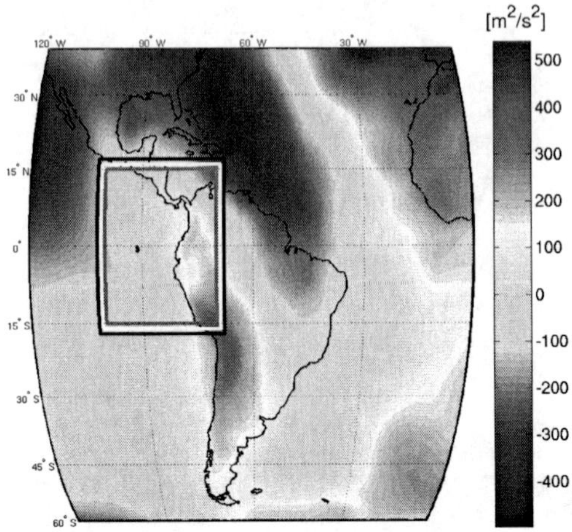

FIGURE 7.12
Low-pass filtered version ${}^{l}T^{\rho_6}$ of the disturbing potential in $[\frac{m^2}{s^2}]$ of the inner-bordered subregion in Figure 7.10 computed by the sum of ${}^{l}T^{\rho_4}$ (Figure 7.10), ${}^{l}WT^{\rho_4}$ (Figure 7.11(a)), and ${}^{l}WT^{\rho_5}$ (Figure 7.11(b)) in this subregion.

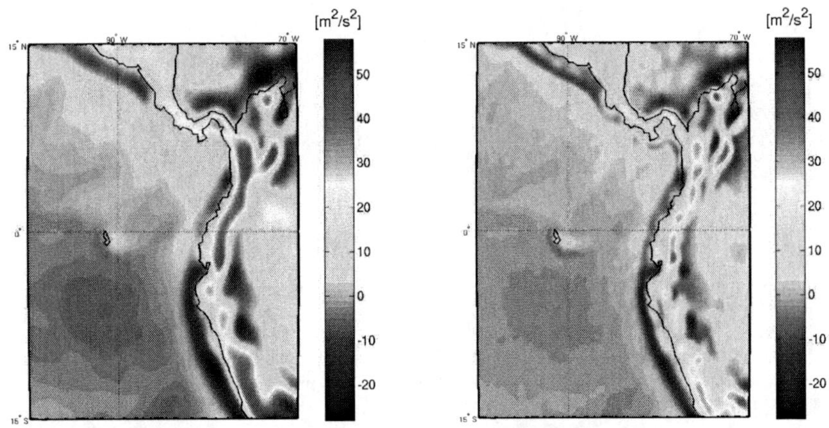

(a) details at scale 6 from 71 253 data points distributed within the outer-bordered subregion in Figure 7.12

(b) details at scale 7 from 63 190 data points distributed within the mid-bordered subregion in Figure 7.12

FIGURE 7.13
Band-pass filtered versions ${}^{l}WT^{\rho_j}$ of the disturbing potential in $[\frac{m^2}{s^2}]$ for subregions in Figure 7.12 for scales $j = 6, 7$ calculated from different numbers of data points (both data sets would correspond to a data set with $4,000,000$ data points distributed over the whole sphere Ω_R).

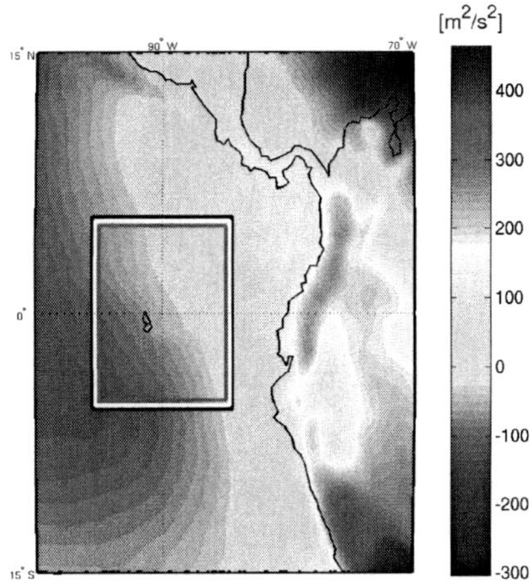

FIGURE 7.14
Low-pass filtered version ${}^{l}T^{\rho_8}$ of the disturbing potential in $[\frac{m^2}{s^2}]$ of the inner-bordered subregion in Figure 7.12 computed by the sum of ${}^{l}T^{\rho_6}$ (Figure 7.12), ${}^{l}WT^{\rho_6}$ (Figure 7.13(a)), and ${}^{l}WT^{\rho_7}$ (Figure 7.13(b)) in this subregion.

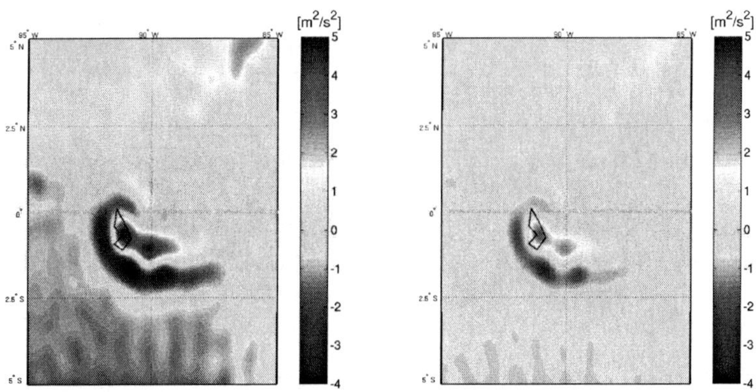

(a) details at scale 8 from 90 951 data points distributed within the outer-bordered subregion in Figure 7.14

(b) details at scale 9 from 85 491 data points distributed within the mid-bordered region in Figure 7.14

FIGURE 7.15
Band-pass filtered versions ${}^{l}WT^{\rho_j}$ of the disturbing potential in $[\frac{m^2}{s^2}]$ for subregions in Figure 7.14 for scales $j = 8, 9$ calculated from different numbers of data points (note that both data sets would correspond to a data set with $49,000,000$ data points distributed over the whole sphere Ω_R).

7.3.2 Disturbing Potential from Deflections of the Vertical

As already known from (7.55), the solution of the surface differential equation

$$\nabla_\xi^* T(R\xi) = -\frac{GM}{R}\Theta(R\xi), \xi \in \Omega, \tag{7.92}$$

determining the disturbing potential T from prescribed deflections of the vertical Θ under the conditions (7.40) is given by

$$T(R\xi) = \frac{R}{4\pi}\int_\Omega \Theta(R\eta)\cdot g\left(\Delta^*;\xi,\eta\right)\, d\omega(\eta), \qquad \xi \in \Omega, \tag{7.93}$$

where the vector kernel $g\left(\Delta^*;\cdot,\cdot\right)$, $1-\xi\cdot\eta > 0$, reads as follows

$$\begin{aligned} g\left(\Delta^*;\xi,\eta\right) &= -\frac{1}{2}\frac{GM}{R^2}\frac{2}{1-\xi\cdot\eta}(\xi-(\xi\cdot\eta)\eta) \\ &= -\frac{1}{2}\frac{GM}{R^2}(S(\xi\cdot\eta))^2(\xi-(\xi\cdot\eta)\eta). \end{aligned} \tag{7.94}$$

Analogous to the determination of the disturbing potential from known gravity disturbances (i.e., the Neumann problem (ENPPG)), the numerical problems of the improper integral in (7.93) can be circumvented by replacing the zonal kernel $S(\xi\cdot\eta)$ with the regularized kernel ${}^lS^\rho(\xi\cdot\eta)$. This process leads to space-regularized representations ${}^lT^\rho$ of the disturbing potential T calculated from deflections of the vertical within a multiscale zooming-in procedure analogous to the approach from Subsection 7.3.1, where gravity disturbances were used as input data. To be more specific, the kernel function $g(\Delta^*;\cdot,\cdot)$ is replaced by the space-regularized function

$$\begin{aligned} &{}^lg^\rho\left(\Delta^*;\xi,\eta\right) \\ &= -\frac{GM}{2R^2}\left({}^lS^\rho(\xi\cdot\eta)\right)^2(\xi-(\xi\cdot\eta)\eta), \\ &= \begin{cases} -\frac{GM}{2R^2}\left(\frac{9R^2}{\rho^2}-\frac{12R^4}{\rho^4}(1-\xi\cdot\eta)\right. \\ \qquad \left.+\frac{4R^6}{\rho^6}(1-\xi\cdot\eta)^2\right)(\xi-(\xi\cdot\eta)\eta), & 0\le 1-\xi\cdot\eta\le\frac{\rho^2}{2R^2}, \\ -\frac{GM}{2R^2}\frac{2}{1-\xi\cdot\eta}(\xi-(\xi\cdot\eta)\eta), & \frac{\rho^2}{2R^2}<1-\xi\cdot\eta\le 2, \end{cases} \end{aligned} \tag{7.95}$$

for $\rho \in (0, 2R]$. This leads to the following approximate representation of the disturbing potential T:

$${}^lT^\rho(R\xi) = \frac{R}{4\pi}\int_\Omega \Theta(R\eta)\cdot {}^lg^\rho\left(\Delta^*;\xi,\eta\right)\, d\omega(\eta), \qquad \xi\in\Omega, \tag{7.96}$$

with ${}^lg^\rho(\Delta^*;\cdot,\cdot)$ given by (7.95).

Theorem 7.13. *Suppose that T is the solution (7.93) of the differential equation (7.92), with Θ being of class $\mathrm{c}^{(0)}(\Omega_R)$. Let ${}^lT^\rho$, $\rho\in(0,2R]$, represent its regularized solution of the form (7.96). Then*

$$\lim_{\rho\to 0+}\sup_{\xi\in\Omega}\left|T(R\xi)-{}^lT^\rho(R\xi)\right| = 0. \tag{7.97}$$

Proof. By inserting (7.93) and (7.96) into the limit relation we obtain

$$\begin{aligned}
&\lim_{\rho\to 0+}\sup_{\xi\in\Omega}\left|T(R\xi) - {}^{l}T^{\rho}(R\xi)\right| \qquad (7.98)\\
&= \lim_{\rho\to 0+}\sup_{\xi\in\Omega}\frac{R}{4\pi}\left|\int_{\Omega}\Theta(R\eta)\cdot\left(g\left(\Delta^*;\xi,\eta\right) - {}^{l}g^{\rho}\left(\Delta^*;\xi,\eta\right)\right)\,d\omega(\eta)\right|.
\end{aligned}$$

The integrands of $T(R\xi)$ and ${}^{l}T^{\rho}(R\xi)$ only differ on the cap $\Gamma_{\rho^2/2R^2}(\xi)$, where

$$\begin{aligned}
&g\left(\Delta^*;\xi,\eta\right) - {}^{l}g^{\rho}\left(\Delta^*;\xi,\eta\right) \qquad (7.99)\\
&= -\frac{GM}{2R^2}\left((S(\xi\cdot\eta))^2 - \left({}^{l}S^{\rho}(\xi\cdot\eta)\right)^2\right)(\xi-(\xi\cdot\eta)\eta)\\
&= -\frac{GM}{2R^2}\left((S(\xi\cdot\eta))^2 - \left({}^{l}S^{\rho}(\xi\cdot\eta)\right)^2\right)\sqrt{1-(\xi\cdot\eta)^2}\,\frac{\xi-(\xi\cdot\eta)\eta}{|\xi-(\xi\cdot\eta)\eta|}.
\end{aligned}$$

Thus, from Lemma 7.6, we obtain the desired result by

$$\lim_{\rho\to 0+}\sup_{\xi\in\Omega}\int_{\Omega}\left|(S(\xi\cdot\eta))^2 - \left({}^{l}S^{\rho}(\xi\cdot\eta)\right)^2\right|\sqrt{1-(\xi\cdot\eta)^2}\,d\omega(\eta) = 0. \qquad (7.100)$$

This completes the proof. □

By restricting $\{{}^{l}g^{\rho}\left(\Delta^*;\cdot,\cdot\right)\}_{\rho\in(0,2R]}$ to the sequence $\{{}^{l}g^{\rho_j}\left(\Delta^*;\cdot,\cdot\right)\}_{j\in\mathbb{N}_0}$, corresponding to a set of scaling parameters $\{\rho_j\}_{j\in\mathbb{N}_0}$ satisfying $\rho_j\in(0,2R]$ and $\lim_{j\to\infty}\rho_j = 0$, we are canonically led to regularized vector scaling functions such that a scale-discrete solution method for the differential equation (7.51) can be formulated. The vector scaling function ${}^{l}g^{\rho_{j+1}}(\Delta^*;\cdot,\cdot)$ at scale $j+1$ is constituted by the sum of the vector scaling function ${}^{l}g^{\rho_j}(\Delta^*;\cdot,\cdot)$ and the corresponding discretized vector wavelet ${}^{l}wg^{\rho_j}(\Delta^*;\cdot,\cdot)$, given by

$$\begin{aligned}
&{}^{l}wg^{\rho_j}\left(\Delta^*;\xi,\eta\right) \qquad (7.101)\\
&= {}^{l}g^{\rho_{j+1}}\left(\Delta^*;\xi,\eta\right) - {}^{l}g^{\rho_j}\left(\Delta^*;\xi,\eta\right)\\
&= \begin{cases}
-\frac{GM}{2R^2}C_{1,j}(\xi\cdot\eta)\;(\xi-(\xi\cdot\eta)\eta), & 0\le 1-\xi\cdot\eta\le\frac{\rho_{j+1}^2}{2R^2},\\
-\frac{GM}{2R^2}C_{2,j}(\xi\cdot\eta)\;(\xi-(\xi\cdot\eta)\eta), & \frac{\rho_{j+1}^2}{2R^2}<1-\xi\cdot\eta\le\frac{\rho_j^2}{2R^2},\\
0, & \frac{\rho_j^2}{2R^2}<1-\xi\cdot\eta\le 2,
\end{cases}
\end{aligned}$$

where we have used the abbreviations

$$\begin{aligned}
C_{1,j}(\xi\cdot\eta) &= \left(\frac{9R^2}{\rho_{j+1}^2}-\frac{9R^2}{\rho_j^2}\right) - \left(\frac{12R^4}{\rho_{j+1}^4}-\frac{12R^4}{\rho_j^4}\right)(1-\xi\cdot\eta) \qquad (7.102)\\
&\quad + \left(\frac{4R^6}{\rho_{j+1}^6}-\frac{4R^6}{\rho_j^6}\right)(1-\xi\cdot\eta)^2
\end{aligned}$$

and

$$C_{2,j}(\xi\cdot\eta) = \frac{2}{1-\xi\cdot\eta} - \frac{9R^2}{\rho_j^2} + \frac{12R^4}{\rho_j^4}(1-\xi\cdot\eta) - \frac{4R^6}{\rho_j^6}(1-\xi\cdot\eta)^2. \qquad (7.103)$$

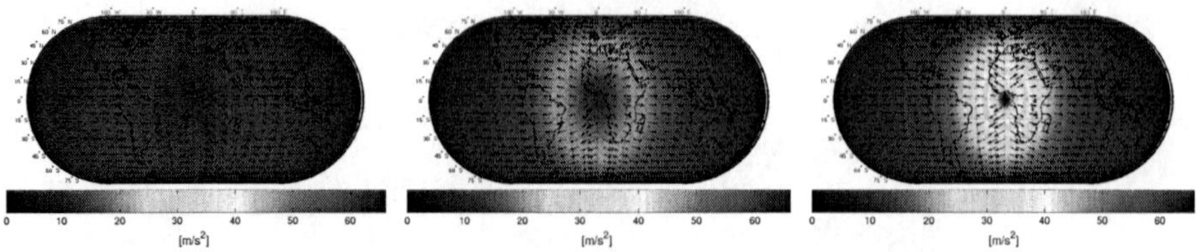

(a) scale $j = 0$ (min $= 0\,\mathrm{m/s^2}$, max $\approx 8\,\mathrm{m/s^2}$) (b) scale $j = 1$ (min $= 0\,\mathrm{m/s^2}$, max $\approx 20\,\mathrm{m/s^2}$) (c) scale $j = 2$ (min $= 0\,\mathrm{m/s^2}$, max $\approx 43\,\mathrm{m/s^2}$)

FIGURE 7.16
Absolute values and the directions of the linearly regularized vector scaling function $\eta \mapsto {}^l g^{\rho_j}(\Delta^*; \xi, \eta)$, $\xi, \eta \in \Omega$ using the single-layer kernel with ξ fixed, for different scales.

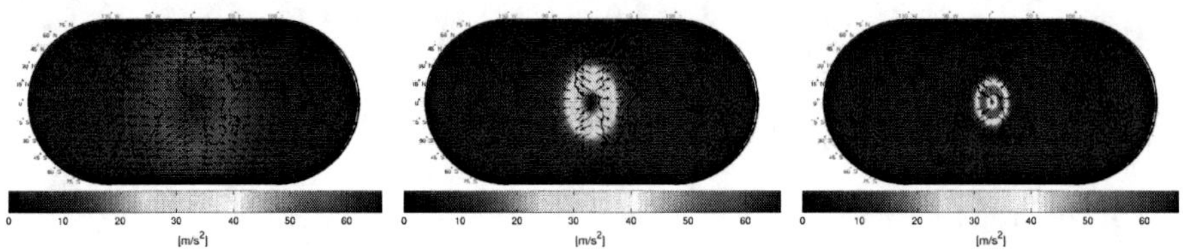

(a) scale $j = 0$ (min $= 0\,\mathrm{m/s^2}$, max $\approx 14\,\mathrm{m/s^2}$) (b) scale $j = 1$ (min $= 0\,\mathrm{m/s^2}$, max $\approx 29\,\mathrm{m/s^2}$) (c) scale $j = 2$ (min $= 0\,\mathrm{m/s^2}$, max $\approx 58\,\mathrm{m/s^2}$)

FIGURE 7.17
Absolute values and the directions of the linearly regularized vector wavelet function $\eta \mapsto {}^l wg^{\rho_j}(\Delta^*; \xi, \eta)$, $\xi, \eta \in \Omega$ using the single-layer kernel with ξ fixed, for different scales.

The vector scaling and wavelet kernels are graphically illustrated in Figures 7.16 and 7.17. All in all, we can formulate the following multiscale reconstruction formula for the disturbing potential.

Theorem 7.14. *Suppose that T is the solution (7.93) of the differential equation (7.92), with Θ being of class $\mathrm{c}^{(0)}(\Omega_R)$. Let ${}^l T^{\rho_{J_0}}$ be the regularized version of the disturbing potential at an initial scale $J_0 \in \mathbb{N}_0$. Moreover, let the wavelet contribution ${}^l WT^{\rho_j}$, $j = 0, 1, \ldots$, be given by*

$$ {}^l WT^{\rho_j}(R\xi) = \frac{R}{4\pi} \int\limits_{\Gamma_{\rho_j^2/2R^2}(\xi)} \Theta(R\eta) \cdot {}^l wg^{\rho_j}(\Delta^*; \xi, \eta)\, d\omega(\eta), \quad \xi \in \Omega, \tag{7.104} $$

where ${}^l wg^{\rho_j}$ is of the form (7.101). Then

$$ \lim_{N\to\infty} \sup_{\xi\in\Omega} \left| T(R\xi) - \left({}^l T^{\rho_{J_0}}(R\xi) + \sum_{j=0}^{N} {}^l WT^{\rho_{J_0+j}}(R\xi) \right) \right| = 0. \tag{7.105} $$

7.3.3 Gravitational Signatures of Mantle Plumes

Nowadays, the concept of mantle plumes is widely accepted in the geoscientific community. Mantle plumes are understood to be approximately cylindrical concentrated upflows of hot mantle material with a common diameter of about 100–200 km. They are an upwelling of abnormally hot rock within the Earth's mantle. As the heads of mantle plumes can partly melt when they reach shallow depths, they are thought to be the cause of volcanic centers known as *hotspots*. Hotspots were first explained by J. Wilson [1963] as long-term sources of volcanism that are fixed relative to the tectonic plate overriding them. Following W.J. Morgan [1971], characteristic surface signatures of hotspots are due to the rise and melting of hot plumes from deep areas in the mantle. Special cases occur as chains of volcanic edifices whose age progresses with increasing distance to the plume, like the islands of Hawaii. They are the result of a pressure-release melting near the bottom of the lithosphere that produces magma rising to the surface and by plate motion relative to the plume. The term *hotspot* is used rather loosely. It is often applied to any long-lived volcanic center that is not part of the global network of mid-ocean ridges and island arcs, like Hawaii, which serves as a classical example. Anomalous regions of thick crust on ocean ridges are also considered to be hotspots, like Iceland.

Hawaii: J.R.R. Ritter, U.R. Christensen [2007] believe that a stationary mantle plume located beneath the Hawaiian Islands created the Hawaii-Emperor seamount chain while the oceanic lithosphere continuously passed over it. The Hawaii-Emperor chain consists of about 100 volcanic islands, atolls, and seamounts that spread nearly 6000 km from the active volcanic island of Hawaii to the 75–80 million year old Emperor seamounts nearby the Aleutian trench. Moving farther southeast along the island chain, the geological age decreases. The interesting area is the relatively young southeastern part of the chain, situated on the Hawaiian swell, a 1200-km-broad anomalously shallow region of the ocean floor, extending from the island of Hawaii to the Midway atoll. Here, a distinct gravity disturbance and geoid anomaly occur that has its maximum around the youngest island that coincides with the maximum topography and both decrease in a northwestern direction. The progressive decrease in terms of the geological age is believed to result from the continuous motion of the underlying plate (cf. W.J. Morgan [1971], J. Wilson [1963]).

With seismic tomography, several features of the Hawaiian mantle plume are gained (cf. J.R.R. Ritter, U.R. Christensen [2007] and the references specified therein). They result in a *low-velocity zone* (LVZ) beneath the lithosphere, starting at a depth of about 130–140 km beneath the central part of the island of Hawaii. So far, plumes have just been identified as low-seismic-velocity anomalies in the upper mantle and the transition zone, which is a fairly new achievement. Because plumes are relatively thin according to their diameter, they are hard to detect in global tomography models. Hence, despite novel advances, there is still no general agreement on the fundamental questions concerning mantle plumes, like their depth of origin, their morphology, their longevity, and even their existence is still controversial. This is due to the fact that many geophysical as well as geochemical observations can be explained by different plume models and even by models that do not include plumes at all (e.g., G. Foulger et al. [2005]). With our space-localized multiscale method of deriving gravitational signatures (more specifically, the disturbing potential) from the deflections of the vertical, we add a new component in specifying essential features of plumes. The deflections of the vertical of the plume in the region of Hawaii are visualized in Figures 7.18 and 7.19. From the band-pass filtered detail approximation we are able to conclude that the Hawaii plume has an oblique layer structure. As can be seen in the lower scale (for which numerical evidence suggests that they reflect the higher depths), the strongest signal is located in the ocean west of Hawaii. With increasing scale, i.e., lower depths, it moves more and more to the Big Island of Hawaii, i.e., in an eastward direction.

Iceland: The plume beneath Iceland is a typical example of a ridge-centered mantle plume. An interaction between the North Atlantic ridge and the mantle plume is believed to be the reason for the existence of Iceland, resulting in melt production and crust generation since the continental breakup in the late Palaeocene and early Eocene. Nevertheless, there is still no agreement on the location of the plume before rifting started in the East. Controversial discussions concerning whether it was located under central or eastern Greenland about 62–64 million years ago are still in progress (cf. D. Schubert et al. [2001] and the references therein). Iceland itself represents the top of a nearly circular rise topography, with a maximum of about 2.8 km above the surrounding seafloor in the south of the glacier Vatnajökull. Beneath this glacier, several active volcanoes are located, which are supposed to be fed by a mantle plume. The surrounding oceanic crust consists of three different types involving a crust thickness that is more than three times as thick as average oceanic crusts. Seismic tomography provides evidence of the existence of a mantle plume beneath Iceland, resulting in low-velocity zones in the upper mantle and the transition zone, but also hints of anomalies in the deeper mantle seem to exist. The low-velocity anomalies have been detected in depths ranging from at least 400 km up to about 150 km. Above 150 km, ambiguous seismic-velocity structures were obtained involving regions of low velocities covered by regions of high seismic velocities. For a deeper access into the theory of the Iceland plume, the interested reader may be referred to J.R.R. Ritter, U.R. Christensen [2007] and the references therein.

From Figures 7.20 and 7.21 it can be seen that the mantle plume in lower scales, i.e., in higher depths, starts in the north of Iceland and with increasing scale, i.e., lower depths, it moves to the south. It is remarkable that from scale 13 on, the plume seems to divide into two sectors. Since it is known that the disturbing potential of the Earth is influenced by its topography, a

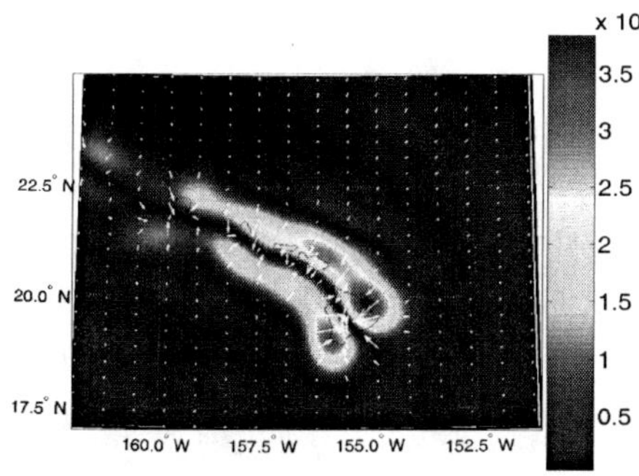

FIGURE 7.18
Illustration of the deflections of the vertical in the region of Hawaii.

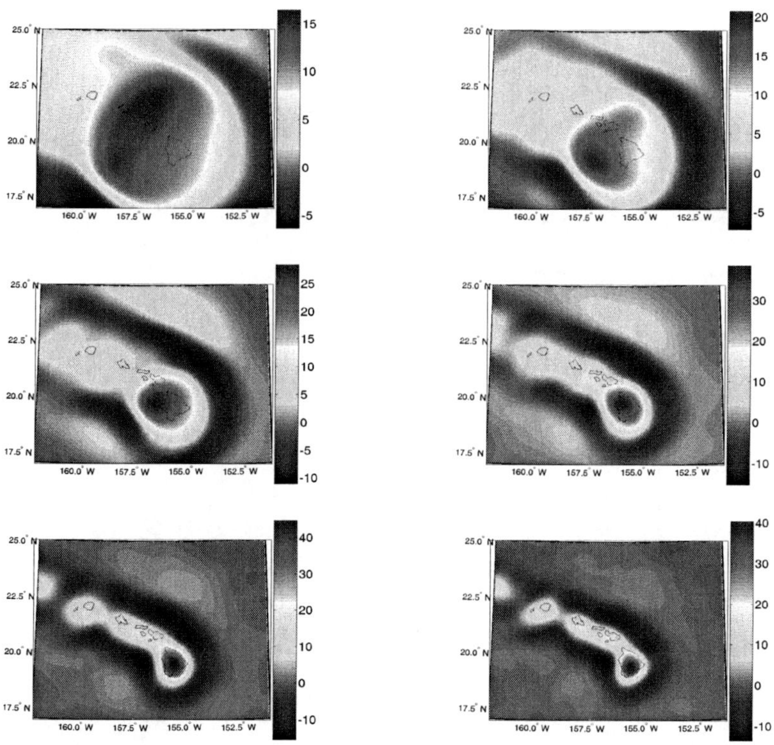

FIGURE 7.19
Band-pass filtered details ${}^{l}WT^{\rho_j}$ of the disturbing potential in $[\frac{m^2}{s^2}]$ in the region of Hawaii with respect to scales $j = 9, 11, 13$ (left, top to bottom) and $j = 10, 12, 14$ (right, top to bottom).

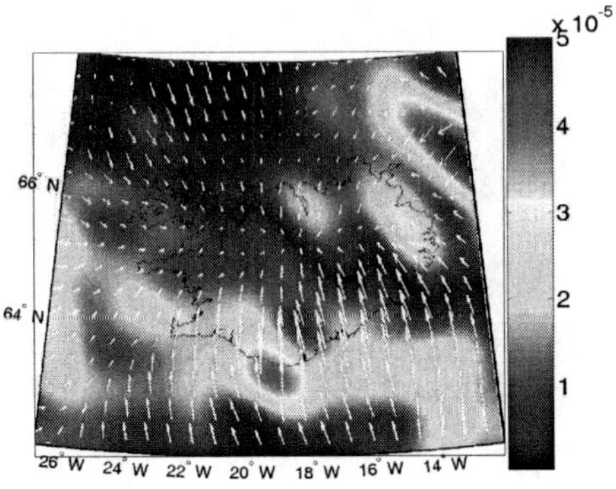

FIGURE 7.20
Illustration of the deflections of the vertical in the region of Iceland.

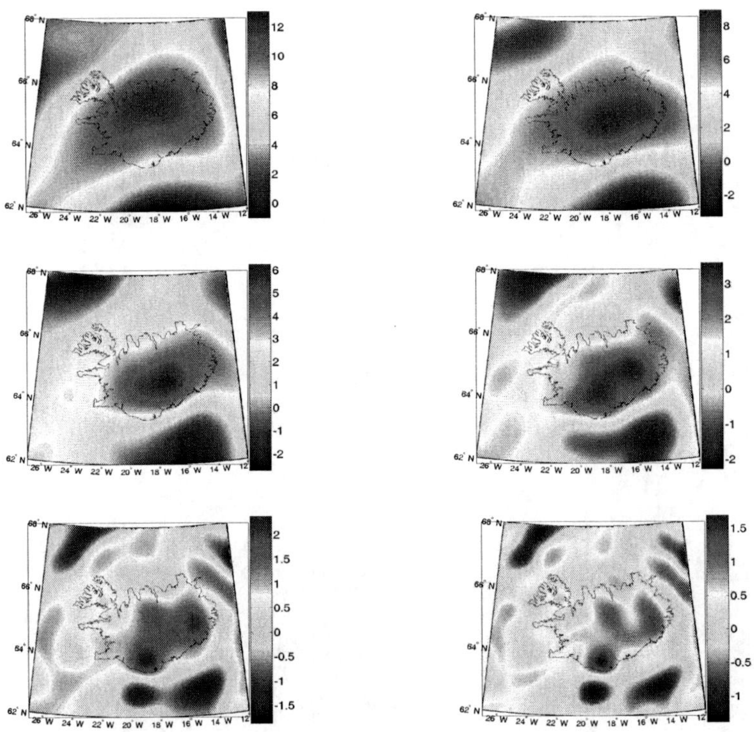

FIGURE 7.21
Band-pass filtered details ${}^{l}WT^{\rho_j}$ of the disturbing potential in $[\frac{m^2}{s^2}]$ in the region of Iceland with respect to the scales $j = 10, 12, 14$ (left, top to bottom) and $j = 11, 13, 15$ (right, top to bottom).

look at a topographic map shows that the sector located farther to the east is (probably) caused by the Vatnajökull glacier (the biggest glacier in Europe).

Altogether, our multiscale method offers an additional component in characterizing mantle plumes with respect to their position and horizontal size. A serious description of their depth, however, cannot be realized. The detailed detection of the depths remains a great challenge for future work (for more details the reader is referred to the case studies due to T. Fehlinger [2009]), in which also gravimetric investigations should be taken into account. It seems that only a combined investigation of all available techniques is well-promising.

7.4 Exercises

Exercise 7.1. Prove the following limit relation: For $F \in \mathrm{C}^{(0)}(\Omega)$ and ${}^{l}S^{\rho}$ as defined by (7.61), we have

$$\lim_{\rho \to 0+} \sup_{\xi \in \Omega} \left| \int_{\Omega} \ln(1 + S(\xi \cdot \eta))F(\eta)\, d\omega(\eta) - \int_{\Omega} \ln(1 + {}^{l}S^{\rho}(\xi \cdot \eta))F(\eta)\, d\omega(\eta) \right| = 0. \tag{7.106}$$

Exercise 7.2. The *Stokes boundary-value problem of physical geodesy* is concerned with the determination of the Earth's disturbing potential T from known gravity anomalies A (see (7.35)). Its potential theoretic formulation reads as follows: Given $A \in \mathrm{C}^{(0)}(\Omega_R)$ with

$$A^{\wedge_{\mathrm{L}^2(\Omega_R)}}(n,k) = \int_{\Omega_R} A(y) H^R_{-n-1,k}(y)\, d\omega(y) = 0, \tag{7.107}$$

$n = 0, 1,\ k = 1, \ldots, 2n+1$. Find $T \in \mathrm{Pot}^{(1)}(\mathbb{R}^3 \backslash \mathcal{B}_R(0))$ corresponding to the boundary data

$$\frac{x}{|x|} \cdot (\nabla T)(x) + \frac{2}{|x|} T(x) = -A(x), \qquad x \in \Omega_R, \tag{7.108}$$

where

$$T^{\wedge_{\mathrm{L}^2(\Omega_R)}}(n,k) = \int_{\Omega_R} T(y) H^R_{-n-1,k}(y)\, d\omega(y) = 0, \tag{7.109}$$

$n = 0, 1,\ k = 1, \ldots, 2n+1$.

(a) Show that the Stokes-boundary value problem is uniquely solvable.

(b) Verify that the unique solution allows the outer harmonic expansion

$$T(x) = R \sum_{n=2}^{\infty} \sum_{k=1}^{2n+1} \frac{1}{n-1} A^{\wedge_{\mathrm{L}^2(\Omega_R)}}(n,k) H^R_{-n-1,k}(x), \quad x \in \mathbb{R}^3 \backslash \mathcal{B}_R(0). \tag{7.110}$$

(c) For $t \in [-1,1]$, $s \in (0,1)$, we may assume that the Legendre expansions

$$\sum_{n=2}^{\infty} s^n P_n(t) = \frac{1}{\sqrt{1+s^2-2st}} - 1 - st \tag{7.111}$$

and

$$\begin{aligned}\sum_{n=2}^{\infty} \frac{1}{n-1} s^{n-1} P_n(t) \\ = \frac{1}{s}\left(1 - st - \sqrt{1+s^2-2st}\right) - t \ln\left(\frac{\sqrt{1+s^2-2st}+1-st}{2}\right)\end{aligned} \tag{7.112}$$

are valid. Use (7.111) and (7.112) to derive the following integral representation for the solution of the Stokes boundary-value problem on Ω_R:

$$T(x) = \frac{1}{4\pi R} \int_{\Omega_R} A(y) St(x,y)\, d\omega(y), \quad x \in \mathbb{R}^3 \backslash \mathcal{B}_R(0), \tag{7.113}$$

where the so-called *Stokes kernel* $St(\cdot,\cdot)$ is given by

$$\begin{aligned} St(x,y) &= \frac{R}{|x|} + \frac{2R}{|x-y|} - \frac{5R^2}{|x|^2}\frac{x}{|x|}\cdot\frac{y}{|y|} - \frac{3R}{|x|^2}|x-y| \\ &\quad -3\frac{R^2}{|x|^2}\frac{x}{|x|}\cdot\frac{y}{|y|} \ln\left(\frac{|x| - R\frac{x}{|x|}\cdot\frac{y}{|y|} + |x-y|}{2|x|}\right). \end{aligned} \tag{7.114}$$

Exercise 7.3. Use the integral representation (7.113) on Ω_R to establish the following integral representation on Ω

$$T(R\xi) = \frac{R}{4\pi} \int_{\Omega} A(R\eta) St(\xi\cdot\eta)\, d\omega(\eta), \quad \xi \in \Omega, \tag{7.115}$$

where the Stokes kernel on Ω is given by

$$\begin{aligned} St(\xi\cdot\eta) &= 1 + \frac{\sqrt{2}}{\sqrt{1-\xi\cdot\eta}} - 5\xi\cdot\eta - 6\frac{\sqrt{1-\xi\cdot\eta}}{\sqrt{2}} \\ &\quad -3\xi\cdot\eta \ln\left(\frac{1-\xi\cdot\eta}{2} + \frac{\sqrt{1-\xi\cdot\eta}}{\sqrt{2}}\right) \end{aligned} \tag{7.116}$$

(note that, in this case, we have $St(R\xi, R\eta) = St(\xi\cdot\eta)$, $\xi,\eta \in \Omega$, $\xi \neq \eta$, i.e., the Stokes kernel is a zonal function).

Exercise 7.4. Show that the (Taylor) linearized Stokes kernel can be represented in the form

$$
\begin{aligned}
&{}^{l}St^{\rho}(\xi\cdot\eta) && (7.117)\\
&= 1-5\xi\cdot\eta-6\frac{1}{S(\xi\cdot\eta)}+{}^{l}S^{\rho}(\xi\cdot\eta)-3\xi\cdot\eta\ln\left(\frac{1}{({}^{l}S^{\rho}(\xi\cdot\eta))^{2}}+\frac{1}{{}^{l}S^{\rho}(\xi\cdot\eta)}\right)\\
&= 1-5\xi\cdot\eta-6\frac{\sqrt{1-\xi\cdot\eta}}{\sqrt{2}}\\
&\quad+\begin{cases}
\frac{R}{\rho}\left(3-\frac{2R^2}{\rho^2}(1-\xi\cdot\eta)\right) & \\
-3\xi\cdot\eta\ln\left(1+\frac{R}{\rho}\left(3-\frac{2R^2}{\rho^2}(1-\xi\cdot\eta)\right)\right) & \\
+6\xi\cdot\eta\ln\left(\frac{R}{\rho}\left(3-\frac{2R^2}{\rho^2}(1-\xi\cdot\eta)\right)\right), & 0\le 1-\xi\cdot\eta\le\frac{\rho^2}{2R^2},\\
\frac{\sqrt{2}}{\sqrt{1-\xi\cdot\eta}}-3\xi\cdot\eta\ln\left(\frac{1-\xi\cdot\eta}{2}+\frac{\sqrt{1-\xi\cdot\eta}}{\sqrt{2}}\right), & \frac{\rho^2}{2R^2}<1-\xi\cdot\eta\le 2,
\end{cases}
\end{aligned}
$$

for $\xi,\eta\in\Omega$ and $\rho\in(0,2R]$.

Exercise 7.5. Let

$$
{}^{l}T^{\rho}(R\xi)=\frac{R}{4\pi}\int_{\Omega}A(R\eta)\,{}^{l}St^{\rho}(\xi\cdot\eta)d\omega(\eta) \tag{7.118}
$$

be the linear regularization of T. Verify that the multiscale reconstruction formula

$$
\lim_{N\to\infty}\sup_{\xi\in\Omega}\left|T(R\xi)-\left({}^{l}T^{\rho_{J_0}}(R\xi)+\sum_{j=0}^{N}{}^{l}WT^{\rho_{J_0+j}}(R\xi)\right)\right|=0 \tag{7.119}
$$

holds true for the solution T of the Stokes boundary-value problem. The wavelet contributions ${}^{l}WT^{\rho_{J_0+j}}(R\xi)$ are defined canonically as in the the case of the Neumann boundary-value problem (ENPPG).

8

Geomagnetism

CONTENTS

8.1 Mie and Helmholtz Decomposition

We revisit both the Helmholtz and the Mie decomposition from Section 5.2 in a spherical framework. While the Helmholtz decomposition can be transferred to a representation intrinsic to spherical vector fields, the Mie decomposition remains a decomposition in the Euclidean space $\mathbb{R}^3$. Yet, the poloidal and toroidal scalars allow a representation in terms of spherical tools like the fundamental solution for the Beltrami operator. Furthermore, we distinguish the two cases of global decompositions on the entire sphere Ω and spherical shells $\mathcal{B}_{R_0,R_1}(0)$, respectively, as well as local decompositions on regular surfaces $\Gamma \subset \Omega$ and conical shells $\mathcal{C}_{R_0,R_1}(\Gamma) \subset \mathcal{B}_{R_0,R_1}(0)$, respectively. With regard to the good data coverage by satellites, the global decompositions are often sufficient for the modeling aspects under consideration. The local framework is of interest whenever we are concerned with a small modeling region and/or if data is only locally available. One should be aware of the fact that the local decompositions usually require the availability of boundary data.

8.1.1 Global Decompositions

The classical Helmholtz decomposition

$$f(x) = \nabla U(x) + \nabla \wedge v(x) \tag{8.1}$$

of a continuously differentiable vector field $f : \mathbb{R}^3 \to \mathbb{R}^3$ with sufficient asymptotic decay at infinity can be reformulated as

$$\begin{aligned} f(x) &= \nabla_x U(x) + \nabla_x \wedge (xV(x)) + \nabla_x \wedge \nabla_x \wedge (xW(x)) \\ &= -x\Delta_x W(x) + \nabla_x \left(U(x) + \frac{\partial}{\partial |x|} (|x|W(x)) \right) - \mathrm{L}_x V(x), \end{aligned} \tag{8.2}$$

for adequate functions $U, V, W : \mathbb{R}^3 \to \mathbb{R}$ (see, e.g., D.P. Stern [1976]). Remembering the connection of the spherical operators L^* and ∇^* to the Euclidean operators L and ∇, Equation (8.2) turns into

$$f(x) = \xi F_1(r\xi) + \nabla_\xi^* F_2(r\xi) + \mathrm{L}_\xi^* F_3(r\xi), \tag{8.3}$$

for $r = |x|$, $\xi = \frac{x}{|x|}$, with F_1, F_2, F_3 depending on U, V, W. In other words, the vector field f can be decomposed into a radial component and two tangential components that only require the use of the spherical operators ∇^* and L^*. Thus, the decomposition (8.3) can be formulated for spherical vector fields $f : \Omega \to \mathbb{R}^3$ and is then called the (spherical) Helmholtz decomposition (see, e.g., G.E. Backus et al. [1996] and W. Freeden, M. Schreiner [2009]).

Theorem 8.1 (Spherical Helmholtz Decomposition). *Let the vector field f be of class $c^{(1)}(\Omega)$. Then there exist scalar fields F_1 of class $\mathrm{C}^{(1)}(\Omega)$ and F_2, F_3 of class $\mathrm{C}^{(2)}(\Omega)$ such that*

$$f(\xi) = o^{(1)} F_1(\xi) + o^{(2)} F_2(\xi) + o^{(3)} F_3(\xi), \quad \xi \in \Omega. \tag{8.4}$$

The scalars F_1, F_2, F_3 are uniquely determined by the properties

$$\frac{1}{4\pi} \int_\Omega F_2(\eta) d\omega(\eta) = \frac{1}{4\pi} \int_\Omega F_3(\eta) d\omega(\eta) = 0. \tag{8.5}$$

Proof. Setting $F_1(\xi) = \xi \cdot f(\xi)$, $\xi \in \Omega$, it is clear that $f_{\mathrm{tan}} = f - o^{(1)} F_1$ is a tangential vector field. Since $\mathrm{L}^* \cdot f_{\mathrm{tan}}$ has a vanishing integral mean value on Ω, Remark 6.15 tells us that there exists a uniquely determined solution F_3 of class $\mathrm{C}^{(2)}(\Omega)$ such that

$$\Delta^* F_3(\xi) = \mathrm{L}^* \cdot f_{\mathrm{tan}}(\xi), \quad \xi \in \Omega, \tag{8.6}$$

and $\frac{1}{4\pi} \int_\Omega F_3(\eta) d\omega(\eta) = 0$. Rewriting Equation (8.6), we obtain

$$\mathrm{L}^* \cdot (f_{\mathrm{tan}}(\xi) - \mathrm{L}^* F_3(\xi)) = 0, \quad \xi \in \Omega. \tag{8.7}$$

Then we know from Theorem 6.62 that a function F_2 of class $\mathrm{C}^{(2)}(\Omega)$ exists such that

$$\nabla^* F_2(\xi) = f_{\tan}(\xi) - \mathrm{L}^* F_3(\xi), \quad \xi \in \Omega. \tag{8.8}$$

The Helmholtz scalar F_2 is uniquely determined under the assumption of a vanishing mean value $\frac{1}{4\pi}\int_\Omega F_2(\eta)d\omega(\eta) = 0$. This completes the proof of the decomposition. □

Remembering the original definition of the vector spherical harmonics $y^{(i)}_{n,k}$, $i = 1, 2, 3$, in Section 2.5, we notice that a Fourier expansion in terms of this orthonormal basis of $\mathrm{l}^2(\Omega)$ corresponds to the spherical Helmholtz decomposition (8.4). More precisely, the part of the Fourier expansion generated by $y^{(i)}_{n,k}$, $n \in \mathbb{N}_{0_i}$, $k = 1, \ldots, 2n+1$, simply denotes the quantity $o^{(i)}F_i$. By use of (8.6), (8.8), and the integral formulas from Theorems 6.2, 6.4, we find integral representations of the Helmholtz scalars F_1, F_2, F_3, which are of particular interest for numerical evaluations.

Lemma 8.2. *Let f be of class $\mathrm{c}^{(1)}(\Omega)$, and F_1, F_2, F_3 be the Helmholtz scalars known from Theorem 8.1. Then*

$$F_2(\xi) = -\int_\Omega \left(\nabla^*_\eta G(\Delta^*; \xi\cdot\eta)\right) \cdot f(\eta)d\omega(\eta), \quad \xi \in \Omega, \tag{8.9}$$

$$F_3(\xi) = -\int_\Omega \left(\mathrm{L}^*_\eta G(\Delta^*; \xi\cdot\eta)\right) \cdot f(\eta)d\omega(\eta), \quad \xi \in \Omega. \tag{8.10}$$

Additionally, if $\frac{1}{4\pi}\int_\Omega F_1(\eta)d\omega(\eta) = 0$ (as it is the case for functions f satisfying the pre-Maxwell equations), then

$$F_1(\xi) = \xi\cdot f(\xi) = \Delta^*_\xi \int_\Omega G(\Delta^*; \xi\cdot\eta)\,\eta\cdot f(\eta)d\omega(\eta), \quad \xi \in \Omega. \tag{8.11}$$

The above representations of the Helmholtz scalars require knowledge of the function f itself. A simple application of Green's formulas as described in Section 2.2 allows representations from the knowledge of $\nabla^*\cdot f$, $\mathrm{L}^*\cdot f$, and the radial part of f only. For tangential vector fields, it results in the well-known fact that such vector fields can be reconstructed from their surface divergence and surface curl divergence. This observation has immediate applications in the modeling of tangential ionospheric current systems. Our approach is treated in some more detail in Section 8.4.

Corollary 8.3. *Let $f \in \mathrm{c}^{(1)}(\Omega)$ be tangential, i.e., $\xi\cdot f(\xi) = 0$, $\xi \in \Omega$. Then the Helmholtz scalars F_2, F_3 known from Theorem 8.1 can be represented in the form*

$$F_2(\xi) = \int_\Omega G(\Delta^*; \xi\cdot\eta)\nabla^*_\eta \cdot f(\eta)d\omega(\eta), \quad \xi \in \Omega, \tag{8.12}$$

$$F_3(\xi) = \int_\Omega G(\Delta^*; \xi\cdot\eta)\mathrm{L}^*_\eta \cdot f(\eta)d\omega(\eta), \quad \xi \in \Omega. \tag{8.13}$$

Next, we go back to the Mie decomposition stated in Theorem 5.10 since we now have the spherical tools for a rigorous proof at hand (see also G.E. Backus et al. [1996]).

Theorem 8.4 (Mie Decomposition). *Let $f : \mathcal{B}_{R_0,R_1}(0) \to \mathbb{R}^3$ be a solenoidal vector field. Then there exist scalar fields P, Q of class $\mathrm{C}^{(1)}(\mathcal{B}_{R_0,R_1}(0))$ such that*

$$f(x) = \nabla \wedge \mathrm{L}P(x) + \mathrm{L}Q(x), \quad x \in \mathcal{B}_{R_0,R_1}(0). \tag{8.14}$$

P, Q are determined uniquely by the additional conditions

$$\frac{1}{4\pi r^2}\int_{\Omega_r} P(y)d\omega(y) = \frac{1}{4\pi r^2}\int_{\Omega_r} Q(y)d\omega(y) = 0, \tag{8.15}$$

for every $r \in (R_0, R_1)$.

Proof. From Theorem 8.1, we obtain the Helmholtz decomposition $f(x) = \xi F_1(r\xi) + \nabla^*_\xi F_2(r\xi) + \mathrm{L}^*_\xi F_3(r\xi)$, where $\xi = \frac{x}{|x|}$, $r = |x| \in (R_0, R_1)$. Using that f is divergence-free (since it is solenoidal), we can deduce that

$$\frac{\partial}{\partial r}(r^2 F_1(r\xi)) + r\Delta^*_\xi F_2(r\xi) = 0, \tag{8.16}$$

for $\xi \in \Omega$, $r \in (R_0, R_1)$. Let us assume for a second that the desired scalars P, Q exist, such that $f(x) = \nabla \wedge \mathrm{L}P(x) + \mathrm{L}Q(x)$, $x \in \mathcal{B}_{R_0,R_1}(0)$. Then we get (as we see later on in Lemma 8.6) that

$$f(x) = \xi\left(\frac{\Delta^*_\xi P(r\xi)}{r}\right) - \nabla^*_\xi\left(\frac{1}{r}\frac{\partial}{\partial r}(rP(r\xi))\right) + \mathrm{L}^*_\xi Q(r\xi), \tag{8.17}$$

for $\xi = \frac{x}{|x|}$, $r = |x| \in (R_0, R_1)$.

It remains to actually construct P, Q by use of (8.17) and the known existence of the Helmholtz decomposition. Setting $Q(x) = F_3(x)$, $x \in \mathcal{B}_{R_0,R_1}(0)$, it is clear that Q satisfies the desired conditions for the toroidal scalar. The function P is assumed to be a solution of the differential equation

$$\Delta^*_\xi P(r\xi) = rF_1(r\xi), \quad \xi \in \Omega, \ r \in (R_0, R_1), \tag{8.18}$$

which we know to exist from Remark 6.15 (because $\int_\Omega rF_1(r\eta)d\omega(\eta) = 0$ by the definition of solenoidality). P is uniquely determined if it is chosen to satisfy $\frac{1}{4\pi}\int_{\Omega_r} P(y)d\omega(y) = 0$. Observing (8.17), we see that the proof is completed if we can show that the previously chosen P satisfies the condition

$$-\frac{\partial}{\partial r}(rP(r\xi)) = rF_2(r\xi), \tag{8.19}$$

for $\xi \in \Omega$, $r \in (R_0, R_1)$. Equation (8.16) implies

$$\begin{aligned} \Delta^*_\xi\left(\frac{\partial}{\partial r}(rP(r\xi)) + rF_2(r\xi)\right) &= \frac{\partial}{\partial r}(r\Delta^*_\xi P(r\xi)) + r\Delta^*_\xi F_2(r\xi) \qquad (8.20)\\ &= \frac{\partial}{\partial r}(r^2 F_1(r\xi)) + r\Delta^*_\xi F_2(r\xi) \\ &= 0, \end{aligned}$$

for $\xi \in \Omega$, $r \in (R_0, R_1)$. Furthermore, $\int_\Omega \frac{\partial}{\partial r}(rP(r\eta)) + rF_2(r\eta)d\omega(\eta) = 0$, due to the corresponding conditions for $P(r\cdot)$ and $F_2(r\cdot)$. Thus, we obtain from Remark 6.15 that $\frac{\partial}{\partial r}(rP(r\xi)) + rF_2(r\xi) = 0$, for all $\xi \in \Omega$, $r \in (R_0, R_1)$, which guarantees the desired assertion. □

Remark 8.5. The poloidal scalar P as constructed in the proof of Theorem 8.4 is only of class $\mathrm{C}^{(1)}(\mathcal{B}_{R_0,R_1}(0))$, yet we can apply the differential operator $\nabla \wedge \mathrm{L}$ of order two. This is due to the fact that L only acts as a spherical differential operator L^*, and $P(r\cdot)$ is of class $\mathrm{C}^{(2)}(\Omega)$, for every $r \in (R_0, R_1)$, since it is a solution to (8.18).

There exists an intimate relationship of the Helmholtz scalars and the Mie scalars. Expressing the gradient in terms of the surface gradient and the radial derivative, we obtain the following relations.

Lemma 8.6. *Let $f : \mathcal{B}_{R_0,R_1}(0) \to \mathbb{R}^3$ be a solenoidal vector field. If P, Q are the corresponding Mie scalars from Theorem 8.4, and $F_1(r\cdot)$, $F_2(r\cdot)$, $F_3(r\cdot)$ the Helmholtz scalars of $f(r\cdot) : \Omega \to \mathbb{R}^3$, $r \in (R_0, R_1)$, known from Theorem 8.1, then*

$$F_1(r\xi) = \frac{\Delta_\xi^* P(r\xi)}{r}, \quad \xi \in \Omega, \tag{8.21}$$

$$F_2(r\xi) = -\frac{1}{r}\frac{\partial}{\partial r}\left(rP(r\xi)\right), \quad \xi \in \Omega, \tag{8.22}$$

$$F_3(r\xi) = Q(r\xi), \quad \xi \in \Omega. \tag{8.23}$$

Proof. We know that L and L^* act essentially in the same way, so that only the term $\nabla \wedge \mathrm{L}P$ of the Mie decomposition has to be investigated in more detail. Observing the properties of the spherical operators in Section 2.1.3, we find that

$$\begin{aligned}
\nabla_x \wedge \mathrm{L}_x P(x) &= \left(\xi\frac{\partial}{\partial r} + \frac{1}{r}\nabla_\xi^*\right) \wedge \mathrm{L}_\xi^* P(r\xi) \qquad (8.24)\\
&= \xi \wedge \mathrm{L}_\xi^* \left(\frac{\partial}{\partial r}P(r\xi)\right) + \frac{1}{r}\nabla_\xi^* \wedge \mathrm{L}_\xi^* P(r\xi)\\
&= -\nabla_\xi^* \left(\frac{\partial}{\partial r}P(r\xi)\right) + \frac{1}{r}\xi\Delta_\xi^* P(r\xi) - \frac{1}{r}\nabla_\xi^* P(r\xi)\\
&= \nabla_\xi^* \left(-\frac{1}{r}\frac{\partial}{\partial r}\left(rP(r\xi)\right)\right) + \xi\left(\frac{1}{r}\Delta_\xi^* P(r\xi)\right),
\end{aligned}$$

for $r = |x|$ and $\xi = \frac{x}{|x|}$. Since $\frac{1}{4\pi r^2}\int_{\Omega_r} P(y)d\omega(y) = 0$ for the poloidal scalar, the quantities $-\frac{1}{r}\frac{\partial}{\partial r}\left(rP(r\cdot)\right)$ and $\frac{1}{r}\Delta^* P(r\cdot)$ possess a vanishing integral mean value over Ω as well, so that the uniqueness of the Helmholtz scalars implies the assertions stated in Lemma 8.6. □

In combination with Lemma 8.2, the relations (8.21), (8.23) allow integral representations for the Mie scalars. Q is directly given by $F_3(r\cdot)$, and P follows from the solution of the Beltrami differential equation (8.18) in dependence on $F_1(r\cdot)$. Since (8.22) implies $\frac{\partial}{\partial r}P(r\cdot) = -\frac{1}{r}P(r\cdot) - F_2(r\cdot)$, we also obtain an integral representation for the radial derivative of the poloidal part of f. One should be aware of the fact that the knowledge of f on a sphere Ω_r, $r \in (R_0, R_1)$, only allows statements on its Mie scalars on that same sphere.

Lemma 8.7. *Let $f : \mathcal{B}_{R_0,R_1}(0) \to \mathbb{R}^3$ be a solenoidal vector field. Then the Mie scalars P, Q known from Theorem 8.4 can be represented by*

$$P(x) = r \int_\Omega G(\Delta^*; \xi \cdot \eta)\eta \cdot f(r\eta) d\omega(\eta), \tag{8.25}$$

$$Q(x) = - \int_\Omega \left(\mathrm{L}^*_\eta G(\Delta^*; \xi \cdot \eta)\right) \cdot f(r\eta) d\omega(\eta), \tag{8.26}$$

for $\xi = \frac{x}{|x|} \in \Omega$, $r = |x| \in (R_0, R_1)$. The radial derivative of the poloidal scalar is given by

$$\frac{\partial}{\partial r} P(x) = \int_\Omega \left(\nabla^*_\eta G(\Delta^*; \xi \cdot \eta) - G(\Delta^*; \xi \cdot \eta)\eta\right) \cdot f(r\eta) d\omega(\eta). \tag{8.27}$$

8.1.2 Local Decompositions

The Helmholtz and the Mie Decomposition can be formulated on subregions of the sphere or spherical shells as well (for some more details, the reader is referred to C. Gerhards [2011b]). We start with the Helmholtz decomposition. An essential difference in comparison to the global case concerns the uniqueness: neither the scalars F_2 and F_3 nor their corresponding vector fields $o^{(2)}F_2$ and $o^{(3)}F_3$ are determined uniquely.

Remark 8.8. Let $\Gamma \subset \Omega$ be a regular region. If f of class $\mathrm{c}^{(1)}(\overline{\Gamma})$ has a Helmholtz decomposition

$$f(\xi) = o^{(1)}F_1(\xi) + o^{(2)}F_2(\xi) + o^{(3)}F_3(\xi), \quad \xi \in \Gamma, \tag{8.28}$$

then an alternative decomposition is given by

$$f(\xi) = o^{(1)}F_1(\xi) + o^{(2)}\left(F_2(\xi) + F(\xi)\right) + o^{(3)}\left(F_3(\xi) - U(\xi)\right), \quad \xi \in \Gamma, \tag{8.29}$$

where F can be any function of class $\mathrm{C}^{(2)}(\Gamma)$ satisfying $\Delta^* F(\xi) = 0$, $\xi \in \Gamma$. The function U is chosen such that $\nabla^* F = \mathrm{L}^* U$. Its existence is guaranteed by the harmonicity of F in connection with Theorem 6.63.

Since the only functions harmonic (with respect to the Beltrami operator) on the entire sphere Ω are the constant functions, we are led to the uniqueness conditions $\frac{1}{4\pi}\int_\Omega F_2(\eta) d\omega(\eta) = \frac{1}{4\pi}\int_\Omega F_3(\eta) d\omega(\eta) = 0$ for the global Helmholtz decomposition as stated in the previous subsection. On subregions of the sphere, the choice of harmonic functions requires more constraints to obtain uniqueness.

Theorem 8.9 (Spherical Helmholtz Decomposition). *Let $\Gamma \subset \Omega$ be a regular region, and the vector field f be of class $c^{(2)}(\overline{\Gamma})$. Then there exist scalar fields F_1 of class $C^{(2)}(\overline{\Gamma})$, F_2, F_3 of class $C^{(2)}(\Gamma)$ such that*

$$f(\xi) = o^{(1)}F_1(\xi) + o^{(2)}F_2(\xi) + o^{(3)}F_3(\xi), \quad \xi \in \Gamma. \tag{8.30}$$

The scalars F_1, F_2, F_3 are uniquely determined by the properties

$$\frac{1}{\|\Gamma\|}\int_\Gamma F_2(\eta)d\omega(\eta) = 0 \tag{8.31}$$

and

$$F_3(\xi) = F(\xi), \quad \xi \in \partial\Gamma, \tag{8.32}$$

for a fixed function F of class $C^{(0)}(\partial\Gamma)$.

Proof. The general idea is the same as presented in the proof of the global version of Theorem 8.1. From Theorem 6.33, we know that there exists a unique solution F_3 of class $C^{(2)}(\Gamma)$ to the problem

$$\Delta^* F_3(\xi) = L^* \cdot f_{\tan}(\xi), \quad \xi \in \Gamma, \tag{8.33}$$

$$F_3(\xi) = F(\xi), \quad \xi \in \partial\Gamma. \tag{8.34}$$

Furthermore, by virtue of Theorem 6.62

$$\nabla^* F_2(\xi) = f_{\tan}(\xi) - L^* F_3(\xi), \quad \xi \in \Gamma, \tag{8.35}$$

is uniquely solvable up to a constant. □

Remark 8.10. Clearly, the type of boundary conditions that have to be prescribed to obtain uniqueness of the Helmholtz decomposition can be varied. They can be imposed on F_2 instead of F_3, or Dirichlet boundary conditions can be substituted by Neumann boundary conditions. Neumann boundary conditions are more advantageous in this respect as they allow imposition of boundary information on the normal and tangential direction of the vectorial quantities $o^{(2)}F_2$ and $o^{(3)}F_3$, respectively, which are often better accessible from the given data than the scalars F_2 or F_3. A comparable formulation of the Neumann case is left to the reader as an exercise.

Using Dirichlet or Neumann Green functions, it is possible to obtain integral representations for the Helmholtz scalars similar to those presented in Lemma 8.2 for the global case. However, in order to apply Green's formulas, we need to require that the boundary value F is of higher regularity. Any choice F of class $C^{(2)}(\partial\Gamma)$ yields a solution of (8.33), (8.34) that is of the required class $C^{(2)}(\Gamma) \cap C^{(1)}(\overline{\Gamma})$ (as already mentioned in Remark 6.46, the reader is referred to D. Gilbart, N.S Trudinger [1977] for more details).

Lemma 8.11. *Let $\Gamma \subset \Omega$ be a regular region. Furthermore, assume that f is of class $c^{(2)}(\overline{\Gamma})$ and F of class $C^{(2)}(\partial\Gamma)$. If F_1, F_2, F_3 are the Helmholtz scalars known from Theorem 8.9, then F_2, F_3 read as follows*

$$\begin{aligned} F_2(\xi) &= -\int_\Gamma \left(\nabla^*_\eta G^{(N)}(\Delta^*;\xi,\eta)\right)\cdot f(\eta)d\omega(\eta) \\ &\quad + \int_{\partial\Gamma} F(\eta)\,\tau_\eta\cdot\nabla^*_\eta G^{(N)}(\Delta^*;\xi,\eta)d\sigma(\eta), \end{aligned} \tag{8.36}$$

$$\begin{aligned} F_3(\xi) &= -\int_\Gamma \left(\mathrm{L}^*_\eta G^{(D)}(\Delta^*;\xi,\eta)\right)\cdot f(\eta)d\omega(\eta) \\ &\quad + \int_{\partial\Gamma} G^{(D)}(\Delta^*;\xi,\eta)\tau_\eta\cdot f(\eta)d\sigma(\eta) \\ &\quad + \int_{\partial\Gamma} F(\eta)\frac{\partial}{\partial\nu_\eta}G^{(D)}(\Delta^*;\xi,\eta)d\sigma(\eta), \end{aligned} \tag{8.37}$$

for $\xi\in\Gamma$. Additionally, if $\frac{1}{\|\Gamma\|}\int_\Gamma F_1(\eta)d\omega(\eta)=0$, then

$$F_1(\xi)=\xi\cdot f(\xi)=\Delta^*_\xi\int_\Gamma G(\Delta^*;\xi\cdot\eta)\,\eta\cdot f(\eta)d\omega(\eta),\quad \xi\in\Gamma. \tag{8.38}$$

Proof. For the solution of (8.33), (8.34), we obtain from Theorem 6.45 and Green's formulas that

$$\begin{aligned} F_3(\xi) &= \int_\Gamma G^{(D)}(\Delta^*;\xi,\eta)\mathrm{L}^*_\eta\cdot f_{\tan}(\eta)d\omega(\eta) \\ &\quad + \int_{\partial\Gamma} F(\eta)\frac{\partial}{\partial\nu_\eta}G^{(D)}(\Delta^*;\xi,\eta)d\sigma(\eta) \\ &= -\int_\Gamma \left(\mathrm{L}^*_\eta G^{(D)}(\Delta^*;\xi,\eta)\right)\cdot f(\eta)d\omega(\eta) \\ &\quad + \int_{\partial\Gamma} G^{(D)}(\Delta^*;\xi,\eta)\,\tau_\eta\cdot f(\eta)d\sigma(\eta) \\ &\quad + \int_{\partial\Gamma} F(\eta)\frac{\partial}{\partial\nu_\eta}G^{(D)}(\Delta^*;\xi,\eta)d\sigma(\eta), \end{aligned} \tag{8.39}$$

for $\xi\in\Gamma$. Inserting (8.39) into (8.35), Theorem 6.64 shows us, in connection with Green's formulas, that

$$\begin{aligned} F_2(\xi) &= \frac{1}{\|\Gamma\|}\int_\Gamma F_2(\eta)d\omega(\eta) \\ &\quad + \int_\Gamma \left(\nabla^*_\eta G^{(N)}(\Delta^*;\xi,\eta)\right)\cdot\left(\mathrm{L}^*_\eta F_3(\eta)-f_{\tan}(\eta)\right)d\omega(\eta) \\ &= \frac{1}{\|\Gamma\|}\int_\Gamma F_2(\eta)d\omega(\eta) - \int_\Gamma \left(\nabla^*_\eta G^{(N)}(\Delta^*;\xi,\eta)\right)\cdot f(\eta)d\omega(\eta) \\ &\quad + \int_{\partial\Gamma} F(\eta)\tau_\eta\cdot\nabla^*_\eta G^{(N)}(\Delta^*;\xi,\eta)d\sigma(\eta), \end{aligned} \tag{8.40}$$

for $\xi \in \Gamma$. Thus, we have found the desired expressions by observing the vanishing integral mean value $\frac{1}{\|\Gamma\|}\int_\Gamma F_2(\eta)d\omega(\eta) = 0$ due to the uniqueness condition on F_2. Moreover, the representation for F_1 is a direct consequence of Theorem 6.14. □

Corollary 8.12. *Let $\Gamma \subset \Omega$ be a regular region. Furthermore, assume that f of class $\mathrm{c}^{(2)}(\overline{\Gamma})$ is tangential and F is of class $\mathrm{C}^{(2)}(\partial\Gamma)$. Then the Helmholtz scalars F_2, F_3 from Theorem 8.9 can be represented in the form*

$$F_2(\xi) = \int_\Gamma G^{(N)}(\Delta^*;\xi,\eta)\nabla^*_\eta \cdot f(\eta)d\omega(\eta) - \int_{\partial\Gamma} G^{(N)}(\Delta^*;\xi,\eta)\nu_\eta \cdot f(\eta)d\sigma(\eta) + \int_{\partial\Gamma} F(\eta)\,\tau_\eta \cdot \nabla^*_\eta G^{(N)}(\Delta^*;\xi,\eta)d\sigma(\eta), \quad \xi \in \Gamma, \tag{8.41}$$

$$F_3(\xi) = \int_\Gamma G^{(D)}(\Delta^*;\xi,\eta)\mathrm{L}^*_\eta \cdot f(\eta)d\omega(\eta) + \int_{\partial\Gamma} F(\eta)\frac{\partial}{\partial\nu_\eta}G^{(D)}(\Delta^*;\xi,\eta)d\sigma(\eta), \quad \xi \in \Gamma. \tag{8.42}$$

A formulation of the representation of the Helmholtz scalars if Neumann boundary conditions are imposed is left as a task for the reader (see Exercise 8.4).

Next, we turn to a local variant of the Mie decomposition. Here as well, the general idea is the same as in the global case, only that one has to take care of additional boundary values to guarantee uniqueness.

Theorem 8.13 (Mie Decomposition). *Let $\Gamma \subset \Omega$ be a regular region, and suppose that the function $f \in \mathrm{c}^{(2)}(\mathcal{C}_{R_0,R_1}(\overline{\Gamma}))$ is divergence-free, i.e., $\nabla \cdot f(x) = 0$, $x \in \mathcal{C}_{R_0,R_1}(\Gamma)$. Then there exist $P \in \mathrm{C}^{(2)}(\mathcal{C}_{R_0,R_1}(\Gamma)) \cap \mathrm{C}^{(0)}(\mathcal{C}_{R_0,R_1}(\overline{\Gamma}))$ and $Q \in \mathrm{C}^{(2)}(\mathcal{C}_{R_0,R_1}(\Gamma)) \cap \mathrm{C}^{(1)}(\mathcal{C}_{R_0,R_1}(\overline{\Gamma}))$ such that*

$$f(x) = \nabla \wedge \mathrm{L}P(x) + \mathrm{L}Q(x), \quad x \in \mathcal{C}_{R_0,R_1}(\Gamma). \tag{8.43}$$

P, Q are uniquely determined by the additional conditions

$$P(R\xi) = F_P(\xi), \quad \xi \in \partial\Gamma, \tag{8.44}$$

$$Q(x) = F_Q(x), \quad x \in \mathcal{C}_{R_0,R_1}(\partial\Gamma), \tag{8.45}$$

for a fixed $R \in (R_0, R_1)$, as well as boundary values F_P of class $\mathrm{C}^{(0)}(\partial\Gamma)$ and F_Q of class $\mathrm{C}^{(2)}(\mathcal{C}_{R_0,R_1}(\partial\Gamma))$.

Proof. Up to Equation (8.18), the proof essentially coincides with the global case treated in Theorem 8.4. The uniqueness of Q follows from the uniqueness of the Helmholtz scalar $F_3(r\cdot)$, $r \in (R_0, R_1)$, if boundary values $F_Q(r\cdot)$ are imposed. For the determination of P, we solve the boundary-value problem

$$\Delta^*_\xi P(r\xi) = rF_1(r\xi), \quad \xi \in \Gamma, r \in (R_0, R_1), \tag{8.46}$$

$$P(r,\xi) = \tilde{F}_P(r\xi), \quad \xi \in \partial\Gamma, r \in (R_0, R_1), \tag{8.47}$$

where $\tilde{F}_P(r\xi) = \frac{R}{r}F_P(\xi) + \frac{1}{r}\int_R^r sF_2(s\xi)ds$, $\xi \in \partial\Gamma$ and $r \in (R_0, R_1)$. The regularity of the boundary value F_Q assures the desired regularity for Q (i.e., the Helmholtz scalar F_3) up to the boundary. Thus, the Helmholtz scalar F_2 is continuous up to the boundary, and finally, $\tilde{F}_P(r\cdot)$ is of class $\mathrm{C}^{(0)}(\partial\Gamma)$. From Theorem 6.33, we now know the existence of a unique solution P of (8.46), (8.47). Due to the choice of $\tilde{F}_P$, this solution satisfies

$$\frac{\partial}{\partial r}(rP(r\xi)) + rF_2(r\xi) = 0, \quad \xi \in \partial\Gamma, r \in (R_0, R_1). \tag{8.48}$$

Furthermore, we already know that

$$\begin{aligned}\Delta^*_\xi\Big(\frac{\partial}{\partial r}(rP(r\xi)) + rF_2(r\xi)\Big) &= \frac{\partial}{\partial r}(r\Delta^*_\xi P(r\xi)) + r\Delta^*_\xi F_2(r\xi) \quad (8.49)\\ &= \frac{\partial}{\partial r}(r^2F_1(r\xi)) + r\Delta^*_\xi F_2(r\xi)\\ &= 0,\end{aligned}$$

for $\xi \in \Gamma$, $r \in (R_0, R_1)$, where (8.16) was used in the last step. With the corresponding boundary values, this leads to $\frac{\partial}{\partial r}(rP(r\xi)) + rF_2(r\xi) = 0$, for all $\xi \in \Gamma$, $r \in (R_0, R_1)$, which completes the proof. □

The integral representations of the local Mie scalars are derivable in an analogous fashion as for the global case. However, one has to observe the boundary values appearing in the proof of Theorem 8.13. Consequently, Lemma 8.6 holds as well for the local setting. The higher order regularity of the boundary values as used in the next lemma simply takes into account the applicability of Green's formulas.

Lemma 8.14. *Let $\Gamma \subset \Omega$ be a regular region, and f of class $\mathrm{c}^{(2)}(\mathcal{C}_{R_0,R_1}(\overline{\Gamma}))$ be divergence-free. For the boundary values, we require F_P to be of class $\mathrm{C}^{(2)}(\partial\Gamma)$ and F_Q of class $\mathrm{C}^{(3)}(\mathcal{C}_{R_0,R_1}(\partial\Gamma))$. Then the Mie scalars P, Q known from Theorem 8.13 can be represented as*

$$\begin{aligned}P(x) &= r\int_\Gamma G^{(D)}(\Delta^*;\xi,\eta)\eta\cdot f(r\eta)d\omega(\eta) \quad (8.50)\\ &\quad + \int_{\partial\Gamma}\tilde{F}_P(r\eta)\frac{\partial}{\partial\nu_\eta}G^{(D)}(\Delta^*;\xi,\eta)d\sigma(\eta),\\ Q(x) &= \int_\Gamma\Big(\mathrm{L}^*_\eta G^{(D)}(\Delta^*;\xi,\eta)\Big)\cdot f(r\eta)d\omega(\eta) \quad (8.51)\\ &\quad + \int_{\partial\Gamma}G^{(D)}(\Delta^*;\xi,\eta)\tau_\eta\cdot f(r\eta)d\sigma(\eta)\\ &\quad + \int_{\partial\Gamma}F_Q(r\eta)\frac{\partial}{\partial\nu_\eta}G^{(D)}(\Delta^*;\xi,\eta)d\sigma(\eta),\end{aligned}$$

for $\xi = \frac{x}{|x|} \in \Gamma$, $r = |x| \in (R_0, R_1)$. The radial derivative of the poloidal scalar is given by

$$\begin{aligned} \frac{\partial}{\partial r}P(x) &= \int_{\Gamma}\left(\nabla^*_\eta G^{(D)}(\Delta^*;\xi,\eta) - G^{(D)}(\Delta^*;\xi,\eta)\eta\right)\cdot f(r\eta)d\omega(\eta) \quad (8.52)\\ &\quad -\frac{1}{r}\int_{\partial\Gamma}\tilde{F}_P(r\eta)\frac{\partial}{\partial\nu_\eta}G^{(D)}(\Delta^*;\xi,\eta)d\sigma(\eta)\\ &\quad -\int_{\partial\Gamma}F_Q(r\eta)\tau_\eta\cdot\nabla^*_\eta G^{(D)}(\Delta^*;\xi,\eta)d\sigma(\eta). \end{aligned}$$

The function $\tilde{F}_P$ is given by $\tilde{F}_P(r\xi) = \frac{R}{r}F_P(\xi) + \frac{1}{r}\int_R^r sF_2(s\xi)ds$, for $\xi \in \partial\Gamma$ and $r \in (R_0, R_1)$, where $F_2(s\cdot)$ is the known local Helmholtz scalar of $f(s\cdot)$.

Remark 8.15. The type of boundary conditions to obtain uniqueness of the Mie scalars can again be varied. Dirichlet boundary values can be entirely substituted by Neumann boundary values or both types of boundary values can be combined. However, one has to note that Neumann boundary conditions additionally require the corresponding Mie scalar to have a vanishing integral mean value.

8.2 Higher-Order Regularization Methods

We have already introduced a linear regularization method in Section 7.2. For applications in geomagnetic modeling and gravity gradiometry, however, it is sometimes necessary to be equipped with regularizations of higher order, so we introduce some more general methods in this section. Yet, the overall idea is essentially the same as for the linear case.

8.2.1 Regularization of the Fundamental Solution

We begin with the regularization of the fundamental solution for the Beltrami operator.

Definition 8.16. For $\rho \in (0,2)$ and $n \in \mathbb{N}$, let R^ρ be of class $\mathrm{C}^{(n)}([-1,1])$, satisfying

$$\lim_{\rho\to 0+}\rho^{\frac{k}{2}}\int_{1-\rho}^{1}\left|\left(\frac{d}{dt}\right)^k R^\rho(t)\right|dt = 0, \quad k = 0,1, \tag{8.53}$$

and

$$\left[\left(\frac{d}{dt}\right)^k R^\rho(t)\right]_{t=1-\rho} = \left[\left(\frac{d}{dt}\right)^k G(\Delta^*;t)\right]_{t=1-\rho}, \quad k = 0,1,\ldots,n. \tag{8.54}$$

Then

$$G^\rho(\Delta^*;\xi\cdot\eta) = \begin{cases} G(\Delta^*;\xi\cdot\eta), & 1-\xi\cdot\eta \geq \rho,\\ R^\rho(\xi\cdot\eta), & 1-\xi\cdot\eta < \rho,\end{cases} \tag{8.55}$$

is called a *regularized fundamental solution (of the Beltrami operator) of order* n. R^ρ is called a regularization function of order n.

A typical regularization function R^ρ of order n would be the Taylor polynomial of degree n of $G(\Delta^*;\cdot)$ with respect to the expansion point $1-\rho$. A regularization using a Taylor polynomial of degree one has been used, e.g., in the proof of Theorem 6.13. The regularization ${}^lG^\rho(\Delta^*;\cdot)$ presented in (7.73) is different from this regularization of order 1 since it relies on a regularization of the single-layer kernel. However, the general properties of the kernel ${}^lG^\rho(\Delta^*;\cdot)$ from (7.73) and the kernel $G^\rho(\Delta^*;\cdot)$ of order 1 from Definition 8.16 are similar (note that the parameter ρ of ${}^lG^\rho(\Delta^*;\cdot)$ actually means a regularization on a spherical cap of radius $\frac{\rho^2}{2R^2}$, a choice that has simply been made to be consistent with the numerical examples discussed later on in Section 7.2). That a regularization of the single-layer kernel implies a regularization of the fundamental solution, and vice versa, can be seen from (7.56). A direct regularization of the fundamental solution as introduced in Definition 8.16 just simplifies some calculations in the geomagnetic framework.

Lemma 8.17. *Let* $G^\rho(\Delta^*;\cdot)\in \mathrm{C}^{(1)}([-1,1])$ *be a regularization of order* 1. *Assume that* F *is of class* $\mathrm{C}^{(0)}(\Omega)$. *Then*

$$\lim_{\rho\to 0+}\sup_{\xi\in\Omega}\left|\int_\Omega G^\rho(\Delta^*;\xi\cdot\eta)F(\eta)d\omega(\eta)-\int_\Omega G(\Delta^*;\xi\cdot\eta)F(\eta)d\omega(\eta)\right|=0. \tag{8.56}$$

Proof. The existence of the integrals is clear. For $\xi\in\Omega$ we get

$$\begin{aligned}
&\left|\int_\Omega G^\rho(\Delta^*;\xi\cdot\eta)F(\eta)d\omega(\eta)-\int_\Omega G(\Delta^*;\xi\cdot\eta)F(\eta)d\omega(\eta)\right| \qquad (8.57)\\
&=\left|\int_{\substack{\eta\in\Omega\\1-\xi\cdot\eta\le\rho}} G^\rho(\Delta^*;\xi\cdot\eta)F(\eta)d\omega(\eta)-\int_{\substack{\eta\in\Omega\\1-\xi\cdot\eta\le\rho}} G(\Delta^*;\xi\cdot\eta)F(\eta)d\omega(\eta)\right|\\
&\le \sup_{\eta\in\Omega}|F(\eta)|\left(\int_{\substack{\eta\in\Omega\\1-\xi\cdot\eta\le\rho}}|R^\rho(\xi\cdot\eta)|\,d\omega(\eta)+\int_{\substack{\eta\in\Omega\\1-\xi\cdot\eta\le\rho}}|G(\Delta^*;\xi\cdot\eta)|\right)d\omega(\eta).
\end{aligned}$$

From the property (8.53) we are immediately able to determine that the first integral of the right side of (8.57) vanishes as ρ tends to zero. The second integral vanishes as ρ tends to zero since the logarithmic singularity of $G(\Delta^*;\cdot)$ is integrable on the sphere Ω. Due to the rotational invariance of the integrands, the convergence is uniform with respect to $\xi\in\Omega$. Altogether, this yields the desired statement. □

Lemma 8.18. *Let F be of class $\mathrm{C}^{(0)}(\Omega)$, and assume that $G^\rho(\Delta^*;\cdot) \in \mathrm{C}^{(1)}([-1,1])$ is a regularization of order 1. Then*

$$\lim_{\rho\to 0+}\sup_{\xi\in\Omega}\left|\int_\Omega \nabla^*_\xi G^\rho(\Delta^*;\xi\cdot\eta)F(\eta)d\omega(\eta) - \nabla^*_\xi\int_\Omega G(\Delta^*;\xi\cdot\eta)F(\eta)d\omega(\eta)\right| = 0, \tag{8.58}$$

$$\lim_{\rho\to 0+}\sup_{\xi\in\Omega}\left|\int_\Omega \mathrm{L}^*_\xi G^\rho(\Delta^*;\xi\cdot\eta)F(\eta)d\omega(\eta) - \mathrm{L}^*_\xi\int_\Omega G(\Delta^*;\xi\cdot\eta)F(\eta)d\omega(\eta)\right| = 0. \tag{8.59}$$

Proof. For $\xi \in \Omega$, similar arguments as in the linear case lead to

$$\begin{aligned}
&\left|\int_\Omega \nabla^*_\xi G^\rho(\Delta^*;\xi\cdot\eta)F(\eta)d\omega(\eta) - \int_\Omega \nabla^*_\xi G(\Delta^*;\xi\cdot\eta)F(\eta)d\omega(\eta)\right| \qquad (8.60)\\
\leq\ & \sup_{\eta\in\Omega}|F(\eta)| \int\limits_{\substack{\eta\in\Omega\\ 1-\xi\cdot\eta\leq\rho}} \left|\left[\frac{d}{dt}R^\rho(t)\right]_{t=\xi\cdot\eta}\right| \left(1-(\xi\cdot\eta)^2\right)^{\frac{1}{2}} d\omega(\eta)\\
&+\sup_{\eta\in\Omega}|F(\eta)| \int\limits_{\substack{\eta\in\Omega\\ 1-\xi\cdot\eta\leq\rho}} \left|\nabla^*_\xi\left(G(\Delta^*;\xi\cdot\eta) - \frac{1}{4\pi}\ln(1-\xi\cdot\eta)\right)\right| d\omega(\eta)\\
&+\sup_{\eta\in\Omega}|F(\eta)| \int\limits_{\substack{\eta\in\Omega\\ 1-\xi\cdot\eta\leq\rho}} \left|\frac{1}{4\pi(1-\xi\cdot\eta)}(\eta-(\xi\cdot\eta)\xi)\right| d\omega(\eta).
\end{aligned}$$

The first integral on the right-hand side vanishes as ρ tends to zero due to Condition (8.53), and the second one due to the continuity and the integrability of the integrand on Ω. The third integral vanishes by the same arguments as used in the proof of Theorem 6.13, which concludes the case of the surface gradient. The results for the surface curl gradient follow in a very similar fashion. □

Remark 8.19. The choice of R^ρ also admits an explicit statement on the convergence rates of the previous integral relations. More precisely, if

$$\int_{1-\rho}^1 |R^\rho(t)|\, dt = \mathcal{O}(\rho) \tag{8.61}$$

and

$$\int_{1-\rho}^1 \left|\frac{d}{dt}R^\rho(t)\right| dt = \mathcal{O}(1), \tag{8.62}$$

then it follows from the proofs of Lemma 8.17 and 8.18 that

$$\left|\int_\Omega G^\rho(\Delta^*;\xi\cdot\eta)F(\eta)d\omega(\eta) - \int_\Omega G(\Delta^*;\xi\cdot\eta)F(\eta)d\omega(\eta)\right| = \mathcal{O}(\rho\ln(\rho)), \tag{8.63}$$

and

$$\left| \int_\Omega \nabla_\xi^* G^\rho(\Delta^*;\xi\cdot\eta)F(\eta)d\omega(\eta) - \nabla_\xi^* \int_\Omega G(\Delta^*;\xi\cdot\eta)F(\eta)d\omega(\eta)\right| = \mathcal{O}(\rho^{\frac{1}{2}}), \tag{8.64}$$

whenever F is of class $\mathrm{C}^{(0)}(\Omega)$. Even more,

$$\left| \int_\Omega \nabla_\xi^* G^\rho(\Delta^*;\xi\cdot\eta)F(\eta)d\omega(\eta) - \nabla_\xi^* \int_\Omega G(\Delta^*;\xi\cdot\eta)F(\eta)d\omega(\eta)\right| = \mathcal{O}(\rho\ln(\rho)), \tag{8.65}$$

provided that F is of class $\mathrm{C}^{(1)}(\Omega)$. The same assertions hold true if we substitute ∇^* by L^*. A regularization via Taylor polynomials typically satisfies the required conditions on R^ρ.

Next we turn to higher-order derivatives. We only discuss the operators Δ^*, $\nabla^*\otimes\nabla^*$, $\mathrm{L}^*\otimes\mathrm{L}^*$, $\mathrm{L}^*\otimes\nabla^*$ as well as $\nabla^*\otimes\mathrm{L}^*$, which are useful for our work later on.

Corollary 8.20. *Let $G^\rho(\Delta^*;\cdot)\in\mathrm{C}^{(2)}([-1,1])$ be a regularization of order 2, and suppose that F is of class $\mathrm{C}^{(1)}(\Omega)$. Then*

$$\lim_{\rho\to 0}\sup_{\xi\in\Omega}\left| \int_\Omega \Delta_\xi^* G^\rho(\Delta^*;\xi\cdot\eta)F(\eta)d\omega(\eta) - \Delta_\xi^* \int_\Omega G(\Delta^*;\xi\cdot\eta)F(\eta)d\omega(\eta)\right| = 0. \tag{8.66}$$

Proof. Since F is assumed to be continuously differentiable, Green's formulas on the unit sphere Ω imply

$$\begin{aligned} & \Delta_\xi^* \int_\Omega G(\Delta^*;\xi\cdot\eta)F(\eta)\,d\omega(\eta) \qquad (8.67)\\ &= \mathrm{L}_\xi^*\cdot\int_\Omega \mathrm{L}_\xi^* G(\Delta^*;\xi\cdot\eta)F(\eta)\,d\omega(\eta)\\ &= \mathrm{L}_\xi^*\cdot\int_\Omega G(\Delta^*;\xi\cdot\eta)\mathrm{L}_\eta^* F(\eta))\,d\omega(\eta)\\ &= \int_\Omega \mathrm{L}_\xi^* G(\Delta^*;\xi\cdot\eta)\cdot\mathrm{L}_\eta^* F(\eta)\,d\omega(\eta). \end{aligned}$$

Thus, we are able to verify the desired assertion by arguments similar to Lemma 8.18. □

Analogously, using Green's formulas and Lemma 8.18, we obtain

Corollary 8.21. *Let $G^\rho(\Delta^*;\cdot)\in\mathrm{C}^{(2)}([-1,1])$ be a regularization of order 2, and suppose that the vector field f is of class $\mathrm{c}^{(1)}(\Omega)$. Then*

$$\begin{aligned} \lim_{\rho\to 0+}\sup_{\xi\in\Omega} \Bigg| & \int_\Omega \left(\left(\Lambda_1^*\right)_\xi \otimes \left(\Lambda_2^*\right)_\eta G^\rho(\Delta^*;\xi\cdot\eta)\right) f(\eta)d\omega(\eta) \qquad (8.68)\\ & -\left(\Lambda_1^*\right)_\xi \int_\Omega \left(\Lambda_2^*\right)_\eta G(\Delta^*;\xi\cdot\eta)\cdot f(\eta)d\omega(\eta)\Bigg| = 0, \end{aligned}$$

where Λ_1^* and Λ_2^* can be chosen to be one of the operators ∇^*, L^*.

Remark 8.22. A regularization of the fundamental solution in the frequency domain, i.e., a truncation of its bilinear series in terms of spherical harmonics, is treated in T. Fehlinger [2009] and W. Freeden, M. Schreiner [2009], where a closed representation for the frequency regularization is also developed. Moreover, a bilinear expansion in terms of spherical harmonics is given for $G^\rho(\Delta^*;\cdot)$ (as introduced in Definition 8.16), where the regularization function R^ρ is chosen to be the Taylor polynomial of degree 1 with respect to the expansion point $1-\rho$.

8.2.2 Regularization of the Single-Layer Kernel

In complete analogy to Definition 8.16, we are able to introduce a regularization of the single-layer kernel. As always, the original single-layer kernel is given by $S(t) = \frac{\sqrt{2}}{\sqrt{1-t}}$, $t \in [-1,1)$.

Definition 8.23. For $\rho > 0$ and $n \in \mathbb{N}$, let R^ρ be a non-negative function of class $\mathrm{C}^{(n)}([-1,1])$, satisfying

$$\lim_{\rho \to 0+} \rho^k \int_{1-\rho}^{1} \left| \left(\frac{d}{dt}\right)^k R^\rho(t) \right| dt = 0, \quad k = 0,1, \tag{8.69}$$

and

$$\left[\left(\frac{d}{dt}\right)^k R^\rho(t)\right]_{t=1-\rho} = \left[\left(\frac{d}{dt}\right)^k S(t)\right]_{t=1-\rho}, \quad k = 0,1,\ldots,n. \tag{8.70}$$

Then the function

$$S^\rho(\xi\cdot\eta) = \begin{cases} S(\xi\cdot\eta), & 1-\xi\cdot\eta \geq \rho, \\ R^\rho(\xi\cdot\eta), & 1-\xi\cdot\eta < \rho, \end{cases} \tag{8.71}$$

is called the *regularized single-layer kernel of order* n. R^ρ is called the regularization function of order n.

A typical regularization function R^ρ of order n is the Taylor polynomial of degree n of S with respect to the expansion point $1-\rho$ (see Figure 8.1 for an illustration). The linear regularization ${}^lS^\rho$ given in Subsection 7.2.1 represents exactly such a regularization of order 1 (again, note that the parameter ρ of ${}^lS^\rho$ actually relates to a parameter $\frac{\rho^2}{2R^2}$ in the sense of Definition 8.23).

Similar to the case of the fundamental solution, it follows that the convolution against the regularized single-layer kernel converges to the convolution against the original kernel. The same holds true if we apply the surface gradient or the surface curl gradient.

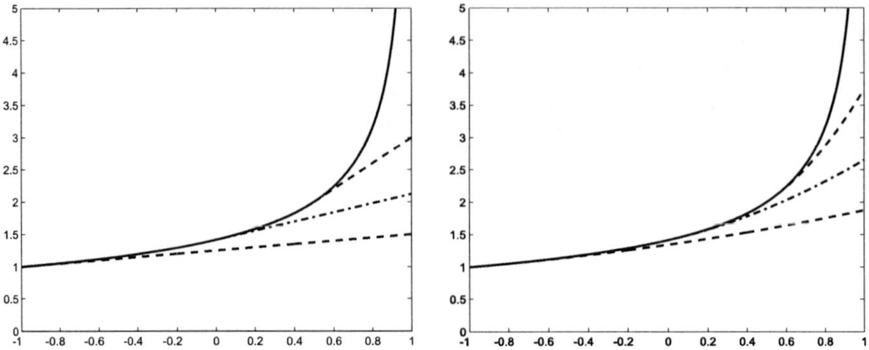

FIGURE 8.1
Plot of a regularized single-layer kernel $S^\rho(t)$ of order 1 (left) and order 2 (right), for $\rho = \frac{1}{2}, 1, 2$ (dotted lines). The original single-layer kernel $S(t)$ is denoted by the continuous black line.

Lemma 8.24. *Let F be of class $\mathrm{C}^{(0)}(\Omega)$. Suppose that $S^\rho \in \mathrm{C}^{(1)}([-1,1])$ is a regularization of order 1. Then*

$$\lim_{\rho\to 0+} \sup_{\xi\in\Omega} \left| \int_\Omega S^\rho(\xi\cdot\eta)F(\eta)d\omega(\eta) - \int_\Omega S(\xi\cdot\eta)F(\eta)d\omega(\eta) \right| = 0. \tag{8.72}$$

Lemma 8.25. *Let F be of class $\mathrm{C}^{(1)}(\Omega)$, and suppose that $S^\rho \in \mathrm{C}^{(1)}([-1,1])$ is a regularization of order 1. Then*

$$\lim_{\rho\to 0+} \sup_{\xi\in\Omega} \left| \int_\Omega \nabla^*_\xi S^\rho(\xi\cdot\eta)F(\eta)d\omega(\eta) - \nabla^*_\xi \int_\Omega S(\xi\cdot\eta)F(\eta)d\omega(\eta) \right| = 0, \tag{8.73}$$

$$\lim_{\rho\to 0+} \sup_{\xi\in\Omega} \left| \int_\Omega \mathrm{L}^*_\xi S^\rho(\xi\cdot\eta)F(\eta)d\omega(\eta) - \mathrm{L}^*_\xi \int_\Omega S(\xi\cdot\eta)F(\eta)d\omega(\eta) \right| = 0. \tag{8.74}$$

Next, we turn to applications of the single-layer operator D^{-1} to the fundamental solution for the Beltrami operator. On the one hand, analogous to our already treated results, we would be able to derive a relation such as

$$\lim_{\rho\to 0+} \sup_{\xi\in\Omega} \left| \int_\Omega \left(\nabla^*_\xi D^{-1}_\xi G^\rho(\Delta^*;\xi\cdot\eta)\right) F(\eta)\, d\omega(\eta) - \nabla^*_\xi D^{-1}_\xi \int_\Omega G(\Delta^*;\xi\cdot\eta)F(\eta)\, d\omega(\eta) \right| = 0, \tag{8.75}$$

where D^{-1} is given as in Lemma 2.49. On the other hand, it is difficult to calculate $D^{-1}_\xi G^\rho(\Delta^*;\xi\cdot\eta)$ in an explicit manner as would be required for our later applications. However, the expression $D^{-1}_\xi G(\Delta^*;\xi\cdot\eta)$ can be calculated and a regularization afterward yields a comparable limit relation.

Lemma 8.26. *For $\xi, \eta \in \Omega$, we have*

$$D_\xi^{-1} G(\Delta^*; \xi \cdot \eta) = \frac{1}{2\pi} \ln \left((1 + \xi \cdot \eta) \left(\frac{1}{2} - \frac{1}{1 - S(\xi \cdot \eta)} \right) \right) - \frac{1}{2\pi}. \tag{8.76}$$

Proof. The bilinear expansion of $G(\Delta^*; \xi \cdot \eta)$, the pseudodifferential representation of D^{-1} (cf. (2.211)), and the generating series of the Legendre polynomials known from Lemma 2.17 imply

$$\begin{aligned} D_\xi^{-1} G(\Delta^*; \xi \cdot \eta) &= \sum_{n=1}^{\infty} \frac{1}{n + \frac{1}{2}} \frac{2n+1}{4\pi} \frac{1}{-n(n+1)} P_n(\xi \cdot \eta) \qquad (8.77) \\ &= \frac{1}{2\pi} \sum_{n=1}^{\infty} \frac{1}{n+1} P_n(\xi \cdot \eta) - \frac{1}{2\pi} \sum_{n=1}^{\infty} \frac{1}{n} P_n(\xi \cdot \eta) \\ &= \frac{1}{2\pi} \ln \left(1 + \frac{\sqrt{2}}{\sqrt{1 - \xi \cdot \eta}} \right) \\ &\quad - \frac{1}{2\pi} \ln \left(\frac{\sqrt{2}\sqrt{1 - \xi \cdot \eta} - 1 + \xi \cdot \eta}{1 - (\xi \cdot \eta)^2} \right) - \frac{1}{2\pi} (1 + \ln(2)) \\ &= \frac{1}{2\pi} \ln \left((1 + \xi \cdot \eta) \left(\frac{1}{2} - \frac{1}{1 - S(\xi \cdot \eta)} \right) \right) - \frac{1}{2\pi}. \end{aligned}$$

The identity (8.77) is well-defined for all $\xi, \eta \in \Omega$. Closed representations of the occurring series involving Legendre polynomials can also be found, e.g., in E.R. Hansen [1975]. □

The representation (8.76) shows that $(\eta, \xi) \mapsto D_\xi^{-1} G(\Delta^*; \xi \cdot \eta)$ is ξ-zonal for fixed $\xi \in \Omega$. Moreover, it is of class $\mathrm{C}^{(1)}(\Omega \times \Omega)$ and, due to the zonality, may as well be regarded as a one-dimensional function of class $\mathrm{C}^{(1)}([-1, 1])$. Some lengthy but elementary computations yield

$$\nabla_\xi^* D_\xi^{-1} G(\Delta^*; \xi \cdot \eta) = \frac{1}{4\pi} \left(1 - S(\xi \cdot \eta) - \frac{1}{1 + S(\xi \cdot \eta)} \right) (\eta - (\xi \cdot \eta)\xi), \tag{8.78}$$

for $\xi, \eta \in \Omega$. A further application of the surface gradient causes a singularity of type $O((1 - \xi \cdot \eta)^{-\frac{1}{2}})$. Therefore, for $\rho > 0$, we are canonically led to a regularization of the form

$$s^\rho_{\nabla^*}(\xi, \eta) = \frac{1}{4\pi} \left(1 - S^\rho(\xi \cdot \eta) - \frac{1}{1 + S^\rho(\xi \cdot \eta)} \right) (\eta - (\xi \cdot \eta)\xi), \tag{8.79}$$

for $\xi, \eta \in \Omega$. Consequently, we find

$$\begin{aligned} &\nabla_\xi^* \otimes s^\rho_{\nabla^*}(\eta, \xi) \qquad (8.80) \\ &= \frac{1}{4\pi} \left(1 - S^\rho(\xi \cdot \eta) - \frac{1}{1 + S^\rho(\xi \cdot \eta)} \right) \nabla_\xi^* \otimes (\xi - (\xi \cdot \eta)\eta) \\ &\quad + \frac{1}{4\pi} \left(-(S^\rho)'(\xi \cdot \eta) + \frac{(S^\rho)'(\xi \cdot \eta)}{(1 + S^\rho(\xi \cdot \eta))^2} \right) (\eta - (\xi \cdot \eta)\xi) \otimes (\xi - (\xi \cdot \eta)\eta), \end{aligned}$$

where, as always, $\left(S^\rho\right)'$ denotes the one-dimensional derivative of S^ρ. The analogous procedure also works for the surface curl gradient. Thus, we obtain for $\rho > 0$,

$$s^\rho_{\mathrm{L}^*}(\xi,\eta) = \frac{1}{4\pi}\left(1 - S^\rho(\xi\cdot\eta) - \frac{1}{1+S^\rho(\xi\cdot\eta)}\right)(\xi\wedge\eta), \tag{8.81}$$

and

$$\begin{aligned} \mathrm{L}^*_\xi \otimes s^\rho_{\mathrm{L}^*}(\eta,\xi) &= \frac{1}{4\pi}\left(1 - S^\rho(\xi\cdot\eta) - \frac{1}{1+S^\rho(\xi\cdot\eta)}\right)\mathrm{L}^*_\xi\otimes(\eta\wedge\xi) \\ &\quad + \frac{1}{4\pi}\left(-\left(S^\rho\right)'(\xi\cdot\eta) + \frac{\left(S^\rho\right)'(\xi\cdot\eta)}{(1+S^\rho(\xi\cdot\eta))^2}\right)(\xi\wedge\eta)\otimes(\eta\wedge\xi). \end{aligned} \tag{8.82}$$

Finally, the relation (8.75) can be rewritten in the following more advantageous way.

Lemma 8.27. *Let F be of class $\mathrm{C}^{(0)}(\Omega)$, and assume that $S^\rho \in \mathrm{C}^{(1)}([-1,1])$ is a regularization of order* 1. *Then, $s^\rho_{\nabla^*}(\cdot,\cdot)$ and $s^\rho_{\mathrm{L}^*}(\cdot,\cdot)$ as defined by (8.79) and (8.81), respectively, satisfy*

$$\lim_{\rho\to 0+}\sup_{\xi\in\Omega}\left|\int_\Omega s^\rho_{\nabla^*}(\xi,\eta)F(\eta)\,d\omega(\eta) - \nabla^*_\xi D_\xi^{-1}\int_\Omega G(\Delta^*;\xi\cdot\eta)F(\eta)\,d\omega(\eta)\right| = 0, \tag{8.83}$$

$$\lim_{\rho\to 0+}\sup_{\xi\in\Omega}\left|\int_\Omega s^\rho_{\mathrm{L}^*}(\xi,\eta)F(\eta)\,d\omega(\eta) - \mathrm{L}^*_\xi D_\xi^{-1}\int_\Omega G(\Delta^*;\xi\cdot\eta)F(\eta)\,d\omega(\eta)\right| = 0. \tag{8.84}$$

In fact, the properties of S^ρ known from Definition 8.23 enable us to see that $\eta\mapsto\nabla^*_\xi D_\xi^{-1}G(\Delta^*;\xi\cdot\eta)$ as well as $\eta\mapsto \mathrm{L}^*_\xi D_\xi^{-1}G(\Delta^*;\xi\cdot\eta)$ are actually continuous on the entire sphere Ω. Hence, the proof of Lemma 8.27 follows analogously to that of Lemma 8.17. However, the relation we are actually aiming at, and which is demanded in later applications, is the upcoming tensorial one.

Lemma 8.28. *Let f be of class $\mathrm{c}^{(1)}(\Omega)$ and $S^\rho\in\mathrm{C}^{(1)}([-1,1])$ be a regularization of order* 1. *Then we get for $s^\rho_{\nabla^*}(\cdot,\cdot)$ and $s^\rho_{L^*}(\cdot,\cdot)$ as defined in (8.79) and (8.81), respectively, that*

$$\begin{aligned} \lim_{\rho\to 0+}\sup_{\xi\in\Omega}\Bigg| &\int_\Omega \left(\nabla^*_\xi\otimes s^\rho_{\nabla^*}(\eta,\xi)\right)f(\eta)\,d\omega(\eta) \\ &-\nabla^*_\xi\int_\Omega\left(\nabla^*_\eta D_\xi^{-1}G(\Delta^*;\xi\cdot\eta)\right)\cdot f(\eta)\,d\omega(\eta)\Bigg| = 0, \end{aligned} \tag{8.85}$$

$$\begin{aligned} \lim_{\rho\to 0+}\sup_{\xi\in\Omega}\Bigg| &\int_\Omega \left(\mathrm{L}^*_\xi\otimes s^\rho_{\mathrm{L}^*}(\eta,\xi)\right)f(\eta)\,d\omega(\eta) \\ &-\mathrm{L}^*_\xi\int_\Omega\left(\mathrm{L}^*_\eta D_\xi^{-1}G(\Delta^*;\xi\cdot\eta)\right)\cdot f(\eta)\,d\omega(\eta)\Bigg| = 0. \end{aligned} \tag{8.86}$$

Proof. Since $|\nabla^*_\xi \otimes (\xi - (\xi\cdot\eta)\eta)|$ and $|f(\eta)|$ are uniformly bounded by some constant $M > 0$, we get the following estimate for $\xi \in \Omega$ and $\rho > 0$,

$$\begin{aligned}
&\left|\int_\Omega \left(\nabla^*_\xi \otimes s^\rho_{\nabla^*}(\eta,\xi)\right) f(\eta)\, d\omega(\eta) - \int_\Omega \left(\nabla^*_\xi \otimes \nabla^*_\eta D_\xi^{-1} G(\Delta^*;\xi\cdot\eta)\right) f(\eta)\, d\omega(\eta)\right| \\
\leq\ & \int\limits_{\substack{\eta\in\Omega\\ 1-\xi\cdot\eta\leq\rho}} \left| S(\xi\cdot\eta) + \frac{1}{1+S(\xi\cdot\eta)} - S^\rho(\xi\cdot\eta) - \frac{1}{1+S^\rho(\xi\cdot\eta)}\right| \qquad (8.87)\\
&\qquad \times \left|\nabla^*_\xi \otimes (\xi - (\xi\cdot\eta)\eta)\right| |f(\eta)|\, d\omega(\eta) \\
&+ \int\limits_{\substack{\eta\in\Omega\\ 1-\xi\cdot\eta\leq\rho}} \left| S(\xi\cdot\eta)^3 - \frac{S(\xi\cdot\eta)^3}{(1+S(\xi\cdot\eta))^2} - 4\left(S^\rho\right)'(\xi\cdot\eta) + \frac{4\left(S^\rho\right)'(\xi\cdot\eta)}{(1+S^\rho(\xi\cdot\eta))^2}\right| \\
&\qquad \times |(\eta - (\xi\cdot\eta)\xi) \otimes (\xi - (\xi\cdot\eta)\eta)|\, |f(\eta)|\, d\omega(\eta) \\
\leq\ & M^2 \int\limits_{\substack{\eta\in\Omega\\ 1-\xi\cdot\eta\leq\rho}} \left| S(\xi\cdot\eta) + \frac{1}{1+S(\xi\cdot\eta)} - S^\rho(\xi\cdot\eta) - \frac{1}{1+S^\rho(\xi\cdot\eta)}\right| d\omega(\eta) \\
&+ M \int\limits_{\substack{\eta\in\Omega\\ 1-\xi\cdot\eta\leq\rho}} \left| S(\xi\cdot\eta)^3 - \frac{S(\xi\cdot\eta)^3}{(1+S(\xi\cdot\eta))^2} - 4\left(S^\rho\right)'(\xi\cdot\eta) + \frac{4\left(S^\rho\right)'(\xi\cdot\eta)}{(1+S^\rho(\xi\cdot\eta))^2}\right| \\
&\qquad \times |\eta - (\xi\cdot\eta)\xi|\, |\xi - (\xi\cdot\eta)\eta| d\omega(\eta).
\end{aligned}$$

Observing the identity

$$|\eta - (\xi\cdot\eta)\xi|\, |\xi - (\xi\cdot\eta)\eta| = \frac{2}{S(\xi\cdot\eta)^2}(1+\xi\cdot\eta) \qquad (8.88)$$

and the integrability of $\eta \mapsto S(\xi\cdot\eta)$ on the sphere Ω, together with the property (8.69) for S^ρ, we see that the integrals on the right-hand side of (8.87) vanish as ρ tends to zero. Due to the zonality of the kernels, this convergence is uniform with respect to $\xi \in \Omega$. Finally, the uniform convergence of (8.87) to zero leads us to

$$\begin{aligned}
&\int_\Omega \left(\nabla^*_\xi \otimes \nabla^*_\eta D_\xi^{-1} G(\Delta^*;\xi\cdot\eta)\right) f(\eta)\, d\omega(\eta) \qquad (8.89)\\
=\ & \nabla^*_\xi \int_\Omega \left(\nabla^*_\eta D_\xi^{-1} G(\Delta^*;\xi\cdot\eta)\right) \cdot f(\eta)\, d\omega(\eta).
\end{aligned}$$

The assertion for the surface curl gradient follows in the same manner. □

Remark 8.29. Analogous to the case of the fundamental solution for the Beltrami operator, an adequate choice of R^ρ admits a statement on the convergence rate. More precisely, if

$$\int_{1-\rho}^1 |R^\rho(t)|\, dt = \mathcal{O}(\rho) \qquad (8.90)$$

and

$$\int_{1-\rho}^1 \left|\frac{d}{dt} R^\rho(t)\right| dt = \mathcal{O}(1), \qquad (8.91)$$

a detailed analysis of the previous proofs gives us

$$\left| \int_{\Omega} \nabla_{\xi}^{*} S^{\rho}(\xi \cdot \eta) F(\eta) d\omega(\eta) - \nabla_{\xi}^{*} \int_{\Omega} S(\xi \cdot \eta) F(\eta) d\omega(\eta) \right| = \mathcal{O}(\rho^{\frac{1}{2}}), \tag{8.92}$$

and

$$\left| \int_{\Omega} \left(\nabla_{\xi}^{*} \otimes s_{\nabla^{*}}^{\rho}(\eta, \xi) \right) f(\eta)\, d\omega(\eta) \right. \tag{8.93}$$

$$\left. -\nabla_{\xi}^{*} \int_{\Omega} \left(\nabla_{\eta}^{*} D_{\xi}^{-1} G(\Delta^{*}; \xi \cdot \eta) \right) \cdot f(\eta)\, d\omega(\eta) \right| = \mathcal{O}(\rho^{\frac{1}{2}}),$$

for all $F \in \mathrm{C}^{(1)}(\Omega)$ and $f \in \mathrm{c}^{(1)}(\Omega)$. The conditions on the regularization function are satisfied, e.g., by the choice of R^{ρ} as a Taylor polynomial of S with expansion point $1 - \rho$. Finally, it should be noted that a higher-order regularization does not per se lead to a better convergence. However, numerical tests have shown that it often does.

8.3 Separation of Sources

In Section 5.4, we saw that the magnetic field b restricted to some sphere Ω_R can be decomposed into

$$b = p_b^{\mathrm{int}} + p_b^{\mathrm{ext}} + q_b, \tag{8.94}$$

where p_b^{int} denotes the poloidal part due to sources inside the sphere, p_b^{ext} the poloidal part due to sources outside the sphere, and q_b the toroidal part due to source currents crossing the sphere. Furthermore, we know that $p_b^{\mathrm{int}} = \nabla U^{\mathrm{int}}$ in $\mathbb{R}^3 \setminus \overline{\mathcal{B}_R(0)}$, for a harmonic potential U^{int} that can be expanded as a Fourier series in terms of outer harmonics. Since these representations are continuous up to the boundary Ω_R, the relation of the outer harmonics to the vector spherical harmonics $\tilde{y}_{n,k}^{(1)}$ (cf. Lemma 2.54) implies

$$p_b^{\mathrm{int}}(x) = \sum_{n=0}^{\infty} \sum_{k=1}^{2n+1} \left(\tilde{b}^{(1)}(R \cdot) \right)^{\wedge}(n,k)\, \tilde{y}_{n,k}^{(1)}(\xi), \tag{8.95}$$

for $\xi = \frac{x}{|x|}$, $R = |x|$. By $\left(\tilde{b}^{(1)}(R \cdot) \right)^{\wedge}(n,k) = \int_{\Omega} b(R\eta) \cdot \tilde{y}_{n,k}^{(1)}(\eta) d\omega(\eta)$ we denote the associated Fourier coefficients. Analogously, we obtain

$$p_b^{\mathrm{ext}}(x) = \sum_{n=1}^{\infty} \sum_{k=1}^{2n+1} \left(\tilde{b}^{(2)}(R \cdot) \right)^{\wedge}(n,k)\, \tilde{y}_{n,k}^{(2)}(\xi), \tag{8.96}$$

$$q_b(x) = \sum_{n=1}^{\infty} \sum_{k=1}^{2n+1} \left(\tilde{b}^{(3)}(R \cdot) \right)^{\wedge}(n,k)\, \tilde{y}_{n,k}^{(3)}(\xi). \tag{8.97}$$

Since the vector spherical harmonics $\tilde{y}_{n,k}^{(i)}$, $i = 1, 2, 3$, are by definition obtained via the spherical operators $\tilde{o}^{(i)}$, $i = 1, 2, 3$, a decomposition

$$b = \tilde{o}^{(1)}\tilde{B}_1 + \tilde{o}^{(2)}\tilde{B}_2 + \tilde{o}^{(3)}\tilde{B}_3 \tag{8.98}$$

provides a separation into internal and external contributions as well. In the following, we want to state this separation in a mathematically rigorous way.

8.3.1 Decomposition with Respect to the Operators $\tilde{o}^{(i)}$

The defining equations (2.198)–(2.200) of the operators $\tilde{o}^{(i)}$, $i = 1, 2, 3$, can be rewritten in the form

$$o^{(1)} = \frac{1}{2}\tilde{o}^{(1)}\mathrm{D}^{-1} + \frac{1}{2}\tilde{o}^{(2)}\mathrm{D}^{-1}, \tag{8.99}$$

$$o^{(2)} = \frac{1}{2}\tilde{o}^{(1)}\left(\frac{1}{2}\mathrm{D}^{-1} - 1\right) + \frac{1}{2}\tilde{o}^{(2)}\left(\frac{1}{2}\mathrm{D}^{-1} + 1\right), \tag{8.100}$$

$$o^{(3)} = \tilde{o}^{(3)}. \tag{8.101}$$

Having this representation at hand, we are able to apply the already known Helmholtz decomposition to derive the following decomposition with respect to the operators $\tilde{o}^{(i)}$, $i = 1, 2, 3$.

Theorem 8.30. *Let f be of class $\mathrm{c}^{(1)}(\Omega)$. Then there exist scalar fields $\tilde{F}_1, \tilde{F}_2$ of class $\mathrm{C}^{(1)}(\Omega)$ and $\tilde{F}_3$ of class $\mathrm{C}^{(2)}(\Omega)$ such that*

$$f(\xi) = \tilde{o}^{(1)}\tilde{F}_1(\xi) + \tilde{o}^{(2)}\tilde{F}_2(\xi) + \tilde{o}^{(3)}\tilde{F}_3(\xi), \quad \xi \in \Omega. \tag{8.102}$$

The scalars $\tilde{F}_1, \tilde{F}_2$, and $\tilde{F}_3$ are uniquely determined by the conditions

$$\frac{1}{4\pi}\int_\Omega \tilde{F}_3(\eta)d\omega(\eta) = 0, \tag{8.103}$$

$$\frac{1}{4\pi}\int_\Omega \tilde{F}_1(\eta) - \tilde{F}_2(\eta)d\omega(\eta) = 0. \tag{8.104}$$

Furthermore, $\tilde{F}_1, \tilde{F}_2, \tilde{F}_3$ admit the representations

$$\tilde{F}_1 = \frac{1}{2}\mathrm{D}^{-1}F_1 + \frac{1}{4}\mathrm{D}^{-1}F_2 - \frac{1}{2}F_2, \tag{8.105}$$

$$\tilde{F}_2 = \frac{1}{2}\mathrm{D}^{-1}F_1 + \frac{1}{4}\mathrm{D}^{-1}F_2 + \frac{1}{2}F_2, \tag{8.106}$$

$$\tilde{F}_3 = F_3, \tag{8.107}$$

with F_1, F_2, and F_3 being the uniquely determined functions of the Helmholtz decomposition in Theorem 8.1.

Proof. Applying the Helmholtz decomposition from Theorem 8.1 to f and observing the identities (8.99)–(8.101), we get

$$\begin{aligned} f &= o^{(1)}F_1 + o^{(2)}F_2 + o^{(3)}F_3 \\ &= \frac{1}{2}\tilde{o}^{(1)}\mathrm{D}^{-1}F_1 + \frac{1}{2}\tilde{o}^{(2)}\mathrm{D}^{-1}F_1 + \frac{1}{2}\tilde{o}^{(1)}\left(\frac{1}{2}\mathrm{D}^{-1} - 1\right)F_2 \\ &\quad + \frac{1}{2}\tilde{o}^{(2)}\left(\frac{1}{2}\mathrm{D}^{-1} + 1\right)F_2 + \tilde{o}^{(3)}F_3 \\ &= \tilde{o}^{(1)}\left(\frac{1}{2}\mathrm{D}^{-1}F_1 + \frac{1}{4}\mathrm{D}^{-1}F_2 - \frac{1}{2}F_2\right) \\ &\quad + \tilde{o}^{(2)}\left(\frac{1}{2}\mathrm{D}^{-1}F_1 + \frac{1}{4}\mathrm{D}^{-1}F_2 + \frac{1}{2}F_2\right) + \tilde{o}^{(3)}F_3. \end{aligned} \tag{8.108}$$

This implies a decomposition as stated in Theorem 8.30. Because of the uniqueness of the Helmholtz representation, $\tilde{F}_3$ is determined uniquely by assuming a vanishing integral mean value. For the uniqueness of $\tilde{F}_1$ and $\tilde{F}_2$ it is sufficient to show that the zero function $f(\xi) = 0$, $\xi \in \Omega$, only has the trivial decomposition with respect to the alternative Helmholtz operators. If

$$\tilde{o}^{(1)}\tilde{F}_1(\xi) + \tilde{o}^{(2)}\tilde{F}_2(\xi) + \tilde{o}^{(3)}\tilde{F}_3(\xi) = 0, \quad \xi \in \Omega, \tag{8.109}$$

we obtain from (2.198)–(2.200) that

$$o^{(1)}\left(\left(\mathrm{D} + \tfrac{1}{2}\right)\tilde{F}_1(\xi) + \left(\mathrm{D} - \tfrac{1}{2}\right)\tilde{F}_2(\xi)\right) + o^{(2)}\left(\tilde{F}_2(\xi) - \tilde{F}_1(\xi)\right) + o^{(3)}\tilde{F}_3(\xi) = 0, \tag{8.110}$$

for $\xi \in \Omega$. The uniqueness of the Helmholtz decomposition then implies

$$\tilde{F}_2(\xi) - \tilde{F}_1(\xi) = 0, \quad \xi \in \Omega, \tag{8.111}$$

$$\left(\mathrm{D} + \tfrac{1}{2}\right)\tilde{F}_1(\xi) + \left(\mathrm{D} - \tfrac{1}{2}\right)\tilde{F}_2(\xi) = 0, \quad \xi \in \Omega, \tag{8.112}$$

provided that $\frac{1}{4\pi}\int_\Omega \tilde{F}_1(\eta) - \tilde{F}_2(\eta)\,d\omega(\eta) = 0$. Inserting $\tilde{F}_1 = \tilde{F}_2$ into the identity (8.112) leads to

$$\mathrm{D}\tilde{F}_1(\xi) = 0, \quad \xi \in \Omega. \tag{8.113}$$

Thus, $\tilde{F}_1(\xi) = 0$, $\xi \in \Omega$, since D is injective, and it follows $\tilde{F}_2(\xi) = 0$, $\xi \in \Omega$, so that uniqueness is assured for this alternative decomposition. □

To obtain a representation of the scalars $\tilde{F}_1, \tilde{F}_2, \tilde{F}_3$, we simply use (8.105)–(8.107), Lemma 2.49, and the known representations of the Helmholtz scalars F_1, F_2, F_3 that are available from Lemma 8.2.

Lemma 8.31. *Let f be of class $\mathrm{c}^{(1)}(\Omega)$, and $\tilde{F}_1, \tilde{F}_2, \tilde{F}_3$ be the scalars known from Theorem 8.30. Then*

$$\begin{aligned}\tilde{F}_1(\xi) &= \int_\Omega \left(\frac{1}{8\pi} S(\xi\cdot\eta)\eta - \frac{1}{4}\nabla^*_\eta \mathrm{D}^{-1}_\xi G(\Delta^*;\xi\cdot\eta) + \frac{1}{2}\nabla^*_\eta G(\Delta^*;\xi\cdot\eta)\right) \\ &\quad \cdot f(\eta)d\omega(\eta), \qquad (8.114)\\ \tilde{F}_2(\xi) &= \int_\Omega \left(\frac{1}{8\pi} S(\xi\cdot\eta)\eta - \frac{1}{4}\nabla^*_\eta \mathrm{D}^{-1}_\xi G(\Delta^*;\xi\cdot\eta) - \frac{1}{2}\nabla^*_\eta G(\Delta^*;\xi\cdot\eta)\right) \\ &\quad \cdot f(\eta)d\omega(\eta), \qquad (8.115)\\ \tilde{F}_3(\xi) &= -\int_\Omega \mathrm{L}^*_\eta G(\Delta^*;\xi\cdot\eta)\cdot f(\eta)d\omega(\eta), \qquad (8.116)\end{aligned}$$

for $\xi\in\Omega$. As usual, S denotes the single-layer kernel, and $\mathrm{D}^{-1}_\xi G(\Delta^*;\xi\cdot\eta)$ allows an explicit representation as given in Lemma 8.26.

For practical applications, our interest is not in the scalars $\tilde{F}_1, \tilde{F}_2, \tilde{F}_3$ but in the vectorial quantities $\tilde{o}^{(i)}\tilde{F}_i$, $i=1,2,3$. In order to find adequate integral representations, we have to interchange the operators $\tilde{o}^{(i)}$ with the integrations occurring in the representations of Lemma 8.31. Unfortunately, due to the singularities of the single-layer kernel and the fundamental solution of the Beltrami operator, this is not possible. As a remedy, we use regularizations as defined in Subsections 8.2.1 and 8.2.2.

Lemma 8.32. *Let f be of class $\mathrm{c}^{(2)}(\Omega)$, and assume $\frac{1}{4\pi}\int_\Omega F_1(\eta)d\omega(\eta) = 0$ for the radial Helmholtz scalar F_1 (note that this holds true if f satisfies the pre-Maxwell equations). Furthermore, let $S^\rho \in \mathrm{C}^{(1)}([-1,1])$ and $G^\rho(\Delta^*;\cdot)\in \mathrm{C}^{(2)}([-1,1])$ be regularizations of order 1 and 2, respectively, as introduced in Definitions 8.23, 8.16. If we introduce*

$$\begin{aligned}\mathbf{g}^{(1)}_\rho(\xi,\eta) &= \xi\otimes\eta\left(\frac{1}{2}\Delta^*_\eta G^\rho(\Delta^*;\xi\cdot\eta) + \frac{1}{16\pi}S^\rho(\xi\cdot\eta)\right) \qquad (8.117)\\ &\quad -\frac{1}{8\pi}\xi\otimes\nabla^*_\eta S^\rho(\xi\cdot\eta) + \frac{1}{4}\nabla^*_\xi\otimes s^\rho_{\nabla^*}(\eta,\xi)\\ &\quad -\frac{1}{8\pi}\nabla^*_\xi S^\rho(\xi,\eta)\otimes\eta - \frac{1}{2}\nabla^*_\xi\otimes\nabla^*_\eta G^\rho(\Delta^*;\xi\cdot\eta),\\ \mathbf{g}^{(2)}_\rho(\xi,\eta) &= \xi\otimes\eta\left(\frac{1}{2}\Delta^*_\xi G^\rho(\Delta^*;\xi\cdot\eta) - \frac{1}{16\pi}S^\rho(\xi\cdot\eta)\right) \qquad (8.118)\\ &\quad +\frac{1}{8\pi}\xi\otimes\nabla^*_\eta S^\rho(\xi\cdot\eta) - \frac{1}{4}\nabla^*_\xi\otimes s^\rho_{\nabla^*}(\eta,\xi)\\ &\quad +\frac{1}{8\pi}\nabla^*_\xi S^\rho(\xi,\eta)\otimes\eta - \frac{1}{2}\nabla^*_\xi\otimes\nabla^*_\eta G^\rho(\Delta^*;\xi\cdot\eta),\\ \mathbf{g}^{(3)}_\rho(\xi,\eta) &= -\mathrm{L}^*_\xi\otimes\mathrm{L}^*_\eta G^\rho(\Delta^*;\xi\cdot\eta), \qquad (8.119)\end{aligned}$$

for $\xi,\eta\in\Omega$ and $\rho\in(0,2)$, then we have

$$\lim_{\rho\to 0+}\sup_{\xi\in\Omega}\left|\tilde{o}^{(i)}\tilde{F}_i(\xi) - \int_\Omega \mathbf{g}^{(i)}_\rho(\xi,\eta)f(\eta)d\omega(\eta)\right| = 0, \quad i=1,2,3, \qquad (8.120)$$

where $\tilde{F}_1, \tilde{F}_2, \tilde{F}_1$ *denote the scalars known from Theorem 8.30, and the kernel* $s^{\rho}_{\nabla^*}(\cdot,\cdot)$ *is given by* (8.79).

Proof. By aid of (8.105) and Lemma 8.2, we obtain

$$\begin{aligned} \tilde{o}^{(1)}\tilde{F}_1(\xi) \; = \; & \frac{1}{2}\xi(\xi\cdot f(\xi)) + \frac{1}{4}\xi \mathrm{D}_\xi^{-1}(\xi\cdot f(\xi)) \\ & -\frac{1}{8}\xi \mathrm{D}_\xi^{-1}\int_\Omega \nabla^*_\eta G(\Delta^*;\xi\cdot\eta)\cdot f(\eta)d\omega(\eta) \\ & +\frac{1}{2}\xi \mathrm{D}_\xi\int_\Omega \nabla^*_\eta G(\Delta^*;\xi\cdot\eta)\cdot f(\eta)d\omega(\eta) \\ & -\frac{1}{2}\nabla^*_\xi \mathrm{D}_\xi^{-1}(\xi\cdot f(\xi)) + \frac{1}{4}\nabla^*_\xi \mathrm{D}_\xi^{-1}\int_\Omega \nabla^*_\eta G(\Delta^*;\xi\cdot\eta)\cdot f(\eta)d\omega(\eta) \\ & -\frac{1}{2}\nabla^*_\xi\int_\Omega \nabla^*_\eta G(\Delta^*;\xi\cdot\eta)\cdot f(\eta)d\omega(\eta), \end{aligned} \tag{8.121}$$

for $\xi\in\Omega$. The expression in the third row on the right-hand side of (8.121) cannot be easily handled since we have no explicit representation for the corresponding regularized convolution kernel. However, observing $\mathrm{D} = \mathrm{D}^{-1}\left(-\Delta^* + \frac{1}{4}\right)$, this inconvenience can be circumvented by writing

$$\begin{aligned} & \mathrm{D}_\xi \int_\Omega \nabla^*_\eta G(\Delta^*;\xi\cdot\eta)\cdot f(\eta)d\omega(\eta) \\ = \; & \frac{1}{4}\mathrm{D}_\xi^{-1}\int_\Omega \nabla^*_\eta G(\Delta^*;\xi\cdot\eta)\cdot f(\eta)d\omega(\eta) \\ & +\mathrm{D}_\xi^{-1}\Delta^*_\xi\int_\Omega G(\Delta^*;\xi\cdot\eta)\nabla^*_\eta\cdot f_{\tan}(\eta)d\omega(\eta) \\ = \; & \frac{1}{4}\mathrm{D}_\xi^{-1}\int_\Omega \nabla^*_\eta G(\Delta^*;\xi\cdot\eta)\cdot f(\eta)d\omega(\eta) + \mathrm{D}_\xi^{-1}\nabla^*_\xi\cdot f_{\tan}(\xi), \end{aligned} \tag{8.122}$$

where Theorem 6.14 has been used in the last step. The application of (8.122) to (8.121) then leads to the more favorable expression

$$\begin{aligned} \tilde{o}^{(1)}\tilde{F}_1(\xi) \; = \; & \frac{1}{2}\xi(\xi\cdot f(\xi)) + \frac{1}{4}\xi \mathrm{D}_\xi^{-1}(\xi\cdot f(\xi)) \\ & +\frac{1}{2}\xi \mathrm{D}_\xi^{-1}\nabla^*_\xi\cdot f_{\tan}(\xi) - \frac{1}{2}\nabla^*_\xi \mathrm{D}_\xi^{-1}(\xi\cdot f(\xi)) \\ & +\frac{1}{4}\nabla^*_\xi\int_\Omega \nabla^*_\eta \mathrm{D}_\xi^{-1} G(\Delta^*;\xi\cdot\eta)\cdot f(\eta)d\omega(\eta) \\ & -\frac{1}{2}\nabla^*_\xi\int_\Omega \nabla^*_\eta G(\Delta^*;\xi\cdot\eta)\cdot f(\eta)d\omega(\eta). \end{aligned} \tag{8.123}$$

Substituting $G(\Delta^*;\cdot)$ by its regularized version and the operator D^{-1} by a convolution with the regularized single-layer kernel $\frac{1}{4\pi}S^\rho$, we set

$$\begin{aligned}\left(\tilde{o}^{(1)}\tilde{F}_1\right)^\rho(\xi) &= \frac{1}{2}\xi(\xi\cdot f(\xi)) + \frac{1}{16\pi}\xi\int_\Omega S^\rho(\xi\cdot\eta)\,\eta\cdot f(\eta)d\omega(\eta) \quad (8.124)\\ &\quad -\frac{1}{8\pi}\xi\int_\Omega \nabla^*_\eta S^\rho(\xi\cdot\eta)\cdot f(\eta)d\omega(\eta)\\ &\quad -\frac{1}{8\pi}\int_\Omega \nabla^*_\xi S^\rho(\xi\cdot\eta)\,\eta\cdot f(\eta)d\omega(\eta)\\ &\quad +\frac{1}{4}\int_\Omega \left(\nabla^*_\xi\otimes s^\rho_{\nabla^*}(\eta,\xi)\right)f(\eta)d\omega(\eta)\\ &\quad -\frac{1}{2}\int_\Omega \left(\nabla^*_\xi\otimes\nabla^*_\eta G^\rho(\Delta^*;\xi\cdot\eta)\right)f(\eta)d\omega(\eta),\end{aligned}$$

for $\xi\in\Omega$ and $\rho\in(0,2)$. Since $G^\rho(\Delta^*;\cdot)$ is of class $\mathrm{C}^{(2)}([-1,1])$ and S^ρ of class $\mathrm{C}^{(1)}([-1,1])$, the considerations in Subsections 8.2.1 and 8.2.2 imply

$$\lim_{\rho\to0+}\sup_{\xi\in\Omega}\left|\tilde{o}^{(1)}\tilde{F}_1(\xi)-\left(\tilde{o}^{(1)}\tilde{F}_1\right)^\rho(\xi)\right| = 0. \quad (8.125)$$

That $\left(\tilde{o}^{(1)}\tilde{F}_1\right)^\rho$ is expressible as a convolution with the kernel $\mathbf{g}^{(1)}_\rho(\cdot,\cdot)$ from (8.117) can directly be seen by a reformulation of (8.124) together with

$$\xi\cdot f(\xi) = \Delta^*_\xi\int_\Omega G(\Delta^*;\xi\cdot\eta)\,\eta\cdot f(\eta)d\omega(\eta), \quad \xi\in\Omega, \quad (8.126)$$

since $\frac{1}{4\pi}\int_\Omega F_1(\eta)d\omega(\eta) = 0$ by assumption (cf. Lemma 8.2). The remaining cases $\tilde{o}^{(i)}\tilde{F}_i(\xi)$, $i=2,3$, follow by similar conclusions. □

Remark 8.33. We forgo treating a local variant of the decomposition with respect to the operators $\tilde{o}^{(i)}$. This would require a detailed study of boundary-value problems for the operator D and is beyond the scope of this book.

8.3.2 Numerical Application: The Earth's Crustal Field

The crustal magnetic field of the Earth is significantly weaker than the main field. Yet, it is of particular interest since it reflects the structure of the Earth's lithosphere. Additionally, as we pointed out in the introductory remarks of Chapter 5, there are iono- and magnetospheric sources that contribute to the Earth's magnetic field and mask the crustal field in which we are interested. Thus, in order to separate the crustal field contributions from the magnetic field data collected by satellites, the data sets have to be corrected for these undesired contributions (e.g., by subtracting main field and iono-/magnetospheric field models like IGRF (International Geomagnetic Reference Field), the CHAOS-model, or POMME (Potsdam Magnetic Model of the Earth)). However, the residual field might still contain contributions not belonging to the crustal field that have not been entirely taken care of by the subtracted models. To correct this, we can use the separation described in the beginning of this section. In doing so, we obtain the part of the residual that is due to sources in the interior of the satellite's orbit, and thus, filter out contributions that can obviously not originate from the Earth's lithosphere. As a result, we get an improved approximation of the crustal field.

One possibility to numerically implement the improvement of the crustal field approximation is to calculate a Fourier expansion of the input data set b^{input} in terms of the vector spherical harmonics $\tilde{y}^{(i)}_{n,k}$, $i = 1, 2, 3$, as described in (8.95)–(8.97). Considering only the contributions for $i = 1$ leads to the interior part p_b^{int} of the magnetic field. An alternative is to use the decomposition provided by Theorem 8.30 and Lemma 8.32. To determine p_b^{int}, we only consider the regularized approximation

$$\left(p_b^{\text{int}}\right)^\rho(x) = \left(\tilde{o}^{(1)}\tilde{B}_1\right)^\rho(x) = \int_\Omega \mathbf{g}_\rho^{(1)}(\xi,\eta)\, b^{\text{input}}(S\eta)d\omega(\eta), \tag{8.127}$$

for $\xi = \frac{x}{|x|}$, $S = |x|$. The latter method has been applied successfully to a set of CHAMP satellite data from 2001 that has been pre-processed at the GFZ in Potsdam. The data set is assumed to be collected on an orbital sphere Ω_S with radius $S = R_0 + 450$ km (where R_0 denotes the mean Earth radius of 6371.2 km). The results for the radial component and the north–south component (i.e., the component pointing in the direction of the vector ε^t of the spherical basis system ε^r, ε^φ, ε^t) of $\left(p_b^{\text{int}}\right)^\rho$ are shown in Figure 8.2. Particularly strong crustal field anomalies can be found in Central Africa, North America, and Northern Europe.

The difference between the calculated approximation $\left(p_b^{\text{int}}\right)^\rho$ of the crustal field and the pre-processed input data set b^{input} (see Figure 8.3) reveals polar contributions that are clearly not due to the crustal field, and thus, have been filtered out successfully. Furthermore, one can detect a positive trend of the radial component in the northern hemisphere and a negative trend in the southern hemisphere. This is a typical signature of a magnetospheric ring current, and is therefore also not part of the crustal field. For more detailed information on pre-processing procedures from a geophysical point of view, the reader is referred to, e.g., R.A. Langel, W.J. Hinze [1998], S. Maus et al. [2006], N. Olsen et al. [2010a], and N. Olsen et al. [2010b].

For the numerical evaluation of the occurring integral in (8.127), we have again used the rule due to J.R. Driscoll, R.M. Healy [1994]. The regularizations $G^\rho(\Delta^*;\cdot)$ and S^ρ, which are required to build up the kernel $\mathbf{g}_\rho^{(1)}(\cdot,\cdot)$ as stated in Lemma 8.32, have been constructed using Taylor polynomials of degree 8. Similar to Section 7.3, we are also able to realize a multiscale method with locally supported wavelets using the regularized kernel $\mathbf{g}_\rho^{(1)}(\cdot,\cdot)$ as a scaling function (see C. Gerhards [2011a], C. Gerhards [2012] for more details). A frequency-based multiscale method using spatially localizing wavelets that are constructed by superposition of the vector spherical harmonics $\tilde{y}^{(1)}_{n,k}$ can be found in C. Mayer [2006].

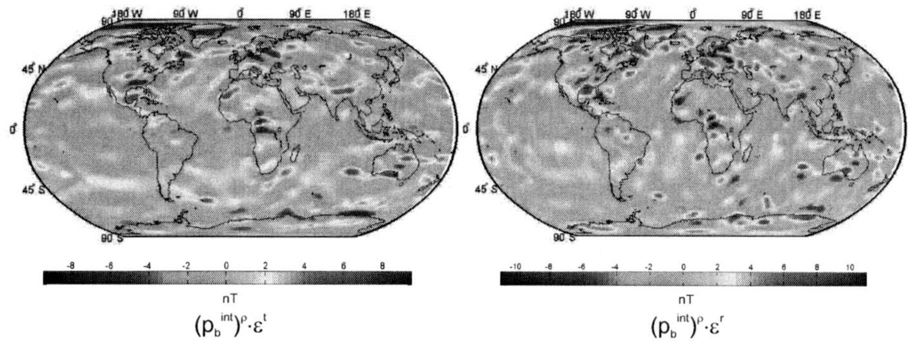

FIGURE 8.2
Approximation of the crustal field $\left(p_b^{\mathrm{int}}\right)^\rho$ for $\rho = 2^{-9}$: north–south component (left) and radial component (right).

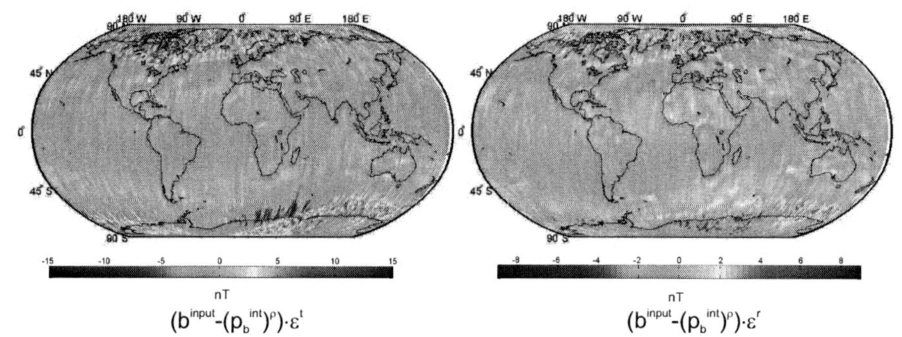

FIGURE 8.3
Difference between the input data b^{input} and the approximation of the crustal field $\left(p_b^{\mathrm{int}}\right)^\rho$: north–south component (left) and radial component (right).

8.4 Ionospheric Current Systems

In Section 5.1, we provided some details on the current systems of the Earth's iono- and magnetosphere. Here, we focus on the mathematical modeling of the field-aligned currents (FAC) in polar regions, coupling magnetospheric current systems with polar near-Earth ionospheric current systems, and horizontal (tangential) current systems in the lower ionosphere (at altitudes R_1 of approximately 100–150 km above the Earth's surface). Field-aligned currents cause magnetic fields that can be detected as non-potential fields at satellite altitude. They are nearly radial in polar regions and can be described by the Helmholtz scalar J_1 of the general current system j. In fact, J_1 at satellite altitude can be determined from knowledge of the magnetic field b on the (spherical) satellite orbit. This is different from the tangential current systems, where one additionally needs information on the radial derivative of b. The derivatives are generally not available from satellite measurements, such that additional assumptions on the current systems become necessary, e.g., that all tangential currents are located on the fixed sphere Ω_{R_1}. The determination of ionospheric current systems from real satellite magnetic field data is usually preceded by a careful selection of the data sets and the subtraction of main and crustal field models (e.g., IGRF, CHAOS, POMME, or the MF(Magnetic Field)-model) to obtain the magnetic field contributions that are due to the current systems under consideration. The interested reader is referred to, e.g., R.A. Langel, W.J. Hinze [1998] for more detailed information on data selection and pre-processing.

8.4.1 Radial Current Systems

Following the standard nomenclature in geomagnetism, we denote by b the twice continuously differentiable solenoidal magnetic field and by j the continuously differentiable current densities, such that the pre-Maxwell equations

$$\nabla \wedge b(x) = j(x), \tag{8.128}$$

$$\nabla \cdot b(x) = 0, \tag{8.129}$$

are satisfied in some spherical shell $\mathcal{B}_{R_0,R_1}(0)$ that contains the satellite orbit (for the sake of brevity, we have chosen the vacuum permeability μ_0 to be equal to one). From Lemma 5.12, we remember the relationships

$$P_j(x) = Q_b(x), \tag{8.130}$$

$$Q_j(x) = -\Delta P_b(x), \tag{8.131}$$

for the Mie scalars P_b, Q_b of the magnetic field, and P_j, Q_j of the current density. Letting $J_1(r\cdot)$, $J_2(r\cdot)$, and $J_3(r\cdot)$ be the Helmholtz scalars of $j(r\cdot)$, we are only interested in the radial contribution $J_1(r\cdot)$. From (8.130) and Lemma 8.6 we find

$$J_1(r\xi) = \frac{\Delta^*_\xi P_j(r\xi)}{r} = \frac{\Delta^*_\xi Q_b(r\xi)}{r}, \quad \xi \in \Omega. \tag{8.132}$$

Since we assume that the magnetic field b is known on the sphere Ω_r, $r \in (R_0, R_1)$, we can determine Q_b from the relation $q_b = \mathrm{L}^* Q_b$, i.e., $\Delta^* Q_b = \mathrm{L}^* \cdot q_b = \mathrm{L}^* \cdot b$. Observing that $\mathrm{L}^* \cdot b(r\cdot)$ has a vanishing integral mean value on Ω, the considerations from Remark 6.15 imply

$$Q_b(x) = -\int_\Omega \mathrm{L}^*_\eta G(\Delta^*; \xi \cdot \eta) \cdot b(r\eta) d\omega(\eta), \tag{8.133}$$

for $r = |x|$ and $\xi = \frac{x}{|x|}$. Application of Δ^* to (8.133) yields $J_1(r\cdot)$ by use of (8.132). It would be desirable to apply the Beltrami operator Δ^*_ξ directly to the kernel $\mathrm{L}^*_\eta G(\Delta^*;\xi\cdot\eta)$, so that numerical differentiation becomes unnecessary in the evaluation process from discrete data. However, the weak singularity of Green's function does not allow us to interchange the differentiation and the integration. But the Green function can be regularized around its singularity (as described in Definition 8.16) so that an interchange becomes possible. Consequently, an integral approximation of the radial current density is obtainable.

Lemma 8.34. *Let $b \in \mathrm{c}^{(2)}(\mathcal{B}_{R_0,R_1}(0))$ be solenoidal, and j be of class $\mathrm{c}^{(1)}(\mathcal{B}_{R_0,R_1}(0))$ such that the pre-Maxwell equations* (8.128), (8.129) *are satisfied. Furthermore, we set*

$$g^\rho_{\mathrm{rad}}(\xi,\eta) = -\mathrm{L}^*_\eta \Delta^*_\xi G^\rho(\Delta^*;\xi\cdot\eta), \tag{8.134}$$

for $\xi,\eta \in \Omega$ and $\rho \in (0,2)$, where $G^\rho(\Delta^*;\cdot) \in \mathrm{C}^{(3)}([-1,1])$ is a regularization of order 3 *as introduced in Definition 8.16. Then*

$$\lim_{\rho\to 0+}\sup_{\xi\in\Omega}\left| J_1(r\xi) - \frac{1}{r}\int_\Omega g^\rho_{\mathrm{rad}}(\xi,\eta)\cdot b(r\eta)\,d\omega(\eta)\right| = 0, \tag{8.135}$$

for $r \in (R_0,R_1)$. As always, $J_1(r\cdot)$ denotes the radial Helmholtz scalar of $j(r\cdot)$.

Proof. The assumption that $G^\rho(\Delta^*;\cdot)$ is of class $\mathrm{C}^{(3)}([-1,1])$ is necessary because $\mathrm{L}^*_\eta\Delta^*_\xi$ is a differential operator of order three. Observing (8.132) and (8.133), it remains to show that

$$\begin{aligned} & J_1(r\xi) - \frac{1}{r}\int_\Omega g^\rho_{\mathrm{rad}}(\xi,\eta)\cdot b(r\eta)\,d\omega(\eta) \qquad (8.136)\\ &= \frac{1}{r}\int_\Omega \Delta^*_\xi \mathrm{L}^*_\eta G^\rho(\Delta^*;\xi\cdot\eta)\cdot b(r\eta)\,d\omega(\eta) - \frac{1}{r}\Delta^*_\xi\int_\Omega \mathrm{L}^*_\eta G(\Delta^*;\xi\cdot\eta)\cdot b(r\eta)\,d\omega(\eta) \end{aligned}$$

converges to zero, uniformly with respect to $\xi \in \Omega$, as ρ tends to zero. Since $\Delta^* = \mathrm{L}^*\cdot\mathrm{L}^*$, we can use Green's formulas to shift two of the three differential operators to b. We obtain

$$\begin{aligned} & \Delta^*_\xi \int_\Omega \mathrm{L}^*_\eta G(\Delta^*;\xi\cdot\eta)\cdot b(r\eta)\,d\omega(\eta) \qquad (8.137)\\ &= -\Delta^*_\xi\int_\Omega G(\Delta^*;\xi\cdot\eta)\mathrm{L}^*_\eta\cdot b(r\eta)\,d\omega(\eta)\\ &= \mathrm{L}^*_\xi\cdot\int_\Omega \left(\mathrm{L}^*_\eta G(\Delta^*;\xi\cdot\eta)\right)\mathrm{L}^*_\eta\cdot b(r\eta)\,d\omega(\eta)\\ &= -\int_\Omega \mathrm{L}^*_\xi G(\Delta^*;\xi\cdot\eta)\cdot\mathrm{L}^*_\eta\left(\mathrm{L}^*_\eta\cdot b(r\eta)\right)d\omega(\eta) \end{aligned}$$

and, for the regularized expression,

$$\begin{aligned} &\int_\Omega \Delta^*_\xi \mathrm{L}^*_\eta G^\rho(\Delta^*;\xi\cdot\eta)\cdot b(r\eta)d\omega(\eta) \\ &= -\int_\Omega \mathrm{L}^*_\xi G^\rho(\Delta^*;\xi\cdot\eta)\cdot \mathrm{L}^*_\eta\left(\mathrm{L}^*_\eta\cdot b(r\eta)\right)d\omega(\eta). \end{aligned} \tag{8.138}$$

In connection with (8.137) and (8.138), the convergence of (8.136) follows from Lemma 8.18. □

Remark 8.35. Our approach to determine $J_1(r\xi)$, $\xi\in\Omega$, has the advantageous property that the kernel $g^\rho_{\mathrm{rad}}(\xi,\cdot)$ is locally supported. More precisely,

$$g^\rho_{\mathrm{rad}}(\xi,\eta) = -\mathrm{L}^*_\eta\Delta^*_\xi G^\rho(\Delta^*;\xi\cdot\eta) = 0, \tag{8.139}$$

for $\eta\in\Omega$ with $1-\xi\cdot\eta>\rho$, i.e., for $\eta\in\left(\Gamma_\rho(\xi)\right)^c$. In other words, it is sufficient to know the magnetic field on the spherical cap $\Gamma_\rho(\xi)$ in order to approximate $J_1(r\xi)$. This fact is also the reason why it is not necessary to treat an explicit local approach in the manner of Subsection 8.1.2 for the modeling from only locally available data.

The classical approach of determining the radial current densities is by use of Fourier expansions in terms of the spherical harmonics from Sections 2.3 and 2.5, and is described next. Due to the global nature of the spherical harmonics, this approach is less suited for modeling from only locally available data, but it works well for global satellite data. Since $q_b(r\cdot)=\mathrm{L}^*Q_b(r\cdot)$, the vector spherical harmonics $y^{(3)}_{n,k}$ suffice to expand the toroidal part of the magnetic field, i.e.,

$$\begin{aligned} q_b(x) &= \sum_{n=1}^{\infty}\sum_{k=1}^{2n+1}\left(b^{(3)}(r\cdot)\right)^\wedge(n,k)\,y^{(3)}_{n,k}(\xi) \\ &= \sum_{n=1}^{\infty}\sum_{k=1}^{2n+1}\left(\mu^{(3)}_n\right)^{-\frac{1}{2}}\left(b^{(3)}(r\cdot)\right)^\wedge(n,k)\,\mathrm{L}^*_\xi Y_{n,k}(\xi) \\ &= \mathrm{L}^*_\xi\sum_{n=1}^{\infty}\sum_{k=1}^{2n+1}\left(\mu^{(3)}_n\right)^{-\frac{1}{2}}\left(b^{(3)}(r\cdot)\right)^\wedge(n,k)\,Y_{n,k}(\xi), \end{aligned} \tag{8.140}$$

for $\xi=\frac{x}{|x|}$ and $r=|x|\in(R_0,R_1)$. By

$$\left(b^{(3)}(r\cdot)\right)^\wedge(n,k) = \int_\Omega b(r\eta)\cdot y^{(3)}_{n,k}(\eta)d\omega(\eta) \tag{8.141}$$

we denote the Fourier coefficient of degree n and order k, and by $\mu^{(3)}_n$ the normalization constant of the vector spherical harmonic. Thus, we find for the toroidal scalar:

$$Q_b(x) = \sum_{n=1}^{\infty}\sum_{k=1}^{2n+1}\left(\mu^{(3)}_n\right)^{-\frac{1}{2}}\left(b^{(3)}(r\cdot)\right)^\wedge(n,k)\,Y_{n,k}(\xi). \tag{8.142}$$

Remembering (8.132), the application of the Beltrami operator to the toroidal scalar from (8.142) yields the representation of the radial current density stated in the following lemma.

Lemma 8.36. *Let* $b \in \mathrm{c}^{(2)}(\mathcal{B}_{R_0,R_1}(0))$ *be solenoidal, and* j *be of class* $\mathrm{c}^{(1)}(\mathcal{B}_{R_0,R_1}(0))$ *such that the pre-Maxwell equations* (8.128), (8.129) *are satisfied. Then*

$$J_1(r\xi) = -\frac{1}{r}\sum_{n=1}^{\infty}\sum_{k=1}^{2n+1} n(n+1)\left(\mu_n^{(3)}\right)^{-\frac{1}{2}}\left(b^{(3)}(r\cdot)\right)^{\wedge}(n,k)\, Y_{n,k}(\xi),$$

for $r \in (R_0, R_1)$, $\xi \in \Omega$, *with* $\left(b^{(3)}(r\cdot)\right)^{\wedge}(n,k)$ *given by* (8.141).

The approach of Lemma 8.36 has been proposed, e.g., by N. Olsen [1997], where some additional geophysical background on the current systems can be found. To avoid the global character of spherical harmonics for radial current determination, superposition of spherical harmonics has been applied in M. Bayer et al. [2001] and T. Maier [2005] to achieve wavelet kernels with a better spatial localization. These superpositions can be regarded as a frequency-oriented regularization of the current systems, while Lemma 8.34 states a spatially oriented regularization. For a multiscale method with locally supported wavelets similar to Section 7.3, the reader is referred to W. Freeden, C. Gerhards [2010].

8.4.2 Tangential Current Systems

Different from the radial current systems, tangential current systems cannot be calculated from the knowledge of the magnetic field b on one fixed sphere only. Indeed, Exercise 8.6 shows that the radial derivatives $\frac{\partial}{\partial r}b$ are also required. These, however, are typically not available from satellite measurements. To circumvent this problem, we assume that all tangential currents are located on a fixed sphere Ω_C in the ionosphere. Since the current-carrying ionosphere is relatively thin in comparison to the entire iono-/magnetosphere structure, this is a legitimate assumption. In Section 5.5, an example with MAGSAT data showed that the surface divergence-free contribution j_{df}, i.e., the toroidal contribution L^*J_3, of the tangential current system $j_{\tan}$ can be reconstructed by inversion of the spherical Biot–Savart operator. The surface curl-free contribution j_{cf}, i.e., the contribution that is represented by ∇^*J_2 (where J_2, J_3 denote the corresponding Helmholtz scalars of $j_{\tan}$), can be obtained, e.g., by its connection to the radial contribution J_1 of the current density $j = j_{\tan} + j_{\mathrm{rad}}$: Ohm's Law tells us that

$$j(x) = \boldsymbol{\sigma}(x)e(x), \quad x \in \mathcal{B}_{R_0,R_1}(0), \tag{8.143}$$

where e denotes the electric field and $\boldsymbol{\sigma}$ the conductivity tensor. Using spherical coordinates, i.e., $j = (J_t, J_\varphi, J_r)^T = J_t\varepsilon^t + J_\varphi\varepsilon^\varphi + J_r\varepsilon^r$ and

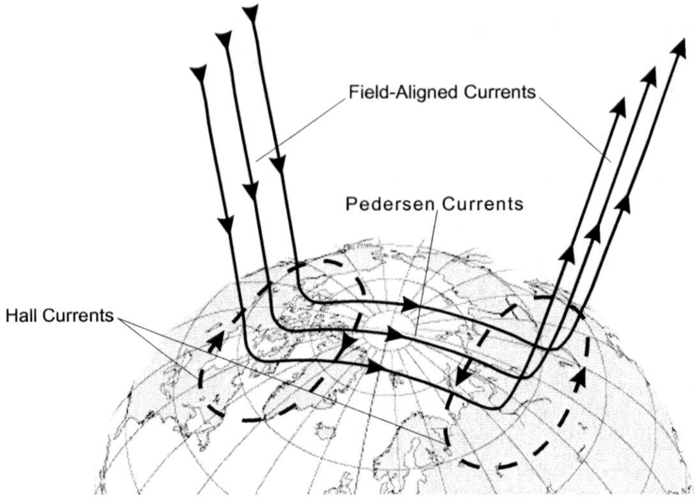

FIGURE 8.4
Simplified illustration of ionospheric Hall and Pedersen currents at high latitudes.

$e = (E_t, E_\varphi, E_r)^T = E_t\varepsilon^t + E_\varphi\varepsilon^\varphi + E_r\varepsilon^r$, the conductivity tensor at high latitudes (where it is valid to assume that the magnetic field lines are radial with respect to the spherical Earth) can be expressed by

$$\boldsymbol{\sigma} = \begin{pmatrix} \sigma_P & -\sigma_H & 0 \\ \sigma_H & \sigma_P & 0 \\ 0 & 0 & \sigma_r \end{pmatrix}. \tag{8.144}$$

σ_r denotes the direct ionospheric conductivity, while σ_P and σ_H mean the Pedersen conductivity and the Hall conductivity, respectively. The corresponding Pedersen currents form those ionospheric currents that flow perpendicular to the magnetic field and in direction of the electric field. Hall currents are perpendicular to the magnetic field and the electric field. In polar regions, where the magnetic field lines are oriented radially, the Pedersen currents coincide with j_{cf} and the Hall currents are equal to j_{df}. Moreover, the Pedersen currents are those currents compensating in- and outgoing field-aligned currents in the ionospheric current sheet (see Figure 8.4).

Since we assume that the tangential currents are located on a sphere Ω_C, the conductivities σ_H and σ_P are substituted by their height-integrated counterparts, i.e.,

$$\Sigma_P(\xi) = \int_{R_0}^{R_1} \sigma_P(r\xi)dr, \quad \Sigma_H(\xi) = \int_{R_0}^{R_1} \sigma_H(r\xi)dr, \quad \xi \in \Omega. \tag{8.145}$$

Here, $\mathcal{B}_{R_0,R_1}(0)$ denotes a spherical shell describing the current-carrying ionosphere. The sphere Ω_C is contained in this shell. Thus, we can express the height-integrated tangential current densities by the modified Ohm Law

$$\tilde{j}_{\tan}(C\xi) = \mathbf{\Sigma}(\xi)\, \tilde{e}_{\tan}(C\xi), \quad \xi \in \Omega, \tag{8.146}$$

with

$$\mathbf{\Sigma} = \begin{pmatrix} \Sigma_P & \Sigma_H \\ \Sigma_H & -\Sigma_P \end{pmatrix} \tag{8.147}$$

and $\tilde{j}_{\tan} = (\tilde{J}_t, \tilde{J}_\varphi)^T$, $\tilde{e}_{\tan} = (\tilde{E}_t, \tilde{E}_\varphi)^T$. For convenience, we keep writing $j_{\tan} : \Omega_C \to \mathbb{R}^3$ in the remainder of this subsection although we mean the height-integrated tangential current densities in the sense of $\tilde{j}_{\tan}$. Current continuity then implies the following connection of the radial current systems to the tangential current systems:

$$\nabla^* \cdot j_{\tan}(C\xi) \;=\; CJ_r(C\xi), \quad \xi \in \Omega. \tag{8.148}$$

Finally, observe that J_r and the Helmholtz scalar J_1 are just two different notations for the radial contribution of the current density j. More details on the geophysical background can be found, e.g., in O. Amm [1997], S.W.H. Cowley [2000], N. Fukushima [1976], and references therein.

The differential equation (8.148) determines the surface curl-free contribution j_{cf}. Since $\nabla^* \cdot \mathrm{L}^* = 0$, the surface divergence-free contribution j_{df} is not obtainable from (8.148). In order to obtain $j_{\tan} = j_{\mathrm{cf}} + j_{\mathrm{df}}$ in total, we need to solve the equations

$$\nabla^* \cdot j_{\tan}(R\xi) \;=\; RJ_1(R\xi), \quad \xi \in \Omega, \tag{8.149}$$

$$\mathrm{L}^* \cdot j_{\tan}(R\xi) \;=\; H(R\xi), \quad \xi \in \Omega, \tag{8.150}$$

(note that J_r is equal to the radial Helmholtz scalar J_1 and that the radius C of the current-carrying sphere has been changed to R in order to return to a mathematically more intuitive notation for the next assertions.) However, the function H is not known from the available data sets, so that (8.149), (8.150) only determine j_{df} in a theoretical sense. But as we have already seen in Section 5.5, there is a way out, since j_{df} can be determined from real data sets by the inversion of the spherical Biot–Savart operator. The surface curl-free part can be obtained from (8.149) since the radial current density J_1 is an available quantity (cf. Subsection 8.4.1). The local version of (8.149), (8.150) reads as

$$\nabla^* \cdot j_{\tan}(R\xi) \;=\; RJ_1(R\xi), \quad \xi \in \Gamma, \tag{8.151}$$

$$\mathrm{L}^* \cdot j_{\tan}(R\xi) \;=\; H(R\xi), \quad \xi \in \Gamma, \tag{8.152}$$

$$\nu(\xi) \cdot j_{\mathrm{cf}}(R\xi) \;=\; G_{\mathrm{cf}}(R\xi), \quad \xi \in \partial\Gamma, \tag{8.153}$$

$$\tau(\xi) \cdot j_{\mathrm{df}}(R\xi) \;=\; G_{\mathrm{df}}(R\xi), \quad \xi \in \partial\Gamma, \tag{8.154}$$

where Γ is a regular region on Ω. Equations (8.153), (8.154) constitute Neumann boundary values for the Helmholtz scalars J_2 and J_3 of $j_{\tan}$.

From Corollary 8.3 and Lemma 8.18, we obtain the following approximation of a solution to (8.149), (8.150).

Lemma 8.37. *Let J_1, H be of class $\mathrm{C}^{(0)}(\Omega_R)$, and assume that $j_{\mathrm{tan}} = j_{\mathrm{cf}} + j_{\mathrm{df}} \in \mathrm{c}^{(1)}(\Omega_R)$ is tangential and satisfies the differential equations (8.149), (8.150). By $j_{\mathrm{cf}} = \nabla^* J_2$, we mean the (surface) curl-free part of j_{tan}, and by $j_{\mathrm{df}} = \mathrm{L}^* J_3$, the (surface) divergence-free part (J_2, J_3 are the corresponding Helmholtz scalars of j_{tan}). Furthermore, we set*

$$g^{\rho}_{\mathrm{cf}}(\xi,\eta) = \nabla^*_{\xi} G^{\rho}(\Delta^*; \xi\cdot\eta), \tag{8.155}$$

$$g^{\rho}_{\mathrm{df}}(\xi,\eta) = \mathrm{L}^*_{\xi} G^{\rho}(\Delta^*; \xi\cdot\eta), \tag{8.156}$$

for $\xi,\eta \in \Omega$ and $\rho \in (0,2)$, where $G^{\rho}(\Delta^*;\cdot) \in \mathrm{C}^{(1)}([-1,1])$ is a regularization of order 1 as introduced in Definition 8.16. Then

$$\lim_{\rho\to 0+} \sup_{\xi\in\Omega} \left| j_{\mathrm{cf}}(R\xi) - R\int_{\Omega} g^{\rho}_{\mathrm{cf}}(\xi,\eta) J_1(R\eta)\, d\omega(\eta)\right| = 0, \tag{8.157}$$

$$\lim_{\rho\to 0+} \sup_{\xi\in\Omega} \left| j_{\mathrm{df}}(R\xi) - \int_{\Omega} g^{\rho}_{\mathrm{df}}(\xi,\eta) H(R\eta)\, d\omega(\eta)\right| = 0. \tag{8.158}$$

Remark 8.38. In order to guarantee the existence of a solution to (8.149), (8.150), it is necessary to claim that

$$\frac{1}{4\pi}\int_{\Omega} J_1(R\eta)\, d\omega(\eta) = \frac{1}{4\pi}\int_{\Omega} H(R\eta)\, d\omega(\eta) = 0. \tag{8.159}$$

For J_1, this condition is satisfied as soon as the current density j satisfies the pre-Maxwell equations.

A local version of Lemma 8.37 can be derived using Exercise 8.4.

Lemma 8.39. *Let $\Gamma \subset \Omega$, and suppose that J_1, H are of class $\mathrm{C}^{(1)}(\overline{\Gamma_R})$ and that $G_{\mathrm{cf}}, G_{\mathrm{df}}$ are of class $\mathrm{C}^{(0)}(\partial\Gamma_R)$, where $\Gamma_R = \{x \in \Omega_R : \frac{x}{|x|} \in \Gamma\}$. Furthermore, the function $j_{\mathrm{tan}} = j_{\mathrm{cf}} + j_{\mathrm{df}} \in \mathrm{c}^{(2)}(\overline{\Gamma_R})$ is assumed to be tangential and satisfy the boundary-value problem (8.151)–(8.154). Finally, we set*

$$g^{\rho}_{\mathrm{cf},\Gamma}(\xi,\eta) = \nabla^*_{\xi} \left(G^{(N)}\right)^{\rho}(\Delta^*; \xi,\eta), \tag{8.160}$$

$$g^{\rho}_{\mathrm{df},\Gamma}(\xi,\eta) = \mathrm{L}^*_{\xi} \left(G^{(N)}\right)^{\rho}(\Delta^*; \xi,\eta), \tag{8.161}$$

for $\xi,\eta \in \Omega$ and $\rho \in (0,2)$. By $G^{\rho}(\Delta^*;\cdot) \in \mathrm{C}^{(1)}([-1,1])$, we mean a regularization of order 1 as introduced in Definition 8.16, and $\left(G^{(N)}\right)^{\rho}(\xi,\eta) = G^{\rho}(\Delta^*;\xi\cdot\eta) - \Phi^{(N)}(\xi,\eta)$ denotes the corresponding regularized Neumann Green function (where $\Phi^{(N)}$ is determined by the conditions from Definition 6.47). Then

$$\lim_{\rho\to 0+} \sup_{\xi\in\Sigma} \left| j_{\mathrm{cf}}(R\xi) - R\int_{\Gamma} g^{\rho}_{\mathrm{cf},\Gamma}(\xi,\eta) J_1(R\eta)\, d\omega(\eta) \right. \tag{8.162}$$

$$\left. - \int_{\partial\Gamma} g^{\rho}_{\mathrm{cf},\Gamma}(\xi,\eta) G_{\mathrm{cf}}(R\eta)\, d\sigma(\eta)\right| = 0,$$

$$\lim_{\rho\to 0+}\sup_{\xi\in\Sigma}\left| j_{\mathrm{df}}(R\xi) - \int_\Gamma g^{\rho}_{\mathrm{df},\Gamma}(\xi,\eta)H(R\eta)d\omega(\eta)\right. \tag{8.163}$$
$$\left. - \int_{\partial\Gamma} g^{\rho}_{\mathrm{df},\Gamma}(\xi,\eta)G_{\mathrm{df}}(R\eta)d\sigma(\eta)\right| = 0,$$

for any subregion $\Sigma \Subset \Gamma$.

Remark 8.40. Since (8.153), (8.154) are just different variants of Neumann boundary values, we need to assume

$$\int_{\partial\Gamma} G_{\mathrm{cf}}(R\eta)d\sigma(\eta) - \int_\Gamma J_1(R\eta)d\omega(\eta) = \int_{\partial\Gamma} G_{\mathrm{df}}(R\eta)d\sigma(\eta) - \int_\Gamma H(R\eta)d\omega(\eta) = 0 \tag{8.164}$$

in order to guarantee the existence of a solution to (8.151)–(8.154) in the first place.

The kernels (8.155), (8.156) can again be regarded as spatial regularizations. However, different from the kernel (8.134) for the reconstruction of radial current densities, these kernels are not locally supported. Thus, it is useful to supply the local version from Lemma 8.39 for reconstructions from only locally available data.

Returning to the global framework, Fourier expansions of j_{cf}, j_{df} in terms of vector spherical harmonics can be derived (compare Lemma 8.36 for radial current densities). Regarding (8.149), we first observe that the known radial current density J_1 can be expanded in terms of (scalar) spherical harmonics:

$$J_1(R\xi) = \sum_{n=1}^{\infty}\sum_{k=1}^{2n+1} (J_1(R\cdot))^\wedge(n,k)Y_{n,k}(\xi), \quad \xi\in\Omega \tag{8.165}$$

with Fourier coefficients $(J_1(R\cdot))^\wedge(n,k) = \int_\Omega J_1(R\eta)Y_{n,k}(\eta)d\omega(\eta)$. Note that $(J_1(R\cdot))^\wedge(0,1) = 0$ because $J_1 = \nabla^*\cdot j_{\mathrm{tan}}$ implies $\int_\Omega J_1(R\eta)d\omega(\eta) = 0$ by virtue of the (surface) theorem of Gauss. Under sufficient differentiability assumptions on the functions, it can be easily seen that

$$\begin{aligned}
\nabla^*_\xi\cdot\sum_{n=1}^{\infty}\sum_{k=1}^{2n+1}\frac{1}{-n(n+1)}(J_1(R\cdot))^\wedge(n,k)\left(\mu_n^{(2)}\right)^{\frac12} y^{(2)}_{n,k}(\xi) \quad & \quad (8.166)\\
= \sum_{n=1}^{\infty}\sum_{k=1}^{2n+1}\frac{1}{-n(n+1)}(J_1(R\cdot))^\wedge(n,k)\Delta^*_\xi Y_{n,k}(\xi) & \\
= \sum_{n=1}^{\infty}\sum_{k=1}^{2n+1}(J_1(R\cdot))^\wedge(n,k)Y_{n,k}(\xi) & \\
= J_1(R\xi), &
\end{aligned}$$

for $\xi\in\Omega$. Similar results hold true for the differential equation (8.150), so that we can state the following lemma.

Lemma 8.41. *Let J_1, H be of class $\mathrm{C}^{(1)}(\Omega_R)$. Assume that the current system $j_{\tan} = j_{\mathrm{cf}} + j_{\mathrm{df}} \in \mathrm{c}^{(2)}(\Omega_R)$ is tangential and satisfies the differential equations (8.149), (8.150). Then we have*

$$j_{\mathrm{cf}}(R\xi) = -\sum_{n=1}^{\infty}\sum_{k=1}^{2n+1} \frac{R}{n(n+1)} (J_1(R\cdot))^{\wedge}(n,k) \left(\mu_n^{(2)}\right)^{\frac{1}{2}} y_{n,k}^{(2)}(\xi), \quad \xi \in \Omega, \tag{8.167}$$

$$j_{\mathrm{df}}(R\xi) = -\sum_{n=1}^{\infty}\sum_{k=1}^{2n+1} \frac{1}{n(n+1)} (H(R\cdot))^{\wedge}(n,k) \left(\mu_n^{(3)}\right)^{\frac{1}{2}} y_{n,k}^{(3)}(\xi), \quad \xi \in \Omega. \tag{8.168}$$

The Fourier coefficients are given by

$$\left(J_1(R\cdot)\right)^{\wedge}(n,k) = \int_{\Omega} J_1(R\eta) Y_{n,k}(\eta)\, d\omega(\eta), \tag{8.169}$$

$$\left(H(R\cdot)\right)^{\wedge}(n,k) = \int_{\Omega} H(R\eta) Y_{n,k}(\eta)\, d\omega(\eta). \tag{8.170}$$

8.4.3 Numerical Application: FACs and Tangential Currents

8.4.3.1 MAGSAT (field-aligned currents)

The FACs can be determined from Lemma 8.34. We have applied this approximation method to a set of MAGSAT magnetic field measurements collected during March 1980 (essentially the same measurements that we used in Subsection 5.5.2). The data set has been pre-processed at DTU Space, the geomagnetic reference field GSFC(12/83) up to degree 12 has been subtracted to obtain the contributions due to ionospheric current systems. The numerically evaluated quantity is given by

$$J_1^{\rho}(x) = \frac{1}{S}\int_{\Omega} g_{\mathrm{rad}}^{\rho}(\xi,\eta) \cdot b^{\mathrm{input}}(R\eta), \tag{8.171}$$

for $\xi = \frac{x}{|x|}$ and $S = |x|$, with $S = R_0 + 450$ km being the radius of the satellite's orbit and R_0 the mean Earth radius. The results for different parameters $\rho \in (0,2)$ are shown in Figure 8.5. Dominating features are strong polar current systems and weaker currents along the dip equator (indicating the radial component of the equatorial electrojet). It has to be noted that the chosen input data set contains only data collected during evening local time (i.e., during times when the satellite was in the dusk sector with respect to the position of the sun). This choice pays tribute to the solar dependence of the ionospheric currents. In a data set collected during morning local time, the equatorial electrojet can hardly be detected (see, e.g., N. Olsen [1997] and T. Maier [2005]).

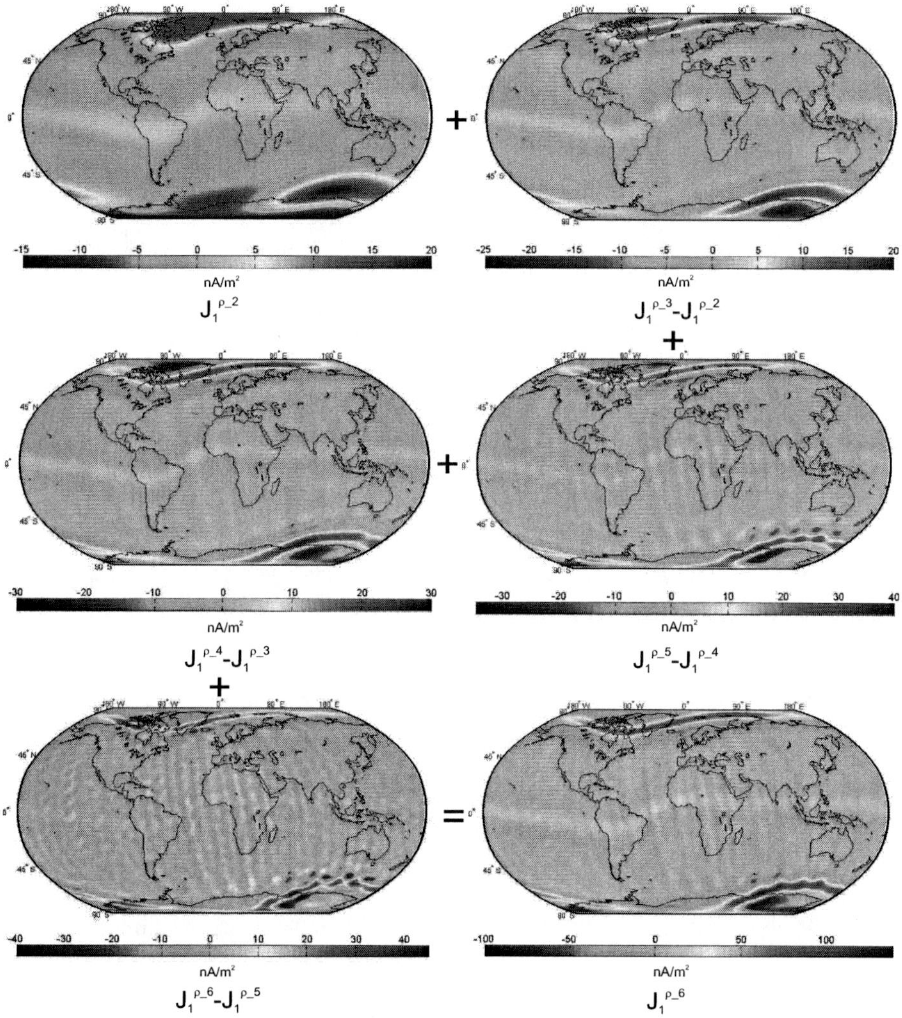

FIGURE 8.5
Multiscale approximation of the radial current densities $J_1^{\rho_k}(R\cdot)$ and their differences from MAGSAT data, for $\rho_k = 2^{-k}$, $k = 2, \ldots, 6$.

In Figure 8.5, we can clearly see that smaller parameters ρ lead to a better spatial resolution of the radial current systems. The differences $J_1^{\rho_{k+1}} - J_1^{\rho_k}$ resolve even finer structures. For small ρ_k, they actually resolve features along the satellite tracks that are not caused by physical processes but originate in measurement inaccuracies of the satellite. In the case $J_1^{\rho_6} - J_1^{\rho_5}$, these features nearly mask the equatorial electrojet entirely. This observation gives an indication of which parameter ρ is a good choice for the approximation J_1^ρ in order to avoid too strong an influence of measurement errors.

For the numerical evaluation of the occurring integral in (8.171), we have used the integration rule due to J.R. Driscoll, R.M. Healy [1994], and the regularization $G^\rho(\Delta^*;\cdot)$ has been constructed using a Taylor polynomial of degree eight (as we have already done for the applications in Subsection 8.3.2). The differences $J_1^{\rho_{k+1}} - J_1^{\rho_k}$ actually represent wavelet contributions analogous to those discussed in Section 7.3. We omit a formal mathematical description due to the similarities, more details can be found in C. Gerhards [2011a].

8.4.3.2 CHAMP (field-aligned currents)

More recent satellite missions supply us with a large amount of accurate magnetic field data over a longer time period than MAGSAT. A study of J_1^ρ with seven years of CHAMP data (2001–2007) is shown in Figure 8.6 (for brevity, we only show the final approximations for a fixed parameter $\rho = 2^{-6}$; see C. Gerhards [2011a] for more details).

The data processing has been more sophisticated than for the previously indicated MAGSAT example. We distinguish four cases, depending on the sign of the y- and z-components of the interplanetary magnetic field (IMF) at the time of measurement, i.e., we are confronted with IMF $B_y \gtrless 0$ and IMF $B_z \gtrless 0$ (note that, in the nomenclature of geomagnetics, the x-, y- and z-components in the IMF coordinate system are chosen such that the z-axis is directed northward, perpendicular to the ecliptic plane of the Earth, while the x-axis lies in the ecliptic plane pointing from the Earth toward the Sun, and the y-axis completes a right-handed orthogonal set). Furthermore, the data has been split with respect to the season: northern winter (November to February), northern summer (from May to August), and equinox (March, April, September, October). Finally, the main field and magnetospheric contributions from the CHAOS-3 model (cf. N. Olsen et al. [2010c]) have been subtracted at DTU Space to obtain the part of the magnetic field data that is due to ionospheric contributions. This forms our input data set b^{input}.

The described pre-selection takes into account the strong solar dependence of the ionospheric current systems. In Figure 8.6, we can see that on the summer hemispheres, as a result of the increased solar heating, FACs are clearly stronger than on the winter hemispheres. Furthermore, we can observe that during times with a negative z-component of the IMF, the currents are stronger than during times with a positive z-component. The reason for this is that during times with IMF $B_z < 0$, the orientation of the IMF is such that

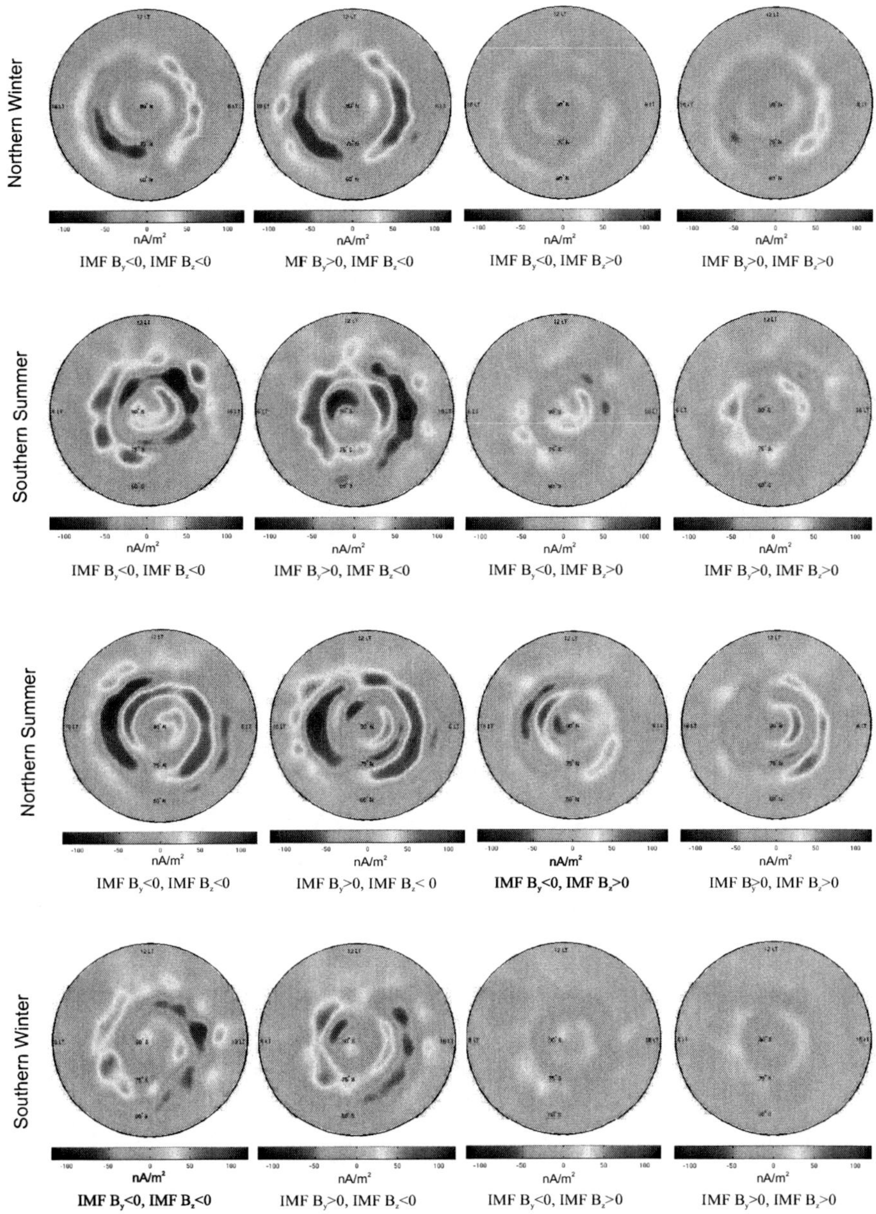

FIGURE 8.6

Approximation of the radial current densities $J_1^\rho(R\cdot)$ from CHAMP data, for $\rho = 2^{-6}$, in SM coordinates. The first and third rows show polar caps on the northern hemisphere, and the second and last rows, polar caps on the southern hemisphere.

the interaction with the Earth's main field simplifies the penetration of the solar wind through the shielding Earth's magnetic field. More detailed geophysical studies with a similar data pre-selection and some more geophysical background can be found in F. Christiansen et al. [2002] and D.R. Weimer [2001].

It should be noted that Figure 8.6 uses a different coordinate system than Figure 8.5. While we have used an Earth-fixed coordinate system (based on latitude and longitude with respect to the geographic North Pole) for MAGSAT data, solar-magnetic (SM) coordinates are used for the present example. This means that the dipole North Pole takes over the role of the geographic North Pole, that latitude is given with respect to the dipole North Pole, and that geographic longitude is substituted by local time (LT). Local time (in hours) describes the position of the satellite with respect to the sun at the time of measurement: during noon (12LT) the satellite is on the dayside of the Earth, while it is on the nightside during midnight (00LT; in Figure 8.6, noon is always located at the top of each plot, and midnight at the bottom). This Sun-fixed SM coordinate system is better suited to represent the solar dependence of the ionospheric current systems than an Earth-fixed geographic coordinate system. In the previous example that is based on MAGSAT data, the solar dependence has been taken into account by a separation into evening and morning data sets. The data coverage of MAGSAT with respect to SM coordinates simply is not sufficient to use this better suited coordinate system.

8.4.3.3 MAGSAT (tangential currents):

We present an example for the (surface) curl-free part j_{cf} of the tangential current system j_{tan}. (Surface) divergence-free current systems have already been treated in Section 5.5 by inversion of the spherical Biot–Savart operator. We now try to obtain the (surface) curl-free contribution by solving (8.149) for the global case and (8.151), (8.153) for the local case. For the approximation from global data (cf. Lemma 8.37), we use

$$j^{\rho}_{\mathrm{cf}}(C\xi) = C\int_{\Omega} g^{\rho}_{\mathrm{cf}}(\xi,\eta)J_1(C\eta)\,d\omega(\eta), \quad \xi\in\Omega, \tag{8.172}$$

where C denotes the radius of the current-carrying sphere Ω_C in the ionosphere. For the local case (cf. Lemma 8.39), we apply

$$\begin{aligned} j^{\rho}_{\mathrm{cf},\Gamma}(C\xi) &= C\int_{\Gamma} g^{\rho}_{\mathrm{cf},\Gamma}(\xi,\eta)J_1(C\eta)\,d\omega(\eta) \qquad (8.173)\\ &\quad -\int_{\partial\Gamma} g^{\rho}_{\mathrm{cf},\Gamma}(\xi,\eta)G_{\mathrm{cf}}(C\eta)\,d\sigma(\eta), \quad \xi\in\Gamma\\ j^{\rho}_{\mathrm{cf},\partial\Gamma}(C\xi) &= -\int_{\partial\Gamma} g^{\rho}_{\mathrm{cf},\Gamma}(\xi,\eta)G_{\mathrm{cf}}(C\eta)\,d\sigma(\eta), \quad \xi\in\Gamma. \qquad (8.174)\end{aligned}$$

The term $j^{\rho}_{\mathrm{cf},\partial\Gamma}$ is discussed separately to illustrate in Figure 8.8 that the boundary contributions cannot be neglected for reconstructions from local data. In our example, we choose Γ to be a spherical cap around the North Pole with a radius of 40° in latitude. The input data $J_1(C\cdot)$ are the radial current densities obtained from MAGSAT data at the beginning of this subsection. However, we have to observe that we only know $J_1(S\cdot)$ at satellite altitude $S = R_0 + 450$ km (R_0 being the mean Earth radius). In order to obtain data $J_1(C\cdot)$ on a sphere of radius $C = R_0 + 110$ km located in the current-carrying ionosphere, we assume current continuity to get

$$S^2 J_1(S\xi) = C^2 J_1(C\xi), \quad \xi \in \Omega. \tag{8.175}$$

The boundary values G_{cf} for the local reconstruction are obtained from j_{cf} calculated by the global approach. (This seems somewhat contradictory to the idea of a local reconstruction. But on the one hand, this example is just meant to illustrate the algorithm (one might as well have used already existing models for the boundary information), and on the other hand, it can be faster to obtain boundary values via the global approach and apply the local scheme afterward than using the global approach for every point in the region of interest.) The numerical integration method as well as the regularized fundamental solution $G^\rho(\Delta^*;\cdot)$ required for g_{cf}^ρ and $g_{\mathrm{cf},\Gamma}^\rho$ are chosen to be the same as for the previous examples.

The results for the (surface) curl-free currents are shown in Figures 8.7 and 8.8. As can be expected from the physical intuition, the patterns of these currents are very similar to the radial currents that have been taken as input

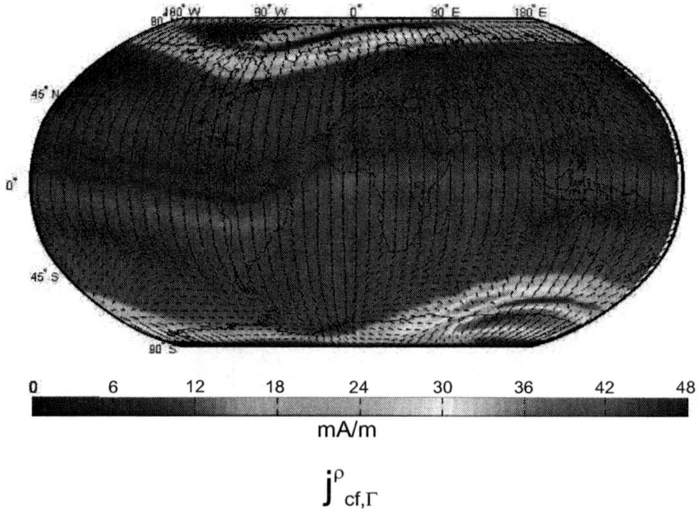

FIGURE 8.7
Approximation of the absolute values and the orientation of the surface curl-free current densities j_{cf}^ρ for $\rho = 2^{-6}$.

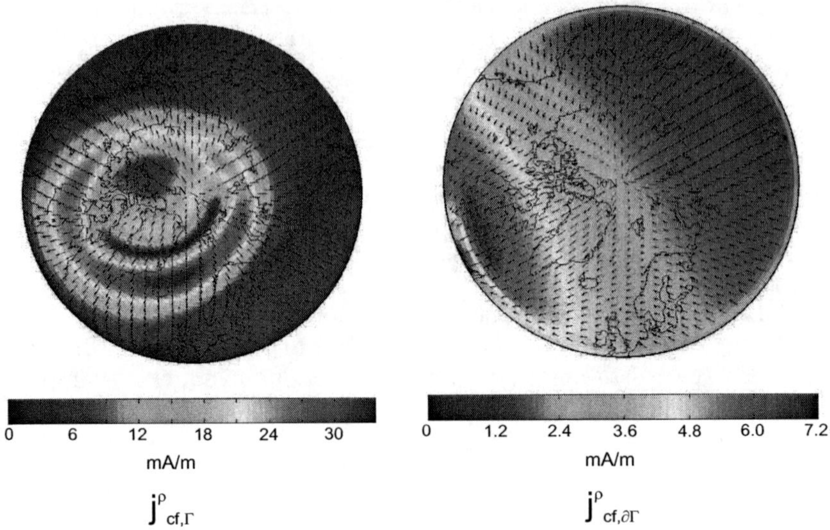

FIGURE 8.8
Local approximation of the absolute values and the orientation of the surface curl-free current densities $j^{\rho}_{\mathrm{cf},\Gamma}$ for $\rho = 2^{-6}$ (left) on a spherical cap around the North Pole. The boundary contribution $j^{\rho}_{\mathrm{cf},\partial\Gamma}$ for $\rho = 2^{-6}$ is shown on the right.

data. We observe sources and sinks located along the polar current systems and the equatorial electrojet. The (surface) curl-free current systems are those that connect the in- and outgoing radial (field-aligned) current systems (as reflected in the continuity equation $\nabla^* \cdot j_{\tan} = J_1$).

8.5 Exercises

Exercise 8.1. Derive explicit representations of the functions F, G, H appearing in the decomposition (8.3) in dependence of the functions U, V, W from decomposition (8.2).

Exercise 8.2. Let $\Gamma \subset \Omega$ be a regular region.

(a) Show that if f is of class $\mathrm{c}^{(2)}(\Gamma)$, then

$$p_{\tan}\left[\Delta^* f(\xi)\right] = \nabla^* \left(\nabla^* \cdot f(\xi)\right) + \mathrm{L}^* \left(\mathrm{L}^* \cdot f(\xi)\right), \quad \xi \in \Gamma. \tag{8.176}$$

(b) Use (a) to prove Theorem 8.9 for the weaker condition of f being of class

$c^{(1)}(\overline{\Gamma})$. More precisely, show the following assertion: If f is of class $c^{(1)}(\overline{\Gamma})$, then there exist scalar fields F_1, F_2, F_3 of class $C^{(1)}(\Gamma)$ such that

$$f(\xi) = o^{(1)} F_1(\xi) + o^{(2)} F_2(\xi) + o^{(3)} F_3(\xi), \quad \xi \in \Gamma. \tag{8.177}$$

Exercise 8.3.

(a) Derive representations for the Helmholtz scalars as in Theorem 8.11 if Dirichlet boundary values are imposed on F_2 instead of F_3.

(b) Derive representations for the Helmholtz scalars as in Theorem 8.11 if Neumann boundary values are imposed on F_3 instead of Dirichlet boundary values.

Exercise 8.4. Let $\Gamma \subset \Omega$ be a regular region. Furthermore, assume that f of class $c^{(2)}(\overline{\Gamma})$ is tangential and G_2, G_3 are of class $C^{(0)}(\partial\Gamma)$, satisfying $\int_{\partial\Gamma} G_2(\eta) d\sigma(\eta) - \int_\Gamma \nabla^*_\eta \cdot f(\eta) d\omega(\eta) = \int_{\partial\Gamma} G_3(\eta) d\sigma(\eta) - \int_\Gamma \mathrm{L}^*_\eta \cdot f(\eta) d\omega(\eta) = 0$. Show that the unique Helmholtz scalars F_2, F_3 satisfying $\frac{\partial}{\partial\nu(\xi)} F_i(\xi) = G_i(\xi)$, $\xi \in \partial\Gamma$, and $\int_\Gamma F_i(\eta) d\omega(\eta) = 0$, $i = 2, 3$, can be represented by

$$F_2(\xi) = \int_\Gamma G^{(N)}(\Delta^*; \xi, \eta) \nabla^*_\eta \cdot f(\eta) d\omega(\eta) - \int_{\partial\Gamma} G^{(N)}(\Delta^*; \xi, \eta) G_2(\eta) d\sigma(\eta), \tag{8.178}$$

$$F_3(\xi) = \int_\Gamma G^{(N)}(\Delta^*; \xi, \eta) \mathrm{L}^*_\eta \cdot f(\eta) d\omega(\eta) - \int_{\partial\Gamma} G^{(N)}(\Delta^*; \xi, \eta) G_3(\eta) d\sigma(\eta), \tag{8.179}$$

for $\xi \in \Gamma$.

Exercise 8.5. Let a regularization $G^\rho(\Delta^*; \cdot)$, using the Taylor polynomial of $G(\Delta^*; \cdot)$ of degree 3 with expansion point $1 - \rho$, be given by

$$G^\rho(\Delta^*; t) = \begin{cases} \frac{1}{4\pi} \ln(1-t) + \frac{1}{4\pi\rho}(1 - \ln(2)), & 1 - r \geq \rho, \\ \frac{1}{12\pi\rho^3}(1-t)^3 - \frac{3}{8\pi\rho^2}(1-t)^2 \\ \quad + \frac{3}{4\pi\rho}(1-t) + \frac{1}{4\pi}(\ln(\rho) - \frac{5}{6} - \ln(2)), & 1 - t < \rho. \end{cases} \tag{8.180}$$

Use this regularization to derive an explicit representation of the kernel $g^\rho_{\mathrm{rad}}(\cdot, \cdot)$ from Lemma 8.34.

Exercise 8.6. Lemma 8.34 yields an approximation of the radial contribution $J_1(r\cdot)$ from knowledge of the magnetic field b. Derive similar approximations for $J_2(r\cdot)$ and $J_3(r\cdot)$. More precisely, prove the following assertion:

Let b of class $c^{(2)}(\mathcal{B}_{R_0,R_1}(0))$ be solenoidal, and j of class $c^{(1)}(\mathcal{B}_{R_0,R_1}(0))$ is assumed to satisfy the pre-Maxwell equations (8.128), (8.129). If $J_2(r\cdot)$ and

$J_3(r\cdot)$ are the Helmholtz scalars of the tangential contributions to j, then

$$J_2(r\xi) = \frac{1}{r}\int_\Omega (\mathrm{L}^*_\eta G(\Delta^*;\xi\cdot\eta))\cdot\left(b(r\eta)+r\frac{\partial}{\partial r}b(r\eta)\right)d\omega(\eta), \tag{8.181}$$

$$\begin{aligned} J_3(r\xi) = & -\frac{1}{r}\Delta^*_\xi\int_\Omega G(\Delta^*;\xi\cdot\eta)\eta\cdot b(r\eta)d\omega(\eta) \\ & -\frac{1}{r}\int_\Omega (\nabla^*_\eta G(\Delta^*;\xi\cdot\eta)-\eta G(\Delta^*;\xi\cdot\eta))\cdot\left(2b(r\eta)+r\frac{\partial}{\partial r}b(r\eta)\right)d\omega(\eta), \end{aligned} \tag{8.182}$$

for $\xi\in\Omega$ and $r\in(R_0,R_1)$.

Bibliography

Aardalan, A.A., Grafarend, E.W., Finn, G.: Ellipsoidal Vertical Deflections and Ellipsoidal Gravity Disturbances: Case Studies, *Stud. Geophys. Geod.* 50, 1–57, 2006.

Alberto, P., Oliveira, O., Pais, M.A.: On the Non-Uniqueness of Main Geomagnetic Field Determined by Surface Intensity Measurements: The Backus Problem, *Geophys. J. Int.* 159, 548–554, 2004.

Alt, H.W.: *Lineare Funktionalanalysis*, 5th edition, Springer, Berlin, 2006.

Amm, O.: Elementary Currents for Ionospheric Fields, *J. Geomagn. Geoelectr.* 49, 947–955, 1997.

Anger, G.: A Characterization of the Inverse Gravimetric Source Problem through Extremal Measures, *Rev. Geophys. Space Phys.* 19, 299–306, 1981.

Anger, G.: *Inverse Problems in Differential Equations*, Akademie–Verlag, Berlin, 1990.

Aronszjain, N.: Theory of Reproducing Kernels, *Trans. Am. Math. Soc.* 68, 337–404, 1950.

Backus, G.E.: Potentials for Tangent Tensor Fields on Spheroids, *Arch. Ration. Mech. Anal.* 22, 210–252, 1966.

Backus, G.E.: Converting Vector and Tensor Equations to Scalar Equations in Spherical Coordinates, *Geophys. J.R. Astron. Soc.* 13, 61–101, 1967.

Backus, G.E.: Application of a Non-Linear Boundary-Value Problem for Laplace's Equation to Gravity and Geomagnetic Intensity Measurements, *Quart. J. Mech. Appl. Math.* 21, 195–221, 1968.

Backus, G.E.: Non-Uniqueness of the External Geomagnetic Field Determined by Surface Intensity Measurements, *J. Geophys. Res.* 75, 6339–6341, 1970.

Backus, G.E., Parker, R., Constable, C.: *Foundations of Geomagnetism*, Cambridge University Press, Cambridge, 1996.

Ballani, L.: Solving the Inverse Gravimetric Problem: On the Benefit of Wavelets, in: *Geodetic Theory Today, Proceedings of the 3rd Hotine–Marussi Symposium on Mathematical Geodesy 1994* (ed. Sansò, F.), 151–161, Springer, Berlin, 1995.

Ballani, L., Böttger, B., Fanselau, G.: Some Remarks to the Approximative Solution of the Inverse Gravimetric Problem, in: *Die moderne Potentialtheorie als Grundlage des Inversen Problems in der Geophysik, Geod. Geophys. Veröff.*, R. III, H. 45, 120–134, Freiberg, 1980.

Ballani, L., Engels, J., Grafarend, E.W.: Global Base Functions for the Mass Density in the Interior of a Massive Body (Earth), *Manuscr. Geod.* 18, 99–114, 1993.

Ballani, L., Stromeyer, D.: The Inverse Gravimetric Problem: A Hilbert Space Approach, in: *Proceedings of the International Symposium Figure of the Earth, the Moon, and Other Planets 1982* (ed. Holota, P.), 359–373, Prague, 1983.

Ballani, L., Stromeyer, D.: On the Structure of Uniqueness in Linear Inverse Source Problems, in: *Theory and Practice of Geophysical Data Inversion* (eds. Vogel, A., Sarwar, A.K.M., Gorenflo, R., Kounchev, O.I.), 85–98, Vieweg, Braunschweig, 1990.

Ballani, L., Stromeyer, D., Barthelmes, F.: Decomposition Principles for Linear Source Problems, in: *Inverse Problems: Principles and Applications in Geophysics, Technology, and Medicine, Math. Res.* 47 (eds. Anger, G., Gorenflo, R., Jochmann, H., Moritz, H., Webers, W.), Akademie–Verlag, Berlin, 1993.

Barzaghi, R., Sansò, F.: Remarks on the Inverse Gravimetric Problem, *Boll. Geod. Scienze Affini* 45, 203–216, 1986.

Barzaghi, R., Sansò, F.: The Integrated Inverse Gravimetric-Tomographic Problem: A Continuous Approach, *Inverse Problems* 14, 499–520, 1998.

Bayer, M., Freeden, W., Maier, T.: A Vector Wavelet Approach in Iono- and Magnetospheric Geomagnetic Satellite Data, *J. Atm. Solar-Terr. Phys.* 63, 581–597, 2001.

Bers, L., John, F., Schechter, M.: *Partial Differential Equations*, Wiley, New York, 1964.

Bertero, M., Brianzi, P., Pike, E.R., Rebolia, L.: Linear Regularizing Algorithms for Positive Solutions of Linear Inverse Problems, *Proc. R. Soc. Lond. A* 415, 257–275, 1988.

Bitzadse, A.V.: *Boundary-Value Problems for Second-Order Elliptic Equations*, North-Holland, Amsterdam, 1968.

Blakely, R.J.: *Potential Theory in Gravity and Magnetic Applications*, Cambridge University Press, Cambridge, 1996.

Blatt, J., Weisskopf, V.: *Theoretical Nuclear Physics*, Wiley, New York, 1952.

Blumenthal, O.: Über die Zerlegung unendlicher Vektorfelder, *Math. Ann.* 61, 235–250, 1905.

Bruns, E.H.: *Die Figur der Erde.* Publikation Königl. Preussisch. Geodätisches Institut, P. Stankiewicz Buchdruckerei, Berlin, 1878.

Buchheim, W.: Zur Geophysikalischen Inversionsproblematik, Seismology and Solid–Earth–Physics, in: *Proceedings of the International Symposium on the Occasion of 50 Years of Seismological Research and 75 Years of Seismic Registration at Jena 1974, Part 2* (ed. Maaz, R.), 305–310, Potsdam, 1975.

Christiansen, F., Papitashvili, V.O., Neubert, T.: Seasonal Variations of High-latitude Field-Aligned Currents Inferred from Orsted and MAGSAT Observations, *J. Geophys. Res.* 107, doi:10.1029/2001JA900104, 2002.

Colton, D., Kress, R.: *Integral Equation Methods in Scattering Theory*, Wiley, New York, 1983.

Courant, R., Hilbert, D.: *Methoden der Mathematischen Physik I + II*, Springer, Berlin, 1924.

Cowley, S.W.H.: Magnetosphere-Ionosphere Interactions: A Tutorial Review, in: *Magnetospheric Current Systems* (eds. Ohtami, S., Fuji, R., Hesse, M., Lysak, R.L.), 91–106, AGU, Washington, 2000.

Deuflhard, P.: On Algorithms for the Summation of Certain Special Functions, *Computing* 17, 37–48, 1975.

Davis, P.J., *Interpolation and Approximation*, Blaisdell Publishing Company, Waltham, MA, 1963.

DiBenedetto, E.: *Partial Differential Equations*, Birkhäuser, Boston, 1995.

Driscoll, J.R., Healy, R.M.: Computing Fourier Transforms and Convolutions on the 2–Sphere, *Advances in Applied Mathematics* 15, 202–250, 1994.

Dufour, H.M.: Fonctions orthogonales dans la sphère résolution théorique du problème du potentiel terrestre, *Bull. Geod.* 51, 227–23, 1977.

Dziewonski, A., Anderson, D.L.: The Preliminary Reference Earth Model, *Phys. Earth Planet. Inter.* 25, 297–356, 1981.

Edmonds, A.R.: *Angular Momentum in Quantum Mechanics.* Princeton University Press, Princeton, 1957.

Engl, H.: *Integralgleichungen*, Springer, Berlin, 1997.

Engl, H., Louis, A.K., Rundell, W. (eds.) *Inverse Problems in Geophysical Applications*, SIAM, Philadelphia, 1997.

ESA (European Space Agency): *The Nine Candidate Earth Explorer Missions*, Publications Division ESTEC, Noordwijk, SP–1196(1), 1996.

ESA (European Space Agency): *European Views on Dedicated Gravity Field Missions: GRACE and GOCE*, ESD–MAG–REP–CON–001, 1998.

ESA (European Space Agency): *Gravity Field and Steady-State Ocean Circulation Mission*, ESTEC, Noordwijk, ESA (European Space Agency) SP—1233(1), 1999.

Èskin, G.I..: *Boundary-Value Problems for Elliptic Pseudodifferential Equations*, Translations of Mathematical Monographs, AMS, Vol. 52, Providence, Rhode Island, 1981.

Fehlinger, T.: *Multiscale Formulations for the Disturbing Potential and the Deflections of the Vertical in Locally Reflected Physical Geodesy.* PhD thesis, Geomathematics Group, TU Kaiserslautern, 2009.

Fehlinger, T., Freeden, W., Gramsch, S., Mayer, C., Michel, D., Schreiner, M.: Local Modelling of Sea Surface Topography from (Geostrophic) Ocean Flow, *ZAMM* 87, 775–791, 2007.

Fehlinger, T., Freeden, W., Mayer, C., Schreiner, M.: On the Local Multiscale Determination of the Earth's Disturbing Potential from Discrete Deflections of the Vertical. *Comput. Geosc.* 4, 473–490, 2008.

Fengler, M.J., Michel, D., Michel, V.: Harmonic Spline-Wavelets on the 3-Dimensional Ball and Their Application to the Reconstruction of the Earth's Density Distribution from Gravitational Data at Arbitrarily Shaped Satellite Orbits, *ZAMM* 86, 856–873, 2006.

Foulger, G., Natland, J., Presnall, D., Anderson, D. (eds.): *Plates, Plumes, and Paradigms*, Geological Society of America, 2005.

Fredholm, I.: Sur une nouvelle méthode pour la résolution du problème de Dirichlet, *Översigt Kongl. Vetenskaps-Akademiens Förhandlingar* 57, 39–46, 1900.

Freeden, W.: On the Approximation of External Gravitational Potential with Closed Systems of (Trial) Functions, *Bull. Géod.* 54, 1–20, 1980.

Freeden, W.: On Approximation by Harmonic Splines, *Manuscr. Geod.* 6, 193–244, 1981.

Freeden, W.: Interpolation and Best Approximation by Harmonic Spline Functions: Theoretical and Computational Aspects, *Boll. Geod. Scienze Affini* 41, 106–120, 1982.

Freeden, W.: *Least Squares Approximation by Linear Combinations of (Multi)-Poles*, Report 344, Department of Geodetic Science and Surveying, The Ohio State University, Columbus, 1983.

Freeden, W.: A Spline Interpolation Method for Solving Boundary Value Problems of Potential Theory from Discretely Given Data, *Math. Part. Diff. Equations* 3, 375–398, 1987.

Freeden, W.: *Multiscale Modelling of Spaceborne Geodata*, B.G. Teubner, Leipzig, 1999.

Freeden, W.: Geomathematik, was ist das überhaupt?, *Jahresbericht der DMV* 111, 125–152, 2009.

Freeden, W.: Geomathematics: Its Role, Its Aim, and Its Potential, in: *Handbook of Geomathematics, Vol. 1* (eds. Freeden, W., Nashed, M.Z., Sonar, T.), 4–42, Springer, Heidelberg, 2010.

Freeden, W.: *Metaharmonic Lattice Point Theory*, Chapman & Hall/CRC, Boca Raton, 2011.

Freeden, W., Gerhards, C.: Poloidal and Toroidal Modeling in Terms of Locally Supported Vector Wavelets, *Math. Geosc.* 42, 817–838, 2010.

Freeden, W., Hesse, K.: On the Multiscale Solution of Satellite Problems by Use of Locally Supported Kernel Functions Corresponding to Equidistributed Data on Spherical Orbits, *Studia Scient. Mathemat. Hungarica* 39, 37–74, 2002.

Freeden, W., Kersten, H.: *The Geodetic Boundary Value Problem Using the Known Surface of the Earth*, Veröff. Geod. Inst. RWTH Aachen, No. 29, 1980.

Freeden, W., Kersten, H.: A Constructive Approximation Theorem for the Oblique Derivative Problem in Potential Theory, *Math. Meth. Appl. Sci.* 3, 104–114, 1981.

Freeden, W., Kersten, H.: An Extended Version of Runge's Theorem, *Manuscr. Geod.* 7, 267–278, 1982.

Freeden, W., Mayer, C.: Wavelets Generated by Layer Potentials, *Appl. Comput. Harm. Anal.* 14, 195–237, 2003.

Freeden, W., Mayer, C.: Modeling Tangential Vector Fields on Regular Surfaces by Means of Mie Potentials, *Int. J. Wavel. Multires. Inf. Process.* 5, 417–449, 2007.

Freeden, W., Michel, V.: *Multiscale Potential Theory (With Applications to Geoscience)*, Birkhäuser, Boston, 2004.

Freeden, W., Nutz, H.: Satellite Gravity Gradiometry as Tensorial Inverse Problem, *Int. J. Geomath.* 2, 177–218, 2011.

Freeden, W., Reuter, R.: A Constructive Method for Solving the Displacement Boundary-Value Problem of Elastostatics by Use of Global Basis Systems, *Math. Meth. Appl. Sci.* 12, 105–128, 1989.

Freeden, W., Schneider, F.: Regularization Wavelets and Multiresolution, *Inverse Problems* 14, 493—515, 1998.

Freeden, W., Schreiner, M.: Local Multiscale Modelling of Geoid Undulations from Deflections of the Vertical, *J. Geodesy* 79, 641–651, 2006.

Freeden, W., Schreiner, M.: *Spherical Functions of Mathematical Geosciences (A Scalar, Vectorial, and Tensorial Setup)*, Springer, Heidelberg, 2009.

Freeden, W., Schreiner, M.: Satellite Gravity Gradiometry (SGG): From Scalar to Tensorial Solution, in: *Handbook of Geomathematics, Vol. 1* (eds. Freeden, W., Nashed, M.Z., Sonar,T.), 269–302, Springer, Heidelberg, 2010.

Freeden, W., Wolf, K.: Klassische Erdschwerefeldbestimmung aus der Sicht moderner Geomathematik, *Math. Semesterb.* 56, 53–77, 2008.

Freeden, W., Gervens, T., Schreiner, M.: *Constructive Approximation on the Sphere (with Applications to Geomathematics)*, Oxford Science Publications, Clarendon, Oxford, 1998.

Freeden, W., Glockner, O., Litzenberger, R.: A General Hilbert Space Approach to Wavelets and Its Application in Geopotential Determination, *Numer. Func. Anal. Opt.* 20, 853–879, 1999.

Freeden, W., Glockner, O., Schreiner, M.: Spherical Panel Clustering and Its Numerical Aspects. *J. Geodesy* 72, 586–599, 1998.

Freeden, W., Mayer, C., Schreiner, M.: Tree Algorithms in Wavelet Approximation of Helmholtz Operators. *Numer. Funct. Anal. Optim.,* 24, 747–782, 2003.

Freeden, W., Michel, V., Nutz, H.: Satellite-to-Satellite Tracking and Satellite Gravity Gradiometry (Advanced Techniques for High-Resolution Geopotential Field Determination), *J. Eng. Math.* 43, 19–56, 2002.

Freeden, W., T. Fehlinger, M. Klug, D. Mathar, K. Wolf.: Classical Globally Reflected Gravity Field Determination in Modern Locally Oriented Multiscale Framework. *J. Geodesy* 83, 1171–1191, 2009.

Fukushima, N.: Generalized Theorem for No Ground Magnetic Effect of Vertical Currents Connected with Pedersen Currents in the Uniform-Conductivity Ionosphere, *Rep. Ion. Space Res. Jap.* 30 35–40, 1976.

Gauss, C.F.: *Allgemeine Theorie des Erdmagnetismus*, Resultate aus den Beobachtungen des magnetischen Vereins, Göttingen, 1838.

Gemmrich, S., Nigma, N., Steinbach, O.: Boundary Integral Equations for the Laplace-Beltrami Operator, in: *Mathematics and Computation, a Contemporary View, Proceedings of the 2006 Abel Symposium* (eds. Munthe-Kaas, H., Owren, B.), 21–37, Springer, Heidelberg, 2008.

Gerhards, C.: *Spherical Multiscale Methods in Terms of Locally Supported Wavelets: Theory and Application to Geomagnetic Modeling.* PhD thesis, Geomathematics Group, TU Kaiserslautern, 2011a.

Gerhards, C.: Spherical Decompositions in a Global and Local Framework: Theory and an Application to Geomagnetic Modelling, *Int. J. Geomath.* 1, 205–256, 2011b.

Gerhards, C.: Locally Supported Wavelets for the Separation of Vector Fields with Respect to Their Sources, *Int. J. Wavel. Multires. Inf. Process.* 10, doi: 10.1142/S0219691312500348, 2012.

Gilbart, D., Trudinger, N.S.: *Elliptic Partial Differential Equations of Second Order.* Springer, Berlin, 1977.

Golberg, M.: The Method of Fundamental Solutions for Poisson's Equation. *Eng. Anal. Bound. Elem.* 16, 205–213, 1995.

Golberg, M., Chen, C: The Method of Fundamental Solutions for Potential Helmholtz and Diffusion Problems. In: *Boundary Integral Methods: Numerical and Mathematical Aspects* (ed. Goldberg, M.), 103–176, WIT Press, Computational Mechanics Publications, Southampton, 1998.

Grafarend, E.W.: Six Lectures on Geodesy and Global Geodynamics, in: *Proceedings of the Third International Summer School in the Mountains* (eds. Moritz, H., Sünkel, H.), 531–685, 1982.

Grafarend, E.W., Klapp, M., Martinec. Z.: Spacetime Modelling of the Earth's Gravity Field by Ellipsoidal Harmonics, in: *Handbook of Geomathematics, Vol. 1* (eds. Freeden, W., Nashed, M.Z., Sonar, T.), 159–253, Springer, Heidelberg, 2010.

Gramsch, S.: *Integral Formulae and Wavelets on Regular Regions of the Sphere*, PhD thesis, Geomathematics Group, TU Kaiserslautern, 2006.

Groten, E.: *Geodesy and the Earth's Gravity Field I + II*, Dümmler, Bonn, 1979.

Grothaus, M., Raskop, T.: Oblique Stochastic Boundary-Value Problem. in: *Handbook of Geomathematics, Vol. 2* (eds. Freeden, W., Nashed, M.Z., Sonar, T.), 1051–1076, Springer, Heidelberg, 2010a.

Grothaus, M., Raskop, T.: Limit Formulae and Jump Relations of Potential Theory in Sobolev Spaces, *Int. J. Geomath.* 1, 51–100, 2010b.

Gubbins, D., Herrero-Bervera, E. (eds.): *Encyclopedia of Geomagnetism and Paleomagnetism*, Springer, Dordrecht, 2007.

Gui, Y.F., Dou, W.B.: A Rigorous and Completed Statement on Helmholtz Theorem, *Prog. Elect. Res. (PIER)* 69, 287–304, 2007.

Günter, N.M.: *Die Potentialtheorie und ihre Anwendung auf Grundaufgaben der Mathematischen Physik*, B.G. Teubner, Leipzig, 1957.

Gurtin, M.E.: *The Linear Theory of Elasticity*, Handbuch der Physik, Vol. VI, 2nd edition, Springer, Heidelberg, 1972.

Gutkin, E., Newton, K.P.: The Method of Images and Green's Functions for Spherical Domains, *J. Phys. A: Math. Gen.* 37, 11989–12003, 2004.

Gutting, M.: *Fast Multipole Methods for Oblique Derivative Problems.* PhD thesis, Geomathematics Group, TU Kaiserslautern, 2007.

Gutting, M.: Fast Multipole Accelerated Solution of the Oblique Derivative Boundary Value Problem, *Int. J. Geomath.*, doi: 10.1007/s13137-012-0038–1, 2012.

Haar, A.: Zur Theorie der orthogonalen Funktionensysteme, *Math. Ann.* 69, 331—371, 1910.

Hansen, E.R.: *A Table of Series and Products*, Prentice Hall, Englewood Cliffs, 1975.

Hein, G., Sansò, F., Strykowsky, G., Tscherning, C.C.: On the Choice of Norm and Base Functions for the Solution of the Inverse Gravimetric Problem, *Ricerche di Geodesia Topografia Fotogrammetria, Vol. 5*, 121–138, CLUP, Milano, 1989.

Heine, E.: *Handbuch der Kugelfunktionen*, Verlag G. Reimer, Berlin, 1878.

Heiskanen, W.A., Moritz, H.: *Physical Geodesy*, Freeman, San Francisco, 1967.

Helmert, F.: *Die Mathematischen und Physikalischen Theorien der Höheren Geodäsie 1*, B.G. Teubner, Leipzig, 1880.

Helmert, F.: *Die Mathematischen und Physikalischen Theorien der Höheren Geodäsie 2*, B.G. Teubner, Leipzig, 1884.

Helms, L.L.: *Introduction to Potential Theory*, Wiley-Interscience, New York, 1969.

Hesse, K.: *Domain Decomposition Methods in Multiscale Geopotential Determination From SST and SGG.* PhD thesis, Geomathematics Group, TU Kaiserslautern, Shaker, 2002.

Hesse, K., Sloan, I.H., Womersley, R.S.: Numerical Integration on the Sphere, in: *Handbook of Geomathematics, Vol. 2* (eds. Freeden, W., Nashed, M.Z., Sonar, T.), 1187–1220, Springer, Heidelberg, 2010.

Hobson, E.W.: *The Theory of Spherical and Ellipsoidal Harmonics*, Reprint Chelsea Publishing Company, New York, 1955.

Hofmann–Wellenhof, B., Moritz, H.: *Physical Geodesy*, Springer, Wien, New York, 2005.

Hörmander, L.: *The Boundary Problems of Physical Geodesy*, The Royal Institute of Technology, Division of Geodesy, Stockholm, Report 9, 1975.

Hulot, G., Finlay, C.C., Constable, C., Olsen, N., Mandea, M.: The Magnetic Field of Planet Earth, *Space Sci. Rev.* 152, 159–222, 2010.

Ilk, K., Flury, J., Rummel, R., Schwintzer, P., Bosch, W., Haas, C., Schröter, J., Stammer, D., Zahel, W., Miller, H., Dietrich, R., Huybrechts, P., Schmeling, H., D.Wolf, H.G., Riegger, J., Bardossy, A., Güntner, A., Gruber, T.: *Mass Transport and Mass Distribution in the Earth System, Contribution of the New Generation of Satellite Gravity and Altimetry Missions to Geosciences*, Proposal for a German Priority Research Program, 2005.

Jackson, D.D.: The Use of A Priori Data to Resolve Non-Uniqueness in Linear Inversion, *Geophys. J. R. Astr. Soc.* 57, 137–157, 1979.

Jackson, J.: *Classical Electrodynamics*, John Wiley & Sons Ltd., New York, 1998.

Jacobs, F., Meyer, H.: *Geophysik-Signale aus der Erde*, B.G. Teubner, Leipzig, and VDF Verlag, Zürich, 1992.

Jantscher, L.: *Distributionen*, Walter de Gruyter, Berlin, New York, 1971.

Jaswon, M.A., Symm, G.T.: *Integral Equation Methods in Potential Theory and Elastostatics*, Academic Press, New York, 1977.

Kantorowitsch, L.W., Akilow, G.: *Funktionalanalysis in normierten Räumen*, Akademie-Verlag, Berlin, 1964.

Kellogg, O.D.: *Foundations of Potential Theory* (reprint), Springer, Berlin, Heidelberg, New York, 1967.

Khokhlov, A., Hulot, G., LeMouel, J.-L.: On the Backus Effect – I, *Geophys. J. Int.* 130, 701–703, 1997.

Kita, E., Kamiya, N.: Trefftz Method: An Overview, *Adv. Eng. Softw.* 24, 3–13, 1995.

Koch, K.R., Pope, A.J.: Uniqueness and Existence for the Geodetic Boundary Value Problem Using the Known Surface of the Earth, *Bull. Géod.* 106, 467–476, 1972.

Kono, M. (ed.): *Geomagnetism: Treatise on Geophysics, Vol. 5*, Elsevier, Amsterdam, 2009.

Kotevska, E.: *Real Earth Oriented Gravitational Potential Determination*, PhD thesis, Geomathematics Group, TU Kaiserslautern, 2011.

Krarup, T.: *A Contribution to the Mathematical Foundation of Physical Geodesy*, Danish Geodetic Institute, Report No. 44, Copenhagen, 1969.

Kupradze, V.D.: *Potential Methods in the Theory of Elasticity*, Israel Program for Scientific Translations, Jerusalem, 1965.

Kupradze, V., Aleksidze M.: The Method of Functional Equations for the Approximate Solution of Certain Boundary Value Problems, *USSR Comp. Math. Math. Phys.* 4, 82–126, 1964.

Kusche, J.: Time-Variable Gravity Field and Global Deformation of the Earth, in: *Handbook of Geomathematics, Vol. 1* (eds. Freeden, W., Nashed, M.Z., Sonar, T.), 253–268, 2010.

Langel, R.A., Hinze, W.J.: *The Magnetic Field of the Earth's Lithosphere: The Satellite Perspective*, Cambridge University Press, Cambridge, 1998.

Laplace, P.S. de: *Theorie des attractions des sphéroides et de la figure des planètes*, Mèm. de l'Acad., Paris, 1785.

Last, B.J., Kubik, K.: Compact Gravity Inversion, *Geophysics* 48, 713–721, 1983.

Laur, H., Liebig, V.: Earth Observation Satellite Missions and Data Access, in: *Handbook of Geomathematics, Vol. 1* (eds. Freeden, W., Nashed, M.Z., Sonar, T.), 71–92, 2010.

Lauricella, G.: Sulla distribuzione della massa nell'interno dei pianeti, *Rend. Acc. Lincei* 21, 18–26, 1912.

Lavrentiev, M.M.: *Some Improperly Posed Problems of Mathematical Physics*, Springer Tracts in Natural Philosophy, Vol. 11, Springer, Berlin, Heidelberg, New York, 1967.

Lax, P.D.: Symmetrizable Linear Transformations, *Comm. Pure Appl. Math.* 7, 633–647, 1954.

Lebedev, N.N.: *Spezielle Funktionen und ihre Anwendungen*, Bibliographisches Institut, Mannheim, 1973.

Leis, R.: *Vorlesungen über partielle Differentialgleichnungen zweiter Ordnung*, BI–Hochschultaschenbücher, Vol. 165/165a, Bibliographisches Institut, Mannheim, 1967.

Legendre, A.M.: Recherches sur l'attraction des sphèroides homogènes, *Mèm. math. phys. près. à l'Acad. Aci. par. divers savantes* 10, 411–434, 1785.

Lemoine, F.G., Kenyon, S.C., Factor, J.K., Trimmer, R.G., Pavlis, N.K., Shinn, D.S., Cox, C.M., Klosko, S.M., Luthcke, S.B., Torrence, M.H., Wang, Y.M., Williamson, R.G., Pavlis, E.C., Rapp, R.H., Olson, T.R.: *The Development of the Joint NASA GSFC and NIMA Geopotential Model EGM96*, NASA/TP–1998–206861, NASA Goddard Space Flight Center, Greenbelt, 1998.

Lewi, E.: *Modelling and Inversion of High Precision Gravity Data*, PhD thesis, University of Darmstadt, Bayerischen Akademie der Wissenschaften, Munich, 1997.

Listing J.B.: *Über unsere jetzige Kenntnis der Gestalt und Größe der Erde*, Dietrichsche Verlagsbuchhandlung, Göttingen, 1873.

Lowes, F.J.: Spatial Power Spectrum of the Main Geomagnetic Field, and Extrapolation to the Core, *Geophys. J. R. Astr. Soc.* 36, 717–730, 1974.

Magnus, W., Oberhettinger, F., Soni, R.P.: *Formulas and Theorems for the Special Functions of Mathematical Physics*, Die Grundlehren der mathematischen Wissenschaften in Einzeldarstellungen, Vol. 52, Springer, Berlin, 1966.

Maier, T.: Wavelet Mie Representation for Solenoidal Vector Fields with Applications to Ionospheric Geomagnetic Data, *SIAM J. Appl. Math.* 65, 1888–1912, 2005.

Martensen, E.: *Potentialtheorie*, Leitfäden der Angewandten Mathematik und Mechanik, Vol. 12, Teubner, Leipzig, 1968.

Martinec, Z., Pec, K.: The Influence of the Core–Mantle Boundary Irregularities on the Mass Density Distribution Inside the Earth, in: *Geophysical Data Inversion Methods and Applications, Proceedings of the 7th International Mathematical Geophysics Seminar Held at the Free University of Berlin* (eds. Vogel, A., Ofoegbu, C.O., Gorenflo, R., Ursin, B.), 233–256, 1989.

Marussi, A.: On the Density Distribution in Bodies of Assigned Outer Newtonian Attraction, *Boll. Geofis. Teorica Appl.* 22, 83–94, 1980.

Maus, S., Roth., Hemant, K., Stolle, C., Lühr, H., Kuvshinov, A., Olsen, N.: Earth's Lithospheric Magnetic Field Determined to Spherical Harmonic Degree 90 from CHAMP Satellite Measurements, *Geophys. J. Int.* 164, 319–330, 2006.

Maxwell, J.C.: *A Treatise on Electricity and Magnetism I + II*, unshortened reprint of the last edition 1891, Dover, 1954.

Mayer, C.: Wavelet Modelling of the Spherical Inverse Source Problem with Application to Geomagnetism, *Inverse Problems* 20, 1713–1728, 2004.

Mayer, C.: Wavelet Decomposition of Spherical Vector Fields with Respect to Sources, *J. Fourier Anal. Appl.* 12, 345–369, 2006.

Mayer, C.: *A Wavelet Approach to the Stokes Problem*, Habilitation thesis, University of Kaiserslautern, Geomathematics Group, 2007.

Meissl, P.A.: *A Study of Covariance Functions Related to the Earth's Disturbing Potential*, Scientific Report, No. 151, The Ohio State University, Department of Geodetic Science, Columbus, 1971.

Meschkowski, H.: *Hilbertsche Räume mit Kernfunktion*, Die Grundlehren der mathematischen Wissenschaften in Einzeldarstellungen, Vol. 113, Springer, Berlin, 1962.

Michel, V.: *A Multiscale Method for the Gravimetry Problem: Theoretical and Numerical Aspects of Harmonic and Anharmonic Modelling*, PhD thesis, Geomathematics Group, TU Kaiserslautern, Shaker, Aachen, 1999.

Michel, V.: *A Multiscale Approximation for Operator Equations in Separable Hilbert Spaces – Case Study: Reconstruction and Description of the Earth's Interior*, Habilitation thesis, TU Kaiserslautern, Geomathematics Group, Shaker, Aachen, 2002a.

Michel, V.: Scale Continuous, Scale Discretized and Scale Discrete Harmonic Wavelets for the Outer and the Inner Space of a Sphere and their Application to an Inverse Problem in Geomathematics, *Appl. Comp. Harm. Anal.* 12, 77–99, 2002b.

Michel, V.: Wavelets on the 3-Dimensional Ball, *Proc. Appl. Math. Mech.* 5, 775–776, 2005.

Michel, V: Regularized Wavelet–Based Multiresolution Recovery of the Harmonic Mass Density Distribution from Data of the Earth's Gravitational Field at Satellite Height, *Inverse Problems* 21, 997–1025, 2005.

Michel, V., Fokas, A.S.: A Unified Approach to Various Techniques for the Non-Uniqueness of the Inverse Gravimetric Problem and Wavelet-Based Methods, *Inverse Problems* 24, doi:10.1088/0266–5611/24/4/045019, 2008.

Michel, V., Wolf, K.: Numerical Aspects of a Spline-Based Multiresolution Recovery of the Harmonic Mass Density out of Gravity Functionals, *Geophys. J. Int.* 173, 1–16, 2008.

Michlin, S.G.: *Lehrgang der Mathematischen Physik*, 2nd edition, Akademie Verlag, Berlin, 1975.

Miranda, C.: *Partial Differential Equations of Elliptic Type*, Springer, Berlin, 1970.

Misner, C.W., Thorne, K.S., Wheeler, J.A.: *Gravitation*, W.H. Freeman, San Francisco, 1973.

Morgan, W.J.: Convection Plumes in the Lower Mantle, *Nature* 230, 42–43, 1971.

Moritz, H.: *Advanced Physical Geodesy*, Wichmann Verlag, Karlsruhe, 1980.

Moritz, H.: *The Figure of the Earth. Theoretical Geodesy of the Earth's Interior*, Wichmann Verlag, Karlsruhe, 1990.

Moritz, H.: Classical Physical Geodesy, in: *Handbook of Geomathematics, Vol. 1* (eds. Freeden, W., Nashed, M.Z., Sonar, T.), 127–158, Springer Verlag, Heidelberg, 2010.

Moritz, H., Sünkel, H. (eds.): *Approximation Methods in Geodesy: Lectures Delivered at 2nd Intern. Summer School in the Mountains on Math. Meth. in Physical Geodesy*, Wichmann Verlag, Karlsruhe, 1978.

Morse, P.M., Feshbach, H.: *Methods of Theoretical Physics*, McGraw-Hill, New York, 1953.

Müller, C.: *Spherical Harmonics*, Lecture Notes in Mathematics, Vol. 17, Springer, Berlin, 1966.

Müller, C.: *Foundations of the Mathematical Theory of Electromagnetic Waves*, Springer, Berlin, 1969.

Müller, C.: *Analysis of Spherical Symmetries in Euclidean Spaces*, Springer, Berlin, 1998.

Nardini, D., Brebbia, C.: A New Approach for Free Vibration Analysis Using Boundary Elements, in: *Boundary Element Methods in Engineering Proceedings* (ed. Brebbia, C.), 312–326, Springer, Berlin, 1982.

Nataf, H.: Seismic Imaging of Mantle Plumes, *Ann. Rev. Earth Planet* 28, 391–417, 2000.

Neumann, F.: *Vorlesungen über die Theorie des Potentials und der Kugelfunktionen*, B.G. Teubner, Leipzig, 1887.

Novikoff, P.: Sur le problème inverse du potentiel, *Comptes Rendus de l'Académie des Sciences de l'URSS* 18, 165–168, 1938.

Nutz, H.: *A Unified Setup of Gravitational Field Observables*, PhD thesis, Geomathematics Group, TU Kaiserslautern, Shaker, Aachen, 2002.

Olsen, N.: Ionospheric F-Region Currents at Middle and Low Lattitudes Estimated from MAGSAT Data, *J. Geophys. Res.* 102, 4563–4576, 1997.

Olsen, N., Glassmeier, K.-H., Jia, X.: Separation of the Magnetic Field into External and Internal Parts, *Space Sci. Rev.* 152, 159–222, 2010a.

Olsen, N., Hulot, G., Sabaka, T.J.: Sources of the Geomagnetic Field and the Modern Data That Enable Their Investigation, in: *Handbook of Geomathematics, Vol. 1* (eds. Freeden, W., Nashed, M.Z., Sonar, T.), 106–124, Springer, Heidelberg, 2010b.

Olsen, N., Mandea, M., Sabaka, T.J., Tøffner-Clausen, L.: The CHAOS–3 Geomagnetic Field Model and Candidates for the 11th Generation IGRF, *Earth Planets Space* 62, 719–727, 2010c.

Parker, R.L.: The Theory of Ideal Bodies for Gravity Interpretation, *Geophys. J. R. Astr. Soc.* 42, 315–334, 1975.

Petrini, H.: Sur l'existence des derivées secondes du potentiel, *C.R. Acad. Sci. Paris* 130, 233–235, 1900.

Pizzetti, P.: Geodesia: sulla espressione della gravita alla superficie del geoide, supposto ellissoidico, *Att. R. Acad. Lincei* 3, 331–350, 1894.

Pizzetti, P.: Corpi equivalenti rispetto alla attrazione newtoniana esterna, *Rom. Acc. L. Rend.* 18, 211–215, 1909.

Pizzetti, P.: Intorno alle possibili distribuzioni della massa nell'interno della terra, *Ann. Mat. Milano* 17, 225–258, 1910a.

Pizzetti, P.: Sopra il calcoba tesrico delle deviazioni del geoide dall' ellisoide, *Att. R. Acad. Sci. Torino* 46, 331–350, 1910b.

Ritter, J.R.R., Christensen, U.R.: *Mantle Plumes, A Multidisciplinary Approach*, Springer, Berlin, Heidelberg, 2007.

Ritz, W.: Über eine neue Methode zur Lösung gewisser Variationsprobleme der mathematischen Physik, *J. Reine Angew. Math.* 135, 1–6, 1909.

Robin, L.: *Fonctions Sphériques de Legendre et Fonctions Sphéroïdales*, Gauthier-Villars, Paris, 1957.

Rubincam, D.P.: Gravitational Potential Energy of the Earth: A Spherical Harmonics Approach, *J. Geophys. R.* 84, 6219–6225, 1979.

Rummel, R.: Spherical Spectral Properties of the Earth's Gravitational Potential and Its First and Second Derivatives, in: *Geodetic Boundary Value Problems in View of the One Centimeter Geoid* (eds. Sansò, F., Rummel, R.), 359–404, Lecture Notes in Earth Science, Vol. 65, Springer, Berlin, Heidelberg, 1997.

Rummel, R.: GOCE: Gravitational Gradiometry in a Satellite, in: *Handbook of Geomathematics, Vol. 1* (eds. Freeden, W., Nashed, M.Z., Sonar,T.), 98–105, Springer Verlag, Heidelberg, 2010.

Rummel R., Balmino, G., Johannessen, J., Visser, P., Woodworth P.: Dedicated Gravity Field Missions - Principles and Aims, *J. Geodyn.* 33, 3–20, 2002.

Rummel, R., van Gelderen, M.: Spectral Analysis of the Full Gravity Tensor, *Geophys. J. Int.* 111, 159–169, 1992.

Runge, C.: Zur Theorie der eindeutigen analytischen Funktionen, *Acta Math.* 6, 229–234, 1885.

Russel, S., Lay, T., Garnero, E.: Seismic Evidence for Small-Scale Dynamics in the Lower Most Mantle at the Root of the Hawaiian Hotspot, *Nature* 396, 255–258, 1998.

Sabaka, T.J., Hulot, G., Olsen, N.: Mathematical Properties Relevant to Geomagnetic Field Modelling, in: *Handbook of Geomathematics, Vol. 1* (eds. Freeden, W., Nashed, M.Z., Sonar, T.), 504–538, Springer, Heidelberg, 2010.

Sansò, F., Barzaghi, R., Tscherning, C.C.: Choice of Norm for the Density Distribution of the Earth, *Geophys. J. R. Astr. Soc.* 87, 123–141, 1986.

Sansò, F., Rummel, R. (eds.): *Geodetic Boundary Value Problems in View of the One Centimeter Geoid*, Lecture Notes in Earth Sciences, Vol. 65, Springer, Berlin, Heidelberg, 1997.

Sansò, F., Tscherning, C.C.: Mixed Collocation: A Proposal, *Quaterniones Geodaesiae* 3, 1–15, 1982.

Sansò, F., Tscherning, C.C.: The Inverse Gravimetric Problem in Gravity Modelling, in: *Festschrift to Torben Krarup* (eds. Kejlsø, E., Poder, K., Tscherning, C.C.), 299–334, Geodatisk Institute, Copenhagen, 1989.

Schauder, J.: Potentialtheoretische Untersuchungen, *Math. Z.* 35, 536–538, 1931.

Schneider, F.: *Inverse Problems in Satellite Geodesy and Their Approximate Solution by Splines and Wavelets*, PhD thesis, Geomathematics Group, TU Kaiserslautern, Shaker, Aachen, 1997.

Schreiner, M.: *Tensor Spherical Harmonics and Their Application in Satellite Gradiometry*, PhD thesis, Geomathematics Group, TU Kaiserslautern, 1994.

Schubert, D., Turcotte, D., Olson, P.: *Mantle Convection in the Earth and Planets*, Cambridge University Press, Cambridge, 2001.

Schulze, B.-W., Wildenhain, G.: *Methoden der Potentialtheorie für elliptische Differentialgleichungen beliebiger Ordnung*, Birkhäuser, Basel, Stuttgart, 1977.

Schwiderski, E.W.: *The Deep Structure of the Earth Inferred from a Satellite's Orbit, Part I: The Density Anomaly*, Technical Report, U.S. Naval Weapons Laboratory, Dahlgren, 1967.

Shure, L., Parker, R.L., Backus, G.E.: Harmonic Splines for Geomagnetic Modelling, *Phys. Earth Planet. Inter.* 28, 215–229, 1982.

Skorvanek, M.: The Inverse Gravimetric Problem for the Earth, in: *Proceedings of the 4th International Symposium on Geodesy and Physics of the Earth 1980*, 464–475, Veröff. Zentralinst. Physik der Erde, Vol. 63, 1981.

Sommerfeld, A.: *Vorlesungen der Theoretischen Physik II, 6th edition*, Verlag Harri Deutsch, Frankfurt, 1978.

Sprössig, W.: On Helmholtz Decompositions and Their Generalizations: An Overview, *Math. Meth. Appl. Sci.* 33, 374–383, 2010.

Stern, D.P.: Representation of Magnetic Fields in Space, *Rev. Geophys.* 14, 199–214, 1976.

Stokes, G.G.: On the Internal Distribution of Matter Which Shall Produce a Given Potential at the Surface of a Gravitating Mass, *Proc. Royal Soc. London* 15, 482–486, 1867.

Stokes, G.G.: On the Variation of Gravity at the Surface of the Earth, *Trans. Cambr. Phil. Soc.* 148, 672–712, 1849.

Stromeyer, D., Ballani, L.: Uniqueness of the Inverse Gravimetric Problem for Point Mass Models, *Manuscr. Geod.* 9, 125–136, 1984.

Strubecker, K.: *Differentialgeometrie II: Theorie der Flächenmetrik*, 2nd edition, Sammlung Göschen, Vol. 1179/1179a, de Gruyter, Berlin, 1969.

Szegö, G.: *Orthogonal Polynomials*, AMS Colloquium Publications, Vol. 23, Providence, Rhode Island, 1939.

Tabley, B.D., Bettadpur, S., Watkins, M.M., Reigber, C.: The Gravity Recovery and Climate Experiment: Mission Overview and Early Results, *Geophys. Res. Lett.* 31, doi:10.1029/2004GL019920, 2004.

Thalhammer, M., Ricard, Y., Rummel, R., Ilk, K.H.: *Application of Spaceborne Gravimetry to Research on the Interior of the Earth*, ESA (European Space Agency) study, CIGAR 4, final report, 1996.

Torge, W.: *Geodesy*, de Gruyter, Berlin, 1991.

Trefftz, E.: Ein Gegenstück zum Ritzschen Verfahren, in: *Verh. d. 2. Intern. Kongr. f. Techn. Mech.*, 131–137, Zürich, 1926.

Tscherning, C.C.: *Some Simple Methods for the Unique Assignment of a Density Distribution to a Harmonic Function*, Report 213, The Ohio State University, Department of Geodetic Science and Surveying, Columbus, 1974.

Tscherning, C.C.: Analytical and Discrete Inversion Applied to Gravity Data, in: *Proceedings of the Interdisciplinary Inversion Workshop 1, Methodology and Application Perspectives in Geophysics, Astronomy and Geodesy* (ed. Holm Jacobsen, B.), 5–8, Aarhus, 1992.

Tscherning, C.C.: Potential Field Collocation and Density Modelling, in: *(IAG) Section IV Bulletin, No. 1*, 165–175, Delft University of Technology, 1995.

Tscherning, C.C.: Isotropic Reproducing Kernels for the Inner of a Sphere or Spherical Shell and Their Use as Density Covariance Functions, *Math. Geol.* 28, 161–168, 1996.

Tscherning, C.C., Rapp, R.H.: *Closed Covariance Expressions for Gravity Anomalies, Geoid Undulations, and Deflections of the Vertical Implied by Anomaly Degree Variance Models*, Scientific Report, No. 208, The Ohio State University, Department of Geodetic Science, Columbus, 1974.

Tscherning, C.C., Strykowski, G.: Quasi-Harmonic Inversion of Gravity Field Data, Model Optimization in Exploration Geophysics 2, in: *Proceedings of the 5th International Mathematical Geophysics Seminar* (ed.: Vogel, A.), 137–154, Vieweg, Braunschweig, Wiesbaden, 1987.

Tscherning, C.C., Sünkel, H.: A Method for the Construction of Spheroidal Mass Distributions Consistent with the Harmonic Part of the Earth's Gravity Potential, *Manuscr. Geod.* 6, 131–156, 1981.

Tykhonov, A.N.: Solution of Incorrectly Formulated Problems and the Regularization Method, *Sov. Math.* 5, 1035–1038, 1963; translation from *Dokl. Akad. Nauk* 151, 501–504, 1963.

Vekua, I.N.: Über die Vollständigkeit des Systems harmonischer Polynome im Raum, *Dokl. Akad. Nauk* 90, 495–498, 1953.

Vening Meinesz, F.A.: A Formula Expressing the Deflection of the Plumb Line in the Gravity Anomalies and Some Formulas for the Gravity Field and the Gravity Potential Outside the Geoid, *Proc. Koninklijke Akad. Wet. Amsterdam* 31, 315–322, 1928.

Walter, W.: *Einführung in die Potentialtheorie*, BI Hochschulskripten, 765/765a, 1971.

Walter, W.: *Einführung in die Theorie der Distributionen*, 3rd edition, BI-Wissenschaftsverlag, Mannheim, 1994.

Walsh, J.L.: The Approximation of Harmonic Functions by Harmonic Polynomials and by Harmonic Rational Functions, *Bull. Amer. Math. Soc.* 35, 499–544, 1929.

Wangerin, A.: *Theorie des Potentials und der Kugelfunktionen*, de Gruyter, Leipzig, 1921.

Weck, N.: Zwei inverse Probleme in der Potentialtheorie, in: *Mitt. Inst. Theor. Geodäsie, Universität Bonn, Vol. 4*, 27–36, 1972.

Weimer, D.R.: Maps of Ionospheric Field-Aligned Currents as a Function of the Interplanetary Magnetic Field Derived from Dynamic Explorer 2 Data, *J. Geophys. Res.* 106, 889–902, 2001.

Wermer, J.: *Potential Theory*, Springer, Berlin, Heidelberg, New York, 1974.

Weyl, H.: Über die Gleichverteilung von Zahlen mod Eins, *Math. Ann.* 77, 313–352, 1916.

Wienholtz, E., Kalf, H., Kriecherbauer, T.: *Elliptische Differentialgleichnungen zweiter Ordnung*, Springer, Heidelberg, 2009.

Wilson, J.: A Possible Origin of the Hawaiian Island, *Can. J. Phys.* 41, 863–868, 1963.

Wolf, K.: *Multiscale Modeling of Classical Boundary Value Problems in Physical Geodesy by Locally Supported Wavelets.* PhD thesis, Geomathematics Group, TU Kaiserslautern, 2009.

Yamabe, H.: On an Extension of Helly's Theorem, *Osaka Math. J.* 2, 15–22, 1950.

Yoshida, K.: *Functional Analysis*, 5th edition, Springer, Berlin, 1980.

Zidarov, D.P.: Presentation of Gravitational Fields with Fields of Set of Multipoles and Solution of the Inverse Gravimetric Problem, *Comptes rendus de l'Académie bulgare des Sciences* 27, 1351–1354, 1974.

Zidarov, D.P.: Some Uniqueness Conditions for the Solution of the Inverse Gravimetric Problem, *Comptes rendus de l'Académie bulgare des Sciences* 33, 909–912, 1980.

Zidarov, D.P.: Conditions for Uniqueness of Self–Limiting Solutions of the Inverse Problems, *Comptes rendus de l'Académie bulgare des Sciences* 39, 57–60, 1986.

Zidarov, D.P.: *Inverse Gravimetric Problem in Geoprospecting and Geodesy*, Developments in Solid Earth Geophysics, Vol. 19, Elsevier, Amsterdam, 1990.

Index